SEISMIC DESIGN OF REINFORCED CONCRETE AND MASONRY BUILDINGS

SEISMIC DESIGN OF REINFORCED CONCRETE AND MASONRY BUILDINGS

T. Paulay
Department of Civil Engineering
University of Canterbury
Christchurch
New Zealand

M. J. N. Priestley
Department of Applied Mechanics
and Engineering Sciences
University of California
San Diego, USA

A WILEY INTERSCIENCE PUBLICATION

JOHN WILEY & SONS, INC.

New York · Chichester · Brisbane · Toronto · Singapore

Library of Congress Cataloging in Publication Data:

Paulay, T., 1923–
   Seismic design of reinforced concrete and masonry buildings/T. Paulay, M. J. N. Priestley.

     p. cm.
   Includes bibliographical references and index.
   ISBN 0-471-54915-0
   1. Earthquake resistant design. 2. Reinforced concrete construction. 3. Buildings, Reinforced concrete—Earthquake effects. 4. Masonry I. Priestley, M. J. N. II. Title.

TA658.44.P38 1992
624.1′762—dc20   91-34862
                                         CIP

Printed in the United States of America

10 9

PREFACE

Involvement over many years in the teaching of structural engineering, the design of structures, and extensive research relevant to reinforced concrete and masonry buildings motivated the preparation of this book. Because of significant seismic activity in New Zealand and California, our interest has naturally focused primarily on the response of structures during severe earthquakes. A continuing dialogue with practicing structural designers has facilitated the translation of research findings into relatively simple design recommendations, many of which have been in use in New Zealand for a number of years.

We address ourselves not only to structural engineers in seismic regions but also to students who, having completed an introductory course of reinforced concrete theory, would like to gain an understanding of seismic design principles and practice. Emphasis is on design rather than analysis, since considerable uncertainty associated with describing expected ground motion characteristics make detailed and sophisticated analyses of doubtful value, and indicate the scope and promise in "telling" the structure how it must respond under potentially wide range of earthquake characteristics, by application of judicious design principles.

The three introductory chapters present basic concepts of seismic design, review the causes and effects of earthquakes and current procedures to quantify seismicity, structural response, and seismic actions to be considered in design, and summarize established principles of reinforced concrete and masonry member design. The remaining six chapters cover in considerable detail the design of typical building structures, such as reinforced concrete ductile frames, structural walls, dual systems, reinforced masonry structures, buildings with restricted ductility, and foundation systems.

Because with few exceptions seismic structural systems must possess significant ductility capacity, the importance of establishing a rational hierarchy in the formation of uniquely defined and admissible plastic mechanisms is emphasized. A deterministic capacity design philosophy embodies this feature and it serves as a unifying guide throughout the book. Numerous examples, some quite detailed and extensive, illustrate applications, including recommended detailing of the reinforcement, to ensure the attainment of intended levels of ductility where required. Design approaches are based on first principles and rationale without adherence to building codes. However, references are made to common codified approaches, particularly those in the United States and New Zealand, which are very similar. Observed structural damage in earthquakes consistently exposes the predominant sources of weakness: insufficient, poorly executed structural details which received little or no attention within the design process. For this reason, great emphasis is placed in this book on the rational quantification of appropriate detailing.

We gratefully acknowledge the support and encouragement received from our colleagues at the University of Canterbury and the University of California–San Diego, and the research contributions of graduate students and technicians. Our special thanks are extended to Professor Robert Park, an inspiring member of our harmonious research team for more than 20 years, for his unfailing support during the preparation of this manuscript, which also made extensive use of his voluminous contributions to the design of structures for earthquake resistance. The constructive comments offered by our colleagues, especially those of Professor Hugo Bachmann and Konrad Moser of the Swiss Federal Institute of Technology in Zürich, improved the text. Further, we wish to acknowledge the effective support of the New Zealand National Society for Earthquake Engineering, which acted as a catalyst and coordinator of relevant contributions from all sections of the engineering profession, thus providing a significant source for this work.

We are most grateful to Jo Johns, Joan Welte, Maria Martin, and especially Denise Forbes, who typed various chapters and their revisions, and to Valerie Grey for her careful preparation of almost all the illustrations.

In the hope that our families will forgive us for the many hours which, instead of writing, we should have spent with them, we thank our wives for their support, care, patience, and above all their love, without which this book could not have been written.

Tom Paulay
Nigel Priestley

Christchurch and San Diego
March 1991

CONTENTS

1 Introduction: Concepts of Seismic Design

1.1 SEISMIC DESIGN AND SEISMIC PERFORMANCE: A REVIEW

Design philosophy is a somewhat grandiose term that we use for the fundamental basis of design. It covers reasons underlying our choice of design loads, and forces, our analytical techniques and design procedures, our preferences for particular structural configuration and materials, and our aims for economic optimization. The importance of a rational design philosophy becomes paramount when seismic considerations dominate design. This is because we typically accept higher risks of damage under seismic design forces than under other comparable extreme loads, such as maximum live load or wind forces. For example, modern building codes typically specify an intensity of design earthquakes corresponding to a return period of 100 to 500 years for ordinary structures, such as office buildings. The corresponding design forces are generally too high to be resisted within the elastic range of material response, and it is common to design for strengths which are a fraction, perhaps as low as 15 to 25%, of that corresponding to elastic response, and to expect the structures to survive an earthquake by large inelastic deformations and energy dissipation corresponding to material distress. The consequence is that the full strength of the building can be developed while resisting forces resulting from very much smaller earthquakes, which occur much more frequently than the design-level earthquake. The annual probability of developing the full strength of the building in seismic response can thus be as high as 1 to 3%. This compares with accepted annual probabilities for achieving ultimate capacity under gravity loads of perhaps 0.01%. It follows that the consequences resulting from the lack of a rational seismic design philosophy are likely to be severe.

The incorporation of seismic design procedures in building design was first adopted in a general sense in the 1920s and 1930s, when the importance of inertial loadings of buildings began to be appreciated. In the absence of reliable measurements of ground accelerations and as a consequence of the lack of detailed knowledge of the dynamic response of structures, the magnitude of seismic inertia forces could not be estimated with any reliability. Typically, design for lateral forces corresponding to about 10% of the building weight was adopted. Since elastic design to permissible stress levels

1

was invariably used, actual building strengths for lateral forces were generally somewhat larger.

By the 1960s accelerograms giving detailed information on the ground acceleration occurring in earthquakes were becoming more generally available. The advent of strength design philosophies, and development of sophisticated computer-based analytical procedures, facilitated a much closer examination of the seismic response of multi-degree-of-freedom structures. It quickly became apparent that in many cases, seismic design to existing lateral force levels specified in codes was inadequate to ensure that the structural strength provided was not exceeded by the demands of strong ground shaking. At the same time, observations of building responses in actual earthquakes indicated that this lack of strength did not always result in failure, or even necessarily in severe damage. Provided that the structural strength could be maintained without excessive degradation as inelastic deformations developed, the structures could survive the earthquake, and frequently could be repaired economically. However, when inelastic deformation resulted in severe reduction in strength, as, for example, often occurred in conjunction with shear failure of concrete or masonry elements, severe damage or collapse was common.

With increased awareness that excessive strength is not essential or even necessarily desirable, the emphasis in design has shifted from the resistance of large seismic forces to the "evasion" of these forces. Inelastic structural response has emerged from the obscurity of hypotheses, and become an essential reality in the assessment of structural design for earthquake forces. The reality that all inelastic modes of deformation are not equally viable has become accepted. As noted above, some lead to failure and others provide ductility, which can be considered the essential attribute of maintaining strength while the structure is subjected to reversals of inelastic deformations under seismic response.

More recently, then, it has become accepted that seismic design should encourage structural forms that are more likely to possess ductility than those that do not. Generally, this relates to aspects of structural regularity and careful choice of the locations, often termed plastic hinges, where inelastic deformations may occur. In conjunction with the careful selection of structural configuration, required strengths for undesirable inelastic deformation modes are deliberately amplified in comparison with those for desired inelastic modes. Thus for concrete and masonry structures, the shear strength provided must exceed the actual flexural strength to ensure that inelastic shear deformations, associated with large deterioration of stiffness and strength, which could lead to failure, cannot occur. These simple concepts, namely (1) selection of a suitable structural configuration for inelastic response, (2) selection of suitable and appropriately detailed locations (plastic hinges) for inelastic deformations to be concentrated, and (3) insurance, through suitable strength differentials that inelastic deformation does not occur at undesirable locations or by undesirable structural

Fig. 1.1 Soft-story sway mechanism, 1990 Philippine earthquake. (Courtesy of EQE Engineering Inc.)

modes—are the bases for the capacity design philosophy, which is developed further in this chapter, and described and implemented in detail in subsequent chapters.

Despite the increased awareness and understanding of factors influencing the seismic behavior of structures, significant disparity between earthquake engineering theory, as reported, for example, in recent proceedings of the World Conferences on Earthquake Engineering [1968–88], and its application in design and construction still prevails in many countries. The damage in, and even collapse of, many relatively modern buildings in seismically active regions, shown in Figs. 1.1 to 1.7, underscores this disparity.

Figure 1.1 illustrates one of the most common causes of failure in earthquakes, the "soft story mechanism." Where one level, typically the lowest, is weaker than upper levels, a column sway mechanism can develop with high local ductility demand. In taller buildings than that depicted in Fig. 1.1, this often results from a functional desire to open the lowest level to the maximum extent possible for retail shopping or parking requirements.

Figure 1.2, also from the July 1990 Philippine's earthquake, shows a confinement failure at the base of a first-story column. Under ductile response to earthquakes, high compression strains should be expected from the combined effects of axial force and bending moment. Unless adequate, closely spaced, well-detailed transverse reinforcement is placed in the potential plastic hinge region, spalling of concrete followed by instability of the compression reinforcement will follow. In the example of Fig. 1.2, there is

Fig. 1.2   Confinement failure of column base of 10-story building. (Courtesy of EQE Engineering, Inc.)

clearly inadequate transverse reinforcement to confine the core concrete and restrain the bundled flexural reinforcement against buckling. It must be recognized that even with a weak beam/strong column design philosophy which seeks to dissipate seismic energy primarily in well-confined beam plastic hinges, a column plastic hinge must still form at the base of the column. Many structures have collapsed as a result of inadequate confinement of this hinge.

The shear failure of a column of a building in the 1985 Chilean earthquake, shown in Fig. 1.3, demonstrates the consequences of ignoring the stiffening effects of so-called nonstructural partial height masonry or concrete infill built hard up against the column. The column is stiffened in comparison with other columns at the same level, which may not have adjacent infill (e.g., interior columns) attracting high shears to the shorter columns, often with

Fig. 1.3 Influence of partial height infill increasing column shear force (1985 Chilean earthquake). (Courtesy of *Earthquake Spectra* and the Earthquake Engineering Research Institute.)

Fig. 1.4 Failure of structural wall resulting from inadequate flexural and shear strength (1990 Philippine earthquake). (Courtesy of EQE Engineering Inc.)

Fig. 1.5 Failure of coupling beams between shear walls (1964 Alaskan earthquake). (Courtesy of the American Iron and Steel Institute.)

disastrous effects. This common structural defect can easily be avoided by providing adequate separation between the column and infill for the column to deform freely during seismic response without restraint from the infill.

Unless adequately designed for the levels of flexural ductility, and shear force expected under strong ground shaking, flexural or shear failures may develop in structural walls forming the primary lateral force resistance of buildings. An example from the 1990 Philippine earthquake is shown in Fig. 1.4, where failure has occurred at the level corresponding to a significant reduction in the stiffness and strength of the lateral force resisting system—a common location for concentration of damage.

Spandrel beams coupling structural walls are often subjected to high ductility demands and high shear forces as a consequence of their short length. It is very difficult to avoid excessive strength degradation in such elements, as shown in the failure of the McKinley Building during the 1964 Alaskan earthquake, depicted in Fig. 1.5, unless special detailing measures are adopted involving diagonal reinforcement in the spandrel beams.

Figure 1.6 shows another common failure resulting from "nonstructural" masonry infills in a reinforced concrete frame. The stiffening effect of the infill attracts higher seismic forces to the infilled frame, resulting in shear

Fig. 1.6   Failure of lower level of masonry-infilled reinforced concrete frame (1990 Philippine earthquake). (Courtesy of EQE Engineering Inc.)

failure of the infill, followed by damage or failure to the columns. As with the partial height infill of Fig. 1.3, the effect of the nonstructural infill is to modify the lateral force resistance in a way not anticipated by the design.

The final example in Fig. 1.7 shows the failure of a beam–column connection in a reinforced concrete frame. The joint was not intended to become the weak link between the four components. Such elements are usually subjected to very high shear forces during seismic activity, and if inadequately reinforced this will result in excessive loss in strength and stiffness of the frame, and even collapse.

While there is something new to be learned from each earthquake, it may be said that the majority of structural lessons should have been learned. Patterns in observed earthquake damage have been identified and reported [B10, J2, M15, P20, S4, S10, U2, W4] for some time. Yet many conceptual, design, and construction mistakes, that are responsible for structural damage

Fig. 1.7 Beam–column connection failure (1990 Philippine earthquake). (Courtesy of T. Minami.)

in buildings are being repeated. Many of these originate from traditional building configurations and construction practices, the abandonment of which societies or the building industry of the locality are reluctant to accept. There is still widespread lack of appreciation of the predictable and quantifiable effects of earthquakes on buildings and the impact of seismic phenomena on the philosophy of structural design.

Well-established techniques, used to determine the safe resistance of structures with respect to various static loads, including wind forces, cannot simply be extended and applied to conditions that arise during earthquakes. Although many designers prefer to assess earthquake-induced structural actions in terms of static equivalent loads or forces, it must be appreciated that actual seismic response is dynamic and is related primarily to imposed deformation rather than forces. To accommodate large seismically induced deformations, most structures need to be ductile. Thus in the design of structures for earthquake resistance, it is preferable to consider forces generated by earthquake-induced displacements rather than traditional loads. Because the magnitudes of the largest seismic-displacement-generated forces in a ductile structure will depend on its capacity or strength, the evaluation of the latter is of importance.

1.1.1 Seismic Design Limit States

It is customary to consider various levels of protection, each of which emphasizes a different aspect to be considered by the designer. Broadly, these relate to preservation of functionality, different degrees of efforts to minimize damage that may be caused by a significant seismic event, and the prevention of loss of life.

The degree to which levels of protection can be afforded will depend on the willingness of society to make sacrifices and on economic constraints within which society must exist. While regions of seismicity are now reasonably well defined, the prediction of a seismic event within the projected lifespan of a building is extremely crude. Nevertheless, estimates must be made for potential seismic hazards in affected regions in an attempt to optimize between the degree of protection sought and its cost. These aspects are discussed further in Chapter 2.

(a) Serviceability Limit State Relatively frequent earthquakes inducing comparatively minor intensity of ground shaking should not interfere with functionality, such as the normal operation of a building or the plant it contains. This means that no damage needing repair should occur to the structure or to nonstructural components, including contents. The appropriate design effort will need to concentrate on the control and limitation of displacements that could occur during the anticipated earthquake, and to ensure adequate strengths in all components of the structure to resist the earthquake-induced forces while remaining essentially elastic. Reinforced concrete and masonry structures may develop considerable cracking at the serviceability limit state, but no significant yielding of reinforcement, resulting in large cracks, nor crushing of concrete or masonry should result. The frequency with which the occurrence of an earthquake corresponding to the serviceability limit state may be anticipated will depend on the importance of preserving functionality of the building. Thus, for office buildings, the serviceability limit state may be chosen to correspond to a level of shaking likely to occur, on average, once every 50 years (i.e., a 50-year-return-period earthquake). For a hospital, fire station, or telecommunications center, which require a high degree of protection to preserve functionality during an emergency, an earthquake with a much longer return period will be appropriate.

(b) Damage Control Limit State For ground shaking of intensity greater than that corresponding to the serviceability limit state, some damage may occur. Yielding of reinforcement may result in wide cracks that require repair measures, such as injection grouting, to avoid later corrosion problems. Also, crushing or spalling of concrete may occur, necessitating replacement of unsound concrete. A second limit state may be defined which marks the boundary between economically repairable damage and damage that is irreparable or which cannot be repaired economically. Ground shaking of intensity likely to induce response corresponding to the damage control limit state should have a low probability of occurrence during the expected life of the building. It is expected that after an earthquake causes this or lesser

intensity of ground shaking, the building can be successfully repaired and reinstated to full service.

(c) Survival Limit State In the development of modern seismic design strategies, very strong emphasis is placed on the criterion that loss of life should be prevented even during the strongest ground shaking feasible for the site. For this reason, particular attention must be given to those aspects of structural behavior that are relevant to this single most important design consideration: survival. For most buildings, extensive damage to both the structure and building contents, resulting from such severe but rare events, will have to be accepted. In some cases this damage will be irreparable, but collapse must not occur. Unless structures are proportioned to possess exceptionally large strength with respect to lateral forces, usually involving significant cost increase, inelastic deformations during large seismic events are to be expected. Therefore, the designer will need to concentrate on structural qualities which will ensure that for the expected duration of an earthquake, relatively large displacements can be accommodated without significant loss in lateral force resistance, and that integrity of the structure to support gravity loads is maintained. In this book, particular emphasis is placed on these aspects of structural response.

It must be appreciated that the boundaries between different intensities of ground shaking, requiring each of the foregoing three levels of protection to be provided cannot be defined precisely. A much larger degree of uncertainty is involved in the recommendations of building codes to determine the intensities of lateral seismic design forces than for any other kind of loading to which a building might be exposed. The capacity design process, which is developed in this book, aims to accommodate this uncertainty. To achieve this, structural systems must be conceived which are tolerant to the crudeness in seismological predictions.

1.1.2 Structural Properties

The specific structural properties that need to be considered in conjunction with the three levels of seismic protection described in the preceding section are described below.

(a) Stiffness If deformations under the action of lateral forces are to be reliably quantified and subsequently controlled, designers must make a realistic estimate of the relevant property—stiffness. This quantity relates loads or forces to the ensuing structural deformations. Familiar relationships are readily established from first principles of structural mechanics, using geometric properties of members and the modulus of elasticity for the material. In reinforced concrete and masonry structures these relationships are, however, not quite as simple as an introductory text on the subject may

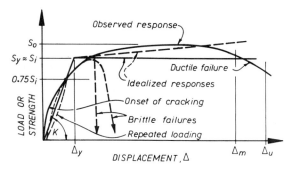

Fig. 1.8 Typical load–displacement relationship for a reinforced concrete element.

suggest. If serviceability criteria are to be satisfied with a reasonable degree of confidence, the extent and influence of cracking in members and the contribution of concrete or masonry in tension must be considered, in conjunction with the traditionally considered aspects of section and element geometry, and material properties.

A typical nonlinear relationship between induced forces or loads and displacements, describing the response of a reinforced concrete component subjected to monotonically increasing displacements, is shown in Fig. 1.8. For purposes of routine design computations, one of the two bilinear approximations may be used, where S_y defines the yield or ideal strength S_i of the member. The slope of the idealized linear elastic response, $K = S_y/\Delta_y$ is used to quantify stiffness. This should be based on the effective secant stiffness to the real load–displacement curve at a load of about $0.75S_y$, as shown in Fig. 1.8, as it is effective stiffness at close to yield strength that will be of concern when estimating response for the serviceability limit state. Under cyclic loading at high "elastic" response levels, the initial curved load–displacement characteristic will modify to close to the linear relationship of the idealized response. An early task within the design process will be the checking of typical interstory deflections (drift), using realistic stiffness values to satisfy local requirements for serviceability [Section 1.1.1(a)].

(b) Strength If a concrete or masonry structure is to be protected against damage during a selected or specified seismic event, inelastic excursions during its dynamic response should be prevented. This means that the structure must have adequate strength to resist internal actions generated during the elastic dynamic response of the structure. Therefore, the appropriate technique for the evaluation of earthquake-induced actions is an elastic analysis, based on stiffness properties described in the preceding section. These seismic actions, combined with those due to other loads on the structure, such as gravity, will lead, perhaps with minor modifications, to

the proportioning of structural members. Thereby the designer can provide the desired strength, shown as S_i in Fig. 1.8, in terms of resistance to lateral forces envisaged.

(c) **Ductility** To minimize major damage and to ensure the survival of buildings with moderate resistance with respect to lateral forces, structures must be capable of sustaining a high proportion of their initial strength when a major earthquake imposes large deformations. These deformations may be well beyond the elastic limit. This ability of the structure or its components, or of the materials used to offer resistance in the inelastic domain of response, is described by the general term *ductility*. It includes the ability to sustain large deformations, and a capacity to absorb energy by hysteretic behavior, as discussed in Section 2.3.3. For this reason it is the single most important property sought by the designer of buildings located in regions of significant seismicity.

The limit to ductility, as shown for example in Fig. 1.8 by the displacement of Δ_u, typically corresponds to a specified limit to strength degradation. Although attaining this limit is sometimes termed *failure*, significant additional inelastic deformations may still be possible without structural collapse. Hence a ductile failure must be contrasted with brittle failure, represented in Fig. 1.8 by the dashed curves. Brittle failure implies near-complete loss of resistance, often complete disintegration, and the absence of adequate warning. For obvious reasons, brittle failure, which may be said to be the overwhelming cause for the collapse of buildings in earthquakes, and the consequent loss of lives, must be avoided. More precise definitions for the essential characteristics of ductility are given in Section 3.5.

Ductility is defined by the ratio of the total imposed displacements Δ at any instant to that at the onset of yield Δ_y. Using the idealizations of Fig. 1.8, this is

$$\mu = \Delta/\Delta_y > 1 \qquad (1.1)$$

The displacements Δ_y and Δ in Eq. (1.1) and Fig. 1.8 may represent strain, curvature, rotation, or deflection. The ductility developed when failure is imminent is, from Fig. 1.8, $\mu_u = \Delta_u/\Delta_y$. Ductility is the structural property that will need to be relied on in most buildings if satisfactory behavior under damage control and survival limit state is to be achieved. An important consideration in the determination of the required seismic resistance will be that the estimated maximum ductility demand during shaking, $\mu_m = \Delta_m/\Delta_y$ (Fig. 1.8), does not exceed the ductility potential μ_u.

The roles of both stiffness and strength, as well as their quantification, are well established. The sources, development, quantification, and utilization of ductility, to serve best the designer's intent, are generally less well understood. For this reason many aspects of ductile structural response are examined in considerable detail in this book.

Structures which, because of their nature or importance to society, are to be designed to respond elastically, even during the largest expected seismic event, may be readily designed with the well-established tools of structural mechanics. Hence these structures will receive only superficial mention in the remainder of the book.

Ductility in structural members can be developed only if the constituent material itself is ductile. Thus it is relatively easy to achieve the desired ductility if resistance is to be provided by steel in tension. However, precautions need to be taken when steel is subject to compression, to ensure that premature buckling does not interfere with the development of the desired large inelastic strains in compression.

Concrete and masonry are inherently brittle materials. Although their tensile strength cannot be relied on as a primary source of resistance, they are eminently suited to carry compression stresses. However, the maximum strains developed in compression are rather limited unless special precautions are taken. The primary aim of the detailing of composite structures consisting of concrete or masonry and steel is to combine these materials in such a way as to produce ductile members, which are capable of meeting the inelastic deformation demands imposed by severe earthquakes.

In the context of reinforced concrete and masonry structures, *detailing* refers to the preparation of placing drawings, reinforcing bar configurations, and bar lists that are used for fabrication and placement of reinforcement in structures. But detailing also incorporates a design process by which the designer ensures that each part of the structure can perform safely under service load conditions and also when specially selected critical regions are to accommodate large inelastic deformations. Particularly, it covers such aspects as the choice of bar sizes, the distribution of bars, curtailment and splice details of flexural reinforcement, and the size, spacing, configuration, and anchorage of transverse reinforcement, intended to provide shear strength and ductility to critical regions. Detailing based on an understanding of and a feeling for structural behavior, on the appreciation of changing demands of economy, and on the limitations of construction practices is at least as important as other attributes in the art of structural design [P1].

1.2 ESSENTIALS OF STRUCTURAL SYSTEMS FOR SEISMIC RESISTANCE

All structural systems are not created equal when response to earthquake-induced forces is of concern. Aspects of structural configuration, symmetry, mass distribution, and vertical regularity must be considered, and the importance of strength, stiffness, and ductility in relation to acceptable response appreciated. The first task of the designer will be to select a structural system most conducive to satisfactory seismic performance within the constraints dictated by architectural requirements. Where possible, architect and structural engineer should discuss alternative structural configurations at the earliest stage of concept development to ensure that undesirable geometry is not locked-in to the system before structural design begins.

Irregularities, often unavoidable, contribute to the complexity of structural behavior. When not recognized, they may result in unexpected damage and even collapse. There are many sources of structural irregularities. Drastic changes in geometry, interruptions in load paths, discontinuities in both strength and stiffness, disruptions in critical regions by openings, unusual proportions of members, reentrant corners, lack of redundancy, and interference with intended or assumed structural deformations are only a few of the possibilities. The recognition of many of these irregularities and of conceptions for remedial measures for the avoidance or mitigation of their undesired effects rely on sound understanding of structural behavior. Awareness to search for undesired structural features and design experience are invaluable attributes. The relative importance of some irregularities may be quantified. In this respect some codes provide limited guidance. Examples for estimating the criticality of vertical and horizontal irregularities in framed buildings are given in Section 4.2.6. Before the more detailed discussion of these aspects later in this chapter, it is, however, necessary to review some general aspects of seismic forces and structural systems.

1.2.1 Structural Systems for Seismic Forces

The primary purpose of all structures used for building is to support gravity loads. However, buildings may also be subjected to lateral forces due to wind or earthquakes. The taller a building, the more significant the effects of lateral forces will be. It is assumed here that seismic criteria rather than wind or blast forces govern the design for lateral resistance of buildings. Three types of structures, most commonly used for buildings, are considered in this book.

(a) Structural Frame Systems   Structures of multistory reinforced concrete buildings often consist of frames. Beams, supporting floors, and columns are continuous and meet at nodes, often called "rigid" joints. Such frames can readily carry gravity loads while providing adequate resistance to horizontal forces, acting in any direction [B4]. Chapter 4 deals with the design of reinforced concrete ductile frames.

(b) Structural Wall Systems   When functional requirements permit it, resistance to lateral forces may be assigned entirely to structural walls, using reinforced concrete or masonry [B4]. Gravity load effects on such walls are seldom significant and they do not control the design. Usually, there are also other elements within such a building, which are assigned to carry only gravity loads. Their contribution to lateral force resistance, if any, is often neglected. In Chapter 5 we present various structural aspects of buildings in

which the resistance to lateral forces is assigned entirely to structural walls. The special features of reinforced masonry, particularly suited for the construction of walls that resist both gravity loads and lateral forces, are presented in Chapter 7.

(c) Dual Systems Finally, dual building systems are studied briefly in Chapter 6. In these, reinforced concrete frames interacting with reinforced concrete or masonry walls together provide the necessary resistance to lateral forces, while each system carries its appropriate share of the gravity load. These types of structures are variously known as dual, hybrid, or wall–frame structures.

The selection of structural systems for buildings is influenced primarily by the intended function, architectural considerations, internal traffic flow, height and aspect ratio, and to a lesser extent, the intensity of loading. The selection of a building's configuration, one of the most important aspects of the overall design [A4], may impose severe limitations on the structure in its role to provide seismic protection. Because the intent is to present design concepts and principles, rather than a set of solutions, various alternatives within each of these three groups of distinct structural systems, listed above, will not be considered. Some structural forms are, however, deliberately omitted. For example, construction consisting of flat slabs supported by columns is considered to be unsuitable on its own to provide satisfactory performance under seismic actions because of excessive lateral displacements and the difficulty to providing the adequate and dependable shear transfer between columns and slabs, necessary to sustain lateral forces, in addition to gravity loads.

Sufficient information for both the design and detailing of components of the principal structural systems is provided in subsequent chapters to allow easy adaptation of any principle to other structural forms, for example those using precast concrete components, which may occur in buildings. Valuable information in this respect may be obtained from a report by ACI-ASCE Committee 442 on the response of concrete buildings to lateral forces [A14].

1.2.2 Gross Seismic Response

(a) Response in Elevation: The Building as a Vertical Cantilever When subjected to lateral forces only, a building will act as a vertical cantilever. The resulting total horizontal force and the overturning moment will be transmitted at the level of the foundations. Once the lateral forces, such as may act at each level of the building, are known, the story shear forces, as well as the magnitude of overturning moments at any level, shown in Fig. 1.9, can readily be derived from usual equilibrium relationships. For example, in Fig. 1.9(*a*), the sum V_j of all floor forces acting on the shaded portion of the building

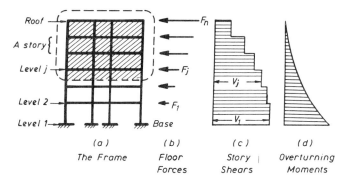

Fig. 1.9   Effects of lateral forces on a building.

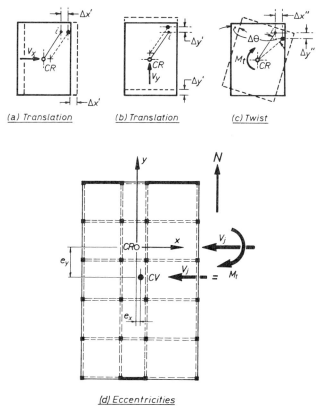

Fig. 1.10   Relative floor displacement.

must be resisted by shear and axial forces and bending moments in the vertical elements in the third story.

In the description of multistory buildings in this book, the following terminology is used. All structures are assumed to be founded at the base or level 1. The position of a floor will be identified by its level above the base. Roof level is identical with the top level. The space or vertical distance between adjacent levels is defined as a story. Thus the first story is between levels 1 and 2, and the top story is that below roof level (Fig. 1.9).

(b)   Response in Plan: Centers of Mass and Rigidity   The structural system may consist of a number of frames, as shown in Fig. 1.9(a), or walls, or a combination of these, as described in Section 1.2.1 [Fig. 1.10(d)]. The position of the resultant force V_j in the horizontal plane will depend on the plan distribution of vertical elements, and it must also be considered. As a consequence, two important concepts must be defined. These will enable the effects of building configurations on the response of structural systems to lateral forces to be better appreciated. The evaluation of the effects of lateral forces, such as shown in Fig. 1.9(a), on the structural systems described in Section 1.2.1 is given in Chapters 4 through 6.

(i)   Center of Mass: During an earthquake, acceleration-induced inertia forces will be generated at each floor level, where the mass of an entire story may be assumed to be concentrated. Hence the location of a force at a particular level will be determined by the center of the accelerated mass at that level. In regular buildings, such as shown in Fig. 1.10(d), the positions of the centers of floor masses will differ very little from level to level. However, irregular mass distribution over the height of a building may result in variations in centers of masses, which will need to be evaluated. The summation of all the floor forces, F_j in Fig. 1.9(a), above a given story, with due allowance for the in-plane position of each, will then locate the position of the resultant force V_j within that story. For example, the position of the shear force V_j within the third story is determined by point CV in Fig. 1.10(d), where this shear force is shown to act in the east–west direction. Depending on the direction of an earthquake-induced acceleration at any instant, the force V_j passing through this point may act in any direction. For a building of the type shown in Fig. 1.10(d), it is sufficient, however, to consider seismic attacks only along the two principal axes of the plan.

(ii)   Center of Rigidity: If, as a result of lateral forces, one floor of the building in Fig. 1.9 translates horizontally as a rigid body relative to the floor below, as shown in Fig. 1.10(a), a constant interstory displacement $\Delta x'$ will be imposed on all frames and walls in that story. Therefore, the induced forces in these elastic frames and walls, in the relevant east–west planes, will be proportional to the respective stiffnesses. The resultant total force, $V_j = V_x$, induced by the translational displacements $\Delta x'$, will pass through the center

of rigidity (CR) in Fig. 1.10(d). Similarly, a relative floor translation to the north, shown as $\Delta y'$ in Fig. 1.10(b), will induce corresponding forces in each of the four frames [Fig. 1.10(d)], the resultant of which, V_y, will also pass through point CR. This point, defined as the center of rigidity or center of stiffness, locates the position of a story shear force V_j, which will cause only relative floor translations.

The position of the center of rigidity may be different in each story. It is relevant to story shear forces applied in any direction in a horizontal plane. Such a force may be resolved into components, such as V_x and V_y shown in Fig. 1.10(a) and (b), which will cause simultaneous story translations $\Delta x'$ and $\Delta y'$, respectively.

Since the story shear force V_j in Fig. 1.10(d) acts through point CV rather than the center of rigidity CR, it will cause floor rotation as well as relative floor translation. For convenience, V_j may be replaced by an equal force acting through CR, thus inducing pure translation, and a moment $M_t = e_y V_j$ about CR, leading to rigid floor rotation, as shown in Fig. 1.10(c). The angular rotation $\Delta \theta$ is termed *story twist*. It will cause additional interstory displacements $\Delta x''$ and $\Delta y''$ in lateral force resisting elements in both principal directions, x and y. The displacements due to story twist are proportional to the distance of the element from the center of rotation, [i.e., the center of rigidity (CR)].

Displacements due to story twist, when combined with those resulting from floor translations, can result in total element interstory displacements that may be difficult to accommodate. For this reason the designer should attempt to minimize the magnitude of story torsion M_t. This may be achieved by a deliberate assignment of stiffnesses to lateral force-resisting components, such as frames or walls, in such a way as to minimize the distance between the center of rigidity (CR) and the line of action of the story shear force (CV). To achieve this in terms of floor forces, the distance between the center of rigidity and the center of mass should be minimized.

1.2.3 Influence of Building Configuration on Seismic Response

An aspect of seismic design of equal if not greater importance than structural analysis is the choice of building configuration [A4]. By observing the following fundamental principles, relevant to seismic response, more suitable structural systems may be adopted.

1. Simple, regular plans are preferable. Building with articulated plans such as T and L shapes should be avoided or be subdivided into simpler forms (Fig. 1.11).

2. Symmetry in plan should be provided where possible. Gross lack of symmetry may lead to significant torsional response, the reliable prediction of which is often difficult. Much greater damage due to earth-

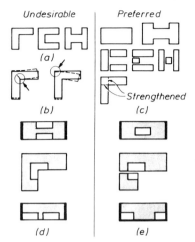

Undesirable   *Preferred*

(a)

(b)

(c)   Strengthened

(d)   (e)

Fig. 1.11   Plan configurations in buildings.

quakes has been observed in buildings situated at street corners, where structural symmetry is more difficult to achieve, than in those along streets, where a more simple rectangular and often symmetrical structural plan could be utilized.

3. An integrated foundation system should tie together all vertical structural elements in both principal directions. Foundations resting partly on rock and partly on soils should preferably be avoided.

4. Lateral-force-resisting systems within one building, with significantly different stiffnesses such as structural walls and frames, should be arranged in such a way that at every level symmetry in lateral stiffness is not grossly violated. Thereby undesirable torsional effects will be minimized.

5. Regularity should prevail in elevation, in both the geometry and the variation of story stiffnesses.

The principles described above are examined in more detail in the following sections.

(a) Role of the Floor Diaphragm   Simple and preferably symmetrical building plans hold the promise of more efficient and predictable seismic response of each of the structural components. A prerequisite for the desirable interaction within a building of all lateral-force-resisting vertical components of the structural system is an effective and relatively rigid interconnection of these components at suitable levels. This is usually achieved with the use of floor systems, which generally possess large in-plane stiffness. Vertical elements will thus contribute to the total lateral force resistance, in proportion to their own stiffness. With large in-plane stiffness, floors can act as di-

aphragms. Hence a close to linear relationship between the horizontal displacements of the various lateral-force-resisting vertical structural elements will exist at every level. From rigid-body translations and rotations, shown in Fig. 1.10, the relative displacements of vertical elements can readily be derived. This is shown for frames in Appendix A.

Another function of a floor system, acting as a diaphragm, is to transmit inertia forces generated by earthquake accelerations of the floor mass at a given level to all horizontal-force-resisting elements. At certain levels, particularly in lower storys, significant horizontal forces from one element, such as a frame, may need to be transferred to another, usually stiffer element, such as a wall. These actions may generate significant shear forces and bending moments within a diaphragm. In squat rectangular diaphragms, the resulting stresses will be generally insignificant. However, this may not be the case when long or articulated floor plans, such as shown in Fig. 1.11(a) have to be used. The correlation between horizontal displacements of vertical elements [Fig. 1.11(b)] will be more difficult to establish in such cases. Reentrant corners, inviting stress concentrations, may suffer premature damage. When such configurations are necessary, it is preferable to provide structural separations. This may lead to a number of simple, compact, and independent plans, as shown in Fig. 1.11(c). Gaps separating adjacent structures must be large enough to ensure that even during a major seismic event, no hammering of adjacent structures will occur due to out-of-phase relative motions of the independent substructures. Inelastic deflections, resulting from ductile dynamic response, must be allowed for.

Diaphragm action may be jeopardized if openings, necessary for vertical traffic within a multistory building or other purposes, significantly reduce the ability of the diaphragm to resist in-plane flexure of shear, as seen in examples in Fig. 1.11(d). The relative importance of openings may be estimated readily from a simple evaluation of the flow of forces within the diaphragm, necessary to satisfy equilibrium criteria. Preferred locations for such openings are suggested in Fig. 1.11(e).

As a general rule, diaphragms should be designed to respond elastically, as they are not suitable to dissipate energy through the formation of plastic regions. Using capacity design principles, to be examined subsequently, it is relatively easy to estimate the magnitudes of the largest forces that might be introduced to diaphragms. These are usually found to be easily accommodated. Other aspects of diaphragms, including flexibility, are discussed in Section 6.5.3.

(b)   Amelioration of Torsional Effects   It was emphasized in Section 1.2.2 that to avoid excessive displacements in lateral-force-resisting components that are located in adverse positions within the building plan, torsional effects should be minimized. This is achieved by reducing the distance between the center of mass (CM), where horizontal seismic floor forces are applied, and the center of rigidity (CR) (Fig. 1.10). A number of examples for both

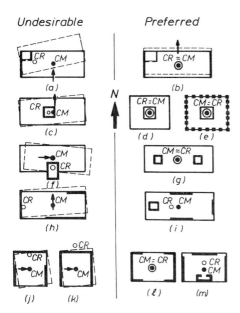

Undesirable   *Preferred*

Fig. 1.12 Mass and lateral stiffness relationship with floor plans. (The grid of frames in each plan, required primarily for gravity loads, is not shown.)

undesirable positioning of major lateral-force-resisting elements, consisting of structural walls and frames, and for the purpose of comparison, preferred locations, are given in Fig. 1.12. For the sake of clarity the positioning of frames required solely for gravity load resistance within each floor plan is generally not shown. While the primary role of the frames in these examples will be the support of gravity load, it must be appreciated that frames will also contribute to both lateral force resistance and torsional stiffness.

Figure 1.12(*a*) shows that because of the location of a stiff wall at the west end of a building, very large displacements, as a result of floor translations and rotations (Fig. 1.10), will occur at the east end. As a consequence, members of a frame located at the east end may be subjected to excessive inelastic deformations (ductility). Excessive ductility demands at such a location may cause significant degradation of the stiffness of a frame. This will lead to further shift of the center of rigidity and consequently to an amplification of torsional effects. A much improved solution, shown in Fig. 1.12(*b*), where the service core has been made nonstructural and a structural wall added at the east end will ensure that the centers of mass and stiffness virtually coincide. Hence only dominant floor translations, imposing similar ductility demands on all lateral force resisting frames or walls, are to be expected.

Analysis may show that in some buildings torsional effects [Fig. 1.12(*c*)] may be negligible. However, as a result of normal variations in material properties and section geometry, and also due to the effects of torsional components of ground motion, torsion may arise also in theoretically perfectly symmetrical buildings. Hence codes require that allowance be made in all buildings for so-called "accidental" torsional effects.

Although a reinforced concrete or masonry core, such as shown in Fig. 1.12(c), may exhibit good torsional strength, its torsional stiffness, particularly after the onset of diagonal cracking, may be too small to prevent excessive deformations at the east and west ends of the building. Similar twists may lead, however, to acceptable displacements at the perimeter of square plans with relatively large cores, seen in Fig. 1.12(d). Closely placed columns, interconnected by relatively stiff beams around the perimeter of such buildings [Fig. 1.12(e)], can provide excellent control of torsional response. The eccentrically placed service core, shown in Fig. 1.12(f), may lead to excessive torsional effects under seismic attack in the east–west direction unless perimeter lateral force resisting elements are present to limit torsional displacements.

The advantages of the arrangement, shown in Fig. 1.12(g), in terms of response to horizontal forces are obvious. While the locations of the walls in Fig. 1.12(h), to resist lateral forces, it satisfactory, the large eccentricity of the center of mass with respect to the center of rigidity will result in large torsion when lateral forces are applied in the north–south direction. The placing of at least one stiff element at or close to each of the four sides of the buildings, as shown in Fig. 1.12(i), provides a particularly desirable structural arrangement. Further examples, showing wall arrangements with large eccentricities and preferred alternative solutions, are given in Fig. 1.12(j) to (m). Although large eccentricities are indicated in the examples of Fig. 1.12(j) and (k), both stiffness and the strength of these walls may well be adequate to accommodate torsional effects.

The examples of Fig. 1.12 apply to structures where walls provide the primary lateral load resistance. The principles also apply to framed systems, although it is less common for excessive torsional effects to develop in frame structures.

(c) Vertical Configurations   A selection of undesirable and preferred configurations is illustrated in Fig. 1.13. Tall and slender buildings [Fig. 1.13(a)] may require large foundations to enable large overturning moments to be transmitted in a stable manner. When subjected to seismic accelerations, concentration of masses at the top of a building [Fig. 1.13(b)] will similarly impose heavy demands on both the lower stories and the foundations of the structure. In comparison, the advantages of building elevations as shown in Fig. 1.13(c) and (d) are obvious.

An abrupt change in elevation, such as shown in Fig. 1.13(e), also called a *setback*, may result in the concentration of structural actions at and near the level of discontinuity. The magnitudes of such actions, developed during the dynamic response of the building, are difficult to predict without sophisticated analytical methods. The separation into two simple, regular structural systems, with adequate separation [Fig. 1.13(f)] between them, is a prefer-

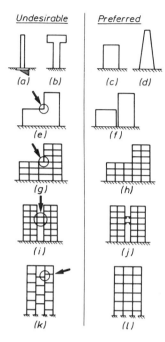

Fig. 1.13 Vertical configurations.

able alternative. Irregularities within the framing system, such as a drastic interference with the natural flow of gravity loads and that of lateral-force-induced column loads at the center of the frame in Fig. 1.13(g), must be avoided.

Although two adjacent buildings may appear to be identical, there is no assurance that their response to ground shaking will be in phase. Hence any connections (bridging) between the two that may be desired [Fig. 1.13(i)] should be such as to prevent horizontal force transfer between the two structures [Fig. 1.13(j)].

Staggered floor arrangements, as seen in Fig. 1.13(k), may invalidate the rigid interconnection of all vertical lateral-force-resisting units, the importance of which was emphasized in Section 1.2.3(a). Horizontal inertia forces, developed during dynamic response, may impose severe demands, particularly on the short interior columns. While such frames [Fig. 1.13(k)] may be readily analyzed for horizontal static forces, results of analyses of their inelastic dynamic response to realistic ground shaking should be treated with suspicion.

Major deviations from a continuous variation with height of both stiffness and strength are likely to invite poor and often dangerous structural response. Because of the abrupt changes of story stiffnesses, suggested in Fig. 1.14(a) and (b), the dynamic response of the corresponding structures [Fig. 1.14(e) and (f)] may be dominated by the flexible stories. Reduced story

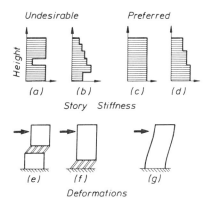

Fig. 1.14 Interacting frames and walls.

stiffness is likely to be accompanied by reduced strength, and this may result in the concentration of extremely large inelastic deformations [Fig. 1.14(*e*) and (*f*)] in such a story. This feature accounts for the majority of collapsed buildings during recent earthquakes. Constant or gradually reducing story stiffness and strength with height [Fig. 1.14(*c*), (*d*), and (*g*)] reduce the likelihood of concentrations of plastic deformations during severe seismic events beyond the capacities of affected members.

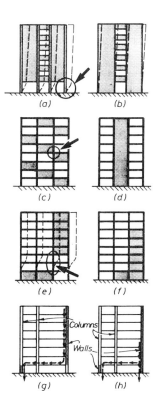

Fig. 1.15 Variation of story stiffness with height.

Examples of vertical irregularities in buildings using structural walls as primary lateral-force-resisting elements are shown in Fig. 1.15, together with suggested improvements. When a large open space is to be provided in the first story, designers are often tempted to terminate structural walls, which may extend over the full height of the building, at level 2 [Fig. 1.15(a)]. Unless other parallel walls, perhaps at the boundaries of the floor plan, are provided, a so-called *soft story* will develop. This is likely to impose ductility demands on columns which may well be beyond their ductility capacity. A continuation of the walls, interconnected by coupling beams at each floor, down to the foundations, shown in Fig. 1.15(b), will, on the other hand, result in one of the most desirable structural configurations. This system is examined in some detail in Section 5.6.

Staggered wall panels, shown in Fig. 1.15(c), may provide a stiff load path for lateral earthquake forces. However, the transmission of these forces at corners will make detailing of reinforcement, required for adequate ductility, extremely difficult. The assembly of all panels into one single cantilever [Fig. 1.15(d)] with or without interacting frames will, however, result in an excellent lateral force resisting system. The structural system of Fig. 1.15(d) is discussed in Chapter 6.

An interruption of walls over one or more intermediate stories [Fig. 1.15(e)] will invite concentrations of drift in those stories, as suggested in Fig. 1.14(a) and (e). Discontinuities of the type shown in Fig. 1.15(f), are, on the other hand, acceptable, as strength and stiffness distribution with height is compatible with the expected forces and displacements.

The side view of the structure shown in Fig. 1.15(a) may be as shown in Fig. 1.15(g). It shows that a major portion of the accumulated earthquake forces from upper levels, resulting in large shear at level 2 of the wall extending above that level, will need to be transferred to a very stiff short wall at the opposite side of the building. The arrows in Fig. 1.15(g) indicate the gross deviation of the path of internal forces leading to the foundation, which may impose excessive demands in both torsion in the first story and actions within the floor diaphragm. Both of these undesired effects will be alleviated if the tall wall terminates in the foundations [Fig. 1.15(b)], while sharing the base shear with a short wall, as shown in Fig. 1.15(h).

Another source of major damage, particularly in columns, repeatedly observed in earthquakes, is the interference with the natural deformations of members by rigid nonstructural elements, such as infill walls. As Fig. 1.16 shows, the top edge of a brick wall will reduce the effective length of one of the columns, thereby increasing its stiffness in terms of lateral forces. Since seismic forces are attracted in proportion to element stiffness, the column may thus attract larger horizontal shear forces than it would be capable of resisting. Moreover, a relatively brittle flexural failure may occur at a location (midheight) where no provision for the appropriate detailing of plastic regions would have been made. The unexpected failure of such major gravity-load-carrying elements may lead to the collapse of the entire building. Therefore, a very important task of the designer is to ensure, both in the

Fig. 1.16 Unintended interference with structural deformations.

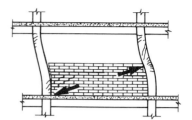

design and during construction, that intended deformations, including those of primary lateral-force-resisting components in the inelastic range of seismic response, can take place without interference.

A wide range of irregularities, together with considerations of numerous other very important issues, relevant to the overall planning of buildings and the selection of suitable structural forms within common architectural constraints are examined in depth by Arnold and Reitherman [A4]. In the context of seismic design the observance of principles relevant to configurations is at least as important as those of structural analysis, the art in detailing for ductilities of critical regions, and the assurance of high quality in workmanship during construction.

1.2.4 Structural Classification in Terms of Design Ductility Level

It is possible to satisfy the performance criteria of the damage control and survival limit states of Section 1.1.1 by any one of three distinct design approaches, related to the level of ductility permitted of the structure. A qualitative illustration of these approaches is shown in Fig. 1.17, where

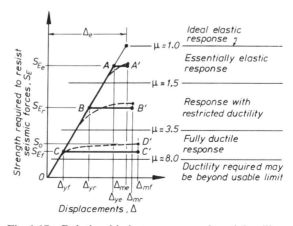

Fig. 1.17 Relationship between strength and ductility.

strength S_E, required to resist earthquake-induced forces, and structural displacements Δ at the development at different levels of strength are related to each other.

(a) Elastic Response Because of their great importance, certain buildings will need to possess adequate strength to ensure that they remain essentially elastic. Other structures, perhaps of lesser importance, may nevertheless possess a level of inherent strength such that elastic response is assured. The analysis and design of both categories of structures can be carried out with conventional procedures. Although the determination of the required resistance at each critical section will usually be based on the principles of strength design, implying that a plastic state is achieved at these sections, it is unlikely that inelastic deformations of any significance will be developed in the structure when the intensity of lateral design forces is attained. This is because specified (nominal) material strength properties and various strength reduction factors, considered in Sections 1.3.3 and 1.3.4, are used when members are being proportioned. The extra protection thus provided ensures that, at most, only insignificant inelastic deformations during an earthquake are expected. Hence the need for special detailing of potential plastic regions does not arise. Even with detailing practices established for structures in reinforced concrete and masonry, subjected only to gravity loads and wind forces, a certain amount of ductility can always be developed. As no special features arise in the design of these structures, even when subjected to design earthquake forces of the highest intensity, no further attention will be given them in this book.

The idealized response of such a structure is shown in Fig. 1.17 by the bilinear strength–displacement path OAA'. The maximum displacement Δ_{me} is very close to the displacement of the ideal elastic structure Δ_e and the displacement of the real structure Δ_{ye} at the onset of yielding.

(b) Ductile Response Most ordinary buildings are designed to resist lateral seismic forces which are smaller than those that would be developed in an elastically responding structure, implying, as Fig. 1.17 shows, that inelastic deformations and hence ductility will be required of the structure. Depending on the force level adopted for strength design, the level of ductility required may vary from insignificant, requiring no special detailing, to considerable, requiring most careful consideration of detailing. It is convenient to divide ductile responding structures into the two subcategories discussed below.

(i) Fully Ductile Structures: These are designed to possess the maximum ductility potential that can reasonably be achieved at carefully identified and detailed inelastic regions. A full consideration must be given to the effects of

dynamic response, using simplified design procedures, to ensure that nonductile modes or undesirable location of inelastic deformation cannot occur.

The idealized bilinear response of this type of structure is shown in Fig. 1.17 by the path OCC'. The magnitude of ductility implied is $\mu_f = \Delta_{mf}/\Delta_{yf}$, where Δ_{mf} and Δ_{yf} are the maximum expected and yield displacements, respectively, and S_{Ef} is the required strength of the fully ductile system. Details of these relationships are also shown in Fig. 1.8. A more realistic strength–displacement path is the curve OD', which shows that the strength S_o, developed at maximum displacement Δ_{mf}, may in fact be larger than the required strength S_{Ef}, used for design purposes. The principal aim of this book is to describe design procedures and detailing techniques for such fully ductile structures.

(ii) Structures with Restricted Ductility: Certain structures inherently possess significant strength with respect to lateral forces as a consequence, for example, of the presence of large areas of structural walls. It may well be that very little strength, if any, in addition to that obtained for the resistance of gravity loads and wind forces would need to be provided to achieve seismic resistance corresponding with elastic response. In other buildings, because of less than ideal structural configurations, it may be difficult to develop large ductilities, which would allow the use of low-intensity seismic design forces. Instead, it may be possible to provide greater resistance to lateral forces with relative ease, to reduce ductility demands. These are structures with *restricted ductility*, sometimes termed *limited ductility*.

An example of the response of a structure with restricted ductility is shown in Fig. 1.17 by the curve OBB'. It shows that the displacement ductility demand is $1 < \mu_r = \Delta_{mr}/\Delta_{yr} < \mu_f$. A more realistic response is shown by the dashed curve. Because requirements for the seismic resistance of such structures are seldom critical, the use of less onerous and simpler design procedures, combined with simpler detailing requirements for components, are recommended by some codes [A1, X3].

It should be appreciated that precise limits cannot be set for structures with full and reduced ductility. The transition from one system to the other will be gradual. Figure 1.17 shows approximate values of ductility factors $\mu = \Delta_m/\Delta_y$ which may be used as guides for the limits of the categories discussed. Although displacement ductilities in excess of 8 can be developed in some well-detailed reinforced concrete structures, the associated maximum displacements Δ_{mf} are likely to be beyond limits set by other design criteria, such as structural stability. Elastically responding structures, implying no or negligible ductility demands, represent the other limit. This suggests that a procedure, incorporating transitions in its requirements for fully ductile to those of elastically responding structures, may be used when designing structures with reduced ductility. This is examined in Chapter 8.

1.3   DEFINITION OF DESIGN QUANTITIES

1.3.1   Design Loads and Forces

In this book the following distinction between loads and forces is made:

1. Loads results from the effect of gravity. Dead loads, live loads, and snow loads are typical examples.
2. Wind, earthquakes, or restraints against deformations, such as shrinkage or creep, lead to forces. Although the term *load* is often also used to describe these forces, it will be avoided herein.

Most codes give characteristic values for design loads or forces, when these cannot be derived simply from the weights of materials to be sustained or from first principles.

(a)   Dead Loads (D)   Dead loads result from the weight of the structure and all other permanently attached materials. Average or typical values for superimposed materials and loads are readily obtained from manuals. The characteristic feature of dead loads is that they are permanent. Codes prescribe certain variable loads, such as movable partition walls, to be considered as permanent. For the sake of safety, dead loads are customarily overestimated. This fact should be considered in seismic design, whenever gravity load effects enhance strength when combined with effects of seismic forces, such as may occur when estimating moment capacity of columns.

(b)   Live Loads (L)   Live loads result from expected usage. They may be movable and their intensity may vary. Maximum intensities specified by codes are based on probablistic estimates. In most cases they are simulated by uniformly distributed loads placed over the entire area of the floor. However, for certain areas with special use, point loads may also be specified.

The probability of an area being subjected to the maximum specified intensity of live load diminishes as the size of the loaded floor area increases. Floors used for offices are typical examples. While an element of floor slab must be designed to sustain the full intensity of the live load, a beam or a column, receiving live load from a considerable tributary floor area A, may be assumed to receive live loads with smaller intensity. A typical code [X8] recommendation is that

$$L_r = rL \qquad (1.2)$$

where

$$r = 0.3 + 3/\sqrt{A} \le 1.0 \qquad (1.3)$$

and L = code specified live load, usually expressed in kPa (or psf)

L_r = reduced live load, assumed to be uniformly distributed over the area tributary to a beam, or to a column extending over the height of one of more stories

r = live-load reduction factor

A = total tributary area in m², not to be taken less than 20 m²

Only the symbol L is used in this book, whenever reference to live load is made, but this will imply that reduced live load L_r will be substituted where appropriate. Equation (1.3) is appropriate for floor slabs of office buildings, but not for storage areas, which have a high probability of being subjected to close to their full design live load over extensive areas.

Dead and live loads need also be considered when an estimate is made for the equivalent mass of a building. For the purpose of estimating seismic acceleration-induced horizontal inertia forces, it is sufficient to assume that the mass of the floor system, including finishes, partitions, beams, and columns one-half story above and below a floor, and with a fraction of the live load acting over the entire floor of a story, is concentrated at the floor level of the structural model (i.e., at the center of the mass) [Section 1.2.2(b)]. In assessing inertia forces some codes ignore some types of live loads [X10], while others [X8] specify that a certain fraction, typically one-third of the code-specified intensity, be converted into an equivalent mass.

(c) Earthquake Forces (E)   In Chapter 2 the various techniques of simulating the effects of earthquakes on structures for buildings are described. The design quantities and procedures used in this book are based on effects resulting from the application of equivalent static horizontal earthquake forces, the determination of which is given in Section 2.4.3. This technique is preferred for its simplicity, its reliability for most common regular building structures, and because most designers are familiar with its use. It will be seen, however, that the method of capacity design for ductile structures, the principal subject of this book, does not depend on the technique with which design earthquake effects have been derived.

(d) Wind Forces (W)   It was shown in Section 1.2.4 that the design intensity of the horizontal earthquake forces, duly adjusted for potential ductility capacity, may be a small fraction of that which would be generated in the elastic structure by the motions of the design earthquake. Thus it may well be, particularly in the case of tall or rather flexible buildings, that code-specified wind, rather than earthquake forces, when combined with appropriate gravity loads, will control strength requirements for many or all components of the structure. While ductility requirements do not arise, or are negligible, in the case when wind forces dominate structural strength, they are of paramount importance if the satisfactory response of the building during a strong ground motion is to be assured. The intensity of horizontal forces, corresponding with the true elastic response of the building to design

seismic excitation, may in fact be many times that of design wind force. For this reason the application of capacity design procedures is still relevant to the majority of multistory buildings, even though wind rather than earthquake forces may control strength requirements.

(e)   Other Forces   Other force effects, including those due to shrinkage, creep, and temperature, must be considered in conventional strength design. Although these have significance when considering the strength design of buildings during elastic response, they have little, if any relevance to structures responding with full or reduced ductility. This is because this category of force effects occurs as a result of small imposed displacements on the structure. The magnitude of forces induced in responding to these effects depends on the incremental stiffness. When fully ductile, or reduced ductile response is assured, the incremental stiffness for imposed displacements is negligible. The structural effect is no longer that of significant increase in force, but of insignificant decrease in ductility. For example, strains induced by thermal or shrinkage effects, when compensated for creep relaxation, will rarely exceed 0.0002. This is less than 7% of the dependable unconfined compression strain capacity of concrete and masonry, and a much smaller proportional of the compression strain capacity when confinement, as discussed in Section 3.2.2, is provided. Although the elastic compression force corresponding to a strain of 0.0002 may be as high as 20% of the compression strength, the significance for a ductile system decreases as ultimate strain capacity increases. As a consequence, it is unnecessary to consider strain-induced forces in conjunction with the ductile response of building systems, and such forces will be ignored in the following.

1.3.2   Design Combinations of Load and Force Effects

Design bending moments, shears, and axial forces are effects developed as a consequence of loads and forces in appropriate combinations. To this end an appropriate model of the real structure must be established which lends itself to rational analyses. Appropriate analytical models are discussed in the relevant chapters on structural types: in particular, Chapters 4 to 6. Loads and forces may be superimposed for structural analysis, or if the structure can reasonably be represented by elastic response, load effects may be superimposed.

Critical combinations of load and force effects to be considered for member design are based on the limit state for strength. The required strength, S_u, defined in Section 1.3.3(a), to be provided at any section of a member is thus

$$S_u = \gamma_D S_D + \gamma_L S_L + \gamma_E S_E \qquad (1.4)$$

TABLE 1.1   Commonly Used Load Factors[a]

Country	γ_D	γ_L	γ_E
United States [A1]	1.40	1.70	—
	1.05	1.28	1.40^b
	0.90	—	1.43^b
New Zealand[X8]	1.40	1.70	—
	1.00	1.30	1.00^b
	0.90	—	1.00^b
	1.00^c	1.00^c	1.00^c
	0.90^c	—	1.00^c

[a]Applicable also to effects due to reduced live load L_r [Eq. (1.1)].
[b]The intensities of specified earthquake forces in the United States and New Zealand, when factored, are similar.
[c]These factors are applicable only when actions due to earthquake effects are derived from capacity design considerations.

where S = denotes strength in general
    D, L, E = denotes the causative load or force (i.e., dead load, live load, etc.)
    $\gamma_D, \gamma_L, \gamma_E$ = specified load factors relevant to dead load, live load, and earthquake forces, respectively

Relevant codes specify values for load factors to be used for different load combinations or combinations of load effects. Some typical values, relevant to building structures studied in this book, are listed in Table 1.1. Load factors are intended to ensure adequate safety against increase in service loads beyond intensities specified, so that failure is extremely unlikely. Load factors also help to ensure that deformations at service load are not excessive.

Although load factors for dead and live loads are very similar in the United States and New Zealand, it will be seen that a substantial discrepancy exits in the treatment of the factors applicable to earthquake forces. In the United States, factors of approximately 1.40 are adopted, while in New Zealand, the appropriate factor is unity.

The original intention of load factors, when first implemented in strength design of structural elements in the 1960s, was to avoid the development of the resistance capacity of elements under maximum loads likely to occur during the building's economic life. With a seismic design philosophy based on ductility this approach is inappropriate, since development of strength, or resistance capacity, is expected under the design-level ground shaking. Applying load factors to a force level that has already been reduced from the level corresponding to elastic response merely implies a reduction to the expected ductility requirement. Unfortunately, this obscures the true level of ductility required. As a consequence, examples of seismic design developed later in the book will be based on load factors of unity for seismic forces.

Thus typical combinations [X8] of bending moments M for a beam, leading to the determination of its required flexural strength, would be

$$M_u = 1.4M_D + 1.7M_L \qquad (1.5a)$$

or

$$M_u = 1.0M_D + 1.3M_L + M_E \qquad (1.5b)$$

The design axial load P on a column would be obtained, for example, from

$$P_u = P_D + 1.3P_L + P_E \qquad (1.6a)$$

or

$$P_u = 0.9P_D + P_E \qquad (1.6b)$$

The latter combination is often critical when the seismic and gravity axial forces counteract each other. For example, a column, under compression due to dead load, may be subjected to axial tension due to earthquake forces.

Where gravity load effects are to be combined with effects resulting from the ductile response of the structure, with overstrength being developed at plastic hinges, as defined in Section 1.3.3(d) and Eq. (1.10), little if any reserve strength is necessary. Hence, where using capacity design procedures to satisfy the limit state for survival, the following combinations of actions [X3] may be used:

$$S_u = S_D + S_L + S_{E_o} \qquad (1.7a)$$

and

$$S_u = 0.9S_D + S_{E_o} \qquad (1.7b)$$

where S_{E_o} denotes an action derived from considerations of earthquake-induced overstrengths of relevant plastic regions, examined in detail in Section 1.3.3(d) and (f) and Chapters 4 and 5.

1.3.3 Strength Definitions and Relationships

Definitions of the strengths of a structure or its members, made in subsequent sections, correspond in general with the intent of most codes. In conformance with general usage, the term *strength* will be used to express the resistance of a structure, or a member, or a particular section. In terms of design practice, strength, however, is not an absolute. Material strengths and section dimensions are not known precisely but vary between probable limits. Choices of these properties should be made dependent on the purpose of application of the computed strength. The meaning of strengths developed at different levels and their relationships as used in this book are given in the following paragraphs.

(a)   Required Strength (S_u)   The strength demand arising from the application of prescribed loads and forces, in accordance with Section 1.3.2, defines the required strength, S_u. The principal aim of the design is to provide *resistance*, also termed *design strength* [A1] or *dependable strength* [X3], to meet this demand.

(b)   Ideal Strength (S_i)   The ideal or nominal strength of a section of a member, S_i, the most commonly used term, is based on established theory predicting a prescribed limit state with respect to failure of that section. It is derived from the dimensions, reinforcing content, and details of the section designed, and code-specified nominal material strength properties. The definition of nominal material strengths differs from country to country. In some cases it is a specified minimum strength, which suppliers guarantee to exceed; in others a characteristic strength is adopted, typically corresponding to the lower 5 percentile limit of measured strengths. A summary of established procedures for the determination of the ideal strengths of sections, subjected to different kinds of actions, is reviewed in Chapter 3. The ideal strength to be provided is related to the required strength by

$$\phi S_i \geq S_u \tag{1.8}$$

where ϕ is a strength reduction factor, typical values of which are given in Section 3.4.1. The designer will aim to proportion member sections so that the relationship $S_i \geq S_u/\phi$ is satisfied. Because of the necessity to round off various quantities in practice, the equality $S_i = S_u/\phi$ will seldom be achieved. Because the design philosophy, pursued in this book, relies on the hierarchy of capacities (i.e., strengths provided in various members), it is important to remember that as a general rule, the ideal strength S_i is not the optimum strength desired, but it is the nominal strength that will be provided in the construction. It will be seen that often the ideal strength of a section may well be in excess of that which is required (i.e., $S_i > S_u/\phi$).

(c)   Probable Strength (S_p)   The probable strength, S_p, takes into account the fact that material strengths, which can be utilized in a member, are generally greater than nominal strengths specified by codes. The probable strength of materials can be established from routine testing, normally conducted during construction. Alternatively, it may be based on previous experience with the relevant materials. The probable strength, or mean resistance, can be related to the ideal strength by

$$S_p = \phi_p S_i \tag{1.9}$$

where ϕ_p is the probable strength factor allowing for materials being stronger than specified, and is thus greater than 1.

Probable strengths are often used when the strength of existing structures are estimated or when time-history dynamic analysis, to predict the likely behavior of a structure when it is exposed to a selected earthquake record (Section 2.4.1), is undertaken. Developments to adopt probable strength as a basis for design, replacing the ideal strength, were known to the authors when preparing this book. Strength reduction factors would then relate dependable strength to probable rather than ideal strength.

(d) Overstrength (S_o)   The overstrength of a section, S_o, takes into account all possible factors that may contribute to strength exceeding the nominal or ideal value. These include steel strength greater than the specified yield strength, additional strength enhancement of steel due to strain hardening at large deformations, concrete of masonry strength at a given age of the structure being higher than specified, unaccounted-for compression strength enhancement of the concrete due to its confinement, and strain rate effects. The overstrength of a section can be related to the ideal strength of the same section by

$$S_o = \lambda_o S_i \tag{1.10}$$

where λ_o is the overstrength factor due to strength enhancement of the constituent materials. This is an important property that must be accounted for in the design when large ductility demands are imposed on the structure, since brittle elements must possess strengths exceeding the maximum feasible strength of ductile elements. Typical values of λ_o for both reinforcing steel and concrete are given in Sections 3.2.4(*e*). Similar strength enhancement in confined concrete is given by Eq. (3.10) and in members subjected to moment and axial compression by Eq. (3.28).

(e)   Relationships Between Strengths   Because strengths to be considered in design are most conveniently expressed in terms of the ideal strength S_i of a section, as constructed, the following simple relationships exist:

$$S_i \geq S_u/\phi \tag{1.8a}$$

$$S_p \geq \phi_p S_i \geq \phi_p S_u/\phi \tag{1.9a}$$

$$S_o \geq \lambda_o S_i \geq \lambda_o S_u/\phi \tag{1.10a}$$

For example, with typical values of $\phi = 0.9$ and $\lambda_o = 1.25$, $S_o \geq 1.39 S_u$, i.e., the overstrength of the section designed "exactly" to match the required strength would be 39% larger than that required by Eq. (1.4).

(f)   Flexural Overstrength Factor ϕ_o   To quantify the hierarchy of strength in the design of ductile structures, it is convenient to express the overstrength of a member in flexure $S_o = M_o$ at a specific section, such as a node point of

the analytical model, in terms of the required flexural strength $S_E = M_E$ at the same section, derived by an elastic analysis for earthquake forces alone. The ratio so formed,

$$\phi_o = S_o/S_E = M_o/M_E \tag{1.11}$$

is defined as the flexural overstrength factor. When the two factors ϕ_o and λ_o, given by Eq. (1.10), are compared, it should be noted that apart from the overstrengths of materials, the following additional sources of flexural overstrength are also included in Eq. (1.11):

1. The strength reduction factor ϕ [Eq. (1.8)] used to relate ideal to required strength
2. More severe strength requirements, if any, due to gravity loads and wind forces
3. Changes in design moments due to any redistribution of these, which the designer may have undertaken (see Section 4.3)
4. Deviations from the optimum ideal strength due to the choice of the amount of reinforcement as dictated by practicality (availability of bar sizes and numbers)

Moreover, the overstrength factor, λ_o, is relevant to the critical section of a potential plastic hinge which may be located anywhere along a member, while the flexural overstrength factor, ϕ_o, expresses strengths ratios at node points. Where the critical section coincides with a node point, as may be the case of a cantilever member, such as a structural wall resisting seismic forces, the relationships from Eqs. (1.8), (1.10), and (1.11) reveal that $M_o = \lambda_o M_i = \phi_o M_E$ and hence that

$$\phi_o = \frac{M_o}{M_E} = \frac{\lambda_o M_i}{M_E} \gtrless \frac{\lambda_o (M_E/\phi)}{M_E} = \frac{\lambda_o}{\phi} \tag{1.12}$$

An equality means that the dependable strength ϕM_i provided, for example, at the base of a cantilever wall is exactly that (M_E) required to resist seismic forces. With typical values of $\lambda_o = 1.25$ and $\phi = 0.9$, the flexural overstrength factor in this case becomes $\phi_o = 1.39$. Values of ϕ_o larger than λ_o/ϕ indicate that the dependable flexural strength of the base section of the wall is in excess of required strength (i.e., the overturning moment M_E resulting from design earthquake forces only). When $\phi_o < \lambda_o/\phi$, a deficiency of required strength is indicated, the sources of which should be identified.

Any of, or the combined sources (2) to (4) of overstrength listed above, as well as computational errors, may be causes of ϕ_o being more or less than the ratio λ_o/ϕ. The flexural overstrength factor ϕ_o is thus a very convenient parameter in the application of capacity design procedures. It is a useful indicator, evaluated as the design of members progresses, to measure first, the extent to which gravity loads or earthquakes forces dominate strength requirements, and second, the relative magnitudes of any over- or under-design by choice or as a result of an error made.

(g) System Overstrength Factor ψ_o The flexural overstrength factor ϕ_o measures the flexural overstrength in terms of the required strength for *earthquake forces alone* at one node point of the structural model. In certain situations it is equally important to compare the sum of the overstrengths of a number of interrelated members with the total demand made on the same members by the specified earthquake forces alone. For example, the sum of the flexural overstrengths of all column sections of a framed building at the bottom and the top of a story may be compared with the total story moment demand due to the total story shear force, such as V_j in Fig. 1.9.

The system or overall overstrength factor, ψ_o, may then be defined:

$$\psi_o = \frac{\Sigma S_o}{\Sigma S_E} = \frac{\Sigma \phi_o S_E}{\Sigma S_E} \tag{1.13}$$

As an example, consider a multistory building in which the entire seismic resistance in a particular direction is provided by n reinforced concrete structural walls. The relevant material overstrength and strength reduction factors in this example are $\lambda_o = 1.4$ and $\phi = 0.9$. The flexural overstrength factor, relevant to the base moments of each of these walls, may well be more or less than $\phi_o = \lambda_o/\phi = 1.56$ [Eq. (1.12)]. Deviations would indicate that some walls have been designed for larger or smaller moments than required for seismic resistance (i.e., $M_i < M_E/\phi$). However, if the value of the system overstrength factor

$$\psi_o = \sum_i^n M_o / \sum_i^n M_E = \sum_i^n (\phi_o M_E) / \sum_i^n M_E$$

is less than 1.56, this indicates that the strength requirement for the structure as a whole, to resist seismic design forces, has been violated. On the other hand, a value much larger than 1.56 in this example will warn the designer that for reasons which should be identified, strength well in excess of that required to resist the specified earthquake forces has been provided. This is of importance when the design of the foundation structure, examined in Chapter 9, is to be considered.

1.3.4 Strength Reduction Factors

Strength reduction factors ϕ, introduced in Section 1.3.3(b), are provided in codes [A1, X3] to allow approximations in the calculations and variations in material strengths, workmanship, and dimensions. In addition, consideration has been given to the seriousness and consequences of failure of a member in respect to the whole structure and the degree of warning involved in the mode of failure [P1].

Thus the overall safety factor for a structure, subjected to dead and live loads only, from Eqs. (1.4) and (1.8) may be expressed by the ratio

$$\frac{S_i}{S_D + S_L} \geq \frac{S_u}{\phi(S_D + S_L)} = \frac{\gamma_D S_D + \gamma_L S_L}{\phi(S_D + S_L)} \tag{1.14}$$

For example, with $L = D$ and $\gamma_D = 1.4$, $\gamma_L = 1.7$ and $\phi = 0.9$, the overall safety factor with respect to the ideal strength S_i to be reached, is 1.72. Commonly used values of strength reduction factors are given in Section 3.4.1.

In some recent codes, strength reduction factors, applicable to a specific action such as flexure or shear, have been replaced by resistance factors with values specified for each of the constituent materials, such as concrete and different types of steels [X5]. Values of load and load combinations factors have been adjusted accordingly. With the use of resistance factors, relationships between ideal, required, probable and overstrength are similar to those described in the previous sections. However, instead of using the equations given here, these relationships have to be established for each case using elementary first principles.

1.4 PHILOSOPHY OF CAPACITY DESIGN

1.4.1 Main Features

Procedures for the application of capacity design to ductile structures, which may be subjected to large earthquakes, have been developed primarily in New Zealand over the last 20 years [P1, P3, P4, P6, P17], where they have been used extensively [X3, X8]. With some modification the philosophy has also been adopted in other countries [A5, C11, X5]. However, for specific situations, the application of capacity design principles was already implied in earlier editions of some codes [X4, X10]. The seismic design strategy adopted in this book is based on this philosophy. It is a rational, deterministic, and relatively simple approach.

In the capacity design of structures for earthquake resistance, distinct elements of the primary lateral force resisting system are chosen and suitably designed and detailed for energy dissipation under severe imposed deformations. The critical regions of these members, often termed *plastic hinges*, are detailed for inelastic flexural action, and shear failure is inhibited by a suitable strength differential. All other structural elements are then protected against actions that could cause failure, by providing them with strength greater than that corresponding to development of maximum feasible strength in the potential plastic hinge regions.

It must be recognized that in an element subjected to full or reduced ductility demands, the strength developed is considerably less than that corresponding to elastic response, as shown, for example, in Fig. 1.17. It follows that it is the actual strength, not the nominal or ideal strengths, that will be developed, and at maximum displacement, overstrength S_o response is expected. Nonductile elements, resisting actions originating from plastic hinges, must thus be designed for strength based on the overstrength S_o rather than the code-specified strength S_u, which is used for determining required dependable strengths of hinge regions. This "capacity" design procedure ensures that the chosen means of energy dissipation can be maintained.

The following features characterize the procedure:

1. Potential plastic hinge regions within the structure are clearly defined. These are designed to have dependable flexural strengths as close as practicable to the required strength S_u. Subsequently, these regions are carefully detailed to ensure that estimated ductility demands in these regions can be reliably accommodated. This is achieved primarily by close-spaced and well-anchored transverse reinforcement.

2. Undesirable modes of inelastic deformation, such as may originate from shear or anchorage failures and instability, within members containing plastic hinges, are inhibited by ensuring that the strengths of these modes exceeds the capacity of the plastic hinges at overstrength.

3. Potentially brittle regions, or those components not suited for stable energy dissipation, are protected by ensuring that their strength exceeds the demands originating from the overstrength of the plastic hinges. Therefore, these regions are designed to remain elastic irrespective of the intensity of the ground shaking or the magnitudes of inelastic deformations that may occur. This approach enables traditional or conventional detailing of these elements, such as used for structures designed to resist only gravity loads and wind forces, to be employed during construction.

An example illustrating the application of these concepts to a simple structure is given in Section 1.4.4.

As discussed in Chapter 2, the intensity of design earthquake forces and structural actions which result from these are rather crude estimates, irrespective of the degree of sophistication on which analyses may be based. Provided that the intended lateral force resistance of the structure is assured, approximations in both analysis and design can be used, within reason, without affecting in any way the seismic performance of the structure. The area of greatest uncertainty of response of capacity-designed structures is the level of inelastic deformations that might occur under strong ground motion. However, the high quality of the detailing of potential plastic regions, the subject addressed in a considerable part of this book, will ensure that significant variations in ductility demands from the expected value can be accommodated without loss of resistance to lateral forces. Hence capacity-designed ductile structures are extremely tolerant with respect to imposed seismic deformations.

It is emphasized that capacity design is not an analysis technique but a powerful design tool. It enables the designer to "tell the structure what to do" and to desensitize it to the characteristics of the earthquake, which are, after all, unknown. Subsequent judicious detailing of all potential plastic regions will enable the structure to fulfill the designer's intentions.

1.4.2 Illustrative Analogy

To highlight the simple concepts of capacity design philosophy, the chain shown in Fig. 1.18 will be considered. Using the adage that the strength of a chain is the strength of its weakest link, a very ductile link may be used to achieve adequate ductility for the entire chain. The ideal or nominal tensile strength [Section 1.3.3(a)] of this ductile steel link is P_i, but the actual strength is subject to the normal uncertainties of material strength and strain hardening effects at high strains. The other links are presumed to be brittle. Note that if they were designed to have the same nominal strength as the ductile link, the randomness of strength variation between all links, including the ductile link, would imply a high probability that failure would occur in a brittle link and the chain would have no ductility. Failure of all other links

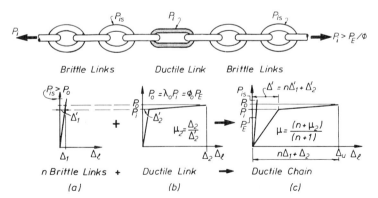

Fig. 1.18 Principle of strength limitation illustrated with ductile chain.

can, however, be prevented if their strength is in excess of the maximum feasible strength of the weak link, corresponding to the level of ductility envisaged. Using the terminology defined in Section 1.3.3, the dependable strength ϕP_{is} of the strong links should therefore not be less than the overstrength P_o of the ductile link $\lambda_o P_i$. As ductility demands on the strong links do not arise, they may be brittle (cast iron).

The chain is to be designed to carry an earthquake-induced tensile force $P_u = P_E$. Hence the ideal strength of the weak link needs to be $P_i \geq P_E/\phi$ [Eq. (1.8)]. Having chosen an appropriate ductile link, its overstrength can be readily established (i.e., $P_o = \lambda_o P_i = \phi_o P_E$) [Eqs. (1.10) and (1.11)], which becomes the design force P_{us}, and hence required strength, for the strong and brittle links. Therefore, the ideal or nominal strength of the strong link needs to be

$$P_{is} > P_{us}/\phi_s = P_o/\phi_s = \phi_o P_E/\phi_s$$

where quantities with subscript s refer to the strong links. For example, when $\phi = 0.9$, $\lambda_o = 1.3$, and $\phi_s = 1.0$, we find that $P_i > 1.11 P_E$ and $P_{is} \geq 1.3 (1.11 P_E)/1.0 = 1.44 P_E$.

The example of Fig. 1.18 may also be used to draw attention to an important relationship between the ductility potential of the entire chain and the corresponding ductility demand of the single ductile link. Linear and bilinear force–elongation relationships, as shown in Fig. 1.18(b) and (c), are assumed for all links. Inelastic elongations can develop in the ductile link only. As Fig. 1.18 shows, elongations at the onset of yielding of the brittle and ductile links are Δ_1' and Δ_2', respectively. Subsequent and significant yielding of the weak link will increase its elongation from Δ_2' and Δ_2, while its resistance increases from $P_y = P_i$ to P_o due to strain hardening. The weak link will thus exhibit a ductility of $\mu_2 = \Delta_2/\Delta_2'$. As Fig. 1.18(c) shows, the total elongation of the chain, comprising the weak link and n strong links, at the onset of yielding in the weak link will be $\Delta' = n\Delta_1' + \Delta_2'$. At the development of the overstrength of the chain (i.e., that of the weak link), the elongation of the strong link will increase only slightly from Δ_1' to Δ_1. Thus at ultimate the elongation of the entire chain becomes $\Delta_u = n\Delta_1 + \Delta_2$. As Fig. 1.18 shows, the ductility of the chain is then

$$\mu = \Delta_u/\Delta' = (n\Delta_1 + \Delta_2)/(n\Delta_1' + \Delta_2')$$

With the approximation that $\Delta_1 \simeq \Delta_1' \simeq \Delta_2' = \Delta_y$, it is found that the relationship between the ductility of the chain μ and that of weak link is μ_2 is

$$\mu = (n + \mu_2)/(n + 1)$$

If the example chain of Fig. 1.18 consists of eight strong links and the maximum elongation of the weak link Δ_2 is to be limited to 10 times its elongation at yield, Δ'_2 (i.e., $\mu_2 = 10$), we find that the ductility of the chain is limited to $\mu = (8 + 10)/(8 + 1) = 2$. Conversely, if the chain is expected to develop a ductility of $\mu = 3$, the ductility demand on the weak link will increase to $\mu_2 = 19$. The example was used to illustrate the very large differences in the magnitudes of overall ductilities and local ductilities that may occur in certain types of structures. In some structures the overall ductility to be considered in the design will need to be limited to ensure that ductility demands at a critical locality do not become excessive. Specific ductility relationships are reviewed in Section 3.4.

1.4.3   Capacity Design of Structures

The principles outlined for the example chain in Fig. 1.18 can be extended to encompass the more complex design of a large structure (e.g., a multistory building). The procedure uses the following major steps:

1. A kinematically admissible plastic mechanism is chosen.
2. The mechanism chosen should be such that the necessary overall displacement ductility can be developed with the smallest inelastic rotation demands in the plastic hinges (Fig. 1.19a).
3. Once a suitable plastic mechanism is selected, the regions for energy dissipation (i.e., plastic hinges) are determined with a relatively high degree of precision.
4. Parts of a structure intended to remain elastic in all events are designed so that under maximum feasible actions corresponding to overstrength in the plastic hinges, no inelastic deformations should occur in those regions. Therefore, it is immaterial whether the failure of regions, intended to remain elastic, would be ductile or brittle (Fig. 1.8). The actions originating from plastic hinges are those associated with the overstrength [Section 1.3.3(d)] of these regions. The required strength of all other regions [Section 1.3.3(b)] is then in excess of the strength demand corresponding to the overstrength of relevant plastic hinges.

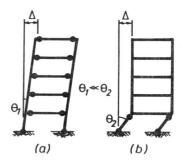

Fig. 1.19 Comparison of energy-dissipating mechanisms.   *(a)*   *(b)*

5. A clear distinction is made with respect to the nature and quality of detailing for potentially plastic regions and those which are to remain elastic in all events.

A comparison of the two example frames in Fig. 1.19 shows that for the same maximum displacement Δ at roof level, plastic hinge rotations θ_1 in case (a) are much smaller than those in case (b), θ_2. Therefore, the overall ductility demand, in terms of the large deflection Δ, is much more readily achieved when plastic hinges develop in all the beams instead of only in the first-story column. The column hinge mechanism, shown in Fig. 1.19(b), also referred to as a soft-story, may impose plastic hinge rotations, which even with good detailing of the affected regions, would be difficult to accommodate. This mechanism accounts for numerous collapses of framed buildings in recent earthquakes. In the case of the example given in Fig. 1.19, the primary aim of capacity design will be to prohibit formation of a soft story and, as a corollary, to ensure that only the mechanism shown in Fig. 1.19(a), can develop.

A capacity design approach is likely to assure predictable and satisfactory inelastic response under conditions for which even sophisticated dynamic analyses techniques can yield no more than crude estimates. This is because the capacity-designed structure cannot develop undesirable hinge mechanisms or modes of inelastic deformation, and is, as a consequence, insensitive to the earthquake characteristics, except insofar as the magnitude of inelastic flexural deformations are concerned. When combined with appropriate detailing for ductility, capacity design will enable optimum energy dissipation by rationally selected plastic mechanisms to be achieved. Moreover, as stated earlier, structures so designed will be extremely tolerant with respect to the magnitudes of ductility demands that future large earthquakes might impose.

1.4.4 Illustrative Example

The structure of Fig. 1.20(a) consists of two reinforced concrete portal frames connected by a monolithic slab. The slab supports computer equipment that applies a gravity load of W to each portal as shown in Fig. 1.20(b). The computer equipment is located at level 2 of a large light industrial two-story building whose design strength, with respect to lateral forces, is governed by wind forces. As a consequence of the required strength for wind forces, the building as a whole can sustain lateral accelerations of at least $0.6g$ elastically (g is the acceleration due to gravity). However, the computer equipment, which must remain functional even after a major earthquake, cannot sustain lateral accelerations greater than $0.35g$. As a consequence, the equipment is supported by the structure of Fig. 1.20(a), which itself is isolated from the rest of the building by adequate structural separation. The design requirements are that the beams should remain elastic and that the

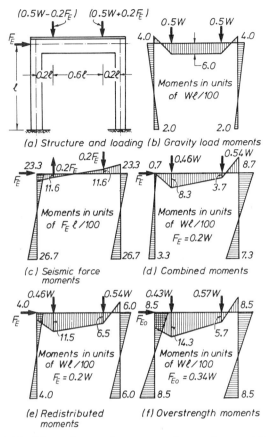

Fig. 1.20   Dimensions and moments of example portal frame.

columns should provide the required ductility by plastic hinging. Columns are to be symmetrically reinforced over the full height.

Elastic response for the platform under the design earthquake is estimated to be 1.0g, and a dependable ductility capacity of $\mu_\Delta = 5$ is assessed to be reasonable. Correspondingly, the design seismic force at level 2 is based on 0.2g lateral acceleration. Design combinations of loads and forces are

$$U \geq 1.4D + 1.7L$$
$$U \geq D + 1.3L + E$$

(i) Gravity Load: The computer is permanent equipment of accurately known weight, so is considered as superimposed dead load. The weight of the platform and the beam and live load are small in comparison with the weight of the equipment, and hence for design purposes the total gravity load W,

corresponding to $U = D + 1.3L$, can be simulated as shown in Fig. 1.20(b), where the induced moments are also shown.

(ii)   Seismic Forces: The center of mass of the computer is some distance above the platform; hence vertical seismic forces of $\pm 0.2 F_E$ as well as the lateral force F_E are induced, as shown in Fig. 1.20(a) and (c). Moments induced by this system of forces are shown in Fig. 1.20(c).

(iii)   Combined Actions: Figure 1.20(d) shows the moments under W and F_E, where the seismic moments are based on the ductile design level of $F = 0.2W$ (i.e., $0.2g$ lateral acceleration) and W is based on $D + 1.3L$. It will be noted that the bending moments generated in the two columns are very different. If column strength is based on the computed moments for the right-hand column, and if the influence of variable axial loads between the two columns is considered to be small enough to ignore, the dependable shear strength of each column will be $V = (0.073 \, Wl + 0.087Wl)/l = 0.16W$, since both columns will have the same reinforcement. (The seismic forces could, of course, act from right to left, making the left-hand column critical.)

The moment pattern of Fig. 1.20(d) corresponds to the elastic distribution of moments applicable at the yield displacement Δ_y. Since the design ductility factor is $\mu_\Delta = 5$, the structure is expected to deform to five times this displacement. Under the large displacements, the left-hand column could also develop its strength at top and bottom, and both columns could be expected to develop overstrength moments. Thus if a strength reduction factor of $\phi = 0.9$ was used and $\lambda_o = 1.25$ is applicable at overstrength, the maximum horizontal force F_o that could be sustained would be, from Eqs. (1.8) and (1.10),

$$F_o = (1/0.9)1.25(2 \times 0.160)W = 0.444W$$

corresponding to a lateral acceleration of $0.444g$. This exceeds the permissible limit of $0.35g$.

(iv)   Redistribution of Design Forces: A solution is to redistribute the moments in Fig. 1.20(d) so as to reduce the shear developed in the right column and increase the shear carried by the left column, as shown in Fig. 1.20(e). Here the maximum moments at the top and bottom of the right column have been decreased to $0.060Wl$, and those of the left column have been increased to $0.040Wl$. The total lateral force carried at this stage by the portal, which is the sum of the shear forces in the two columns, is $0.2W$. It would be possible to make all column end moments equal at $0.05Wl$, but that would make the column top moment significantly less than that required for gravity load alone ($U \geq 1.4D$; $M_u \geq 1.4 \times 0.04Wl = 0.056Wl$). Also, to reduce elastic design moments by 43% might be considered excessive.

(v) Evaluation of Overstrengths: After selection of suitable reinforcement, the dependable column moment strengths are found to be, for both columns,

$$\phi M_i = 0.061Wl$$

The dependable lateral strength of the frame is thus

$$\phi F_i = (4 \times 0.061)Wl/l = 0.244W = 1.22F_E$$

The column overstrength moment capacity is, from Eq. (1.10), $M_o = \lambda_o M_i = 1.25 \times 0.0061Wl/\phi = 0.085Wl$, and the system overstrength factor is, from Eq. (1.13) and Fig. 1.20(f) and (c), $\psi_o = (4 \times 0.085)/[2 \times (0.233 + 0.267) \times 0.2] = 1.70$. Thus the maximum feasible strength will correspond to $1.7 \times 0.2g = 0.34g$, which is less than the design limit of $0.35g$. Moments at overstrength response are shown in Fig. 1.20(f).

To ensure ductile response ($\mu_\Delta \simeq 1.0/0.244 = 4.1$ required), shear failure must be avoided. Consequently, the required shear strength of each column must be based on flexural overstrength, that is,

$$V_u = 2 \times 0.085Wl/l = 0.17W = 0.5\psi_o F_E$$

The beam supporting the computer platform must be designed for a maximum required positive moment capacity of $0.143Wl$ to ensure that the platform remains elastic. Note that this is 1.72 times the initial value given in Fig. 1.20(d). In a more realistic example, however, it might be necessary to design for even larger flexural strength because of higher mode response, corresponding to the computer equipment "bouncing" on the vertically flexible beam.

2 Causes and Effects of Earthquakes: Seismicity → Structural Response → Seismic Action

2.1 ASPECTS OF SEISMICITY

2.1.1 Introduction: Causes and Effects

Although it is beyond the scope of this work to discuss in detail seismicity and basic structural dynamics, a brief review of salient features is warranted in order to provide a basis for assessing seismic risk and for estimating structural response. Earthquakes may result from a number of natural and human-induced phenomena, including meteoric impact, volcanic activity, underground nuclear explosion, and rock stress changes induced by the filling of large human-made reservoirs. However, the vast majority of damaging earthquakes originate at, or adjacent to, the boundaries of crustal tectonic plates, due to relative deformations at the boundaries. Because of the nature of the rough interface between adjacent plates, a stick–slip phenomenon generally occurs, rather than smooth continuous relative deformation, or creep. The relative deformation at the adjacent plates is resisted at the rough interface by friction, inducing shear stresses in the plates adjacent to the boundary. When the induced stresses exceed the frictional capacity of the interface, or the inherent material strength, slip occurs, releasing the elastic energy stored in the rock primarily in the form of shock waves propagating through the medium at the ground-wave velocity.

Relative deformations in the vicinity of the plate boundaries may reach several meters before faulting occurs, resulting in substantial physical expression of the earthquake activity at the ground surface in the form of fault traces with considerable horizontal or vertical offsets. Clearly, structures built on a foundation within which faulting occurs can be subjected to extreme physical distress. However, it has been noted that buildings constructed on strong integral foundation structures, such as rafts or footings interconnected by basement walls, cause the fault trace to deviate around the boundaries rather than through a strong foundation. The problem of structural dislocation caused by relative ground movement at a fault trace is potentially more serious for bridges or for low-rise buildings of considerable length where the footings of adjacent supports may be unconnected. Although physical ground dislocation is the most immediately apparent structural threat, it affects only a very restricted surface area and hence does not generally constitute significant seismic risk. Of much greater significance is the inertial response of structures to the ground accelerations resulting from the energy released during fault slip, and it is this aspect that is of primary interest to the structural designer.

Typically, the boundaries between plates do not consist of simple single-fault surfaces. Frequently, the relative movement is spread between a number of essentially parallel faults, and earthquakes may occur not only along these faults, but along faults transverse to the plate boundaries, formed by the high shearing strains and deformations in the plate boundaries. Figure 2.1 shows a distribution of the known fault lines capable of generating significant earthquakes in southern California. It is worth noting that even with the intensive mapping efforts that have taken place recently, such maps are incomplete, with new faults being discovered continuously, often only after an unexpected earthquake makes their presence abundantly obvious, as

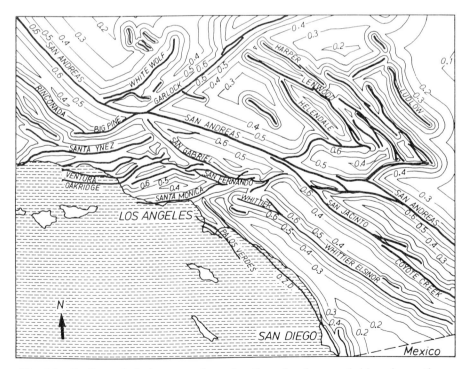

Fig. 2.1 Faults and design ground accelerations for freeway bridges in southern California [X14].

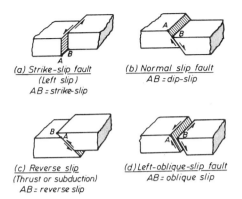

(a) Strike-slip fault
(Left slip)
AB = strike-slip

(b) Normal slip fault
AB = dip-slip

(c) Reverse slip
(Thrust or subduction)
AB = reverse slip

(d) Left-oblique-slip fault
AB = oblique slip

Fig. 2.2   Categories of fault movement.

with the 1987 Whittier earthquake, which has lead to the discovery of several previously unknown significant faults in the Los Angeles basin area. It has been said, somewhat facetiously, that the extent of detail in seismic zoning maps is more a reflection of the density of geologists in the area than of actual fault locations. This point is worth noting when construction is planned in an undeveloped region with apparently few faults.

Figure 2.2 describes the basic categories of fault movement. Strike-slip faults [Fig. 2.2(a)] display primarily lateral movement, with the direction of movement identified as left-slip or right-slip, depending on the direction of movement of one side of the fault as viewed from the other side. Note that this direction of movement is independent of which side is chosen as the reference. Normal-slip faults [Fig. 2.2(b)] display movement normal to the fault, but no lateral relative displacement. The movement is associated with extension of distance between points on opposite sides of the fault, and hence the term *tension fault* is sometimes used to characterize this type of movement. Reverse-slip faults [Fig. 2.2(c)] also involve normal movement, but with compression between points on opposite sides of the fault. These are sometimes termed *thrust* or *subduction faults*. Generally, fault movement is a combination of strike and normal components, involving oblique movement, resulting in compound names as illustrated, for example, in Fig. 2.2(d).

Rates of average relative displacement along faults can vary from a few millimeters a year to a maximum of about 100 mm/year (4 in./year). The magnitude of dislocation caused by an earthquake may be from less than 100 mm (4 in.) up to several meters, with 10 m (33 ft) being an approximate upper bound.

The major faults shown in Fig. 2.1 are characteristic of a strike-slip system. In some regions of plate boundaries, subduction occurs, generally at an angle acute to the ground surface, as is the case with the great Chilean earthquakes, where the Nazca plate is subducting under the South American plate, as shown in Fig. 2.3. Subducting plate boundaries are thought to be

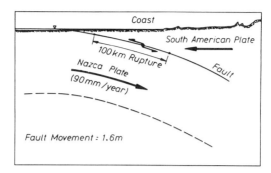

Fig. 2.3   Origin of Chilean earthquake of March 3, 1985.

capable of generating larger earthquakes than plate boundaries with essentially lateral deformation, and appear to subject larger surface areas to strong ground motion. Conversely, normal-slip movements are thought to generate less intense shaking because the tensile force component across the fault implies lower stress drop associated with fracture.

Despite the clear preponderance of earthquakes associated with plate boundaries, earthquakes can occur, within plates, at considerable distance from boundaries with devastating effects. The largest earthquake in the contiguous 48 states of the United State in recorded history did not occur in California but in New Madrid, Illinois, in 1811. Frequently, the lack of recent earthquake activity in an area results in a false sense of security and a tendency to ignore seismic effects in building design. Although the annual risk of significant earthquake activity may be low in intraplate regions, the consequences can be disastrous, and should be assessed, particularly for important or hazardous structures.

2.1.2 Seismic Waves

The rupture point within the earth's crust represents the source of emission of energy. It is variously known as the *hypocenter*, *focus*, or *source*. For a small earthquake, it is reasonable to consider the hypocenter as a point source, but for very large earthquakes, where rupture may occur over hundreds or even thousands of square kilometers of fault surface, a point surface does not adequately represent the rupture zone. In such cases the hypocenter is generally taken as that point where rupture first initiated, since the rupture requires a finite time to spread over the entire fracture surface.

The *epicenter* is the point on the earth's surface immediately above the hypocenter, and the *focal depth* is the depth of the hypocenter below the epicenter. *Focal distance* is the distance from the hypocenter to a given reference point.

The energy released by earthquakes is propagated by different types of waves. Body waves, originating at the rupture zone, include P waves (primary or dilatation waves), which involve particle movement parallel to the direction of propagation of the wave, and S waves (secondary or shear waves), which involve particle movement perpendicular to the direction of propagation.

When body waves reach the ground surface they are reflected, but also generate surface waves which include Rayleigh and Love waves (R and L waves). Love waves produce horizontal motion transverse to the direction of propagation; Rayleigh waves produce a circular motion analogous to the motion of ocean waves. In both cases the amplitude of these waves reduces with depth from the surface.

P and S waves have different characteristic velocities v_p and v_s. For an elastic medium these velocities are frequency independent and in the ratio $v_p/v_s \approx \sqrt{3}$. As a consequence, the time interval ΔT between the arrival of P and S waves at a given site is thus proportional to the focal distance x_f. Hence

$$x_f = v_p \Delta T / (\sqrt{3} - 1) \tag{2.1}$$

Recordings of the P–S time interval at three or more noncollinear sites thus enables the epicentral position to be estimated, as shown in Fig. 2.4. Generally, sites at substantial distance are chosen so the epicentral and hypocentral distances are essentially identical.

As distance from the epicenter increases, the duration of shaking at a given site increases and becomes more complex, as illustrated in Fig. 2.5. This is because of the increase in time between the arrival of P and S waves, and also due to scattering effects resulting from reflection of P and S waves from the surface. In Fig. 2.5, PP and PPP refer to the first and second reflections of seismic waves at the surface. As noted above, the incident P and S waves at the surface also induce surface waves. These travel at speeds

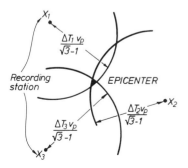

Fig. 2.4   Locating epicenter from P–S time intervals ΔT at three noncollinear recording stations.

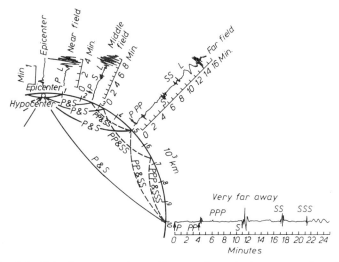

Fig. 2.5   Seismic waves at large distances from the hypocenter [M16].

that are frequency dependent, thus further confusing the motion at distance from the epicenter.

The above is a very brief description of seismic wave motion. The interested reader is referred elsewhere for more complete coverage [R1, R4].

2.1.3   Earthquake Magnitude and Intensity

Earthquake magnitude is a measure of the energy released during the earthquake and hence defines the size of the seismic event. Intensity is a subject assessment of the effect of the earthquake at a given location and is not directly related to magnitude.

(a) Magnitude   The accepted measure of magnitude is the Richter scale [R1]. The magnitude is related to the maximum trace deformation of the surface-wave portion of seismograms recorded by a standard Wood–Anderson seismograph at a distance of 100 km from the epicenter. As such, it can be sensitive to the focal depth of the earthquake, and magnitudes computed from the body wave portions of seismograms are often used to refine estimates of the magnitude. However, the result is generally converted back to equivalent Richter magnitude for reporting purposes. The accepted relationship between energy released, E, and Richter magnitude, M, is

$$\log_{10} E = 11.4 + 1.5M \tag{2.2}$$

where E is in ergs.

Earthquakes of Richter magnitude less than 5 rarely cause significant structural damage, particularly when deep seated. Earthquakes in the $M5$ to $M6$ range can cause damage close to the epicenter. A recent example is the 1986, magnitude $M5.4$, San Salvador earthquake, which was located at a depth of 7 km below the city [X14] and caused damage estimated at U.S. $1.5 billion. The surface area subjected to strong ground shaking was approximately 100 km^2 and corresponded closely to the city limits.

In the $M6$ to $M7$ range, the area of potential damage is considerably larger. The 1971 San Fernando earthquake ($M6.4$) caused structural damage over an area of approximately 2000 km^2. In the large $M7$ to $M8$ range, structural damage may occur over an area up to 10,000 km^2. Recent examples are the Tangshan earthquake (China, 1976, $M7+$) [B21], which destroyed the city and left more than 250,000 people dead, and the Chilean earthquake of 1985 ($M7.8$) [X1]. Earthquakes of magnitude $M8$ or greater, often termed *great earthquakes*, are capable of causing widespread structural damage over areas greater than 100,000 km^2. The Alaskan earthquake of 1964 ($M8+$) [B22] and the Chilean earthquake of 1960 ($M8+$) [X1], each of which caused widespread damage to engineered structures, are in this category.

The logarithmic scale of Eq. (2.2) implies that for each unit increase in the Richter magnitude, the energy released increases by $10^{1.5}$. Thus a magnitude $M8$ earthquake releases 1000 times the energy of an $M6$ earthquake. Primarily, the increased energy of larger earthquakes comes from an increase in the fault surface area over which slip occurs. A magnitude $5+$ earthquake may result from fault movement over a length of a few kilometers, while a magnitude 8 event will have fault movement over a length as much as 400 km (250 miles), with corresponding increase in the fault surface area. Other factors influencing the amount of energy released include the stress drop in the rock adjacent to fault slip. The seismic moment is a measure of the earthquake size based on integration of the parameters above, and can be related to Richter magnitude [R1].

A secondary but important effect of the increased size of the fault surface of large earthquakes is the duration of strong ground shaking. In a moderate earthquake the source may reasonably be considered as a point source, and the duration may be only a few seconds. In a large earthquake, shock waves reach a given site from parts of the fault surface which may be hundreds of kilometers apart. The arrival times of the shock waves will clearly differ, extending the duration of shaking.

(b) ***Intensity*** Earthquake intensity is a subjective estimate of the perceived local effects of an earthquake and is dependent on peak acceleration, velocity, and duration. The most widely used scale is the modified Mercalli scale, MM, which was originally developed by Mercalli in 1902, modified by Wood and Neumann in 1931, and refined by Richter in 1958 [R1]. The effective range is from $MM2$, which is felt by persons at rest on upper floors

of buildings, to *MM* 12, where damage is nearly total. A listing of the complete scale is given in Appendix B. As a measure of structural damage potential the value of this scale has diminished over the years, as it is strongly related to the performance of unreinforced masonry structures (Appendix B). The expected performance of well-designed modern buildings, of masonry or other materials, cannot be directly related to modified Mercalli intensity. However, it is still of value as a means for recording seismic effects in regions where instrumental values for ground shaking are sparse or nonexistent.

It is important to realize that the relationship between maximum intensity (whether assessed by subjective methods such as *MM* intensity, or measured by peak effective ground acceleration) and size or magnitude is at best tenuous and probably nonexistent. A shallow seated magnitude 5 + earthquake may induce local peak ground accelerations almost as high as those occurring during a magnitude 8 + earthquake, despite the $10^{4.5}$ difference in energy release. For example, peak horizontal ground accelerations recorded in both the 1986 *M* 5.4 San Salvador earthquake and the 1985 *M* 7.8 Chilean earthquake were approximately $0.7g$.

The most important differences between moderate and large (or great) earthquakes are thus the area subjected to, and the duration of, strong ground motions. A secondary difference is the frequency composition of the ground motion, with accelerograms [Section 2.1.4(*a*)] from large earthquakes typically being richer in long-period components.

2.1.4 Characteristics of Earthquake Accelerograms

(*a*) *Accelerograms* Our understanding of seismically induced forces and deformations in structures has developed, to a considerable extent, as a consequence of earthquake accelerograms recorded by strong-motion accelerographs. These accelerographs record ground acceleration in optical or digital form as a time-history record. When mounted in upper floors of buildings, they record the structural response to the earthquake and provide means for assessing the accuracy of analytical models in predicting seismic response. Integration of the records enables velocities and displacements to be estimated. Although many useful data have recently been recorded in major earthquakes, there is still a paucity of information about the characteristics of strong ground motion, particularly for large or great earthquakes. Of particular concern is the lack of definitive information on attenuation of shaking with distance from the epicenter.

As new data are recorded, seismologists are constantly revising estimates of seismic characteristics. For example, earlier theoretical predictions that peak ground acceleration could not exceed $0.5g$, which were widely accepted in the 1960s and early 1970s, have recently been consistently proven to be low. Several accelerograms with peak acceleration components exceeding $1.0g$ have now been recorded. Figures 2.6 and 2.7 show examples of accelerograms recorded during the 1986 San Salvador earthquake and the 1985

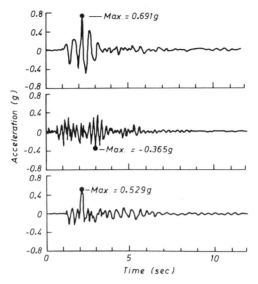

Fig. 2.6   Uncorrected accelerograph, San Salvador, 1985 [X11].

Chilean earthquake. It will be noted that the San Salvador record is of shorter duration and appears to have a reduced range of frequency components compared to the Chilean record. This is typical of near-field records (Fig. 2.5) of small-to-moderate earthquakes. Peak accelerations of the two earthquakes are, however, very similar.

(b)   Vertical Acceleration   Vertical accelerations recorded by accelerographs are generally lower than corresponding horizontal components and frequently are richer in low-period components. It is often assumed that peak vertical accelerations are approximately two-thirds of peak horizontal values. Although this appears reasonable for accelerograms recorded at some distance from the epicenter there is increasing evidence that it is nonconservative for near-field records, where peak horizontal and vertical components

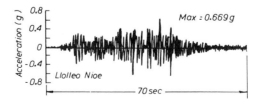

Fig. 2.7   Uncorrected accelerograph, Chile, 1985 [X1].

are often of similar magnitude. In general, the vertical components of earthquakes, discussed in Section 2.3, are not of great significance to structural design.

(c) Influence of Soil Stiffness   It is generally accepted that soft soils modify the characteristics of strong ground motion transmitted to the surface from the underlying bedrock. The extent and characteristics of this modification are, however, still not fully understood. Amplification of long-period components occurs, and generally peak accelerations in the short-period range are reduced, as a result of strength limitations of the soil. It also appears that amplification of ground motion is dependent on the intensity of ground shaking. The high levels of site amplification that have been measured on soft soils in microtremors and aftershocks are probably not applicable for stronger levels of excitation because of increased damping and limited strength of the soil.

Simple linear elastic models simulating soil amplification of vertically propagating shear waves are now known to give a poor representation of actual response and ignore the influence of surface waves. Nevertheless, soil amplification of response is extremely significant in many cases. A classic example is the response of the soft lake bed deposits under Mexico City. These deposits are elastic to high shearing strain, resulting in unusually high amplification of bedrock response. Figure 2.8 compares acceleration recorded at adjacent sites on rock and on medium depth lake deposits in the 1985 Mexico earthquake. Mexico City was some 400 km from the epicenter of the earthquake, and peak bedrock accelerations were about $0.05g$. These were amplified about five times by the elastic characteristics of the old lake bed deposits and generated modified ground motion with energy predominantly in the period range 2 to 3 s. As a consequence, buildings with natural periods

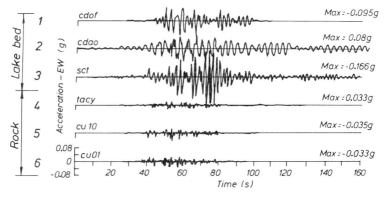

Fig. 2.8   Comparison of lake bed (1–3) and rock (4–6) accelerographs, Mexico City, 1985.

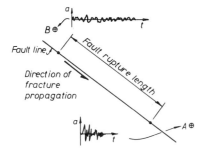

Fig. 2.9 Influence of fracture directionality on site response.

in this range were subjected to extremely violent response, with many failures resulting. There is still controversy as to the extent of site amplification that can be expected from deep alluvial deposits in very large earthquakes.

(d) Directionality Effects   Energy is not released instantaneously along the fault surface. Rather, fracture initiates at some point and propagates in one or both directions along the fault. There is evidence that in many cases, the fracture develops predominantly in one direction. In this case the location of a site with respect to the direction of rupture propagation can influence the local ground motion characteristics, as shown in Fig. 2.9. Station A, "down-stream" of the rupture propagation, is likely to experience enhanced peak accelerations due to reinforcement interaction between the traveling shock waves and new waves released downstream as the fault propagates. High-frequency components should be enhanced by a kind of Doppler shift, and the duration of shaking should be reduced. Station B, "upstream," should see reduced intensity of ground motion, but with an increased duration. Energy should be shifted toward the long-period range.

(e) Geographical Amplification   Geographical features may have a signifi-cant influence on local intensity of ground motion. In particular, steep ridges may amplify the base rock accelerations by resonance effects in a similar fashion to structural resonance of buildings. A structure built on top of a ridge may thus be subjected to intensified shaking. This was graphically illustrated during the 1985 M7.8 Chilean earthquake. At the Canal Beagle site near Viña del Mar, planned housing development resulted in identical four- and five-story reinforced concrete frame apartment buildings with masonry-infill panels being constructed by the same contractor along two ridges and in a valley immediately adjacent to one of the ridges, as shown schematically in Fig. 2.10. While the earthquake caused extensive damage to the buildings along both ridges, the buildings in the valley site escaped unscathed. Simultaneous recordings of aftershock activity [C5] at the ridge and valley sites indicated intensive and consistent amplification of motion at the ridge site. Figure 2.11 shows a typical transfer function for aftershock

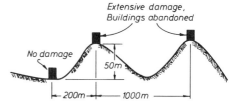

Fig. 2.10 Influence of geographical amplification on structural damage, Canal Beagle, Viña del Mar, Chile, 1985.

activity found by dividing the ridge acceleration response by the valley response. Although the geographical amplification clearly resulted in the increased damage at the ridge sites, it is probable that the transfer function of Fig. 2.11 overestimates the actual amplification that occurs during the main event, as a result of increased material damping and other nonlinear effects at the higher levels of excitation.

2.1.5 Attenuation Relationships

A key element in the prediction of seismic risk at a given site is the attenuation relationship giving the reduction in peak ground acceleration with distance from the epicenter. Three major factors contribute to the attentuation. First, the energy released from an earthquake may be considered to be radiated away from the source as a combination of spherical and cylindrical waves. The increase in surface area of the wave fronts as they move away from the source implies that accelerations will decrease with distance as the sum of a number of terms proportional to $R_e^{-1/2}$, R_e^{-1}, R_e^{-2}, and $\ln R_e$ [N1], where R_e is the distance to the point source or cylindrical axis. Second, the total energy transmitted is reduced with distance due to

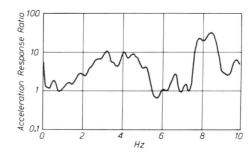

Fig. 2.11 Ridge/valley site acceleration transfer function, Canal Beagle, Chile, 1985 [C5].

material attenuation or damping of the transmitting medium. Third, attenuation may result from wave scattering at interfaces between different layers of material.

It would appear that for small-to-moderate earthquakes, the source could reasonably be considered as a point source, and spherically radiating waves would characterize attenuation. For large earthquakes, with fault movement over several hundred kilometers, cylindrical waves might seem more appropriate, although this assumes instantaneous release of energy along the entire fault surface. Hence attenuation relationships for small and large earthquakes might be expected to exhibit different characteristics.

Despite this argument, most existing attenuation relationships have been developed from analyses of records obtained from small-to-moderate earthquakes, as a result of the paucity of information on large earthquakes. The relationships are then extrapolated for seismic risk purposes to predict the response under larger earthquakes.

Typical attenuation relationships take the form

$$a_o = C_1 e^{C_2 M} (R_e + C_3)^{C_4} \tag{2.3}$$

where a_o is the peak ground acceleration, M the Richter magnitude of the earthquake, R_e the epicentral distance, and $C_1 \cdots C_4$ are constants. Many relationships of the general form of Eq. (2.3) have been proposed [N1], resulting in a rather wide scatter of predicted values. This is illustrated in Fig. 2.12(a) for Richter magnitude $M6.5$ earthquakes and in Fig. 2.12(b) for

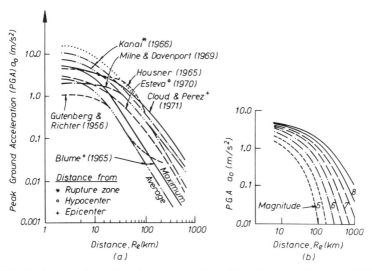

Fig. 2.12 Peak ground acceleration attenuation with epicentral distance: (a) for an M6.5 earthquake; (b) average values for different magnitudes [P52].

earthquakes with different magnitudes. At small focal distances the scatter can be partially attributed to the measurement of distance from the epicenter, hypocenter, or edge of fracture zone in the different models, but the scatter is uniform at more than an order of magnitude over the full range of distance. Esteva and Villaverde [X3] recommend the following form of Eq. (2.3):

$$a_o = 5829e^{0.8M}(R_e + 40)^{-2} \qquad (2.4)$$

where the peak ground acceleration a_o is in cm/s^2 and R_e is in kilometers. The coefficient of variation of Eq. (2.4) for the data set used to develop it was about 0.7. It is a measure of the difficulty in producing reliable attenuation relationships that other researchers, using essentially the same data sets, have developed expressions resulting in predictions that differ by up to ±50% from the prediction of Eq. (2.4), without significantly worse coefficients of variation.

It would appear that expressions of the form of Eqs. (2.3) and (2.4) can never adequately predict attenuation. The constant coefficient C_4 does not agree with theoretical observations of the propagation of spherical and cylindrical waves, which indicate that as distance from the epicenter increases, terms containing different powers of R_e dominate attenuation. Also, the prediction that peak ground acceleration in the near field differ according to (say) $e^{0.8M}$ are not supported by observations. Equation (2.4) would predict that 10 km from the epicenter, peak ground acceleration resulting from magnitude 6 or 8 earthquakes would be 283 or 1733 cm/s^2, respectively. Part of the reason for this behavior is the assumption in equations of the form of Eq. (2.4) that earthquakes can be considered as point sources. This is clearly inappropriate for large earthquakes.

The assumption that the rate of attenuation is independent of magnitude is also highly suspect. An example of the difficulty in predicting ground motion in large earthquakes from smaller earthquakes is provided by the 1985 M7.8 Chilean earthquake. On the basis of records obtained from three earthquakes, ranging in magnitude from 5.5 to 6.5 prior to 1982, Saragoni et al. [X1] proposed an attenuation relationship:

$$a_o = 2300e^{0.71M}(R_e + 60)^{-1.6} \qquad (2.5)$$

It will be seen that this is of the form of Eq. (2.3). This expression [Eq. (2.5)] is plotted in Fig. 2.13 together with peak ground accelerations recorded by various seismographs during the 1985 earthquake. Data plotted in Fig. 2.13 are the maxima of the two recorded components, but are not necessarily the peak ground accelerations found from resolving the two components along different axes. Attenuation curves are typically plotted to log/log scales. However, this tends to disguise the scatter, and Fig. 2.13 is plotted to a natural scale. It will be seen that Eq. (2.5) underestimates the ground

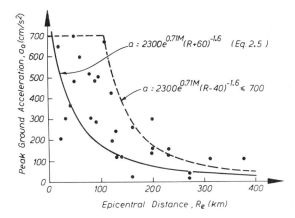

Fig. 2.13   Attenuation of peak ground acceleration, 1985 Chilean earthquake.

acceleration seriously. For the purpose of seismic risk assessment, it might be considered appropriate to use an attenuation relationship that provides predictions with, say, 5% probability of exceedence. It is of interest that by adding 100 km to the epicentral distance R_e, and adopting a cutoff of the peak ground acceleration at $0.7g$, as shown by the modified equation in Fig. 2.13, an approximate upper 5% bound to the data is provided.

The problem of ground motion attenuation has been discussed here at some length because of the current trend toward site-specific seismic studies for important structures and facilities. It must be realized that although site-specific studies provide a valuable means for refining estimates of local seismicity, the results are often based on incomplete data. Uncertainty in predicting seismicity should be appreciated by the designer.

2.2   CHOICE OF DESIGN EARTHQUAKE

2.2.1   Intensity and Ground Acceleration Relationships

Structures are designed to withstand a specified intensity of ground shaking. In earlier times this was expressed in terms of a design Modified Mercalli (*MM*) level, and in some parts of the world this approach is still adopted. Nowadays intensity is generally expressed as a design peak ground acceleration, since this is more directly usable by a structural engineer in computing inertia forces.

The relationship between *MM* intensity and peak ground acceleration (PGA), based on a number of studies [M16], is shown in Fig. 2.14. It will be observed that considerable scatter is exhibited by the data, with PGA typically varying by an order of magnitude for low *MM* values, and about half

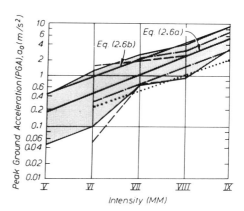

Fig. 2.14 Relationships between intensity and peak ground acceleration [M16].

an order of magnitude for high *MM* levels. To some extent this is a result of the subjective nature of the *MM* scale but is also in part due to the inadequacy of PGA to characterize earthquake intensity.

Peak ground acceleration is only one factor that affects intensity. Other factors include duration and frequency content of the strong motion. Different earthquakes with the same PGA can thus have different destructive power and are perceived, correctly, to have different intensities. To some extent the scatter exhibited by Fig. 2.14 can be reduced by use of effective peak ground acceleration (EPA), which is related to the peak response acceleration of short-period elastic oscillation rather than the actual maximum ground acceleration [X7].

The damage potential of an earthquake is as much related to peak ground velocity as to acceleration, particularly for more flexible structures. For this class of structure, *MM* intensity or some measure of ground acceleration, both of which are mainly relevant to the response of stiff structures, provides a poor estimate of damage potential.

Despite the wide scatter in Fig. 2.14, there is an approximate linear relationship between the logarithm of PGA and *MM* intensity, *I*. An average relationship may be written as

$$PGA_{ave} = 10^{-2.4 + 0.34I} \qquad (2.6a)$$

However, if design to a given intensity *I* is required, a conservative estimate for PGA should be adopted, and the following expression is more appropriate:

$$PGA_{des} = 10^{-1.95 - 0.32I} \qquad (2.6b)$$

These equations are plotted in Fig. 2.14.

2.2.2 Return Periods: Probability of Occurrence

To assess the seismic risk associated with a given site, it is necessary to know not only the characteristics of strong ground shaking that are feasible for a given site, but also the frequency with which such events are expected. It is common to express this by the *return period* of an earthquake of given magnitude, which is the average recurrence interval for earthquakes of equal or larger magnitude.

Large earthquakes occur less frequently than small ones. Over much of the range of possible earthquake magnitudes the probability of occurrence (effectively, the inverse of the return period) of earthquakes of different magnitude M are well represented by a Gumbel extreme type 1 distribution, implying that

$$\lambda(M) = \alpha V e^{-\beta M} \qquad (2.7)$$

where $\lambda(M)$ is the probability of an earthquake of magnitude M or greater occurring in a given volume V of the earth's crust per unit time, and α and β are constants related to the location of the given volume. Figure 2.15 shows data for different tectonic zones compared with predictions of Eq. (2.7) calibrated to the data by Esteva [E2]. It is seen that the form of the recurrence relationships does indeed agree well with extreme type 1 distributions, except at high magnitudes, where the probability of occurrence is overpredicted. Equation (2.7) gives poor agreement for small earthquakes, since it predicts effectively continuous slip for very small intensities.

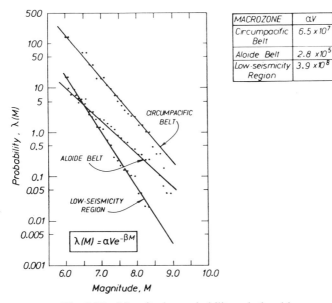

MACROZONE	αV	β
Circumpacific Belt	6.5×10^7	2.16
Aloide Belt	2.8×10^5	1.71
Low-seismicity Region	3.9×10^8	2.82

Fig. 2.15 Magnitude–probability relationships.

TABLE 2.1   **Historical Seismicity of Valparaiso**

Year	Interval (yrs)	Magnitude (approx.)
1575		7.0–7.5
	72	
1647		8.0–8.5
	83	
1730		8.7
	92	
1822		8.5
	84	
1906		8.2–8.6
	79	
1985		7.8
	Ave: 82 ± 10	Ave: 8.1 ± 0.6

Equation (2.7) assumes a stochastic process, where the value of $\lambda(M)$ is constant with time regardless of recent earthquake activity. Thus it is assumed that the occurrence of a $M8$ earthquake in one year does not reduce the probability of a similar event occurring in the next few years. However, it is clear that slip at portions of major tectonic boundaries occurs at comparatively regular intervals and generates earthquakes of comparatively uniform size. An example is the region of the San Andreas fault east of Los Angeles. Cross-fault trenching and carbon dating of organic deposits have enabled the year and magnitude of successive earthquakes to be estimated. This research indicates that $M7.5+$ earthquakes are generated with a return period between 130 and 200 years. Even more regular behavior has been noted on the Nazca/South American plate boundary at Valparaiso in Chile. Table 2.1 lists data and approximate magnitudes of earthquakes in this region since the arrival of the Spanish in the early sixteenth century led to reliable records being kept. It will be seen that the average return period of 82 years has a standard deviation of only 7 years. In an area such as Valparaiso, the annual risk of strong ground motion is low immediately after a major earthquake and its associated aftershocks, but increases to very high levels 70 years after the last major shake. This information can be relevant to design of structures with limited design life.

2.2.3   Seismic Risk

As has already been discussed, small earthquakes occur more frequently than large earthquakes. They can generate peak ground accelerations of similar magnitudes to those of much larger earthquakes, but over a much smaller area. The quantification of seismic risk at a site thus involves assessing the probability of occurrence of ground shaking of a given intensity as a result of the combined effects of frequent moderate earthquakes occurring close to the site, and infrequent larger earthquakes occurring at greater distances. Mathematical models based on the probability of occurrence of earthquakes of given magnitude per unit volume, such as Eq. (2.7) and attenuation relationships such as Eq. (2.4), can be used to generate site-specific seismic risk, and the relationship between risk, generally expressed in terms of annual probability of exceedance of a given level of peak ground acceleration and that level of peak ground acceleration.

2.2.4 Factors Affecting Design Intensity

(a) *Design Limit States* The intensity of ground motion adopted for seismic design will clearly depend on the seismicity of the area. It will also depend on the level of structural response contemplated and the acceptable risk associated with that level of response. Three levels, or limit states, were identified in Chapter 1:

1. A *serviceability limit state* where building operations are not disrupted by the level of ground shaking. At the limit state, cracking of concrete and the onset of yield of flexural reinforcement might be acceptable provided that these would not result in the need for repairs.
2. A *damage control limit* state where repairable damage to the building may occur. Such damage might include spalling of cover concrete and the formation of wide cracks in plastic hinge regions.
3. A *survival limit state* under an extreme event earthquake, where severe and possibly irreparable damage might occur but collapse and loss of life are avoided.

The acceptable risk for each level of response being exceeded will depend on the social and economic importance of the building. Clearly, hospitals must be designed for much lower risk values than office buildings. This aspect is considered in more detail in Section 2.4.3(*b*). Values of annual probability p which are often considered appropriate for office buildings are:

$$\begin{aligned}
&\text{Serviceability limit state:} && p \approx 0.02/\text{year} \\
&\text{Damage control limit state:} && p \approx 0.002/\text{year} \\
&\text{Survival limit state:} && p \approx 0.0002/\text{year}
\end{aligned}$$

In general, only one of the three limit states will govern design at a given site, dependent on the seismicity of the region. There is growing evidence that for the very long return periods events appropriate for survival limit states, the PGA is rather independent of seismicity or proximity to a major

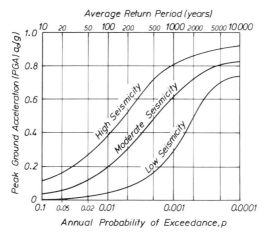

Fig. 2.16 Relationship between peak ground acceleration and annual probability of exceedance for different seismic regions.

fault. For example, large regions of the east coast and midwest of the United States are thought to be susceptible to similar maximum levels of ground shaking as the much more seismically active west coast, when return periods are measured in thousands of years. However, at short return periods, expected PGAs in moderate- to high-seismicity regions may be an order of magnitude greater than for low-seismicity regions. These trends are included in Fig. 2.16, which plots PGA against annual probability of exceedance for three levels of site seismicity. If we consider a typical design of a ductile frame building, the serviceability limit state might be taken as the onset of yield in beam members, the damage control limit state as that corresponding to a displacement ductility of $\mu_\Delta = 4$, and survival limit state as $\mu_\Delta = 8$. Table 2.2 compares risk and resistance for the three levels of seismicity of Fig. 2.16, using these levels of ductility and a relative risk on resistance of 1.0

TABLE 2.2   Seismic Risk and Resistance Compared for Different Seismic Regions

Limit State	Annual Probability	Relative Resistance	Relative Risk[a]			Risk/Resistance[a]		
			Hi	Mod.	Low	Hi	Mod.	Low
Serviceability	0.0200	0.125	0.30	0.15	0.03	2.50	1.20	0.25
Damage control	0.0020	0.500	0.80	0.60	0.30	1.60	1.20	0.40
Survival	0.0002	1.000	1.00	1.00	1.00	1.00	1.00	1.00

[a]Hi, high seismicity; Mod., moderate seismicity; Low, low-seismicity region.

at $p = 0.0002$. Resistance is based on the equal-displacement concept of equivalent elastic response, described in Section 2.3.4.

By dividing risk by resistance for the three limit states, the highest resulting number identifies the critical state. For moderate seismicity, the risk/resistance ratio is seen to be reasonably uniform, but the damage control limit state appears somewhat more critical. For high-seismicity regions, it is clear that the serviceability state is significantly more critical, while for regions of low seismicity, the survival limit state dominates.

The approach to risk presented above creates considerable difficulty when related to current design practice, where regions of low seismicity have traditionally been assigned low seismic design intensities. It is, however, becoming increasingly accepted that such a design approach could result in catastrophic damage and loss of life, at a level that would be socially unacceptable. It should be emphasized that the numeric values of Fig. 2.16 and Table 2.2 are intended only to indicate trends and are subject to considerable uncertainty. Caution should therefore be exercised in adopting them in practice.

(b)   Economic Considerations   Economics are, of course, another factor influencing the choice of design intensity. The extent to which economics become the overriding consideration depends on a number of factors: some quantifiable, other apparently not. The main factor that can readily be quantified is the cost of providing a given level of seismic protection, since this is objective. The key unquantifiable factor is the value of human life, which is subjective and controversial. To make a valid economic assessment of the cost of providing increased seismic resistance, the following factors must be considered:

Initial cost of providing increased seismic resistance

Reduced cost of repair and replacement, both structural and nonstructural, as a result of damage or collapse

Reduced loss of revenue resulting from loss of serviceability

Reduced costs caused by third-party consequences of collapse

Possible reduced insurance costs

Reduced costs arising from injury or loss of life

The extent to which the initial cost is balanced by the latter factors depends on circumstance. The relationship will be different if the building is designed for a specific client or as a speculative venture where the initial owner's commitment to the building is short-lived. Hence, to some extent, it depends on the social system of the country where the building is constructed. Clearly, the economic state of the country will also affect the economic equation, although in which sense is not always clear. For example, it can be argued that a country with limited financial reserves cannot afford increased seismic protection. On the other hand, it can be argued with at least equal logic that such a country cannot afford the risk associated with recovering from the effects of severe and widespread damage under an earthquake of relatively high probability of occurrence, since this will require a massive commitment of financial resources beyond the capabilities of a poor country.

In fact, the cost of providing increased seismic resistance is generally significantly less than believed by uninformed critics, particularly when the increased resistance is provided by improved detailing rather than increased strength. Typical studies [A7] comparing the cost of doubling strength of frame buildings from resisting a total lateral force corresponding to $0.05g$ to $0.10g$ indicate increased structural costs of about 6 to 10%. When it is considered that structural costs are typically only 20 to 25% of total building costs, it is apparent that the increase in total building costs is likely to be only a few percent. Costs associated with providing increased ductility by improving detailing are typically less.

2.3 DYNAMIC RESPONSE OF STRUCTURES

The challenge in seismic design of building structures is primarily to conceive and detail a structural system that is capable of surviving a given level of lateral ground shaking with an acceptable level of damage and a low probability of collapse. Assessment of the design level of ground motion was discussed in Section 2.2.

Two other aspects of earthquake activity have not been included in the definition above: vertical acceleration and ground dislocation. The problem of sustaining vertical accelerations resulting from earthquake activity is almost always a lesser problem than response to lateral acceleration because the vertical accelerations are typically less than horizontal accelerations and because of the characteristically high reserve strength provided as a result of design for gravity load. For example, a typical beam of a multistory office building may be designed to support live loads equal to 70% of dead loads. Assuming load factors of 1.4 and 1.7 for dead and live load, respectively [Eq. (1.5a)], and a strength reduction factor of $\phi = 0.9$ (Section 1.3.4), the ratio of the ideal strength of the beam to the strength demand due to dead and live loads is

$$\frac{1}{0.9}\left(\frac{1.4 + 0.7 \times 1.7}{1.0 + 0.7}\right) = 1.69$$

Thus under full dead plus live load, a vertical response acceleration of $0.69g$ would be required to develop the strength of the beam. In fact, significantly higher accelerations would be required, since probable strength will normally exceed ideal strength.

Although ground dislocation by faulting directly under a building could have potentially disastrous consequences, the probability of occurrence is extremely low. Where fault locations are identified it is common to legislate against building directly over the fault. As discussed earlier, buildings with strong foundations tend to deflect the path of faulting around the building perimeter rather than through the relative strong structural foundation, particularly when the building is supported by other than rock foundation material. Subsequent discussions are thus limited to establishing design forces and actions for response to lateral ground excitation. To achieve this aim it is first necessary to review some basic principles of structural dynamics.

2.3.1   Response of Single-Degree-of-Freedom Systems to Lateral Ground Acceleration

Figure 2.17 represents a simple weightless vertical cantilever supporting a concentrated mass M at a height H above its rigid base. An absolute coordinate system XY is defined, and a second system xy defined relative to the base of the cantilever. Thus the xy frame of reference moves with the structure as the ground is subjected to ground motion X_b relative to the absolute frame. Inertial response of the mass will induce it to displace an amount x_r relative to the base of the cantilever. D'Alembert's principle of dynamic equilibrium requires that the inertial force associated with the acceleration of the mass is always balanced by equal or opposite forces induced (in this case) by the flexing of the cantilever and any damping forces.

The inertial force of response is $M(\ddot{x}_r + \ddot{X}_b)$, where $(x_r + X_b)$ is the absolute lateral displacement of the mass. The force due to flexing of the cantilever is Kx_r, where K is the lateral stiffness of the cantilever. The force due to damping is $c\dot{x}_r$, assuming viscous damping, where c is a damping coefficient with units of force per unit velocity. Thus D'Alembert's principle requires that

$$M\left(\ddot{x}_r + \ddot{X}_b\right) + c\dot{x} + Kx_r = 0$$
$$M\ddot{x}_r + c\dot{x}_r + Kx_r = -M\ddot{X}_b \qquad (2.8)$$

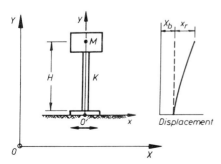

Fig. 2.17 Response of a single-degree-of-freedom structure.

Equation (2.8) is the characteristic equation solved in structural response to lateral earthquake motion. Some aspects of this equation deserve further examination.

(a) *Stiffness*   If the cantilever in Fig. 2.17 responds linear elastically, the stiffness (force per unit displacement of mass relative to base) is

$$K = \frac{1}{H^3/3E_c I + H/A_{ev}G} \tag{2.9}$$

where E_c is the modulus of elasticity, I the moment of inertia, A_{ev} the effective shear area, and G the shear modulus. The two terms in the denominator of Eq. (2.9) represent the flexural and shear flexibility, respectively. It should, however, be noted that concrete and masonry structures cannot strictly be considered linear elastic systems. I and A_{ev} in Eq. (2.9) depend on the extent of cracking, and hence the lateral force levels, E_c and G are dependent on the stress level, and unloading stiffness is different from elastic stiffness. Thus even at levels of force less than at yield strength of the cantilever, K is a variable. If the cantilever response is in the postyield state, clearly K must be considered as a variable in Eq. (2.8).

(b) *Damping*   It is traditional to use the form of D'Alembert's principle given in Eq. (2.8), which assumes viscous damping. This is primarily a matter of mathematical convenience rather than structural accuracy. Other forms of damping, such as Coulomb damping, are possible where the damping force is constant at $\pm c_1$, independent of displacement or velocity, with the $+$ or $-$ sign being selected dependent on the direction of motion. Coulomb forces thus reasonably represent frictional damping.

Viscous damping is applicable to displacement of oil in ideal dashpots. It is rather difficult to accept that it is equally applicable to concrete or masonry structural elements. In fact, the predicted influence of viscous damping on a linear elastic system appears to produce behavior opposite to that observed in reinforced concrete and masonry structural elements. Consider the ideal damped linear elastic system of Fig. 2.18(a) subjected to sinusoidal displacements of the form

$$x = x_m \sin \omega t \tag{2.10}$$

where ω is the circular frequency and t represents time. The damping force

$$F_c = c\dot{x} = \omega c x_m \cos \omega t$$

has a maximum value of

$$F_{cm} = \omega x_m c$$

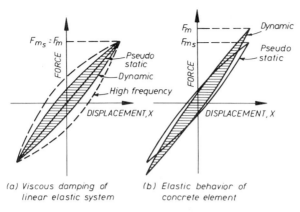

(a) Viscous damping of
   linear elastic system

(b) Elastic behavior of
   concrete element

Fig. 2.18 Dynamic response of linear systems to sinusoidal displacements.

If the frequency of the applied displacements is very small (i.e., $\omega \to 0$), the damping force is effectively zero, as shown by the pseudostatic straight line in Fig. 2.18(a). As the frequency ω increases, the maximum damping force increases in proportion, and hence so does the width of the hysteretic loop as shown in Fig. 2.18(a). However, since the velocity when the system reaches x_m is always zero, the same peak force F_m is always developed.

Actual concrete or masonry behavior is represented in Fig. 2.18(b). At very low frequencies of displacement ("pseudostatic"), the width of the hysteretic force displacement loop is large, as a result of creep effects, which are significant at high "elastic" levels of structural response. Under dynamic loading, however, element stiffness is typically higher, and the width of the hysteretic loop is typically less than at pseudostatic cycling rates, primarily due to inhibition of tensile cracking, which is very dependent on strain rate. Peak resistance attained is higher than that of the equivalent pseudostatic system.

Thus viscous damping does not represent actual behavior, although the errors are typically not large at the levels of damping (2 to 7%) normally assumed for elastic response of structural concrete. It is probable that more realistic representation can be achieved by ignoring damping in Eq. (2.8) and treating the stiffness K as a function of displacement, effective strain rate, and direction. This approach is commonly adopted for inelastic analyses, by the specification of hysteretic rules with the appropriate characteristics.

(c) Period   For the elastic system of Fig. 2.17 the natural period of vibration T is given approximately by

$$T = 2\pi\sqrt{M/K} \qquad (2.11)$$

2.3.2   Elastic Response Spectra

A response spectrum defines the frequency dependence of peak response to a given dynamic event. In earthquake engineering, response spectra for a defined level of strong ground shaking are commonly used to define peak structural response in terms of peak acceleration $|\ddot{x}_r|_{max}$, velocity $|\dot{x}_r|_{max}$, and displacement $|x_r|_{max}$. The defined level of ground shaking may be an actual earthquake accelerogram or it may represent a smoothed response curve corresponding to a design level of ground motion.

The value of response spectra lies in their condensation of the complex time-dependent dynamic response to a single key parameter, most likely to be needed by the designer: namely, the peak response. This information can then generally be treated in terms of equivalent static response, simplifying design calculations. It is, however, important to recognize that the response spectra approach omits important information, particularly relating to duration effects. Survivability of a structure depends not only on peak response levels, but also on the duration of strong ground shaking and the number of cycles where response approaches the peak response level. For example, the severe structural damage in the 1985 Mexico earthquake is at least partly attributable to the high number of cycles of response at large displacements demanded of structures by the earthquake characteristics.

Elastic response spectra are derived by dynamic analyses of a large number of single-degree-of-freedom oscillators to the specified earthquake motion. Variables in the analysis are the natural period of the oscillator T and the equivalent viscous damping. Typically, a period range from about 0.5 to 3.0 s is adopted, corresponding to the typical period range of structures, and damping levels of 0, 2, 5, 10 and 20% of critical damping are considered. Measurements of the dynamic response of actual structures in the elastic range close to yield strength indicate that equivalent viscous damping levels of 5 to 7% for reinforced concrete and 7 to 10% for reinforced masonry are appropriate. These are significantly higher than for steel structures, where 2 to 3% is more appropriate and results primarily from the nonlinear elastic behavior of concrete and masonry systems. Where foundation deformation contributes significantly to structural deformations, higher equivalent viscous damping levels are often felt to be appropriate.

For elastic response, peak response acceleration, velocity, and displacement are approximately interrelated by the equations of sinusoidal steady-state motion, namely:

Velocity: $$|\dot{x}_r|max = (T/2\pi)|\ddot{x}_r|max \qquad (2.12a)$$

Displacement: $$|x_r|max = (T^2/4\pi^2)|\ddot{x}_r|max \qquad (2.12b)$$

The interrelations of Eqs. (2.12a) and (2.12b) enable peak velocity and displacement to be calculated from peak acceleration. Tripartite response

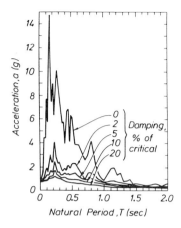

Fig. 2.19 Elastic acceleration response spectra for Llolleo accelerogram, Chile, 1985.

spectra include acceleration, displacement, and velocity information on the one logarithmic graph, but are of limited practical use because of difficulty in extracting values with any reasonable accuracy from them.

Figure 2.19 shows typical acceleration response spectra for a moderate earthquake. Structural period rather than the inverse form, natural frequency, is traditionally used as the horizontal axis, as this provides a better expansion of the scale over the range of greatest interest to the structural designer. A period of $T = 0$ represents an infinitely rigid structure. For this case the maximum acceleration response is equal to the peak ground acceleration (i.e., $a = a_o$).

The shapes in Fig. 2.19 indicate that peak accelerations are irregularly distributed over the period range but decrease very significantly at long structural periods. In the low- to middle-period range, response shows significant amplification above peak ground acceleration. The period at which peak elastic response occurs depends on the earthquake characteristics and the ground conditions. Moderate earthquakes recorded on firm ground typically result in peak response for periods in the range 0.15 to 0.4 s. On soft ground, peak response may occur at much longer periods. The case of the 1985 Mexico earthquake has been mentioned earlier. The soft lake bed deposits amplified long-period motion and resulted in peak response at the unusually long period of about 2 to 2.5 s. The significant influence of damping in reducing peak response, apparent in Fig. 2.19, should be noted.

2.3.3   Response of Inelastic Single-Degree-of-Freedom Systems

It is generally uneconomic, often unnecessarily, and arguably undesirable to design structures to respond to design-level earthquakes in the elastic range. In regions of high seismicity, elastic response may imply lateral accelerations as high as $1.0g$. The cost of providing the strength necessary to resist forces associated with this level of response is often prohibitive, and the choice of structural system capable of resisting it may be severely restricted. For tall buildings, the task of providing stability against the overturning moments generated would become extremely difficult.

If the strength of the building's lateral force resisting structural system is developed at a level of seismic response less than that corresponding to the design earthquake, inelastic deformation must result, involving yield of reinforcement and possibly crushing of concrete or masonry. Provided that the strength does not degrade as a result of inelastic action, acceptable response can be obtained. Displacements and damage must, however, be controlled at acceptable levels.

An advantage of inelastic response, in addition to the obvious one of reduced cost, is that the lower level of peak response acceleration results in reduced damage potential for building contents. Since these contents (including mechanical and electrical services) are frequently much more valuable than the structural framework, it is advisable to consider the effect of the level of seismic response not only on the structure but also on the building contents.

When the structure is able to respond inelastically to the design-level earthquake without significant strength degradation, it is said to possess ductility. Ductility must be provided for the full duration of the earthquake, possibly implying many inelastic excursions in each direction.

Perfect ductility is defined by the ideal elastic/perfectly plastic (often also called elastoplastic) model shown in Fig. 2.20(a), which describes typical response in terms of inertia force (mass × acceleration) versus displacement at the center of mass. Diagrams of this form are generally termed *hysteresis loops*.

The structural response represented by Fig. 2.20(a) is a structural ideal, seldom if ever achieved in the real world, even for steel structures which may exhibit close to ideal elastoplastic material behavior under monotonic loading if strain hardening effects are ignored. Hysteresis loops more typical of reinforced concrete and masonry structures are shown in Fig. 2.20(b) to (d). In reinforced concrete frame structures it is desirable to concentrate the inelastic deformation in plastic hinges occurring in the beams, generally adjacent to column faces. Under ideal conditions, hysteresis loops of the form of Fig. 2.20(b) result, where the energy absorbed is perhaps 70 to 80% of that of an equivalent elastoplastic loop. When energy is dissipated in plastic hinges located in columns with moderate to high axial load levels, the loops diverge further from the ideal elastoplastic shape, as illustrated in Fig. 2.20(c).

However, many structural elements exhibit dependable ductile behavior with loops very different from elastoplastic. Figure 2.20(d) is typical of squat structural walls with low axial load. The low stiffness at low displacements results from sliding of the wall on a base-level crack opened up during previous inelastic excursions. T-section structural walls typically exhibit dif-

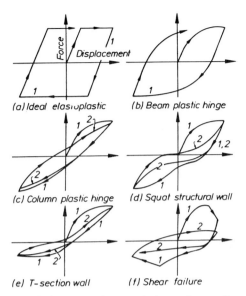

Fig. 2.20 Typical force-displacement hysteresis loop shapes for concrete and masonry structural elements.

ferent strengths and stiffnesses in opposite directions of loading, parallel to the web. The hysteresis loops are thus asymmetric, as shown in Fig. 2.20(e), and may be very narrow, particularly if the flange is wide, indicating low energy absorption.

All the loops of Fig. 2.2(a) to (e) represent essentially ductile behavior, in that they do not indicate excessive strength degradation with increasing displacement or with successive cycling to the same deflection. The loop shapes of Fig. 2.20(a) and (b) are to be preferred to those of Fig. 2.20(c), (d), or (e) since the area inside the loop is a measure of the energy that can be dissipated by the plastic hinge. For short-period structures, the maximum displacement response levels are very sensitive to the hysteretic damping, as measured by the area inside the hysteresis loops. For long-period structures, analyses have shown [G3] that hysteretic damping is less important. In each case the loops of Fig. 2.20(a) to (e) result primarily from inelastic flexural action. Inelastic shear deformation typically results in strength degradation, as shown in Fig. 2.20(f). This behavior is unsuitable for seismic resistance.

As discussed in Sections 1.1.2 and 1.2.4, ductility is generally defined by the ductility ratio μ, relating peak deformation Δ_m to the yield deformation Δ_y. Thus the displacement and curvature ductility ratios are, respectively,

$$\mu_\Delta = \Delta_m/\Delta_y \qquad (2.13a)$$

$$\mu_\phi = \phi_m/\phi_y \qquad (2.13b)$$

Although the definition of yield deformation is clear for hysteretic characteristics similar to Fig. 2.20(a), it is less obvious for the other cases of Fig. 2.20. Aspects relating to types and quantification of ductilities are considered in detail in Section 3.5.

The response of inelastic single-degree-of-freedom systems to seismic attack can be found by modifying Eq. (2.8), replacing the constant stiffness K with a variable $K(x)$ which is dependent not only on the displacement x, but also on the current direction of change in x (loading or unloading) and the previous history of x. To describe behavior represented by Fig. 2.20 requires carefully defined hysteretic rules. Equation (2.8) is then solved in stepwise fashion in the time domain, changing the stiffness at each time interval, if necessary.

2.3.4   Inelastic Response Spectra

Inelastic time-history analyses of single-degree-of-freedom systems with strength less than that corresponding to elastic response force levels by a factor R, and with hysteretic characteristics represented by Fig. 2.20(a) and (b), indicate consistent behavior dependent on the structural natural period. For structures with natural periods greater than that corresponding to peak elastic spectral response T_m (see Fig. 2.21) for the earthquake under consideration, it is observed that maximum displacements achieved by the inelastic system are very similar to those obtained from an elastic system with the same stiffness as the initial elastic stiffness of the inelastic system, but with unlimited strength, as illustrated in Fig. 2.22(a). The geometry of Fig. 2.22(a) thus implies that the ductility achieved by the inelastic system is approxi-

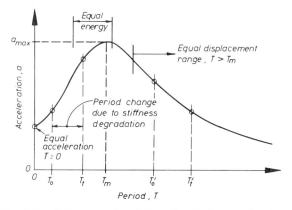

Fig. 2.21   Influence of period on ductile force reduction.

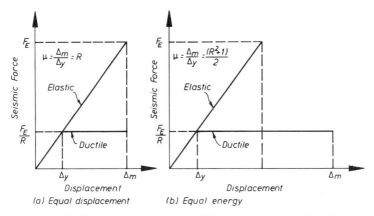

Fig. 2.22 Relationship between ductility and force reduction factor.

mately equal to the force reduction factor. That is,

$$\mu = R \qquad\qquad (2.14a)$$

This observation is sometimes referred to as the *equal-displacement principle*, although it does not enjoy the theoretical support or general applicability to warrant being called a principle.

For shorter-period structures, particularly those whose natural period is equal to or shorter than the peak spectral response period, Eq. (2.14a) is nonconservative. That is, the displacement ductility demand is greater than the force reduction factor. For many such systems it is found that the peak displacement ductility factor achieved can be estimated reasonably well by equating the area under the inelastic force–deflection curve and the area under the elastic relationship with equal initial stiffness as shown in Fig. 2.22(b). Since the areas represent the total energy absorbed by the two systems under a monotonic run to maximum displacement, Δ_m, this is sometimes termed the *equal-energy principle*. Again, the elevation of the observation to the status of "principle" is unwarranted.

From Fig. 2.22(b), the relationship between displacement ductility factor and force reduction factor can be expressed as

$$\mu = (R^2 + 1)/2 \qquad\qquad (2.14b)$$

For very-short-period structures (say $T < 0.2$ s) the force reduction factor given by Eq. (2.14b) has still been found to be unconservative. Gulkan and Sozen [G3] report displacement ductility factors of 28 to 30 resulting from a $T = 0.15$ s structure designed for a force reduction factor of about 3.3 and analyzed under different earthquake records. The equal-displacement and equal-energy approaches would result in expected ductility demands of $\mu = 3.33$ and 5.95, respectively. This inadequacy of the equal-energy principle for short-period structures results from a tendency for the period to lengthen from T_0 to a period range of higher response T_1, as a result of inelastic action and consequent stiffness degradation, as shown in Fig. 2.21. For medium- and long-period structures, the period lengthening due to inelastic action causes a shift away from the period range of maximum response.

In the limit when the period approaches $T = 0$, even small force reduction factors imply very large ductility, since the structural deformations become insignificant compared with ground motion deformations. Consequently, the structure experiences the actual ground accelerations, regardless of relative displacements, and hence ductility. If the structure cannot sustain the peak ground acceleration, failure will occur. The corollary of this is that very-short-period structures should not be designed for force levels less than peak ground acceleration. The behavior above is theoretically consistent and may be reasonably termed the *equal-acceleration principle*.

The information above may be used to generate inelastic response spectra from given elastic response spectra, for specific levels of displacement ductility factor. The force reduction factors for a given value of μ are thus

For long-period structures: $\qquad R = \mu$ $\qquad\qquad\qquad\qquad$ (2.15a)

For short-period structures: $\qquad R = \sqrt{2\mu - 1}$ $\qquad\qquad\qquad$ (2.15b)

For zero-period structures: $\qquad R = 1$ (regardless of μ) $\qquad$ (2.15c)

Figure 2.23 shows a typical inelastic acceleration response spectra based on the foregoing principles. The elastic 5% damped spectra has peak re-

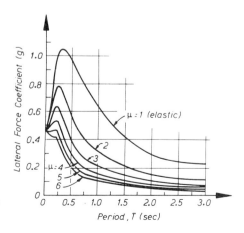

Fig. 2.23 Typical inelastic acceleration response spectra.

sponse at approximately 0.35 s. Equation (2.15*a*) is assumed to be applicable for $T > 0.70$ s. At $T = 0$, Eq. (2.15*c*) is applicable. Between $T = 0$ and $T = 0.70$ s, a linear increase in R is assumed with T according to the relationship

$$R = 1 + (\mu - 1)T/0.7 \qquad (2.15d)$$

For moderately large ductility values (say, $\mu = 6$), Eqs. (2.15*b*) and (2.15*d*) are equal at about $T = 0.3$ s.

It should be noted that currently very few codes or design recommendations adopt force reduction values that depend on natural period, although the behavior described above has been accepted for many years [N4]. An exception is the New Zealand seismic design recommendations for bridges [B10]. In design to reduced or inelastic spectra, the designer must be aware of the paramount importance of providing ductility capacity at least equal to that corresponding to the assumed force reduction factor.

It was mentioned earlier that many perfectly sound ductile structural elements of reinforced concrete or masonry exhibit hysteresis loops of very different shapes from the elastoplastic shape of Fig. 2.20(*a*), which is generally adopted for dynamic inelastic analysis. For long-period structures, the equal displacement observation indicates that the ductility level will be insensitive to the shape of the hysteresis loop. However, for short-period structures (say, $T < 0.5$ s) where the equal-energy approach is more realistic, reduction in the energy dissipated, as represented by the thinner hysteresis loops of Fig. 2.20(*c*) to (*e*), will imply a corresponding increase in ductility demand. Thus the inelastic response spectra of Fig. 2.23 are likely to be nonconservative for short-period systems with poor hysteretic loop shapes.

2.3.5 Response of Multistory Buildings

Thus far, discussion has been limited to single-degree-of-freedom systems, and hence, by implication, to single-story structures. It is now necessary to expand this to multistory structures, which must be represented by multi-degree-of-freedom systems. Equation (2.8) can be generalized for multi-degree-of-freedom systems by writing in matrix form. The distributed mass system of the building is generally lumped at nodes joining the structural elements (beams and columns, structural walls, spandrels and slabs, etc.) and then solved by standard methods of matrix structural analysis. It is beyond the scope of this book to provide consideration of such solution techniques, and the reader is referred to specialized texts [C4, C8].

2.4   DETERMINATION OF DESIGN FORCES

Three levels of analysis are available to enable the designer to estimate design-level forces generated by seismic forces in multistory buildings [X12].

They are discussed below in decreasing order of complexity and increasing order of utility.

2.4.1   Dynamic Inelastic Time-History Analysis

The most sophisticated level of analysis available to the designer for the purpose of predicting design forces and displacements under seismic attack is dynamic inelastic time-history analysis. This involves stepwise solution in the time domain of the multi-degree-of-freedom equations of motion representing a multistory building response. It requires one or more design accelerograms representing the design earthquake. These are normally generated as artificial earthquakes analytically or by "massaging" recorded accelerograms to provide the requisite elastic spectral response. Since structural response will depend on the strengths and stiffnesses of the various structural elements of the building, which will not generally be known at the preliminary stages of a design, it is unsuitable for defining design force levels. It is worth noting that the level of sophistication of the analytical technique may engender a false sense of confidence in the precision of the results in the inexperienced designer. It must be recognized that assumptions made as to the earthquake characteristics and the structural properties imply considerable uncertainty in the predicted response.

The main value of dynamic inelastic analysis is as a research tool, investigating generic rather than specific response. It may also be of considerable value in verifying anticipated response of important structures after detailed design to forces and displacements defined by less precise analytical methods. Probably the best known commercially available program is Drain-2D [P62]. As with most other dynamic inelastic programs, it is limited to investigation of simultaneous response of planar structures to vertical accelerations and lateral accelerations in the plane of the structure.

2.4.2   Modal Superposition Techniques

Modal superposition is an elastic dynamic analysis approach that relies on the assumption that the dynamic response of a structure may be found by considering the independent response of each natural mode of vibration and then combining the responses in some way. Its advantage lies in the fact that generally only a few of the lowest modes of vibration have significance when calculating moments, shears, and deflections at different levels of the building. In its purest form, the response to a given accelerogram in each significant mode of vibration is calculated as a time history of forces and displacements, and these responses are combined to provide a complete time history of the structural response. In practice, it is used in conjunction with an elastic response spectrum to estimate the peak response in each mode. These peak responses, which will not necessarily occur simultaneously in real structures, are then combined in accordance with one of several combination schemes.

Typically, for analysis, the mass of the structure is lumped at the floor levels. Thus for planar systems, only one degree of freedom per floor results. For eccentric structures where torsional response must be considered, a three-dimensional analysis is necessary, and the rotational inertia of the floor mass must also be considered. Two degrees of freedom—lateral displacement and angle of twist around the vertical axis—as shown in Fig. 1.10, thus result for each floor.

Let the modal displacements at the ith floor in the nth mode of vibration be Δ_{in}, and let W_i be the weight of the ith floor. The effective total weight W_n of the building participating in the nth mode is then [R4]

$$W_n = \frac{\left(\Sigma_{i=1}^N W_i \, \Delta_{in}\right)^2}{\Sigma_{i=1}^N W_i \, \Delta_{in}^2} \qquad (2.16)$$

where N is the number of stories. Note that as the mode number n increases, W_n decreases. The maximum base shear in the nth mode of response is then

$$V_{on} = C_{E,n} W_n \qquad (2.17)$$

where $C_{E,n}$ is the ordinate of the elastic design acceleration response spectrum at a period corresponding to the nth mode, expressed as a fraction of g, the acceleration due to gravity [see Fig. 2.24(a)].

The lateral force at the mth-floor level in the nth mode of vibration is then

$$F_{mn} = V_{on} \frac{W_m \, \Delta_{mn}}{\Sigma_{i=1}^N W_i \, \Delta_{in}} \qquad (2.18)$$

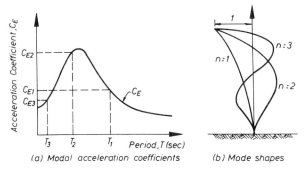

(a) Modal acceleration coefficients   (b) Mode shapes

Fig. 2.24   Modal superposition for design force levels.

The design lateral force at level m is found by combining the modal forces $F_{m1} \cdots F_{mi} \cdots F_{mn}$ of the n modes considered to provide significant response. To combine them by direct addition of their numerical value would imply simultaneous peak response in each mode and would clearly be overly conservative and inconsistent. For example, considering only the first two modes, numerical addition of first and second modes at the top is inconsistent with simultaneous numerical addition of first and second modes at midheight of the building where the modal contributions will be out of phase [Fig. 2.24(b)] if they are in phase at roof level. The most common modal combination scheme is the square-root-sum-of-squares (SRSS) scheme, where the total lateral force at level m is

$$F_m = \sqrt{\sum_{i=1}^{n} F_{mi}^2} \tag{2.19}$$

It must be emphasized that this produces only an estimate of the appropriate force level. Forces and displacements in the structure may be found by calculating the individual modal components in accordance with Eq. (2.19), or may be found from a static elastic analysis under the equivalent lateral forces F_1 to F_N.

It has been noted [C8] that the SRSS method of combination can lead to significant errors when adjacent modal frequencies are close together. This is often the case when buildings with symmetrical floor plans are subjected to torsional response as a result of eccentric mass. In such cases it has been shown that a more correct combination is provided by the complete quadratic combination (CQC) method [C8]. In this method Eq. (2.19) is replaced by

$$F_m = \sqrt{\sum_i \sum_j F_{mi} \rho_{ij} F_{mj}} \tag{2.20}$$

where the cross-modal coefficients ρ_{ij} are functions of duration and frequency content of the earthquake and of modal frequencies and damping of the structure. If the duration of the earthquake is long compared to structural periods and if the earthquake spectrum is reasonably uniform, then for constant modal damping ζ, the cross-modal coefficients are given by

$$\rho_{ij} = \frac{8\zeta^2(1 + r)r^{1.5}}{(1 - r^2)^2 + 4\zeta^2 r(1 + r)^2} \tag{2.21}$$

where $r = T_i/T_j$ is the ratio of the modal periods. If different modes are assigned different damping levels, a more complex form of Eq. (2.21) must be adopted [C8].

2.4.3   Equivalent Lateral Force Procedures

Despite the comparative elegance and simplicity of modal superposition approach for establishing seismic design forces, it has some drawbacks as a preliminary to seismic design. First, it is based on elastic response. As noted earlier, economic seismic designs for buildings will generally be based on ductile response, and the applicability of the modal superposition decreases as reliance on ductility increases. Second, even for elastic response it provides only an approximate solution for the maximum seismic design forces, particularly when the real nonlinear nature of concrete response in the elastic stage is considered. Third, it implies knowledge that is often not available at the start of the seismic design process. Member sizes and stiffness will only be estimates at this stage of the design. Fourth, it implies a knowledge of the seismic input. Although this may be provided by a design response spectra, this is at best an indication of the probable characteristics of the design-level earthquake. There is a tendency for designers utilizing modal superposition techniques as a means for defining seismic design forces to get carried away by the elegance of the mathematics involved and forget the uncertainty associated with the design seismic input.

For these reasons and others, the simple equivalent lateral force method for defining seismic resistance is still the most useful of the three methods described herein. When combined with a capacity design philosophy, ensuring that ductility can occur only in carefully selected and detailed plastic regions, and undesirable modes of inelastic deformation such as due to shear are suppressed, it can be shown, by dynamic inelastic time-history analyses, that structures which are relatively insensitive to earthquake characteristics can be designed satisfactorily. This is the essence of the design approach suggested in this book. The remainder of this section describes the process for determining lateral design forces. Subsequent chapters describe the application of the capacity design process to different common structural systems.

The equivalent lateral force procedure consists of the following steps:

1. Estimate the first-mode natural period.
2. Choose the appropriate seismic base shear coefficient.
3. Calculate the seismic design base shear.
4. Distribute the base shear as component forces acting at different levels of the structure.
5. Analyze the structure under the design lateral forces to obtain design actions, such as moments and shears.
6. Estimate structural displacements and particularly, story drifts.

Each step is now examined in more detail.

(a) First-Mode Period   Preliminary estimates may be made from empirical equations, computer matrix inversion of the stiffness matrix, or from Rayleigh's method. In all cases, the period should not be based on properties of uncracked concrete or masonry sections, as it is the period associated with elastic response at just below flexural yield which is of relevance. Sections 4.1.3 and 5.3.1 give information on cracked-section member stiffnesses. The following empirical methods have often been used for first estimates [X10].

(i) Concrete Frames

$$T_1 = 0.061 H^{0.75} \quad (H \text{ in } m)$$

$$T_1 = 0.025 H^{0.75} \quad (H \text{ in ft}) \tag{2.22}$$

where H is the building height. Alternatively, $T_1 = 0.08n$ to $0.13n$ may be used, where n is the number of stories. These estimates are likely to be conservative for multistory frames, in as far as they are likely to predict a shorter natural period and as a consequence increased response [Fig 2.24(a)].

(ii) Concrete and Masonry Structural Wall Buildings

$$T_1 = 0.09 H / \sqrt{L} \quad (H \text{ in m})$$

$$T_1 = 0.05 H / \sqrt{L} \quad (H \text{ in ft}) \tag{2.23}$$

where L is the length of the building in the direction of earthquake attack. Alternatively, $T_1 = 0.06n$ to $0.09n$ may be used.

The empirical equations listed above are very crude and should only be used where initial estimates of member sizes cannot easily be made. A much preferable approach is to estimate the natural period using Rayleigh's method [R4], in which the period is calculated from lateral displacements induced by a system of lateral forces applied at floor levels. Although the period so calculated is relatively insensitive to the distribution of lateral forces chosen, these will normally correspond to the code distribution of seismic forces, so that final seismic design member forces can be scaled directly from the results once the building period, and hence the seismic coefficient, has been established. Using Rayleigh's method, the natural period is given as

$$T_1 = 2\pi \sqrt{\frac{\sum_1^N W_i \Delta_i^2}{g \sum_1^N F_i \Delta_i}} \tag{2.24}$$

where F_i is the lateral force applied at levels $i = 1$ to N, Δ_i are the corresponding lateral displacements, and W_i are the floor weights. Note that for the evaluation of Eq. (2.24) the magnitude of forces F_i chosen is irrelevant and may be based on an initial crude estimate of period. If an initial period is assessed on the basis of the empirical approaches of Eq. (2.23) or (2.24), it is strongly recommended that a refined estimate be made, based on Eq. (2.24), once member sizes have been selected.

(b) *Factors Affecting the Seismic Base Shear Force*   The value of the seismic base shear selected will depend on the response spectra, the level of ductility capacity assumed to be appropriate for the building type, and the acceptable probability of exceedance of the design earthquake. As mentioned above, inelastic response spectra should be used to assess the influence of ductility and period, since the assumption, common to many building codes, of a constant force reduction factor to be applied to an elastic response spectrum to allow for ductility is unconservative for short-period structures.

The acceptable annual probability of exceedance of the design earthquake is a measure of acceptable risk. Many codes take this into account by "importance" or "risk" factors which relate to the need for facilities to remain functional after a major earthquake (e.g., hospitals, fire control headquarters) or the consequences of damage (i.e., release of toxic fumes). Generally, these factors have been arrived at somewhat subjectively. A consistent probabilistic approach, which relates earthquake magnitude to annual probability of occurrence, is to be preferred.

Using the nomenclature adopted above and the approach advocated earlier in this chapter, the consistent method for expressing the base shear would be

$$V_b = C_{T,S,\mu,p} \sum_1^N W_{tr} \tag{2.25}$$

where $C_{T,S,\mu,p}$ = inelastic seismic coefficient appropriate to the seismic zone, building period T, soil type S, assigned ductility capacity μ, acceptable probability of exceedance p, and the design limit state

W_{tr} = total floor weight at level r, which should include dead load W_D plus the probable value of live load W_L occurring in conjunction with the design-level earthquake

Most seismic building codes adopt a simplified representation of Eq. (2.25), where $C_{T,S,\mu,p}$ is expressed as a compound coefficient comprised of a number of variables that are individually assessed for the building and site. This has the advantage of convenience, but the disadvantage that the period-dependent coefficient does not account for nonlinear effects of ductility and return period.

TABLE 2.3   Comparison of Effective Peak Accelerations (EPA) for United States, Japan, and New Zealand (g)

	Limit State		
	Serviceability	Damage Control	Survival
United States	—	0.075–0.40	0.85[a]
Japan	0.056–0.080	0.280–0.40	—
New Zealand	0.027–0.053	0.160–0.32	—

[a]NEHRP Recommendations [X7].

The general form of Eq. (2.25), adopted by seismic provisions of building codes, may be expressed as

$$V_b = ZC_{T,S}\frac{I}{R}\sum_1^N W_{tr} \qquad (2.26)$$

where the variables are defined and discussed below. Although many base shear coefficient equations in codes appear very different from Eq. (2.26), they can be manipulated to this form [X9].

(i)   Zone Factor (Z): Z expresses the zone seismicity, generally in terms of effective peak ground acceleration. Most codes specify a single value of Z for a specific site, generally applicable to the damage control limit state. An exception is Japanese practice [A7], which specifies a two-stage design process with Z factors listed for both serviceability and damage control limit states. A two-stage approach is also advocated in the 1988 NEHRP [X7] recommendations.

A direct comparison between Z values recommended by different countries is difficult, since in some countries the Z value includes spectral amplification for short-period structures and is essentially the peak response acceleration rather than peak ground acceleration. Table 2.3 compares the range of effective peak accelerations (EPA) relevant to U.S. [X10], Japanese [A7], and New Zealand [X8] codes. In the New Zealand code and the U.S. Uniform Building Code, EPAs are listed directly. The Japanese code lists peak response accelerations, and these have been divided by 2.5 to obtain EPA values.

It will be seen that similar levels of zone coefficients are implied for the damage control limit state for the three countries. In most cases countries are divided geographically into different seismic zones, often with significant changes in Z value as a boundary between zones is crossed. There is a current tendency, however, to develop contour maps for Z, providing more gradual transition in Z values.

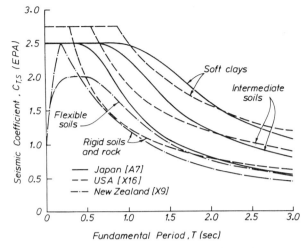

Fig. 2.25   Period-dependent coefficients $C_{T,S}$ for different countries.

(ii)   Period-Dependent Coefficient ($C_{T,S}$): The period-dependent seismic coefficient generally includes the influence of soil type by specifying different curves for different soil stiffness or by multiplying the basic coefficient for rock by an amplification factor for softer soils. Figure 2.25 compares $C_{T,S}$ coefficients implied by Japanese, U.S., and New Zealand codes for different soil stiffnesses, where the coefficients are normalized to EPA, using the assumptions of the preceding section. For rigid soils this implies maximum coefficients of 2.5 for Japanese and New Zealand coefficients, but a value of 2.75 is appropriate for U.S. conditions, as this value is specifically listed in the requirements [X10]. For flexible soils a maximum value of 2.0 is adopted by the New Zealand code.

It will be seen that the U.S. and Japanese codes adopt a constant coefficient from zero period to about 0.3 to 0.8 s, depending on the soil stiffness. Only the New Zealand code adopts the reduced coefficients for short-period structures which are clearly apparent in all recorded accelerograms. In the U.S. and Japanese codes this compensates to some extent for the use of a constant force reduction factor R, which is nonconservative at short periods, as stated earlier, if a realistic spectral shape is adopted. Coefficients used in Japan and the United States agree reasonable well, but the coefficients adopted in New Zealand do not show such a significant influence of soils, and they appear nonconservative in comparison with U.S. and Japanese coefficients.

(iii)   Importance Factor (I): The importance factor is applied in U.S. and New Zealand codes to reflect the need to protect essential facilities that must

TABLE 2.4   Force Reduction Factors, R, for Ductility, and Base Shear Coefficients $(ZC_{T,S}/R)$ Different Structures and Materials

		R Factor			$ZC_{T,S}/R$		
		U.S.	Japan	N.Z.[a]	U.S.	Japan	N.Z.
Elastically responding structure stiff ground	$T = 0.4$ s	(1.0)	1.0	1.25	(0.92)	1.00	0.550
Ductile frame with restricted ductility stiff ground	$T = 1.0$ s	5.0	2.2	3.00	0.10	0.30	0.160
Slender concrete structural wall stiff ground	$T = 0.6$ s	4.3	2.5	5.00	0.16	0.38	0.140
Slender masonry structural wall stiff ground	$T = 0.6$ s	4.3	2.0	3.50	0.16	0.47	0.210
Ductile concrete frame stiff ground	$T = 1.8$ s	8.6	3.3	6.00	0.04	0.11	0.038
Ductile concrete frame soft clay	$T = 1.2$ s	8.6	3.3	6.00	0.10	0.28	0.046

[a] Ductility factor.

operate after earthquakes, such as hospitals, fire stations, and civil defense headquarters, and is also applied to buildings whose collapse could cause unusual hazard to the public, such as facilities storing toxic chemicals. The implication is to reduce the acceptable probability of occurrence of the design earthquake. In the United States this factor only varies within the range 1.0 to 1.25, but in New Zealand values as high as 1.6 are adopted for specially important structures at the serviceability limit state. A maximum value of 1.3 applies at the damage control limit state. Japanese design philosophy does not specify an importance factor, but limits construction of important facilities to structural types that are perceived to be less at risk than others.

(iv)   Force Reduction Factor (R) (Structural Ductility): The largest variation between approaches adopted to define seismic base shear by different countries lies in the value of the force reduction coefficient assigned to different structural and material types. Effectively, the R factor can be thought of as reflecting the perceived available ductility of the different structural systems. A comparative summary is listed in Table 2.4. In preparing Table 2.4, coefficients have been adjusted to obtain a uniform approach based on the form of Eq. (2.26). For example, the Uniform Building Code in the United States [X10] adopts a load factor of 1.4 in conjunction with Eq. (2.26), implying that the real force reduction factor is $R/1.4$, where R is the listed value. New Zealand does not specify force reduction factors but directly lists

the maximum design displacement ductility factor. Design spectra include curves for ductility factors from $\mu_\Delta = 1$ to 6. For concrete and masonry structures $\mu_\Delta = 1.25$ is taken as appropriate for elastically responding structures. As the New Zealand inelastic spectra exhibit some of the characteristics of Fig. 2.23, the ductility factor and effective force reduction factor are not equal for short-period structures, but are equal for $T \geq 1.0$ s.

Considerable scatter is apparent in Table 2.4 between R factors recommended by the three countries, with U.S. values generally highest and Japanese values invariably lowest. It should be noted that strength reduction factors, which are used in both U.S. and New Zealand practice, are not used in Japanese design, and hence the Japanese coefficients could be increased by approximately 11% to obtain a more realistic comparison. In Table 2.4 a value of 1.0 for the U.S. coefficient for elastically responding structures has been assumed, although no such category exists [X10].

Table 2.4 also compares the compound effects of $ZC_{T,S}$ and R for the same categories of structures. These final coefficients, related to $I = 1.0$ in all cases indicate that the variability of the R factors between countries dominates the final base shear coefficients adopted to determine required seismic resistance. Japanese coefficients are typically between two and three times those of U.S. or New Zealand values. There is no consistent relationship between U.S. and New Zealand coefficients, which are generally in reasonable agreement, except, as noted above, for the low emphasis given in the New Zealand code to influence of soil type, particularly for long-period structures.

(c) Distribution of Base Shear over the Height of a Building The building is typically considered to respond in a simplified first-mode shape. For buildings less than 10 stories high, the mode shape is often assumed to be linear, as shown in Fig. 2.26(*b*). Assuming sinusoidal response, the peak accelerations at the floor levels are thus also distributed approximately linearly with height.

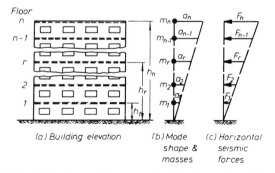

Fig. 2.26 Equivalent lateral force method for base shear distribution.

The acceleration a_r of the rth floor is related to the acceleration a_n of the nth floor:

$$a_r = (h_r/h_n)a_n \qquad (2.27)$$

The floor inertial forces are thus

$$F_r = m_r a_r = m_r h_r(a_n/h_n) = K_n W_{tr} h_r \qquad (2.28)$$

where

$$K_n = a_n/h_n$$

Now the total base shear is

$$V_b = \sum_1^n F_r = \sum_1^n K_n W_{tr} h_r = C \sum_1^n W_{tr} \qquad (2.29)$$

Hence

$$K_n = \frac{C \sum_1^n W_{tr}}{\sum_1^n W_{tr} h_r} \qquad (2.30)$$

Therefore, from Eqs. (2.28) and (2.30),

$$F_r = \left(C \sum_1^n W_{tr} \right) \frac{W_{tr} h_r}{\sum_1^n W_{tr} h_r} \qquad \text{that is,} \qquad F_r = V_b \frac{W_{tr} h_r}{\sum_1^n W_{tr} h_r} \qquad (2.31)$$

For structures higher than about 10 stories, it is common to apply a roof-level force of $0.1 V_b$ in addition to the "inverted triangle" component of $0.9 V_b$ to account for the influence of higher modes in increasing moments and shears in upper-level members. In this case,

$$F_n = 0.1 V_b + 0.9 V_b \frac{W_{tn} h_n}{\sum_1^n W_{tn} h_r} \qquad (2.32a)$$

$$F_r = 0.9 V_b \frac{W_{tr} h_r}{\sum_1^n W_{tr} h_r} \qquad (2.32b)$$

The approach represented by Eqs. (2.31) and (2.32) is rather crude, and recent building codes have sought to provide a refined distribution of forces. In the 1988 NEHRP recommendations [X7], Eqs. (2.31) and (2.32) are replaced by the following equation:

$$F_r = \frac{W_{tr}h_r^k}{\sum_1^n W_{tr}h_r^k} \tag{2.33}$$

where k is an exponent related to the building period as follows:

$$T \le 0.5 \text{ s} \qquad k = 1.0$$
$$0.5 \le 2.5 \text{ s} \qquad k = 1.0 + 0.5(T - 0.5)$$
$$T \ge 2.5 \text{ s} \qquad k = 2.0$$

Japanese practice [A7] similarly adopts vertical distributions of lateral seismic forces that are period dependent. Lateral force distributions resulting from the NEHRP [X7] and Japanese [A7] provisions are compared in Fig. 2.27 for 10-story buildings with equal floor mass at each level and with different periods. The NEHRP provisions provide larger variations in distribution shape between $T = 0.5$ and $T = 2.5$ than is apparent in the Japanese recommendations. The Japanese provisions have a more pronounced increase in lateral force at higher levels, although story shears below level 9 would in fact be lower than those from the NEHRP recommendations. For periods less than 0.5 s, the familiar triangular distribution of force, as in New Zealand, is adopted by NEHRP, but the Japanese provisions show continuous variation of floor forces down to $T = 0$. At this period the distribution is constant with height, which is theoretically correct. However, it should be recognized that $T = 0$ is an impractical case, particularly for a 10-story building.

(d)   Lateral Force Analysis   Member actions (moments, axial forces, and shear forces) may be calculated from the design-level lateral forces by many methods. Where parallel load-resisting elements act together to resist the

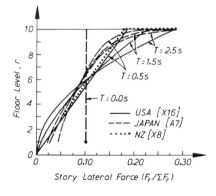

Fig. 2.27   Seismic lateral force distributions for 10-story building with uniformly distributed mass.

seismic forces, some additional distribution of horizontal forces between different elements at the same level will need to be carried out in proportion to their relative stiffnesses, prior to the structural analyses of these elements. Such force–distribution methods are discussed in chapters for different structural systems. Structural actions derived from such analyses, where appropriately combined with load effects due to gravity, are then used to determine the required strength of each component of the structure and to check that the limit state of serviceability or damage control [Section 2.2.4(*a*)] are also satisfied.

(e) Estimates of Deflection and Drift   The analysis carried out under the design-level forces will produce estimates of lateral deflections, Δ_e. However, if these forces were calculated assuming a structure ductility of (say) μ_Δ, it must be realized that the actual displacements achieved will be greater than the elastic values predicted by analysis. A reasonable approximation is

$$\Delta_m = \mu_\Delta \Delta_e \tag{2.34}$$

provided that the force levels corresponding to ductility μ_Δ have been based on consistent inelastic spectra, as discussed above. Using current code approaches, it is reasonable to approximate $\mu_\Delta = R$.

Equation (2.34) can only be applied to the displacement at the center of seismic force, which is typically at about two-thirds of the building height. Story drifts, particularly in the lower floors of frame buildings, may be substantially higher than estimated by multiplying elastic drifts by the structure ductility factor. Detailed guidance on this matter is given in Section 4.7.

(f) P–Δ Effects in Frame Structures   When flexible structures, such as reinforced concrete frames are subjected to lateral forces, the resulting horizontal displacements lead to additional overturning moments because the gravity load P_g is also displaced. Thus in the simple cantilever model of Fig. 2.28(*a*), the total base moment is

$$M_b = F_E H + P_g \Delta \tag{2.35}$$

Therefore, in addition to the overturning moments produced by lateral force F_E, the secondary moment $P_g \Delta$ must also be resisted. This moment increment in turn will produce additional lateral displacement, and hence Δ will increase further. In very flexible structures, instability, resulting in collapse, may occur. Provided that the structural response is elastic, the nonlinear response, including P–Δ effects, can be assessed by different "second-order" analysis techniques.

When the inelastic response of ductile structural systems under seismic actions is considered, the displacement Δ is obtained from rather crude

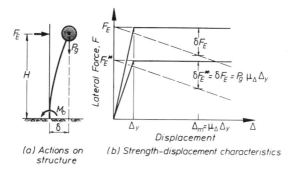

(a) Actions on
structure

(b) Strength-displacement characteristics

Fig. 2.28   Influence of P–Δ effects on resistance to lateral forces.

estimates, and hence exact assessment of P–Δ effects cannot be made, and approximate methods must be used. This is discussed in detail in Section 4.7.

It is, however, necessary to recognize when assessing seismic design forces that the importance of P–Δ effects will generally be more significant for structures in regions of low-to-moderate seismicity than for structures in regions of high seismicity, where design lateral forces will be correspondingly higher. Consider the characteristic inelastic load–deformation response of two structures, one designed for a region of high seismicity with a strength under lateral forces of F_E, and the other, in a less seismically active region, with a lateral strength of F_E^*, as shown in Fig. 2.28(*b*). As a result of the different required strengths, the stiffnesses will also be different, and for simplicity we shall assume that the stiffnesses of the two buildings are proportional to their required strengths (a reasonable approximation), and hence the yield displacements of the two structures, Δ_y, are the same.

If P–Δ effects are insignificant, the load–deformation characteristic may be approximated by the elastoplastic response shown in Fig. 2.28(*b*). The P–Δ effect on a ductile system is to reduce the lateral forces that can be resisted. For example, since the base moment capacity M_b is not affected, from Eq. (2.35) it follows that

$$F_E' = \frac{M_b - P_g \Delta}{H} \tag{2.36}$$

The effective strength available to resist lateral forces thus reduces as the displacement increases to its maximum Δ_m, as shown in Fig. 2.28(*b*). Since the weight P_g of the two structural systems is presumed the same, the reduction in strength δF for a specified displacement Δ_m is the same. That is, $\delta F_E = \delta F_E^* = P_g \Delta_m$. If both structures have the same yield displacement Δ_y, and the same expected structural ductility μ_Δ, the reduction in strength available to resist lateral forces at maximum expected displacement is $\delta F_E = \delta F_E^* = P_g \mu_\Delta \Delta_y$.

It is clear from Fig. 2.28(b) that this reduction in strength is more significant for the weaker system, since the ratio $\delta F_E / F_E^* > \delta F_e / F_E$. As will be shown in Section 4.7, the design approach will be to compensate for the effective strength loss by provision of extra initial strength above the level F_E or F_E^* corresponding to code forces when the strength loss δF_E exceeds a certain fraction of required strength F_E. It is important to recognize this at an early stage of the design process to avoid the necessity for redesign because of a late check of $P{-}\Delta$ effects.

(g) Torsion Effects   In Section 1.2.2(b) it was shown that lateral forces F_j, applied at the center of floor mass at a level, may result in eccentricity of the story shear force V_j with respect to the center of rigidity (CR) [Fig. 1.10(d)] in that story. However, additional torsion may also be introduced during the dynamic response of the structure because of the torsional ground motions, deviations of stiffness from values assumed, and different degrees of stiffness degradation of lateral-force-resisting components of the structure during the inelastic response of the building. For this reason allowance is made in codes [X4, X10] for additional eccentricity due to *accidental torsion*, so that the design eccentricities become

$$e_d = e_s + \alpha B_b \tag{2.37}$$

where e_s is the computed static eccentricity and B_b is the overall plan dimension of the building measured at right angles to the earthquake action considered. Typical values of α range from 0.05 to 0.10. The design eccentricities e_{dx} and e_{dy} are determined independently for each of the principal directions y and x of the framing plan [Fig. 1.10(d)]. Lateral force resisting elements located on each side of the center of rigidity are designed for the most adverse combination of lateral forces, due to story translations and story twists. When torsional irregularity exists, further amplification of the eccentricity due to accidental torsion is recommended [X10]. The treatment of torsional effects is discussed for frames in Sections 4.2.5 and 4.2.6 and for structural walls in Section 5.3.2(a).

The designer's aim should be to minimize the imposition of excessive inelastic lateral displacements on those vertical framing elements within a building that are at maximum distance from the shear center, (i.e., the center of rigidity). The choice of torsionally balanced framing systems, the preferred systems, such as seen in Fig. 1.12 and the example in Fig. 5.50(a), are likely to achieve this aim with greater promise than meeting the requirements of the relevant equations of building codes.

3 Principles of Member Design

3.1 INTRODUCTION

In this chapter we summarize and review the established principles used in the design of reinforced concrete or masonry members. We assume that the reader is familiar with the elementary concepts of the relevant theory and that some experience with design, at least related to gravity-load created situations, exists. Standard texts used in undergraduate courses in structural engineering may need to be consulted.

In presenting procedures for section analysis to provide adequate resistance of various structural actions, inevitably some reference needs to be made to design codes. The guidance provided in codes of various countries are generally based on accepted first principles, on the finding of extensive research, on engineering judgments, and on consideration of traditional or local construction practices. As stated, in its effort to emphasis important principles and to promote the understanding of structural behavior, this book does not follow any specific concrete design code. However, to emphasize generally accepted sound design guides and to enable numerical examples to be presented, illustrating the application of design procedures and details, some reference to quantified limits had to be made. To this end, particularly for the sake of examples, recommendations of the code used in New Zealand [X3], the drafting of which was strongly influenced by concepts presented in this book, are frequently followed. These recommendations are similar to those of the American Concrete Institute [A1]. No difficulties should arise in ensuring that the principles and techniques presented in subsequent chapters conform with basic requirements of modern seismic codes in any country.

While this chapter summarizes basic design requirements in general, in subsequent chapters we concentrate on the specific issues that arise from seismic demands. Frequent references in the remaining chapters, particularly in the extensive design examples, will be made to sections and equations of this chapter.

3.2 MATERIALS

3.2.1 Unconfined Concrete

(a) Stress–Strain Curves for Unconfined Concrete The stress–strain relationship for unconfined concrete under uniaxial stress is dealt with in numerous

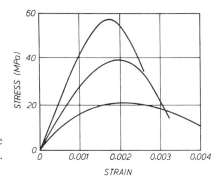

Fig. 3.1 Stress–strain curves for concrete cylinders loaded in unaxial compression. (1 MPa = 145 psi.)

texts [N5, P1] and it is assumed that the reader is familiar with standard formulations. Consequently, only a very brief description of the behavior of unconfined concrete under compression is presented here. Figure 3.1 shows typical curves for different-strength concretes, where the compression strength, as obtained from the test of standard cylinders at an age of 28 days, is defined as f'_c.

It should be noted that as the compression strength f'_c increases, the strain at peak stress and at first crushing decreases. This apparent brittleness in high-strength concrete is of serious concern and must be considered when ductility requirements result in high concrete compression strains.

The modulus of elasticity, E_c, used for design is generally based on secant measurement under slowly applied compression load to a maximum stress of $0.5f'_c$. Design expressions relate compression modulus of elasticity to compression strength by equations of the form

$$E_c = 0.043w^{1.5}\sqrt{f'_c} \quad (\text{MPa}) \tag{3.1}$$

for values of concrete unit weight, w, between 1400 and 2500 kg/m^3 (88 to 156 lb/ft^3). For normal-weight concrete,

$$E_c = 4700\sqrt{f'_c} \quad (\text{MPa})$$
$$= 57{,}000\sqrt{f'_c} \quad (\text{psi}) \tag{3.2}$$

is often used [X3].

It should be noted that Eqs. (3.1) and (3.2) have been formulated primarily with the intent of providing conservative (i.e., large) estimates of lateral deflections, particularly of beams and slabs, and hence tend to underestimate average values of E_c obtained from cylinder tests. On the basis of full-scale tests of concrete structures, it has been observed that cylinders give a low estimate of insitu modulus of elasticity [X3]. Also, actual concrete strength in

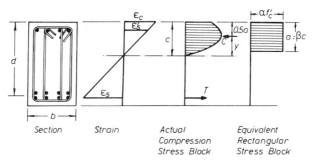

Fig. 3.2 Concrete stress block design parameters for flexural strength calculations.

a structure will tend to exceed the specified, or nominal, 28-day strength. Finally, moduli of elasticity under the dynamic rates of loading characteristic of seismic loading are higher than the values given by low-strain-rate tests [Section 3.2.2(d)].

As a consequence of these points, values of the modulus of elasticity based on Eqs. (3.1) or (3.2) and using the specified design compression strength can be as much as 30 to 40% below actual values. While this is conservative and perhaps desirable for static deflection calculations, it has different significance for seismic design. Calculated building periods based on low E_c values will exceed true values. Generally, this will mean seismic base shear coefficients lower than those corresponding to the true E_c value. If a conservative seismic design philosophy is to be achieved, there is a case for amplifying E_c given by Eqs. (3.1) and (3.2) by, say, 30% when calculating the stiffness of lateral-force-resisting elements.

(b) Compression Stress Block Design Parameters for Unconfined Concrete   In this book the flexural strength of reinforced concrete elements subjected to flexure with or without axial load will be based on the widely accepted ACI concept of an equivalent rectangular stress block for concrete in compression [A1]. As shown in Fig. 3.2, for convenience of calculation, the actual compression stress block is replaced by an equivalent rectangular block of average stress $\alpha f_c'$ and extent βc from the extreme compression fiber, where c is the distance from the extreme compression fiber to the neutral axis. The essential attributes of the equivalent rectangular stress block are that it should have the same area and centroidal height as those of the actual stress block. Thus, referring to the rectangular section in Fig. 3.2, we have

$$\alpha\beta bc f_c' = C \tag{3.3}$$

$$\alpha\beta(1 - 0.5\beta)bc^2 f_c' = Cy \tag{3.4}$$

where C is the resultant force of the compression stress block and is located a distance y from the neutral axis. For unconfined concrete the values for α and β commonly adopted [A1] are

$$\alpha = 0.85 \quad \text{for all values of } f'_c$$

and
$$0.85 \geq \beta = 0.85 - 0.008(f'_c - 30) \geq 0.65 \quad \text{(MPa)} \qquad (3.5)$$
$$= 0.85 - 0.055(f'_c - 4.35) \geq 0.65 \quad \text{(ksi)}$$

The other design parameter required for strength and ductility calculations is the ultimate compression strain, ϵ_{cu}. The normally accepted value for unconfined concrete is 0.003. However, this value is based on experiments on concrete elements subjected to uniform compression, or to constant moment. The critical regions of concrete members under seismic loading are generally subjected to significant moment gradients. This is particularly the case for members of seismic resistant frames, and tests on such elements invariably indicate that the onset of visible crushing is delayed until strains well in excess of 0.003 and sometimes as high as 0.006 to 0.008. For such members it is recommended that an ultimate compression strain of 0.004 can be conservatively adopted.

(c) Tensile Strength of Concrete The contribution of the tensile strength of concrete to the dependable strength of members under seismic action must be ignored, because of its variable nature, and the possible influence of shrinkage- or movement-induced cracking. However, it may be necessary to estimate member tension or flexural behavior at onset of cracking to ensure in certain cases that the capacity of the reinforced section in tension is not exceeded. For this purpose, the following conservatively high values for tensile strength may be assumed:

Concrete in direct tension: $\quad f'_t = 0.5\sqrt{f'_c} \quad \text{(MPa)} \qquad = 6\sqrt{f'_c} \quad \text{(psi)}$

$$(3.6a)$$

Concrete in flexural tension: $f'_t = 0.75\sqrt{f'_c} \quad \text{(MPa)} \qquad = 9\sqrt{f'_c} \quad \text{(psi)}$

$$(3.6b)$$

At high strain rates, tension strength may considerably exceed these values. It must be emphasized that although concrete tension strength is ignored in flexural strength calculations, it has a crucial role in the successful resistance to actions induced by shear, bond, and anchorage.

3.2.2 Confined Concrete

(a) Confining Effect of Transverse Reinforcement In many cases the ultimate compression strain of unconfined concrete is inadequate to allow the struc-

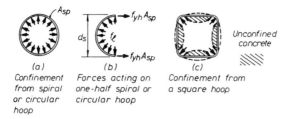

(a)
Confinement
from spiral
or circular
hoop

(b)
Forces acting on
one-half spiral or
circular hoop

(c)
Confinement from
a square hoop

Unconfined
concrete

Fig. 3.3 Confinement of concrete by circular and square hoops [P43].

ture to achieve the design level of ductility without extensive spalling of the cover concrete. Unless adequate transverse reinforcement is provided to confine the compressed concrete within the core region, and to prevent buckling of the longitudinal compression reinforcement, failure may occur. Particularly susceptible are potential plastic hinge regions in members that support significant axial load, such as columns at the base of building frames, where inelastic deformations must occur to develop a full hinging mechanism, even when the design is based on the weak beam/strong column philosophy (Section 1.4.3).

When unconfined concrete is subjected to compression stress levels approaching the crushing strength, high lateral tensile strains develop as a result of the formation and propagation of longitudinal microcracks. This results in instability of the compression zone, and failure. Close-spaced transverse reinforcement in conjunction with longitudinal reinforcement acts to restrain the lateral expansion of the concrete, enabling higher compression stresses and more important, much higher compression strains to be sustained by the compression zone before failure occurs.

Spirals or circular hoops, because of their shape, are placed in hoop tension by the expanding concrete, and thus provide a continuous confining line load around the circumference, as illustrated in Fig. 3.3(a). The maximum effective lateral pressure f_l that can be induced in the concrete occurs when the spirals or hoops are stressed to their yield strength f_{yh}. Referring to the free body of Fig. 3.3(b), equilibrium requires that

$$f_l = 2f_{yh}A_{sp}/(d_s s_h) \tag{3.7}$$

where d_s is the diameter of the hoop or spiral, which has a bar area of A_{sp}, and s_h is the longitudinal spacing of the spiral.

Square hoops, however, can only apply full confining reactions near the corners of the hoops because the pressure of the concrete against the sides of the hoops tends to bend the sides outward [as illustrated by dashed lines in Fig. 3.3(c)]. The confinement provided by square or rectangular hoops can be significantly improved by the use of overlapping hoops or hoops with cross-ties, which results in several legs crossing the section. The better confinement

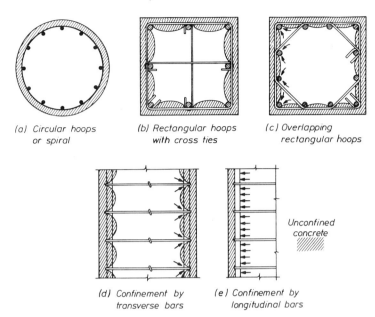

(a) Circular hoops or spiral

(b) Rectangular hoops with cross ties

(c) Overlapping rectangular hoops

Unconfined concrete

(d) Confinement by transverse bars

(e) Confinement by longitudinal bars

Fig. 3.4 Confinement of column sections by transverse and longitudinal reinforcement.

resulting from the presence of a number of transverse bar legs is illustrated in Fig. 3.4(b) and (c). The arching is more efficient since the arches are shallower, and hence more of the concrete area is effectively confined. For this reason, recommendations for the minimum spacing of vertical bars in columns are made in Section 3.6.1.

The presence of a number of longitudinal bars well distributed around the perimeter of the section, tied across the section, will also aid the confinement of the concrete. The concrete bears against the longitudinal bars and the transverse reinforcement provides the confining reactions to the longitudinal bars [see Fig. 3.4(d) and (e)].

Clearly, confinement of the concrete is improved if transverse reinforcement layers are placed relatively close together along the longitudinal axis. There will be some critical spacing of transverse reinforcement layers above which the section midway between the transverse sets will be ineffectively confined, and the averaging implied by Eq. (3.7) will be inappropriate. However, it is generally found that a more significant limitation on longitudinal spacing of transverse reinforcement s_h is imposed by the need to avoid buckling of longitudinal reinforcement under compression load. Experiments have indicated [M5, P38] that within potential hinging regions this spacing should not exceed six times the diameter of the longitudinal bar to be restrained.

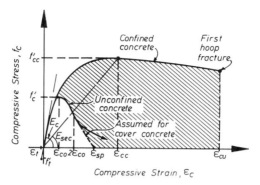

Fig. 3.5 Stress–strain model for monotonic loading of confined and unconfined concrete in compression.

(b) Compression Stress–Strain Relationships for Confined Concrete   The effect of confinement is to increase the compression strength and ultimate strain of concrete as noted above and illustrated in Fig. 3.5. Many different stress–strain relationships have been developed [B20, K2, M5, S16, V2] for confined concrete. For the designer, the significant parameters are the compression strength, the ultimate compression strain (needed for ductility calculations), and the equivalent stress block parameters.

(i) Compression Strength of Confined Concrete: The compression strength of confined concrete [M5] is directly related to the effective confining stress f_l' that can be developed at yield of the transverse reinforcement, which for circular sections is given by

$$f_l' = K_e f_l \tag{3.8}$$

and for rectangular sections is given by

$$f_{lx}' = K_e \rho_x f_{yh} \tag{3.9a}$$

$$f_{ly}' = K_e \rho_y f_{yh} \tag{3.9b}$$

in the x and y directions, respectively, where f_l' for circular sections is given by Eq. (3.7), ρ_x and ρ_y are the effective section area ratios of transverse reinforcement to core concrete cut by planes perpendicular to the x and y directions, as shown in Fig. 3.4(b) and (c), and K_e is a confinement effectiveness coefficient, relating the minimum area of the effectively confined core (see Fig. 3.4) to the nominal core area bounded by the centerline

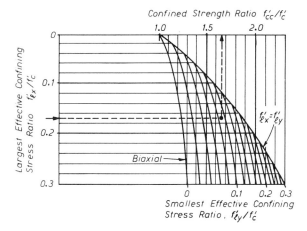

Fig. 3.6 Compression strength determination of confined concrete from lateral confining stresses for rectangular sections [P43].

of the peripheral hoops. Typical values of K_e are 0.95 for circular section, 0.75 for rectangular column sections, and 0.6 for rectangular wall sections.

The compression strength f'_{cc} of confined circular sections, or rectangular sections with equal effective confining stress f'_l in the orthogonal x and y directions is related to the unconfined strength by the relationship [M5, M13]

$$K = \frac{f'_{cc}}{f'_c} = \left(-1.254 + 2.254\sqrt{1 + \frac{7.94 f'_l}{f'_c}} - \frac{2 f'_l}{f'_c} \right) \qquad (3.10)$$

For a rectangular section with unequal effective confining stresses f'_{lx} and f'_{ly}, $K = f'_{cc}/f'_c$ may be found from Fig. 3.6, where $f'_{ly} > f'_{lx}$. The peak stress is attained (Fig. 3.5), at a strain of

$$\epsilon_{cc} = 0.002 \left[1 + 5(f'_{cc}/f'_c - 1) \right] \qquad (3.11)$$

(ii) Ultimate Compression Strain: The strain at peak stress given by Eq. (3.11) does not represent the maximum useful strain for design purposes, as high compression stresses can be maintained at strains several times larger (Fig. 3.5). The useful limit occurs when transverse confining steel fractures, which may be estimated by equating the strain-energy capacity of the transverse steel at fracture to the increase in energy absorbed by the concrete, shown shaded in Fig. 3.5 [M5]. A conservative estimate for ultimate

compression strain is given by

$$\epsilon_{cu} = 0.004 + 1.4\rho_s f_{yh}\epsilon_{sm}/f'_{cc} \tag{3.12}$$

where ϵ_{sm} is the steel strain at maximum tensile stress and ρ_s is the volumetric ratio of confining steel. For rectangular sections $\rho_s = \rho_x + \rho_y$. Typical values for ϵ_{cu} range from 0.012 to 0.05, a 4- to 16-fold increase over the traditionally assumed value for unconfined concrete.

(c) Influence of Cyclic Loading on Concrete Stress–Strain Relationship   Experiments on unconfined [P46] and confined [S15] concrete under cyclic loading have shown the monotonic loading stress–strain curve to form an envelope to the cyclic loading stress–strain response. As a consequence, no modification to the stress–strain curve is required when calculating the flexural strength of concrete elements subjected to the stress reversals typical of seismic loading.

(d) Effect of Strain Rate on Concrete Stress–Strain Relationship   Concrete exhibits a significant increase in both strength and stiffness when loaded at increased strain rates. Response to seismic loading is dynamic and compression strain rates in the critical plastic hinge regions may exceed 0.05 s^{-1}. The actual maximum strain rate is a function of the peak strain within the plastic hinge region and the effective period of the critical inelastic response displacement pulse of the structure.

Figure 3.7 shows typical increases in strength and initial stiffness of compressed concrete with strain rate, and indicate that for a strain rate of 0.05 s^{-1} and a typical strength of $f'_c = 30$ MPa (4350 psi), the compression strength would be enhanced by about 27% and the stiffness by about 16% compared with quasi-static strain rates. Under cyclic straining, this increase

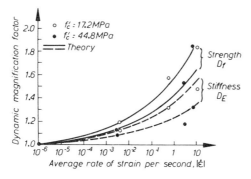

Fig. 3.7   Dynamic magnification factors D_f and D_E to allow for strain rate effects on strength and stiffness [P43]. (1 MPa = 145 psi.)

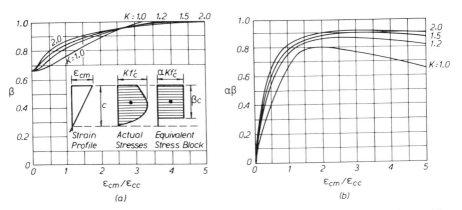

Fig. 3.8 Concrete compressive stress block parameters for rectangular sections with concrete confined by rectangular hoops for use with Eqs. (3.3) and (3.4) [P46].

dissipates, and an effective strain rate from start of testing to the current time is appropriate.

(e) Compression Stress Block Design Parameters for Confined Concrete   The approach taken for defining equivalent rectangular compression stress block parameters for unconfined concrete can be extended to confined concrete, provided that the average stress $\alpha f'_c$ shown in Fig. 3.2 is redefined as $\alpha K f'_c$, where K given by Eq. (3.10) or Fig. 3.6, is found from the assessed strength of the confined concrete. The appropriate values of α and β depend on the value of K and on the extreme compression fiber strain [P46]. Design values of β and the product $\alpha\beta$ are included in Fig. 3.8 for different values of peak compression strain ϵ_{cm}, expressed as the ratio $\epsilon_{cm}/\epsilon_{cc}$.

Values of α and β from Fig. 3.8 may be used in conjunction with the calculated value of K to predict the flexural strength of confined rectangular sections. However, it must be realized that the parameters apply only to the confined core. At high levels of compression strain, the concrete cover will have spalled and become ineffective, so the core dimensions, measured to the centerline of transverse confining reinforcement, should be used when calculating flexural strength.

Example 3.1: Design Parameters for a Rectangular Confined Column Section
The column section of Fig. 3.9 is confined by transverse sets of R16 hoops on $s_h = 90$-mm (0.63 in. diameter at 3.5 in.) centers, with $f_{yh} = 300$ MPa (43.5 ksi). The compression strength of the unconfined concrete is $f'_c = 30$ MPa (4350 psi). As shown in Fig. 3.9, the cross section dimensions are 500×400 (19.7 in. × 15.7 in.), and the confined core dimensions are 440×340 (17.3 in. × 13.4 in.). The steel strain corresponding to maximum stress may be

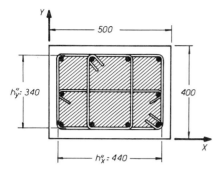

Fig. 3.9 Column section for Example 3.1.
(1 mm = 0.0394 in.)

taken as $\epsilon_{sm} = 0.15$. Calculate the strength of the confined core concrete, the ultimate compression strain, and the design parameters for the equivalent rectangular stress block.

SOLUTION: In the Y direction, there are four D16 mm (0.63-in.-diameter) legs. Consequently, the reinforcement ratio ρ_y is

$$\rho_y = \frac{4A_b}{sh''_x} = \frac{4 \times 201}{90 \times 440} = 0.0203$$

In the X direction, the central one-third of the section is confined by five legs, as a result of the additional central hoop with the remainder of the core confined by three legs. Taking an average value of 3.67 effective legs, we have

$$\rho_x = \frac{3.67A_b}{sh''_y} = \frac{3.67 \times 201}{90 \times 340} = 0.0241$$

Assuming an effectiveness coefficient of $K_e = 0.75$, then from Eq. (3.9),

$$f'_{lx}/f'_c = 0.75 \times 0.0241 \times 300/30 = 0.181$$
$$f'_{ly}/f'_c = 0.75 \times 0.0230 \times 300/30 = 0.152$$

From Fig. 3.6, entering on the left axis with 0.181 and interpolating between curves $f'_{lx}/f'_c = 0.14$ and 0.16 gives $K = f'_{cc}/f'_c = 1.82$ (follow the dashed line). The strength of the confined core is thus

$$f'_{cc} = 1.82 \times 30 = 54.6 \text{ MPa } (7920 \text{ psi})$$

Equation (3.12) gives the ultimate concrete compression strain

$$\epsilon_{cu} = 0.004 + 1.4\rho_s f_{yh}\epsilon_{sm}/f'_{cc}$$

where $\rho_s = \rho_x + \rho_y$ for rectangular confinement. Therefore, with $\rho_s = 0.0444$,

$$\epsilon_{cu} = 0.004 + 1.4 \times 0.0444 \times 300 \times 0.15/54.6 = 0.055$$

From Eq. (3.11),

$$\epsilon_{cc} = 0.002[1 + 5(54.6/30 - 1)] = 0.0102$$

Therefore, $\epsilon_{cu}/\epsilon_{cc} = 5.4$.

From Fig. 3.8(a), the appropriate design parameters for the equivalent rectangular stress block (extrapolating to $\epsilon_{cm}/\epsilon_{cc} = 5.4$) are

$$\beta = 1.0; \qquad \alpha\beta = 0.88$$

Thus the average strength to use for the equivalent rectangular stress block is $0.88 \times 54.6 = 48.0$ MPa (6960 psi). Note that using the full ultimate strain will be conservative for assessing the stress block parameters even if this strain is not reached during the design-level earthquake. However, for strains less than $\epsilon_{cu} = \epsilon_{cc}$, the values of α and β above will be nonconservative.

In this example it was assumed that the compression zone extended over virtually the complete core area. For a more realistic case where the neutral axis parallel to x was, say, close to middepth of the section, the additional confinement in the X direction provided by the central hoop should be ignored.

3.2.3 Masonry

Surprisingly, until recently, little information was available concerning the complete compression stress–strain behavior of masonry. To a large extent this is because most masonry design codes have insisted on elastic design to specified stress levels, even for seismic forces. As was established in Section 1.1.1, this approach is uneconomical if strictly applied, and potentially unsafe if applied to seismic forces reduced from the full elastic response level on the assumption of ductile behavior.

Elastic design requires knowledge of the initial modulus of elasticity E_m and the crushing strength f'_m. Other compression stress–strain parameters, such as ultimate compression strain and shape of the stress–strain curve, are superfluous to the needs of elastic design. As a consequence, the great majority of masonry compression tests have been designed to measure only E_m and f'_m.

More recently, considerable research effort has been directed toward establishing the necessary information for ductile strength design: namely, compression strength, ultimate strain, and compression stress block parameters for flexural strength design. These are discussed in the following sec-

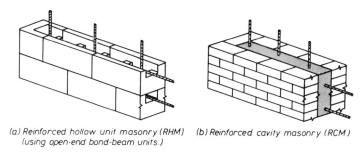

(a) Reinforced hollow unit masonry (RHM)   (b) Reinforced cavity masonry (RCM)
(using open-end bond-beam units.)

Fig. 3.10   Common forms of masonry construction for seismic regions.

tions. As with reinforced concrete, the tension strength of masonry is ignored in strength calculations.

The two main forms of masonry constructions are illustrated in Fig. 3.10. Hollow-block masonry consists of masonry units, most commonly with two vertical flues or cells to allow vertical reinforcement and grout to be placed. In Fig. 3.10(a) wall construction using bond beam units with depressed webs to facilitate placement of horizontal reinforcement is shown. Bond beam units also provide a passage for grout to flow horizontally, and it is recommended that these be used at all levels, regardless of whether horizontal reinforcement is placed at a given level, to provide a full interconnecting lattice of grout that will improve the integrity of the masonry. For the same reason, masonry units with one open end should be used, to avoid the weak header joint between end shells of adjacent blocks. The aim of masonry construction should be to make it structurally as close as possible to monolithic concrete construction. Hollow masonry units are typically constructed to a nominal 200×400 (8 in. $\times$ 16 in.) module size in elevation, with nominal widths of 100, 150, 200, 250, or 300 mm (4, 6, 8, 10, or 12 in.). Actual dimensions are typically 10 mm ($\frac{3}{8}$ in.) less than the nominal size to allow the placement of mortar beds. Grouted cavity masonry, a less common alternative form of construction, is depicted in Fig. 3.10(b). Two skins, or wythes, of solid masonry units (clay brick, concrete, or stone) are separated by a gap in which a two-way layer of reinforcement is placed, and which is subsequently filled with grout. The grouted cavity is typically 50 to 100 mm (2 to 4 in.) wide.

A further form of construction common in seismic regions of Central and South America and in China, generally termed *confined masonry construction*, involves a form of masonry-infilled frame in which the infill is first constructed out of unreinforced solid clay brick units. A reinforced concrete frame is then cast against the masonry panel, providing better integrity of the frame and panel than occurs with conventional infill construction where the panel is built after the frame. Vertical loads from the floor system are transmitted to the masonry panel, improving shear strength of the panel. The panel is sometimes built with castellated vertical edges, resulting naturally from the stretcher bonding of masonry units to improve the connection between column and panel.

(a) *Compression Strength of the Composite Material*   There are well-established and dependable methods for predicting the compression strength of concrete, given details on aggregate grading and aggregate/cement/water ratios. The compression strength of masonry (Figs. 3.11, 7.10, and 7.18), depending as it does on the properties of the masonry units (clay brick, concrete block, or stone), the mortar and the grout, is less easy to predict. As a consequence, most masonry design codes specify low design values for compression strength f'_m unless prism tests are carried out to confirm higher values. Because of the bulk of typical masonry prisms, the tests are difficult and expensive to perform, and most designers use the low-strength "default" option.

There is also a wide variation in the compression strength of the various constituents of masonry. The compression strength of the masonry units may vary from as low as 5 MPa (725 psi) for low-quality limestone blocks to over 100 MPa (14,500 psi) for high-fired ceramic clay units. Concrete masonry units vary in compression strength from 12 MPa (1750 psi) for some lightweight (pumice aggregate) blocks to over 30 MPa (4350 psi) for blocks made with strong aggregates. A minimum strength of about 12.5 MPa (1800 psi) is typically required by design codes.

The mortar strength depends on the proportion of cement, lime, and sand used in the mix, and the amount of water, which is generally added "by eye" by the mason to achieve a workable mix. It is also known that the strength of mortar test specimens bears little relationship to strength in the wall, because of absorption of moisture from the mortar by the masonry units.

The grout used for reinforced masonry construction can be characterized as a small–aggregate sized, high-slump concrete. High water/cement ratios are necessary to enable the grout to flow freely under vibration to all parts of the masonry flues or grout gaps. Free water loss to the masonry unit face shells can drastically reduce the workability of the grout. High cement contents are typically used to satisfy the contradictory requirements of specified minimum compression strength and flow ability. The result is often a grout that suffers excessive shrinkage in the masonry, causing the formation of voids since the slumping of mortar under the shrinkage is restrained by the sides of the flue or grout gap. To avoid this it is desirable to add an expansive agent to the mortar to compensate for expected high shrinkage. Compression strength of grout will again depend on the method adopted for sampling. More realistic values are obtained from cylinders molded directly from the casting against molds formed from masonry units lined with absorbent paper. A minimum compression strength of 17.5 MPa (2500 psi) is typically required for grout, but strengths up to 30 MPa (4350 psi) can be obtained without difficulty.

It is beyond the scope of this book to discuss masonry material properties in detail. For further information, the reader is referred to specialist texts [M5, S15].

(b) Ungrouted Masonry   The compression strength f'_p of a stack-bonded prism of masonry units of compression strength f'_{cb} bonded with mortar beds of compression strength $f'_j < f'_{cb}$ is invariably higher than the strength of the weaker mortar, as shown in Fig. 3.11. Further, failure appears to be precipitated by vertical splitting of the masonry unit rather than by crushing of the mortar. This behavior can be explained as a consequence of the mismatch of material properties of the masonry unit and mortar. Because of the lower strength and hence generally lower modulus of elasticity of the mortar compared with that of the masonry units, both axial and transverse (Poisson's ratio) strains in the mortar are higher than in the masonry units. As the axial stress approaches the crushing strength f'_j of the unconfined mortar, lateral mortar expansions increase markedly, unless restrained, as shown in Fig. 3.12(a) The combined effects of lower modulus of elasticity and higher Poisson's ratio result in a tendency for lateral mortar tensile strains to greatly exceed the lateral masonry unit strains. Since friction and adhesion at the mortar–masonry interface constrains the lateral strains of mortar and masonry unit to be equal, self-equilibrating lateral compression forces in the mortar and lateral tension forces in the masonry unit are set up [Fig. 3.12(b) and (c)]. The resulting triaxial compression stress state in the mortar enhances its crushing strength, while the combination of longitudinal compression and lateral biaxial tension in the masonry unit reduces its crushing strength and induces a propensity for vertical splitting.

The strength of the confined mortar may be approximated by

$$f'_{cj} = f'_{cb} + 4.1 f_l \tag{3.13}$$

where f'_{cj} is the compression strength of the confined mortar, and f_l is the

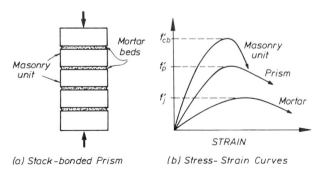

(a) Stack-bonded Prism   (b) Stress-Strain Curves

Fig. 3.11   Stress–strain behavior of masonry prism.

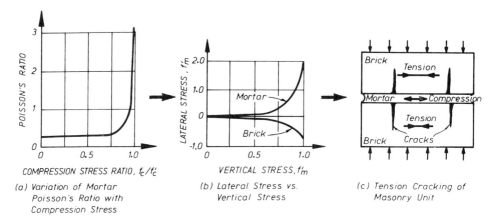

Fig. 3.12   Failure mechanism for masonry prisms.

lateral compression stress developed in the mortar. Hilsdorf [H1] proposed a linear failure criterion for the masonry unit as shown by the failure envelope of Fig. 3.13. This failure criterion may be written as

$$f_x/f'_{tb} + f_y/f'_{cb} = 1 \qquad (3.14)$$

where f'_{cb} and f'_{tb} are uniaxial compression and biaxial tension strengths of the masonry unit, and f_y is the axial compression stress occurring in conjunction with lateral tension stress f_x, at failure. Figure 3.13 also shows the stress path to failure of the masonry unit and a simplified path that ignores the stresses induced by different elastic lateral tension strains between mortar and masonry unit as being insignificant compared with the increase in Poisson's ratio at high mortar strains, indicated in Fig. 3.12.

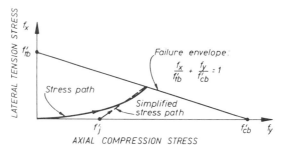

Fig. 3.13   Mohr's failure criterion for a masonry unit.

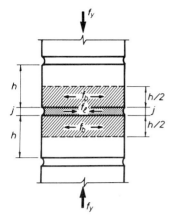

Fig. 3.14 Transverse equilibrium of masonry unit and mortar in prism.

By considering the transverse equilibrium requirements of a mortar joint of thickness j, and a tributary height of masonry unit equal to one-half a masonry unit above and below the joint, as shown in Fig. 3.14, in conjunction with Eqs. (3.13) and (3.14), the longitudinal stress f'_p to cause failure is found to be

$$f'_p = f_y = \frac{f'_{cb}(f'_{tb} + \alpha f'_j)}{U_u(f'_{tb} + \alpha f'_{cb})} \tag{3.15}$$

where

$$\alpha = j/4.1h \tag{3.16}$$

and h is the height of the masonry unit. The stress nonuniformity coefficient U_u is 1.5 [H1].

(c)   Grouted Concrete Masonry   For grouted masonry, either of hollow unit masonry or grouted cavity masonry, the shell strength (masonry unit and mortar) can be expected to be given by Eq. (3.15), where the strength f'_p will apply to the net area of the masonry units. The grout strength f'_g will apply to the area of grout in flues or cavities. Since the strength f'_p of a concrete masonry shell is typically reached at a strain less than that applicable to the grout, direct addition of strengths is not appropriate. Further, parameter studies of the influence of the masonry unit biaxial tensile strength f'_{tb} and the mortar strength f'_j and shell strength f'_p given by Eq. (3.15) indicate comparative insensitivity to these variables, and the composite compression strength can be conservatively approximated by

$$f'_m = \phi\left[0.59xf'_{cb} + 0.90(1 - x)f'_g\right] \tag{3.17}$$

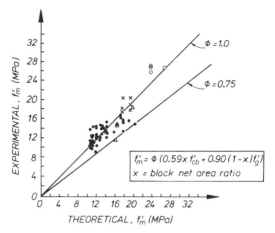

Fig. 3.15 Comparison of measured prism strength with predictions for grouted concrete masonry by Eq. (3.17) [P53]. (1 MPa = 145 psi.)

where x is the ratio of net block area to gross area. A value of $\phi = 1.0$ provides a good average agreement between Eq. (3.17) and test results as shown in Fig. 3.15 [P53]. A value of $\phi = 0.75$ provides a lower bound to the data and is useful when adopting a suitable design value for concrete masonry.

(d)   Grouted Brick Masonry   For clay brick masonry, the influence of mortar strength on prism strength is more significant. The compression strength may be expressed as

$$f'_m = \phi\left[xf'_p + (1 - x)f'_g\right] \tag{3.18}$$

where f'_p is given by Eq. (3.15).

Example 3.2   Predict the compression strength of a grouted brick cavity masonry wall [see Fig. 3.10(b)] given that the brick size is 65 mm high × 240 mm long × 95 mm (2.56 in. × 9.45 in. × 3.75 in.) wide, the grout gap between the two wythes of brick is 100 mm (3.94 in.), the mortar joint thickness is 10 mm (0.39 in.), and the material strengths are $f'_{cb} = 40$ MPa (5800 psi), $f'_g = 20$ MPa (2900 psi), $f'_j = 9.6$ MPa (1400 psi), and the stress nonuniformity coefficient is $U_u = 1.5$.

SOLUTION:

$$\text{Net area ratio } x = 2 \times 95/(2 \times 95 + 100) = 0.655$$
$$\text{Height factor } \alpha = j/4.1h = 10/(4.1 \times 65) = 0.0375$$

From Eq. (3.15), assuming that $(f'_{tb}/f'_{cb}) = 0.1$,

$$f'_p = 40(4 + 0.0375 \times 9.6)/[1.5(4 + 0.0375 \times 40)] = 21.1 \text{ MPa (3060 psi)}$$

From Eq. (3.18), taking $\phi = 1.0$,

$$f'_m = xf'_p + (1 - x)f'_g = 0.655 \times 21.1 + 0.345 \times 20 = 20.7 \text{ MPa (3000 psi)}$$

Note that doubling the biaxial tension strength ratio to $(f'_{tb}/f'_{cb}) = 0.2$ would result in $f'_p = 23.5$ MPa (3410 psi) and $f'_m = 22.3$ MPa (3230 psi). It will be seen that the compression strength of brick masonry is not very sensitive to the biaxial tension strength.

As with concrete masonry, a strength reduction of $\phi = 0.75$ should be used in conjunction with Eq. (3.18) for design calculations.

(e) *Modulus of Elasticity* There is still a lack of consensus as to the appropriate relationship between modulus of elasticity E_m and compression strength f'_m of masonry. In part this stems from the considerable variability inherent in a material with widely ranging constituent material properties, but it is also related to different methods adopted to measure strain in compression tests. Bearing in mind that it is advisable to adopt conservatively high values for E_m to ensure seismic lateral design forces are not underestimated, the following values are recommended:

$$\text{Concrete masonry:} \quad E_m = 1000 f'_m$$
$$\text{Clay brick masonry:} \quad E_m = 750 f'_m$$

(f) *Compression Stress–Strain Relationships for Unconfined and Confined Masonry* Compression stress–strain curves for masonry are similar to those for concrete and may be represented by similar equations, provided that allowance is made for a somewhat reduced strain corresponding to the development of peak compression stress [P61]. An increased tendency for splitting failure noted in Fig. 3.12 means that the ultimate compression strain of masonry, at 0.0025 to 0.003, is lower than that appropriate to concrete.

It was noted in Section 3.2.2 that the compression stress–strain characteristics of concrete can be greatly improved by close-spaced transverse reinforcement in the form of ties, hoops, or spirals, which increase the strength and ductility of the concrete. A degree of confinement to masonry can be provided by thin galvanized steel or stainless steel plates placed within the mortar beds in critical regions of masonry elements. Tests on prisms with 3-mm-thick confining plates, cut to a pattern slightly smaller than the block net area, enabling the plates to be placed in the mortar beds without impeding grout continuity, show increased strength and ductility compared with results from unconfined prisms, as shown in Fig. 3.16. This was despite the wide vertical spacing of 200 mm (8 in.) dictated by the block module size.

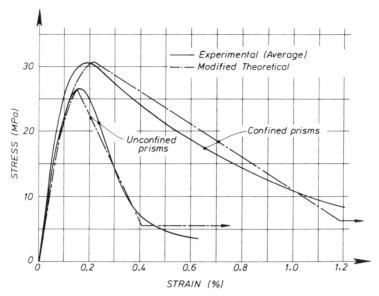

Fig. 3.16 Compression stress–strain curves for concrete masonry [P61]. (1 MPa = 145 psi.)

Figure 3.16 also compares the stress–strain curves with theoretical predictions based on a modification of equations for confined concrete [P61].

(g) *Compression Stress Block Design Parameters for Masonry* The approach taken for defining equivalent rectangular stress block parameters for concrete in Section 3.2.1 can also be extended to masonry. Based on experimental results and theoretical stress–strain curves for unconfined and confined masonry, the following parameters are recommended:

Unconfined masonry:
$$\alpha = 0.85, \quad \beta = 0.85, \quad \epsilon_{cu} = 0.003$$

Confined masonry (with confining plates equivalent to $\rho_s = 0.0077$):
$$\alpha = 0.9K, \quad \beta = 0.96, \quad \epsilon_{cu} = 0.008$$

where the strength enhancement factor [Eq. (3.10)] for confined masonry K can be approximated by

$$K = 1 + \rho_s f_{yh}/f'_m \tag{3.19}$$

where f_{yh} is the yield strength of the confining plate material. Figure 3.17

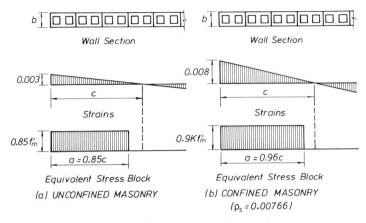

Fig. 3.17 Compression stress block parameters for unconfined and confined masonry.

shows ultimate strain profiles and the corresponding equivalent stress blocks for unconfined and confined masonry.

3.2.4 Reinforcing Steel

(a) Monotonic Characteristics The prime source of ductility of reinforced concrete and masonry structural elements is the ability of reinforcing steel to sustain repeated load cycles to high levels of plastic strain without significant reduction in stress. Figure 3.18 shows representative curves for different strengths of reinforcing steel commonly used in concrete and masonry construction. Behavior is characterized by an initial linearly elastic portion of the stress–strain relationship, with a modulus of elasticity of approximately $E_s = 200$ GPa (29,000 ksi), up to the yield stress f_y, followed by a yield plateau of variable length and a subsequent region of strain hardening. After maximum stress is reached, typically at about $f_{su} \simeq 1.5 f_y$, strain softening occurs, with deformation concentrating at a localized weak spot. In terms of structural response the strain at peak stress may be considered the "ultimate" strain, since the effective strain at fracture depends on the gauge length over which measurement is made. In structural elements, the length of reinforcement subjected to effectively constant stress may be considerable.

Figure 3.18 indicates that typically, ultimate strain and the length of the yield plateau decrease as the yield strength increases. This trend is, however, not an essential attribute. It is possible, without undue effort, and desirable from a structural viewpoint, for steel manufacturers to produce reinforcing steel with a yield strength of 400 MPa (58 ksi) or 500 MPa (72.5 ksi) while

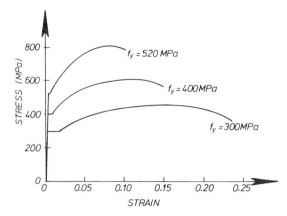

Fig. 3.18   Typical stress–strain curves for reinforcing steel. (1 MPa = 145 psi.)

retaining the ductility of lower strength reinforcement, as has recently been demonstrated by New Zealand steel producers [A15].

The desirable characteristics of reinforcing steel are a long yield plateau followed by gradual strain hardening and low variability of actual yield strength from the specified nominal value. The desire for these properties stems from the requirements of capacity design, namely that the shear strength of all elements and flexural strength of sections not detailed as intended plastic hinges should exceed the forces corresponding to development of flexural overstrength at the chosen plastic hinge locations. If the reinforcing steel exhibits early and rapid strain hardening, the steel stress at a section with high ductility may exceed the yield stress by an excessive margin. Similarly, if the steel for a specified grade of reinforcement is subject to considerable variation in yield strength, the actual flexural strength of a plastic hinge may greatly exceed the nominal specified value. In both cases, the result will be a need to adopt high overstrength factors (Section 1.3.3) to protect against shear failures or unexpected flexural hinging.

Of particular concern in the United States is the practice of designating reinforcing steel that has failed the acceptance level for grade 60 (415 MPa) strength as grade 40 (275 MPa) reinforcement for structural usage. This provides reinforcement with a nominal yield strength of 275 MPa (40 ksi), but typical yield strength in the range 380 to 400 MPa (55 to 58 ksi). For the reason discussed above, such reinforcement should not be used for construction in seismic regions.

(b). Inelastic Cyclic Response   When reinforcing steel is subject to cyclic loading in the inelastic range, the yield plateau is suppressed and the stress–strain curve exhibits the Bauschinger effect, in which nonlinear re-

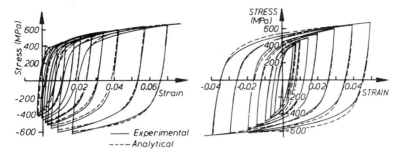

Fig. 3.19   Cyclic straining of reinforcing steel [f_y = 380 MPa (55 ksi)] [L3].

sponse develops at a strain much lower than the yield strain. Figure 3.19 shows results of two different types of cyclic testing of reinforcing steel. In Fig. 3.19(a) the cyclic inelastic excursions are predominantly in the tensile strain range, while in Fig. 3.19(b) the excursions are symmetrically tensile and compressive. The former case is typical of reinforcement in a beam plastic hinge that is unlikely to experience large inelastic compression strain. For such a response the monotonic stress-strain curve provides an envelope to the cyclic response.

Symmetrical strain behavior such as that shown in Fig. 3.19(b) can result during cyclic response of columns with moderate to high axial load levels. As the amplitude of response increases, the stress level for a given strain also increases and can substantially exceed the stress indicated by the monotonic stress–strain curve.

(c) Strain Rate Effects   At strain rates characteristic of seismic response (0.01 to 0.10 s^{-1}), reinforcing steel exhibits a significant increase in yield strength above static test values. For normal-strength steel [300 MPa (43.5 ksi) $\leq f_y \leq$ 400 MPa (58 ksi)], yield strength is increased by about 10 and 20% for strain rates of 0.01 and 0.10 s^{-1}, respectively [M13]. Under cyclic straining the effective strain rate decreases, minimizing this effect.

(d) Temperature and Strain Aging Effects   Below a certain temperature (typically about $-20°$C), the ductility of reinforcing steel is lost and it behaves in a brittle fashion on reaching the yield stress. Care is thus needed when designing structures for ductile response in cold climates. A related effect relevant to warmer climates is the gradual increase, with time, of the threshold temperature between brittle and ductile steel behavior subsequent to plastic straining of reinforcement [P47]. The threshold temperature may rise to as high as $+20°$C in time. Thus reinforcing steel that has been plastically strained to form a bend or standard hook will eventually exhibit brittle characteristics at the region of the bend. It is thus essential to ensure,

by appropriate detailing, that the steel stress in such regions can never approach the yield stress.

It would appear that structures which have responded inelastically during earthquake response may be subject to the effects of strain aging and could behave in brittle fashion in a subsequent earthquake. This possible effect deserves more research attention.

(e)   Overstrength Factor (λ_o)   As mentioned in Section 1.3.3(f), it is necessary to assess maximum feasible flexural overstrength of sections in the capacity design of structures. This overstrength results primarily from variability of reinforcement actual yield strength above the specified nominal value, and from strain hardening of reinforcement at high ductility levels. Thus the overstrength factor λ_o can be expressed as

$$\lambda_o = \lambda_1 + \lambda_2 \tag{3.20}$$

where λ_1 represents the ratio of actual to specified yield strength and λ_2 represents the potential increase resulting from strain hardening.

λ_1 will depend on where the local supply of reinforcing steel comes from, and considerable variability is common, as noted earlier in this section. With tight control of steel manufacture, values of $\lambda_1 = 1.15$ are appropriate. It is recommended that designers make the effort to establish the local variation in yield strength, and where this is excessive, to specify in construction specifications the acceptable limits to yield strength. Since steel suppliers keep records of yield strength of all steel in stock, this does not cause any difficulties with supply.

λ_2 depends primarily on yield strength and steel composition, and again should be locally verified. If the steel exhibits trends as shown in Fig. 3.18, the appropriate values may be taken as

$$\text{for } f_y = 275 \text{ MPa (40 ksi)} \qquad \lambda_2 = 1.10$$
$$\text{for } f_y = 400 \text{ MPa (58 ksi)} \qquad \lambda_2 = 1.25$$

For $\lambda_1 = 1.15$, these result in $\lambda_o = 1.25$ and 1.40 for $f_y = 275$ and 400 MPa (40 and 60 ksi), respectively.

3.3   ANALYSIS OF MEMBER SECTIONS

3.3.1   Flexural Strength Equations for Concrete and Masonry Sections

It is assumed that the reader is familiar with the analysis of reinforced concrete elements for flexural strength, and only a very brief review will be included in this book. More complete coverage is given in basic texts on reinforced concrete [P1]. Simplified design methods are covered in subsequent chapters.

(a) Assumptions   The normal assumptions made in assessing flexural strength of concrete and masonry elements are:

1. Initially plane sections remain plane after bending. As ideal strength is approached, and particularly where diagonal cracks occur as a result of high shear stresses, significant departure from a linear strain profile, implied by this assumption, may occur. However, this has negligible effect on accuracy of estimating flexural strength.

2. Perfect bond exists between reinforcement and concrete. Whenever shear forces are to be resisted, bond stresses are generated, and relative displacements between a bar and its surrounding concrete, termed *slip*, are inevitable. Generally, this effect is negligible. At location of very high shear, such as beam–column joints, bar slip may be large enough to significantly affect predictions of both steel and concrete or masonry forces within the section. In such locations the phenomenon requires special attention.

3. Concrete tension strength at the critical section is ignored after cracking.

4. Equivalent stress block parameters are used to describe the magnitude and centroidal position of the internal concrete or masonry compression force.

5. The steel stress–strain characteristic is idealized by an elastoplastic approximation. That is, strain hardening is ignored. This assumption is not necessary if the full stress–strain characteristics are known, but is conservative and convenient in assessing flexural strength.

6. Flexural strength is attained when the extreme concrete compression fiber reaches the ultimate compression strain, ϵ_{cu}.

(b) Flexural Strength of Beam Sections   Figure 3.20 shows stress resultants of a doubly reinforced beam section at flexural strength. With the use of the

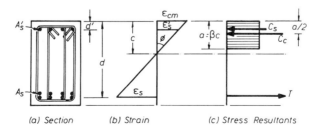

Fig. 3.20   Equilibrium of a beam section at flexural strength.

stress block parameters defined in Sections 3.2.1(b), 3.2.2(e), and 3.2.3(g), two equilibrium equations and a strain compatibility relationship enable the ideal flexural strength to be determined, as follows:

1. Force equilibrium requires that

$$C_c + C_s = T \tag{3.21}$$

that is,

$$\alpha f'_c ab + A'_s f'_s = A_s f_s$$

2. Moment equilibrium (about centroid of tension steel) requires that

$$M_i = C_c(d - a/2) + C_s(d - d') \tag{3.22}$$

3. Strain compatibility is satisfied when

$$\epsilon'_s = \epsilon_{cu}\frac{c - d'}{c} \tag{3.23a}$$

$$\epsilon_s = \epsilon_{cu}\frac{d - c}{c} \tag{3.23b}$$

If the steel strains ϵ'_s and ϵ_s exceed the yield strain ϵ_y, the steel stress is put equal to the yield stress f_y; otherwise, $f'_s = E_s \epsilon'_s$, $f_s = E_s \epsilon_s$.

In Eqs. (3.21) to (3.23), f_s and f'_s are the steel stress at the centroid of tension and compression steel, respectively, and $a = \beta c$. For beams designed for seismic response, ϵ_s will always exceed ϵ_y, even if the section is not required to exhibit ductility. Hence $f_s = f_y$. However, since substantial areas of compression reinforcement are commonly required for beam sections because they are subjected to moment reversal under seismic response [P51], the compression reinforcement will generally not yield. The compression steel stress can thus be expressed as

$$f'_s = E_s \epsilon_{cu}(a - \beta d')/a \le f_y \tag{3.24}$$

Substitution into Eq. (3.21) produces an expression where $a = \beta c$ is the only unknown, and hence solution for a is possible. Substitution of this value in Eq. (3.22) then enables the ideal flexural strength to be computed. A simplified approach applicable when beam sections are subjected to high ductility levels, sufficient to cause spalling of cover concrete, is presented in Section 4.5.1(a).

When large amounts of tension reinforcement (A_s) are used without, or with small amounts of compression reinforcement (A'_s), the corresponding large compression force C_c in Fig. 3.20(c) will necessitate an increase of the neutral axis depth c. It is then possible that the ultimate compression strain in the unconfined concrete is reached before the tension steel yields. Where ductile response of a beam is desired, clearly this situation must be avoided by limiting the maximum amount of tension reinforcement. This is examined and discussed in Sections 3.4.2 and 4.5.1, where corresponding recommendations are made.

(c) Flexural Strength of Column and Wall Sections   The addition of axial force and a distribution of reinforcement that cannot readily be represented by lumped compression and tension areas at corresponding centroidal positions make the calculation of flexural strength of columns and walls more tedious than for beams. As a consequence, flexural strength of columns and walls are often estimated using design charts or computer programs, often developed in-house by designers. However, the principles are straightforward and are summarized below.

Consider the column section of Fig. 3.21(a) subjected to axial force P_i, resulting from gravity and seismic actions. The same three conditions as for beams (two equilibrium equations and one strain compatability relationship) are used to define the response.

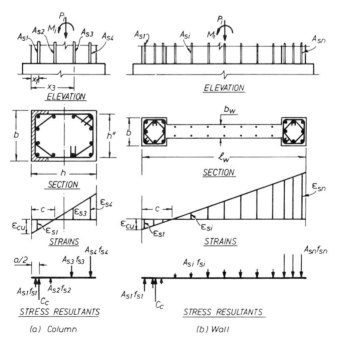

Fig. 3.21 Equilibrium of column and wall sections at flexural strength.

1. Force equilibrium:

$$C_c + \sum_1^4 A_{si} f_{si} = P_i \qquad (3.25)$$

where

$$C_c = \alpha f_c' ab$$

2. Moment equilibrium is expressed with respect to the neutral axis, for convenience:

$$M_i = C_c\left(c - \frac{a}{2}\right) + \sum_1^4 A_{si} f_{si}(c - x_i) + P_i\left(\frac{h}{2} - c\right) \qquad (3.26)$$

3. Strain compatibility:

$$\epsilon_{si} = \epsilon_{cu} \frac{c - x_i}{c}$$

and

$$f_y \geq E_s \epsilon_{si} \geq -f_y \qquad (3.27)$$

Equation (3.26) can be solved after a trial-and-error approach based on successive approximation for c (or $a = \beta c$) is used to solve Eqs. (3.25) and (3.27) simultaneously.

The approach for structural walls, represented in Fig. 3.21(b), is identical and can be obtained by generalizing Eqs. (3.25) and (3.26) from 4 to n layers of steel. However, if, as is the case in Fig. 3.21(b), the neutral axis is in the web of the wall rather than the boundary element, the center of compression of C_c will not be at middepth of the equivalent rectangular compression stress block. The necessary corrections are obvious.

It should be noted that all the web vertical reinforcement has been included in the flexural strength assessment. This should be the case even if only nominal reinforcement is placed in accordance with code minimum requirements, since an accurate assessment of strength is essential when capacity design procedures are used to determine maximum feasible shear force to be resisted by the wall and its foundation.

When implementing the procedure outlined in Eqs. (3.25) to (3.27), some care needs to be exercised in the relationship between ultimate compression strain and section dimensions. If the ultimate compression strain ϵ_{cu} corresponds to the crushing strain (0.004 to 0.005), the full section dimensions may be used. However, if much higher compression strain are required, as will be the case for columns required to exhibit significant ductility, that portion of cover concrete where strains exceed 0.004, shown shaded in Fig. 3.21(a), should be ignored in the analysis. In conjunction with the reduced compression area, stress block parameters appropriate to confined concrete, defined in Fig. 3.8, should be used.

The influence of confinement on the flexural strength of beam and column members is typically not significant for low levels of axial load. However, at high levels of axial load, the significance of enhanced concrete compression strength becomes considerable, as illustrated by Fig. 3.22, which compares experimental flexural strengths of columns of circular, square, and rectangular section with predictions based on conventional flexural strength theory using measured material strengths, an ultimate compression strain of 0.003, the full section dimension, including cover, and a strength reduction factor of $\phi = 1.0$. The increased influence of confined compression strengths results from the increase in compression zone depth, c, with axial load, and hence the greater importance of the term $C_c(c - a/2)$ to the total flexural strength in Eq. (3.26). At low levels of axial load, the average ratio of experimental strength to code-predicted strength based on measured material strength properties is 1.13, resulting primarily from the effects of strain hardening of flexural reinforcement at high ductility factors. At higher axial loads, particularly for $P/f_c' A_g \geq 0.3$, the strength enhancement factor increases rapidly. As an alternative to predicting the flexural strength of column sections using stress block parameters based on confined concrete and the core concrete

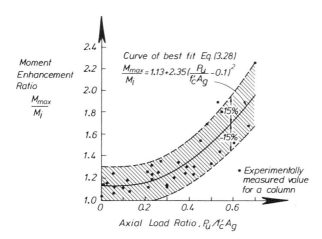

Fig. 3.22 Flexural strength enhancement of confined columns at different axial force levels [A13].

dimensions, the average value of the experimentally obtained strength enhancement factor may be used. As shown in Fig. 3.22, this is given by

$$\frac{M_{\max}}{M_i} = 1.13 + 2.35\left(\frac{P_i}{f_c' A_g} - 0.1\right)^2 \tag{3.28}$$

The experimental data fall within $\pm 15\%$ of this equation. When the design is based on the specified yield strength of the steel, 1.13 in Eq. (3.28) should be replaced by λ_o [Section 3.2.4(e)].

For wall sections, the depth of compression will not normally be great enough for significant strength enhancement to result from the increased compression strength of confined concrete. The reduction in section area resulting from cover spalling will typically more than compensate for the increased compression strength. Consequently, the flexural strength corresponding to development of the extreme fiber crushing strain (0.004) is likely to be a good estimate of ideal strength, with any strength enhancement resulting from the steel yield stress exceeding the nominal value, and from strain hardening of the steel at high ductilities.

3.3.2 Shear Strength

Under actions due to gravity loads and earthquake forces, the development of cracks in reinforced concrete and masonry structures must be considered to be inevitable. Therefore, the shear strength of concrete alone cannot be relied on, and hence all members of frames or structural walls must be provided with sufficient shear reinforcement.

(a) Control of Diagonal Tension and Compression Failures The design for shear resistance is based on well-established models of shear mechanisms [C3, P1]. The use of truss models in particular is generally accepted. In these the tension chords consist of flexural tension reinforcement, the compression chords are assumed to consist of concrete alone, and shear resistance is assigned to the web, which comprises a diagonal concrete compression field combined with web tension reinforcement. In some models and in most code specifications [A1, X3] a part of the shear resistance is assigned to mechanisms other than the model truss. Aggregate interlock along crack interfaces, dowel action of chord reinforcement, shear transfer by concrete in the flexural compression regions, arch action, and the tensile strength of uncracked concrete are typical components of such complex mechanisms. The combined strength of these mechanisms is commonly termed the *contribution of the concrete* to shear strength. Its magnitude, in conjunction with a truss model consisting of diagonal struts at 45° to the axis of a member, is based on empirical relationships.

Under gravity loading the sense of the shear force at a given section does not change, or if it does, as a result of different live-load patterns, the effect of the change is insignificant. However, under the actions due to earthquake forces, reversal of shear forces over significant parts or over the entire length of members will be common. Response of web reinforcement to cyclic straining imposed by seismic response does not involve significant cyclic degradation, provided that the reinforcement remains in the elastic range. However, the concrete of the web, subjected to diagonal compression, may be seriously affected by seismic actions. This is because at each shear reversal the direction of diagonal concrete struts of the truss model changes by approximately 90°. Moreover, as a result of similar changes in the directions of principal tensile strains, diagonal cracks, crossing each other at approximately 90°, will also develop (Fig. 5.37). Thus the compression response of diagonally cracked concrete, with alternating opening and closure of the cracks, needs to be considered. Shear transfer along diagonal cracks in turn will depend on the strains developed in the web reinforcement, which crosses these cracks at some angle.

If the web reinforcement is permitted to yield, significant shear deformations will result. The closure of wide diagonal cracks upon force reversal is associated with insignificant shear and hence seismic resistance. As a consequence, marked reduction of energy dissipation will also occur during the corresponding hysteretic response of affected members. Typical "pinching" effects on hysteretic response due to inelastic shear deformations of this type is seen in Fig. 2.20(d). For this reason the prevention of yielding of the web reinforcement during the progressive development of full collapse mechanisms within structural systems is one of the aims of the capacity design procedure. Mechanisms associated with yielding stirrups, leading to large shear deformations, are illustrated in Fig. 3.23. The approach presented here and used in subsequent design examples is extracted from one code [X3], but it follows the general trends embodied in many other codes [A1].

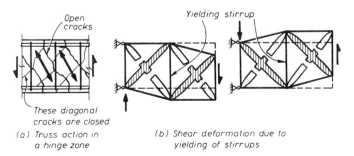

Fig. 3.23   Mechanisms of shear transfer in plastic hinges [F3].

(i) Nominal Shear Stress: For convenience in routine design, shear strength is commonly quantified in terms of a nominal shear stress v_i, defined as

$$v_i = V_i/b_w d \tag{3.29}$$

where V_i is the ideal shear strength at a particular section of the member, and b_w and d are the width of the web and the effective depth, respectively, of the member at the same section. No physical meaning, in terms of real stresses, should be attached to v_i. It should be used as an index only, measuring the magnitude of a shear force relative to the cross section of the member. Note that in terms of the definitions of strength in Section 1.3.3, $V_i \geq \phi V_u$. In members affected by seismic forces V_i will generally be derived from capacity design consideration (Section 1.4.3), details of which are given in subsequent chapters for various types of structures and members.

(ii) Limitations on Nominal Shear Stress: To ensure that premature diagonal compression failure will not occur in the web before the onset of yielding of the web shear reinforcement, the nominal shear stress needs to be limited. Recommended limitations are:

1. In general

$$v_i \leq 0.2 f'_c \leq 6 \text{ MPa (870 psi)} \tag{3.30}$$

2. In plastic hinge regions of beams, columns, and walls:

$$v_i \leq 0.16 f'_c \leq 6 \text{ MPa (870 psi)} \tag{3.31}$$

3. In structural walls, in accordance with Section 5.4.4, where dynamic effects and expected ductility demands are also taken into account.
4. In diagonally reinforced coupling beams, no limitation need be applied because negligible reliance is placed on the contribution of concrete to shear resistance (i.e., on its diagonal compression strength) [Section 5.4.5(*b*)].

When the computed shear stress exceeds the values given above, the dimensions of the member should be increased.

(iii) Shear Strength: The shear strength at a section of a member is derived from

$$V_i = V_c + V_s \tag{3.32}$$

where $V_c = v_c b_w d$ is the contribution of the concrete to shear strength, and V_s is the contribution of shear reinforcement.

(iv) *Contribution of the Concrete, v_c:* The contribution of the concrete to shear strength, expressed in terms of nominal shear stress, may be taken as:

1. In all regions except potential plastic hinges
    In cases of flexure only:

$$v_c = v_b = (0.07 + 10\rho_w)\sqrt{f_c'} \leq 0.2\sqrt{f_c'} \quad \text{(MPa)}$$

$$= (0.85 + 120\rho_w)\sqrt{f_c'} \leq 2.4\sqrt{f_c'} \quad \text{(psi)} \tag{3.33}$$

where the ratio of the flexural tension reinforcement ρ_w is expressed in terms of the web width b_w.
In cases of flexure with axial compression P_u:

$$v_c = (1 + 3P_u/A_g f_c')v_b \tag{3.34}$$

In cases of flexure with axial tension when $P_u < 0$:

$$v_c = (1 + 12P_u/A_g f_c')v_b \tag{3.35}$$

In structural walls:

$$v_c = 0.27\sqrt{f_c'} + P_u/4A_g \quad \text{(MPa)}$$

$$3.3\sqrt{f_c'} + P_u/4A_g \quad \text{(psi)} \tag{3.36}$$

When the force load P_u produces tension, its value in Eqs. (3.35) and (3.36) must be taken negative.

2. In regions of plastic hinges:
    In beams:

$$v_c = 0 \tag{3.37}$$

In columns:

$$v_c = 4v_b\sqrt{P_u/A_g f_c'} \tag{3.38}$$

In walls:

$$v_c = 0.6\sqrt{P_u/A_g} \quad \text{(MPa)}; \qquad 7.2\sqrt{P_u/A_g} \quad \text{(psi)} \tag{3.39}$$

Equations (3.38) and (3.39) apply when the axial load P_u results in compression. When P_u represents tension, $v_c = 0$. A_g is the area of the gross concrete section. Regions of plastic hinges, over which the equations above are applicable, are given in Sections 4.5.1(*d*) and 4.6.11(*e*) for beams and columns, respectively, and in Section 5.4.3(*e*)(iii) for walls.

(v) *Contribution of Shear Reinforcement*: To prevent a shear failure result-ing from diagonal tension, shear reinforcement, generally in the form of stirrups, placed at right angles to the axis of a member, is to be provided to resist the difference between the total shear force V_i and the contribution of the concrete V_c. Accordingly, the area of a set of stirrups A_v with spacing s along a member is

$$A_v \geq \frac{(v_i - v_c)b_w s}{f_y} \qquad (3.40)$$

The contribution of shear reinforcement to the total shear strength V_i, based here on truss models with 45° diagonal struts, is thus

$$V_s = A_v f_y (d/s) \qquad (3.41)$$

Because of the reversal of shear forces in members affected by earthquakes, the placing of stirrups at angles other than 90° to the axis of such members is generally impractical.

The choice of the angle (45°) for the plane of the diagonal tension failure in the region of potential plastic is a compromise. Some codes [X5] have adopted the more rational variable-angle truss analogy to calculate the required amount of web reinforcement. However, as a result of bond deterio-ration along yielding flexural bars, the adoption of assumptions used in the derivation of the angle of the diagonal compression field [C3] may be unsafe. Under repeated reversed cyclic loading, failure planes in plastic hinges at angles to the axis of the member larger than 45° have been observed [F3], leading to eventual yielding of stirrups that have been provided in accordance with these recommendations based on 45° failure planes.

(vi) *Minimum Shear Reinforcement*: Current codes require the provision of a minimum amount of shear reinforcement in the range $0.0015 \leq A_v/b_w s \leq 0.0020$ in members affected by earthquake forces.

(vii) *Spacing of Stirrups*: To ensure that potential diagonal tension failure planes are crossed by sufficient sets of stirrups, spacing limitations, such as set out below, have been widely used. The spacing s should not exceed:

1. In beams
    In general: $0.50d$ or 600 mm (24 in.)
    When $(v_i - v_c) > 0.07f_c'$: $0.25d$ or 300 mm (12 in.)
2. In columns
    When $P_u/A_g \leq 0.12f_c'$: as in beams
    When $P_u/A_g > 0.12f_c'$: $0.75h$ or 600 mm (24 in.)
3. In walls,
    2.5 times the wall thickness or 450 mm (18 in.)

Spacing limitations to satisfy requirements for the confinement of compressed concrete and the stabilizing of compression bars in potential plastic hinge regions are likely to be more restrictive.

(b) Sliding Shear   The possibility of failure by sliding shear is a special feature of structures subjected to earthquakes. Construction joints across members, particularly when poorly prepared, present special hazards. Flexural cracks, interconnected during reversed cyclic loading of members, especially in plastic hinge regions, may also become potential sliding planes. The treatment of the relevant issues, presented in Sections 5.4.4(c) and 5.7.4, is somewhat different from that for members of frames.

(i) Sliding Shear in Walls: Shear transfer across potential sliding planes across walls, where construction joints occur or where wide flexural cracks originating from each of the two edges interconnect, may be based on mechanisms of aggregate interlock [P1], also termed *shear friction*. Accordingly, the transferable shear force must rely on a dependable coefficient of friction μ, and on a force, transverse to the sliding plane, provided by compression load on the wall P_u, and the clamping forces generated in reinforcement with area A_{vf}. This consideration leads to the required amount of reinforcement transverse to the potential sliding plane.

$$A_{vf} = (V_u - \phi\mu P_u)/\phi\mu f_y \qquad (3.42)$$

where ϕ is the strength reduction factor in accordance with Section 1.3.4 with a value of 0.85, except when V_u is derived from capacity design considerations, when $\phi = 1.0$, and μ is the friction coefficient with a value of 1.4 when concrete is placed against previously hardened concrete where the interface for shear transfer is clean, free of laitance, and intentionally roughened to an amplitude of not less than 5 mm (0.2 in.), or $\mu = 1.0$ when under the same conditions the surface is roughened to an amplitude less than 5 mm (0.2 in.) but more than 2 mm (0.08 in.). When concrete is placed against hardened concrete free of laitance but without special roughening, $\mu = 0.7$ should be assumed [X3].

Direct tension across the assumed plane of sliding should be transmitted by additional reinforcement. Shear-friction reinforcement in accordance with Eq. (3.42) should be well distributed across the assumed plane. It must be adequately anchored on both sides by embedment, hooks, welding, or special devices. All reinforcement within the effective section, resisting flexure and axial load on the cross section and crossing the potential sliding plane, may be included in determining A_{vf}. Sliding shear presents a common hazard in squat structural walls when their inelastic behavior is controlled by flexure. In such cases diagonal reinforcement may be used to control shear sliding. The design of these walls is examined in considerable detail in Section 5.7.

Fig. 3.24   Large shear displacements along interconnecting flexural cracks across a plastic hinge of a beam.

(ii)   Sliding Shear in Beams: Sliding displacements along interconnected flexural and diagonal cracks in regions of plastic hinges can significantly reduce energy dissipation in beams [B2]. With reversed cyclic high-intensity shear load, eventually a sliding shear failure may develop. The order of the magnitude of shear sliding displacement along approximately vertical cracks of a test beam [P41] can be gauged from Fig. 3.24.

To prevent such a failure and to improve the hysteretic response of beams, diagonal reinforcement in one or both directions—for example, in the form shown in Fig. 3.25 or Fig. 4.17(a)—should be provided within the plastic hinge region whenever

$$v_i \geq 0.25(2 + r)\sqrt{f_c'} \quad \text{(MPa)}; \qquad \geq 3.0(2 + r)\sqrt{f_c'} \quad \text{(psi)} \quad (3.43)$$

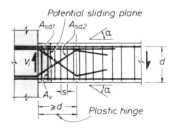

Fig. 3.25   Control of sliding shear in potential plastic hinge regions.

where r defines the ratio of design shear forces associated with moment reversals at the critical section of the plastic hinge, that is,

$$r = V_{un}/V_{um} \leq 0 \qquad (3.44)$$

Its value is always negative. In Eq. (3.44) V_{un} denotes the smaller and V_{um} the larger of the shear forces applied in opposite directions.

When Eq. (3.43) is applicable, the diagonal reinforcement provided should resist a shear force of not less than

$$V_{di} \geq 0.7\left(\frac{v_i}{\sqrt{f_c'}} + 0.4\right)(-r)V_i \quad \text{(MPa)}$$

$$\geq 0.7\left(\frac{v_i}{12\sqrt{f_c'}} + 0.4\right)(-r)V_i \quad \text{(psi)} \qquad (3.45)$$

whenever the ratio of shear forces r is in the range of $-1 < r < -0.2$. When $r > -0.2$, no diagonal reinforcement is considered to be necessary.

As an example, consider an extreme case, when $r = -1$, $f_c' = 30$ MPa (4350 psi) and $v_i = 0.16f_c'$ [Eq. (3.31)]. It is found that according to Eq. (3.45), 89% of the total shear V_i will need to be resisted by diagonal reinforcement. Because there will always be some shear force due to gravity loads, r will seldom approach -1. Using, as an example, more representative values, such as $r = -0.6$ and $v_i = 0.4\sqrt{f_c'}$ ($4.8\sqrt{f_c'}$ psi), typical for a short span spandrel beam, Eq. (3.45) will require that $V_{di} = 0.34V_i$.

As Fig. 3.25 shows, the diagonal bars crossing a potential vertical sliding plane may be utilized simultaneously in both tension and compression. Thus the required total area of diagonal reinforcement shown in Fig. 3.25, to control sliding shear in accordance with Eq. (3.45), is

$$A_{sd1} + A_{sd2} \geq V_{di}/(f_y \sin \alpha) \qquad (3.46)$$

For the purpose of these requirements, the plastic hinge should be assumed to extend to a distance not less than d from the face of the support (Fig. 3.25) or from a similar cross section where maximum yielding due to reversed seismic attack can be expected (Fig. 4.17).

When diagonal reinforcement of the type shown in Fig. 3.25 is used, the bars in tension (A_{sd1}) may also be utilized to contribute to shear strength associated with potential diagonal tension failure. Thereby the amount of stirrup reinforcement in the plastic hinge zone may be reduced. For the example, as shown in Fig. 3.25,

$$V_i = A_{sd1}f_y \sin \alpha + A_v f_y d/s$$

(iii) *Sliding Shear in Columns*: Columns with axial compression less than $P_i = 0.1 f'_c A_g$ and with the great majority of bars placed in two opposite faces only, as in beams, should be treated in regions of potential plastic hinges, as beams. However, when the vertical reinforcement is evenly distributed around the periphery of the column section, as seen in Figs. 4.30 and 4.31, more reliance may be placed on the dowel resistance against sliding of the vertical bars. Any axial compression on the column that may be present will of course greatly increase resistance against sliding shear. Therefore, no consideration need be given to sliding shear in columns constructed in the usual form.

(c) **Shear in Beam–Column Joints** Shear resistance is often the critical feature of the design of beam–column joints of ductile frames under earthquake attack. Relevant and detailed aspects of joint behavior are examined in Section 4.8.

3.3.3 Torsion

As a general rule, torsion in beams and columns is not utilized in the development of earthquake resistance. Mechanisms of torsion with force reversals, like those of shear resistance, are not suitable for stable energy dissipation desired in ductile systems used in a seismic environment. For these reasons no further consideration is given in this book to design for torsion.

Because of deformation compatibility, significant twisting may, however, be imposed on members even in the elastic range of response. This is particularly the case with beams of two-way frames, which support cast-in-place floor slabs. The aim in such cases should be to restrain the opening of diagonal cracks by providing closed hoops in affected members and in particular in plastic hinge regions, rather than to attempt to resist any torsion that may arise. Whenever reinforcement is necessary to meet the requirements of equilibrium or continuity torsion arising from gravity loads or wind forces, existing code requirements [X5] and appropriate design procedures [C7] should be followed.

3.4 SECTION DESIGN

In the preceding section, methods for analyzing sections to determine ideal strength were discussed. Section design is related to analysis by the imposing of additional constraints; typically:

1. Member strength must exceed required strength.
2. Maxima and minima reinforcement contents must be satisfied.
3. Member proportions must satisfy certain limits. These aspects are discussed briefly in the following sections.

3.4.1 Strength Reduction Factors

In Section 1.3.3 the relationship between ideal and required strength was identified as

$$\phi S_i \geq S_u$$

where ϕ is a strength reduction factor taking into account the possibilities of substandard materials, errors in methods of analyses, and normal tolerances in section dimensions. Values for flexural strength reduction factors specified in codes [A1, X5] typically are dependent in the axial compression force level, decreasing from 0.9 for members in pure flexure to 0.7 or 0.75 for members with axial force levels exceeding $0.1 f'_c A_g$, where A_g is the gross section area. This approach is very conservative when applied to column sections with well-confined cores, as may be seen from Fig. 3.22 and Eq. (3.28), where it is apparent that flexural strength increases above values computed by conventional theory as axial compression force increases. Since in this book we always recommend significant levels of confinement for columns, even when they are not required to be ductile, the variation in ϕ with axial load is not seen to be relevant, and a constant value of $\phi = 0.9$ will be adopted [X3].

When the required moment is based on the maximum feasible actions developing at flexural overstrength of plastic hinges, in accordance with the principles of capacity design, it would be unnecessarily conservative to reduce the ideal strength by the use of a strength reduction factor less than unity.

Shear strength is governed by the same arguments presented above for flexural strength. Thus a constant value will be adopted regardless of axial load, and when the required shear strength is based on flexural overstrength in plastic hinges, a value of unity will be assumed.

Masonry walls and columns cannot be confined to the extent envisaged for concrete walls, even when confining plates are used, and strength degradation is likely to be more severe than in concrete walls or columns. Consequently, a sliding scale for the masonry flexural strength reduction factor ϕ is recommended as follows:

$$0.85 \geq \phi = 0.85 - 2(P_u/f'_m A_g) \geq 0.65 \qquad (3.47)$$

The values given by Eq. (3.47) are 0.05 lower than those specified in many codes [A1, X3, X5] for reinforced concrete columns. Similar reasoning leads to the suggestion that the shear strength reduction factor should be $\phi = 0.80$ for masonry structures. The values of strength reduction factor recommended for design, based on the consideration above, are summarized in Table 3.1.

Simplified methods for designing sections, once the required strength has been determined, are covered in the appropriate chapters for frames, walls, masonry, and so on.

TABLE 3.1 Strength Reduction Factors (ϕ)

	Concrete		Masonry	
	Code Required Strength[a]	Capacity Required Strength[b]	Code Required Strength[a]	Capacity Required Strength[b]
Flexure, with or without axial load	0.9	1.0	$0.65 \leq \phi \leq 0.85^c$	1.0
Shear	0.85	1.0	0.80	1.0

[a]Required strength M_u or V_u derived from factored loads and forces [Section 1.3.3(a)].
[b]Required strength M_u or V_u derived from overstrength development of member or adjacent members [Section 1.3.3(d)].
[c]See Eq. (3.47).

3.4.2 Reinforcement Limits

Most codes [A1, X3, X5] impose minimum and maximum limits on reinforcement content for beams, columns, and walls. When sections are required to exhibit ductility, these limits assume special importance and frequently need narrowing from the wide limits typically specified.

If a section has insufficient reinforcement, there is a danger that the cracking moment may be close to, or even exceed, the flexural strength, particularly since the modulus of rupture at seismic strain rates is significantly higher than the static value. Coupled with the fact that moments typically reduce with distance from the critical section of the plastic hinge, the consequence is that only one flexural crack may form in the plastic hinge region. This will result in an undesirable concentration of inelastic deformations over a greatly reduced plastic hinge length and hence very high local curvature ductility requirement. Fracture of the longitudinal reinforcement could result.

On the other hand, if the section contains too much flexural reinforcement, the ultimate curvature will be limited as a result of increased depth of the concrete compression zone at flexural strength necessary to balance the large internal tension force. Further, when beams in ductile frames contain excessive levels of reinforcement, it will be found that the resulting high levels of shear forces in adjacent beam–column joint regions require impractical amounts of joint shear reinforcement. In masonry walls limited grout space in vertical flues can result in bond and anchorage problems if too much reinforcement is used.

For reinforced concrete beams, practical reinforcement limits for tensile reinforcement will be in the range $0.0035 \leq \rho = A_s/bd \leq 0.015$. For columns, total reinforcement should be limited to $0.007 \leq \rho_t = A_{st}/A_g \leq 0.04$. More complete details on suggested reinforcement limits are included for frame members and walls in the appropriate chapters.

3.4.3 Member Proportions

It is important that there be some relationship between the depth, width, and clear length between faces of lateral support of members designed for seismic response, particularly if the member is expected to exhibit ductile response to the design-level earthquake. If the member is too slender, lateral buckling of the compression edge may result. If it is too squat, it may be difficult to control strength and stiffness degradation resulting from shear effects.

It is recommended that the limits to slenderness for rectangular reinforced concrete frame members be set by [X3]

$$l_n/b_w \leq 25 \qquad \text{and} \qquad l_n h/b_w^2 \leq 100 \qquad (3.48)$$

where l_n is the clear span between positions of lateral support, b_w the web width, and h the overall section depth. If the section is a T or L beam, the limits of Eq. (3.48) may be increased by 50%, because of the restraint to lateral buckling provided by the flange. There does not seem to be any justification to setting absolute limits to beam and column dimensions [i.e., $b_w \geq 200$ mm (8 in.)] or proportions (i.e., $b_w \geq 0.33h$) as required by some codes [X10]. It is clear that the division of members into categories termed *columns* or *walls*, for example, is largely a matter of terminological convenience. Member proportions that would be unacceptable by codes as a column can frequently be solved by renaming the member a *wall*. It is doubtful whether the element appreciates the semantic subtleties, and it is better to preclude schizophrenia on the part of the structural element (and the designer) by avoiding inconsistent limitations.

Limitations on squatness of elements are best handled in terms of the shear stress levels induced during earthquake response and are covered in the appropriate sections, as are special requirements for slenderness of walls.

3.5 DUCTILITY RELATIONSHIPS

It was emphasized in Section 1.1.2 that ductility is an essential property of structures responding inelastically during severe shaking. The term *ductility* defines the ability of a structure and selected structural components to deform beyond elastic limits without excessive strength or stiffness degradation. A general definition of ductility was given by Eq. (1.1) and a comparison of ductile and brittle responses was illustrated in Fig. 1.8. It is now necessary to trace briefly specific sources of ductility and to establish the relationship between different kinds of ductilities. As the term *ductility* is not specific enough, and because misunderstandings in this respect are not uncommon, the various ways of quantifying ductilities are reviewed here in some detail.

3.5.1 Strain Ductility

The fundamental source of ductility is the ability of the constituent materials to sustain plastic strains without significant reduction of stress. By similarity to the response shown in Fig. 1.8, strain ductility is simply defined as

$$\mu_\epsilon = \epsilon / \epsilon_y \qquad\qquad (3.49)$$

where ϵ is the total strain imposed and ϵ_y is the yield strain. The strain imposed should not exceed the dependable maximum strain capacity, ϵ_m.

It is evident (Fig. 3.1) that unconfined concrete exhibits very limited strain ductility in compression. However, this can be significantly increased if compressed concrete is appropriately confined [Section 3.2.2(b)], as seen in Fig. 3.5. A strain ductility of $\mu_\epsilon = \epsilon_m / \epsilon_y \geq 20$, and if necessary more can be readily attained in reinforcing bars (Fig. 3.18).

Significant ductility in a structural member can be achieved only if inelastic strains can be developed over a reasonable length of that member. If inelastic strains are restricted to a very small length, extremely large strain ductility demands may arise, even during moderate inelastic structural response. An example of this, using the analogy of a chain, was given in Section 1.4.2.

3.5.2 Curvature Ductility

The most common and desirable sources of inelastic structural deformations are rotations in potential plastic hinges. Therefore, it is useful to relate section rotations per unit length (i.e., curvature) to causative bending moments. By similarity to Eq. (1.1), the maximum curvature ductility is expressed thus:

$$\mu_\phi = \phi_m / \phi_y \qquad\qquad (3.50)$$

where ϕ_m is the maximum curvature expected to be attained or relied on and ϕ_y is the yield curvature.

(a) Yield Curvature Estimates of required ductility are based on assumed relationships between ductility and force reduction factors as discussed in some detail in Section 2.3.4. Such relationships are invariably based on an elastoplastic or bilinear approximation to the structural force–displacement response. Consequently, it is essential that when assessing ductility capacity the actual structural response idealized in Fig. 1.8 and presented in terms of the moment–curvature characteristic in Fig. 3.26(a) be similarly approxi-

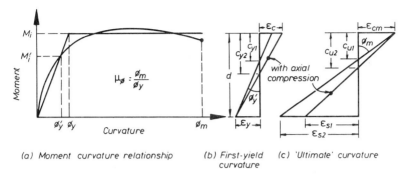

(a) Moment curvature relationship   *(b) First-yield curvature*   *(c) 'Ultimate' curvature*

Fig. 3.26   Definition of curvature ductility.

mated by an elastoplastic or bilinear relationship. This means that the yield curvature ϕ_y will not necessarily coincide with the first yield of tensile reinforcement, which will generally occur at a somewhat lower curvature ϕ'_y [see Fig. 3.26(a)], particularly if the reinforcement is distributed around the section as would be the case for a column. As discussed in Section 1.1.2, it is appropriate to define the slope of the elastic portion of the equivalent elastoplastic response by the secant stiffness at first yield.

For this typical case, the first-yield curvature ϕ'_y is given from Fig. 3.26(b) as

$$\phi'_y = \epsilon_y/(d - c_y) \tag{3.51}$$

where $\epsilon_y = f_y/E_s$ and c_y is the corresponding neutral-axis depth. Extrapolating linearly to the ideal moment M_i, as shown in Fig. 3.26(a), the yield curvature ϕ_y is given by

$$\phi_y = \frac{M_i}{M'_i}\phi'_y \tag{3.52}$$

If the section has a very high reinforcement ratio, or is subjected to high axial load, high concrete compression strain may develop before the first yield of reinforcement occurs. For such cases the yield curvature should be based on the compression strains

$$\phi'_y = \epsilon_c/c_y$$

where ϵ_c is taken as 0.0015.

An acceptable approximation for beam sections is to calculate steel and concrete extreme fiber strains, and hence the curvature ϕ'_y, based on conventional elastic section analyses [P1] at a moment of $M_i' = 0.75M_i$, thus providing an equivalent yield curvature of $\phi_y = 1.33\phi_y'$.

(b) Maximum Curvature   The maximum attainable curvature of a section, or *ultimate curvature* as it is generally termed, is normally controlled by the maximum compression strain ϵ_{cm} at the extreme fiber, since steel strain ductility capacity is typically high. With reference to Fig. 3.26(c), this curvature may be expressed as

$$\phi_m = \epsilon_{cm}/c_u$$

where c_u is the neutral-axis depth at ultimate curvature.

For the purpose of estimating curvature, the maximum dependable concrete compression strain in the extreme fiber of unconfined beam, column, or wall sections may be assumed to be 0.004, when normal-strength concrete [$f_c' \leq 45$ MPa (6.5 ksi)] is used. However, as was shown in Sections 3.2.2(b) and 3.2.3(f), much larger compression strains may be attained when the compressed concrete or masonry is adequately confined. In such situations the contribution of any concrete outside a confined core, which may be subjected to compression strains in excess of 0.004, should be neglected. This generally implies spalling of the cover concrete.

(c) Factors Affecting Curvature Ductility   A detailed quantitative treatment of parameters affecting curvature ductility is beyond the scope of this book, and reference should be made to the extensive literature on the subject [P38, P48] for further details. However, a brief qualitative examination will indicate typical trends. The most critical parameter is the ultimate compression strain ϵ_{cm}, which has been considered in some detail in this chapter. Other important parameters are axial force, compression strength, and reinforcement yield strength.

(i) Axial Force: As shown in Fig. 3.26(b) and (c), the presence of axial compression will increase the depth of the compression zone at both first yield (c_{y2}) and at ultimate (c_{u2}). By comparison with conditions without axial force (c_{y1} and c_{u1}) it is apparent that the presence of axial compression increases the yield curvature, ϕ_y, and decreases the ultimate curvature, ϕ_u. Consequently, axial compression can greatly reduce the available curvature ductility capacity of a section. As a result, spalling of cover concrete is expected at an earlier stage with ductile columns than with beams, and the need for greater emphasis on confinement is obvious. Conversely, the presence of axial tension force greatly increases ductility capacity.

(ii) Compression Strength of Concrete or Masonry: Increased compression strength of concrete or masonry has exactly the opposite effect to axial compression force: the neutral axis depth at yield and ultimate are both

reduced, hence reducing yield curvature and increasing ultimate curvature. Thus increasing compression strength is an effective means for increasing section curvature ductility capacity.

(iii) Yield Strength of Reinforcement: If the required tensile yield force is provided by a reduced area of reinforcement of higher yield strength, the ultimate curvature will not be affected unless the associated steel strain exceeds the lower ultimate tensile strain of the steel [Section 3.2.4(*a*)]. However, the increased yield strain ϵ_y means that the yield curvature will be increased. Hence the curvature ductility ratio given by Eq. (3.49) will be less for high-strength steel.

3.5.3 Displacement Ductility

The most convenient quantity to evaluate either the ductility imposed on a structure by an earthquake μ_m, or the structure's capacity to develop ductility μ_u, is displacement [Section 1.1.2(*c*)]. Thus for the example cantilever in Fig. 3.27, the displacement ductility is

$$\mu_\Delta = \Delta/\Delta_y \qquad (3.53)$$

where $\Delta = \Delta_y + \Delta_p$. The yield ($\Delta_y$) and fully plastic ($\Delta_p$) components of the total lateral tip deflection Δ are defined in Fig. 3.27(*f*).

For frames the total deflection used is commonly that at roof level, as seen in Fig. 1.19. Although an approach consistent with the use of force reduction factors as defined in Section 2.3.4 would investigate structure ductility at the height of the resultant lateral seismic force, the error in assessing μ_Δ at roof level will normally be negligible in comparison with other approximations made. Of particular interest in design is the ductility associated with the

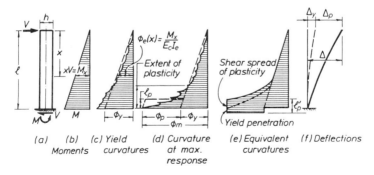

Fig. 3.27 Moment, curvature, and deflection relationships for a prismatic reinforced concrete or masonry cantilever.

maximum anticipated displacement $\Delta = \Delta_m$ (Fig. 1.8). Equally, if not more important are displacement ductility factors μ_Δ which relate interstory deflections (story drifts) to each other. This may also be seen in Fig. 1.19. It is evident that while displacements ductilities in terms of the roof deflection Δ of the two frames shown may be comparable, dramatically different results are obtained when displacements relevant only to the first story are compared. Figure 1.19 also suggests that the displacement ductility capacity of such a frame μ_Δ will be largely governed by the ability of plastic hinges at the ends of beams and/or columns to be sufficiently ductile, as measured by individual member ductilities.

The yield deflection of the cantilever Δ_y, as defined in Fig. 3.27(f) for most reinforced concrete and masonry structures, is assumed to occur simultaneously with the yield curvature ϕ_y at the base. Its realistic estimate is very important because absolute values of maximum deflections $\Delta_m = \mu_\Delta \Delta_y \leq \Delta_u$ will also need to be evaluated and related to the height of the structure over which this displacement occurs.

3.5.4   Relationship Between Curvature and Displacement Ductilities

For a simple structural element, such as the vertical cantilever of Fig. 3.27, the relationship between curvature and displacement ductilities can simply be expressed by integrating the curvatures along the height. Thus

$$\mu_\Delta = \frac{\Delta_m}{\Delta_y} = \frac{\int \phi(x) x \, dx}{\int \phi_e(x) x \, dx} = \frac{K_1 \phi_m}{K_2 \phi_y} = K \mu_\phi \qquad (3.54)$$

where $\phi(x)$ and $\phi_e(x)$ are the curvature distributions at maximum response and at yield respectively, K, K_1, and K_2 are constants, and x is measured down from the cantilever tip. In practice, the integrations of Eq. (3.54) are tedious and some approximations are in order.

(a)   *Yield Displacement*   The actual curvature distribution at yield, $\phi_e(x)$, will be nonlinear as a result of the basic nonlinear moment curvature relationship as shown in Fig. 3.26(a) and because of local tension stiffening between cracks. However, adopting the linear approximation suggested in Fig. 3.26(a) and shown in Fig. 3.27(c), the yield displacement may be estimated as

$$\Delta_y = \phi_y l^2 / 3 \qquad (3.55)$$

(b)   *Maximum Displacement*   The curvature distribution at maximum displacement Δ_y is represented by Fig. 3.27(d), corresponding to a maximum curvature ϕ_m at the base of the cantilever. For convenience of calculation, an equivalent plastic hinge length l_p is defined over which the plastic curvature

$\phi_p = \phi - \phi_e$ is assumed equal to the maximum plastic curvature $\phi_m - \phi_y$ [see Fig. 3.27(d)]. The length l_p is chosen such that the plastic displacement at the top of the cantilever Δ_p, predicted by the simplified approach is the same as that derived from the actual curvature distribution.

The plastic rotation occurring in the equivalent plastic hinge length l_p is given by

$$\theta_p = \phi_p l_p = (\phi_m - \phi_y) l_p \qquad (3.56)$$

This rotation is an extremely important indicator of the capacity of a section to sustain inelastic deformation. Assuming the plastic rotation to be concentrated at midheight of the plastic hinge, the plastic displacement at the cantilever tip is thus

$$\Delta_p = \theta_p(l - 0.5l_p) = (\phi_m - \phi_y) l_p(l - 0.5l_p) \qquad (3.57)$$

The displacement ductility factor [Eq. (3.53)] is thus

$$\mu_\Delta = \frac{\Delta}{\Delta_y} = \frac{\Delta_y + \Delta_p}{\Delta_y} = 1 + \frac{\Delta_p}{\Delta_y}$$

Substituting from Eqs. (3.55) and (3.57) and rearranging yields the relationship between displacement and curvature ductility:

$$\mu_\Delta = 1 + 3(\mu_\phi - 1)\frac{l_p}{l}\left(1 - 0.5\frac{l_p}{l}\right) \qquad (3.58)$$

or conversely,

$$\mu_\phi = 1 + \frac{(\mu_\Delta - 1)}{3(l_p/l)\left[1 - 0.5(l_p/l)\right]} \qquad (3.59)$$

(c) *Plastic Hinge Length*   Theoretical values for the equivalent plastic hinge length l_p based on integration of the curvature distribution for typical members would make l_p directly proportional to l. Such values do not, however, agree well with experimentally measured lengths. This is because, as Fig. 3.27(c) and (d) show, the theoretical curvature distribution ends abruptly at the base of the cantilever, while steel tensile strains continue, due to finite bond stress, for some depth into the footing. The elongation of bars beyond the theoretical base leads to additional rotation and deflection. The phenomenon is referred to as *tensile strain penetration*. It is evident that the extent of strain penetration will be related to the reinforcing bar diameter, since large-diameter bars will require greater development lengths. A second reason for discrepancy between theory and experiment is the increased spread of plasticity resulting from inclined flexure-shear cracking. As is shown in Section 3.6.3, inclined cracks result in steel strains some distance above the base being higher than predicted by the bending moment at that level.

A good estimate of the effective plastic hinge length may be obtained from the expression

$$l_p = 0.08l + 0.022d_bf_y \quad \text{(MPa)}$$
$$= 0.08l + 0.15d_bf_y \quad \text{(ksi)} \tag{3.60}$$

For typical beam and column proportion, Eq. (3.60) results in values of $l_p \simeq 0.5h$, where h is the section depth. This value may often be used with adequate accuracy.

A distinction must be made between the equivalent plastic hinge length l_p, defined above, and the region of plasticity over which special detailing requirements must be provided to ensure dependable inelastic rotation capacity. This distinction is clarified in Fig. 3.27(d). Guidance on the region of plasticity requiring special detailing is given in the specialized chapters on frames, walls, and so on.

Example 3.3   The cantilever of Fig. 3.27 has $l = 4$ m (13.1 ft), $h = 0.8$ m (31.5 in.) and is reinforced with D28 (1.1-in.-diameter) bars of $f_y = 300$ MPa (43.5 ksi). Given that the required structure displacement ductility is $\mu_\Delta = 6$, what is the required curvature ductility μ_ϕ?

SOLUTION: From Eq. (3.60), the equivalent plastic hinge length is

$$l_p = 0.08 \times 4 + 0.022 \times 0.028 \times 300 = 0.505 \text{ m } (1.66 \text{ ft})$$

Note that $l_p/l = 0.126$ and $l_p/h = 0.63$. Substituting into Eq. (3.59) yields

$$\mu_\phi = 1 + (6 - 1)/[3 \times 0.126(1 - 0.5 \times 0.126)] = 15.1$$

As will usually be the case, the curvature ductility factor is much larger than the displacement ductility factor.

3.5.5   Member and System Ductilities

The simple relationships between curvature and displacement ductilities of Eqs. (3.58) and (3.59) depended on the assumption that total transverse tip deflections $\Delta = \Delta_y + \Delta_p$ originated solely from flexural deformations within the rigidly based cantilever members. If foundation rotation occurred the yield displacement would be increased by the displacement due to foundation rotation. However, the plastic displacement Δ_p originates only from plastic rotation in the cantilever member and would remain unchanged. Hence the displacement ductility factor would be reduced. If lateral forces acting on the cantilever produced moment patterns different from that shown in Fig. 3.27(b), Eqs. (3.58) and (3.59) would not apply.

In multistory frames or even in relatively simple subassemblages, these relationships become more complex. Three important features of such relationships are briefly reviewed here. For this purpose, another example, the portal frame studied in Fig. 1.20, will be used.

(a) Simultaneity in the Formation of Several Plastic Hinges   In Section 1.4.4 it was explained how bending moment patterns for combined gravity loads and seismic forces, such as shown in Fig. 1.20(d), were derived. In the eventual inelastic response of the structure, redistribution of design moments was carried out to arrive at more desirable distribution of required resistances, as shown in Fig. 1.20(e). In this example it was decided to restrict plastic hinge formation to the columns. The frame will be reinforced symmetrically because of the need to design, also for the reversed loading direction.

From the study of the moment diagram, reproduced in Fig. 3.28(b), and the positions of the potential plastic hinges, shown in Fig. 3.28(c), it is evident that yielding will first set in at C, closely followed by yielding at D. As the seismic force F_E increases further, plastic rotations will begin at A and end at B. Thus the nonlinear seismic force F_E–lateral deflection Δ relationship during the full inelastic response of the frame will be similar to that shown in Fig. 1.8. At the application of 75% of the seismic force F_E, corresponding with the ideal strength of the frame, the moment at the corner near C will be $0.75 \times 8.7 \approx 6.5$ moment units. As the ideal strength available at that location will be least $6/0.9 = 6.7$, the frame will still be elastic at the application of $0.75F_E$. Elastic analysis under this condition will enable, with use of the bilinear relationship in accordance with Fig. 1.8, the defined yield displacement Δ_y and hence the maximum expected deflection $\Delta_m = \mu_\Delta \Delta_y$ to be determined. As noted earlier, the assumption of equivalent elastoplastic

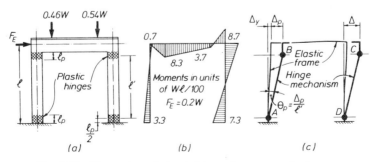

Fig. 3.28   Formation of plastic hinges in an example frame.

response means that the equivalent displacement will be 1.33 times the elastic displacement under $0.75F_E$.

Because yield curvatures in the four potential plastic hinge regions will be attained at different instants of the application of the gradually increasing seismic force F_E, it is evident that for a given displacement ductility μ_Δ, the curvature ductility demands μ_ϕ at these four locations will be different. The maximum curvature ductility will arise as location C, where yielding first occurs, while the minimum will occur at B.

(b)   Kinematic Relationships   The relationships between the plastic hinge rotation $\theta_p = \phi_p l_p$ and the inelastic transverse deflection Δ_p of the cantilever of Fig. 3.27 was readily established. However, it is seen in Fig. 3.28(b) that for this frame, in comparison with those in the cantilever, hinge rotations are magnified to a greater extent because $\theta_p = \Delta_p/l' > \Delta_p(l - 0.5l_p)$. The farther away from node points are the plastic hinges, the larger is the magnification of plastic hinge rotation due to a given inelastic frame displacement Δ_p. Another example is shown in Fig. 4.14.

(c)   Sources of Yield Displacements and Plastic Displacements   The yield displacement of the frame in Figs. 1.20 and 3.28, Δ_y, is derived from routine elastic analysis. This elastic yield deformation contains components resulting from beam flexibility Δ_b, joint flexibility Δ_j, column flexibility Δ_c, and possibly foundation horizontal and rotational flexibility Δ_f. Therefore, $\Delta_y = \Delta_b + \Delta_j + \Delta_c + \Delta_f$. Moreover, elastic column deflections (Δ_c) may result from flexural (Δ_{cm}) and shear (Δ_{cv}) deformations within the column, so that $\Delta_c = \Delta_{cm} + \Delta_{cv}$.

In the example frame shown in Fig. 3.28, inelastic deflections Δ_p originate solely from plastic hinge rotations (i.e., curvature ductility) in the columns. Thus simple geometric relationships, such as those given by Eqs. (3.58) and (3.59), between displacement and curvature ductility will underestimate the required curvature ductility μ_ϕ. A more realistic estimate of the ductility required, for example, in the columns of the frame, shown in Fig. 3.28, may be obtained by relating the total plastic deflection Δ_p to the frame deflection due to only the flexural deformations in the columns:

$$\frac{\Delta_p}{\Delta_y} \approx (\mu_\Delta - 1)\frac{l}{l'} = \frac{(\mu_{c\Delta} - 1)\Delta_{cm}}{\Delta_b + \Delta_j + (\Delta_{cm} + \Delta_{cv}) + \Delta_f} \qquad (3.61a)$$

where the sources of elastic frame deflection from the frame components have been defined. The column ductility $\mu_{c\Delta}$ originating from rotations in the column plastic hinges can now be expressed in terms of lateral deflections:

$$\mu_{c\Delta} = (\mu_\Delta - 1)\frac{l}{l'}\left(\frac{\Delta_b + \Delta_j + \Delta_{cv} + \Delta_f}{\Delta_{cm}} + 1\right) + 1 \qquad (3.61b)$$

For a frame with an infinitely rigid beam, joints, and foundation, so that $\Delta_b = \Delta_j = \Delta_f = 0$, we find that $\mu_{\Delta c} = \mu_\Delta$ if shear deformations in the columns are ignored and $l = l'$. However, when elastic beam, joint, and foundation deformations occur, the ductility demand on the column may be increased significantly. For example, when Δ_b, Δ_j, Δ_c, and Δ_f represent 33, 17, 35, and 15%, respectively of the total lateral deflection at yield Δ_y, and 20% of the column deformation is due to shear, it is found that by assuming that $l/l' = 1.2$,

$$\mu_{c\Delta} = 1 + 1.2(\mu_\Delta - 1)\left[\frac{0.33 + 0.17 + 0.2 \times 0.35 + 0.15}{(1 - 0.2)0.35} + 1\right]$$

Thus when $\mu_\Delta = 4$, the column displacement ductility demand to be supplied by inelastic rotations in the plastic hinges increases to $\mu_{c\Delta} = 1 + 1.2(4 - 1) \times (2.57 + 1) = 13.9$.

If this displacement ductility demand, requiring plastic hinge rotations $\theta_p = \Delta_p/l'$, as shown in Fig. 3.28(c) is translated into curvature ductility demand, in accordance with Eq. (3.59), it is found that using typical values, such as $l_p/l = 0.2$,

$$\mu_\phi = 1 + \frac{13.9 - 1}{3 \times 0.2(1 - 0.5 \times 0.2)} = 24.9$$

The computed curvature ductility in this example relates to the average behavior of the four potential plastic hinges. However, it was pointed out above that inelastic rotations in the four hinges do not commence simultaneously. In the worst case, hinge C begins at $0.77F_E$. Thus the effective yield displacement relevant to this hinge alone is proportionally reduced and the effective system displacement ductility increases to $4/0.77 = 5.2$. Correspondingly, the column displacement ductility increases to $\mu_{c\Delta} = 19.0$, and the curvature ductility for the hinge at C to $\mu_\phi = 34.3$.

The ductility relationship within members of frames, such as seen in Fig. 3.28, are similar to those between the links of a chain, illustrated in the example of Fig. 1.18. There it was shown that the numerical value of the ductility demand on one ductile link may be much greater than that applicable to the entire chain.

Masonry structures are particularly sensitive to ductility demands. Therefore, ductility relationships, taking several variables into account, are examined in greater detail in Sections 7.2.4 and 7.3.6.

3.5.6 Confirmation of Ductility Capacity by Testing

The example of the preceding section (Figs. 1.20 and 3.28) illustrated relationships between different ductilities within one system. The overall inelastic response of this simple structure was characterized by the displacement ductility factor $\mu_\Delta = \Delta/\Delta_y$, as seen in Fig. 3.28(c). In relation to a more complex structure, this is referred to as the *system ductility*. It originates from the ductility of all inelastic regions of components of the system.

It was also seen that numerical values of various ductility ratios can be very different. Whereas the choice of the lateral earthquake design forces depends on the system ductility potential (i.e., ductility capacity μ_Δ), detailing requirements of potential plastic regions must be based on the curvature ductility demand, relevant to these regions.

Considerations of ductility were so far based on monotonic response of components or sections, as seen in Figs. 1.8, 1.17, 2.22, 2.28, and 3.28. During intense ground shaking, however, cyclic displacement of variable amplitudes, often multidirectional, are imposed on the structures. Comparable response of structures and their components to simulated seismic motions in experiments are gauged by familiar hysteretic response curves, such as shown in Figs. 2.20, 5.26, and 7.35. A significant reduction of stiffness and some reduction in strength, as a result of such hysteretic response, is inevitable. Usually, it is not possible to quantify in a conveniently simple way the associated loss of ability to dissipate seismic energy. A shift in the fundamental period of vibration due to reduced stiffness and the likely duration of the earthquake are only two of the parameters that should be considered.

It is for such reasons that a simple criterion was introduced in New Zealand [X8], to enable the ductility capacity of structures or their components to be confirmed, either by testing or by interpretation of existing test results. This rather severe test of the adequacy of detailing for ductility is based largely on engineering judgment. The criterion is defined as follows: The reduction of the strength of the structure with respect to horizontal forces, when subjected to four complete cycles of displacements in the required directions of the potential earthquake attack with an amplitude of $\Delta_u = \mu_\Delta\Delta_y$, shall not exceed 20% of its ideal strength [Section 1.3.3(a)], where μ_Δ is the system ductility factor intended to be used in the derivation of the design earthquake forces, and the yield displacement Δ_y is as defined in Fig. 1.8. The confirmation of the adequacy of components should be based on the same principle except that the four cycles of displacements (eight load reversals) should be applied with an amplitude that corresponds with the location of that component within the system.

The majority of recommendation with respect to detailing for ductility, given in subsequent chapters, is based on laboratory testing of components in accordance with the foregoing performance criterion or a very similar one.

3.6 ASPECTS OF DETAILING

It is reemphasised that judicious detailing of the reinforcement is of paramount importance if reliance is to be placed during a severe earthquake on the ductile response of reinforced concrete and masonry structures [P14].

One of the aims of detailing is to ensure that the full strength of reinforcing bars, serving either as principal flexural or as transverse reinforcement, can be developed under the most adverse conditions that an earthquake may impose. Well-known principles, most of which have been codified, are summarized in this section, while other aspects of detailing relevant to a particular structural action are systematically brought to the designer's attention in subsequent chapters.

3.6.1   Detailing of Columns for Ductility

Recommended details of enforcement for potential plastic hinge regions are covered in the relevant chapters of this book. These details will normally be adequate to ensure that typical curvature ductility demand associated with expected inelastic response can be safely met, particularly for beam and wall section. For such cases, calculation of ductility capacity will not be necessary.

(a)   Transverse Reinforcement for Confinement   Columns subjected to high axial compression need special consideration, as noted above. Most codes [A1], include provisions specifying the amount of confinement needed for columns. Generally, this has been made independent of the axial force level. Recent theoretical and experimental research has shown that the amount of confining steel required for a given curvature ductility factor is in fact strongly dependent on the axial force level [P11, S1, Z1]. A simplified conservative representation of the recommendations of this research for required confining reinforcement area is given for rectangular sections by the following relationship:

$$\frac{A_{sh}}{s_h h''} = k \frac{f_c'}{f_{yh}} \frac{A_g}{A_c} \left(\frac{P_u}{f_c' A_g} - 0.08 \right) \tag{3.62}$$

where $k = 0.35$ for a required curvature ductility of $\mu_\phi = 20$, and $k = 0.25$ when $\mu_\phi = 10$. Other values may be found by interpolation or extrapolation. In Eq. (3.62) A_{sh} is the total area of confining transverse reinforcement in the direction perpendicular to the concrete core width h'' and at vertical spacing s_h; f_{yh} is the yield strength of the hoop reinforcement, A_g the gross concrete section area, and A_c the core concrete area measured to the center of the hoops.

For an example column section, Eq. (3.62) is compared with various code requirements for transverse reinforcement in Fig. 3.29. It will be seen that existing code equations [A1, X10] tend to be very conservative for low axial compression force levels but may be considerably nonconservative at high axial force levels. Equation (3.62) will be up to 40% conservative when the section contains high longitudinal reinforcement ratios [S3].

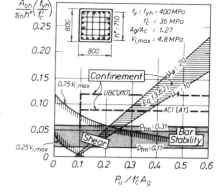

Fig. 3.29 Confinement reinforcement for columns from Eq. (3.62), and comparison with typical requirements for bar stability and shear resistance (A_g/A_c = 1.27). (1 MPa = 145 psi; 1 mm = 0.0394 in.)

The amount of reinforcement indicated by Eq. (3.62) should be provided in each of the two orthogonal principal section directions. Equation (3.62) may also be used to estimate the required volumetric ratio of confinement for circular columns, $\rho_s = 4A_{sp}/s_h d_c$, taking $k_1 = 0.50$ and 0.35 for $\mu_\phi = 20$ and 10, respectively, where A_{sp} is the cross-sectional area of the spiral or circular hoop reinforcement, and d_c is the diameter of the confined core. A more refined estimate may be obtained from the design charts of reference [P38], or from first principles, estimating ultimate curvature from the ultimate compression strain given by Eq. (3.12). Applications of this approach to rectangular columns are given in Section 4.11.8.

For low axial compression loads requirements other than that of confining the concrete will dictate the necessary amount of transverse reinforcement in the plastic hinge region. Typical values of relative transverse reinforcement to stabilize compression bars in accordance with Eq. (4.19) are shown for common values of $\rho_{tm} = \rho_t(f_{yh}/f_c')$ in Fig. 3.29. It is also seen that where axial compression is low, shear strength requirements (Section 3.3.2) are most likely to represent critical design criteria for transverse reinforcement.

(b) Spacing of Column Vertical Reinforcement The important role of column vertical reinforcement in confining the concrete core was emphasized in Section 3.2.2(*a*). To ensure adequate integrity of the confined core, it is recommended that when possible, at least four bars be placed in each side of the column. Because of bar size limitations, this will pose no problems for larger columns, but for small columns, or compression boundary elements in walls, it may be impractical to meet this recommendation. In such cases three bars per side is acceptable. There does not seem to be any logical justification for an arbitrary upper limit on bar spacing [say, 200 mm (8 in.)], as is commonly specified in many codes.

3.6.2 Bond and Anchorage

Efficient interaction of the two constituent components of reinforced concrete and masonry structures requires reliable bond between reinforcement and concrete to exist. In certain regions, particularly where inelastic and reversible strains occur, heavy demand may be imposed on stress transfer by bond. The most severe locations are beam–column joints, to be examined in Section 4.8.

Established recommendations, embodied in various codes, aim to ensure that reinforcement bars are adequately embedded in well-compacted concrete so that their yield strength can be developed reliably without associated deformations, such as slip or pullout, becoming excessive. Important code recommendations, relevant only to the design and detailing of structures covered in the following chapters, are reviewed briefly here. A detailed examination of the mechanisms of bond transfer [P1] is beyond the scope of this book. It should be noted, however, that the conditions of the concrete surrounding embedded bars, particularly in plastic regions and where extensive multidirectional cracking may occur (Fig. 3.24) as a result of inelastic seismic response, are often inferior to those which prevailed in test specimens from which empirical code-specified rules for bar anchorages have been derived.

Only bars with appropriately deformed surfaces are considered here. Plain round bars are not suitable when seismic actions would require bar development by means of a bond to the plain surface (i.e., in beams and columns). Plain bars can be and are used efficiently, however, as transverse reinforcement where anchorage relies on bends and suitable hooks engaging longitudinal bars which distribute by means of bearing stresses concentrated forces from bent plain bars to the concrete.

(a) Development of Bar Strength   The length of a deformed bar required to develop its strength, whether it is straight (l_d) or hooked (l_{dh}), is affected by a number of principal parameters, such as concrete tensile strength, yield strength of steel, thickness of cover concrete, and the degree of confinement afforded by transverse reinforcement or transverse compression stresses.

Code provisions are generally such that an adequately anchored bar, when overloaded in tension, will fracture rather than pull out from its anchorage. Development lengths used in the design examples, which are given in subsequent chapters, are based on the following semiempirical rules [X3].

(i) Development of Straight Deformed Bars in Tension: A bar should extend beyond the section at which it may be required to develop its strength f_y by at least a distance

$$l_d = m_{db} l_{db} \qquad (3.63)$$

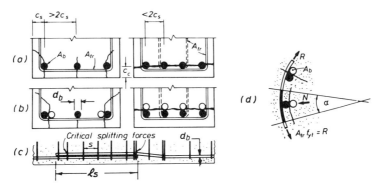

Fig. 3.30   Bar force transmission by shear friction at a lapped splice.

where the basic development length is

$$l_{db} = \frac{1.38 A_b f_y}{c\sqrt{f_c'}} \quad \text{(MPa)}; \qquad l_{db} = \frac{0.114 A_b f_y}{c\sqrt{f_c'}} \quad \text{(psi)} \qquad (3.64)$$

where A_b = cross-sectional area of bar, mm² (in.²)

c = lesser of the following distances [Fig. 3.30(a)]

   = three times the bar diameter d_b

   = center of the bar from the adjacent concrete surface

   = one-half of the distance between centers of adjacent bars in a layer

m_{db} = modification factor with values of 1.3 for horizontal top reinforcement where more than 300 mm (12 in.) fresh concrete is cast in the member below the bar

   = $c/(c + k_{tr}) \leq 1.0$ (3.65)

   when reinforcement, transverse to the bar being developed and outside it [Fig. 3.30(a)], consisting of at least three tie legs, each with area A_{tr} and yield strength f_{yt} and distance s between transverse ties, are provided along l_d, and where

$$k_{tr} = \frac{A_{tr} f_{yt}}{10 s} \leq d_b \quad \text{(MPa)}; \qquad k_{tr} = \frac{A_{tr} f_{yt}}{1450 s} \leq d_b \quad \text{(psi)} \qquad (3.66)$$

When applying Eqs. (3.63), (3.65), and (3.66), the following limitations apply: $k_{tr} \leq d_b$, $c \leq c + k_{tr} \leq 3d_b$.

Transverse reinforcement crossing a potential splitting crack [Fig. 3.30(a)] and provided because of other requirement (shear, temperature, confinement, etc.) may be included in A_{tr}. To simplify calculations it may always be assumed that k_{tr} in Eq. (3.65) is zero.

The interpretation of the distance c is also shown in Fig. 3.30(a). The area A_{tr} refers to that of one tie adjacent to the bar to be developed. It is similar to the area A_{te}, as shown in Fig. 4.20.

(ii) Development of Deformed Bars in Tension Using Standard Hooks: The following limitations apply to the development l_{dh} of tension bars with hooks:

$$150 \text{ mm (6 in.)} < l_{dh} = m_{hb}l_{hb} > 8_{db} \qquad (3.67)$$

where the basic development length is

$$l_{hb} = 0.24d_b f_y/\sqrt{f_c'} \text{ (MPa)}; \qquad l_{hb} = 0.02d_b f_y/\sqrt{f_c'} \text{ (psi)} \quad (3.68)$$

and where m_{hb} is a modification factor with values of:

1. 0.7 when side cover for 32-mm (1.26-in.) bars or smaller, normal to the plane of the hooked bar, is not less than 60 mm (2.4 in.) and cover to the tail extension of 90° hooks is not less than 40 mm (1.58 in.)
2. 0.8 when confinement by closed stirrups or hoops with area A_{tr} and spacings s not less than 6_{db} is provided so that

$$\frac{A_{tr}}{s} \geq \frac{A_b}{1000} \frac{f_y}{f_{yt}} \text{ (mm}^2/\text{mm)}; \qquad \frac{A_{tr}}{s} \geq \frac{A_b}{40} \frac{f_y}{f_{yt}} \text{ (in.}^2/\text{in.)} \quad (3.69)$$

The specific geometry of a hook with bends equal or larger than 90°, such as tail end and bend radii and other restrictions, should be obtained from relevant code specifications. The development length l_{dh} is measured from the outer edge of the bent-up part of the hook.

(b) Lapped Splices   By necessity, reinforcing bars placed in structural member need often to be spliced. This is commonly achieved by overlapping parallel bars, as shown in Figs. 3.30 and 3.31. Force transmission relies on bond between bars and the surrounding concrete and the response of the

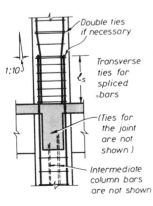

1:10

Double ties
if necessary

Transverse
ties for
spliced
bars

l_s

(Ties for
the joint
are not
shown)

Intermediate
column bars
are not shown   **Fig. 3.31**   Splice details at the end region of a column.

concrete in between adjacent bars. Therefore, the length of the splice l_s, as shown in Fig. 3.30(c), is usually the same as the development length l_d, described in Section 3.6.2(a). However, when large steel forces are to be transmitted by bond, cracks due to splitting of the concrete can develop. Typical cracks at single or lap-spliced bars are shown in Fig. 3.30(a) and (b). To enable bar forces to be transmitted across continuous splitting cracks between lapped bars, as seen in Fig. 3.30, a shear friction mechanism needs to be mobilized [P19]. To control splitting forces, particularly at the end of splices [Fig. 3.30(c)], clamping forces developed in transverse ties are required. In regions where high-intensity reverse cyclic steel stresses need to be transferred, an increase of splice length beyond l_d, without adequate transverse clamping reinforcement, is not likely to assure satisfactory performance [P19]. Under such force demands, lapped splices tend to progressively unzip. Conservatively, it may be assumed that the clamping force along the distance $l_s \geq l_d$ should be equal to the tension force to be transmitted from one spliced bar to the other. Thereby a diagonal compression field at approximately 45° can develop.

The application of these concepts is particularly relevant to columns, where it is desirable to splice all longitudinal bars at the same level. Despite the high intensity of reversed stresses in bars, such splices are possible at the end regions (i.e., at the bottom end, of columns), provided that yielding of spliced bars, even under severe seismic attack, is not expected. Such conditions can be achieved with the application of capacity design principles, details of which for columns are examined in some detail in Section 4.6. By considering that the maximum force to be transmitted across the splice by a column bar is that which occurs at the end of the splice at distance l_s away from the critical (bottom) end of the column (Fig. 3.31), it is found [P19, X3] that the area of transverse clamping reinforcement relevant to each spliced bar with diameter d_b per unit length is

$$\frac{A_{tr}}{s} \geq \frac{d_b}{50} \frac{f_y}{f_{yt}} \tag{3.70}$$

where the symbols are as defined previously. Typical splices in columns are shown in Figs. 3.31 and 4.28(b) and (c).

When a tie leg is required to provide clamping force for more than one bar, the area A_{tr} from Eq. (3.70) should be increased by an amount proportional to the tributary area of the unclamped bar. Lapped splices without a tie should not be farther than 100 mm (4 in.) from either of the two adjacent ties on which the splice relies for clamping.

It is emphasized that the splices should not be placed in potential plastic hinge regions. While the transverse reinforcement in accordance with Eq. (3.70) will ensure strength development of a splice after the application of many cycles of stress reversals close to but below yield level (f_y), it will not ensure satisfactory performance with ductility demands. Typical splices that failed in this way are shown in Fig. 4.29. Additional detailing requirements of column splices are discussed in Section 4.6.10.

In circular columns the necessary clamping force across potential splitting cracks, which may develop between lapped bars, is usually provided by spiral reinforcement or by circular hoops. As shown in Fig. 3.30(d), two possibilities for contact laps may arise. When lapped bars are arranged around the periphery, as seen in the top of Fig. 3.30(d) and in Figs. 4.28(c) and 4.29(b), radial cracks can develop. Therefore, the circumferential (spiral) reinforcement must satisfy the requirement of Eq. (3.70). In its role to provide a clamping force for each pair of lapped bars, circular transverse reinforcement is very efficient, as it can secure an unlimited number of splices.

When lapped bars are arranged as shown in the lower part of Fig. 3.30(d), circumferential splitting cracks may develop. Thus a clamping force $N = \alpha A_{tr} f_{yt}$ is necessary, where α is the angle of the segment relevant to one pair of lapped bars. A comparison of the two arrangements in Fig. 3.30(d) shows that to develop a clamping force $N = A_{tr} f_{yt}$, only $n = 6$ splices ($\alpha = 60°$) could be placed around the circumference. When the number of uniformly spaced splices around the circumference with bars aligned radially is larger than 6, the required area of spiral on circular hoop A_{tr}, given by Eq. (3.70), needs to be increased by a factor of $n/6$, where n is the number of splices.

Splices such as shown in Figs 4.28(b) and 4.29(a) and in the lower part of Fig. 3.30(d), where longitudinal bars are offset by cranking, as seen in Fig. 3.31, require special attention. To minimize radial forces at the bends, the inclination of the crack should not be more than $1:10$. To ensure that the resulting radial steel force can be resisted by transverse reinforcement, stirrup ties or circular hoops should be capable of resisting without yielding radial forces on the order of $0.15 A_b f_y$. This may require the use of double ties at these critical localities, as suggested in Figs. 3.30(c) and 3.31.

(c) Additional Considerations for Anchorages The preceding sections summarized some established and common procedures to ensure that the strength of an individual reinforcing bar, even at the stage of strain hardening, can be developed by means of bond forces. When detailing the reinforcement, attention should also be paid to the bond paths or stress field necessary to enable bond forces to be equilibrated. This is particularly important when a number of closely spaced or bundled bars are required to transfer a significant force to the surrounding concrete.

Figure 3.32 shows the anchorage within a structural wall of a group of diagonal bars from a coupling beam, such as seen in Fig. 5.56. The components of the tension force T, developed in this group of bars, may exceed the tensile strength of the concrete, so that diagonal cracks may form. Clearly, the free body, shown shaded in Fig. 3.32, needs to be tied with a suitable mesh of reinforcement to the remainder of the wall. To increase the bound-

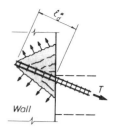

Fig. 3.32 Anchorage of a group of tension bars.

aries of the free body, formed by diagonal cracks, the anchorage of the group of bars l_d^* should be larger than the development length l_d specified for individual bars. Corresponding recommendations are made in Section 5.4.5(b).

Another example (Fig. 3.33) shows two columns, one transmitting predominantly tension and the other a compression force to a foundation wall. It is evident that concurrent vertical and diagonal concrete compression forces can readily equilibrate each other at the node point at B. Hence a development length l_d, required for individual bars, should also be sufficient for the entire group of bars in that column. However, the internal forces at the exterior column A necessitate a node point near the bottom of the foundation wall. The horizontal force shown there results from the anchorage of the flexural reinforcement at the bottom of the wall. Thus the vertical column bars must be effectively anchored at the bottom of the foundation wall at a distance from the top edge significantly larger than l_d^*, required for a group of bars. Alternatively, extra web reinforcement in the wall, close to column A, must be designed, using the concept implied by Fig. 3.32, to enable the tension force P_t to be transferred from column A to the bottom of the wall.

No detailed rules need be formulated for cases such as those illustrated in these two examples, as only first principles are involved. Once a feasible load path is chosen for the transmission of anchorage forces to the remainder of the structural members, elementary calculations will indicate the approximate quantity of additional reinforcement, often only nominal, that may be

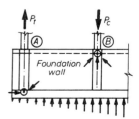

Fig. 3.33 Anchorage of a tension column in a foundation wall.

required and the increased anchorage length of groups of bars necessary to introduce tensile forces at appropriate node points.

3.6.3 Curtailment of Flexural Reinforcement

To economize, to reduce possible congestion of bars, and to accommodate splices, the flexural reinforcement along a member may be curtailed whenever reduced moment demands allow this to be done and when it is practicable. Beams and structural walls are examples. It is seldom practicable to curtail column bars.

Clearly, a bar must extend by a distance not less than the development length l_d beyond a section at which it is required at full strength (f_y). Such sections may be determined from bending moment envelopes with due allowance for tension shift due to shear [P1]. This phenomenon is reviewed briefly with the aid of Fig. 3.34.

The internal forces, such as flexural concrete compression C_1, flexural tension T_2, vertical tension generated in stirrups V_s, shear transmitted across the flexural compression zone V_{co} and by aggregate interlock V_a, transmitting a total moment and shear of M and V_b, respectively, at an approximately 45° diagonal section across a beam, are shown in Fig. 3.34. It is seen that at section 1,

$$M_1 = z_b T_2 + 0.5 z_b V_s \tag{3.71}$$

where z_b is the internal lever arm.

Because $M_1 = M_2 + z_b V$, we find that the flexural tension force T_2 at section 2 is not proportional to the moment at M_2 at this section but is larger, that is,

$$T_2 = \frac{1}{z_b}(M_2 + z_b V - 0.5 z_b V_s)$$

$$= \frac{M_2}{z_b} + (1 - 0.5\eta)V \tag{3.72}$$

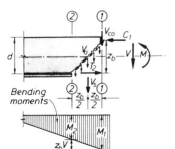

Bending moments

Fig. 3.34 Internal forces in a diagonally cracked reinforced concrete member.

where η is the ratio of the shear resisted by stirrups to the total applied shear (i.e., V_s/V). Thus the flexural tension force at section 2 is proportional to a moment $[M_2 + (1 - 0.5\eta)Vz_b]$ that would occur a distance

$$e_v = (1 - 0.5\eta)z_b \qquad (3.73)$$

to the right of section 2. The distance e_v is termed the *tension shift*. When the entire shear V_b in Fig. 3.34 is resisted by web reinforcement, we find that $e_v = 0.5z_b$.

In routine design it is seldom justified to evaluate accurately the value of the tension shift. Conservatively, it may therefore be assumed that $\eta = 0$ and hence

$$e_v = z_b \approx d \qquad (3.74)$$

In terms of bar curtailment this means that if the moment diagram indicates that a bar is required to develop its full strength (f_y), say, at section 1 in Fig. 3.34, it must extend to the left beyond this section by the development length l_d plus the tension shift $e_v \approx d$. Because the location of the section is not exactly known, bars which according to the bending moment diagram, including tension shift, are theoretically not required to make any contribution to flexural strength should be extended by a small distance, say $0.3d$, beyond that section [X3]. Applications of these principles are presented in Section 4.5.2 and the design examples are given in Section 4.11.7.

When the design of the web reinforcement is based on the use of a diagonal compression field with an inclination to the axis of the member considerably less than 45°, the tension shift [Eq. (3.73)] will be larger, and this may need to be taken into account when curtailing beam bars.

3.6.4 Transverse Reinforcement

The roles of and detailing requirements for transverse reinforcement in regions of beams, columns and walls, which are expected to remain elastic, are well established in various building codes. In structures affected by earthquakes, however, special attention must be given to potential plastic hinge regions. The role of transverse reinforcement in the development of ductile structural response cannot be emphasized enough! Because of its importance, the contributions of transverse reinforcement in resisting shear (Fig. 3.25), preventing premature buckling of compression bars (Fig. 3.35), confining compressed concrete cores (Fig. 3.4) and providing clamping of lapped slices [Figs. 3.30(c) and 3.31)], are examined in considerable detail for beams in Section 4.5.4, for columns in Section 4.6.11, for beam–column joints in Section 4.8.9, and for walls in Section 5.4.3(e). Figure 3.35 points to the need to stabilize each beam bar in the potential plastic hinge zone against buckling. Such bars are subject to the Bauschinger effect and lateral pressure

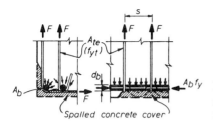

Fig. 3.35 Lateral restraint to prevent premature buckling of compression bars situated in plastic hinge regions.

from an expanding concrete core. Therefore, an estimate of the magnitude of the restraining forces F to be provided at sufficiently close intervals s, as shown in Section 4.5.4, need be made. These will be functions of the expected curvature ductility demand.

The spacing of the transverse reinforcement is as important as the quantity to be provided. For this reason, recommended maximum spacings of sets of transverse ties along a member, required for four specific purposes, are summarized here.

1. *To Provide Shear Resistance*: Except as set out in Section 3.3.2(a)(vii):

   In beams: $s \leq 0.5d$ or 600 mm ($\approx$ 24 in.)
   In columns: $s \leq 0.75h$ or 600 mm ($\approx$ 24 in.)
   In walls: $s \leq 2.5b_w$ or 450 mm ($\approx$ 18 in.)

2. *To Stabilize Compression Bars in Plastic Regions*: As described in Section 4.5.4 for beams, but also applicable to bars with diameters d_b in columns and walls [Section 5.4.3(e)]:

$$s \leq 6d_b, \quad s \leq d/4, \quad s \leq 150 \text{ mm } (\approx 6 \text{ in.})$$

3. *To Provide Confinement of Compressed Concrete in Potential Plastic Regions*: As described in Sections 3.6.1(a), 4.6.11(e), and 5.4.3(e),

$$s_h \leq b_c/3, \quad s_h \leq h_c/3, \quad s_h \leq 6d_b, \quad s_h \leq 180 \text{ mm } (\approx 7 \text{ in.})$$

4. *At Lapped Splices*: As described in Sections 3.6.2(b), 4.6.10, and 4.6.11(f) for the end regions of columns where plastic hinges are not expected to occur:

$$s \leq 8d_b, \quad s \leq 200 \text{ mm } (\approx 8 \text{ in.})$$

4 Reinforced Concrete Ductile Frames

4.1 STRUCTURAL MODELING

It was emphasized in Section 2.4.3 that elastic analyses, under a simplified representation of seismically induced inertia forces by horizontal static forces, remains the most convenient, realistic, and widely used approach for the derivation of member forces in frames.

It is useful to restate and examine relevant common assumptions made in analyses, recognizing that the approximate nature of the applied forces makes precise evaluation of member properties unwarranted. It will be assumed that analysis for lateral forces is carried out using a computer frame analysis program. It should be recognized, however, that simple hand analysis techniques can produce member forces of adequate accuracy with comparative speed and may result in a more accurate representation of the effects, for example of torsion, when the computer analysis is limited to planar frame analysis. The derivation of an efficient approximate hand analysis [M1] is given in Appendix A. It is applied to an example design in Section 4.11.

4.1.1 General Assumptions

1. Elastic analyses based on member stiffnesses applying at approximately 75% of member yield strength, as discussed in Section 1.1.2(a) and shown in Fig. 1.8, adequately represent the distribution of member forces under the design-level loading. Since the true response to design-level seismic attack will generally be in the inelastic range of member behavior, a precise assessment of elastic response is unwarranted.

2. Nonstructural components and cladding do not significantly affect the elastic response of the frame. Provided that nonstructural components are deliberately and properly separated from the structure, their stiffening effect during strong shaking will be small. Their contribution to the resistance of maximum lateral forces is likely to be negligible when inelastic deformations have occurred in the structure during previous response cycles. Thus in routine design, the contribution of nonstructural components to both stiffness

and strength is ignored. The proper separation between nonstructural and structural elements, to ensure the validity of this assumption, is therefore important. Moreover, inadequate separation may result in excessive damage to the nonstructural content of the building during moderate earthquakes. However, it must be recognized that infill panels, either partial or full height within structural frames, such as that discussed in Section 1.2.3(c) and shown in Fig. 1.16, cannot be considered nonstructural, even when the infill is of light weight low-strength masonry.

3. The in-plane stiffness of the floor system, consisting of cast-in-place slabs or prefabricated components with a cast-in-place reinforced concrete topping, is normally considered to be infinitely large. This is a reasonable assumption for framed buildings, with normal length-to-width ratios. Special aspects of the diaphragm action of the floors in distributing the horizontal inertia forces to vertical resisting elements are discussed in Section 6.5.3. The assumption of infinitely rigid diaphragms at each floor allows, with the use of simple linear relationships, the allocation of lateral forces to each bent by taking into account the translational and torsional displacements of floors relative to each other [Section 1.2.2(b) and Fig. 1.10].

4. Regular multistory frames may be subdivided for analysis purposes into a series of vertical two-dimensional frames, which are analyzed separately. The relative displacements of individual bents will be governed by a simple relationship that follows from assumption 3. This is covered in more detail in Section 4.2. Three-dimensional effects, such as torsion from beams framing into beams or columns transverse to the plane of the bent, may in most cases be neglected. For irregular structures, or structures whose plan dimensions are such that assumption 3 is invalid, three-dimensional computer analyses [W1, W2] should be employed to determine member forces.

5. Floor slabs cast monolythically with beams contribute to both the strength and the stiffness of the beams, but floor slabs as independent structural elements may be considered infinitely flexible for out-of-plane bending actions. However, slab flexural stiffness may be considerable in comparison with the torsional stiffness of supporting beams, for example at the edges of the floor. For this reason slabs tend to restrain beam rotation about the longitudinal axis during lateral displacements of frames transverse to the axis of the beam. As a consequence, torsional rotations of the beam are concentrated at the beam ends and may lead to extensive diagonal cracking. Significant damage to beams due to slab restraint has been observed in tests [P1].

6. The effects on the behavior of frames of axial deformations in columns and beams can usually be neglected. The influence of axial deformation of columns increases with the number of stories and when beams with large flexural stiffness are used. Most computer analysis programs, however, consider the elastic axial deformations of members. Computer analyses will not

normally model the effects of inelastic beam extension resulting from high-curvature ductilities at beam ends. This can result in significant changes to column member forces from values predicted by elastic analysis, particularly in the first two stories of frames (Fig. 4.14).

7. Shear deformations in slender members, such as normally used in frames, are small enough to be neglected. When relatively deep beams are used in tube frame structures with close-spaced columns, shear deformations should be accounted for. The torsional stiffness of typical frame members, relative to their flexural stiffness, is also small and thus may also be neglected.

4.1.2 Geometric Idealizations

For the purpose of analysis, beams and columns are replaced by straight bars as shown in Fig. 1.9. The position of such an idealized bar coincides with the centroidal axis of the beam or the column it models. The centroidal axis may be based on the gross concrete section of the member. As a matter of convenience this reference axis will often be taken at the middepth of the member for normal slender T and L beams as well as for rectangular columns, despite the fact that the neutral axis position will vary along the beam length as a result of the effect of moment reversal and variable flange contribution to strength and stiffness along the member length. Figure 4.1 shows an example of how sectional properties may vary along the span of a beam. During an earthquake attack the flange of a T beam abutting against the two opposite faces of a column, will be subjected to tension and compression, respectively, as the moments in Fig. 4.1(*b*) suggest. This feature, examined in some detail in Section 4.8, renders the compression flanges

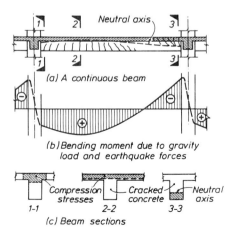

Fig. 4.1 Variation of sectional properties along the span of a beam.

of T beams in the vicinity of columns largely ineffective. For normally proportioned members the consequences of resulting errors are negligible compared with other approximations in the analysis. Haunched reinforced concrete beams may also be modeled with straight bars.

The span length of members is taken as the distance between node points at which reference axes for beams and columns intersect. At these node points a rigid joint is assumed. Accordingly, relative rotations between intersecting members meeting at a joint do not occur in the model structure. The flexural stiffness of members is then based on these span lengths.

It has been common in analysis to consider at least a part of the joint region formed by the intersection of beams and columns as an infinitely rigid end element of the beam or the column. This assumption leads to some increase of member flexural stiffness. However, in normally proportioned frames with reasonably uniform and slender members, the rigid end regions will have very little effect on the relative stiffnesses of beams and columns and will hence not influence computed member forces. Further, under earthquake actions, joints are subjected to high shear stresses, resulting in diagonal cracking and significant shear deformation within the joint region. As a consequence of this and possible bond slip of the flexural reinforcement within the joint, total joint deformations can be considerable, as shown in Section 4.8. Typically, 20% of the interstory deflection due to earthquake forces may originate from joint deformations. For this reason it is strongly recommended that no allowance for rigid end regions be made in the lateral force analysis of ductile frames.

Beams framing into exceptionally wide columns or into walls should, however, be given special consideration, because in such situations joint deformations are likely to be very small. A reasonable modeling for such a situation is shown in Fig. 4.2(*a*). When beams frame into a wall at right

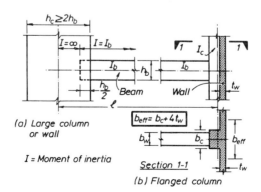

Fig. 4.2   Effective dimensions that may be used for stiffness modeling.

angles to its plane, the effective part of the wall, forming part of a flanged column section, may be modeled for the purpose of analysis, as shown in Fig. 4.2(b), where the meaning of symbols may readily be identified.

4.1.3 Stiffness Modeling

When analyzing concrete frame structures for gravity loads, it is generally considered acceptable to base member stiffnesses on the uncracked section properties and to ignore the stiffening contribution of longitudinal reinforcement. This is because under service-level gravity loads, the extent of cracking will normally be comparatively minor, and relative rather than absolute values of stiffness are all that are needed to obtain accurate member forces.

Under seismic actions, however, it is important that the distribution of member forces be based on realistic stiffness values applying at close to member yield forces, as this will ensure that the hierarchy of formation of member yield conforms to assumed distributions, and that member ductilities are reasonably uniformly distributed through the frame. A reasonably accurate assessment of member stiffnesses will also be required if the building period (Section 2.3.1) and hence seismic forces are to be based on the global frame stiffness resulting from computer analysis [F4].

Clearly, however, under seismic actions, when frame members typically exhibit moment reversal along their length, with flexural cracking at each end, and perhaps an uncracked central region, the moment of inertia (I) will vary along the length (Fig. 4.1). At any section, I will be influenced by the magnitude and sign of the moment, and the amount of flexural reinforcement, as well as by the section geometry and the axial load. Tension stiffening effects will cause further stiffness variations between cracked sections and sections between cracks. For monolythic slab–beam construction, the effective flange width and the stiffening effect of the slab depend on whether the slab is in tension or compression and on the moment pattern along the beam. Diagonal cracking of a member due to shear, intensity and direction of axial load, and reversed cyclic loading are additional phenomena affecting member stiffness.

In terms of design effort, it is impractical to evaluate the properties of several cross sections in each member of a multistory frame, and a reasonable average value should be adopted. As a corollary of this essential lack of precision, it must be recognized that the results of any analysis will be only an approximation to the true condition. The aim of the design process adopted should be to ensure that the lack of precision in the calculated member forces does not affect the safety of the structure when subjected to seismic forces.

Thus, in estimating the flexural stiffness of a member, an average value of EI, applicable to the entire length of a prismatic member, should be assumed. The moment of inertia of the gross concrete section I_g should be modified to take into account the phenomena discussed above, to arrive at an

TABLE 4.1   Effective Member Moment of Inertia[a]

	Range	Recommended Value
Rectangular beams	0.30–$0.50I_g$	$0.40I_g$
T and L beams	0.25–$0.45I_g$	$0.35I_g$
Columns, $P > 0.5f_c'A_g$	0.70–$0.90I_g$	$0.80I_g$
Columns, $P = 0.2f_c'A_g$	0.50–$0.70I_g$	$0.60I_g$
Columns, $P = -0.05f_c'A_g$	0.30–$0.50I_g$	$0.40I_g$

[a]A_g = gross area of section; I_g = moment of inertia of gross concrete section about the centroidal axis, neglecting the reinforcement.

equivalent moment of inertia I_e. Typical ranges and recommended average values for stiffness are listed in Table 4.1.

The column stiffness should be based on an assessment of the axial load that includes permanent gravity load, which may be taken as 1.1 times the dead load on the column, plus the axial load resulting from seismic overturning effects. Unless the span of adjacent beams is very different, the earthquake-induced axial forces will normally affect only the outer columns in a frame, since seismic beam shear forces, which provide the seismic axial force input to columns, will typically balance on opposite sides of an interior joint (Fig. 4.1). Since the axial column forces resulting from seismic actions will not be known at the start of the analysis process, a successive approximation approach may be needed, with the column stiffnesses modified after the initial analysis, based on the axial forces predicted by the first analysis. Alternatively, a satisfactory approximation for seismic axial forces on the outer columns of a regular planar frame based on the assumption of an inverted triangle distribution of lateral forces is given by

$$P_i = \frac{V_{bf}l_c}{jl} \sum_i^n \left[1 - \left(\frac{i}{n} \right)^2 \right] \tag{4.1}$$

where V_{bf} is the frame base shear, and P_i the axial seismic force at level i of an n-story frame with j approximately equal bays, l_c the constant story height, and l the bay length. It is thus assumed that the seismic overturning moment at approximately the midheight of any story is resisted by the outer columns only. It will be necessary to make an initial estimate for the base shear for the frame V_{bf} in using Eq. (4.1).

Flange contribution to stiffness in T and L beams is typically less than the contribution to flexural strength, as a result of the moment reversal occurring across beam–column joints and the low contribution of tension flanges to flexural stiffness (Fig. 4.1). Consequently, it is recommended that for load combinations including seismic actions, the effective flange contribution to

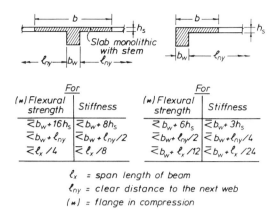

ℓ_x = span length of beam
ℓ_{ny} = clear distance to the next web
$(*)$ = flange in compression

Fig. 4.3   Assumptions for effective width of beam flanges.

stiffness be 50% of that commonly adopted for gravity load strength design (A1). For convenience the assumptions [X3] for effective flange width for the evaluation of both flexural compression strength and stiffness are given in Fig. 4.3.

The stiffness values used for gravity loads, in accordance with Eq. (1.5b), should preferably be the same as those used in the analysis for lateral seismic forces. This is readily achieved when the structure is analyzed for the simultaneous effects of gravity loads and seismic forces. As mentioned, the evaluation of gravity load effects alone, in accordance with Eq. (1.5a), stiffnesses may be based on uncracked section properties. It is often convenient to use results of this analysis, when reduced proportionally to meet the requirements of Eq. (1.5b), and to superimpose subsequently the actions due to seismic forces alone. Although this is strictly not correct, the superposition will usually result in small errors, particularly when seismic actions dominate strength requirements.

Special consideration needs to be given to the modeling of the joint detail at the base of columns in ductile frames. The common assumption of full fixity at the column base may only be valid for columns supported on rigid raft foundations or on individual foundation pads supported by short stiff piles, or by foundation walls in basements. Foundation pads supported on deformable soil may have considerable rotational flexibility, resulting in column forces in the bottom story quite different from those resulting from the assumption of a rigid base. The consequence can be unexpected column hinging at the top of the lower-story columns under seismic lateral forces. In such cases the column base should be modeled by a rotational spring, details of which are given in Section 9.3.2 and Fig. 9.5.

4.2 METHODS OF ANALYSIS

4.2.1 "Exact" Elastic Analyses

The matrix form of stiffness and force methods of analysis, programmed for digital computation, present a systematic approach to the study of rigid jointed multistory frames. Standard programs such as ICES-STRUDL, Drain-2D, TABS, and ETABS are readily available. These require only the specification of material properties, stiffnesses, structural geometry, and the loading. In seismic design the advantages of such analyses is speed rather than accuracy.

Analyses for any load or for any combination of (factored) loads can readily be carried out for the elastic structure. By superposition or directly, the desired combinations of load effects can be determined. However, these values of actions are not necessarily the most suitable ones for the proportioning of components.

4.2.2 Nonlinear Analyses

A more accurate and realistic prediction of the behavior of strength of reinforced concrete structures may be achieved by various methods [C1] of nonlinear analysis. Some of these are rather complex and time consuming. With current available techniques, the computational effort involved in the total nonlinear analysis of a multistory framed building is often prohibitive. A separate analysis would need to be carried for each of the load combinations given in Section 1.3.2. Nonlinear analysis techniques have no particular advantage when earthquake forces, in combination with gravity loads, control the strength of the structure.

4.2.3 Modified Elastic Analyses

With the general acceptance of the principles of strength design for reinforced concrete structures, their nonlinear response has also been more widely recognized. As a first approximation for the pattern of internal actions, such as bending moments, to be considered in determining the required resistance of beams and columns of a frame, an elastic analysis is traditionally used. This satisfies the criteria of equilibrium and, within the limitations of the assumptions made, compatibility of elastic deformations. In many instances the results may be used directly and a practical and economic solution can be achieved. More frequently, however, a more efficient structural design will result when internal actions are adjusted and redistributed in recognition of nonlinear behavior, particularly when the full strength of the structure is being approached. This is done while the laws of equilibrium are strictly preserved. The redistribution of actions predicted by the elastic analysis is kept within certain limits to ensure that serviceability criteria are also satisfied and that the ductility potential of affected regions is not exhausted. Moment redistribution, discussed in Section 4.3, is an example of the application of these concepts. Because potential plastic hinge regions in earthquake-resisting frames are detailed with special care for large possible ductility demand, the advantages of nonlinear behavior can generally be fully utilized.

The results of elastic analyses, subsequently modified to allow for inelastic redistribution of internal actions, will be used here for the design of frame components. Such redistribution can be applied within wide limits. These limits suggest that the results of the elastic analysis need not be particularly accurate provided equilibrium of internal and external forces is maintained. For this reason approximate elastic analyses techniques should not be considered as being inferior to "exact" ones in the design of ductile earthquake resisting structures.

4.2.4   Approximate Elastic Analyses for Gravity Loads

Tributary areas at each floor are assumed to contribute to the loading of each beam. The complex shedding of the load from two-way slabs can be satisfactorily simulated by the subdivision of panel areas as shown in Fig. 4.4(a). As

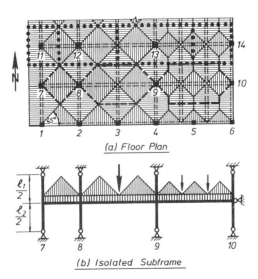

Fig. 4.4   Tributary floor areas and their contributions to the gravity loading of beams and columns.

can be seen, triangular, trapezoidal, and rectangular subareas will result. The total of the tributary areas assigned to each beam, spanning in the north–south direction, for example, can be readily established. The shading at right angles to a beam shows the tributary area for the relevant length of that beam. In this example framing system, there are three secondary beams in the north–south direction. These must be supported by the interior east–west girders. Therefore, secondary beams may be assumed to impose concentrated loads on the girders. The total tributary areas, A, relevant to each of the beam spans in frame 7–10 to be used in evaluating the live-load reduction factor r given by Eq. (1.3), are shown in Fig. 4.4(a) by the heavy dashed boundaries.

When prefabricated flooring systems with cast-in-place reinforced concrete topping are used, the tributary areas are suitably adjusted to recognize the fact that these systems normally carry the loading in one direction only.

With this subdivision of tributary floor areas, the gravity load pattern for each beam is readily determined. Figure 4.4(b) shows, for example, a set of triangular and concentrated loads on the continuous beam 7–8–9–10, resulting from the dead and live loads on the two-way slab. The uniformly distributed load shown in each span represents the weight of the beam below the underside of the slab (i.e., the stem of the T beam).

In the seismic design of reasonably regular framed buildings, it is sufficient to consider an isolated subframe as shown in Fig. 4.4(b). It may be assumed that similar beams above and below the floor, to be considered subsequently, are loaded in the same fashion. Therefore, joint rotations under gravity loading at each floor along one column will be approximately the same. With all spans loaded with dead load or dead and live loads, this simplification allows a point of contraflexure to be assumed at the midstory of each deformed column. With the pin-ended columns, as shown in Fig. 4.4(b), a simple isolated subframe results, which can be readily analyzed even with hand calculations. Moreover, only one such analysis may be applicable to a large number of floors at different levels. When seismic actions dominate strength requirements, alternating loading of spans by live load, to give more adverse beam or column moments, is seldom justified. Subsequent moment redistribution would usually render such refinement meaningless.

The axial load on columns induced by gravity load on the floor system should strictly be derived from the reactions induced at each end of the continuous beams. However, for design purposes it is usually sufficient to consider the approximate tributary area relevant to each column. The boundaries of these rectangular assumed tributary areas for four columns, passing through the midspan of adjacent beams of the structure of Fig. 4.4(a), are shown by pointed lines. This assumption implies that the beam reactions are evaluated as for simply supported beams. It will be seen later, however, that when seismic actions are also considered, the beam shear forces induced by end moments will be accounted for fully and correctly.

4.2.5   Elastic Analysis for Lateral Forces

(a)   Planar Analysis   When the lateral-force-resisting system for a building consists of a number of nonplanar frames, the total seismic lateral force assigned to each floor must be distributed between the frames in accordance with their stiffness. If a full three-dimensional computer analysis is adopted, the lateral force distribution will be an automatic by-product of the structural analysis. However, if a planar analysis is to be used, with the different parallel frames analyzed separately, the lateral forces must be distributed between the frames prior to analysis of each frame.

(b)   Distribution of Lateral Forces Between Frames   Figure 4.5 shows a typical floor plan of a regular multistory two-way frame building. Frames X1 to X4 in the x direction have in-plane stiffnesses K_{x1} to K_{x4}, and frames Y1 to Y4 in the y directions have in-plane stiffnesses K_{y1} to K_{y4}. These stiffnesses will be calculated based on preliminary analyses. At this stage it is relative rather than absolute stiffnesses that are important, and simplifying assumptions regarding the influence of cracking may be made. For regular buildings it may be assumed that the relative frame stiffnesses, based on member properties at the midheight of the building, apply for all levels of load application. For buildings with irregular stiffness distributions [Fig. 1.14(a), (b), and (d)], particularly those resulting from stepped elevations, such as

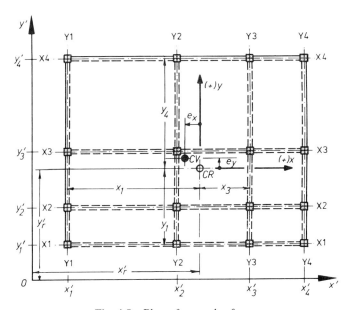

Fig. 4.5   Plan of a regular frame.

those shown in Fig. 1.13, the calculations below will have to be made for several levels of the structure.

The fundamental principles of this analysis are given in detail in Appendix A, where equations suitable for hand calculation are derived. Instead of finding the shear force induced by story shear forces in each column in each floor, as presented in Appendix A, the following equations allow the share of each frame in the total lateral force resistance to be determined. This will then allow a plane frame computer analysis to be applied for each frame.

Seismic floor forces are assumed to act at the center of mass, M, of the floor (Section 1.2.2). The center of rigidity (CR), or center of stiffness of the framing system, defined by point CR in Fig. 1.10(d), is located by the coordinates x'_r and y'_r, where

$$x'_r = \frac{x'_1 K_{y1} + x'_2 K_{y2} + x'_3 K_{y3} + x'_4 K_{y4}}{K_{y1} + K_{y2} + K_{y3} + K_{y4}} \tag{4.2}$$

or more generally,

$$x'_r = \Sigma\,(x'_i K_{yi})/\Sigma\,K_{yi} \tag{4.3a}$$

and similarly,

$$y'_r = \Sigma\,(y'_i K_{xi})/\Sigma\,K_{xi} \tag{4.3b}$$

where x'_i and y'_i are the coordinates of a frame taken from an arbitrary origin, as shown in Fig. 4.5.

When the seismic lateral force $V_n = V_x$ acts through the center of rigidity, shown as CR in Fig. 4.5, only translation of the floors in the x direction [Fig. 1.10(a)] will occur. Thus the total lateral force due to identical translation of all frames in the x or y direction to be resisted by one frame will be

$$V'_{ix} = (K_{xi}/\Sigma\,K_{xi})V_x \tag{4.4a}$$

$$V'_{iy} = (K_{yi}/\Sigma\,K_{yi})V_y \tag{4.4b}$$

respectively.

However, as explained in Section 1.2.2(b), the earthquake-induced lateral force at each floor, F_j in Fig. 1.9, will act through the center of mass at that floor. Hence the resultant total lateral force on the entire structure V_n will, as a general rule, not pass through the center of rigidity but instead through point CV shown in Figs. 1.10(d) and 4.5. As Fig. 1.10(d) shows, a torsional moment $M_t = e_y V_x$ or $M_t = e_x V_y$ will be generated.

Torsional moments, causing floor rotations, as shown in Fig. 1.10(c), will induce horizontal forces in frames in both the x and y directions. It may be shown from first principles, or by similarity to the corresponding expressions given in Appendix A, that the total lateral force induced in a frame by the torsional moment M_t alone is

$$V''_{ix} = (y_i K_{xi}/I_p)M_t \qquad (4.5a)$$

in the x direction and similarly,

$$V''_{iy} = (x_i K_{yi}/I_p)M_t \qquad (4.5b)$$

in the y direction, where the polar moment of inertia of frame stiffnesses is

$$I_p = \Sigma x_i^2 K_{yi} + \Sigma y_i^2 K_{xi} \qquad (4.6)$$

and the coordinates for the frames, x_i and y_i, are to be taken from the center of rigidity (CR), as shown in Fig. 4.5.

Thus the total horizontal force applied to one frame in the x or y direction becomes

$$V_{ix} = V'_{ix} + V''_{ix} \qquad (4.7a)$$

or

$$V_{iy} = V'_{iy} + V''_{iy} \qquad (4.7b)$$

where depending on the sense of the eccentricity and the coordinate for a particular frame, the torsional contributions, V''_{ix} and V''_{iy}, may be positive or negative. Using the sign convention for the example frame in Fig. 4.5, the static eccentricities are such that $e_x < 0$ and $e_y > 0$. For design purposes larger eccentricities must be considered and this was examined in Section 2.4.3(g), where appropriate recommendations have been made.

(c) Corrected Computer Analyses   A plan frame analysis of a space frame of the type shown in Fig. 4.5 implies that the total lateral static force acts through the center of rigidity CR. Hence all frames are subjected to identical lateral displacements. The corresponding modeling for a computer analysis is readily achieved by placing all frames in a single plane while interconnecting them by infinitely rigid pin ended bars at the level of each floor. The total lateral force so derived for a frame, V'_{ix} or V'_{iy}, is a direct measure of the stiffness of those frames, K_{xi} and K_{yi}. Hence Eqs. (4.3) and (4.6) and the static eccentricities can be evaluated.

Once the design eccentricity [Section 2.4.3(g)] is known, the torsional moments M_t, relevant to an earthquake attack in the x or y direction, are determined, and hence from Eq. (4.5) the torsion-induced lateral force, V''_{ix} and V''_{iy}, applied to each frame, is also found. The total lateral force on a frame to be used for the design is thus given by Eq. (4.7). For convenience, all moments and axial and shear forces derived for members of a particular frame by the initial computer plane frame analysis for lateral forces, V'_{ix} or

V'_{iy}, can be magnified by the factors V_{ix}/V'_{ix} and V_{iy}/V'_{iy}, respectively, to arrive at the maximum quantities to be considered for member design.

4.2.6   Regularity in the Framing System

In Chapter 1 it was pointed out that one of the major aims of the designer should be to conceive, at an early stage, a regular structural system. The greater the irregularity, the more difficult it is to predict the likely behavior of the structure during severe earthquakes. As some irregularity is often unavoidable, it is useful to quantify it. When the irregularity is severe, three-dimensional dynamic analysis of the structure is necessary.

(a)   Vertical Regularity   Vertical regularity in a multistory framed building is assured when the story stiffnesses and/or story masses do not deviate significantly from the average value. Examples were given in Figs. 1.13 and 1.14. It is shown in Appendix A that the story stiffness in each principal direction is conveniently expressed for the rth story as the sum of the column stiffnesses: that is, the D values for all the columns in that story (i.e., $\Sigma_r\, D_{ix}$ and $\Sigma_r\, D_{iy}$). The average story stiffnesses for the entire structure, consisting of n stories, are then

$$\frac{1}{n}\Sigma_n\,\Sigma_r\,D_{ix} \quad \text{and} \quad \frac{1}{n}\Sigma_n\,\Sigma_r\,D_{iy}$$

respectively, in the two principal directions of the framing system. The stiffness of any story with respect to story translation, as shown in Fig. 1.10(a) and (b), should not differ significantly from the average value [X10]. The Building Standard Law in Japan [A7] requires, for example, that special checks be employed when the ratio $n\,\Sigma_r\,D_i/\Sigma_n\,\Sigma_r\,D_i$ in the rth story becomes less than 0.6. Such vertical irregularity may arise when the average story height, or when column dimensions in a story, are drastically reduced, as shown in Fig. 1.14(e). Irregularity resulting from the interaction of frames with walls is examined in Section 6.2.5.

(b)   Horizontal Regularity   Horizontal irregularity arises when at any level of the building the distance between the center of rigidity (CR) of the story and the center of the applied story shear (CV), defined in Figs. 1.10(d) and 4.5 as static eccentricity e_x or e_y, become excessive. The torsional stiffness of a story is given by the term I_p [i.e., by Eq. (4.6)]. The ratios of the two kinds of stiffnesses may be conveniently expressed by the *radii of gyration of story stiffnesses* with respect to the principal directions:

$$r_{Dx} = \sqrt{I_p/\Sigma_r\,D_{yi}} \quad \text{and} \quad r_{Dy} = \sqrt{I_p/\Sigma_r\,D_{xi}}$$

Again using Japanese recommendations [A7] as a guide, horizontal irregularity may be considered acceptable when

$$e_x/r_{Dx} < 0.15 \quad \text{and} \quad e_y/r_{Dy} < 0.15$$

An application of this check for regularity is shown in Section 4.11.6(c).

There are other ways to define horizontal irregularity [X10]. One approach [X4] magnifies further the design eccentricity when torsional irregularity exists. Frames with eccentricity exceeding the above limits should be considered irregular. With full consideration of the effects of eccentricity such structures can also be designed satisfactorily.

4.3 DERIVATION OF DESIGN ACTIONS FOR BEAMS

4.3.1 Redistribution of Design Actions

The combined effects of gravity loads and seismic forces often result in frame moment patterns that do not allow efficient design of beam and column members. This is illustrated by Figs. 4.6(c) and 4.7(c), which show typical moment patterns that may develop under the combined effects of gravity loads and seismic forces for one-story subsections of two frames. Figure 4.6

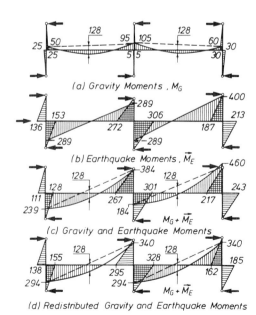

(a) Gravity Moments, M_G

(b) Earthquake Moments, $\tilde{M}_E$

(c) Gravity and Earthquake Moments

(d) Redistributed Gravity and Earthquake Moments

Fig. 4.6 Redistribution of design moments for an earthquake-dominated regular frame.

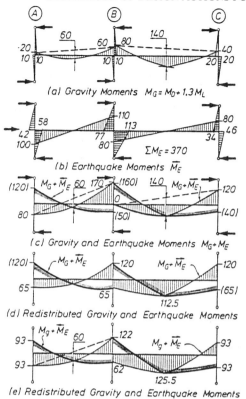

Fig. 4.7 Redistribution of design moments for a gravity-dominated unsymmetrical frame.

represents moments for a tall regular geometrically symmetrical frame, whose design is dominated by seismic forces, and Fig. 4.7 represents moments of a lower nonsymmetrical gravity-dominated frame. The moments of Figs. 4.6 and 4.7 are without units and have magnitudes solely for illustrative purposes in what follows.

Design gravity and seismic actions for the regular frame of Fig. 4.6 are based on different stiffnesses for the outer tension and compression columns, as discussed in Section 4.1.3. Although this creates artifically unsymmetrical gravity load moments when considered in isolation, it results in consistent total moments when combined with seismic moments in Fig. 4.6. As discussed earlier, it is preferable to use a single analysis for the combined effects of gravity loads and seismic forces. The separation into individual components in Figs. 4.6 and 4.7 is merely to facilitate comparison of gravity and seismic moment components.

For the unsymmetrical frame of Fig. 4.7, the variation in column axial loads due to seismic actions was assessed to be minor, and equal stiffnesses were assumed for the tension and compression columns during analysis. The point load on the long beam span results from a secondary transverse beam framing into the midpoint of the long span.

Because of geometric symmetry of the regular frame in Fig. 4.6, moments are shown only under forces from left to right, denoted symbolically as $\vec{E}$. The moment pattern is antisymmetrical under the reversed seismic force direction ($\overleftarrow{E}$). For the nonsymmetrical frame in Fig. 4.7, the combination of seismic and gravity moments results in moment patterns for the two opposite directions that are not antisymmetrical. Hence beam moment profiles for both loading directions ($\vec{E}$ and $\overleftarrow{E}$) are shown in Fig. 4.7(c), and column moments are omitted for clarity.

Examination of the beam moment profiles of Figs. 4.6(c) and 4.7(c) reveals some typical trends. Even in the seismic-dominated regular frame, where gravity load moments are approximately 30% of seismic moments, the resulting combination of gravity and seismic moments results in the maximum negative beam moment (460 units) being about 2.5 times the maximum positive moment of 184 units in that span. Maximum negative beam moments at center and outer columns differ, as do maximum positive moments at the same locations, meaning that different amounts and patterns of reinforcement would be required at different sections. An efficient structural design would aim at equal negative moments at the critical beam sections at an interior column and at peak negative and positive moments of similar magnitude. The situation is rather worse for the gravity-load-dominated frame of Fig. 4.7(c), where maximum negative moments vary between 120 and 170 units and peak positive moments vary between 0 and 80 units.

Section design to the elastic moment patterns of Figs. 4.6(c) and 4.7(c) would result in inefficient structures. Section size would be dictated by the moment demand at the critical negative moment location, and other sections would be comparatively under-reinforced. It may be impossible to avoid undesirable beam flexural overstrength [Section 1.3.3(d)] at critical locations of low moment demand, such as the positive moment value of 0 units at the central column in Fig. 4.7(c) or where the beam moment requirement on either side of a column under $\vec{E}$ and $\overleftarrow{E}$ are different, as in the values of 160 and 170 for negative moments in Fig. 4.7(c), since it is not convenient to terminate excess top reinforcement within an interior beam–column joint.

It should be noted that many building codes require positive moment strength at column faces to be at least 50% of the negative moment strength [Section 4.5.1(c)]. In Fig. 4.6(c) this would require a modest increase in positive moment strength from 184 units to 192 units, but for the gravity dominated frame of Fig. 4.7(c) would require a strength of 85 units, compared with the elastic moment requirement of 0 or 50 for $\vec{E}$ or $\overleftarrow{E}$, respectively. The consequence of this structurally unnecessary beam overstrength has greater influence than just the efficiency of beam flexural design. As has been discussed in detail in Sections 1.2.4(b), 1.4.3, and 2.3.4, the consequences of the philosophy of design for ductility are that design forces are much lower than the true elastic response levels, and it is the actual beam strengths rather than the design levels that will be developed under design-level seismic forces. Since total beam and column moments at a given joint must be in equilibrium, excess beam moment capacity must be matched by additional column moment capacity if undesirable column plastic hinges [Fig. 1.19(*b*)] are to be avoided. Actual beam and column shear forces will be increased similarly.

The moment profiles of Figs. 4.6(*c*) and 4.7(*c*) can be adjusted by moment redistribution to result in more rational and efficient structural solutions without sacrificing structural safety or violating equilibrium under applied loads. The mechanisms by which this is achieved are discussed in the following sections.

4.3.2 Aims of Moment Redistribution

The purpose of moment redistribution in beams of ductile frames is to achieve an efficient structural design by adopting the following measures:

1. Reduce the absolute maximum moment, usually in the negative moment region of the beam, and compensate for this by increasing the moments in noncritical (usually positive) moment regions. This makes possible a better distribution of strength utilization along the beam. Where convenient, the adjustment will be made so that negative and positive design moments at critical sections approach equality. This will result in a simple and often symmetrical arrangement of flexural reinforcement at these beam sections.

2. Equalize the critical moment requirements for beam sections on opposite sides of interior columns resulting from reversed (opposite) directions of applied seismic forces. This will obviate the need to terminate and anchor beam flexural reinforcement at interior beam column joints.

3. Utilize the minimum positive moment capacity required by codes when this exceeds the requirements derived from elastic analyses. As mentioned earlier, most codes require that positive moment capacity be at least equal to 50% of the negative moment capacity at column faces. The intent of this code provision is to ensure that with the presence of flexural compression reinforcement, the required curvature ductility can readily be developed under large negative moments.

4. Reduce moment demands on critical columns, particularly those subject to small axial compression or to axial tension. This will sometimes be necessary to avoid the need to use excessive flexural reinforcement in such columns.

4.3.3 Equilibrium Requirements for Moment Redistribution

The essential requirement of the moment redistribution process is that equilibrium under the applied seismic forces and gravity loads must be maintained. Figure 4.8 represents a typical subframe of a multistory frame,

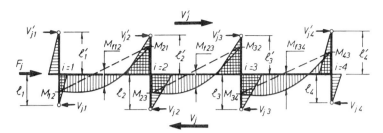

Fig. 4.8   Equilibrium of a subframe.

isolated by cutting columns above and below the beam at column contraflexure points, as was the case for the examples of Figs. 4.6 and 4.7. The moment pattern shown resulted from an elastic analysis for the simultaneous actions of gravity loads and earthquake forces. The total shear forces transmitted by all the columns below and above the floor are V_j and V_j', respectively, for the direction of earthquake forces considered ($\vec{E}$). Note that these column shear forces include gravity load components. However, as no horizontal forces are applied to the framed structure subjected to gravity loads only, no corresponding story shear forces can exist. Therefore, the sum of column shear forces in any story, associated with gravity loading of beams, must be zero. Thus the shear force V_j' and V_j in Fig. 4.8, generally called *story shear forces*, are due entirely to applied lateral seismic design forces, such as F_j.

In considering equilibrium criteria, the simplifying assumption is made that the distance between the two column points of contraflexure, above and below a beam centerline, is the same for all columns of the frame, and this distance does not change while beam moments are redistributed. Using the symbols in Fig. 4.8, this means that $(l_i' + l_i) = l_c$ is a constant length. The assumption implies that the redistribution of moments in beams at sequential levels is essentially similar in nature and extent, a phenomenon normally encountered in the design of frames.

The equilibrium criteria to be satisfied can be established by considering horizontal forces and moments separately. Equilibrium of horizontal forces requires that

$$V_j' + F_j + V_j = \Sigma_i V_{ji}' + F_j + \Sigma_i V_{ji} = 0 \qquad (4.8a)$$

Conservation of story shear forces requires that

$$V_j' = \Sigma_i V_{ji}' \qquad (4.8b)$$

and

$$V_j = \Sigma_i V_{ji} \qquad (4.8c)$$

where V'_{ji} and V_{ji} are the column shears in the ith column above and below the beam, respectively, and F_j is the lateral force assigned to this frame at level j by the elastic analysis for the entire structure, discussed in Section 4.2.5. Equations (4.8b) and (4.8c) imply that whereas some or all of the column shear forces may change in a story during moment redistribution, the total horizontal forces in that story (i.e., $\Sigma_i V_{ji}$ or $\Sigma_i V'_{ji}$) must remain constant.

The requirements of Eq. (4.8) can also be expressed in terms of moment equilibrium, which is more convenient to use in design calculations. As Fig. 4.8 shows, shear forces in any individual column apply a moment $M_{ci} = l'_i V'_{ji} + l_i V_{ji}$ to the continuous beam at joint i. However, during moment redistribution a moment increment or decrement ΔM_i may be introduced at joint i. This will result also in a change of shear force

$$\Delta V_i = \Delta M_i / (l'_i + l_i) = \Delta M_i / l_c \qquad (4.8d)$$

in the affected column both above and below the floor. From Eq. (4.8c) it follows, however, that $\Sigma_i (V_{ji} + \Delta V_i) = \Sigma_i V_{ji} + \Sigma_i \Delta V_i = V_j = $ constant, and this means that the sum of incremental column shear forces $\Sigma_i \Delta V_i$ in the stories above and below level j must be zero. Consequently, from Eq. (4.8d) the sum of moment increments at beam–column joints must also be zero, since $l_c = l_i + l'_i$ is constant (i.e., $\Sigma_i \Delta M_i = 0$). Thereby the requirement of Eq. (4.8a) will also be satisfied after moment redistribution if

$$\Sigma_i M_{ci} + \Sigma_i \Delta M_i = \Sigma_i l'_i (V'_{ji} + \Delta V_i) + \Sigma_i l_i (V_{ji} + \Delta V_i)$$
$$= \Sigma_i l'_i V'_{ji} + \Sigma_i l_i V_{ji} + (l'_i + l_i) \Sigma_i V_\Delta$$
$$= \Sigma_i M_{ci} = \text{constant} \qquad (4.9a)$$

Thus, as stated above, the shear forces in individual columns, V_{ji} or V'_{ji}, may change during moment redistribution, but the total moment input to the beams, $\Sigma_i M_{ci}$, must remain unchanged. Therefore, the moments applied to the ends of the beams, shown in Fig. 4.8, must satisfy the same condition, that is,

$$\Sigma_i (M_{i,i-1} + M_{i,i+1}) = \Sigma_i M_{ci} = \text{story moment} = \text{constant} \quad (4.9b)$$

which in terms of the example shown in Fig. 4.8 is

$$\Sigma_i M_{ci} = M_{12} + (M_{21} + M_{23}) + (M_{32} + M_{34}) + M_{43} = \Sigma_i M_{bi}$$

where M_{bi} refers to the moments introduced to a column by the beams that are joined to the column at node point i. Equation (4.9b) implies that the magnitude of any or all beam end moments may be changed as long as the sum of beam end moments remains unchanged. In practical applications of moment redistribution, only the rule of Eq. (4.9b) needs to be observed, whereby Eq. (4.8) will also be satisfied. This follows from the equilibrium requirement for any joint, whereby at that joint $\Sigma M_b = \Sigma M_c$. It will be shown in Section 4.6.3 that in the derivation of design moments for columns, in accordance with the recommended capacity design procedure, the foregoing equilibrium criterion for joints will be satisfied automatically.

There are two characteristic situations for moment redistribution along continuous beams. The first involves beam moment redistribution across a joint. For example, the moment M_{21} in Fig. 4.8 may be reduced by an amount and the moment M_{23} be increased by the same amount. The total beam moment input to the joint remains unaltered, and hence the moments and shear forces for the relevant column remain the same as before.

The second type of moment redistribution in the beam involves redistribution of actions between columns. For example, when the beam moment M_{43} is reduced by ΔM_4, any or all other beam end moments must be increased correspondingly. As an example, end moments M_{12} and M_{34} may be increased by increments ΔM_1 and ΔM_3, respectively, so that the total change in beam end moments is $\Delta M_4 + \Delta M_1 + \Delta M_3 = 0$. Because the beam moment input to columns 1, 3, and 4 has changed, the shear forces in these columns will also change. Thus redistribution of both moments and shear forces between these three columns will take place. The conditions stipulated by Eqs. (4.8) and (4.9) will not be violated.

With the application of these two cases, any combination of moment reductions or increases is permissible, provided that Eq. (4.9) is satisfied. This makes the process of moment redistribution extremely simple.

In an attempt to achieve a desirable solution, the designer manipulates end moments of beams only. Gravity load equilibrium will be maintained provided that the part of the bending moment which originates from gravity load only and is applicable to a simply supported beam (i.e., the moment superimposed on a straight baseline which extends between the beam end moments), is not changed. Typical moment values that must not change are M_{f12}, M_{f23}, and M_{f34} in Fig. 4.8.

A change of beam end moment in any span will change the beam reactions and hence the forces introduced to individual columns below the beam under consideration. The actions shown in Figs. 4.6 to 4.8 uniquely define axial load increments applied to each column at the level considered. Hence axial forces on columns have not been shown in these figures.

4.3.4   Guidelines for Redistribution

Moment redistribution can be relied upon only if adequate rotational ductility is available at critical beam locations. The consequence of redistribution of design actions will be that members whose design actions are reduced by redistribution will begin to yield at somewhat less than the design intensity of the lateral forces and will need to sustain increased rotational ductility demand, approximately in inverse proportion to the change in moment from the elastic level. However, the global ductility demand on the structure as a whole due to seismic actions remains unchanged. Since the design philosophy implied by Fig. 1.19(a) requires a weak beam/strong column system, where ductility is assigned primarily to beams, special detailing will be provided at critical beam sections to ensure that adequate rotational ductility exists for seismic actions. The much smaller rotations required for redistribution of elastic actions can thus easily be accommodated.

It should be noted that the redistribution process outlined here relies entirely on rotations within plastic hinges in the beams. The apparent redistribution of moments and shears between individual columns also relies on plastic hinge rotations in the beams only.

It is recommended that in any span of a continuous beam in a ductile frame, the maximum moments may be decreased, if so desired, by up to 30% of the absolute maximum moment derived for that span from elastic analysis, for any combination of seismic and gravity loading. This limit is placed to ensure that plastic hinges do not occur prematurely under a moderate earthquake, and that the beam rotational ductility demand is not increased excessively. The inaccuracies inherent in the elastic analyses, noted earlier, may influence the true level of effective redistribution. The impact of the 30% limit is that yielding may begin at 70% of the design level forces. Because of the reduced curvature at yield, the peak curvature ductility demand at such a section may increase by up to 43%. This is considered acceptable in light of conservative detailing requirements for plastic hinge regions, to be defined later. Increased curvature ductility at such sections implies larger steel tensile strains rather than increased concrete compression strains.

Figure 4.9 illustrates this change of ductility demand for a simple redistribution between two potential plastic hinges, A and B, in terms of a characteristic bilinear force–displacement relationship. Assuming that the elastic bending moment diagram indicated equal strength demands at A and B, a corresponding design would ensure that both hinges would begin yielding simultaneously. Thus the elastoplastic response of each hinge, corresponding to the resistance of forces $F_A = F_B = 0.5F_D$, is the same as that of the whole structure. This is shown by the full lines in Fig. 4.9(a). The rotational ductility demand for each hinge (ignoring effects of elastic column and joint deformations) is $\mu_{\theta A} = \mu_{\theta B} = \mu_\theta = \Delta_{\max}/\Delta_y$.

A redistribution from hinge A to hinge B, corresponding to a 30% force decrement of $0.3F_A = 0.15F_D$, would change the characteristic yield displacements to $\Delta_{yA} = 0.7\Delta_y$ and $\Delta_{yB} = 1.3\Delta_y$, assuming that stiffnesses at A and B are as before. The ensuing response is shown by the dashed lines in Fig. 4.9(a). It is also seen that as a result of 30% moment redistribution, the ductility demand increased by 43% at A while it decreased by 23% at B.

The example above is, however, somewhat conservative because it did not take account of a change in incremental stiffness of the regions at A and B.

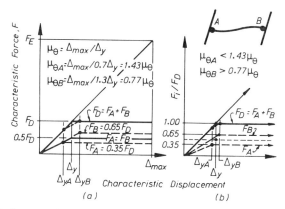

Fig. 4.9 Influence of moment redistribution on rotational ductility demand.

A reduced design moment will result in a reduced amount of flexural reinforcement, and hence a reduced amount of flexural rigidity (EI) of the section at A and a corresponding increase of the characteristic displacement at yield, Δ_{yA}. The dashed lines in Fig. 4.9(b) show the different responses of the two hinges. These indicate that the differences in rotational ductility demands (i.e., $\mu_{\theta A}$ and $\mu_{\theta B}$) are not as large as in the previous case.

It will be found that in most practical situations optimum moment patterns can be obtained with moment changes much less than the recommended 30% limit. Since increasing the moment capacity at a section as a consequence of moment redistribution (say hinge B in Fig. 4.9) delays the onset of yield and reduces the local rotational ductility demand at that section, no limit needs be placed on the amount by which moments along the beam can be increased. However, as mentioned earlier, this should be done within the constraints of the equilibrium requirement of Eq. (4.9) to ensure that unnecessary overstrength does not result.

It has been found in theoretical studies that moment redistribution, within the limit suggested here, had insignificant effects on both the overall elastic–plastic dynamic response of frames and changes in ductility demands in plastic hinges [O3].

4.3.5 Examples of Moment Redistribution

The redistribution process may be outlined with reference to the two examples of Figs. 4.6 and 4.7. Considering the symmetrical frame of Fig. 4.6 first, satisfaction of Eq. (4.9b) requires that

$$\Sigma_i M_{bi} = 239 + 384 + 184 + 460 = 1267$$

The sum must not be reduced after redistribution.

The moment redistribution process will aim to result in beam longitudinal reinforcement requirements at potential plastic hinges locations that are as close as practical to uniform. Generally, plastic hinges will form at column faces, never at column centerlines. The influence of gravity loads in the span is such that beam negative moments reduce more rapidly with distance from the column centerline than will beam positive moments. This can be seen, for example, in Fig. 4.6(c). Hence negative moments after redistribution should exceed positive moments at the column centerline if it is desired to obtain equal moments at the column faces. Since some slab reinforcement parallel to the beam axis will often contribute to the beam negative flexural strength, as discussed in Section 4.5.1(b), a further small increase in negative design moments relative to positive moments is generally advisable if equal top and bottom beam flexural reinforcement is desired at critical hinge locations.

It will be noted that equal moments of $1267/4 \approx 317$ units could be provided at all plastic hinge locations. However, the maximum reduction of the critical moment of 460 units is to $0.7 \times 460 = 322 > 317$. Further, as outlined above, it is desirable to retain slightly higher negative than positive moments at column centerlines. As a consequence, negative beam moments of 340 units and positive moments of 294 are chosen for the column centerlines, as shown in Fig. 4.6(d), giving

$$\Sigma_i M_{bi} = 2 \times 340 + 2 \times 294 = 1268 \approx 1267$$

The frame of Fig. 4.6 has been based on span lengths of 8 m and a column depth of 600 mm in the plane of the frame. With these dimensions, negative moments of 300 units and positive moments of 290 units result at the column faces. Further minor adjustments to the moments may be made during the beam design stage to further improve efficiency.

The unsymmetrical frame of Fig. 4.7 requires more careful consideration. Equation (4.9b) requires that $\Sigma_i M_{bi} = 370$ [Fig. 4.7(b)]. Note that in Fig. 4.7(c) the same total is obtained whether $\overrightarrow{E}$ or $\overleftarrow{E}$ is considered. At the center support, the minimum permissible moment after redistribution is $0.7 \times 170 = 119$. A value of 120 is chosen. The requirement that minimum positive moment at the support is at least 50% of the negative moment capacity requires a positive moment capacity of at least 60 units. Initially, this value is chosen. At the external supports negative moments capacity should exceed positive moment capacity, as discussed earlier. Adopting a differential of 20 units between negative and positive capacities at the outer columns results in $0.5[370 - (120 + 60)] \pm 10 = 105$ and 85 units, respectively, to provide $\Sigma_i M_{bi} = 120 + 60 + 105 + 85 = 370$.

It would be possible to increase the positive moment capacity at the center support above 60 units, and thereby to reduce moments elsewhere (but not the negative moment capacity at the center column). However, because of the gravity load domination of the long span, the critical positive moment section is at midspan with a required moment capacity of 122.5 units, which exceeds the maximum negative moment requirement.

A simpler and better solution results from making all negative moments equal to 120 units (the minimum possible at the center support). Equal positive moments of $(370 - 2 \times 120)/2 = 65$ units would then be required at the two positive moment locations at the column centerlines. This solution is plotted in Fig. 4.7(d). Figure 4.7(e) shows an alternative and equally acceptable solution, also satisfying Eq. (4.9b), whereby positive and negative moment requirements at the outer columns have been made the same.

Before finalizing design moments, such as those derived in Fig. 4.7(d) and (e), gravity load requirements must also be checked. For example, the factored gravity load demands $M_u = 1.4M_D + 1.7M_L$, according to Eq. (1.5a) may well be 40% larger than those shown in Fig. 4.7(a). In this example it is seen that the maximum negative moment demand at the interior column is of the order of $1.4 \times 80 = 112 < 120$, and at the center of the long span it is $1.4[140 - 0.5(80 + 40)] = 112 \approx 112.5$. These moments are very similar to those derived for the combined gravity and seismic requirements. Gravity load in this example is clearly not critical at the outer columns.

Figure 4.7(d) and (e) show that gravity requirements dominate the center region of the long span. For both spans careful consideration of the location of plastic hinges for positive moments, as discussed in Section 4.5.1(a) and shown in Fig. 4.18, will be required.

4.3.6 Moment Redistribution in Inelastic Columns

Moment redistribution between elastic columns has already been discussed as a mechanism for reducing design moments for columns subjected to tension or to low axial compression force. It is also possible to consider moment redistribution from one end of a column to another. In general, no significant advantage will result from this action in upper stories of ductile frame, but there can be advantages in first-story columns.

When columns are relatively stiff, the elastic analysis for lateral forces may indicate that moments at the two ends of columns in the first story are very different. Such a case, by no means extreme, is seen in the first story of the example column in Fig. 4.21. The designer may wish to proportion the column at the base, where a plastic hinge is to be expected, to resist a smaller moment; thus a smaller column may be used. This may be achieved by assigning a larger design moment to the top end of such a column. It is essential, however, to ensure that the resistance to the total story shear force is not altered. This means that in the process of moment redistribution the sum of the design moments at the top and the bottom of all columns in the first story must not be reduced. By assigning larger design moments to the top ends of these first-story columns, the resistance of the first-floor beams must be increased correspondingly. The procedure may allow a column section, suitable to resist the design actions over a number of the lower stories, also to be used in the first story. To safeguard columns against premature yielding at the ground floor, it would be prudent to limit the reduction of design moments at the base to less than 30%, the maximum value suggested for beams.

4.3.7 Graphical Approach to the Determination of Beam Design Moments

The effort involved in producing beam design moment envelopes can be greatly reduced if instead of the traditional horizontal baseline shown in Figs. 4.6 to 4.8, the gravity moment curve for a simply supported beam is used for this purpose. This is shown in Fig. 4.10. All other moments, resulting from moment applications at the ends of the spans only, consist of straight lines. Repeated plotting of moment curves for each direction of earthquake attack and for each improved redistribution is thereby avoided. If a suitable sign convention such as that suggested in Fig. 4.10 is used, moment superposition and subsequent redistribution will consist of the simple adjustment of straight lines only. Accordingly, moment values measured downward from the curved baseline are positive, and those measured upward will be taken negative (i.e., causing tension in the top fibers of a beam). With this convention the effects of both load cases, $\overrightarrow{E}$ and $\overleftarrow{E}$, can readily be assessed in a single diagram.

To illustrate this method, the steps of the previous example, given in Fig. 4.7, will be used, the results of which are shown by numbered straight lines in Fig. 4.10. The construction proceeds as follows:

1. First the baseline is drawn for each span. These are the curves shown (1) in Fig. 4.10 and these correspond to the gravity moment curves shown in Fig. 4.7(a).

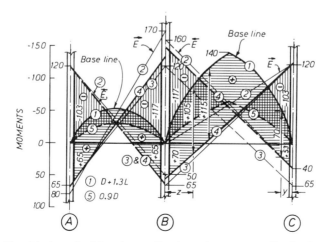

Fig. 4.10 Graphical method for the application of moment redistribution.

2. The end moments obtained from the elastic analysis for gravity loads and earthquake forces, shown in Fig. 4.7(c), are plotted. These moments are connected by the (thin) straight full lines marked (2) for $\vec{E}$ and corresponding dashed lines for $\overleftarrow{E}$. The vertical ordinates between curves (1) and lines (2) record the magnitudes of the moments exactly as in Fig. 4.7(c). Lines marked $\vec{E}$ and $\overleftarrow{E}$ imply the application of clockwise or anticlockwise moments, respectively, at the ends of the beams.

3. The designer may now adjust the straight lines (i.e., end moments) only, to arrive at suitable moment intercepts at critical sections. This is a trial-and-error procedure which, however, converges fast with a little design experience. It readily accommodates decisions that the designer may make in order to arrive at simple and practical solutions. In Fig. 4.10, lines (3) show the specific choice made in Fig. 4.7(d). Accordingly, the magnitudes of the negative moments at column centerlines are 120, and those of the positive moments are 65 units.

4. Although these centerline moments are well balanced, there are significant differences in the magnitudes of critical moments at column faces. With further minor adjustments of the column centerline moments (i.e., the diagonal lines that connect them in a span), while observing the previously stated equilibrium condition ($\sum_i M_{bi}$ = constant), equal moment ordinates at column faces may be obtained. As Fig. 4.10 shows, lines (4) result in scaled (approximate) values of 103 and 117 units for the negative moments at the exterior and interior column faces, respectively. Positive moments of 65 units were achieved in the short span. These critical moment ordinates at column faces are emphasized with heavy lines in Fig. 4.10. Thus the amounts and details of the flexural reinforcement at both ends of the short-span beam may be identical. Generally, more slab reinforcement will contribute [Section 4.8.9(a)] to negative flexural strength enhancement at an interior column than at exterior columns. For this reason the negative design moment at the interior column (117) was made a little larger than at the other ends of the beams (103).

It is emphasized that a high degree of precision in balancing design moments and in determining their magnitudes is unnecessary at this stage of the graphical construction. It is almost certain that the chosen arrangement of reinforcing bars will not exactly match the demand resulting from the design moments shown in Fig. 4.10. If for practical reasons, more or less flexural reinforcement is to be used at a section, a slight adjustment of the moment line (4), to correspond with the strengths in fact provided, will immediately indicate whether the equilibrium condition [Eq. (4.9)] has been violated, or alternatively, where unnecessary excess strength has been provided.

For the long-span beam in Fig. 4.10, it will not be possible to develop positive plastic hinges at the column faces, because these moments, as a result of gravity load dominance in that span, turned out to be smallest over the entire span. For this reason no effort needs to be made to attempt to equalize these noncritical moments. It will be shown subsequently that with suitable curtailment of the bottom bars in such spans, the development of a positive plastic hinge can be enforced a short distance away from the column faces (Fig. 4.18).

5. The moment envelopes constructed in Fig. 4.10, when based on the actual flexural strength provided at the critical sections of potential plastic hinges, will give a good indication for curtailment requirements for the flexural reinforcement along each span. However, for this the influence of the minimum specified gravity load [Eq. (1.6b)] must also be considered. This may be achieved by plotting, with the dashed-line curve (5), a new baseline in each span for the moment which accounts for the presence of only 90% of dead load. The vertically shaded areas so obtained in Fig. 4.10 will provide a good estimate of the beam length over which negative moments could possible develop after the formation of negative plastic hinges. Figure 4.18 illustrates the simple principles involved in the determination of curtailment of beam bars based on moments developed at flexural overstrength of the plastic hinges.

A detailed application of the graphical procedures, outlined above, is presented as part of a design example in Section 4.11.7. The same technique, even more simple, can be used to establish a suitable redistribution of gravity moments alone. An example of this is also given in Section 4.11.

4.4   DESIGN PROCESS

For frame structures designed in accordance with the weak beam/strong column philosophy [P45], the design process involves a well-defined sequence of operations, as summarized below, and described in detail in the following sections.

4.4.1   Capacity Design Sequence

(a)   Beam Flexural Design   Beams are proportioned such that their dependable flexural strength at selected plastic hinge locations is as close as possible to the moment requirements resulting from the redistribution process of Section 4.3. Generally, though not always, the plastic hinge locations are chosen to be at the column faces. Flexural strength of other regions of the beams are chosen to ensure that expected plastic hinges cannot form in regions where special detailing for ductility, described in Section 4.5.4, has not been provided.

(b)   Beam Shear Design   Since inelastic shear deformation does not exhibit the prerequisite characteristics of energy dissipation, shear strength at all sections along the beams is designed to be higher than the shear corresponding to maximum feasible flexural strength at the beam plastic hinges. Conservative estimates of shear strength and special transverse reinforcement details are adopted within potential plastic hinge regions, as shown in Section 4.5.3.

(c)   Column Flexural Strength   Considerations of joint moment equilibrium and possible higher-mode structural response are used to determine the maximum feasible column moments corresponding to beam flexural overstrength. Ideal column moment capacity is matched to these required strengths to ensure that the weak beam/strong column hierarchy is achieved (Section 4.6).

(d)   Transverse Reinforcement for Columns   The determination of the necessary amount of transverse reinforcement, a vital aspect of column design, is based on the more stringent of the requirements for shear strength, confinement of compressed concrete, stability of compression reinforcement, and lapped bar splices. Again an estimate of the maximum feasible shear force in the column is made on the basis of equilibrium considerations at beam flexural overstrength (Section 4.6.7).

(e)   Beam–Column Joint Design   Because beam–column joints are poor sources of energy dissipation, inelastic deformations due to joint shear forces or bond deterioration must be minimized. The ideal strength of joints is matched to the input from adjacent beams when these develop flexural overstrength at the critical sections of plastic hinges. We examine all relevant issues in detail in Section 4.8.

It should be noted that only for the case of beam flexural design will design actions correspond to the code level of lateral seismic forces, and these moments may differ from the elastic analysis results considerably, as a result of moment redistribution. For beam shear and all column design actions, the design forces are calculated on the assumption of beam plastic hinge sections developing maximum feasible flexural strength using simple equilibrium relationships.

4.4.2   Design of Floor Slabs

The proportioning of floor slabs in buildings, for which lateral force resistance is assigned entirely to ductile frames, is very seldom affected by seismic actions. Cast-in-place floor slabs spanning in one or two directions and monolithic with the supporting beams generally provide ample strength for diaphragm action unless penetration by large openings is excessive.

In precast or other types of prefabricated floor systems, sufficiently strong and rigid in-plane connections must be provided to ensure the necessary diaphragm action, discussed in Section 1.2.3(a). This is usually achieved by a relatively thin (typically, 60 mm, 2.5 in.) reinforced concrete slab (topping), cast on top of the prefabricated floor system and the supporting beams and providing the finished-level floor surface. The thickness of this concrete topping may have to be checked to ensure that in-plane shear stresses due to seismic actions are not excessive and that connections to vertical lateral force resisting elements are adequate. A more detailed examination of these aspects is given in Section 6.5.3. Positive attachment of thin topping slabs, by bonding or by suitably spaced mechanical connectors to the structural floor system underneath, must be ensured to prevent separation during critical diaphragm action [X3]. If separation can occur, the topping may buckle when subjected to diagonal compression resulting from shear in the diaphragm. Usually, a light mesh of reinforcement in the topping slab is sufficient to ensure its integrity for diaphragm action. Without an effectively bonded topping slab, precast floor systems will not be capable of transmitting floor inertial forces back to designated lateral-force-resisting elements. This was illustrated, with tragic consequences, by the extremely poor performance of precast frame buildings in the Armenian earthquake of 1988. A contributing factor in these failures was the lack of positive connection between precast floor slab elements, and the poor connection between them and the supporting beam members.

The analysis and design of slabs for gravity loading is beyond the scope of this book. The subject is well covered in the technical literature [P2] and design procedures are suggested in various building codes [A1].

4.5 DESIGN OF BEAMS

4.5.1 Flexural Strength of Beams

The flexural behavior of reinforced concrete sections was reviewed in Section 3.3.1. Therefore, in this section only issues specific to seismic design are considered. Before detailed treatment of a beam begins, the designer must make sure that slenderness criteria, recommended in Section 3.4.3 are satisfied.

(a) Design for Flexural Strength

(i) *Conventionally Reinforced Beam Sections*: The design of a critical section for flexural strength in a potential plastic hinge zone of a beam, such as seen in Fig. 4.15, involves very simple concepts. As shown in Fig. 4.11(a) either because of the reversed nature of moment demands or to satisfy code

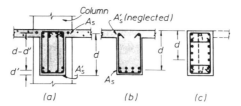

Fig. 4.11 Distribution of flexural reinforcement in beam sections.

requirements to ensure adequate ductility potential as summarized in Section 3.5.2, such beam sections are always doubly reinforced. Due to reversed cycling inelastic displacements, both the top and the bottom beam reinforcement may be expected to yield in tension. When the flexural bars have yielded extensively in one face of the beam, wide cracks in the concrete will occur. These cracks cannot close again upon moment reversal unless the bars in the compression zone yield or slip. Even if the Bauschinger effect is taken into account [P1], the development of the full yield strength of these bars in compression may need to be considered. A prerequisite is full effective anchorage of such bars on either side of the plastic hinge.

This can be achieved when beams frame into walls or massive columns. However, when plastic beam hinges develop at both sides of a beam–column joint, bond deterioration within the joint, examined in detail in Sections 4.8.3(c) and 4.8.8, will to some extent invalidate some of the traditional assumptions [Section 3.3.1(a)] used in the analysis of reinforced concrete sections for flexure. Therefore, after reversed cyclic curvature ductility demands, compression reinforcement may not contribute to the flexural response as effectively as assumed. As a consequence, the contribution to flexural strength of the concrete in compression may be greater than that indicated by section analysis.

This phenomenon will influence mechanisms within beam–column joints (Section 4.8.5), but it will normally have a negligible effect on the accuracy of flexural strength predictions. The center of internal compression force $C = C_c + C_s$ in Fig. 3.20 will be very close to the centroid of the compression steel. The design of a beam section, such as that shown in Fig. 4.11(a), can thereby be greatly simplified. It is evident that at this stage the internal lever arm is simply the distance between centroids of the top and bottom flexural reinforcement, namely, with the notation used in Section 3.3.1(b) and Fig. 3.20, $jd = d - d'$. This distance may be used as an approximation also in cases when the area of compression steel A'_s is less than that of the tension steel A_s. It should be remembered that $A'_s \geq 0.5A_s$ (Section 3.4.2).

The necessary flexural tension reinforcement for the top or the bottom of a conventionally reinforced beam section can therefore be determined very rapidly from

$$A_s = \frac{M_u}{\phi f_y (d - d')} \tag{4.10}$$

wherever reversed moments in potential plastic hinges can occur.

In positive moment areas, such as the midspan of beam B–C in Fig. 4.10, where only very small or no negative moments can occur [Fig. 4.11(b)], the contribution of the concrete compression zone should be taken into account. Because in these localities an effective compression flange (Fig. 4.3) is usually available, the principles of Section 3.3.1(b), utilizing a larger internal lever arm, should be employed.

(ii)   Beam Sections with Vertically Distributed Flexural Reinforcement: An unconventional arrangement of bars, shown in Fig. 4.11(c), leads to equal efficiency in flexural resistance, particularly when the positive and negative moment demands are equal or similar. However, this arrangement offers numerous advantages that designers should consider. It may be shown from first principles [W7] that the flexural strength of a beam with a given amount of total reinforcement ($A_{st} = A_s + A'_s$) is for all practical purposes the same, irrespective whether it is uniformly distributed or placed in two equal lumps in the top and bottom of the section, similar to that shown in Fig. 4.11(a). In the case shown in Fig. 4.11(c), the flexural strength results from a number of bars, significantly larger than one-half of the total, operating in tension with a reduced internal lever arm. This principle was reviewed and emphasized in connection with the analysis of wall sections in Section 3.3.1(a) and Fig. 3.21.

The major advantages include (1) easier access in the top of the beam for placing and vibrating the concrete during construction; (2) better distribution and early closing of flexural cracks on moment reversal compared with conventional sections; (3) reducing the tendency for excessive sliding shear deformations in the plastic hinge region [Section 3.3.2(b)]; (4) increased depth of concrete compression zone in beam plastic hinges, thereby improving shear transfer; (5) a tendency to develop smaller flexural overstrength than those developed by conventionally reinforced sections under large curvature ductility demands; and (6) satisfactory performance of beam–column joints with reduced joint shear reinforcement, an aspect examined in Section 4.8.5 [W7].

(b)   Effective Tension Reinforcement   The effective tension reinforcement to be considered should be that which is likely to be utilized during earthquake motions. When the top bars near the support of a continuous beam are yielding extensively, adjacent parallel bars in the slab, which forms the integrally built tension flange of the beam section, will also yield. Thus during large inelastic displacements, the flexural strength of the section could be increased significantly [C14, Y2].

In the design for gravity loads, the contribution of such slab reinforcement to flexural strength is traditionally neglected. However, in seismic design, the development of strength considerably in excess of that anticipated may lead to undesirable frame behavior. This will become evident subsequently when the shear strength of beams and the strength of columns are examined.

The participation of slab reinforcement in the development of beam flexural strength at interior and exterior beam–column joints has been consistently observed in experiments [C14, K4, M2, Y2]. However, it is difficult to estimate for design purposes the effective amount of slab reinforcement that might participate in moment resistance of a beam section [S5]. There are several reasons for this. First, the extent of mobilization of slab bars depends on the magnitude of earthquake-imposed inelastic deformations. The larger the rotations in plastic hinges adjacent to column faces, the more slab bars, placed farther away from the column, will be engaged. Second, tension forces in slab bars due to flange action need to be transferred via the beams to the beam–column joint. Thus the contribution of any slab bar will depend on its anchorage within the slab acting as a flange. The effectiveness of short bars, placed in the top of a slab to resist negative gravity moments over a transverse beam, will decrease rapidly with distance from the joint. Third, the effectiveness of slab bars will also be affected by the presence or absence of transverse beams. This is of particular importance where a slab is monolithic with an edge beam. The mechanisms of slab action in a tension flange of a T or L beam are presented in Section 4.8.9(a), where the relative importance of the foregoing aspects are examined in some detail.

To be consistent with the philosophy of capacity design, beam flexural strengths at two levels of tension flange participation should be evaluated. The dependable flexural strength of a beam section should be based on deliberate underestimations with respect to the effective tributary width of a tension flange. On the other hand, evaluation of the overstrength of the critical section of a plastic hinge [Section 1.3.3(d)] should consider a larger effective width, recognizing the probable magnitude of imposed plastic hinge rotations.

Under normal circumstances the amount of slab reinforcement, which could contribute to flexural tension in a beam, is a relatively small fraction of the total steel content represented by the top bars in a beam. Hence sophistication in the estimate of effective slab tension reinforcement would seldom be warranted. Therefore, a compromise approach is suggested here, whereby the same effective tension flange width b_e should be assumed for the estimation of both dependable strength and overstrength in flexure. Accordingly, it is recommended [C14] that in T and L beams, built integrally with floor slabs, the longitudinal slab reinforcement placed parallel with the beam, to be considered effective in participating as beam tension (top) reinforcement, in addition to bars placed within the web width of the beam, should include all bars within the effective width in tension b_e, which may be

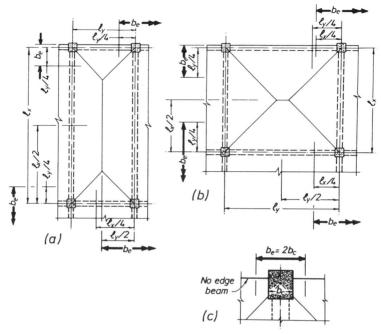

Fig. 4.12   Effective width of tension flanges for cast-in-place floor systems.

assumed to be the smallest of the following [C14]:

1. One-fourth of the span of the beam under consideration, extending each side from the center of the beam section, where a flange exists
2. One-half of the span of a slab, transverse to the beam under consideration, extending each side from the center of the beam section, where a flange exists
3. One-fourth of the span length of a transverse edge beam, extending each side of the center of the section of that beam which frames into an exterior column and is thus perpendicular to the edge of the floor

Effective tension flange widths b_e, determined as above, are illustrated in Fig. 4.12. Within this width b_e, only those bars in the slab that can develop their tensile strength at or beyond a line at 45° from the nearest column should be relied on. At edge beams, effective anchorage of bars, in both the top and bottom of the slab, must also be checked. Where no beam is provided at the edge of a slab, only those slab bars that are effectively anchored in the immediate vicinity of a column as shown in Fig. 4.12(c) should be relied on.

No corresponding provisions exist in current U.S. codes. It has been a recommended in Japan [A10] that b_e be taken as one-tenth of the beam span. In New Zealand the effective width of a slab in tension is taken [X3] up to four times the thickness of the slab, measured from each side of a column face, depending on whether an interior or an exterior joint with or without transverse beams is being considered.

Under seismic actions the end moments in beams are balanced by similar moments in the column above and below the joint rather than by moment in the beam at the opposite side of the column, as in the case of gravity-loaded continuous beams. Therefore, it is desirable to place most of the principal top and bottom flexural bars within the width of the beam web and carry these into or across the cores of supporting columns. For this reason it is also preferable that the width of the beam should not be larger than the width of the column. As an upper limit it is recommended [X3] that the width of the beam b_w should not be more than the width of the column plus a distance on each side of that column equal to one-fourth of the overall depth of the column in the relevant direction, but not more than twice the width of the column. The interpretation of this practical recommendation is shown in Fig. 4.13 and discussed in Section 4.8.7(c).

The transfer of forces from bars outside a column to the joint core by means of anchorage over the depth of the column h_c is doubtful. Therefore, any reinforcement in a beam section which passes through the column outside the column core should be assumed to be ineffective in compression. Particular care should be taken when using wide beams at exterior joints, where transverse reinforcement will need to be provided to ensure that forces developed in tension bars, anchored outside columns, are transferred to the joint core.

The effective interaction of beams with columns under earthquake attack must also be assured when the column is much wider than the beam. Clearly,

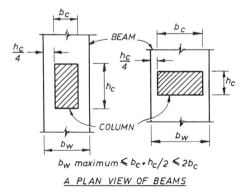

$$b_w \ maximum \leqslant b_c + h_c/2 \leqslant 2b_c$$

A PLAN VIEW OF BEAMS

Fig. 4.13   Recommended maximum widths of beams [X3].

concrete areas or steel bars in the column section a considerable distance away from vertical faces of the beam will not fully participate in resisting moment inputs from that beam. Problems may also arise at eccentric beam–column joints. The approach to these issues (shown in Fig. 4.63) is examined, together with the design of beam–column joints, in Section 4.8.9(c).

(c)   Limitations to the Amounts of Flexural Tension Reinforcement

(i)   Minimum Reinforcement: Unless the amount of flexural tension reinforcement is sufficient to ensure that the flexural strength exceeds the cracking moment by a reasonable margin, there is a real danger that only one crack will form, at the critical section, within the potential plastic hinge region. This is particularly the case for negative moments, which reduce rapidly with distance from the column face. Well-distributed flexural cracking within plastic hinges is needed to avoid the possibility of excessive local curvature ductility demand, particularly if the diameter of beam bars is small, which is likely with low reinforcement ratios.

When tension flanges of T and L beams contribute to flexural tension strength, the cracking moment will be substantially higher than for rectangular sections or for the same section under positive moment. The following expressions for the required minimum amount of tension reinforcement ensure that ideal flexural strength is at least 50% greater than the probable cracking moment.

Rectangular sections, and T beams with flange in compression:

$$\rho_{min} = 0.25\sqrt{f'_c}\big/f_y \ \ (\text{MPa}); \qquad \rho_{min} = 3\sqrt{f'_c}\big/f_y \ \ (\text{psi}) \qquad (4.11)$$

T-beams with flange in tension:

$$\rho_{min} = 0.40\sqrt{f'_c}\big/\big/f_y \ \ (\text{MPa}); \qquad \rho_{min} = 4.8\sqrt{f'_c}\big/f_y \ \ (\text{psi}) \qquad (4.12)$$

where ρ_{min} includes the tension reinforcement in the flange in accordance with Section 4.5.1(*b*).

(ii)   Maximum Reinforcement: To ensure adequate curvature ductility, the maximum flexural reinforcement content should be limited to

$$
\begin{aligned}
\rho_{max} &= \left[\frac{275}{f_y}\right]\frac{1 + 0.17(f'_c/7 - 3)}{100}\left(1 + \frac{\rho'}{\rho}\right) < \frac{7}{f_y} \ \ (\text{MPa}) \\[2mm]
&= \left[\frac{40}{f_y}\right]\frac{1 + 0.17(f'_c - 3)}{100}\left(1 + \frac{\rho'}{\rho}\right) < \frac{1}{f_y} \ \ (\text{ksi})
\end{aligned}
\tag{4.13}
$$

where

$$\rho' \geq 0.5\rho \qquad (4.14)$$

within potential plastic hinge regions, defined in Section 4.5.1(d). The reinforcement ratio ρ is computed using the width of the web b_w.

Equation (4.13) will ensure that a curvature ductility factor of at least 8 can be attained with an extreme fiber compression strain of 0.004 [P1, X3] and recognizes the influence of increasing values of f'_c and ρ'/ρ in reducing the depth of the concrete compression zone when the limiting concrete strain of 0.004 is attained and hence increasing ultimate curvature and the curvature ductility factor. A curvature ductility factor of 8 will ensure that premature spalling of cover concrete does not occur in moderate earthquakes. However, it should be noted that spalling is probable under design-level seismic response.

The upper limit of $7/f_y$ (MPa) $[1/f_y$ (ksi)] is similar, for grade 275 reinforcement, to that required by some well-known codes [A1, X3] ($\rho <$ 0.025), but recognizes that it is the tension force (proportional to ρf_y) rather than reinforcement content that must be controlled. In many cases it will be found necessary to further limit the maximum amount of reinforcement, to avoid excessive shear stresses in beam–column joints (Section 4.8.3), and a practical upper limit of $\rho = 5/f_y$ (MPa) $[0.7/f_y$ (ksi)] is suggested.

When selecting bar sizes, consideration must be given to rather severe bond criteria through beam–column joints, discussed in Section 4.8.6(b).

(d)   Potential Plastic Hinge Zones   Locations of plastic hinges in beams must be clearly identified since special detailing requirements are needed in inelastic regions of beams of frames subjected to earthquake forces. Also, the limitations of Eqs. (4.11) to (4.14) apply only within potential plastic hinges, and may be relaxed elsewhere.

Plastic hinges in beams of ductile frames, the design of which is dominated by seismic actions, commonly develop immediately adjacent to the sides of columns, as shown for the short-span beams of the frame in Fig. 4.14 and in Fig. 4.82. This would also be the case for the beams in Fig. 4.6 and for span A–B of the beams in Fig. 4.10.

When the positive moments in the span become large because of the dominance of gravity loading, particularly in long-span beams, it may be difficult, if not impossible, to develop a plastic hinge at a face of a column. The designer may then decide to allow a plastic hinge to form at some distance away from the column. Typical examples are shown for the long-span beams in Fig. 4.14(a). When earthquake forces act as shown, the positive moment plastic hinge in the top beam of Fig. 4.14(a) will develop close to the inner column, at the location of the maximum moment. If plastic hinges formed at column faces, hinge plastic rotations would be θ, as seen in the short span. However, with a positive hinge forming a distance l_1^* from the right column, the hinge plastic rotations will increase to $\theta' = (l/l_1^*)\theta$.

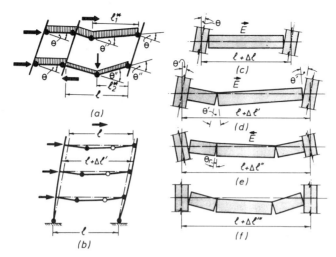

Fig. 4.14   Beam hinge patterns.

It is evident in Fig. 4.14(a) that the farther the positive plastic hinge is from the left-hand column, the larger will be the hinge rotations. In the lower of the long-span beams, the presence of a significant point load at midspan indicates that this will probably be the location of maximum positive moment under the action of gravity load and seismic forces. If the plastic hinge is located there, the plastic rotations would increase to $\theta'' = (l/l_2^*)\theta$.

The desired curvature ductility in plastic hinges is achieved primarily by very large inelastic tensile strains. Therefore, the mean strain over the depth of a beam and along the length of a plastic hinge will be tension, resulting in a lengthening of that part of the beam. Because the neutral-axis depth varies along the span, elongations also occur after cracking in elastic parts of the beam. However, these are negligible in comparison with those developed over plastic hinges. Thus it should be realized that seismic actions, resulting in two plastic hinges in a beam, such as shown in Fig. 4.14(c), will cause beams to become longer. The magnitude of span length increase Δl will be affected by the depth of the beam, the hinge plastic rotations θ or θ', and hence by the location of plastic hinges [Fig. 4.14(a) and (d)].

A particular feature of a plastic hinge, developing away from columns in the positive-moment region of a beam, is that inelastic rotations, and hence axial displacements, may increase in sequential cycles of inelastic displacement to constant-displacement ductility levels. With the dominance of gravity-load-induced positive moments at such a hinge, moment reversal due to seismic actions may never occur, or if it does, negative moments so developed may be very small (Fig. 4.10). Reversed-direction inelastic rotation occurs at a separate hinge location, the column face. Thus a significant residual rotation θ_r may remain, as illustrated in Fig. 4.14(e), contributing to further increase of total beam elongation $\Delta l''$. In the next cycle of inelastic displacement in the initial loading direction, additional inelastic rotation is added to the residual rotation at the positive moment hinge. After several inelastic frame displacements in both directions of earthquake attack, residual plastic hinge rotations may result in both large deflections and elongations in beams that have been designed to develop (positive) plastic hinges in the span, as shown in Fig. 4.14(f). Such beam elongations, also identified in experiments [F3, M11], may impair or even destroy connections to nonstructural components or to elements of precast floor systems, and also alter bending moments in columns of the lower two stories, seen in Fig. 4.14(b). It is evident that because of the progressive accumulation of residual rotations in both positive and negative plastic hinges, the prescribed performance of a frame required to sustain a given amount of cumulative ductility, such as described in Section 1.1.2(c), will be more difficult to achieve.

Unless expected ductility demands are moderate or the number of inelastic displacement reversals are likely to be few, as in the case of frames with long periods T, the plastic beam mechanisms of Fig. 4.14(d) to (f) should be avoided. It should be noted that progressive beam elongations will occur also with mechanisms shown in Fig. 4.14(c). This is because of the increasing misfit between crack faces after repeated opening and incomplete closure of cracks. This lengthening of the plastic hinge zone may be further increased when the shear force to be transmitted is large, leading to sliding, and when cracks do not follow a straight line [Fig. 3.24].

The length of a plastic hinge in a beam, over which special detailing of transverse reinforcement is required, should be twice the depth h of the beam [X3].

1. When the critical section of the plastic hinge is at the face of the supporting column or wall, this length is measured from the critical section toward the span. Examples are shown in Fig. 4.15, where the moment M_A or M_B is to be resisted.

2. Where the critical section of the plastic hinge is not at the face of a column (Fig. 4.16) and is located at a distance not less than the beam depth h or 500 mm (20 in.) away from a column or wall face, the length should be assumed to begin between the column or wall face and the critical section, at least $0.5h$ or 250 mm (10 in.) from the critical section, and to extend at least $1.5h$ past the critical section toward midspan. Examples are shown in Fig. 4.17.

3. At positive plastic hinges where the shear force is zero at the critical section, such as at C in Fig. 4.15, the length should extend by h in both directions from the critical section.

Advantages that stem from the relocation of plastic hinges away from column faces, as shown in Fig. 4.16, are that yield strain in beam bars at

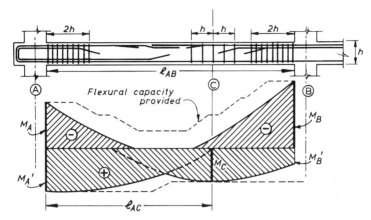

Fig. 4.15   Localities of potential plastic hinges where special detailing is required.

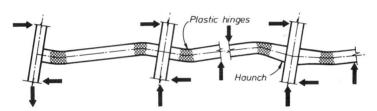

Fig. 4.16   Beams with relocated plastic hinges.

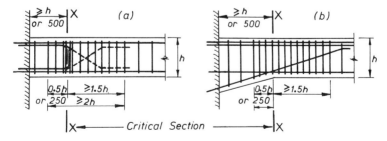

Fig. 4.17   Details of plastic hinges located away from column faces.

column faces may be avoided, and hence the penetration of yielding along beam bars into adjacent joint cores, as a result of bond deterioration, can be prevented. This improves joint behavior, particularly when beams are heavily reinforced, and relevant features are examined in detail in Section 4.8. As a consequence, the designer may choose to relocate plastic hinges away from column faces even when maximum moments occur at column faces. As long as the critical sections for the chosen plastic hinges for positive and negative moments coincide, this solution does not contradict the aims of avoiding plastic hinge formations, such as illustrated in Fig. 4.14(e). In short beams, however, relocated plastic hinges may be too close to each other, resulting in significant increase in curvature ductility demands [Fig. 4.14(a)]. In such cases the solution suggested in Fig. 4.74 may be used. Typical locations are shown in Fig. 4.16. Figure 4.17 shows two methods of achieving this. In Fig. 4.17(a) extra flexural reinforcement is provided for a distance not less than h or 500 mm (20 in.) from the column face and anchored by standard 90° bends into the beam or by bending over some of the top and bottom bars at an angle of 45° or less into the opposite face of the beam, so as to ensure that the critical section occurs at X–X. The amount of extra reinforcement is chosen to be sufficient to ensure that yield will not occur at the column face, despite the higher moment there.

In long-span beams the use of haunches may be advantageous. With the location of a plastic hinge, required to develop negative moments, at the shallow end of a haunch with a carefully selected slope, the length over which the top reinforcement will yield can be increased considerably. For a given plastic hinge rotation this will result in greatly reduced curvature ductility demands and hence reduced local damage. Typical detailing of the reinforcement is shown in Fig. 4.17(b).

When gravity load on a span is significant, as in the cases of beam B–C in Fig. 4.10 and at section C of the beam in Fig. 4.15, it is difficult to detail the bottom tension reinforcement in the beam so as to develop the critical section of a plastic hinge at the face of a column. In such cases the bottom bars must be curtailed so as to produce a (positive) plastic hinge at the locality of the maximum positive moment (Fig. 4.15) or at a short distance away from a column face. Such locations were selected at distances z and y from adjacent columns in Fig. 4.10. The relocation of plastic beam hinges for special cases is discussed further in Sections 4.8.11(e) and 4.10.2.

At the critical section of a positive plastic hinge in the span, such as C in Fig. 4.15, and its vicinity, shear forces will generally be rather small. For this reason and because the bottom beam bars will never be subjected to yield in compression, the requirements for transverse reinforcement over the specified length $2h$ of the plastic hinge can be relaxed (Section 4.5.4). It must be noted, however, that in such situations the position of the critical section of a positive plastic hinge is not unique. The position will depend on the intensity of gravity load present during the earthquake and the relative values of bending moments developed in the beam at the column faces at different instants of seismic response. The limits within which the critical section of such a positive plastic hinge can shift readily be determined, and this is shown in the design examples in Section 4.11. This will result in the less onerous detailing requirements for positive plastic hinges, given in Section 4.5.4, to be applied over a length considerably larger than $2h$, shown in Fig. 4.15.

(e) Flexural Overstrength of Plastic Hinges   In accordance with the philosophy of capacity design, discussed in Chapter 1, the maximum likely actions imposed on the beam during a very large inelastic displacement must be estimated. This is achieved simply by the use of Eq. (1.10). In evaluating the flexural overstrength of the critical beam sections:

1. All the flexural beam reinforcement provided, including the assumed tributary steel area in tension flanges, in accordance with Section 4.5.1(b), is included in the total effective flexural tension steel area.
2. Strength enhancement of the steel, allowing for the yield strength exceeding nominal design values f_y and also for some strain hardening at maximum curvature ductilities, is considered with the use of a magnified yield strength $\lambda_o f_y$ [Section 3.2.4(e)]. Because the influence of concrete strength f'_c on the flexural strength of doubly reinforced beam sections is negligible, strength enhancement due to steel properties only need be considered.

With this, the flexural overstrength of the critical section of a potential plastic hinge at a location X becomes with good approximation

$$M_{ox} = \lambda_o A_s f_y (d - d') \tag{4.15}$$

(f) Beam Overstrength Factors (ϕ_o) As a convenient measure of the flexural overstrength of beam sections developed under large ductility demand, the beam flexural overstrength is expressed in terms of the design moments M_E that resulted from the analysis for code-specified lateral earthquake forces alone. As explained in Section 1.3.3(f), both the moment quantifying flexural overstrength M_o and that resulting from the specified design earthquake forces M_E are expressed for beam node points of the model frame (i.e., at the centerlines of supporting columns). Typical beam moments M_E, to be used as a reference, are those shown in Figs. 4.6(b) and 4.7(b). The flexural overstrength factor, described in general terms in Section 1.3.3(f) for a beam, is thus

$$\phi_o = M_o / M_E \tag{1.12}$$

Generally, for a given direction of applied earthquake forces there will be two values for ϕ_o, one for each end of a beam span. For unsymmetrical situations it is convenient to identify also the direction of the relevant earthquake attack, and in these cases the symbols $\vec{\phi}_o$ and $\overleftarrow{\phi}_o$ will be used.

The flexural overstrength of a beam with reference to a column centerline, may be influenced by factors listed in Section 1.3.3(f). The beam flexural overstrength at the centerline of an exterior column $M_o = \phi_o M_E$ informs the designer about the maximum beam strength developed during large ductility demands without having to search for and identify the various sources of strength enhancement. Clearly, this relationship is very important in the context of capacity design and is simply the extension of the concepts outlined in Section 1.4.

It was shown in Section 1.3.3(f) and with Eq. (1.12) that in the exceptional case when at a node point the dependable strength provided, ϕM_i, matches exactly the strength demand for earthquake forces alone, M_E, the value of the overstrength factor becomes

$$\phi_o = M_o/M_E = \lambda_o M_i/\phi M_i = \lambda_o/\phi \qquad (4.16)$$

This value serves as an indicator of the excess or deficiency of strength that has been provided at one particular node point. The beam overstrength factor ϕ_o, relevant to node points, will be used subsequently in determining design actions for columns of ductile frames. For this reason the value of ϕ_o at the centerline of an interior column will be obtained from the ratio of the sum of the flexural overstrengths, developed by two adjacent beams, to the sum of the required flexural strengths derived from the seismic forces alone (i.e., $\phi_o = \Sigma M_o/\Sigma M_E$).

(g) System Overstrength Factor (ψ_o) To distinguish and quantify the overstrength relevant to the structure as a whole from that applicable to a particular beam or column, the system overstrength factor ψ_o was introduced in Section 1.3.3(g) and quantified by Eq. (1.13). In terms of the flexural strength of the beams of a ductile frame at a particular level above level 1, Eq. (1.13) becomes

$$\psi_o = \frac{\Sigma_1^n M_{o,j}}{\Sigma_1^n M_{E,j}} = \frac{\Sigma_1^n (\phi_o M_E)_j}{\Sigma_1^n M_{E,j}} \qquad (4.17)$$

where $M_{o,j}$ is the flexural overstrength of a beam measured at the column centerline at j, $M_{E,j}$ is the bending moment derived from the application of design earthquake forces for the same beam at the same node point j, and n is the total number of beam node points at that level. Note that there are two node points for each beam span. The summation applies to all beams at one floor level in one frame, or in all frames of the entire framed structure.

(h) Illustration of the Derivation of Overstrength Factors To illustrate the implications of overstrength factors, the beams of the subframe in Fig. 4.7

will be used. Figure 4.7(b) shows that the code-specified lateral forces alone required a moment of 80 units to be resisted by the beam at column C for both directions of seismic actions. After superposition with gravity moments and the application of moment redistribution, we arrived at design terminal beam moments of -120 and $+65$ units at the same column, as seen in Fig. 4.7(d). Let us assume that reinforcement ($f_y = 275$ MPa (40 ksi) and $\lambda_o = 1.25$) at the critical beam sections has been provided in such a way as to resist, at ideal strength, moments of $120/0.9 = 133$ and $65/0.9 = 72$ units with respect to column centerlines, exactly as required. Therefore, the flexural overstrengths at C will be $\overrightarrow{M}_{oc} = 1.25(-133) = -167$ and $\overleftarrow{M}_{oc} = 1.25 \times 72 = 90$ units, respectively. Hence the beam flexural overstrength factors with respect to the two directions of earthquake forces alone become $\overrightarrow{\phi}_{oC} = 167/80 = 2.09$ and $\overleftarrow{\phi}_{oC} = 90/80 = 1.13$, respectively. The apparent large deviation from the value of $\lambda_o/\phi = 1.25/0.9 = 1.39$ results from the influence of gravity-induced moments and the application of moment redistribution, whereby we allocated more or less strength to this section than indicated by the initial approximate elastic analysis for the specified earthquake forces only.

If it is assumed that in the same example structure, top and bottom beam reinforcement passing continuously through the interior column B has been provided exactly to give a flexural overstrengths of $1.25 \times 120/0.9 = 167$ and $1.25 \times 65/0.9 = 90$ units, then from Fig. 4.7(b) and according to the definition above, the beam flexural overstrength factor at column B is $\overrightarrow{\phi}_{oB} = (167 + 90)/(110 + 80) = 1.35$. Similar considerations will lead at column A to $\overrightarrow{\phi}_{oA} = 90/100 = 0.9$ and $\overleftarrow{\phi}_{oA} = 167/100 = 1.67$. It is also evident that with these assumptions the overstrength of the entire subframe, with respect to seismic forces, shown in Fig. 4.7(d), is according to Eq. (4.17), $\overrightarrow{\psi}_o = \overleftarrow{\psi}_o = (90 + 167 + 90 + 167)/(100 + 110 + 2 \times 80) = 514/370 = 1.39$.

It is seen that although the flexural overstrength factors at the ends of beams are significantly different from the "ideal" value of 1.39 [Eq. (1.11)], the system overstrength factor [Eq. (4.17)] $\psi_o = 1.39$ indicates that a "perfect" match of the total seismic strength requirements has been attained. The numerical values chosen for the purpose of illustration are somewhat artificial. For example, as stated earlier, it would be difficult to provide bottom beam reinforcement to the right of the central column [Fig. 4.7(d)] without developing a dependable beam strength in excess of 65 units. Curtailment of bottom bars in the beam span at the column face with a standard hook, similar to that shown in Fig. 4.84, is undesirable because this would result in a concentration of plasticity over a very short length.

More realistic values for both ϕ_o and ψ_o are obtained if instead of the idealized moment patterns of Fig. 4.7(d), those of Fig. 4.10 are used. In the latter an attempt was made to optimize required beam moments at column faces rather than at column centerlines. To assist in the understanding of the details of this technique, the strength of the beams of the structure shown in Fig. 4.7(a) will be reevaluated, considering various aspects that arise in the

TABLE 4.2 Design Quantities for the Beam of the Frame Shown in Figs. 4.7, 4.10, and 4.18 [a]

(1)	(2)	(3)	(4)	(5)	(6)	(7)	(8)	(9)
At	Sense	M_u	ϕM_i	M_o	M_o^c	$\overleftrightarrow{M}_E$	$\overrightarrow{\phi}_o$	$\overleftarrow{\phi}_o$
A–B	−	103	108	150	165	100	—	1.65
	+	65	70	97	100		1.00	—
B–A	−	117	120	167	183	110		
	+	65	70	97	98		1.70	1.52
B–C	−	117	120	167	191	80		
	+	—	—	(146)	(140)			
Span	+	105[b]	105[b]	146[b]	—	—	—	—
C–B	−	103	108	150	175	80	2.19	—
	+	—	—	(97)	90		—	1.13
Span	+	53	70[b]	97[b]	—	—	—	—

[a](2) Positive moments cause tension in the bottom fibers of beam sections; (3) required moments at column faces as shown in Fig. 4.10; (4) dependable flexural strength provided ($\phi = 0.9$); (5) flexural overstrength of critical sections $M_o = \lambda_o M_i = 1.25 M_i$; (6) flexural overstrength at column centerlines, as shown in Fig. 4.18; (7) moments due to earthquake forces alone from Fig. 4.7(b); (8), (9) flexural overstrength factors at column central lines.
[b]At a distance z or y from the face of column B or C, respectively.

real structure but which are not apparent when using a structural model, such as seen in Fig. 4.7(a). Again $f_y = 275$ MPa (40 ksi) will be assumed.

The critical moments at column faces, shown in Fig. 4.10, will be used as a basis of a realistic design. Proportioning of the relevant sections leads to a practical selection of beam bars so that dependable strengths ϕM_i at each beam section match as closely as possible, but not exactly, the required flexural strengths M_u. These strengths are recorded in Table 4.2. It is decided that a (positive) plastic hinge in the long-span B–C should not be located where the maximum positive moment (115 units) occurs but somewhat closer to the support, at a distance z from the central column, as shown in Fig. 4.10.

If the amount of bottom reinforcement in the beam to the right of column B is made 50% larger than that to the left of this column, the resulting dependable strength will be $1.5 \times 70 = 105$ units. With this value the location of the positive plastic hinge section z can be determined. To the right of this section additional bottom bars will be provided so that yielding in the central region of the beam cannot occur.

For similar reasons it is inadvisable to detail the bottom reinforcement in the beam in such a way as to enable a (positive) plastic hinge to form at the face of column C. Therefore, it is decided to provide over a distance y from the center of column C the same dependable strength (70 units) as in the short-span A–B. This enables the distance y to be established. To the left of this section the bottom beam reinforcement will be sufficient to preclude hinge formation. It is evident that plastic hinges can readily develop adjacent to column faces in the short-span beam A–B.

With all plastic hinge locations established, the flexural overstrength at these sections $M_o = \lambda_o M_i$ can be determined and the appropriate values are given in column 5 of Table 4.2. These values are plotted in Fig. 4.18. Straight lines passing through these maximum values of plastic hinge moments lead to values of corresponding moments at column centerlines, which may be scaled off. For the sake of comparison only, they are also listed in column 6 of Table 4.2.

Figure 4.18 also shows that with gravity load only a little less than assumed $(D + L)$, the critical section of the positive plastic hinges in span B–C may move to the column forces. Thus flexural overstrength based on the bottom reinforcement in span B–C can develop anywhere over the lengths z and y shown in Fig. 4.18.

Finally, the flexural overstrength factors, defined in Section 1.3.3(f) and required for the capacity design of the columns, can readily be derived. For example, from Fig. 4.7(a) at column A, $M_E = 100$ units is obtained. Hence from Fig. 4.18, $\overleftarrow{\phi}_{oA} = 165/100 = 1.65$. Similarily, for earthquake action $\overrightarrow{E}$ in

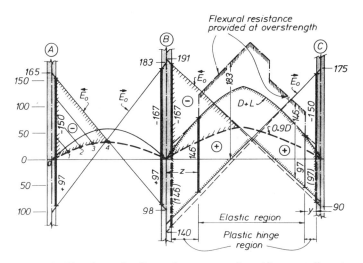

Fig. 4.18   Envelopes for flexural overstrength and bar curtailment.

the other direction at column B, it is found that $\vec{\phi}_{oB} = (183 + 140)/(110 + 80) = 1.70$. All other values are recorded in Table 4.2. It is seen that there are considerable deviations from the "ideal" value of $\lambda_o/\phi = 1.25/0.9 = 1.39$.

The final check for the adequacy of the strength of the entire beam with respect to the total required flexural strength $\Sigma M_E = 370$ for earthquake forces alone [Fig. 4.7(b)], is obtained from the evaluation of the system overstrength factor [Eq. (1.13)] thus:

$$\vec{\psi}_o = (100 + (183 + 140) + 175)/370 = 1.62 > 1.39$$

$$\overleftarrow{\psi}_o = (165 + (98 + 191) + 90)/370 = 1.47 > 1.39$$

It is seen that the strength provided is more than adequate. The 17% excess capacity with respect to the required seismic resistance for $\vec{E}$ could not be avoided. The graphical approach used in Fig. 4.18, reflecting the actual detailing of the beam flexural reinforcement, may conveniently be replaced by a corresponding computer graphic.

4.5.2 Development and Curtailment of the Flexural Reinforcement

The principles of bond, anchorage, and bar curtailment have been summarized in Section 3.6.2, and various aspects are given in greater detail elsewhere [A1, P1]. The principles of bar curtailment shown in Fig. 3.34 can be conveniently applied also to moment envelopes, such as shown in Fig. 4.18. As explained in Section 3.6.3, a flexural bar should extend past the sections at which according to the moment envelope it could be required to develop its full strength, by a distance $(d + l_d)$. Also a bar should extend past the section at which it is no longer required to resist moment, by a distance $1.3d$. These critical sections can readily be located, as illustrated for the example structure in Fig. 4.18, in the vicinity of the exterior column A. In Fig. 4.18 it is assumed that 25% of the total top flexural reinforcement required at the face of column A is curtailed at a time. Thus the division of the moment (150 units) into four equal parts is shown. Therefore, each of the four groups of bars must extend beyond points marked 0, 1, 2, and 3 by the distance $(d + l_d)$, or beyond points 1, 2, 3, and 4 by the distance $1.3d$. In the region of a large moment gradient, the first criterion usually governs. However, at least one fourth of the top bars should be carried right through the adjacent spans [A1, X3]. These bars may then be spliced in the midspan region.

As it is has been decided that (positive) plastic hinges will be restricted to end regions over length z and y in span B–C in Fig. 4.18, additional bottom bars must be provided over the remainder of the span to ensure that no yielding of significance will occur over the elastic region. The moment envelope for positive moments over span $B-C$, plotted for convenience in terms of flexural overstrength $(\lambda_o M_i)$, is shown shaded in Fig. 4.18. The bottom bars near column C provide a resistance of 97 units over the length y, where the reinforcement is abruptly increased by 50% (146 units). This arrangement of bottom bars is used over length z near column B, beyond which additional bars are provided to increase the resistance to $1.25 \times 146 = 183$ units. The envelope of flexural resistance at overstrength so obtained is shown in Fig. 4.18 by the stepped envelope with vertical shading. The detailing of bottom bars for a similar situation is shown in Section 4.11 and Fig. 4.84.

The envelope for positive moments in span $B-C$ also shows that the associated shear forces are very small. Hence only insignificant shear reversal can occur and the need for diagonal reinforcement to control sliding shear displacement in the plastic hinge region, in accordance with Section 3.3.2(b), will not arise.

Lapped splices for bottom bars, preferably staggered, may be placed anywhere along span $B-C$ of the beam in Fig. 4.18, with the exception of a length equal to the depth h of the beam from the column faces. This is to ensure that yielding of the bottom bars in the (positive) plastic hinges is not restricted to a small length, resulting in a concentration of inelastic rotations. Wherever stress reversal, according to moment envelopes, such as shown in Fig. 4.18, can occur at lapped splices, with stresses exceeding $\pm 0.6 f_y$, transverse reinforcement over the splice length, in accordance with Section 3.6.2(b), should be provided.

4.5.3 Shear Strength of Beams

(a) Determination of Design Shear Forces   From considerations of the transverse loading and the simultaneous development of two plastic hinges due to lateral forces, the shear forces in each span are readily found. Because inelastic shear deformations are associated with limited ductility, strength reduction, and significant loss of energy dissipation, they should be avoided. To this end the shear forces developed with the flexural overstrength of the beam at both plastic hinges will need to be considered. This is the simplest example of the application of capacity design philosophy. With respect to the beam, shown in Fig. 4.15, the maximum feasible shear, at the right-hand column face is,

$$V_B = V_{gB} + \frac{M_{oB} + M'_{oA}}{l_{AB}} = V_{gB} + \vec{V}_{Eo} \qquad (4.18a)$$

where M_{oB} and M'_{oA} are the end moments at the development of flexural overstrengths at the plastic hinges, l_{AB} is the clear span (i.e., between column faces), and V_{gB} is the shear force at B due to gravity load placed on the base structure (i.e., as for a simply supported beam). $\vec{V}_{Eo}$ is the earthquake-induced shear force generated during ductile response of the frame and the arrow indicates the direction of relevant horizontal forces. This is constant over the span. Similarly, at the other end of this beam,

$$V_A = V_{gA} + \frac{M_{oA} + M'_{oB}}{l_{AB}} = V_{gA} + \overset{\leftarrow}{V}_{Eo} \qquad (4.18b)$$

In Eq. (4.18b) M'_{oB} is the end moment at B, evaluated from the flexural overstrength M_{oC}. The shear at the positive plastic hinge in the span will be zero. However, as discussed in Section 4.5.1(f), the position of the critical section of this plastic hinge (i.e., the distance l_{AC} in Fig. 4.15) will depend on the intensity of gravity loading, given in Eqs. (1.7a) and (1.7b).

In accordance with Eq. (1.7a) the shear due to gravity loads alone is $V_g = V_D + V_L$. Equation (1.7b) does not result in a critical combination for shear. When comparing Eq. (4.18) with available shear strength, the strength reduction factor is, in accordance with Section 3.4.1, $\phi = 1.0$. The emphasis placed here on principles of capacity design should not detract from the necessity to check other combinations of strength requirements listed in Section 1.3.2.

The design shear force envelopes that would qualitatively correspond with the loading situation of the subframe examined in Figs. 4.7 and 4.10 are shown in Fig. 4.19. Here the shear force diagram for the gravity-loaded base frame $V_g = V_D + V_L$ may be used as a baseline. This enables the effect of the

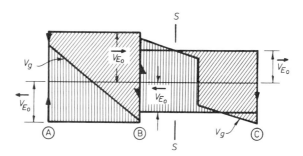

Fig. 4.19   Shear force envelopes for a two-span beam.

lateral-force-induced shear forces at flexural overstrength V_{Eo} to be readily considered for both directions of seismic attack. In the example, shear reversal in plastic hinges of the left-hand span will occur. However, the relevant value of r, defined by Eq. (3.44), is not very large. Shear reversal does not occur in plastic hinges of the long-span beam $B-C$ of this example.

(b) Provisions for Design Shear Strength   Different treatments are required for plastic hinges and regions between hinges, as follows:

1. In potential plastic hinge zones, defined in Section 4.5.1(d), the contribution of the concrete to shear resistance is discounted since aggregate interlock across wide flexural cracks will be ineffective and thus shear reinforcement, as outlined in Section 3.3.2(a), needs to be provided for the entire design shear. If the computed shear stress is in excess of that given by Eq. (3.43) and shear reversal can occur, the use of diagonal shear reinforcement, to resist a fraction of the design shear force, in accordance with Eq. (3.45), should also be considered. When providing stirrup reinforcement and selecting suitable bar sizes and spacings, other detailing requirements, outlined in Section 4.5.4, must also be taken into account.

2. Over parts of the beam situated outside potential plastic hinge regions, the flexural tension reinforcement is not expected to yield under any load conditions. Hence web reinforcement, including minimum requirements, in accordance with standard procedures are used and the contribution of v_c [Eq. (3.33)] may be relied on. The design example of Section 4.11 illustrates application of this approach.

4.5.4 Detailing Requirements

To enable the stable hysteretic response of potential plastic hinge regions to be maintained, compression bars must be prevented from premature buckling. To this end it should be assumed that when severe ductility demands are imposed, the cover concrete in these regions will spall off. Consequently, compression bars must rely on the lateral support provided by transverse stirrup ties only. The following semi-empirical recommendation [X3] will ensure satisfactory performance.

1. Stirrup ties should be arranged so that each longitudinal bar or bundle of bars in the upper and lower faces of the beam is restrained against buckling by a 90° bend of a stirrup tie, except that where two or more bars, at not more than 200-mm (8-in.) centers apart, are so restrained, any bars between them may be exempted from this requirement. Figure 4.20 shows examples of tie arrangements around longitudinal bars in the bottom of beams. Additional notes and dimensions are also provided in this figure to clarify the intents of these rules. It is seen in Fig. 4.20(a) that bars 1 and 2

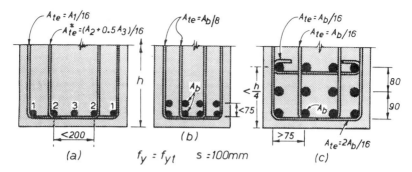

Fig. 4.20 Arrangement and size of stirrup ties in potential plastic hinge zones of beams.

are well restrained against lateral movements. Bar 3 needs not be tied because the distance between adjacent bars is less than 200 mm (8 in.) and bar 3 is assumed to rely on the support provided by the short horizontal leg of the tie extending (and bending) between bars 2. The two vertical legs of the tie around bars 2 are thus expected to support three bars against buckling.

2. The diameter of stirrup ties should not be less than 6 mm (0.25 in.), and the area of one leg of a stirrup tie in the direction of potential buckling of the longitudinal bars should not be less than

$$A_{te} = \frac{\Sigma A_b f_y}{16 f_{yt}} \frac{s}{100} (\text{MPa})$$

$$= \frac{\Sigma A_b f_y}{16 f_{yt}} \frac{s}{4} (\text{ksi}) \qquad (4.19a)$$

where ΣA_b is the sum of the areas of the longitudinal bars reliant on the tie, including the tributary area of any bars exempted from being tied in accordance with the preceding section. Longitudinal bars centered more than 75 mm (3 in.) from the inner face of stirrup ties should not need to be considered in determining the value of ΣA_b. In Eq. (4.19), f_{yt} is the yield strength of the tie leg with area A_{te} and horizontal spacings s.

Equation (4.19) is based on the consideration that the capacity of a tie in tension should not be less than $1/16$ of the force at yield in the bar (with area A_b) or group of bars (with area ΣA_b) it is to restrain, when spaced on 100-mm (4-in.) centers. For example, the area of the tie restraining the corner bar 1 in Fig. 4.20(a) against vertical or horizontal movements, and spaced on 100-mm (4-in.) centers, should be $A_{te} = A_1/16$, assuming that the yield strength of all bars is the same. However, the area of the inner ties around bar 2 must be $A_{te}^* = (A_2 + 0.5 A_3)/16$ because they must also give support to the centrally positioned bar marked 3. In computing the value of ΣA_b, the tributary area of unrestrained bars should be based on their position relative to the two adjacent tie legs.

Figure 4.20(b) shows a beam with eight bottom bars of the same size, A_b. Again assuming that $f_y = f_{yt}$, the area of identical ties will be $A_{te} = 2A_b/16$ because the second layer of bars is centered at less than 75 mm (3 in.) from the inside of the horizontal legs of stirrup ties. The vertical legs of the ties in Fig. 4.20(c) need only support the bottom layer of beam bars. The second layer, being at more than 75 mm (3 in.) from the horizontal tie legs, is assumed sufficiently restrained by the surrounding concrete and hence is not considered to require lateral support.

For design purposes it is convenient to rearrange Eq. (4.19a) in this form:

$$\frac{A_{te}}{s} = \Sigma A_b f_y / 1600 f_{yt} \quad (\text{mm}^2/\text{mm}); \qquad \frac{A_{te}}{s} = \Sigma A_b f_y / 64 f_{yt} \quad (\text{in.}^2/\text{in.})$$

$$(4.19b)$$

which gives the area of tie leg required per millimeter or inch length of beam and enables ready comparison to be made with other requirements for transverse reinforcement.

3. If a layer of longitudinal bars is centered farther than 100 mm (4 in.) from the inner face of the horizontal leg of a stirrup, the outermost bars should also be tied laterally as required by Eq. (4.19), unless this layer is situated farther than $h/4$ from the compression edge of the section.

The reason for this requirement is that the outer bars placed in second or third layers in a beam may buckle horizontally outward if they are situated too far from a horizontal transverse leg of a stirrup tie. This situation is illustrated in Fig. 4.20(c), which shows a single horizontal tie in the third layer, because those outer bars are farther than 100 mm (4 in.) from the horizontal leg of the peripheral stirrup ties at the bottom of the beam section. The inner four bars need not be considered for restraint, as they are situated more than 75 mm (3 in.) from any tie. The outer bars in the second layers shown in Figs. 4.20(b) and (c) are considered satisfactorily restrained against horizontal buckling as long as they are situated no farther than 100 mm (4 in.) from the horizontal bottom tie. However, the horizontal bottom tie should be capable of restraining two outer beam bars, one in each of the two layers. Any layer of bars in a beam situated farther than $h/4$ from the compression edge of the section is not considered to be subjected to compression strains large enough to warrant provisions for lateral restraints. This waiver does not apply to columns.

4. In potential plastic hinge regions, defined in Section 4.5.1(d), conditions 1 and 2, the center-to-center spacing of stirrup ties should not exceed the smaller of $d/4$, or six times the diameter of the longitudinal bar to be restrained in the outer layers, whenever the bar to be restrained may be

subjected to compression stress in excess of $0.6f_y$. The first stirrup tie in a beam should be as close as practicable to column bars and should not be farther than 50 mm (2 in.) from the face of the column (Fig. 4.15).

5. In potential "positive" plastic hinge regions, defined in Section 4.5.1(d), condition 3, and also in regions defined by Section 4.5.1(d) conditions 1 and 2, provided that the bar to be restrained cannot be subjected to compression stress exceeding $0.6f_y$, the center-to-center spacing of stirrup ties should not exceed the smaller of $d/3$ or 12 times the diameter of the longitudinal compression bar to be restrained, nor 200 mm (8 in.) (Fig. 4.15).

The limitations on maximum tie spacing are to ensure that the effective buckling length of inelastic compression bars is not excessive and that the concrete within the stirrup ties has reasonable confinement. The limitations are more severe when yielding of longitudinal bars can occur in both tension and compression. Because of the Bauschinger effect and the reduced tangent modulus of elasticity of the steel, a much smaller effective length must be considered for such flexural compression bars than for those subjected only to compression. Equation (4.19) ensures that the tie area is increased when the spacing s is in excess of 100 mm (4 in.), which is often the case when the limit is set by $s \leq 6d_b$ and bars larger than 16 mm (0.63 in.) are used.

Since tension in vertical stirrup legs will act simultaneously to restrict longitudinal bar buckling and to transfer shear force across diagonal cracks, the steel areas calculated to satisfy the requirement above and those for shear resistance in accordance with Eq. (3.40) need not be additive. Stirrup size and spacing will be governed by the more stringent of the two requirements.

Regions other than those for potential plastic hinges are exempted from these rules for detailing. In those regions the traditional recommendations of building codes [A1] may be considered to be sufficient. Application of these simple rules is shown in Section 4.11. It is emphasized that these recommendations, which have been verified by results of numerous experimental projects, are applicable to beams only. Similar rules, developed for columns, are examined in Section 4.6.11.

4.6 DESIGN OF COLUMNS

4.6.1 Limitations of Existing Procedures

As outlined in Section 1.4, the concept of a desirable hierarchy in the energy-dissipating mechanisms, to be mobilized in ductile multistory frames during very large earthquakes, requires that plastic hinges develop in beams rather than in columns and that "soft-story" column failure mechanisms be avoided. Evaluation of design actions and consideration of the concurrency of such actions along the two principal directions of the building during the inelastic dynamic response of two-way framed structures involves complex and time-consuming computational efforts. Probabilistic modal superposition techniques have been used to estimate likely maxima that may be encountered during the elastic response of the structure. The predominantly inelastic nature of the structural response is not sufficiently recognized, however, by these techniques. Moreover, the designer is still required to use judgment if a quantification of the hierarchy in the development of failure mechanisms is to be established.

Time-history analyses of the inelastic dynamic response of frames to given ground excitations are likely to furnish the most reliable information with respect to structural behavior. Unfortunately, these are analyses rather than design techniques. They are useful in verifying the feasibility of the design. However, the results must be assessed in view of the probable relevance of the chosen (observed or artificial) earthquake record to local seismicity. To overcome some of these difficulties and in an attempt to simplify routine design procedures for ductile frames, a simple deterministic technique has been suggested [P3, P4]. Frames designed using this method have subsequently been subjected to inelastic time-history studies [J1, P5, T1], which resulted in minor modifications [P6, P7] of the technique. This modified designed procedure [X3] is presented in detail in the following sections.

4.6.2 Deterministic Capacity Design Approach

In this procedure, bending moments and shear and axial forces for columns resulting from elastic analysis (static or modal) representing the design earthquake level are magnified in recognition of the expected effects during dynamic response and to ensure the development of only the chosen plastic hinge mechanism. It should ensure that no inelastic deformations of any significance will occur except by flexural action in designated plastic hinge regions, even under extreme earthquake excitations with a wide range of spectral characteristics. The procedure is conservative and simple, yet case studies indicate no increased material costs compared with structures designed with less conservative methods [X8]. It is applicable for regular frames except those with excessively flexible beams, where cantilever action may govern the moment pattern in columns in the lower stories or in low frames where column sway mechanisms are considered acceptable.

When gravity loads rather than lateral forces govern the strength of beams, capacity design philosophy will require columns of ductile frames to be designed for moments that may be much larger than those resulting from code-specified earthquake forces. In such cases the acceptance of column hinging before the development of full beam sway mechanisms, at a lateral force in excess of that stipulated by building codes, may be more appropriate. The approach to such frames is examined in Section 4.9.

4.6.3 Magnification of Column Moments due to Flexural Overstrength of Plastic Hinges in Beams

(a) *Columns Above Level 2*   The primary aim of the capacity design of columns is to eliminate the likelihood of the simultaneous formation of plastic hinges at both ends of all columns of a story (Fig. 1.19(b)). Therefore, columns must be capable of resisting elastically the largest moment input from adjacent beam mechanisms. This moment input with reference to a node point can readily be evaluated as

$$M_c = \phi_o M_E \qquad (4.20)$$

where M_E is the moment derived for the column for the code-specified seismic forces, measured at the centerline of the beam, and ϕ_o is the beam overstrength factor determined in accordance with Eq. (1.11). The required ideal strengths of columns derived from an elastic analysis for typical code-specified lateral forces are shown on the left of Fig. 4.21. The meaning of the beam flexural overstrength factor ϕ_o was discussed in Sections 1.3.3(f) and 4.5.1(f). Because load- or displacement-induced moments in beams and columns must be in equilibrium at a beam–column joint (i.e., node point), any magnification of moments at ends of beams necessitates an identical magnification of column moments. Equation (4.20) performs this simple operation.

In evaluating Eq. (4.20), column moments induced by gravity loading on the frame, such as shown in Figs. 4.6(a) and 4.7(a), need not be considered, since ϕ_o is related to seismic actions alone, while the strength of the beam has been based on considerations of gravity loads and earthquake forces, together with the effects of moment redistribution and the actual arrangement of beam reinforcement.

Equation (4.20) implies that the beam overstrength moment input is shared by the columns, above and below a beam, in the same proportions as determined by the initial elastic frame analysis for lateral forces only. This is unlikely because of dynamic effects; hence Eq. (4.20) will be revised accordingly in Section 4.6.4. The negligible effect of gravity-induced moments on the relative proportions of column moments at a node point was pointed out in Section 4.5.1(f).

The generally accepted [X4, X10] aim to eliminate the possibility of plastic hinges forming simultaneously at the top and bottom of the columns in a story higher than level 2 is achieved when the ideal flexural strength of the critical column section, say at the level of the top of a beam at the nth level of a multistory frame, is

$$M_{i,n} \geq \left(1 - \frac{M_{c,n} + M_{c,n+1}}{2M_{c,n}} \frac{h_b}{l_c}\right) \frac{M_{c,n}}{\phi_c} \qquad (4.21)$$

where $M_{c,n} = \phi_{o,n} M_{E,n}$ from Eq. (4.20) and where the subscript n refers to the level of the floor, h_b is the average depth of the beams framing into the column at levels n and $n + 1$, l_c is the height of the story above level n, and ϕ_c is the strength reduction factor to be used for this column, in accordance with Section 3.4.1. A similar equation will apply to the critical column section at the soffit on the beam at level n but will include moments at levels n and $n - 1$. The first term on the right in Eq. (4.21) reduces the centerline moment $M_{c,n}$ to that relevant to the column section at the top or bottom face of the beam.

This approach results in a column-to-beam required flexural strength [Section 1.3.3(d)] ratio at a node point of

$$\frac{\Sigma M_c}{\Sigma M_E} = \frac{\phi_c \Sigma M_{i,c}}{\phi_b \Sigma M_{i,b}} = \phi_o \qquad (4.22)$$

Hence the ratio of ideal strengths is

$$\frac{\Sigma M_{i,c}}{\Sigma M_{i,b}} = \phi_o \frac{\phi_b}{\phi_c} \qquad (4.23)$$

where $M_{i,c}$ and $M_{i,b}$ at a node are the ideal flexural strengths of the columns and beams, respectively, and ϕ_b and ϕ_c are relevant strength reduction factors according to Section 1.3.4. At an exterior node, where typically two beams and one column are joined, Eq. (4.23) is expressed simply in terms of $\Sigma M_{i,c}/M_{i,b}$.

With typical values of $\lambda_o = 1.25$, $\phi_b = 0.9$, and $\phi_c = 0.75$, the ratio of ideal strengths from Eq. (4.23) becomes 1.50. U.S. practice [A1], whereby the ratio of dependable strengths of columns to those of beams at a node should not be less than 6/5, gives similar ratios of ideal strengths. This is sufficient to ensure that a "soft story" will not develop.

It should be noted that these ratios of flexural strengths are approximate because at this stage no consideration has been given to the axial force level to be resisted by the column in combination with the flexural overstrength. Moreover, during the inelastic dynamic response of a frame, when frame distortions similar to those of higher mode shapes [Fig. 2.24(b)] occur, moments may significantly increase at one or the other end of a column, and hence the formation of a plastic hinge at either ends must be expected. Accordingly, relevant codes [A1, X3, X10] specify that each end of such a column be designed and detailed for adequate rotational ductility. Also, the placing of lapped splices of bars in the end regions of columns, discussed in greater detail in Section 4.6.10, is prohibited, with laps required to be located in the midheight region of columns.

Columns in stories above level 2 may, however, be provided with additional flexural strength in the end regions, so that the likelihood of the development of plastic hinges is eliminated. The behavior of such columns, even during extreme seismic events, can then be expected to remain essentially elastic. If significant curvature ductility demands and high-intensity reversed cyclic steel stresses in the end regions of columns above level 2 cannot arise, adequately detailed lapped splices of column bars may be placed immediately above a floor. This enables easier and speedier erection of reinforcing cages for columns. Moreover, the need to confine end regions, to provide for adequate rotational ductility, does not arise, and some transverse reinforcement in columns may be saved. In New Zealand, where the capacity design procedure has been in use since 1980, these advantages have been found to offset cost increases resulting from the use of either larger column sizes or larger amounts of vertical reinforcement. The following sections set out the details of the design of such "elastic" columns above level 2 of ductile frames.

(b)   Columns of the First Story   At ground floor of the first story (level 1) or at foundation level, where normally, full base fixity is assumed for a column, the formation of a plastic hinge, as part of the chosen sway mechanisms shown in Fig. 1.19(a), is expected. Accordingly, at this level the design moment for the column is that derived from the appropriate combination (Section 1.3.2) of gravity and earthquake effects. Because the moment demand of this level does not depend on the strength of adjacent members, such as components of the foundation system, the beam flexural overstrength factor ϕ_o, is not applicable. The strength hierarchy between foundation and superstructure is examined in Chapter 9.

It may well be, particularly for columns supporting a large number of stories, that the critical moment at the base will result from wind rather than specified earthquake forces. It should not be overlooked, however, that the formation of a plastic hinge during the design earthquake is still to be expected, perhaps with some reduction in ductility demand, and hence detailing the column for ductility is essential. To eliminate the likelihood of a plastic hinge developing at the top end of a column in the first story, the design moment at that level should be derived with the use of ϕ_o, as described in Section 4.6.3(a).

(c)   Columns in the Top Story   At roof level, gravity loads will generally govern the design of beams. Moreover, plastic hinges in columns should be acceptable because ductility demands on columns, arising from a column sway mechanism in the top story, are not excessive. Further, axial compression on such columns are generally small, and hence rotational ductility in plastic hinges can readily be achieved with amounts of transverse reinforcement similar to those used in plastic hinges of beams. Thus at roof level, strength design procedures for flexure are appropriate. The designer may choose to allow plastic hinge formation in either the beams or the columns. Hinge formation at the bottom end of top-story columns is also acceptable.

However, in this case transverse reinforcement in the lower-end region must also be provided to ensure adequate rotational ductility, and lapped splices of column bars should then be located at midstory.

(d) Columns Dominated by Cantilever Action When a column is considerably stiffer than the beams that frame into it, cantilever action may dominate its behavior in the lower stories. In such cases the column moment above a floor, derived from elastic analyses, may be larger than the total beam moment input at such a floor. Because beam moment inputs are not dominant in the moment demands on such columns, the beam overstrength factor ϕ_o is not relevant. A typical moment pattern for such a column is shown in Fig. 4.23. Therefore, at all floors below the lower column contraflexural point indicated by the elastic analysis for code-specified lateral forces, moment magnification due to strength enhancement of beams must be modified. To this end the flexural overstrength factor at the column base ϕ_o^* needs to be evaluated. From Eqs. (1.10a) and (1.12) this is

$$\phi_o^* = M_o^*/M_E^* = \frac{\lambda_o}{\phi_c} \qquad (4.24)$$

where M_E^* is the moment derived for the design earthquake forces at the column base (Fig. 4.23), and M_o^* is the flexural overstrength of the column base section as designed, taking into account the effect of axial force associated with the direction of earthquake attack, in accordance with Section 4.6.5. It needs to be remembered that the base section of the column is proportioned with standard strength design procedure as described in Section 4.6.3(b). In the evaluation of the overstrength of this critical section in accordance with Eq. (1.11), the value of λ_o [Section 3.2.4(e)] should also include strength enhancement due to confinement of the concrete [Eq. (3.28)]. Hence for the purpose of capacity design, all column moments above level 1 over the height where cantilever action dominates should be increased by the overstrength factor ϕ_o^*. This is shown by the outer dashed moment envelope curve in Fig. 4.23. Alternatively, the approach adopted for the design of walls of dual systems (Section 6.4) may be adopted for the design of columns dominated by cantilever action.

4.6.4 Dynamic Magnification of Column Moments

To give columns a reasonably high degree of protection against premature yielding, allowance must be made for the fact that column moments during the inelastic dynamic response of a framed building to a severe earthquake will differ markedly from those preducted by analyses for static forces. This is due to dynamic effects, particularly during responses in the higher modes of vibrations. As an example, Fig. 4.21 shows bending moment patterns for a

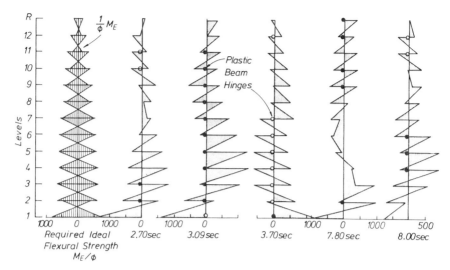

Fig. 4.21 Comparison of column moment patterns due to horizontal static and dynamic forces.

column of a 12-story ductile frame. This structure was designed in accordance with the principles presented in this chapter [P7]. The first diagram shows the demand for column moments, in terms of ideal flexural strength M_E/ϕ, based on the results of the elastic analysis for the specified equivalent lateral static forces. Subsequent moment patterns show critical instants of the response based on an inelastic time-history analysis and show dramatic departures at certain levels from the regular moment pattern traditionally used for design. The circles indicate that at the given instant of the earthquake record, the formation of a plastic hinge in the adjacent beam or at the column base was predicted by the time-history analysis. Circles indicate clockwise, and heavy dots anticlockwise, plastic hinge rotations. The analysis did not allow for beam overstrength, but it took into account that for practical reasons the ideal strength of some beams was slightly in excess of that required. It is seen that at some instants the point of contraflexure for a column in a story disappears, and that sometimes the moments and shear forces change sign over a number of stories. It will also be seen that the peak column moments frequently exceed the values shown at the left of Fig. 4.21, and that further amplification of column moments is needed to ensure that plastic hinges do not occur. The large moments shown at the column base, where, as expected, a plastic hinge did form, were based in the analysis on the probable strength of materials [Section 1.3.3(c)], including strength enhancement of the steel, corresponding with the instantenously imposed large curvature at the critical base section.

The bending moment pattern derived for lateral static force can be considered to give a reasonable representation for moment demands during the first mode of vibration of the frame. Higher mode effects significantly change these moment patterns, particularly in the upper stories of frames with long fundamental periods of vibration. To allow for such dynamic effects, the moments resulting from lateral static forces must be increased further if, as intended, column hinging above level 1 is to be avoided. This is achieved by the dynamic moment magnification factor ω. Hence to ensure that column plastic hinges do not form in frames above level 1, elastic analysis results, M_E, must be amplified in accordance with the relationship

$$M_u = \omega \phi_o M_E$$

where ϕ_o represents the effects of flexural overstrength of plastic hinges in the beams as finally detailed, and ω accounts for the dynamic amplification of column moments. While the factor ϕ_o was examined in detail in Sections 1.3.3(f), 4.5.1(f), and 4.6.3(a), the factor ω is considered in the following sections.

In the evaluation of the suggested values of the dynamic moment magnification factor ω, three points were considered in particular:

1. With the exception of the topmost story, the formation of a column story mechanism, involving column hinge development simultaneously at the top and bottom of columns of the story, is to be prevented.
2. With the exception of the column base at ground floor (level 1 or foundation level) and in the top story, hinge formation in columns should be avoided. If this can be achieved, considerable relaxation in the detailing requirements for the end regions of columns, with respect to confinement, shear strength, and bar splicing, may be made at all other levels.
3. Under extreme circumstances, overloading and hence yielding of any column section during the inelastic dynamic response of the framed building can be tolerated. Column yielding and hinge development are not synonymous in the context of seismic design. The latter involves ductility demands of some significance and usually necessitates hinge development at one end of all columns in a story. As long as some of the columns of a story can be shown to remain elastic, all other columns will be protected against ductility demands of any significance, unless adjacent beam hinges do not develop.

(a) *Columns of One-Way Frames* The dynamic moment magnification for such columns may be estimated with

$$\omega = 0.6T_1 + 0.85 \tag{4.25a}$$

provided that

$$1.3 \leq \omega \leq 1.8 \qquad (4.25b)$$

where T_1 is the computed fundamental period of the framed structure in seconds, as evaluated, for example, by Eq. (2.24).

When earthquake forces in the direction transverse to the plane of the frame is resisted predominantly by structural walls, the columns may be considered as part of a one-way frame.

(b)   Columns of Two-Way Frames   Such frames should be considered under simultaneous attack of earthquake forces along the two principal directions of the system. This normally involves analysis of column sections for biaxial bending and axial load. The concurrent development of plastic hinges in all beams framing into a column, as shown in Fig. 4.46(d), should also be taken into account. It should be noted that this does not imply simultaneity of maximum response in two orthogonal directions, since plastic hinges may form in beams at comparatively small levels of seismic attack, albeit with low ductility demand. Assessment of concurrency effect, however, may become an involved process. For example, at an interior column, supporting four adjacent beams, the interrelationship of the strengths of up to four adjacent plastic hinges, ranging from probable strength to flexural overstrength, and the interdependence of the dynamic magnification of moments at column ends above or below the beam in the two principal directions would need to be estimated. The probability of beam flexural overstrength development with extreme dynamic magnification being present at a section concurrently in both directions is considered to diminish with the increased number of sources for these effects.

To simplify the design process and yet retain sufficient protection against premature yielding in columns of two-way frames, dynamic moment magnification will be increased so as to allow the column section to be designed for unidirectional moment application only. Columns so designed, separately in each of the two principal directions, may then be assumed to possess sufficient flexural strength to resist various combinations of biaxial flexural demands. This may be achieved by use of dynamic moment magnification of

$$\omega = 0.5T_1 + 1.1 \qquad (4.26a)$$

with the limitations of

$$1.5 < \omega < 1.9 \qquad (4.26b)$$

Values for the dynamic moment magnificationfactor are given for both types of frames in Table 4.3. The minimum value of $\omega = 1.5$ for two-way frames results from consideration that a column section should be capable of sustaining simultaneous beam hinge moment inputs at overstrength from two

TABLE 4.3 Dynamic Moment Magnification Factor ω

Type of Frame	Period of Structure, T_1 (s)									
	< 0.7	0.8	0.9	1.0	1.1	1.2	1.3	1.4	1.5	> 1.6
One-way	1.30	1.33	1.39	1.45	1.51	1.57	1.63	1.69	1.75	1.80
Two-way	1.50	1.50	1.55	1.60	1.65	1.70	1.75	1.80	1.85	1.90

directions, corresponding with the moment pattern predicted by the initial elastic analysis. Analysis shows that a square column section, subjected to a moment along its diagonal, is only about 90% as efficient as for moment action along the principal directions. To allow for this, the approximate minimum value of ω, to allow for concurrent seismic action only, becomes $\omega = \sqrt{2}/0.9 \approx 1.5$. The relationship will be somewhat different for other column sections, but this approximation may be considered as being a reasonable average allowance for all columns of a story.

The probability of concurrence of large orthogonal moments at one column section due to responses in the higher mode shapes diminishes with the lengthening of the fundamental period. Therefore, the allowance for concurrent moment attack in Eq. (4.26) gradually reduces with the increase of the fundamental period, in comparison with the values given by Eq. (4.25).

(c) Required Flexural Strength at the Column Base and in the Top Story As discussed in Section 4.6.3(*b*), hinge formation at the base of columns, possibly with significant ductility demand, is to be expected, and hence this region will need to be detailed accordingly. To ensure that the flexural strength of the column sections at the base in two-way frames is adequate to sustain at any angle an attack of code force intensity, the unidirectional moment demand should logically be increased by approximately 10%. Similar considerations apply to the top store. Accordingly, the appropriate value of ω at the column base and for the top story should be.

$$\text{For columns of one-way frames:} \quad \omega = 1.0$$

$$\text{For columns of two-way frames:} \quad \omega = 1.1$$

(d) Higher-Mode Effects of Dynamic Response In terms of moment magnification, higher mode effects are more significant in the upper than in the first few stories above the column bases. To recognize this, Eqs. (4.25) and (4.26) are intended to apply only to levels at and above 0.3 times the height of the frame, measured from the level at which the first-story columns are considered to be effectively restrained against rotations. This level is normally at ground floor or at the foundations, depending on the configuration of the basement. In the lower 30% of the height, a linear variation of ω may be

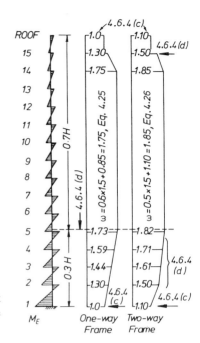

Fig. 4.22 The evaluation of dynamic moment magnification factor ω for two 15-story example columns.

assumed. However, at level 2 ω should not be taken less than the minima required by Eq. (4.25b) or (4.26b), respectively. At the soffit of the beams at the floor immediately below the roof, the value of ω may be taken as 1.3 for one-way frames and 1.5 for two-way frames.

The interpretation of the suggested rules for the estimation of the dynamic moment magnification factor ω, as set out above, is shown in Fig. 4.22 separately for 15-story one- and two-way frames, each with an assumed fundamental period of 1.5 s and a given moment pattern that resulted from the elastic analysis for lateral static forces. The arrows shown in this figure refer to the appropriate section number in the text.

(e) Columns with dominant Cantilever Action Columns with moment patterns such as shown in Fig. 4.23 require special consideration. Over the stories in which, because of dominant cantilever action (i.e., flexible beams), points of contraflexures are not indicated by the elastic analysis, critical moments are not likely to be affected significantly by the higher modes of dynamic response. In such columns the value of ω may be taken as the minima at first-floor level (i.e., 1.3 or 1.5 as applicable) and then linearly increased with height to the value obtained from Eq. (4.25) or (4.26), as appropriate, at the level immediately above the first point of contraflexure indicated by the analysis. This provision is less stringent than that shown for the lower stories in Fig. 4.22, when the first point of contraflexure appears

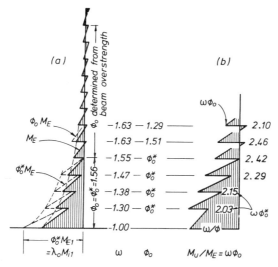

Fig. 4.23 Moment magnifications in the lower stories of a column of a 13-story one-way frame dominated by cantilever action.

above a floor that is further than 0.3 times the height above column base level. The intent of these provisions is to ensure that plastic hinges in cantilever-action-dominated columns will occur at the base and not in one of the lower stories. Specific values of ω, so derived for an example column, are shown in Fig. 4.23, where it was assumed that $\phi_o^* = 1.56$.

4.6.5 Column Design Moments

(a) Column Design Moments at Node Points   The magnified moments at the centers of beam–column joints are obtained simply from $\omega\phi_o M_E$, where ω and ϕ_o are calculated in accordance with Sections 4.6.3 and 4.6.4. Except at level 1 and the roof, the magnification by ϕ_o is intended to apply to column moments M_E, in each story as shown in Fig. 4.22. However, only end moments are magnified by ω. These two steps are illustrated for a column in Fig. 4.24. The design moments at the top and bottom ends so obtained in this column will not occur simultaneously.

The application of this moment magnification in the lower stories of the example column in the 15-story two-way frames, referred to in Fig. 4.22, is shown in Fig. 4.25. The numerical values of ϕ_o are those assumed to have resulted from the overstrengths of the beams as designed.

A similar example for the lower stories of the column of a one-way frame, dominated by cantilever action, as discussed in Sections 4.6.3(d) and 4.6.4(e),

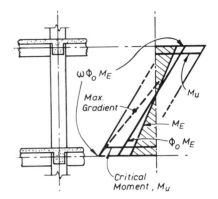

Fig. 4.24 Moment magnifications in an upper-story column.

is shown in Fig. 4.23(*b*). A period $T_1 = 1.3$ s was applicable to this building and hence for the upper levels, from Table 4.3, $\omega = 1.63$.

(b) Critical Column Section The critical column section to be designed is close to the top or the soffit of the beams. Accordingly, the centerline column moments should be reduced when determining the longitudinal reinforcement requirements. However, the gradient of the moment diagram is unknown, because it is not possible to determine what the shear force might be when the locally magnified moment is being approached during an earthquake. To be conservative, it may be assumed that only 60% of the critical shear V_u, to be examined in Section 4.6.7, will act concurrently with the design moment. Hence the centerline moments, such as shown in Fig. 4.24,

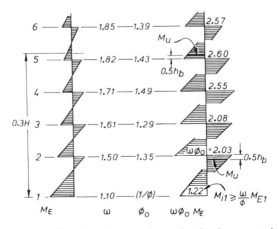

Fig. 4.25 Moment magnifications for a column in the lower stories of a 15-story two-way frame.

can be reduced by $\Delta M = 0.6(0.5h_b V_{col})$, where h_b is the depth of the beam. Consequently, the critical design moment M_u, shown in Fig. 4.24, to be used together with the appropriate axial load P_u and strength reduction factor $\phi = 1.0$, for determination of the ideal strength of sections at the ends of columns is

$$M_u = \phi_o \omega M_E - 0.3 h_b V_u \qquad (4.27)$$

where the value of V_u is that derived in Section 4.6.7 for the particular story. Equation (4.27) will need to be evaluated separately for each of the two principal directions for two-way frames.

(c)   Reduction in Design Moments   When yielding only in a small number of all columns in a story would result, a reduction of design moments should be acceptable. This is particularly relevant to columns that are subjected to low axial compression or to net axial tension, because in such columns the required flexural reinforcement might be rather large. Such columns are like vertical beams, and hence will be very ductile. Therefore, the shedding of moment from critical ends, necessitating some curvature ductility demand, will be associated with moderate concrete strains in the extreme compression fiber of the section affected. The larger the axial tension, the more moment reduction should be acceptable. Also, when design moments are large because of large dynamic magnification, larger local column strength reduction should be accepted. To achieve this, it is suggested that when the total design axial compression P_u on a column section does not exceed $0.1 f_c A_g$, the design moment may be reduced, if desirable for economic reasons, so that

$$M_{u,r} = R_m (\phi_o \omega M_E - 0.3 h_b V_u) \qquad (4.28)$$

where the reduction factor R_m should not be less than that given in Table 4.4, and where P_u is to be taken negative if causing tension, provided that the following criteria are also satisfied:

1. In selecting R_m in Table 4.4, the value of $P_u/f_c' A_g$ should not be taken less than -0.15 nor less than $-0.5 \rho_t f_y/f_c'$. The second requirement is intended to prevent excessive moment reduction in columns with small total steel content $\rho_t = A_{st}/A_g$, when the axial tension exceeds $0.5 f_y A_{st}$.
2. The value of R_m taken for any one column should not be less than 0.3. Thus up to 70% moment reduction is recommended when other criteria do not restrict it.
3. The total moment reduction, summed across all columns of a bent, should not be more than 10% of the sum of the unreduced design moments as obtained from Eq. (4.27) for all columns of the bent and taken at the same level. This is to ensure that no excessive story shear

TABLE 4.4 Moment Reduction Factor R_m

ω^a	$P_u/f_c' A_g$										
	-0.150	-0.125	-0.100	-0.075	-0.050	-0.025	0.000	0.025	0.050	0.075	0.100
1.0	1.00	1.00	1.00	1.00	1.00	1.00	1.00	1.00	1.00	1.00	1.00
1.1	0.85	0.86	0.88	0.89	0.91	0.92	0.94	0.95	0.97	0.98	1.00
1.2	0.72	0.75	0.78	0.81	0.83	0.86	0.89	0.92	0.94	0.97	1.00
1.3	0.62	0.65	0.69	0.73	0.77	0.81	0.85	0.88	0.92	0.96	1.00
1.4	0.52	0.57	0.62	0.67	0.71	0.76	0.81	0.86	0.90	0.95	1.00
1.5	0.44	0.50	0.56	0.61	0.67	0.72	0.76	0.83	0.89	0.94	1.00
1.6	0.37	0.44	0.50	0.56	0.62	0.69	0.75	0.81	0.88	0.94	1.00
1.7	0.31	0.38	0.45	0.52	0.59	0.66	0.73	0.79	0.86	0.93	1.00
1.8	0.30	0.33	0.41	0.46	0.56	0.63	0.70	0.78	0.85	0.93	1.00
1.9	0.30	0.30	0.37	0.45	0.53	0.61	0.68	0.76	0.84	0.92	1.00
	Tension							Compression			

$^a\omega$ is the local value of the dynamic moment magnification factor applicable to the column section considered.

carrying capacity is lost as a consequence of possible excessive moment reduction in columns. Normally, moment from an inelastic column could be transferred to others, in accordance with the principles of moment redistribution. However, column design is being considered when a beam sway mechanism in a subframe, such as shown in Fig. 4.10, has already developed. Moments possibly transferred to a stronger column of the bent can no longer be equilibrated by adjacent beams. Thus the moment reductions, suggested by Eq. (4.28), normally means strength loss. Because at this stage all actions are being considered in the bent, with all possible beam hinges being at overstrength (i.e., at least ϕ_o times code design force level), 10% loss of strength in the bent is acceptable.

The interpretation of this third limitation is shown in Fig. 4.26. If, for example, the design moment M_{u1} in the tension column is to be reduced, the reduction ΔM_{u1} must be such that

$$\Delta M_{u1} = (1 - R_m) M_{u1} \leq 0.1(M_{u1} + M_{u2} + M_{u3} + M_{u4}) \qquad (4.29)$$

where each of these four column moments was determined with Eq. (4.27). Such moment reduction will allow outer columns in symmetrical frames, such as shown in Fig. 4.26, to be designed in such a way that the requirements for reinforcement for the tension case (column 1) will not be much different from that of the compression (column 4) case.

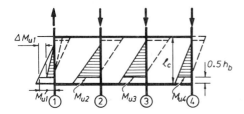

Fig. 4.26 Reduction of design moments in tension columns.

4.6.6 Estimation of Design Axial Forces

To be consistent with the principles of capacity design, the earthquake-induced axial load input at each floor should be V_{Eo} the seismic shear force induced by large seismic motions in the adjacent beam or beams at the development of their flexural overstrengths, as described in Section 4.5.3(a). The summation of such shear forces above the level under consideration, as shown in Fig. 4.27, would give an upper-bound estimate of the earthquake-induced axial column force. It should be recognized, however, that with an increasing number of stories above the level to be considered, the number of beam plastic hinges at which the full flexural overstrength will develop is likely to be reduced, as shown, for example, in the moment patterns of Fig. 4.21. This axial force, used together with the appropriately factored gravity loads and the design moments and shears to determine the strength of the critical column section, is then simply

$$P_{Eo} = R_v \Sigma V_{Eo} > P_E \qquad (4.30)$$

where ΣV_{Eo} is the sum of the earthquake-induced beam shear forces from all floors above the level considered, developed at all sides of the column,

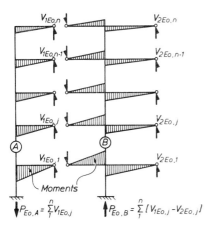

Fig. 4.27 Maximum possible column axial forces due to seismic actions at flexural overstrength of all beams.

TABLE 4.5   Axial Load Reduction Factor R_v

Number of Floors Above the Level Considered	Dynamic Magnification Factor, ω^a					
	1.3 or less	1.5	1.6	1.7	1.8	1.9
2	0.97	0.97	0.96	0.96	0.96	0.95
4	0.94	0.94	0.93	0.92	0.91	0.91
6	0.91	0.90	0.89	0.88	0.87	0.86
8	0.88	0.87	0.86	0.84	0.83	0.81
10	0.85	0.84	0.82	0.80	0.79	0.77
12	0.82	0.81	0.78	0.76	0.74	0.72
14	0.79	0.77	0.75	0.72	0.70	0.67
16	0.76	0.74	0.71	0.68	0.66	0.63
18	0.73	0.71	0.68	0.64	0.61	0.58
20 or more	0.70	0.68	0.64	0.61	0.57	0.54

$^a\omega$ is given by Eqs. (4.25) and (4.26).

taking into account the beam overstrengths and the appropriate sense of the shear forces. Values of R_v are given in Table 4.5.

In summing the beam shear forces at the column faces, strictly, all beams in both directions should be considered. In general, this step may be ignored at interior columns when the beam spans on either side of a column in each orthogonal frame are similar. This is because earthquake-induced axial forces are likely to be very small in such columns compared with the gravity-induced compression. However, for outer columns and corner columns in particular, a significant increase in axial force will result from skew earthquake attack, and this should be considered. When dynamic magnifications in the two principal directions of a structure are different, the larger of the values of ω, relevant to the level under consideration, may be taken in obtaining R_v from Table 4.5 to evaluate the axial force due to concurrent earthquake actions. Note that the higher dynamic amplification factors applicable to two-way frames ensures that the R_v factor will be lower than for equivalent one-way frames. This provides some recognition of the further reduced probability of beam hinging at all levels in both directions for a two-way frame.

4.6.7   Design Column Shear Forces

(a) Typical Column Shear Forces   In all but the first and top stories the shear force can be estimated from the gradient of the bending moment along the column. The minimum shear force to be considered is ϕ_o times the shear

derived from the elastic analysis for code forces, V_E. This is evident from the gradient of the $\phi_o M_E$ diagram in Fig. 4.24. An allowance, however, should also be made for a disproportionate distribution of beam moments between the columns above and below a beam, giving a somewhat larger gradient than implied by the elastic analysis moment pattern. A 20% increase of moment gradient, as shown in Fig. 4.24, is appropriate. Finally, the more serious consequences of a shear failure, also recognized by the magnitudes of strength reduction factors recommended [A1] for flexure ($\phi = 0.9$) and shear ($\phi = 0.85$), should be taken into account. Hence it is recommended [X3] that the ideal shear strength of the column, in conjunction with the design axial load relevant to the direction of earthquake attack, should not be less than

$$V_u = 1.3\phi_o V_E \tag{4.31}$$

It is seen that with a typical average value of $\phi_o = 1.4$, the ideal shear strength of the column with $\phi = 1.0$ will need to be $1.8V_E$.

(b) Design Shear in First-Story Columns This must also be related to the flexural overstrength [Eq. (4.24)] of the potential plastic hinge at the column base, $\phi_o^* M_E^*$. If this overstrength is large, which may be the case when the axial compression load intensity is in excess of $P_i = 0.3f_c' A_g$ [Eq. (3.28) and Fig. 3.22], the moment gradient may well exceed that assumed for Eq. (4.31). Hence for these columns the following design shear force should also be considered:

$$V_u = \frac{\phi_o^* M_E^* + 1.3\phi_o M_{E,\,\text{top}}}{l_n + 0.5h_b} \tag{4.32}$$

where $M_{E,\,\text{top}}$ is the column moment at the centerline of the beam at level 2 with depth h_b, derived from code forces, and l_n is the clear height of the first-story column.

(c) Shear in Columns of Two-Way Frames Additional considerations are necessary because of the possibility of concurrent earthquake attack from two directions. The shear strength of symmetrically reinforced square columns has been found to be the same when subjected to shear load in any direction. If it is assumed that the strengths of the beams framing into such a column from two directions are the same, the principal induced shear force in the column in the direction of the diagonal could be $\sqrt{2}$ times the shear applied under unidirectional earthquake attack. By again considering the reduced probability of concurrence of all critical load conditions, such as measured by ϕ_o, ω, and the 20% increase in moment gradient, it is suggested that for all columns of two-way frames, in which plastic hinges cannot develop, instead of using $\sqrt{2}$ times the value given by Eq. (4.31),

$$V_u = 1.6\phi_o V_E \tag{4.33}$$

and for first-story columns of two-way frames,

$$V_{\text{col}} = \frac{\phi_o^* M_E^* + 1.6\phi_o M_{E,\text{top}}}{l_n + 0.5h_b} \tag{4.34}$$

should be used to estimate the shear strength demand. These shear forces may be used while considering separately only unidirectional lateral forces in each of the two principal directions.

(d) Shear in Top-Story Columns   When column plastic hinges may develop at roof level before the onset of yield in the roof beams, the column shear may be assessed the same way as in first-story columns using either Eq. (4.32) or (4.34), as appropriate. When top-story columns are designed to develop plastic hinges simultaneously at both ends, the design for shear becomes the same as for beams, outlined in Section 4.5.3(*a*).

4.6.8   Design Steps to Determine Column Design Actions: A Summary

To summarize the issues of the deterministic design approach presented in Section 4.6, and to illustrate the simplicity of what might first appear to be a complex procedure, a step-by-step application of the technique, as used in a design office, is given in the following.

Step 1: Derive the bending moments for all members of the frame for the specified lateral earthquake forces only, using an appropriate elastic analysis. M_E refers to moments so derived at the node points of the frame model.

Step 2: Superimpose the beam bending moments resulting from elastic analyses for the lateral forces and those for the appropriately factored gravity loading, or obtain the moments from both sources in a single operation. Subsequently, carry out moment redistribution for all beam spans in each bent in accordance with the principles given in Section 4.3.

Step 3: Design all critical beam sections to provide the necessary ideal flexural strength, and determine and detail the reinforcement for all beams of the frame.

Step 4: Compute the flexural overstrength of each potential plastic hinge, as detailed, for each span of each continuous beam for both directions of the applied lateral forces. Using bending moment diagrams or otherwise, determine the corresponding beam overstrength moments at each column centerline (Fig. 4.18) and subsequently determine the beam seismic shear forces V_{Eo} in each span associated with these end moments, as outlined in Section 4.5.3(a).

Step 5: Determine the beam overstrength factor ϕ_o at the centerline of each column for both directions of the lateral forces acting on the frame, as explained in Section 4.6.3. ϕ_o factors are not applicable where plastic hinges in columns are expected, as at column bases and at roof level.

Step 6: From the fundamental period of vibration of the structure T_1, obtain the value of the dynamic magnification factor ω from Table 4.3. Consider one- and two-way frames separately, and observe the following exceptions:

(a) At the base and at roof level, $\omega = 1.0$ or $\omega = 1.1$ for one-way and two-way frames, respectively.

(b) At the soffit of beams at the level immediately below the roof, $\omega = 1.3$ and $\omega = 1.5$ for one- and two-way frames, respectively.

(c) At floors situated within the lower 30% of the total height H of the frame, the value of ω may be interpolated between the minimum values (1.3 or 1.5) at level 2 and the value obtained from Table 4.3, which is applicable at and above the level of $0.3H$ above the column base hinge level.

(d) For frames in which the analysis for lateral forces does not indicate a column point of contraflexure in a story, the minimum value of ω may be linearly increased to its full value, obtained from Table 4.3 at the floor immediately above the level were the first column point of contraflexure is indicated by analysis.

Step 7: Sum up all the earthquake-induced overstrength beam shear forces V_{Eo} from all floors from roof level down to level 2, as shown in Fig. 4.27, and determine at each floor $P_{Eo} = R_v \Sigma V_{Eo}$, where R_v is obtained from Table 4.5. Determine the design axial forces on the columns P_u at each floor for the appropriate load combinations: for example, $(D + L_R + E_o)$ or $(0.9D + E_o)$.

Step 8: The column design shear force V_u at a typical upper story is generally computed from Eq. (4.31) or (4.33), depending on whether the column is part of a one- or a two-way frame. For first-story columns, Eq. (4.32) or (4.34) need also be considered.

Step 9: The critical design moments for the columns at the top or the soffit of beams, to be considered together with the axial load P_u, obtained in step 7, are found from

$$M_u = R_m(\phi_o\omega M_E - 0.3h_bV_u) \qquad (4.35)$$

For columns under low axial compression or subjected to axial tension, the column design bending moments obtained from Eq. (4.27) may be reduced by the factor R_m. Obtain the value of R_m from Table 4.4, and note that when $P_u/f_c' A_g \geq 0.10$, $R_m = 1.0$.

Note further that for all column sections where design actions have been derived from capacity design consideration (Sections 1.4 and 4.6.1 to 4.6.7) the necessary ideal strength is based on a strength reduction factor of $\phi = 1.0$ (Section 3.4.1). Thus in these regions $M_i \geq M_u$, $P_u \geq P_i$, and $V_i \geq V_u$. On the other hand, at sections where plastic hinges are expected (e.g., at the base of a column), the required strength for moment and axial load, M_u and P_u, is based only on code-specified combination of factored loads and forces (Section 1.3.2). Accordingly, the necessary ideal strength must be evaluated with use of strength reduction factors $\phi < 1.0$ listed in Table 3.1. Moment magnifications by the product $\omega \phi_o$ for columns, to be designed to remain essentially elastic at and above level 2, are compared in Figs. 4.23 and 4.25 with equivalent magnifications derived from strength design principles, applicable at the column base. Moments so magnified then give the minimum ideal flexural strength of critical column sections. The necessary longitudinal reinforcement may now be determined and the special detailing requirements considered.

4.6.9 Choice of Vertical Reinforcement in Columns

Because of drift limits, slender columns cannot be used for seismic-dominated ductile frames. Therefore, it is often found that columns are large enough to resist the specified earthquake forces with steel contents in the range $0.01 < \rho_t < 0.03$.

Either mild or high-strength steel may be used for column reinforcement. Generally, the latter ($f_y = 400$ MPa, U.S. grade 60) will be more economical. Above level 2 of a frame, the characteristics of the steel should not affect column response, because if designed in accordance with the outlined capacity design principles, plastic hinges should never develop. At the base of the first-story column, however, significant ductility demand must be expected. As explained in Sections 3.3.1(*b*), 4.6.3, and 4.6.7, the flexural overstrength of the column section must then be evaluated accordingly, to ensure that the column design shear V_u is not underestimated.

If bond limitations for beam bars passing through interior beam–column joints are to be satisfied, the depth of the column may need to be up to 30 times the diameter of the beam bars. Various issues relevant to the development of the strength of beam bars in beam–column joints are examined in Section 4.8.6. These criteria may govern the choice of column dimensions in medium-rise buildings. For example, approximately 750 mm (30 in.) will be required for the column depth if 22-mm (0.88-in.)-diameter bars [$f_y = 415$ MPa (60 ksi)] are to be used in the top of the beams of a two-way frame when plastic hinges are expected to develop at both faces of that column when $f'_c = 31$ MPa (4.5 ksi) $\lambda_0 = 1.25$ and $P_u = 0.2 f'_c A_g$.

In the choice of the number and size of bars to be used in columns to satisfy strength requirements, apart from aspects of economy and ease in construction, the following points should also be considered:

1. For efficient confinement in end regions, column bars should be reasonably closely spaced around the periphery. The need for this is outlined in Section 3.6.1(a). Bars should not be farther apart than 200 mm (8 in.) center to center, or one-third of the cross-section dimension in the direction considered in the case of rectangular columns, or one-third of the diameter in the case of circular cross sections.

2. Intermediate column bars, such as shown in Fig. 4.31, also serve as vertical joint shear reinforcement. If they are omitted, additional joint shear reinforcement would need to be provided, as outlined in Section 4.8.9(b).

3. The two points above suggest a reasonably uniform spacing of bars, preferably of the same size, around the section periphery. Column design charts are prepared for this common case. Some designers are tempted to utilize bundled column reinforcement, grouped in the four corners, as shown in Fig. 4.28(a), because of greater efficiency in moment resistance. For the reasons mentioned, in seismic design this arrangement is undesirable, and the strength increase obtained is generally not significant.

4. The curtailment of bars to suit a bending moment pattern, as it is done in beams, is impractical in columns. Therefore, all bars are normally carried through the full story. Column bars must therefore be spliced. Either welded or mechanical or lapped splices may be used. The spacing of column bars must therefore be such that there is enough room to splice the bars. Strength considerations of lapped splices were examined in Section 3.6.2(b).

Lapped splices are best arranged in rectangular columns so that each pair of spliced bars is tied so that the tie crosses the plane of potential splitting, as shown in Fig. 3.30. Such an arrangement is also shown in Fig. 4.28(b). Vertical bars are to be suitably offset as shown in Fig. 3.31. A similar arrangement of spliced bars may also be used in circular columns. However,

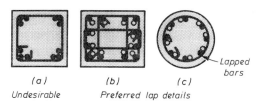

(a)   (b)   (c)

Undesirable   Preferred lap details

Lapped bars

Fig. 4.28 Typical arrangement of column bars in bundles and at lapped splices.

it will be more economical to avoid the use of offsets in such situations, as bars can be safely spliced side by side, as shown in Fig. 4.28(c). Spirals or circular ties can efficiently control splitting cracks that might develop between bars [P19].

5. Because of bond limitations for bars within beam–column joints, bar sizes may need to be limited as recommended in Section 4.8.6(b).

4.6.10   Location of Column Splices

In deciding where to locate column bar splices, the following aspects should be considered:

1. Within the region of a potential plastic hinge, yielding of the reinforcement, both in compression and tension, with possible strain hardening, must be expected. A splice should not be located in such an area. A plastic hinge must be expected to occur at the base of first-story columns. (Section 3.6.2(b)). In most cases it will be possible to carry the column reinforcing cage through the lowest level of beams for splicing above level 2. Alternatively, the splice can be located approximately at midheight of the first-story column.

When transverse confining reinforcement in accordance with requirements of Section 3.6.1(a) is provided in a potential plastic hinge region and column bar diameters are not large, it may be possible to preserve the strength of lapped splices, even during severe cyclic reversed loading. However, in such a case the plastic hinge length, over which yielding of the column bars is expected to spread, will be very short. Yielding and consequent plastic hinge rotations have been found [P19] in such cases to be restricted to the end of the lapped splice at the section of maximum moment. This involves extremely large curvature ductility demands and may lead to fracture of the bars. When larger-diameter bars are lap spliced in plastic hinge regions, splices will fail, with gradual deterioration of bond transfer between spliced bars, after a few excursions into the inelastic range. Figure 4.29 shows typical splice failures in test specimens [P19].

2. One of the aims of the capacity design procedure is to eliminate the possibility of plastic hinges forming above level 2. Therefore, lapped splices in such columns may be placed anywhere within the story height. For ease of erection the preferred location will be immediately above a floor, so that prefabricated reinforcing cages can be supported directly on top of the floor slab. Transverse reinforcement, in accordance with Eq. (3.70), must be provided, however, to ensure adequate strength of the splice.

3. Splicing by welding or special proprietary mechanical connectors (couplers) is popular for column design, particularly for large-diameter bars where lap splice length may be considerable. Such connections should not be located in potential plastic hinge regions unless cyclic inelastic testing of the

Fig. 4.29 Failures in columns with lapped splices in the end region.

connection detail in realistically sized column test units has established that ductility will not be impaired by the detail. Note that a tension test on a single splice, indicating unimpaired tension strength, is inadequate to establish satisfactory performance.

4. Particular care must be taken with site welding of column bar splices. Lap welding must not be used, and butt welds must be carefully prepared and executed. With high-strength reinforcement, as is commonly used for columns, there is a real danger of embrittlement of the steel adjacent to the weld unless preheating procedures are followed rigorously. Welded splices should never be located in potential hinge regions.

4.6.11 Design of Transverse Reinforcement

(a) General Considerations There are four design requirements that control the amount of transverse reinforcement to be provided in columns: shear strength, prevention of buckling of compression bars, confinement of compressed concrete in potential plastic hinge regions or over the full length of columns subjected to very large compression stresses, and the strength of lapped bar splices. The basic requirements for each of these criteria have been given previously, and only specific seismic aspects are examined in the subsections that follow.

Because the configuration of transverse reinforcement, consisting of stirrups, ties, and hoops in various shapes, is usually the same over the full height of a column in a story, variations in required quantities are achieved by changing the spacing s of sets of ties or hoops that make up the transverse reinforcement. However, in certain situations other, generally more stringent, limitations need be imposed on spacing s. In noncritical regions of columns, code-specified limits on spacings are often based on established detailing practice, engineering judgment, and constructability constraints.

Requirements for transverse reinforcement vary according to the criticality of regions along a column. In particular, end regions need to be distinguished from parts of the column in between these end regions. Moreover, a distinction must be made between end regions that are potential plastic hinges and those that are expected to remain essentially elastic at all times. With this distinction, the discussions that follow concentrate on end regions of columns where seismic actions are most critical. The most severe of the four design criteria noted above will control requirements for the quantity, spacing, and configuration of transverse reinforcement.

(b)   Configurations and Shapes of Transverse Reinforcement   The required amount of transverse reinforcement is traditionally made up in the form of stirrups; rectangular, square, or diamond-shaped closed ties or hoops; single leg ties with hooks to provide anchorage; and circular or rectangular spirals. In most rectangular sections a single peripheral hoop will not be sufficient to confine the concrete properly or to provide lateral restraint against buckling for the longitudinal bars. Therefore, an arrangement of overlapping rectangular hoops and/or supplementary cross-ties will be necessary, as shown in Fig. 4.30. Supplementary cross-ties, shown in Fig. 4.30(*b*), are sometimes specified to be fitted tightly around peripheral ties, a rather difficult requirement in

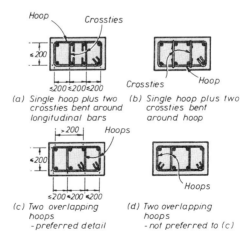

Fig. 4.30   Alternative details using hoops and supplementary cross-ties in columns.

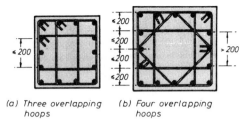

(a) Three overlapping   (b) Four overlapping
    hoops   hoops

Fig. 4.31 Typical column details showing the use of overlapping hoops.

practice. It is likely to be more effective if intermediate cross-ties, like any other form of transverse reinforcement, are bent tightly around a vertical column bar, as seen in Fig. 4.30(a). Another practical solution is to use a number of overlapping rectangular hoops, as seen in Figs. 4.30(c) and 4.31, rather than a single peripheral hoop and supplementary cross-ties. Thereby, the concrete may be confined by vertical arching between sets of hoops, supplementary cross-ties, and also by horizontal arching between longitudinal bars held rigidly in position by transverse legs as illustrated in Fig. 3.4. In a set of overlapping hoops it is preferable to have one peripheral hoop enclosing all the longitudinal bars, together with one or more hoops covering smaller areas of the section, as in Fig. 4.30(c). While the detail in Fig. 4.30(d), which has two hoops each enclosing six bars, is equally effective, it will be more difficult to construct.

Figure 4.31 shows typical details using overlapping hoops for sections with a greater number of longitudinal bars. It is to be noted that the diamond-shaped hoop surrounding the four bars at the center of each face in Fig. 4.31(b) can be counted on, making a contribution to A_{sh} in Eq. (3.62) by determining the equivalent bar area of the component of forces in the required direction. For example, two such hoop legs inclined at 45° to the section sides could be counted as making a contribution of $\sqrt{2}$ times the area of one perpendicular bar in assessing A_{sh} or A_v. That is, in Fig. 4.31(b), A_{sh} or A_v may be taken as $(4 + 1.41)A_{te}$, where A_{te} is the area of each hoop bar having the same diameter.

As in the case of beams, discussed in Section 4.5.4, recommendation 1, not all bars need to be laterally supported by a bend of a transverse hoop or cross-tie. If bars or groups of bars that are laterally supported by bends in the same transverse hoop or cross-tie are less than or equal to 200 mm (8 in.) apart, any bar or bundle of bars between them need not have effective lateral support from a bent transverse bar, as is demonstrated in Fig. 4.31(a).

Figure 4.32 illustrates commonly used configurations for hoops and ties. Several of these would not fulfill intended structural functions. Tie anchorages with 90° overlapping hooks at corners, shown in Fig. 4.32(a), clearly cannot confine a concrete core (shown shaded) after the cover concrete

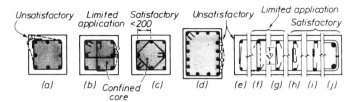

Fig. 4.32 Alternative tie arrangements.

spalls. Also, several column bars will thereby be deprived of lateral support. This has often been observed in earthquake damage. In some countries, J-type intermediate ties, with a 135° hook at one end and 90° hook at the other, shown in Fig. 4.32(b), are preferred because of ease of construction. Such ties, when arranged so that the positions of different hooks alternate, as in Fig. 4.32(f), can effectively stabilize compression bars but can make only limited contribution to the confinement of the concrete core [T2]. When, in the presence of high compression forces and ductility demands, the strength of such a tie is fully mobilized, the 90° hook may open, as illustrated in Fig. 4.32(b). Hence it is recommended [T2] that J ties be used only in members exposed to restricted ductility demands. The use of two types of hoops, with 135° hooks suited for small columns, shown in Fig. 4.32(c), can ensure optimum performance while providing ready access for the placing of fresh concrete. It should be appreciated that even very large amounts of transverse reinforcement, consisting of peripheral ties only as seen in Fig. 4.32(d), will fail to ensure satisfactory performance in terms of both confinement of the concrete core and prevention of bar buckling.

Various solutions are possible when, because of construction difficulties, the use of intermediate ties with a 135° or 180° hook on each end, is not possible. Figure 4.32(g) shows that ties made of deformed bars may be spliced in the compressed core. These may contribute effectively to confinement when ductility demand is restricted [T2]. However, the use of such splices for shear reinforcement in beams, particularly in potential plastic hinge regions, where reversed moments can also occur, must be avoided. More effective splicing of ties, particularly where plain rather than deformed bars are used, is shown in Fig. 4.32(h).

As ties are not subjected to alternating inelastic strains, lap welding as in Fig. 4.32(i), should be acceptable, provided that good-quality weld is assured. When relatively large bars need to be used for ties, as in beam–column joints or at the boundaries of end regions of massive structural walls, the accommodation of overlapping 135° hooks, typically with a 6 to $8d_b$ straight extension beyond the bend, may lead to congestion of reinforcement. In such cases prefabricated hoops with proper butt welds, as shown in Fig. 4.32(j), may facilitate construction. For obvious reasons, ties, lap spliced in the cover concrete, as in Fig. 4.32(e), should not be assumed to make any contribution to strength or stability, and preferably should not be used. Similarly lapped splices of circular hoops or spirals in the cover concrete, known to have resulted in the collapse of bridge piers during earthquakes when the cover concrete spalled, must be avoided. Instead, welding as in Fig. 4.32(i) or hooked splices as in Fig. 4.28(c) should be used. It may be noted that spacing limitations for shear reinforcement, set out in Section 3.3.2(a), are usually less critical than similar limitations for confining reinforcement.

(c) **Shear Resistance**   It will be necessary to ensure that some or all of the design shear force V_u be resisted by transverse reinforcement in the form of column ties or circular hoop or spiral reinforcement. As in beams, the approach to shear design in potential plastic hinge regions is different from that applicable to other (elastic) parts of the column. The details of the requirements of design for shear resistance are set out in Section 3.3.2.

(d) **Lateral Support for Compression Reinforcement**   The instability of compression bars, particularly in the potential plastic hinge zone, must be prevented. Some yielding of the column bars, both in tension and compression, may be expected in the end regions of "elastic" columns above level 2, even though full development of plastic hinges will not occur. Therefore, transverse stirrup ties, sometimes referred to as *antibuckling* reinforcement, should be provided in the end regions of all columns of the frame in the same way as for the end regions of beams, discussed in Section 4.5.4. In particular, Eq. (4.19) and Fig. 4.20 are relevant.

Lateral supports for compression bars in between end regions are as for nonseismic design. Rules for these are given in various codes [A1, X3] and are summarized in Section 3.6.4. Generally, the spacing limitations required by shear strength or confinement of the concrete in the middle region of a column are more stringent and hence will govern tie spacing.

(e) **Confinement of the Concrete**   To ensure adequate rotational ductility in potential plastic hinge regions of columns subjected to significant axial compression force, confinement is essential. The principles of confinement were summarized in Sections 3.2.2 and 3.6.1, and the amounts of transverse reinforcement for this purpose were given by Eq. (3.62).

The full amount of confining reinforcement, in accordance with Eq. (3.62), is to be provided in all end regions of columns where plastic hinges are expected. However, only one-half of this confining reinforcement is required in regions adjacent to potential plastic hinges and in the end regions of columns in which, because of the application of capacity design summarized in Section 4.6.8, plastic hinges will not occur.

This full confinement will always be required at the column base and in the top story, if hinges are also to be expected there. Full amounts of confining reinforcement will also be required in the end regions of any column in any story that has been designed to resist only moments given by Eq. (4.21), that is, moments without the application of the dynamic moment magnification factor ω.

No reinforcement for confinement of the compressed concrete in columns needs to be provided outside end regions or those adjoining end regions, the dimensions of which together with other relevant requirements are given in subsections (i) and (ii), which follow.

(i)   End Region Requirements: These requirements apply to distance l_0 over which special transverse reinforcement is required. Lengths of potential plastic hinges in columns are generally smaller than in beams. This is partly because column moments vary along the story height with a relatively large gradient, so the spread of yielding in the tension reinforcement is limited.

When the axial compression load on the column is high, the necessary confining reinforcement content in the hinge zone will be considerable. As shown in Fig. 3.5, this will increase the strength of the concrete. In the potential plastic hinge zone at the end of the column, the flexural strength of sections may thus be considerably greater than that in the less heavily confined region away from the critical section of the potential plastic hinge. Therefore, the length of the end region of a column to be confined must be greater when the axial load is high. The critical axial force at which this extension should be considered was found from experiments to be approximately $0.3 f'_c A_g$ [G4, P8]. Accordingly, it is recommended [X3] that the length of the end region in columns, shown as l_0 in Fig. 4.33 and measured from the section of maximum end moment, be taken as follows:

1. When $P_e < 0.3 f'_c A_g$, not less than the larger of the longer section dimension in the case of rectangular columns or the diameter in the case of circular columns, or the distance to the section where the moment equals 80% of the maximum moment at that end on the member (Fig. 4.33).
2. When $P_e > 0.3 f'_c A_g$, 1.5 times the dimensions in (1) above.
3. When $P_e > 0.6 f'_c A_g$, twice the dimensions required in (1).

These rules require a knowledge of the shape of the column bending moment diagram. It was shown, for example in Fig. 4.21, that during inelastic response of a frame, moment patterns along columns can be quite different from those predicted by the initial elastic analysis. For this reason the following assumptions regarding moment gradients, to determine the lengths of end regions above, should be made:

1. When a point of contraflexure is predicted in the first story, a linear variation of moments from maximum at the base to zero at the center of the level 2 beam should be taken. The moment pattern marked $M_{\text{earthquake}}$ in Fig. 4.33(b) shows this assumption.

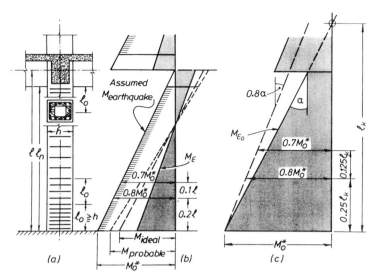

Fig. 4.33 Definition of end regions in a first-story column, where special transverse reinforcement is required.

2. In columns dominated by cantilever action, such as those shown in Fig. 4.23, 80% of the gradient obtained from the analysis for lateral forces may be assumed to determine the level at which the moment is 80% and 70%, respectively, of that at the base. An example is shown in Fig. 4.33(c). In this case a substantial length of the first-story column may need to be fully confined.

(ii) Transverse Reinforcement Adjacent to End Regions: The reinforcement for full confinement can be reduced because in these regions no plastic deformations are expected and the concrete does not need as much confinement. However, the reduction should be achieved by gradual increase of the spacing of the sets of transverse reinforcement. It is therefore recommended that over the length of the column adjacent to the end region, defined in the preceding section, and equal in length to the end region, shown as l_0 in Fig. 4.33, not less than one-half of the amount of confining reinforcement required by Eq. (3.62) be provided. This is conveniently achieved by simply doubling the spacing s_h specified in Section 3.6.4 for plastic hinge regions. As noted above, this reduced amount of reinforcement also applies to end regions of "elastic" columns above level 2.

(f) Transverse Reinforcement at Lapped Splices Splices must be provided with special transverse reinforcement in accordance with the requirements of Section 3.6.2(*b*). When splices in upper stories are placed within the end

region of the column, as seen in Fig. 3.31, it is likely that the transverse reinforcement to satisfy the more critical of the requirements for either the confinement of the concrete core or for shear resistance, will be found adequate to control splitting and hence bond strength loss within splices. It is emphasized again that splices can be placed in end regions of columns only if the columns have been provided with sufficient reserve strength, in accordance with Section 4.6.8, to eliminate the possibility of plastic hinge formation. Otherwise, column splices must be placed in the center quarter of the column height.

4.7 FRAME INSTABILITY

4.7.1 P–Δ Phenomena

Although P–Δ phenomena in elastic structures have been studied and reported extensively [C2], limited guidance with respect to inelastic seismic response is currently available to the designer [M3, P9, X10]. For this reason the examination of some principles here is rather detailed. Fortunately, in most situations, particularly in regions where large seismic design forces need to be considered, P–Δ phenomena will not control the design of frames.

From the review of the issues relevant to the influence of P–Δ phenomena on the dynamic response of frames, given in Chapter 2, two questions of importance arise:

1. Are secondary moments due to $P\Delta$ effects critical in seismic design, and if so, when are they critical?
2. If $P\Delta$ effects prove to be critical, is the remedy to be found in increased stiffness to reduce Δ, or in added strength to accommodate $P\Delta$ moments and to preserve the lateral strength and energy dissipation capacity in frames?

4.7.2 Current Approaches

Most building codes do not give definitive guidance with respect to the quantifying of $P\Delta$ effects. Typically, reference is made to accepted engineering practice [X4]. In this, it is suggested that in assessing Δ, a multiple of the story drift predicted by an elastic analysis, typically 4.5Δ for ductile frames, should be used, to allow for ductility demands.

Elsewhere [X6], it has been suggested that interstory drift to be expected be assessed from

$$\Delta = \frac{\mu_\Delta \Delta_e}{1 - \theta_\Delta} \tag{4.36}$$

where Δ_e is the elastic drift resulting from the application of the code-specified lateral forces, μ_Δ is an estimate of the expected story displacement
ductility ratio, and θ_Δ is a coefficient similar to the stability index subsequently given in Eq. (4.37). It is seen that the elastic drift Δ_e is magnified by
the factor $1/(1 - \theta_\Delta)$ to allow for the additional drift due to $P\Delta$ effect. It is
implied [X6] that if Δ, so computed, is less than 1 to 1.5% of the story height,
no further precautions in the design need be taken.

It has been suggested for steel structures [X7] that if in any frame the $P\Delta$
moment, resulting from the product of factored column load P, excluding
earthquake-induced axial forces, and the interstory deflection Δ at first yield,
exceeds 5% of the plastic moment capacity of the beam framing into the
column, the strength of the frame should be increased to carry the $P\Delta$
moment. This provision concedes 20% loss of beam capacity to resist lateral
forces when the imposed story displacement ductility is 4.

In focusing on the role of compression members, codes tend to divert
designers' attention from the significance that $P\Delta$ effects have on beams of
ductile frames [A2]. The seismic resistance of weak beam–strong column
frame systems, described in previous sections, is controlled by the beams
(weak links). Thus if additional resistance with respect to lateral force effects
is to be provided, it is the beams that need to be strengthened. The strength
of the columns (strong links), affected by P–Δ phenomena only in an indirect
way, is derived from the strength of the beams, as described in Section 4.6.

4.7.3 Stability Index

It is useful to define the stability index Q_r, which compares the magnitude of
$P\Delta$ moments in a story with the story moment generated by lateral earthquake forces. Consider the distorted shape of a multistory frame at an instant
of severe inelastic lateral displacement, as shown in Fig. 4.34. Plastic hinges
that have developed are also shown. While the maximum displacement at

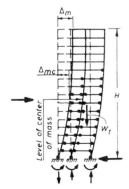

Fig. 4.34 Typical deflection of an inelastic multistory frame
during severe earthquake attack and actions due to $P\Delta$
effects.

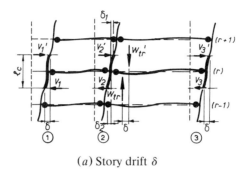

(*a*) Story drift δ

(*b*) Beam moments due to seismic design forces

Fig. 4.35   Actions on a subframe at an intermediate level.

roof level is Δ_m, the center of mass of the building is displaced only by Δ_{mc}. The secondary moment due to the $P\Delta$ effect with respect to the base is therefore $W_t \Delta_{mc}$, where W_t is the total weight of the frame. This overturning moment will need to be resisted by moments and particularly, by axial forces at the base of the columns, as indicated in Fig. 4.34. These forces are additional to those required to equilibrate gravity loads and lateral forces. It is evident that these actions, when applied to plastic hinges, will reduce the share of the member capacities that will be available to resist lateral forces generated by earthquake motions. The axial forces, induced in the columns due to $P\Delta$ moments only, originate from shear forces generated in beams at each floor. Additional moments for the beams result from the product of the sum of the gravity load at that floor and from the frame above and the story drift.

A somewhat idealized configuration of the plastified beams and elastic columns at intermediate stories of the frame of Fig. 4.34 is shown in Fig. 4.35(*a*). The relative displacement of the floors at this stage (i.e., the drift) is δ. Consequently, the secondary moment $W'_{tr}\delta_1 + W_{tr}\delta_2 \approx W_{tr}\delta$ due to the total weight above (W'_{tr}) and below (W_{tr}) level r must be resisted by the subframe, drawn with heavy lines in Fig. 4.35(*a*).

The stability index Q_r is thus given by

$$Q_r = W_{tr}\delta / \Sigma M_E \qquad (4.37)$$

where ΣM_E is the sum of the required beam moments at column center lines resulting from the specified lateral forces. With reference to Fig. 4.35(*a*),

$$\Sigma M_E \approx \Sigma(V_i' + V_i)(l_c/2)$$

and this is also termed the story moment. Equation (4.37) is similar in formulation to stability indices developed by other researchers [A1, M6] but relates the $P\Delta$ effect to maximum expected displacements rather than elastic displacements corresponding to specified lateral force levels.

For a rigorous analysis the evaluation of the story drift δ in each of the stories is required. With reference to Fig. 4.34, it is seen, however, that δ may be estimated from the average slope of the frame, Δ_m/H, and the relevant story height l_c with the introduction of a suitable displacement magnification factor λ. This factor relates the story slope to the average slope of the frame and will vary over the height of the frame. With this substitution Eq. (4.37) becomes

$$Q_r = \lambda \frac{l_c W_{tr} \Delta_m}{H \Sigma M_E} = \lambda \frac{l_c W_{tr} \mu_\Delta \Delta y}{H \Sigma M_E} \tag{4.38}$$

and with this the severity of $P\Delta$ effects with respect to seismic design for strength can be gauged. The evaluation of λ, from considerations of the estimated deflected shape of the inelastic structure, is presented in Section 4.7.5(*b*), where Δ_m is also defined.

Equation (4.38) is a modification of the elastic stability index [A1]

$$Q_e = W_{tr}\delta_e/V_r l_c \tag{4.39}$$

which has been used [M6] to gauge the seriousness of second-order $P\Delta$ moments on column instability and hence the possible need for more precise derivation of column design moments.

4.7.4 Influence of $P\Delta$ Effects on Inelastic Dynamic Response

(*a*) *Energy Dissipation*   During the response of well-behaved subframes of the type shown in Fig. 4.35, energy dissipation will not be reduced by $P\Delta$ effects during inelastic oscillations with equal amplitudes in each direction. This is shown in Fig. 4.36, where curve 1 illustrates an ideal elastoplastic response when unaffected by $P\Delta$ load demands. The contributions of the beam dependable flexural strengths ΣM_E to the resistance of the story moment, given above and shown in Fig. 4.36, are, however, reduced when the moment δW_{tr} is also introduced. Hysteretic energy so lost is shown by the vertically shaded area in Fig. 4.36. This loss is recovered, however, as shown by the horizontally shaded area, when in the next displacement cycle the frame is restored to its original undeformed position. Consequently, during

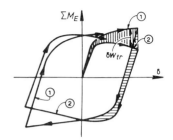

Fig. 4.36 Lateral force–displacement relationship for a ductile subframe with and without $P\Delta$ effects.

the complete cycle of displacements shown in Fig. 4.36, there will be no loss of energy dissipation due to $P\Delta$ effects.

Figure 4.36 shows that the secondary moments reduce both strength and stiffness during the first quadrant of the response, whereas increased resistance is offered against forces that tend to restore the frame to its original perfectly vertical position. There is thus increased probability that after a very large displacement in one direction, resulting from a long velocity pulse, the frame may not be restored to its original undisplaced position. During subsequent ground excitations the frame will exhibit degrading strength characteristics in the loading direction and this could encourage "crawling," leading to incremental collapse, as illustrated in Fig. 4.37. This type of response is more likely to occur in structures in which energy dissipation is concentrated in only one story, by column sway mechanisms, which require inherently large story drifts to provide the necessary system displacement ductility [K1].

(b)   Stiffness of Elastic Frames   Frame stiffness is reduced by $P\Delta$ effects, but only slightly, since lateral displacement during the elastic range of response

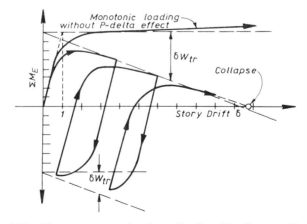

Fig. 4.37 $P\Delta$ moments causing "crawling" and leading to collapse.

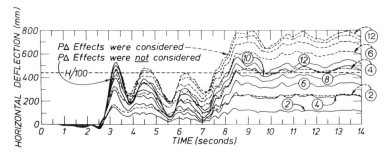

Fig. 4.38   Displacement history of even levels of a 12-story example frame during the Pacoima Dam earthquake record [M4].

will be generally small. The consequent small increase in the period T_1 of the frame has been found to result in slightly reduced frame responses to most earthquake records [M4], as expected.

(c)   Maximum Story Drift   As expected, P–Δ phenomena will increase drift, but several numerical analyses of typical building frames have indicated that effects are small when maximum interstory drift is less than 1% [M3, M4, P10]. However, for greater interstory drifts, $P\Delta$ effects have lead to rapidly increasing augmentation of these drifts. Correspondingly larger inelastic deformations were recorded, particularly in the lower parts of the frames.

Figure 4.38 compares the computed horizontal deflections of even-numbered levels of a 12-story example frame when this was subjected to the 1971 Pacoima Dam record, an extremely severe excitation. It is seen that the displacements were significantly larger after 3.5 s of the excitation, when $P\Delta$ effects were included in the analysis. While after 8 s of excitation, P–Δ considerations indicated some 60% increase in the top-story deflections, the maximum story drift observed in the lower two stories increased by 100%, to 3.7% of the story height. The curves in Fig. 4.38 also show the nonuniform distribution of inelastic drifts, with the dramatic increase of drifts in the lower half of the structure. These drifts correspond to the deflection profile shown qualitatively in Fig. 4.34.

It has been shown that increasing strength of a frame is more effective in controlling drift than is increasing stiffness [M3, M4]. This is to be expected because the more vigorously a frame responds in the inelastic range, the less is the significance of stiffness, a characteristic of elastic behavior.

(d)   Ductility Demand   Inelastic story drifts are directly related to ductility demands in plastic hinges. Thus when drift increases in the lower stories of a frame due to $P\Delta$ effects, the plastic rotational demand increases correspondingly. When analysis ignoring $P\Delta$ effects predicts inelastic drifts significantly in excess of 1.5% of the story height, the influence of P–Δ phenomena on

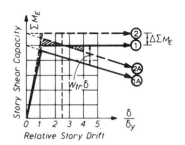

Fig. 4.39 Idealized bilinear response showing the effects of $P\Delta$ moments.

the inelastic dynamic response may be so large that plastic rotational demands in both beams and first-story columns would approach or exceed limits attainable with normal detailing recommended in seismic design.

4.7.5 Strength Compensation

(a) Compensation for Losses in Energy Absorption   Energy absorption may be considered as a basis for assessing the influence of $P\Delta$ effects on the inelastic seismic response of a frame. To illustrate this simple approach, the idealized bilinear force displacement relationships of a beam sway mechanism are compared in Fig. 4.39. Initial relationships without, and including, $P\Delta$ effects are shown by lines 1 and 1A, respectively. To compensate for the loss of absorbed energy, the strength of the beams must be increased so that the bilinear responses shown by curves 2 and 2A in Fig. 4.39 result. The shaded areas for curve 2A show that if a suitable increase in lateral force resistance is made, the reduction in energy absorbed for a given story ductility demand can be compensated. It can be expected that the performance of two frames, one characterized by curve 1 and the other by curve 2A in Fig. 4.39, will be very similar when a story drift $\delta = \mu_\delta \delta_y$ is imposed. This consideration leads to a required strength increase by

$$\Delta \sum M_E = \frac{1 + \mu_\delta}{2} W_{tr} \delta_y \qquad (4.40)$$

where δ_y is the interstory drift in the elastic structure subjected to the design lateral static (code) forces.

(b) Estimate of Story Drift   In evaluating Eq. (4.40), the fact that the distorted shape of the inelastic structure, as shown in Fig. 4.34, will be fundamentally different from that of the elastic structure, must be recognized. The first two curves in Fig. 4.40 show the range of deflection profiles for elastic multistory frames subjected to lateral forces prior to the onset of yielding. When the concept of displacement ductility factor $\mu_\Delta = \Delta_m / \Delta_y$ is

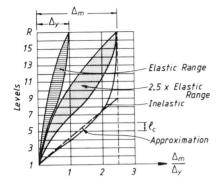

Fig. 4.40 Comparison of deflected shapes of elastic and inelastic frames.

used, it is often erroneously assumed [X4] that all deformations predicted by the elastic analysis are simply magnified by this factor, as shown by the second set of curves in Fig. 4.40. In this it has been assumed that $\Delta_m/\Delta_y =$ 2.5. The critical "inelastic" deformed shape of a frame, resulting in much larger interstory drift in the lower stories, is, however, similar to that of the third curve in Fig. 4.40. It should be appreciated that the larger story drifts will occur in lower stories when significant plastic hinge rotations develop also at the base of the columns. Therefore, it is suggested that the magnitude of the story drift in the lower half of the frame, where $P\Delta$ effects could be critical, be based on twice the average slope of the frames, as shown by the straight dashed line in Fig. 4.40. This means that for the lower half of the frame, the magnitude of the story displacement ductility factor is much larger than the overall displacement ductility for the entire frame, that is,

$$\mu_\delta \approx 2\mu_\Delta \qquad (4.41)$$

This assumption, implying that the value of λ in Eq. (4.38) is a constant equal to 2, may be unduly conservative for stories at midheight of the building. In any story λ should not be taken less than 1.2.

(c) Necessary Story Moment Capacity   The contribution of the beams to the story moment capacity at a level, shown in Fig. 4.35(b), can now readily be derived from the requirement that

$$\Sigma M_E < \phi(\Sigma M_i - \Delta\Sigma M_i) \qquad (4.42a)$$

or

$$\phi\Sigma M_i > \Sigma M_E(1 + \Delta\Sigma M_E/\Sigma M_E) \qquad (4.42b)$$

With the use of Eq. (4.40) and substitution from Eq. (4.41), the need for the increased required story moment strength becomes

$$\phi\Sigma M_i > (1 + Q_r^*)\Sigma M_E \qquad (4.43)$$

where the modified stability index is

$$Q_r^* = (\mu_\Delta + 0.5)\frac{l_c W_{tr}\Delta_y}{H\Sigma M_E} \qquad (4.44)$$

It is seen that in Eq. (4.43) the sum $\phi\Sigma M_i$ of the reduced ideal flexural strengths of all beams, as detailed at a floor, is compared with the corresponding sum of moments ΣM_E (i.e., required strength), derived from the specified seismic forces. Therefore, any excess flexural strength that may have been provided in the beams because of gravity load or construction considerations may be considered to contribute toward the resistance of $P\Delta$ secondary moments.

A small amount of loss in lateral force resistance and hence energy absorption, when maximum displacement ductility is imposed on the frame, is not likely to be detrimental. For this reason it is suggested that compensation for $P\Delta$ effect should be considered only when the value of Q_r [computed from Eq. (4.37) or from Eq. (4.38) with $\lambda = 2$] exceeds 0.15. As a good approximation, the limit $Q_r^* \geq 0.085$ may also be used.

4.7.6 Summary and Design Recommendations

1. When the final member sizes, and hence the stiffness of the frame, have been determined, the elastic frame deflection at roof level, Δ_y, should be checked. This is also required to confirm whether the initial design assumption with respect to period estimation [Section 2.4.3(a)] is acceptable and, if necessary, to ensure sufficient separation from adjacent buildings.

2. Determine whether the $P\Delta$ effect should be considered by evaluating the stability index, Q_r^*, from Eq. (4.44). If $Q_r^* > 0.085$, $P\Delta$ effects should be given further consideration. Typical variation of total weight, W_{tr}, and story moment demands ΣM_E due to lateral design story shear forces of ΣV_E for a 16-story frame are shown in Fig. 4.41(a) and (e). From previous examination of the story moment on a subframe or from Fig. 4.35, it is evident that $\Sigma M_E \approx l_c \Sigma V_E$, where ΣV_E is the design story earthquake shear, the typical distribution of which is shown in Fig. 4.41(b).

3. The sum of the ideal beam flexural strengths with respect to column centerlines, ΣM_i, based on properties of the beams as detailed, should be evaluated. This is readily obtained from the material overstrength factor λ_o and the system flexural overstrength factor ψ_o, given by Eq. (4.17). Accordingly $\Sigma M_i = \Sigma M_{Eo}/\lambda_o = \psi_o \Sigma M_E/\lambda_o$.

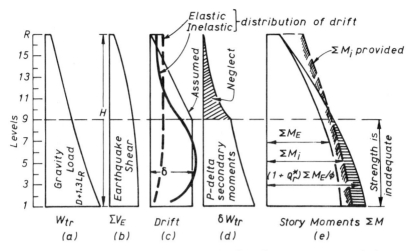

Fig. 4.41 Design quantities relevant to $P\Delta$ effects in a 16-story example frame.

4. At levels in the lower half of the frame only, wherever it was found that $Q_r^* > 0.085$, it should be ascertained that

$$\Sigma\, M_i \geq (1 + Q_r^*)\Sigma\, M_E/\phi \tag{4.43}$$

If this requirement is not satisfied, the flexural reinforcement in some or all of the beams at that floor should be increased. In accordance with the concepts of capacity design philosophy, the columns, supporting these beams, may also need to be checked for correspondingly increased strength. It is advisable to make an estimate for the likely $P\Delta$ moment contribution during the preliminary stages of the design, when deflections are being computed, to establish the fundamental period of the structure. In most cases such estimates may readily be incorporated immediately into the beam design, leading to flexural reinforcement in excess of that required for the appropriate combination of gravity load and lateral force-induced beam moments. As moment redistribution does not change the value of the story moment $\Sigma\, M_E$, it does not affect P–Δ considerations either.

Figure 4.41(d) shows a typical distribution of the secondary moments δW_{tr}. These are based on the assumed uniform distribution of drift in the lower eight stories, and they may be compared in Fig. 4.41(c) with typical drift distributions obtained for elastic and inelastic frames. Figure 4.41(e) compares the required flexural strength $\Sigma\, M_E$ for the specified lateral forces with the ideal flexural strengths that might have been provided $\Sigma\, M_i$, and the total ideal strength required to resist lateral forces plus the $P\Delta$ secondary moments. The shaded area in Fig. 4.41(e) shows, somewhat exaggerated, the magnitudes of the total additional beam flexural strengths that should be provided at different levels in this example frame to compensate for $P\Delta$ moment demands. As Fig. 4.41(e) implies, $P-\Delta$ phenomena are not likely to be critical in the upper half of ductile multistory frames unless column and beam sizes are reduced excessively, whereby story drifts might increase proportionally.

4.8 BEAM-COLUMN JOINTS

4.8.1 General Design Criteria

It is now generally recognized that beam-column joints can be critical regions in reinforced concrete frames designed for inelastic response to severe seismic attack. As a consequence of seismic moments in columns of opposite signs immediately above and below the joint, and similar beam moment reversal across the joint, the joint region is subjected to horizontal and vertical shear forces whose magnitude is typically many times higher than in the adjacent beams and columns. If not designed for, joint shear failure can result. The reversal in moment across the joint also means that the beam reinforcement is required to be in compression on one side of the joint and at tensile yield on the other side of the joint. The high bond stresses required to sustain this force gradient across the joint may cause bond failure and corresponding degradation of moment capacity accompanied by excessive drift. In the following sections, aspects of joint behavior specific to seismic situations only are examined, and subsequently, design recommendations are made. Detailed studies of joints for buildings in seismic regions have been undertaken only in the past 20 years. Therefore, the design of these important components, an aspect which up until a few years ago had been entirely overlooked and which as yet has received no in-depth treatment in standard texts, is discussed in the following sections in considerable detail.

Up until a few years ago there was very little coordination in relevant research work conducted in different countries. This led to design recommendations in some countries which in certain aspects are in conflict with each other. However, between 1984 and 1989 significant efforts, including coordinated experimental work, by researchers from the United States, New Zealand, Japan, and China [A16, K4, K6, K7, P12, S6] were made to identify and to resolve these conflicts.

It is sometimes claimed that the importance of joints in seismic design is overemphasized because there is little evidence from past earthquakes of major damage or collapse that could be attributed to joint failures. This observation is largely due to the inferior standard of design of beams and particularly, the poor detailing of columns. These members thus became the weak links in the structural system. Many failures of framed buildings resulted from soft-story mechanisms in which column failure due to shear or inadequate confinement of the concrete occurred before the development of

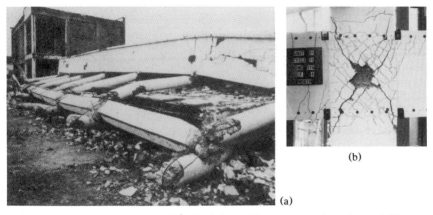

(b)

(a)

Fig. 4.42 Beam–column joint failure in (a) the El Asnam earthquake and (b) a test specimen.

available beam strengths. It has recently been reported, however, that in no other comparable event have as many beam–column joint failures been observed as in the 1980 El Asnam earthquake [B16] [Fig. 4.42(a)]. Shear and anchorage failures, particularly at exterior joints, have also been identified after the 1985 Mexico [M15], the 1986 San Salvador [X14], and the 1989 Loma Prieta [X2] earthquakes.

Criteria for the desirable performance of joints in ductile structures designed for earthquake resistance may be formulated as follows [P18]:

1. The strength of the joint should not be less than the maximum demand corresponding to development of the structural plastic hinge mechanism for the frame. This will eliminate the need for repair in a relatively inaccessible region and for energy dissipation by joint mechanisms, which, as will be seen subsequently, undergo serious stiffness and strength degradation when subjected to cyclic actions in the inelastic range.

2. The capacity of the column should not be jeopardized by possible strength degradation within the joint. The joint should also be considered as an integral part of the column.

3. During moderate seismic disturbances, joints should preferably respond within the elastic range.

4. Joint deformations should not significantly increase story drift.

5. The joint reinforcement necessary to ensure satisfactory performance should not cause undue construction difficulties.

The fulfillment of these criteria may readily be achieved by the application of a capacity philosophy, outlined in previous sections, and the development of practical detailing procedures. These aspects are examined in the following sections.

4.8.2 Performance Criteria

The ductility and associated energy dissipating capacity of a reinforced concrete or masonry frame is anticipated to originate primarily from chosen and appropriately detailed plastic hinges in beams or columns. Because the response of joints is controlled by shear and bond mechanisms, both of which exhibit poor hysteretic properties, joints should be regarded as being unsuitable as major sources of energy dissipation. Hence the response of joints should be restricted essentially to the elastic domain. When exceptionally large system ductility demands (Section 3.5.5) arise, when damage beyond repair to members of a frame may well occur, some inelastic deformations within a joint should be acceptable. Thus the performance criteria for a beam–column joint during testing should conform with those recommended in Section 3.5.6 for confirmation of the ductility capacity of primary ductile components.

It is of particular importance to ensure that joint deformations, associated with shear and particularly bond mechanisms, do not contribute excessively to overall story drifts. When large-diameter beam bars are used, the early breakdown of bond within the joint may lead to story drifts in excess of 1%, even before the yield strength of such bars is attained in adjacent beams. Excessive drift may cause significant damage to nonstructural components of the building, while frames respond within the elastic domain. By appropriate detailing, to be examined subsequently, joint deformations can be controlled. Under the actions of seismic forces, producing moments of the type shown, for example, in Fig. 4.6(d), the component of lateral structural displacement resulting from deformations of well-designed joints will generally be less than 20% of the total displacement [C14]. When estimating the stiffnesses of members, in accordance with Sections 1.1.2(a) and 4.1.3, due allowance should be made for the contribution of joint deformations in order to arrive at realistic estimates of story drifts under the action of lateral forces.

4.8.3 Features of Joint Behavior

Under seismic action large shear forces may be introduced into beam–column joints irrespective of whether plastic hinges develop at column faces or at some other section of beams. These shear forces may cause a failure in the joint core due to the breakdown of shear or bond mechanisms or both.

(a) Equilibrium Criteria As a joint is also part of a column, examination of its function as a column component is instructive. An interior column extending between points of contraflexure, at approximately half-story heights, may be isolated as a free body, as shown in Fig. 4.43(a). Actions introduced by symmetrically reinforced beams to the column are shown in this figure to be internal horizontal tension T_b and compression C_b forces and vertical beam shear V_b forces. Making the approximations that $C_b = T_b$ and that

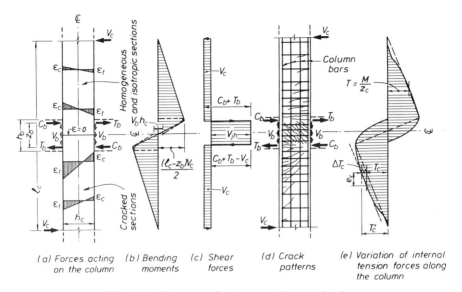

(a) Forces acting on the column   (b) Bending moments   (c) Shear forces   (d) Crack patterns   (e) Variation of internal tension forces along the column

Fig. 4.43 Features of column and joint behavior.

beam shears on opposite sides of the joint are equal, equilibrium of the free body shown requires a horizontal column shear force of

$$V_c = \frac{2T_b z_b + V_b h_c}{l_c} \tag{4.45}$$

where the variables are readily identified in Fig. 4.43(a).

The corresponding moment and shear force diagrams for the column are shown in Fig. 4.43(b) and (c). The large horizontal shear force across the joint region is, from first principles,

$$V_{jh} = C_b + T_b - V_c = \left(\frac{l_c}{z_b} - 1\right)V_c - \frac{h_c}{z_b}V_b \tag{4.46}$$

where the right-hand expression is obtained from consideration of the moment gradient within the joint core. Because the conventionally used full-line moment diagram in Fig. 4.43(b) does not show the moment decrement $h_c V_b$, its slope across the joint does not give the correct value [Eq. (4.46)] of the horizontal joint shear force. The correct moment gradient would be obtained if the moment decrement $h_c V_b$ is, for example, assumed to be introduced at the horizontal centerline of the joint, as shown by the dashed line in Fig. 4.43(b). As Fig. 4.43(c) indicates, the intensity of horizontal shear within the joint V_{jh} is typically four to six times larger than that across the column between adjacent joints V_c.

To appreciate the relative magnitudes of horizontal, vertical, and diagonal forces within the joint regions, some consideration must also be given to strain compatibility. To this end the upper part of Fig. 4.43(a) shows flexural strain distributions across sections of an assumed homogeneous isotropic column, not subjected to axial load. In the lower part of the same figure, strain distributions corresponding to the traditional assumptions of flexurally cracked concrete sections are shown. In both cases strain gradients (i.e., curvatures) at sections are proportional to bending moments at the same levels. Accordingly, strains across the column at the level of the beam centerline should be zero, as shown in Fig. 4.43(a).

The top half of Fig. 4.43(d) shows a typical reinforced concrete column in which a number of approximately horizontal cracks have developed. The total internal tension force T_c, corresponding to a cracked-section analysis, is then $T_c = M/z_c$, as shown in the top half of Fig. 4.43(e), where z_c is the internal lever arm in the column. Where moments are small, uncracked concrete may resist the internal flexural tension force.

The lower part of Fig. 4.43(d) shows a column in which, as a result of the shear force V_c, distinct diagonal cracks have developed. It was shown in Section 3.6.3 that these cracks lead to an increase of internal tension forces with respect to those applied by the bending moment at the relevant section of a member (i.e., $T_c + \Delta T_c$), as shown in the lower half of Fig. 4.43(e). The tension force increment is

$$\Delta T_c \approx \frac{e_v}{z_c} V_c$$

where e_v is the tension shift, defined in Section 3.6.3 and Eq. (3.73). The usual range of e_v/z_c is 0.5 to 1.0.

It is thus seen that the tension force increment ΔT_c is proportional to the shear force across the region considered. It is for this reason that tensile forces in column bars within the joint region, seen in Fig. 4.43(d), are significantly larger than suggested by the bending moment diagram. This increase of tension forces, even at the level of zero moment, as seen in Fig. 4.43(e), leads to a vertical expansion of the joint core. Similar consideration may be used to show that after diagonal cracking of the joint core concrete, simultaneously horizontal expansion will also occur.

(b) Shear Strength   Internal forces transmitted from adjacent members to the joint, as shown in Fig. 4.43(a), result in joint shear forces in both the horizontal and vertical directions. These shear forces lead to diagonal compression and tension stresses in the joint core. The latter will usually result in diagonal cracking of the concrete core, as shown in Figs. 4.42(b) and 4.52. The mechanism of shear resistance at this stage changes drastically.

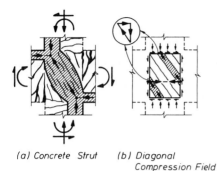

(a) Concrete Strut   (b) Diagonal
                      Compression Field

Fig. 4.44 Mechanisms of shear transfer at an interior joint.

Basic mechanisms of shear transfer are shown in Fig. 4.44. Some of the internal forces, particularly those generated in the concrete, will combine to develop a diagonal strut [Fig. 4.44(a)]. Other forces, transmitted to the joint core from beam and column bars by means of bond, necessitate a truss mechanism, shown in Fig. 4.44(b).

To prevent shear failure by diagonal tension, usually along a potential corner to corner failure plane [Fig. 4.42(b)], both horizontal and vertical shear reinforcement will be required. Such reinforcement will enable a diagonal compression field to be mobilized, as shown in Fig. 4.44(b), which provides a feasible load path for both horizontal and vertical shearing forces. The amount of horizontal joint shear reinforcement required may be significantly more than would normally be provided in columns in the form of ties or hoops, particularly when axial compression on columns is small.

When the joint shear reinforcement is insufficient, yielding of the hoops will occur. Irrespective of the direction of diagonal cracking, horizontal shear reinforcement transmits tension forces only. Thus inelastic steel strains that may result are irreversible. Consequently, during subsequent loading, stirrup ties can make a significant contribution to shear resistance only if the tensile strains imposed are larger than those developed previously. This then leads to drastic loss of stiffness at low shear force levels, particularly immediately after a force or displacement reversal. The consequence is a reduction in the ability of a subassembly to dissipate seismic energy [see Figs. 2.20(d) and (f)].

When sufficient joint shear reinforcement has been provided to ensure that unrestricted yielding of the same cannot occur during repeated development of adjacent plastic hinges in beams, the crushing of the concrete in the joint core due to diagonal compression must also be considered as a potential primary cause of failure. However, this is to be expected only if the average shear and axial compression stresses to be transferred are high. This mode of failure can be avoided if an upper limit is placed on the diagonal compression stress, conveniently expressed in terms of the joint shear stress, that could be developed when the overstrength of the structure is mobilized.

(c) **Bond Strength** The development length specified by building codes for a given size of straight beam bar, as described in Section 3.6.2(*a*), is usually larger than the depth of an adjacent column. At an exterior column the difficulty in anchoring a beam bar for full strength can be overcome readily by providing a standard hook, as seen in Fig. 4.69. At interior columns, however, this is impractical. Some codes [A1] require that beam bars at interior beam–column joints must pass continuously through the joint. It will be shown subsequently that bars may be anchored with equal if not greater efficiency using standard hooks within or immediately behind an interior joint.

The fact that bars passing through interior joints, as shown in Fig. 4.43(*a*), are being "pulled" as well as "pushed" by the adjacent beams, to transmit forces corresponding to steel stresses up to the strain hardening range in tension, has not, as a rule, been taken into account in code specifications until recently. It may be shown easily that in most practical situations bond stresses required to transmit bar forces to the concrete of the joint core, consistent with plastic hinge development at both sides of a joint, would be very large and well beyond limits considered by codes for bar strength development [M14].

Even at moderate ductility demands, a slip of beam bars through the joint can occur. A breakdown of bond within interior joints does not necessarily result in sudden loss of strength. However, bond slip may seriously affect the hysteretic response of ductile frames. As little as 15% reduction in bond strength along a bar may result in 30% reduction in total energy dissipation capacity of a beam–column subassembly [F5]. Because the stiffness of frames is rather sensitive to the bond performance of bars passing through a joint, particularly at interior columns, special precautions should be taken to prevent premature bond deterioration in joints under seismic attack. At exterior joints, anchorage failure of beam bars is unacceptable at any stage because it results in complete loss of beam strength.

The bond performance of bars anchored in a joint profoundly affects the relative contribution to shear strength of the two mechanisms shown in Fig. 4.44, and this is examined in Section 4.8.6.

4.8.4 Joint Types Used in Frames

Joints may be classified in terms of geometric configuration as well as structural behavior. Because of fundamental differences in the mechanisms of beam bar anchorages, it is customary to differentiate between interior and exterior joints. Response in both the elastic and inelastic range due to seismic motions need to be examined. Only cast-in-place joints are considered in this study.

(a) **Joints Affected by the Configuration of Adjacent Members** Various types of exterior beam–column joints are shown in Fig. 4.45. Some of these occur

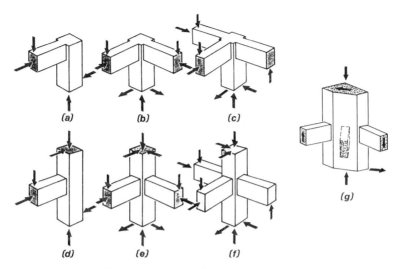

Fig. 4.45   Exterior beam–column joints.

in plane frames [Fig. 4.45(*a*) and (*d*)], and others in two way or space frames. They may occur at the top floor or at intermediate floors. Types (*b*) and (*e*) in Fig. 4.45 are referred to as corner joints. The arrows indicate typical input forces that may be encountered during earthquake attack in a particular direction. Types (*c*) and (*f*) in Fig. 4.45 are referred to as edge joints. Figure 4.46 shows similar variants of interior joints of plane space frames.

The examples above occur in rectilinear frames. It is not uncommon that beams frame into a column at angles other than 90° or 180°. Such an example is shown in Fig. 4.45(*g*), where exterior beams include an angle of 135°. A third beam from the interior of the floor, shown by dashed lines, may also frame into such a column. For the sake of clarity, floor slabs, cast together with the beams, are not shown in these illustrations.

(b)   Elastic and Inelastic Joints   If the design criteria outlined in Section 4.8.1 are to be met, it is preferable to ensure that joints remain essentially in the elastic range throughout the inelastic response of the structure. This may be achieved with relative ease. Mechanisms of shear and bond resistance within the joint are profoundly affected by the behavior of the adjacent beams and columns. When inelastic deformations do not or cannot occur in the beams and columns adjacent to a joint, the joint may readily be reinforced so as to remain elastic even after a very large number of displacement reversals. Under such circumstances, smaller amounts of joint shear reinforcement generally suffice.

As a general rule, when subjected to the design earthquake plastic hinges are expected to develop at the ends of the beams, immediately adjacent to a

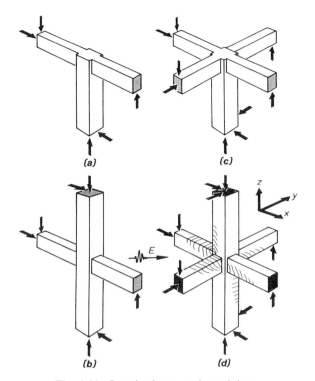

Fig. 4.46   Interior beam–column joints.

joint. In such cases, particularly after a few cycles of excursions into the inelastic range, it is not possible to prevent some inelastic deformation occurring also in parts of the joint. This is due primarily to the penetration of inelastic strains along the reinforcing bars of the beams into the joint. These joints are classified here as "inelastic." They require larger amounts of joint shear reinforcement.

In frames designed in accordance with the weak beam–strong column principle, it is relatively easy to ensure that if desired, the joints remain elastic. This involves the relocation of potential plastic hinges in beams, some distance away from column faces, as shown, for example, in Figs. 4.16 and 4.17. A more detailed study of such joints is presented in Sections 4.8.7(*e*) and 4.8.11(*e*), while typical detailing of the beam reinforcement is shown in Figs. 4.17 and 4.71.

4.8.5   Shear Mechanisms in Interior Joints

It is important, particularly in the case of reinforced concrete beam–column joints, to evaluate, either by reasoning or with carefully planned experimental

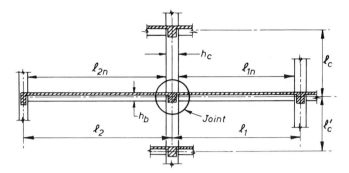

Fig. 4.47   Interior beam–column assemblage.

techniques, the limitations of mathematical modeling. Without a suitable model, designers would be deprived of using rationale in the understanding of joint behavior and innovation in situations where conventional and codified procedures may not offer the best or any solution. In the formulation of mathematical models of joints, by necessity, compromises between simplicity and accuracy must be made. In this endeavor, test observations can furnish invaluable inspiration. The primary role of laboratory tests is, however, to provide evidence for the appropriateness as well as the limits of the mathematical model conceived, so as to enable the designer to use it with confidence in varied situations rather than to provide a wealth of data for the formulation of empirical rules.

It is the aim of this section to postulate models, with the aid of which overall joint behavior, as well as the effects of significant parameters, can be approximated. Some of these models are generally accepted; others are less well known [P1, P18].

(a)   Actions and Disposition of Internal Forces at a Joint   The disposition of forces around and within a joint may be studied by examining a typical interior beam–column assemblage of a plane frame, such as shown in Fig. 4.47. It is assumed that due to gravity loads and earthquake-induced lateral forces on the frame, moments introduced to the joint by the two beams cause rotations in the same sense.

Typical moments and shear and axial forces introduced under such circumstances to an interior joint are shown in Fig. 4.48(a). To enable simple equilibrium requirements, as set out in Section 4.8.3(a), to be satisfied and load paths to be identified, the stress resultants are assembled around the joint core, shown shaded in Fig. 4.48(b). Tensile stress resultants are denoted by T, and the compression stress resultants in the concrete and steel are shown by the symbols C_c and C_s, respectively. Figure 4.48(b) shows a

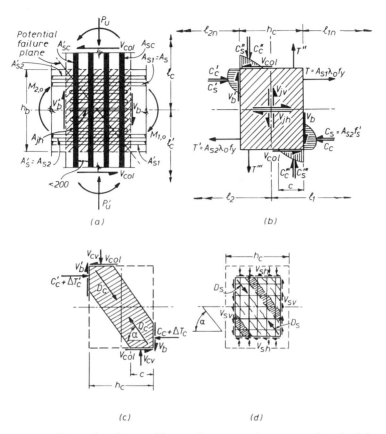

Fig. 4.48 External actions and internal stress resultants at an interior joint.

situation where plastic hinges would have developed in the beams, immediately adjacent to the joint.

(b) Development of Joint Shear Forces By similarity to Eq. (4.46), and by neglecting the horizontal floor inertia force introduced to the joint during an earthquake, the horizontal joint shear force is readily estimated using the notation given in Fig. 4.48(*b*), whereby

$$V_{jh} = T + C'_c + C'_s - V_{\text{col}} = T' + C_c + C_s - V_{\text{col}} \qquad (4.47a)$$

Since the approximation $T' = C'_c + C'_s$ may be made, the joint shear force simplifies to

$$V_{jh} = T + T' - V_{\text{col}} \qquad (4.47b)$$

where V_{col} is the average of column shears above and below the joint estimated by Eq. (4.48).

As the quantities of flexural reinforcement A_{s1} and A_{s2} are known and the tensile steel is assumed to have developed its overstrength $\lambda_o f_y$, as defined by Eq. (1.10) and discussed in Sections 3.2.4(e) and 4.5.1(e), the maximum intensity of the horizontal joint shear force to be used in the design for shear strength is found to be

$$V_{jh} = (A_{s1} + A_{s2})\lambda_o f_y - V_{col} = (1 + \beta)\lambda_o f_y A_{s1} - V_{col} \quad (4.47c)$$

where $\beta = A_{s2}/A_{s1}$.

The column shear force V_{col} is readily derived from the computed beam flexural overstrength at the column faces, shown as $M_{1,o}$ and $M_{2,o}$ in Fig. 4.48(a), or corresponding scaled beam bending moments at column center lines [Figs. 4.10 and 4.18], may be used. With the dimensions shown in Fig. 4.47, it is then found that

$$V_{col} \approx 2\left(\frac{l_1}{l_{1n}}M_{1,o} + \frac{l_2}{l_{2n}}M_{2,o}\right)\bigg/(l_c + l'_c) \quad (4.48)$$

The consideration of the equilibrium of vertical forces at the joint of Fig. 4.48 would lead to expressions for the vertical joint shear forces V_{jv} similar to Eqs. (4.47a) to (4.47c). However, because of the multilayered arrangement of the column reinforcement, the derivation of vertical stress resultants is more cumbersome. By taking into account the distances between stress resultants and the member dimensions shown in Fig. 4.48(b), for common design situations it will be sufficiently accurate to estimate the intensity of the vertical joint shear force thus:

$$V_{jv} = (h_b/h_c)V_{jh} \quad (4.49)$$

It is now necessary to estimate the contribution to shear resistance of each of the basic mechanisms shown in Fig. 4.44.

(c) Contribution to Joint Shear Strength of the Concrete Alone Compression forces introduced to a joint by a beam and a column at diagonally opposite corners of the joint, and combined into a single diagonal compression force D_c, are shown in Fig. 4.48(c). At the upper left-hand corner of the joint, C'_c and C''_c represent the stress resultants of the corresponding concrete stress blocks in Fig. 4.48(b). It is reasonable to assume that the shear forces developed in the beam, V'_b, and the column, V_{col}, are predominantly introduced to the joint via the respective flexural compression zones.

The total horizontal force to be transmitted by the beam top flexural reinforcement to the joint by means of bond is $T + C'_s$. A fraction of this force, $\Delta T_c'$, to be examined more closely, will be transmitted to the diagonal strut in the shaded region seen in Fig. 4.48(c). Similarly, a fraction of the total force, $T''' + C_s''$ (i.e., $\Delta T_c''$), developed in the vertical column bars may be transmitted to the same region of the joint. Similar concrete compression, shear, and bond forces at the lower right-hand corner of the joint [Fig. 4.48(c)] will combine into an equal and opposing diagonal compression force D_c.

Provided that the compression stresses in the diagonal strut are not excessive, which is usually the case, the mechanism in Fig. 4.48(c) is very efficient. The contribution of this strut to joint shear strength may be quantified by

$$V_{ch} = D_c \cos \alpha \qquad (4.50a)$$

and

$$V_{cv} = D_c \sin \alpha \qquad (4.50b)$$

When axial force is not applied to the column, the inclination of the strut in Fig. 4.48(c) is similar to that of the potential failure plane in Fig. 4.48(a). In this case $\alpha \approx \tan^{-1}(h_b/h_c)$. With axial compression in the column, transmitted through the joint, the inclination of this strut will be steeper. The forces V_{ch} and V_{cv} will be referred to subsequently as contributions of the concrete strut mechanism to joint shear strength.

(d) Contribution to Joint Shear Strength of the Joint Shear Reinforcement   It was postulated that a fraction ΔT_c of the total bond force along beam or column bars is transmitted to the diagonal strut, shown in Fig. 4.48(c). The remainder, V_{sh}, of the bond force, for example in the top beam bars where $V_{sh} = T + C_s' - \Delta T_c$, is expected to be introduced to the core concrete in the form of shear flow, as suggested in Fig. 4.48(d). Similar bond forces, introduced to the concrete at the four boundaries of the joint core model in Fig. 4.48(d), being in equilibrium, will generate a total diagonal compression force D_s. The contribution of joint shear reinforcement in the horizontal direction can thus be expressed as

$$V_{sh} = D_s \cos \alpha = V_{jh} - V_{ch} \qquad (4.51)$$

The contribution of the vertical joint shear reinforcement is evident from the model shown in Fig. 4.48(d). In deriving the necessary amount of vertical joint reinforcement vertical axial forces acting at the top and bottom of the panel in Fig. 4.48(d) need also be considered. Based on the assumption that the inclination of diagonal α in both mechanisms is the same, this aspect is considered further in Section 4.8.7(b).

Contributions of the mechanism shown in Fig. 4.48(d), as discussed above, are based on the assumption that in a thoroughly cracked core of a joint, no (diagonal) tensile stresses can be transmitted by the concrete. When beams with very small amounts of flexural reinforcement are used, or when column sections relative to beam sizes are large, joint shear stresses [Eq. (4.73)] may be rather small, and no or very few diagonal cracks may develop. As the concrete core in such cases will resist shear by means of diagonal tensile stresses, the truss mechanism in Fig. 4.48(*d*) will hardly be mobilized.

The assignment of total joint shear force to the two shear resisting mechanisms, shown in Fig. 4.44, is an important design step. The strut mechanism does not rely on steel contributions, but the truss mechanism may require considerable amounts of reinforcement, particularly in the horizontal direction.

4.8.6   Role of Bar Anchorage in Developing Joint Strength

(*a*)   *Factors Affecting Bond Strength*   Some established aspects of bond, as it affects the flexural reinforcement in beams and columns, were discussed in Section 3.6.2. However, our knowledge of the mechanisms of bond in a seismic environment is not as extensive as many other aspects of reinforced concrete behavior. Significant advances have been made in identifying various factors that affect the bond strength and bond–slip relationship for bars subjected to high-intensity reversed cyclic forces [C10, E1, F5, P54]. Such research findings cannot be readily translated into relatively simple design recommendation because it has been found difficult to formulate both mathematical models and experimental simulation which would adequately resemble the severe conditions prevailing in the concrete core of a beam–column joint, such as seen in Fig. 4.52.

It is established that bond deterioration begins as soon as the yield strain in the steel is exceeded at the locality under consideration. Therefore, in elastic joints higher average bond stresses can be maintained. Bond deterioration due to plastic strains in a bar embedded in a joint core can contribute up to 50% of the overall deflections in beam–column subassemblages [S9].

The conditions for bond change along the embedment length of a beam bar passing through a joint. Bond is rapidly reduced in the concrete that is outside the joint core [Figs. 4.49(*b*) and 4.50(*e*)]. Within the column (joint) core, high bond stresses can be developed, because some confinement of the concrete perpendicular to the beam bar is always present. A transition region between these two zones exists [E1]. Therefore, the bond–slip relationship varies along the bar depending on the region of embedment that is considered. For this reason it is difficult, if not impossible, to model in a simple way, suitable for routine design, the global bond–slip response of bars anchored in an interior joint.

Experimental evidence with respect to simulated seismic bond response is usually obtained from tests with conditions as shown in Fig. 4.49(*a*). These are similar to traditional pull-out tests [P1]. Bond stresses in such tests are uniformly distributed around the periphery of a bar, causing uniform tangential and radial stresses in the surrounding concrete, as shown in Fig. 4.49(*a*).

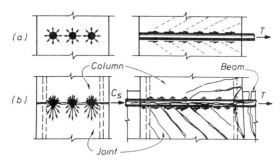

Fig. 4.49 Bond stresses around (*a*) bars simply anchored or (*b*) those passing through an interior joint.

However, a bar situated in the top of a beam as shown in Fig. 4.49(*b*) encounters more unfavorable conditions within a joint. Because of the very large bond force, $\Delta T = T + C_s$, to be transferred and also because the transverse tension at the right hand of the joint imposed by the column flexure, a splitting crack along the bar will usually form. From consideration of the horizontal shear across the joint, given by Eq. (4.47), it is evident from Fig 4.48(*d*) that the total bond force from a top beam bar needs to be transferred predominantly downward into the diagonal compression field of the joint core. Therefore, the bond stress distribution around a bar will not be uniform, as in the case of standard tests. Much larger bond stresses will need to be generated in the side of the bar facing the joint core, as suggested in Fig. 4.49(*b*). Any bond force in excess of about 15% of the total which might be transferred toward the column [i.e., from the top half of a bar shown in Fig. 4.49(*b*)] will have to enter the joint core. This involves shear transfer by shear friction across the horizontal splitting crack shown in Fig. 4.49(*b*). The mechanism is not likely to be efficient unless effective confinement across this crack is present.

The uneven distribution of bond stresses around a bar may adversely affect top beam bars, the underside of which may be embedded in concrete of inferior quality due to sedimentation, water gain, and other causes. Parameters that influence the bond response of bars in joints are as follows:

1. *Confinement*, transverse to the direction of the embedded bar, significantly improves bond performance under seismic conditions. Confinement may be achieved with axial compression on the column and with reinforcement that can exert clamping action across splitting cracks. Therefore, intermediate column bars [Fig. 4.48(*a*)], apart from their role as vertical joint shear reinforcement, are essential to prevent premature bond failure in the joints when columns are subjected to small or no axial compression force.

There is an upper limit to confinement. A fully confined environment for bond exists when additional transverse reinforcement or transverse concrete compression due to column axial force does not result in improvement in the local bond–slip relationship [E1]. At this level of confinement, the maximum bond strength is attained and failure occurs as a consequence of crushing of the concrete in front of bar ribs and the breakdown of the frictional shear resistance around the outer diameter of the deformed bar [P1].

2. *Bar diameter*, d_b, has no significant effect on bond strength in terms of bond stress. A variation of about 10% in local bond strength can be expected for common bars of normal sizes, with smaller bars displaying increased strength [E1]. Therefore, if bond stress limits the maximum bond force $\Delta T = T + C_s$ in Fig. 4.49(b), the ratio of the effective embedment length, l_e, to bar diameter, d_b, for bars with identical force input must remain constant, that is,

$$l_e/d_b = \text{constant}$$

In interior beam–column joints the designer would seldom have the opportunity to specify adequate effective embedment length. The available embedment length is fixed, being approximately equal to the dimension of the confined joint core h_c taken parallel with the bar to be anchored [Fig. 4.50(e)]. Thus the designer is restricted to selecting a bar diameter that satisfies the requirement imposed by an appropriate l_e/d_b ratio.

3. The *compression strength of concrete* is not a significant parameter. It has been reconfirmed [E1], as it is implied by most code provisions [A1, X3], that under revised cyclic loading, local bond strength depends more on the tensile strength of the concrete, which is a function of $\sqrt{f_c'}$. Maximum local bond stresses on the order of $2.5\sqrt{f_c'}$ MPA ($30\sqrt{f_c'}$ psi) have been observed [E1].

4. The *Clear distance between bars* affects moderately the bond strength. A reduction in bond strength was found when this clear distance was less than $4d_b$, but the reduction was not more than 20% [E1].

5. *Bar deformation* (i.e., the area of ribs of deformed bars) is the major source of bond strength. Therefore, the rib area will influence both the bond strength and the bond–slip relationship [E1]. Of even greater importance with regard to slip is the direction of casting the concrete relative to the position of the bar and the sense of the force applied to the bar [P1].

(b) ***Required Average Bond Strength*** Experimental work carried out over the last 20 years has revealed that beam bar anchorages at interior joints, to satisfy seismic design criteria, may be based on theoretical bond stresses significantly higher than those implied by code [A1, X3] requirements, as reviewed in Section 3.6.2(a). Tests carried out in New Zealand [B5, B6, C14,

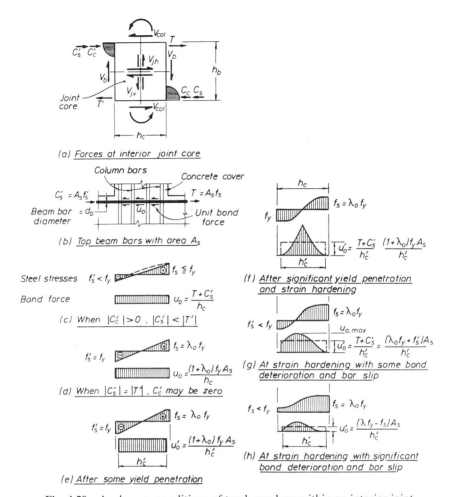

Fig. 4.50 Anchorage conditions of top beam bars within an interior joint.

P13, P18] consistently indicated that ductilities corresponding with a displacement ductility factor of at least $\mu_\Delta = 6$ or interstory drifts of at least 2.5% could be achieved if the bar diameter-to-column depth ratio at an interior joint was limited so that

$$\frac{d_b}{h_c} f_y \leq 11 \quad (\text{MPa}); \qquad \frac{d_b}{h_c} f_y \leq 1600 \quad (\text{psi}) \qquad (4.52)$$

This limit was based on the most severe stress condition, whereby it was assumed that a beam bar passing through a joint was subjected to over-strength $\lambda_o f_y$ simultaneously in tension and compression. In realistic situations this extreme stress condition cannot be expected. As will be shown subsequently and as illustrated for a number of distinct cases in Fig. 4.50, beam bar stresses could approach yield strength only if extremely good anchorage conditions exist. Moreover, this limitation [X3] was based on test observations when the compression strength of the concrete f'_c in the joint was typically 25 MPa (3.6 ksi). Thus, by considering the forces shown in Fig. 4.50 at maximum strength $\lambda_o f_y$ at both ends of a bar, the average unit bond force over the length of the joint h_c is

$$u_o = (T + C'_s)/h_c = 2\lambda_o f_y A_s/h_c \qquad (4.53a)$$

or in terms of average bond stresses,

$$u = \frac{u_o}{\pi d_b} = \frac{d_b}{2h_c}\lambda_o f_y \qquad (4.53b)$$

Based on observed maximum values of local bond stresses that could be developed in tests [E1] simulating average confinement in the core of a beam-column joint, the average usable bond stress can be estimated. Using the stress distribution patterns in Fig. 4.50(g) the average bond force per unit length over the concrete core $0.8h'_c$ is approximately $u'_o \simeq 0.67u_{o,\,\text{max}} = 0.67\pi d_b u_{\text{max}}$. When this is expressed in terms of average bond stress u_a over the length of the joint h_c, the recommended design value is found

$$u_a = 0.67 \times 0.8 \times 2.5\sqrt{f'_c} \approx 1.35\sqrt{f'_c} \text{ (MPa)}; \ 16\sqrt{f'_c} \text{ (psi)} \qquad (4.54)$$

A similar value is obtained from Eqs. (4.52) and (4.53b) when $\lambda_o = 1.25$ when $f'_c \approx 25$ MPa (3.6 ksi).

From first principles the diameter of a bar under the actions shown in Fig. 4.50(a) must be chosen so that when other factors do not affect the quality of bond

$$d_b \geq (T + C'_s)/\pi h_c u_a \qquad (4.55)$$

By limiting the average bond strength u_a to that given by Eq. (4.54) but considering compression steel stresses less than that at overstrength (i.e., $f'_s < \lambda_o f_y$) and other factors that influence bond performance, increased values of d_b/h_c can be estimated.

Further studies [P17] indicated that other parameters, which would allow a relaxation of the severe limitation of Eq. (4.52), should also be taken into account. These are:

1. In most situations stressing of beam bars in compression up to the strain hardening range, gauged by the quantity $\lambda_o f_y$, will not arise [Fig.

4.50(d) and (e)]. This is particularly the case with the top beam reinforcement when its total area A_s is more than that of the bottom reinforcement A'_s (i.e., when $A'_s < A_s$). Clearly, in this case the top bars will develop compression stresses which are less than those in the bottom bars in tension.

2. When gravity loads on beams are significant or dominant in comparison with actions due to earthquake forces, a plastic hinge at the face of the column with the bottom beam bars in tension may never develop. An example of this was given in Fig. 4.7, and further cases are studied in Section 4.9. Maximum tensile steel stresses $f_s < f_y$ can then readily be estimated.

3. Beam bar stress variations within a joint and corresponding bond forces under severe seismic actions are likely to be similar to those shown in Fig. 4.50(g). This implies that limited loss in the efficiency of beam bars developing compression stresses at column faces should be accepted. In view of this, it appears unduly conservative to assume that compression bars will always yield, and that when they do they will enter the strain hardening range.

4. Some allowance for improved bond strength seems justified when the surrounding concrete over a significant length of a beam bar is subjected to transverse compression, such as when the column carries also axial compression load.

5. Precautions should be taken with beam bars passing through joints of two-way frames [Figs. 4.45(e) and (f) and 4.46(d)] that are embedded in concrete that will be subjected to transverse tensile strains. This severe but common situation must be expected when plastic hinges in adjacent beams develop simultaneously at all four faces of the column [Fig. 4.46(d)]. The inferior anchorage of beam bars at interior joints of two-way frames, in comparison with those at joints of one-way frames, has been observed [B7, C14].

By considering all the aspects listed above and assuming that bar stress in compression does not exceed f_y, the basic limitation for the extremely severe case represented by Eq. (4.52) may be modified as follows:

$$\frac{d_b}{h_c} < k_j \frac{\xi_p \xi_t \xi_f}{\xi_m \lambda_o} \frac{\sqrt{f'_c}}{f_y} \qquad (4.56)$$

where $k_j = 5.4$ when stresses are in MPa and 65 when in psi units.

$\xi_m \geq 1.0$, a factor that considers stress levels likely to be developed in an embedded beam bar at each face of the joint as a result of moments, given by Eqs. (4.57c) and (4.57d).

$\xi_p \geq 1.0$, a factor given by Eq. (4.58), recognizing the beneficial effect on bond of confinement by axial compression force on the joint.

$\xi_t = 0.85$ for top beam bars where more than 300-mm (12 in.) of fresh concrete is cast underneath such bars, in recognition of the inferior bond performance in such situations. For other cases $\xi_t = 1.0$.

ξ_f = is a factor that allows for the detrimental effect of simultaneous plastic hinge formation of all four faces of an interior column for which $\xi_f = 0.90$; in all other cases, $\xi_f = 1.0$.

The value of f'_c in Eq. (4.56) should not be taken greater than 45 MPa (6500 psi) unless tests with high strength concrete justifies it.

(i)   Typical Values of ξ_m: By substituting the appropriate values of bar stresses or bar forces, such as shown in Fig. 4.48(*b*), the ratio ξ_m of the maximum bond force $(T + C'_s)_{max}$ to the maximum steel tensile force T can be estimated when $C'_s = \gamma f_y A_b$, thus

$$(T + C'_s)_{max} = \lambda_o f_y A_b + \gamma f_y A_b$$

$$= \left(1 + \frac{\gamma}{\lambda_o}\right)\lambda_o f_y A_b = \xi_m T \qquad (4.57a)$$

$$\text{where } \xi_m = 1 + \frac{\gamma}{\lambda_o} \qquad (4.57b)$$

When the amount of bottom reinforcement $A'_s = A_{s2}$ is less than that of the top beam bars $A_s = A_{s1}$, including any reinforcement in tension flanges which might contribute to the total tension force T in accordance with Section 4.5.1(*b*), expressed by the ratio $\beta = A'_s/A_s$, the compression stress in the top bars can not exceed $f'_s = \beta\lambda_o f_y$.

Moreover, bond stresses and associated slips implied by Fig. 4.50(*g*) are such that after a few cycles of stress reversals the compression stress in the top beam bars is not likely to exceed $0.7f_y$. This has been observed in tests [C14]. Because of the significant slip when top bars are again subject to compression, cracks close before the top bars can yield and thus, even in the case when $\beta = 1.0$, the concrete is contributing to flexural compression. These considerations indicate that the effectiveness of the beam bars in carrying compression stresses may be estimated with the parameter γ where

$$\lambda_o \beta \geq \gamma \leq 0.7$$

The value of ξ_m (Eq. 4.57b) is not overly sensitive to the magnitude of the steel overstrength factor λ_o and hence for its usual range of values ($1.2 \leq \lambda_o \leq 1.4$) and with $\beta = A'_s/A_s > 0.7/\lambda_o$

$$\xi_m = 1.55 \qquad (4.57c)$$

is considered to be an acceptable approximation.

Whenever the bottom beam bars do not yield because a positive plastic hinge can not develop at the column face, $\gamma = f'_s/f_y \leq 0.7$ may be used, if desired, to estimate ξ_m from Eq. (4.57b).

When $\beta \leq 1.0$, the bottom beam bars can be subjected to higher stresses in compression. Hence for the estimation of the total bond force transmitted to the joint core by a bottom bar the following approximation may be used

$$\xi_m = 2.55 - \beta \leq 1.8 \qquad (4.57d)$$

When, for example $\beta \leq 0.75$, using Eq. (4.57d), anchorage for a bottom bar with area A_b should be provided for a total force of $1.8\lambda_o f_y A_b$.

(ii) Value of ξ_p: When under the design moment and the minimum axial design compression load P_u, according to Sections 4.6.5 and 4.6.6, a significant part of the gross column section is subjected to compression stresses, it seems reasonable, when plastic hinges are expected at joint faces, to allow for some enhancement of bond strength due to increased confinement [E1]. Accordingly, it is suggested that

$$1.0 \leq \xi_p = \frac{P_u}{2 f'_c A_g} + 0.95 < 1.25 \qquad (4.58)$$

although little experimental evidence [B5] is currently available to indicate the influence of axial compression on anchorage strength in prototype joints.

Example 4.1 To illustrate the range of limitations on the size of beam bars passing through an interior joint, consider a 610-mm (24-in.)-deep column of a two-way frame subjected to a minimum effective axial compression force of $P_u = 0.25 f'_c A_g$. Hence when evaluating Eq. (4.56) and assuming that $f'_c = 27.5$ MPa (4000 psi), we find from Eq. (4.57c) that $\xi_m = 1.55$, from Eq. (4.58) that $\xi_p = 0.5 \times 0.25 + 0.95 = 1.075$, and for two-way frames that $\xi_f = 0.90$. Hence when using beam bars with $f_y = 275$ MPa (40 ksi) and $\lambda_o = 1.25$, it is found from Eq. (4.56) that for top bars with $\xi_t = 0.85$

$$\frac{d_b}{h_c} = 5.4 \frac{1.075 \times 0.85 \times 0.90}{1.55 \times 1.25} \frac{\sqrt{27.5}}{275} \ (\text{MPa}) \approx 0.044$$

or

$$\frac{d_b}{h_c} = 65 \frac{1.075 \times 0.85 \times 0.90}{1.55 \times 1.25} \frac{\sqrt{4000}}{40,000} \ (\text{psi}) \approx 0.044$$

Hence

$$d_b \leq 0.044 \times 610 = 26.9 \text{ mm } (1.06 \text{ in.})$$

In a one-way frame, 29.9-mm (1.18-in.)-diameter bars could be used.

The limitations are more severe when high-strength steel [f_y = 415 MPa (60 ksi)] with, say, λ_o = 1.25 is used. In this case Eq. (4.56) indicates that beam bars of diameter $d_b \le$ 18 mm (0.71 in.) and $d_b \le$ 20.0 mm (0.80 in.) would need to be used in two- and one-way frames, respectively. For bottom reinforcement bar diameter could be increased by 18%.

(c)   Distribution of Bond Forces Within an Interior Joint   To be able to assign the total joint shear force to the two mechanisms shown in Fig. 4.44, it is necessary to estimate the distribution of bond forces u along bars passing through a joint. As Fig. 4.50 illustrates, the pattern of steel stresses and hence bond forces will change continually as the frame passes through different phases of elastic and inelastic dynamic response. For the purpose of design, a representative, realistic, and not unduly optimistic stress pattern must be selected, and this is attempted in this section.

Figure 4.50(*b*) shows the transfer of unit bond force [Eq. (4.53*a*)] from a beam bar to the column and the joint as a result of the steel forces T and C'_s shown in Fig. 4.50(*a*). In subsequent illustrations, patterns of stresses and bond forces along the same beam bar for various conditions of structural response are shown. For example, Fig. 4.50(*c*) and (*d*) show traditionally assumed distributions of steel stresses and bond forces during elastic response. At the attainment of yield strains and stress reversal the cover concrete over the column bars may not be able to absorb any bond forces, and thus the effective anchorage length of the beam bar, as shown in Fig. 4.50(*e*), is reduced from h_c to h'_c. When reaching the level of strain hardening, further yield penetration and consequent redistribution of bond forces can be expected [Fig. 4.50(*f*)], even if no bar slip occurs at the center of the joint.

In the cases considered so far, perfect anchorage of beam bars was assumed insofar that steel stresses and strains at the center of the joint were very small and no slip of bars at this location would have occurred. In these cases steel tensile and compression strains being approximately equal, the change in the length of a beam bar within the joint h_c is negligible.

Figure 4.50(*g*) shows a more realistic situation, when after some inelastic displacement reversals of a frame, significant tensile yield penetration in the joint core with some deterioration of bond has occurred. As a consequence, compression stresses f'_s in the beam bar are reduced well below yield level. In comparison with the previous cases in Fig. 4.50, a reduction of the total bond force $T + C'_s$ and a significant increase of the bar length within the joint are distinct features of this state. Figure 4.51 shows measured steel strains along a beam bar of a test unit seen in Fig. 4.52. At the development of a displacement ductility of μ = 6, tensile strains of the order of $20\epsilon_y$ where recorded in the beam plastic hinge, while tensile strains at the center of the joint core were in the vicinity of yield strain. Note that on the compression side of the beam bar residual tensile strains are large. The unit behaved very well, with significant bar slip being recorded only at the end of the test.

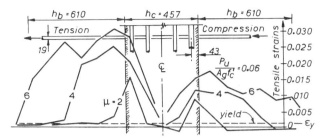

Fig. 4.51   Distribution of measured steel strains along a beam bar passing through plastic hinges and an interior joint [B5].

Fig. 4.52   Crack patterns in an interior beam–column assembly with clearly defined plastic hinges in the beam [B5].

Figure 4.50(g) is considered to be representative of stress distribution when ductilities on the order of 4 to 6 are imposed. Equation (4.56) would allow 19.4 mm (0.76-in.) beam bars to be used in the subframe shown in Fig. 4.52, in which 19-mm (0.75-in.)-diameter bars were used.

Figure 4.50(h) illustrates the case when a major breakdown of bond has occurred. Frame deformations due to significant elongation of the beam bars and slip will be very large [B2]. Following the recommendation presented here should ensure that such a situation will not arise even when during an earthquake extremely large ductility demands are imposed on a frame.

(d) *Anchorage Requirements for Column Bars* Yielding of column bars is not expected when columns are designed in accordance with Section 4.6. Hence bond deterioration resulting from yield penetration is not anticipated. Moreover, because of the greater participation of the concrete in resisting compression forces, bond forces introduced by column bars are smaller than those developed in beam bars. Consequently more advantageous values of the factors in Eq. (4.56) are appropriate. It may be shown that this results in bar diameter to beam depth ratios at least 35% larger than those derived from Eq. (4.56) when the following values are used: $\xi_m = 1.55$, $\xi_p = \xi_t = \xi_f = 1.0$, and $\lambda_o = 1.0$. As a result, in general no special check for usable column bar diameters will need to be made. It is restated, however, that the requirements of Eq. (4.56) do apply when plastic hinges in columns with significant rotational ductility demands are to be expected.

4.8.7 Joint Shear Requirements

Using the model shown in Fig. 4.44, the total shear strength of an interior beam–column joint can be estimated. To this end the actions shown in Fig. 4.48(c) and (d), representing the two postulated components of shear resistance, need be quantified. Accordingly, the joint shear strength can be derived from the superposition of the two mechanisms thus:

$$V_{jh} = V_{ch} + V_{sh} \qquad (4.59a)$$

$$V_{jv} = V_{cv} + V_{sv} \qquad (4.59b)$$

where the subscripts c and s refer to the contribution of the concrete strut and the truss mechanism, respectively. The latter requires joint shear reinforcement [Fig. 4.44(b)] in the horizontal and vertical directions.

The primary purpose of the discussion on the influence of bond performance on shear mechanisms in Section 4.8.6(c) was to enable a rational approach to be developed in assigning the appropriate fractions of the total joint shear to these two mechanisms. In the following, joints with adjacent beam plastic hinges will be considered. Columns designed in accordance with Section 4.6 are assumed to remain elastic. In cases where columns rather than beams are designed to develop plastic hinges, the roles of the two members, in terms of joint design, are simply interchanged.

(a) *Contributions of the Strut Mechanism* (V_{ch} *and* V_{cv}) The horizontal component of the diagonal compression strut D_c in Fig. 4.48(c) consists of a concrete compression force C_c, a steel force ΔT_c transmitted by means of bond to the strut approximately over the depth c of the flexural compression zone in the column, and the shear force from the column V_{col}. Hence

$$V_{ch} = C_c' + \Delta T_c' - V_{col} \qquad (4.60)$$

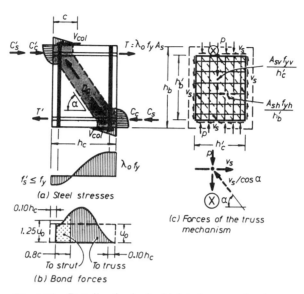

Fig. 4.53 Details of principal joint shear mechanisms.

It is now necessary to make a realistic estimate with respect to the magnitudes of C_c' and $\Delta T_c'$. The fraction of the total bond force at the top beam bars $\Delta T_c'$ that can be transmitted to the strut depends on the distribution of bond forces along these bars (Fig. 4.50). Realistic distributions of steel stresses and bond forces were suggested in Fig. 4.50(g). Such distribution of steel stresses and corresponding bond forces is simulated in Fig. 4.53(a) and (b), where it is assumed that no bond is developed within the cover over an assumed thickness of $0.1h_c$ and that some tension yield penetration into the joint core also occurs. The maximum value of the horizontal bond force introduced over the flexural compression zone of the column in Fig. 4.53(b), is approximately 1.25 times the average unit bond force u_o, expressed by Eq. (4.53a). It may then be assumed that the bond force is introduced to the diagonal strut over an effective distance of only $0.8c$, where c is the depth of the flexural compression zone of the elastic column, which can be approximated by

$$c = \left(0.25 + 0.85\frac{P_u}{f_c' A_g}\right)h_c \qquad (4.61)$$

where P_u is the minimum compression force acting on the column. With these approximations,

$$\Delta T_c' = (1.25u_o)(0.8c) = u_o c = (C_s' + T)\frac{c}{h_c} \qquad (4.62)$$

It is now necessary to estimate what the relative magnitudes of the internal forces C'_c and C'_s in Fig. 4.53 might be. Because bond deterioration along beam bars precludes the development of large steel compression stresses f'_s [Fig. 4.53(a)], it will be assumed that the steel compression force is limited and that it is not exceeding that based on yield strength f_y. The assumptions made here are the same as those used to estimate in Section 4.8.6(b) the maximum usable diameter of beam bars to ensure that premature anchorage failure in the joint core will not occur. Hence from Eq. (4.57a) we find that using the total beam reinforcement rather than just one bar

$$C'_s = \gamma f_y A_b = \frac{\gamma}{\lambda_o} T \qquad (4.63)$$

where from Fig. 4.53 the maximum tension force applied to top beam reinforcement is

$$T = \lambda_o f_y A_b$$

and from Eq. (4.57a)

$$C'_s + T = \left(1 + \frac{\gamma}{\lambda_o}\right) T$$

By considering the range of values of typical steel overstrength factor $1.2 < \lambda_o < 1.4$ and relative compression reinforcement contents in beam sections $0.5 \le \beta = A'_s/A_s \le 1.0$, it was shown in Section 4.8.6(b) and Eq. (4.57c) that $\gamma/\lambda_o = 0.55$ appears to be an acceptable approximation. Hence with this value, we find from Eq. (4.62) that

$$\Delta T'_c = 1.55 \frac{c}{h_c} T \qquad (4.64)$$

and that from Fig. 4.53 the concrete compression force becomes

$$C'_c = T' - C'_s = \beta T - \frac{\gamma}{\lambda_o} T$$

$$= (\beta - 0.55) T \qquad (4.65)$$

Therefore the contribution of the strut mechanism shown in Fig. 4.53 is by substituting into Eq. (4.60)

$$V_{ch} = (\beta - 0.55) T + 1.55 \frac{c}{h_c} T - V_{col}$$

$$= \left(1.55 \frac{c}{h_c} + \beta - 0.55\right) T - V_{col} \qquad (4.66)$$

With the evaluation of Eq. (4.59a) the joint shear resistance to be assigned to the truss mechanism V_{sh} can now be obtained. However, it is more convenient to bypass this intermediate step and obtain V_{sh} directly.

The joint shear force to be resisted by the truss mechanism in Fig. 4.53(c) is found from Eq. (4.59a).

$$V_{sh} = V_{jh} - V_{ch}$$

$$= [(1 + \beta)T - V_{col}] - \left[\left(1.55\frac{c}{h_c} + \beta - 0.55\right)T - V_{col}\right]$$

$$= 1.55\left(1 - \frac{c}{h_c}\right)T \qquad (4.67a)$$

When expressing the ratio c/h_c from Eq. (4.61), the value of V_{sh} becomes simply

$$V_{sh} \approx \left(1.15 - 1.3\frac{P_u}{f'_c A_g}\right)T \qquad (4.67b)$$

When, for example, the axial load on the column produces an average compression stress $f_c = 0.115f'_c$, joint shear reinforcement will need to be provided to resist a shear force equal to the maximum tensile force in the top beam reinforcement at overstrength, that is $T = \lambda_o f_y A_s$.

To appreciate the relative contribution of the two mechanisms to the total joint shear force, relevant forces in typical frames may be compared. In such frames it is found that the beam depth h_b to story height l_c is such that $V_{col} \approx 0.15(1 + \beta)T$ and hence $V_{jh} = 0.85(1 + \beta)T$ and thus

$$\frac{V_{sh}}{V_{jh}} = \frac{1.15 - 1.3(P_u/f'_c A_g)}{0.85(1 + \beta)} \qquad (4.68)$$

This shows that when for example $P_{min} = P_u = 0.1f'_c A_g$ and hence from Eq. (4.67) $V_{sh} = 1.02T$, the truss mechanism will need to resist 60 to 80% of the total horizontal joint shear force when $1.0 \geq \beta = A'_s/A_s \geq 0.5$. However, it is more important to note that the total force V_{sh} is independent of the amount of bottom reinforcement used, expressed by the ratio β, and is simply a fixed fraction of the maximum tensile force developed by the top reinforcement T. As Eq. (4.67b) shows, this fraction diminishes with increased axial compression load P_u on the column.

The contribution of the strut mechanism to vertical joint shear resistance

$$V_{cv} = V_{ch} \tan \alpha \approx \frac{h_b}{h_c} V_{ch} \qquad (4.69)$$

results from considerations of equilibrium [Figs. 4.44 and 4.48(c)]. However, V_{sv}, which could now be obtained from Eq. (4.59b), should not be used to determine the necessary amount of vertical joint reinforcement, because axial forces, shown in Figs. 4.44(b) and 4.53(c), need also be taken into account.

(b)   Contributions of the Truss Mechanism (V_{sh} and V_{sv})

(i)   Horizontal Reinforcement: The horizontal joint shear force V_{sh} to be resisted by the truss mechanism [Figs. 4.44(b) and 4.48(d)] may now be readily obtained from Eq. (4.67b), and thus the required amount of joint shear reinforcement is found from

$$A_{jh} = V_{sh}/f_{yh} \qquad (4.70a)$$

where f_{yh} is the yield strength of the joint shear reinforcement. By expressing T [Fig. 4.53(a)], Equation (4.67b) may be conveniently rearranged so that

$$A_{jh} = \left(1.15 - 1.3 \frac{P_u}{f_c' A_g}\right) \frac{\lambda_o f_y}{f_{yh}} A_s \qquad (4.70b)$$

where A_s is the area of the effective top beam reinforcement passing through the joint. Using the previous example in which $P_u = 0.1 f_c' A_g$ and assuming that $f_y = f_{yh}$ with $\lambda_o = 1.25$, Eq. (4.70b) indicates that $A_{jh} \geq 1.28 A_s$.

This reinforcement should be placed in the space within the outer layers of bars in beams so that the potential corner to corner failure plane across the joint is effectively crossed, as shown in Fig. 4.48(a). The purpose of the joint shear reinforcement is to sustain the assigned shear forces in accordance with the model of Fig. 4.48(d), with restricted yielding.

In evaluating the area of horizontal joint reinforcement, A_{jh}, provided normally by horizontal stirrup ties, it is necessary to:

1. Consider the effective area of legs across the potential failure plane in accordance with the orientation of individual tie legs with respect to that plane. For example, the effective area of the two legs of a square diamond-shaped tie, as seen in Fig. 4.31(b), would be $\sqrt{2} A_b$.

2. Only include ties that are placed within the effective width, b_j, of the joint. For example, tie legs placed outside the densely shaded area in Fig. 4.55(a) should be disregarded.

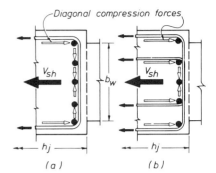

Fig. 4.54 Development of diagonal compression forces necessary to transmit bond forces from column bars in a joint.

3. Place horizontal sets of ties approximately uniformly distributed over the vertical distance between the top and bottom beam bars which pass through the joint. Ties placed very close to the beam flexural reinforcement are inefficient in resisting shear.

4. Consider only those of the shorter ties which in their given position can effectively cross the potential diagonal failure plane. It is advisable to discount those tie legs which, in the direction of the applied horizontal joint shear force, are shorter than one-third of the corresponding joint dimension, h_j. The estimation of effective tie legs, such as in Fig. 4.31(b), is shown in Section 4.11.9(a)(2)(i).

5. Enable bond forces from all column bars in opposite faces of the joint to be effectively transmitted to the diagonal compression field modeled in Fig. 4.48(d). Therefore, multilegged stirrup ties will be more efficient than only perimeter ties. The differences in developing diagonal struts in the joint core, required for each column bar to transmit the necessary bond forces, are shown in Fig. 4.54 for two cases of tie arrangements.

6. The horizontal joint shear reinforcement should not be less than that required in the end regions of adjacent columns, as outlined in Section 4.6.11(d), to provide lateral support to those column bars, typically at the four corners of the column section, which could possibly buckle.

(ii) Vertical Joint Shear Reinforcement: The vertical joint shear reinforcement consists commonly of intermediate column bars, that is, all vertical bars other than those placed in the outermost layers in Fig. 4.48(a). Because columns are expected to remain elastic, the intermediate column bars in particular will not be highly stressed in tension. Thus they can be expected to contribute to the truss mechanism of the joint shear resistance. Hence when four-bar columns, or columns with layers of vertical bars along two opposite faces only, are used, special vertical joint shear reinforcement is required. This may consist of a few larger-diameter bars, similar in size to those used for the main reinforcement of the column, extending at least by development

length beyond the ends of the joint. Alternatively, such short bars may have 90° standard hooks bent toward the column core above and below the horizontal beam reinforcement.

The horizontal spacing of vertical joint shear reinforcement will normally be dictated by the arrangement of column bars above and below the joint to satisfy the requirements of Section 4.6.9. In any case, at least one intermediate bar at each side of the column should be provided.

The derivation of the amount of necessary vertical joint reinforcement may be based on the truss model reproduced in Fig. 4.53(c). By providing horizontal ties with total area A_{jh} [Eq. (4.70)], the shear flow is $v_s = V_{sh}/h'_c$, and by considering the equilibrium of one node of the truss model, marked X in Fig. 4.53(c), with diagonal compression inclined at an angle α, it is found that

$$p + \frac{A_{jv} f_{yv}}{h'_c} - v_s \tan \alpha = 0$$

and hence

$$A_{jv} = \frac{1}{f_{yv}} (V_{sh} \tan \alpha - h'_c p) \tag{4.71}$$

where $\tan \alpha = h'_b/h'_c \approx h_b/h_c$.

The vertical force $h'_c p$ consists of the axial force P_u acting on the column and the difference between the vertical forces $v_s h'_b$ [Fig. 4.53(c)] and T''' [Fig. 4.48(b)] when no axial force is applied. When evaluating these forces, again using equilibrium criteria only, and substituting into Eq. (4.71) with the sole approximation that $V_b \approx V'_b$ [Fig. 4.48(b)], the area of vertical joint shear reinforcement is found:

$$A_{jv} = \frac{1}{f_{yv}} \left[0.5(V_{jv} + V_b) - P_u \right] \tag{4.72a}$$

When Eq. (4.72a) becomes negative, obviously no vertical joint shear reinforcement will be required. This stage represents the limit of the truss model chosen because the inclination of the compression field becomes larger than that in Fig. 4.53(c).

In the presence of some axial compression force on interior columns, the requirements of Eq. (4.72a) are in general readily met with the use of intermediate column bars. Vertical reinforcement in columns designed in accordance with Section 4.6 will have ample reserve strength to absorb the vertical tension forces due to joint shear. Outer elastic column bars may also contribute to restricting the vertical growth of the joint due to shear. In view of this, and since unless very short span beams are used, the earthquake-induced beam shear V_b is very small in comparison with the joint shear V_{jv},

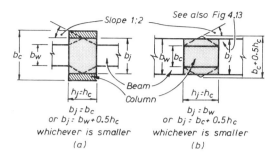

Fig. 4.55   Assumptions for effective joint area.

Eq. $(4.72a)$ may be further simplified to

$$A_{jv} \geq \frac{1}{f_{yv}} (0.5V_{jv} - P_u) \qquad (4.72b)$$

(c)   Joint Shear Stress and Joint Dimensions   To gauge the relative severity of joint shear forces, it is convenient to express this in terms of shear stresses. Because of the different mechanisms engaged in shear transfer after the onset of diagonal cracking in the core concrete of the joint, no physical meaning should be attached to shear stress. It should be considered only as a useful index of the severity of joint shear forces.

The cross-sectional area over which these forces can be transferred cannot be defined uniquely. For convenience, joint shear stresses may be assumed to be developed over the gross concrete area of the column. The accuracy in quantifying the assumed effective joint shear area, used for computing shear stress, is of no importance in design, provided that the assumptions are rational and the limiting shear stresses so derived are appropriately calibrated.

Effective horizontal joint shear area, based largely on engineering judgment, is shown in Fig. 4.55 for two typical cases. The effective width of a joint, b_j, may be taken as the width of the narrower member plus a distance included between lines on a slope of 1 in 2, as shown by the dashed lines in Fig. 4.55. The assumed width of a joint in the case when a narrow beam frames into a wide column is shown in Fig. 4.55(a). The case when the width of a beam is greater than that of the column is shown in Fig. 4.55(b). The length of the joint core, h_j, is taken as the overall depth of the column, h_c, measured parallel to the beam or beams that frame into the column. Reinforcement in both the horizontal and vertical directions located within this arbitrarily defined joint area, $b_j h_j$, may then considered to be effective in contributing to joint strength. Hence the nominal horizontal joint shear stress can be expressed as

$$v_{jh} = V_{jh}/b_j h_j \qquad (4.73)$$

The effective interchange of internal forces within the joint is possible only if the width of the weaker members is not significantly larger than that of the stronger members. Preferably, beam width should be smaller, so that all main beam bars can be anchored in the joint core. Recommendations in this respect are shown in Fig. 4.13.

It should be appreciated that bond forces, to be transferred to the joint core from beam bars that are placed outside the faces of the column, generate horizontal shear stresses in poorly confined concrete. Notwithstanding the limitations of Fig. 4.55, it is therefore suggested that at least 75% of the beam flexural reinforcement should pass through the joint within the column core (i.e., within peripherally placed and tied vertical column bars).

The effective interaction of beams with columns under seismic attack must also be assured when the column is much wider than the beam. Clearly, concrete areas or reinforcing bars in the column section, at a considerable distance away from the vertical faces of the beams, will not participate efficiently in resisting moment input from beams. Hence the effective width of the column, considered to be utilized to resist moments from adjacent beams at a floor, should not be taken larger than b_j, shown in Fig. 4.55(a). This problem may also arise in eccentric beam–column joints (Fig. 4.63) to be examined in Section 4.8.9(c).

(d)   Limitations of Joint Shear   The maximum shear in reinforced concrete joints may be considered to be governed by two criteria.

(i)   Amount of Joint Shear Reinforcement: Large shear forces across interior joints will require very large amounts of joint shear reinforcement. The placing of numerous legs of horizontal stirrup ties may present insurmountable construction difficulties.

(ii)   Diagonal Compression Stresses: As Fig. 4.48 shows, the axial compression load from the columns and the joint shear forces are transmitted by diagonal compression in the core concrete. When the joint shear force is large and extensive diagonal cracking in both directions has occurred in the joint core, as seen for example in Fig. 4.52, the strength of the diagonal compression field rather than the joint reinforcement may control the strength of the joint. To safeguard the joint core against diagonal crushing, it is necessary to limit the magnitude of the horizontal joint shear stress. In determining this limit, the following aspect should be considered:

1. It is well established [C3] that the angle of diagonal compression within the chosen shear mechanism should be kept within certain limits. Typical

inclinations α in beam–column joints, as suggested in Fig. 4.53, are well within these limits. Moreover, it is to be recognized that tensile strains in both the horizontal and vertical directions, as a result of the functioning of the joint shear reinforcement, reduce the diagonal compression strength f'_c of the concrete. In this respect, tensile strains imposed by the horizontal joint ties are of particular importance. If horizontal inelastic strains are permitted to develop, with repeated cycles diagonal compression failure of the core concrete will eventually occur [S18].

2. Diagonal cracks will develop in two directions as a consequence of reversing earthquake forces. The phenomenon is likely to reduce further the compression strength of the concrete [S18].

To avoid brittle diagonal compression failure in joints, it is suggested that the horizontal shear stress computed with Eq. (4.73) be limited in joints of one-way frames to

$$v_{jh} \le 0.25 f'_c \le 9 \text{ MPa (1300 psi)} \tag{4.74a}$$

Shear stresses of this order will generally necessitate large amounts of joint shear reinforcement, the placing of which, rather than Eq. (4.74), is likely to govern the design of joints.

Reinforcement in the joint not less than 50% of that required by Eq. (4.70b), transverse to the horizontal joint shear reinforcement, should also be provided to confine the diagonally compressed concrete.

When small amounts of flexural reinforcement are used in beams, or when columns are much larger than the beams, joint shear stress may be very small. As a consequence, there will be very little or no diagonal cracking within the joint core, even when plastic hinges develop in the beams. For these cases the mechanisms postulated in previous sections would furnish overly conservative predictions for joint shear strength.

(e) Elastic Joints When the design precludes the formation of any plastic hinges at a joint, or when all beams at the joint are detailed so that the critical section of the plastic hinge is located at a distance from the column in accordance with Fig. 4.17, the joint may be considered to remain elastic. For typical interior elastic joints the contribution of the strut mechanism to joint shear strength can be estimated by

$$V_{ch} = 0.5\left(\beta + 1.6\frac{P_u}{f'_c A_g}\right)V_{jh} \tag{4.75}$$

The required joint shear reinforcement is then found using Eqs. (4.59a) and (4.70a).

Elastic joints enable the quantity of joint shear reinforcement to be reduced, particularly when $A'_s/A_s = \beta > 0.7$. Moreover, because bond forces to be transmitted from beam bars to the joint core are reduced, larger-diameter beam bars can be used. In estimating the maximum size of bar with Eq. (4.56),

$$\xi_m = (f_s + f'_s)/(\lambda_o f_y) \le 1.2 \qquad (4.76)$$

may be used, where f_s and f'_s are the estimated tensile and compression stress in the beam bars at the two column faces. The maximum suggested value $\xi_m = 1.2$ will allow bars 30% larger than those required according to Eq. (4.56) to be used. For the determination of vertical joint shear reinforcement Eq. (4.72b) may be used.

4.8.8   Joints in Two-Way Frames

An idealized interior joint in a two-way frame, subjected to the actions of an earthquake attack, is shown in Fig. 4.46(d). For the sake of clarity, the floor slab is not shown. The possibility of plastic hinges developing in beams simultaneously at all four faces of the column during large inelastic frame displacement must be recognized. This is a consequence of the design approach relying on ductility, where strength can be developed at a fraction of the design earthquake intensity in each direction. Obvious computational difficulties arise. To overcome these, a simplified approach to the design of such joints is suggested in the following.

The mechanism of shear resistance in a joint of a space frame is likely to be similar to that described previously for plane frame joints, except that the orientation of critical failure planes is different. Figure 4.56(a) shows compression blocks that could develop at the six faces of a joint core. Thus a

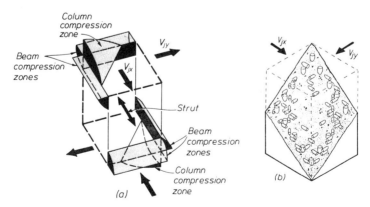

Fig. 4.56 Components of a strut mechanism and potential failure plane in an interior joint of a two-way frame.

diagonal compression strut could develop approximately across diagonally opposite corners. The exact nature of this strut is complex.

When shear reinforcement in the joint is insufficient, a diagonal failure plane, such as shown in Fig. 4.56(b), could develop. If conventional horizontal ties with sides parallel to the faces of the column are used, only one leg of each tie in each direction will be crossed by this plane. Moreover, the legs cross this plane at considerably less than 90°, so that they resist shear with reduced efficiency.

It is more convenient to consider two truss mechanisms acting simultaneously at right angles to each other. This approach utilizes all tie legs in the joint. It enables the joint shear strength to be considered separately in each of the two principal directions. The only modification that needs to be made to the application of the procedures outlined in Section 4.8.7 is in the allowance for the contribution of axial compression on the column assessing joint shear resistance. In two-way frames it would be inappropriate to assume that the full contribution of the axial compression to joint shear strength, given in Eqs. (4.67), (4.68), (4.70b), and (4.72), would exist simultaneously in both diagonal compression fields in perpendicular planes. An approximation may therefore be made by replacing P_u in these equations by $C_j P_u$, where

$$C_j = \frac{V_{jh}}{V_{jx} + V_{jy}} \qquad (4.77)$$

is a factor that apportions beneficial effects of the axial compression P_u in the x and y directions [Fig. 4.46(d)], respectively. V_{jx} and V_{jy} [Fig. 4.56(a)] are the horizontal joint shear forces derived independently from beam moment inputs at overstrengths in the x and y directions. With this reduced effective axial load, the required joint shear reinforcement in the two principal directions may be independently calculated as for one-way frames. Joints so designed performed very satisfactorily when subjected in tests to simulated bidirectional earthquake attack [C14].

It has been suggested that joints of two-way frames benefit significantly from the confinement provided by transverse beams. Accordingly, it has been recommended [A1, A3] that considerably less joint shear reinforcement is required in such joints than in identical joints of one-way frames. These recommendations were justified on the basis of tests in which short stub beams, simulating unloaded transverse beams, were used. However, when plastic beam hinges can develop on all four forces of interior columns, the joint will be dilated simultaneously in both horizontal directions. This is evident from postulated steel stress variations along beam bars [Fig. 4.50(g)], as well as from observed steel strain distributions within the joint core (Fig. 4.51). Therefore, conditions in inelastic joints in two-way frames for both shear and bond strength are more adverse than those in joints of one-way frames [B5, C14].

Since the diagonally compressed concrete in the core of two-way joints is subjected to tensile strains in both horizontal directions, joint shear stresses in the x and y directions should be limited to less than that permitted by Eq. (4.74a) for one-way joints. Hence

$$v_{jx} \text{ or } v_{jh} \leq 0.2 f'_c < 7 \text{ MPa } (1000 \text{ psi}) \qquad (4.74b)$$

4.8.9 Special Features of Interior Joints

(a) Contribution of Floor Slabs   If it is to be assured that a beam–column joint does not become the weakest link, the maximum strengths of the adjoining weakest members, normally the beams, must be assessed. This also involves estimation of the contribution of slab reinforcement, placed parallel with the beam in question to the flexural overstrength of that beam. Corresponding recommendations have been made in Section 4.5.1(b) and Fig. 4.12 as to the effective tension flanges of beams that should be assumed in design.

In this section the perceived mechanism of tension flange contributions and the introduction of corresponding forces to an interior beam–column joint is reviewed briefly. Only expected behavior at the development of large curvature ductilities in beam hinges at column faces is considered. Corresponding crack patterns in the slab of an isolated test beam–column–slab subassemblage [C14] are seen in Fig. 4.57. The test unit has been subjected to multidirectional displacements with progressively increasing displacement ductilities and with applied forces as shown in Fig. 4.46(d) [A16].

Idealized actions on a slab quadrant (Fig. 4.57), considered as a free body, are shown in Fig. 4.58. The tension forces T_x are associated with significant yielding of the slab bars. The resulting cracks may be assumed to be large enough to inhibit significant shear forces in the Y direction to be introduced to the slab edge in Fig. 4.58(a). It is seen that the total tension force ΣT_x, applied to the north–south edge, gives rise to shear force and moment M at the east–west edge of the quadrant. To enable the tensile forces T_x developed in the slab bars after diagonal cracking of the slab (Fig. 4.57) to be transmitted to the top fiber of the east–west beam, a diagonal compression field, as shown in Fig. 4.58(b), needs to be mobilized. Points A and B in Fig. 4.58(b) indicate locations at which steel forces can be transmitted to the concrete by means of bond. It is also evident that this is possible only if tensile force T_y in slab bars in the perpendicular direction are also developed.

Slab reinforcement anchored in the quadrant adjacent to that in Fig. 4.58(a), introducing similar T_x forces, will necessitate the development of a diagonal compression field similar to that shown in Fig. 4.58(b). The flexural and torsional stiffness of a transverse beam, shown by dashed lines in Fig. 4.58(a), relative to the membrane stiffness of the slab, is likely to be dramatically reduced after the development of plastic hinges resulting from north–south earthquake actions. Hence at this stage transmission of slab forces to the joint core by means of transverse beams may be neglected.

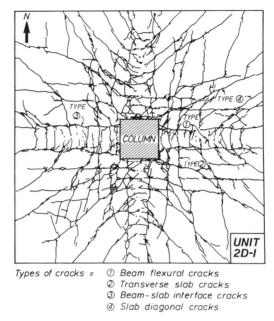

Types of cracks = ① Beam flexural cracks
② Transverse slab cracks
③ Beam-slab interface cracks
④ Slab diagonal cracks

Fig. 4.57   Crack pattern in a two-way slab after simulated multi directional earthquake attack [C14].

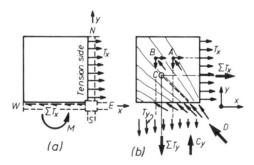

Fig. 4.58   Equilibrium criteria for membrane forces in a slab quadrant acting as a tension flange.

The diagonal membrane forces developed in a cracked slab can then be introduced to the beams as shown in Fig. 4.59. It is seen that no strength enhancement will occur in the east beam, where the slab is in compression. However, an additional force ΔP, due to tension flange contribution, can be applied to the west beam. As a consequence, a moment $\Delta M = z_b C_x$ can be introduced to the column, where z_b is a beam moment arm [Fig. 4.43(a)], and in terms of the forces shown in Fig. 4.58(b), $C_x = 2 \sum T_x$.

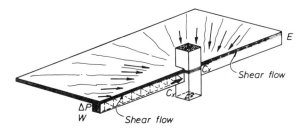

Fig. 4.59   Transfer of flange membrane forces to beams and columns.

As the forces originating in the tension flanges of the beams are introduced primarily by concrete compression forces to the joint, the strut mechanism of Fig. 4.44(a) may be relied on for transmitting these forces to the columns. Hence no joint shear reinforcement should be required on account of tension flange actions.

The mechanism just described is consistent with the observed fact that after the development of plastic hinges, beams will become significantly longer. As Fig. 4.60(b) suggests, for a given inelastic story drift, this elongation will be proportional to the depth of the beams. The previously postulated membrane mechanism is also applicable when continuous beams are to be considered instead of an isolated unit (Figs. 4.57 and 4.59). These are shown in Fig. 4.60, where the beam (C_x) and column forces shown are those necessary to equilibrate only tensile forces in the slabs $\sum T_x$.

Figure 4.60(a) implies that slab reinforcement, developing tensile forces $\sum T_x$, is constant over the four slab panels. When this is not the case, for example when the positive midspan slab reinforcement is less than the

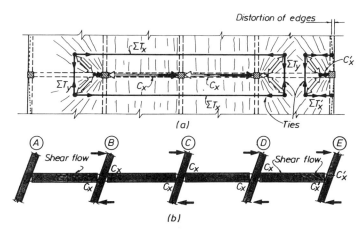

Fig. 4.60   Flange mechanisms in continuous beams.

negative reinforcement over the slab edges, the tensile forces carried over the interior panels by the mechanism shown in Fig. 4.60(a) will reduce to ΣT_{xp}. The differential tension force in the slabs across the transverse beams, $\Sigma \Delta T_x = \Sigma T_x - \Sigma T_{xp}$, will then be transmitted to each column by the mechanism postulated in Fig. 4.59.

Tension forces ΣT_y, transverse to those introducing compression forces (C_x) to the columns, shown in Figs. 4.58 and 4.60, are of similar order. This suggests that under bidirectional earthquake attack, the full contribution of slab reinforcement to beam strength, as shown in Fig. 4.59, cannot be developed simultaneously in both orthogonal directions.

Since flange force transfer at exterior columns, seen in Fig. 4.60, will depend primarily on the flexural strength of edge beams with respect to the vertical axis of the beam section, strength enhancement of the interior beam where it is connected to the exterior column will diminish when plastic hinges develop in the edge beams [C14].

(b) Joints with Unusual Dimensions In previous sections situations were examined that arise in common frames where members with similar dimensions are joined. Experimental evidence to support the suggested design procedures has also been obtained from such test units. Members with significantly different dimensions require additional considerations when joints between them are to be designed.

In low-rise frames with long-span beams, the column depth may be considerably smaller than that of the beam it supports. An oblong joint core, such as shown in Fig. 4.61(a) results. It is likely that during earthquake attack, plastic hinges would develop in the columns, perhaps with no moment reversal in the beams, which is suggested in Fig. 4.61(a). This is common in gravity-load-dominated frames, to be examined in Section 4.9.

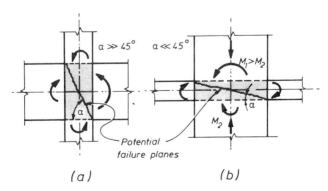

Fig. 4.61 Elongated joint cores.

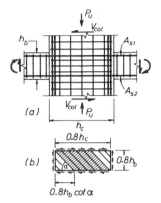

Fig. 4.62 Joint between a shallow beam and a deep column.

A related situation arises when shallow beams frame into deep wall-like columns, as shown in Fig. 4.61(b). The column in this situation is much stronger than the beams. It might principally act as a cantilever, so that the sense of the column moments above and below the floor will remain the same. Also, there is a greater length available to develop the strength of the beam bars. The following discussion will be restricted to the situation of Fig. 4.61(b), but by inversion (beam-column, column-beam) it applies equally to the situation in Fig. 4.61(a).

Typical reinforcing details at and in the vicinity of a joint, such as in Fig. 4.61(b), are shown in Fig. 4.62(a). The vertical reinforcement would have been determined from considerations of flexure and axial load. As explained earlier, the vertical column bars, other than the main flexural bars at the extremities of the column section, may be assumed to contribute to joint strength.

Figure 4.62(b) models the joint core and the shear flow resulting from the beam moment inputs and column shear force. Typically, the core may be assumed to have the dimensions $0.8h_b$ and $0.8h_c$, respectively. The horizontal joint shear force V_{jh} is determined, as discussed in Section 4.8.5(b) and from Eq. (4.47). Because of the favorable bond conditions for the beam bars, steel stress distributions similar to those shown in Fig. 4.50(e) may be assumed. Thus with a close to uniform shear flow and because the amount of required joint reinforcement will not be great, the entire joint shear resistance may conveniently be assigned to the truss mechanism [Fig. 4.53(c)]. Equation (4.72) can be correspondingly modified so that with a suitable selection of the angle α, say $45°$, the vertical joint reinforcement becomes

$$A_{jv} = \frac{1}{f_{yv}}\left(V_{jh}\tan\alpha - P_u\right) \qquad (4.78a)$$

Alternatively, the adequacy of the existing interior vertical bars in Fig. 4.62(a) may be checked, and from Eq. (4.78a) the corresponding angle α may be determined. However, A_{jv} should not be less than derived from Eq. (4.78a) with $\alpha \geq 30°$.

As Fig. 4.62(b) suggests, horizontal joint reinforcement is required only to sustain the diagonal compression field in the unshaded area of the joint core. Using, for example, the concept of shear flow at the boundaries of the core in Fig. 4.62(b) it is found that

$$A_{jh} = \frac{1}{f_{yh}} \frac{h_b}{h_c} V_{jh} \cot \alpha \qquad (4.78b)$$

When unusual dimensions are encountered, some engineering judgment will be required in sensibly applying the principles outlined here. The application of this simple approach is shown in Section 9.5.3(e), where a joint between two walls is considered.

The procedure above may be applied in situations similar to that shown in Fig. 4.61(a). The roles of the beams and the columns are simply interchanged. Instead of the intermediate column bars in Fig. 4.62(a), horizontal stirrup ties may be used in the joint shown in Fig. 4.61(a), with the beneficial effect of axial load being discounted.

(c) Eccentric Joints Beam–column joints, where the axes of the beam and column are not coplanar, will introduce some torsion to the joint. Although this may not be critical, some allowance for it should be made. A simple design approach is suggested here, whereby a smaller concentric joint is assumed to transfer the necessary total internal shearing forces that arise from lateral forces on the frame.

The approximation is illustrated in Fig. 4.63, which shows a relatively deep exterior column supporting continuous spandrel beams that frame eccentrically into it. When due to the forces shown, moments are introduced at the joint by the column and the two beams at opposite faces of the column, these

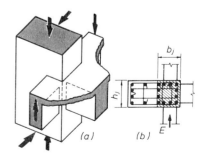

Fig. 4.63 Eccentric beam–column joint.

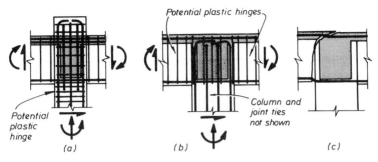

Fig. 4.64 Beam–column joints at roof level.

moments not being coplanar, torsion will be generated in the column, mainly within the depth of the beams.

Instead of attempting to account for this torsion, we may assume that the joint shear forces are transferred with negligible torsion across the effective joint core only, as shown in plan in Fig. 4.63(b). The effective width b_j is as defined in Fig. 4.55(a). It is then only necessary to place all joint shear reinforcement in both the horizontal and vertical directions, within this effective width, shown shaded in Fig. 4.63(b). Also, to ensure effective interaction between the column and the spandrel beams, the column flexural reinforcement, necessary to resist the moments originating from the spandrel beams, should be placed within the effective width b_j.

(d)   Joints with Inelastic Columns   Joints with inelastic columns will commonly occur at roof level, where the beam flexural strengths are likely to be in excess of that of a column. Moreover, at this level no attempt need be made to enforce plastic hinges in the beams. The principles outlined for inelastic joints apply also in this situation, except that the roles of the beams and columns interchange. The prime consideration now is the anchorage of the column bars. As Fig. 4.64 suggests, a column stub at roof level above the beam is preferable, to allow standard hooks to be accommodated. In many situations, however, functional requirements at the roof will not allow this to be done. In such cases the welding of column bars to anchorage plates should be considered. The estimation of the vertical joint shear reinforcement, shown in Fig. 4.64(a) and (b), should be based on that of the horizontal joint shear reinforcement in exterior joints, examined in Section 4.8.11.

If possible plastic hinges in beams should be avoided because it is more difficult to provide good anchorage for the top beam bars within the joint, unless adequate vertical stirrup, as shown in Fig. 4.64(b), are also provided. Congestion of reinforcement in the joint is thereby aggrevated. Moreover, the potential for diagonal tension failure across the flexural compression zone of the beam exists [Fig. 4.64(c)].

4.8.10 Alternative Detailing of Interior Joints

(a)   Beam Bar Anchorage with Welded Anchorage Plates   When the use of large-diameter beam bars passing through small columns is unavoidable and the bond strength requirements described in Section 4.8.6(b) cannot be satisfied, anchorage plates welded to beam bars close to each column face may be used. Beam bar forces can thereby be effectively introduced to the strut mechanism [Fig. 4.44(a)] and the need for large amounts of horizontal joint reinforcement can be reduced significantly. Such anchorage plates should be designed to transmit the combined tension and compression forces, T and C'_s in Fig. 4.53(a), to the diagonal strut. As the bond within the joint is expected to break down, inelastic tensile strains along beam bars within the joint may occur. The distance between anchorage plates may thus increase, leading to a slack connection. To avoid this, the area of beam bars within the joint must be increased, for example by welding smaller-diameter bars to the large-diameter beam bars. Frame assemblages with such joints have been found to exhibit excellent hysteretic response [F1].

(b)   Diagonal Joint Shear Reinforcement   Diagonal joint shear reinforcement, as shown in Fig. 4.65(a), is another alternative whereby the horizontal joint shear reinforcement, consisting of stirrup ties, can be greatly reduced. A fraction of the beam reinforcement can be bent diagonally across the joint core. The detailing of the reinforcement for this weak beam/strong column assembly requires careful planning. Also, there are a number of prerequisites that should be met when this type of detailing is contemplated, such as:

1. The column should have a greater structural depth than the beam, to enable beam bars to be bent within the column, preferably without exceeding a slope of 45°. This is essential to control bearing stresses, both within and at the outside of the bends of the beam bars. Radii of bend should be large and the surrounding concrete should be laterally confined by transverse ties.
2. Beam width must be sufficient to allow diagonal beam bars to be placed in different vertical planes so that they can readily bypass each other where the diagonals would intersect.
3. The arrangement is likely to prove impractical in beam–column joints of two-way frames. Thus this solution is largely restricted to one-way frames. It has been found suitable for the peripheral frames of tube-framed buildings.

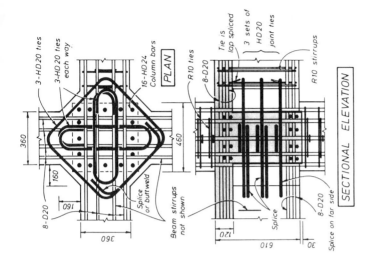

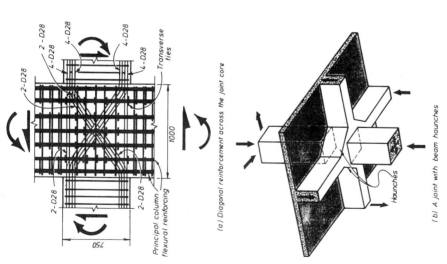

Fig. 4.65 Alternative detailing of interior joints.

293

A study of such an arrangement will reveal that under the earthquake actions indicated, the beam bars bent diagonally across the joint core do not transfer bond forces. Within the joint core they would be subjected to nearly uniform stresses, up to yield if necessary, in tension or in compression. Therefore, the limitations on bar sizes, as discussed in Section 4.8.6, need not be considered. Fewer large-diameter bars may be used in the beam.

(c) Horizontally Haunched Joints   Horizontally haunched joints may find application with two-way frames, where this arrangement could lead to significant relief with respect to the congestion of reinforcement in the beam–column joint region. Horizontal beam haunches, such as shown in Fig. 4.65(*b*), enable large-diameter bars to be used for joint shear reinforcement. Some of these ties can be placed outside the joint core proper. Because of the large-size ties, fewer sets are required and thus more space becomes available, particularly for compacting the fresh concrete within the joint. Typical details of reinforcement for a specific joint are given in Fig. 4.65(*b*). Standard ties used in the columns, immediately above and below the beam, can be omitted altogether within the depth of the beam, where the large-diameter peripheral ties in the haunch provide adequate confinement.

With a suitable choice of the haunch, the critical beam section can be moved away from the column face. The development lengths for beam bars is thereby increased. If this is taken into account, then by considering the principles outlined in Section 4.8.6, larger-diameter beam bars can be used. For example, the dimension $h = 360 + 2 \times 160 = 680$ mm (26.8 in.) from Fig. 4.65(*b*) can be used in place of $h_j = h_c = 460$ mm (18.1 in.), when determining from Eq. (4.56) the maximum usable beam bar size.

4.8.11   Mechanisms in Exterior Joints

(a) Actions at Exterior Joints   Because at an exterior joint only one beam frames into a column, as shown in Fig. 4.66, the joint shear will generally be less than that encountered with interior joints. From the internal stress resultants, shown in Fig. 4.66, it is evident that by similarity to Eq. (4.47*b*), the horizontal joint shear force is

$$V_{jh} = T - V_{col} \qquad (4.79)$$

where the value of the tension force T is either $f_s A_s$ or $\lambda_o f_y A_s$, depending on whether an elastic beam section or the critical section of a beam plastic hinge at the face of the column is being considered. Using first principles, the vertical joint shear force may be found from the appropriate summation of the stress resultants shown in Fig. 4.66, or approximated by using Eq. (4.49).

Various types of exterior joints are shown in Fig. 4.45. For the purpose of examining the behavior of exterior joints, however, only a plane frame joint [Fig. 4.45(*d*)] will be discussed in the following. It was pointed out that as a

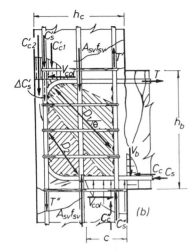

Fig. 4.66   Mechanisms of exterior joints.

general rule, no beneficial contribution to joint strength should be expected from inelastic transverse beams of frames entering a joint, such as shown in Fig. 4.45(*e*) and (*f*).

(b)   Contributions of Joint Shear Mechanisms   As expected, shear transfer within the joint core by mechanisms of a "concrete strut" and a "truss," sustaining a diagonal compression field, will be similar to that postulated for interior joints. Both top and bottom bars in the beam must be bent into the joint as close to the far face of the column as possible, as shown in Fig. 4.66, unless the column is exceptionally deep (i.e., when it is approaching the dimensions of a structural wall).

A diagonal strut similar to that shown in Fig. 4.53(*a*) will develop between the bend of the hooked top tension beam bars and the lower right-hand corner of the joint in Fig. 4.66, where compression forces in both the horizontal and vertical directions are introduced. In developing a diagonal compression force D_c, it is the latter region that is critical. Anchorage forces at the hook of the top bars can readily adjust themselves to balance the diagonal compression force D_c [Y1].

As explained in Section 4.8.7(*a*), the horizontal component of the strut mechanisms is, with the notation used in Fig. 4.66,

$$V_{ch} = C_c + \Delta T_c - V_{col} \qquad (4.80)$$

where ΔT_c is the fraction of the steel compression force C_s developed in the bottom beam reinforcement, introduced to the strut by means of bond over the length of bar subjected to transverse compression from the lower column. By assuming good anchorage conditions for the bottom bars, compression stresses could be developed so that $C_s \leq A'_s f_y = \beta T / \lambda_o$, where $T = \lambda_o f_y A_s$. Hence the flexural concrete compression force in the bottom of the beam is $C_c \geq T - C_s = (1 - \beta/\lambda_o)T$, where $\beta = A'_s/A_s$. Evidence obtained from tests [C14] indicated that steel compression stresses at joints do not exceed yield strength.

As in the case of interior joints presented in Section 4.8.7(a), an assumption needs to be made with respect to realistic distributions of bond forces along the bottom beam bars within the core of the exterior joint. Because the bent-up hook can not be considered to be effective in transferring steel compression forces, parts of the bottom bars close to the beam are likely to provide the bulk of the required anchorage. In comparison with the distribution of bond forces from beam bars within an interior joint, shown in Fig. 4.50(g), the contribution in exterior joints of the strut mechanism to joint shear strength will be underestimated if uniform bond distribution is assumed. By assuming an effective anchorage length of $0.7h_c$ for the bottom beam bars, the unit bond force will be

$$u_o \approx 1.4 C_s / h_c$$

and hence the anchorage force introduced to the diagonal strut is

$$\Delta T_c = u_o c = 1.4 C_s c / h_c = \frac{1.4 c \beta}{\lambda_o h_c} T \tag{4.81}$$

By assuming again that effective bond transfer to the diagonal strut occurs over only 80% of the compression zone c of the column, defined by Eq. (4.61), we can determine from Eqs. (4.80) and (4.81) the magnitude of V_{ch} and hence the joint shear force to be resisted by horizontal shear reinforcement. With minor rounding up of coefficients, this is

$$V_{sh} = V_{jh} - V_{ch} \approx \frac{\beta}{\lambda_o} \left(0.7 - \frac{P_u}{f'_c A_g} \right) T \tag{4.82}$$

Provided that adequate anchorage of the beam flexural tension reinforcement by a standard hook, to be reviewed briefly in the next section, exists, there is no need to investigate the force transfer V_{ch} to that end of the diagonal strut. The anchorage force introduced by the hook [i.e., the force R in Fig. 4.68(b)] and the bond force from the horizontal part of the top beam bars in Fig. 4.66, together with vertical forces from the column, such as C'_{c1} and C'_{c2} in Fig. 4.66, will combine into a diagonal compression force acting at an angle much less than α (Fig. 4.53). This force resolves itself into D_c and another compression force, shown as D_1 in Fig. 4.66, acting at an angle $\theta < \alpha$. The latter force will engage joint ties and hence the diagonal force D_2 in Fig. 4.66.

(c) **Joint Shear Reinforcement**   The required amount of horizontal joint reinforcement, consisting of ties as shown in Fig. 4.66, can readily be determined from Eq. (4.82) (i.e., $A_{jh} = V_{sh}/f_{yh}$). It may also be expressed in terms of the total beam tension reinforcement, that is,

$$A_{jh} = \beta\left(0.7 - \frac{P_u}{f_c'A_g}\right)\frac{f_y}{f_{yh}}A_s \tag{4.83}$$

When plastic hinge formation in the beam develops tension in the top bars, there will always be some compression force acting on the column so that $P_u > 0$. However, under reversed seismic forces, when the bottom beam bars are in tension, axial tension will be introduced to the exterior columns. This seismic action will thus reduce the net axial compression and may even result in net tension acting on the column and the joint. In the latter case, P_u in Eqs. (4.82) and (4.83), which are also applicable when the bottom beam bars are in tension, must be taken negative. This case should be checked because in terms of joint shear reinforcement it is likely to be critical.

Following the reasoning employed in the derivation of the required vertical joint shear reinforcement for interior joints in Section 4.8.7(*b*), it may be shown that the same result is applicable to exterior joints; that is,

$$A_{jv} \geq (1/f_{yv})(0.5V_{jv} - P_u) \tag{4.72b}$$

(d)   **Anchorages in Exterior Joints**

(i)   Forces from the Longitudinal Column Reinforcement: Figure 4.66 suggests that bond forces from bars placed at the outer face of the column must be transferred into the joint core. There are no mechanisms whereby bond stresses transmitted to the cover concrete could be absorbed reliably, particularly after intense reversing earthquake shaking. At this stage, splitting cracks along these outer column bars develop, separating the cover concrete from the core.

When the horizontal joint shear reinforcement is inadequate, premature yielding in the tie legs, seen in Fig. 4.66, may occur. Consequently, the lateral expansion of the joint core in the plane of the frame will be larger than the corresponding expansion of adjoining end regions of the column. Fracture of the cover concrete above and below the joint and its complete separation may result [U1]. Some examples of the phenomenon may be seen in Fig. 4.67. Spalling of the cover concrete over the joint will also reduce the flexural strength of the adjacent column sections [P1, P41, S13] because the compression force C_{c2}' in Fig. 4.66 cannot be transmitted to the joint.

(ii)   Forces from the Beam Flexural Reinforcement: The tension force, T, developed in the top bars of the beam in Fig. 4.66 is introduced to the

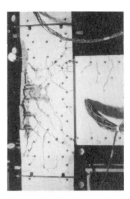

Fig. 4.67 Bond deterioration along outer column bars passing through a joint.

surrounding concrete by bond stresses along the bar and also by bearing stresses in the bend of a hook. Traditionally, the effective development length for these hooked beam bars, l_{hb} [Section 3.6.2(a)], is assumed to begin at the inner face of the column or of the joint core [A1]. This assumption is satisfactory only in elastic joints, where yielding of beam bars at the face of the column is not expected.

When a plastic hinge develops adjacent to the joint, with the beam bars entering the strain hardening range, yield penetration into the joint core and simultaneous bond deterioration, as discussed in Section 4.8.6(c), is inevitable. Consequently, after a few cycles of inelastic loading, anchorage forces for tension will be redistributed progressively to the hook except for very deep columns. Bond loss along a straight bar anchored in an exterior joint would result in complete failure. Therefore, beam bars at exterior joints, which can be subjected to yield in tension during an earthquake, should be anchored with a hook or with other means of positive anchorage [P1].

The penetration of inelastic tensile strains into the joint core and the consequent development of the yield strength of beam bars very close to a 90° hook has been observed, and an example of this is shown in Fig. 4.68(a) [P41]. The numbers in circles in this figure refer to the displacement ductility factor imposed on the test specimen. It is seen that for upward loading, when these bars [d_b = 20 mm (0.79 in.) diameter] carry flexural compression forces, significant inelastic tensile strains remain within the joint. Transmission of compression forces from beam bars to the joint is by means of bond only. After a number of cycles of inelastic displacements, bond deterioration along beam bars is likely to be significant, and hence their contribution as compression reinforcement may be reduced unless special detailing enables the hook to support a compression force. However, as a result of redistribution of compression forces from the steel to concrete (C_s and C_c in Fig. 4.66), the strut mechanism postulated in Section 4.8.11(b) should not be affected.

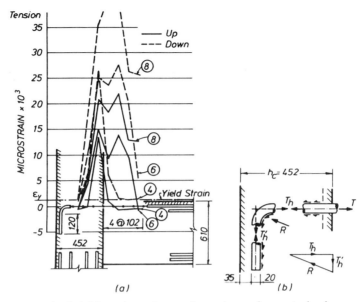

Fig. 4.68 Spread of yielding along the top beam bars of an exterior beam–column joint assembly and forces generated at a right angle hook.

Provided that the surrounding concrete is confined, the 90° standard hook commonly used will ensure that a bar will not pull out. The forces that are applied to various parts of such a beam bar are shown in Fig. 4.68(b). After a number of load reversals, the maximum tensile force, $T = \lambda_o f_y A_b$ at the face of the column may be reduced to $T_h \leq f_y A_b$ at the beginning of the 90° bend. The vertical bond forces along the straight length of the bar following the bend develop the tensile force T_h'. Codes [A1, X3] normally require a straight extension of length $12d_b$ past the hook. Any additional length sometimes provided does not improve bar development [W3]. As Fig. 4.68(b) shows, a force R will be introduced to the concrete core by means of bearing and bond stresses.

Development lengths required for bars with standard hooks were given in Section 3.6.2(a) and Eqs. (3.67) to (3.69). Because the location of the hook is critical, the following aspects should also be considered in the design of exterior joints:

1.   When a plastic hinge is expected to develop at or near the column face, the anchorage of beam bars should be assumed to begin only at a distance from the inner column face, well inside the joint core. It has been suggested [X3] that the section of beam bars at which embedment should be assumed to begin should be at the center of the column but that it need not

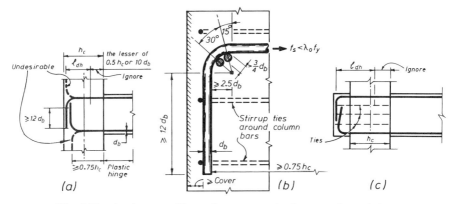

Fig. 4.69 Anchorage of beam bars at exterior beam–column joints.

be more than 10 times the bar diameter d_b from the column face of entry. This requirement is shown in Fig. 4.69(a).

2. The basic development length for a standard hook l_{dh} beyond the section defined above should be as required by codes [A1, X3]. This is given in Section 3.6.2(a). In shallow columns, the length available to accommodate the beam bars satisfactorily may be insufficient. In such cases:

(a) Smaller-diameter bars may need to be used in the beams.

(b) Anchorage plates welded to the ends of a group of beam bars may be provided [P1].

(c) The effective straight length in front of the hook may be reduced, provided that the concrete against which the beam bars bear within the bend is protected against premature crushing or splitting. This may be achieved by placing short bars across and tightly against the beam bars in the bend, as shown in Fig. 4.69(b).

(d) Horizontal ties in the joint, shown in Fig. 4.69(b) and (c), should be placed so as to provide some effective restraint against the hook when the beam bar is subjected to compression.

3. To develop a workable strut mechanism, such as shown in Fig. 4.66, it is vital that beam bars be bent into the joint core. Bending beam bars away from the joint, as shown by the dashed lines in Fig. 4.69(a), not an uncommon practice, is undesirable in a seismic environment [W3, K3, X3].

4. Because the inclination of the principal diagonal strut, D_1 in Fig. 4.66, critically influences the shear transfer mechanisms, it is essential that the hook at the ends of the beam bars be as close to the outer face of the column as possible [K3]. It is suggested that the inner face of the hooked end of beam bars be no closer than $0.75h_c$ to the inner face of the column, as shown in Fig. 4.69(b) [X3].

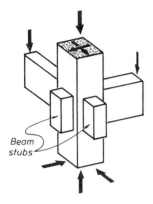

Fig. 4.70   Beam stubs at corner joints to accommodate beam bar anchorages.

5. Whenever architectural considerations will allow, and particularly when shallow columns are joined with relatively deep beams, the beam bars may be terminated in a beam stub at the far face of the column [P1], as shown in Figs. 4.69(c) and 4.70. When this anchorage detail is compared with that used in standard practice, shown in Figs. 4.66 and 4.67, it is seen that greatly improved bond conditions are also provided for the outer column bars. Also, because of the relocation of the bearing stresses developed inside the bend of the beam bars, improved support conditions for the diagonal strut exist. Some column ties, as shown in Fig. 4.69(c), should be extended into the beam stub for crack control. The excellent performance of beam–column joint assemblages with this detail has been observed in tests [M10, P1, P40].

6. To reduce bond stresses, it is always preferable to use the smallest bar diameter that is compatible with practicality. Because at exterior joints reliance of beam bar anchorage is placed primarily on a standard hook rather than the length preceding it, the requirements of Section 4.8.6 limiting bar diameter in relation to column depth need not apply. In general, it is easier to satisfy beam anchorage requirements at exterior than at interior joints.

(e)   *Elastic Exterior Joints*   The improvement of the performance of interior joints resulting from the prevention of yielding of the beam flexural reinforcement at the column faces was discussed in Section 4.8.7(e). To a lesser degree the same principles apply to exterior joints. One way to ensure elastic joint response is to relocate the potential plastic hinge from the face of the column, as suggested in Figs. 4.16 and 4.17. As Fig. 4.71 shows, beam bars may be curtailed so that stresses in the reinforcement will not exceed yield stress at the face of the column, while strain hardening may be developed at the critical section of the plastic hinge. Therefore, the latter section must be a sufficient distance away from the face of the column. A minimum distance of h_b or 500 mm, whichever is less, seems appropriate [X3]. This suggestion was borne out by tests [P16, P40]. In this case the effective development

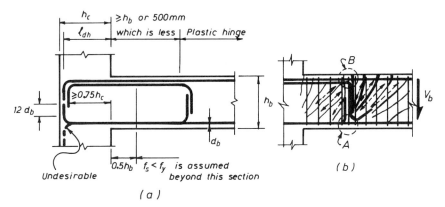

Fig. 4.71 Potential plastic hinge relocated from the face of a column.

length, l_{dh}, for the hooked beam bars may be assumed to begin at the face of the column [Fig. 4.71(a)].

In view of the unusual nature of shear transfer across the critical section of the plastic hinge region, care must be taken with the detailing of such relocated plastic hinges. Figure 4.71(b) indicates that the beam shear force V_b, introduced by diagonal forces to region A, needs to be transferred to the top of the beam at region B. This necessitates stirrups extending between these two regions, perhaps supplemented by the specially bent top beam bars, to carry the entire shear force.

4.8.12 Design Steps: A Summary

Because relevant design procedures are as yet not well established, issues of beam–column joints in reinforced concrete ductile frames have been discussed in disproportionate detail in this chapter. However, the conclusions drawn lead to a relatively simple design process which, in terms of design steps, is summarized here for interior joints.

Step 1: Determination of Design Forces. After evaluation of the overstrength of each of the two potential plastic hinges in each beam, or in exceptional cases those in columns, the forces acting on a joint are readily found. Depending on whether the tensile forces introduced to the joint by means of steel stresses are at or below yield level ($f_s \leq f_y$) or at overstrength ($f_s = \lambda_o f_y$), the joint is considered as being elastic or inelastic. Both the horizontal and vertical joint shear forces, V_{jh} and V_{jv}, are found from Eqs. (4.47) and (4.49).

Step 2: Joint Shear Stress. To safeguard against premature diagonal compression failure of the concrete in the joint core, nominal shear stresses v_{jh}, based on Eq. (4.73), should be limited by Eq. (4.74a) in one-way and by Eq. (4.74b) in two-way frames.

Step 3: Anchorage Requirements.   As the prevention of bond failure in joints is as important as that due to shear, the diameter of beam and column bars in relation to the available anchorage length within a joint must be limited in interior joints in accordance with Eq. (4.56) and in exterior joints in accordance with Eq. (3.67) and Fig. 4.69(a). These issues must be considered early when bars are being selected for the detailing of beams and columns.

Step 4: Assigning Joint Shear Resistance to Two Mechanisms.   With consideration of the relative quantities of reinforcement in adjacent beam sections ($A'_s/A_s = \beta$), the contribution of the strut mechanism to horizontal joint shear resistance V_{ch} may be determined from Eq. (4.66). More conveniently, the shear resistance assigned to the truss mechanism V_{sh}, involving reinforcement, can be obtained directly for inelastic joints from Eq. (4.67b). Similarly, the required vertical joint shear resistance is obtained with the use of Eq. (4.72b).

Step 5: Joint Shear Reinforcement.   As the contribution of the truss mechanism to the total joint shear resistance required has been established in the previous step, the necessary amount of total joint shear reinforcement is obtained from $A_{jh} = V_{sh}/f_{yh}$ or from Eq. (4.70b) and A_{jv} from Eq. (4.72b). To satisfy these requirements, horizontal sets of ties, normally consisting of several legs, and intermediate vertical column bars are utilized satisfying usual anchorage requirements. Particular attention must be paid to the avoidance of congestion of reinforcement in joints.

4.9 GRAVITY-LOAD-DOMINATED FRAMES

4.9.1 Potential Seismic Strength in Excess of That Required

As outlined in previous sections, it is necessary to evaluate the flexural overstrengths of the potential plastic hinges in all beams in order to be able to assess the desirable reserve strength that adjacent columns should have. This is done to ensure that a weak beam/strong column system will result from the design, so that story column mechanisms cannot develop during any seismic excitation. With the diminished likelihood of the development of plastic hinges in the columns at levels other than at the base, considerable relaxation in the detailing of columns can be accepted.

In low-rise ductile reinforced concrete frames, particularly in those with long-span beams, and also in the top stories of multistory frames, often gravity load rather than seismic force requirements will govern the design strength of beams.

When the strength of the beams is substantially in excess of that required by the seismic lateral forces specified, an indiscriminate application of the capacity design philosophy can lead to unnecessary or indeed absurd conservatism, particularly in the design of columns. Excess beam strengths with respect to lateral forces originate in plastic hinge regions, where gravity loads induce large negative moments when lateral forces impose moderate positive moments. It was shown that to complete a story beam mechanism, two plastic hinges need to develop in each span. In gravity-load-dominated frames, considerable additional lateral story shear forces may be required after the formation of the first (negative) hinge in the span to enable the second (positive) hinge to develop also. If the designer insists on the full execution of capacity design to ensure that yielding of columns above level 1 will not occur before the full development of beam plastic hinges, the columns would have to be designed for even larger story shear forces.

Skillfully applied moment redistribution, given in Section 4.3, may reduce considerably the unintended potential lateral-force carrying capacity in such frames. Another means whereby reduction in lateral force resistance may be achieved is the relocation of potential positive (sagging) plastic hinges away from column faces to regions where both gravity loads and lateral forces generate positive moments. Such hinge patterns were shown in Figs. 4.10 and 4.16. However, in many situations the most meticulous allocation of beam hinge strength will not offset the excess potential lateral force resistance.

In such frames the designer might decide to allow the formation of plastic hinges in some columns to enable the complete frame mechanism to develop at lower lateral force resistance. This would be for economic reasons. However, two criteria should be satisfied if such an energy dissipating system is adopted.

1. At least a partial beam mechanism should develop to ensure that no story sway mechanisms (soft stories) can form. This can be achieved if plastic hinges are made to develop in the outer spans of beams close to exterior columns, which in turn must have adequate flexural strength to absorb without yielding the moment input from these outer beam spans. Column hinges above level 1 will thus not develop in the two outer columns of the frame. At the inner column–beam joints, column hinges above and below each floor will need to form to complete the frame mechanisms. As long as the outer columns above level 1 are assured of remaining elastic, plastic hinge development, in both beams and interior columns, should spread over several or all of the stories. Hence story column mechanism will be avoided. Such a mechanism is shown in Fig. 4.72.

2. Because the formation of plastic hinges in columns is less desirable, restricted ductility demand in such frames might be considered. This can be achieved if the resistance of the frame with respect to lateral forces is increased, whereby the inelastic response of the frame to the design earthquake is reduced.

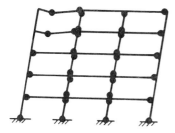

Fig. 4.72 Frame hinge mechanisms involving plastic hinges in interior columns.

4.9.2 Evaluation of the Potential Strength of Story Sway Mechanisms

Whether a frame is dominated by earthquake forces or by gravity load depends on the relation of its potential lateral force resistance to the lateral design earthquake forces. This is best quantified in terms of the strength of story mechanisms. In connection with the principles of moment redistribution in beams, it was explained in Section 4.3.3 with reference to Fig. 4.8 that the lateral force demand for a bent at a floor can be expressed by Eq. (4.9) (i.e., ΣM_{ci} = constant). The resistance to meet this demand was achieved in previous weak beam/strong column frames by the selection and appropriate design of beam plastic hinges. Consequently, the moments that would be induced in the beams at column centerlines while two plastic hinges in predetermined positions develop in each span at flexural overstrength were determined. Such moments are evaluated from those shown for an example frame in Fig. 4.10. With the aid of these terminal moments at column centerlines, the system overstrength factors for the entire bent ψ_o could be determined for each direction of the seismic action [Eq. (1.13)]. Typical minimum values of ψ_o were given in Section 1.3.3(g). In the actual design of earthquake-force-dominated frames, the values of ψ_o will be in general somewhat larger. This is because of routine rounding-up errors. In gravity-load-dominated frames, however, the value of ψ_o so derived may be larger than 2 or even 3.

To illustrate the effect of gravity load dominance, the simple symmetrical three-bay bent of Fig. 4.73 may be considered [P15]. Instead of presenting the procedure in general terms, the principles involved are explained here with the aid of a numerical example. The first three diagrams of Fig. 4.73 show in turn the bending moments for the factored gravity load only ($1.4D + 1.7L_r$), the factored gravity load to be considered together with the lateral forces ($D + 1.3L_r$), and the code-specified lateral earthquake forces ($\vec{E}$). A comparison of these cases will reveal that beam strength is governed by gravity load considerations only [Fig. 4.73(a)]. With modest moment redistribution in the center span, shown by the dashed line, the beam at the interior column may be designed to give a dependable flexural strength of 190 moment units. For the purpose of illustrating the principles involved, moments at column centerlines, rather than those at column faces, are used in this example. The

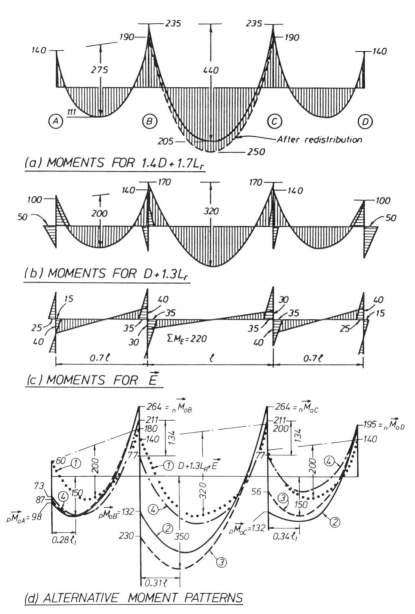

Fig. 4.73 Beam moment and plastic hinge patterns for a three-bay symmetrical gravity-load-dominated frame (*continued next page*).

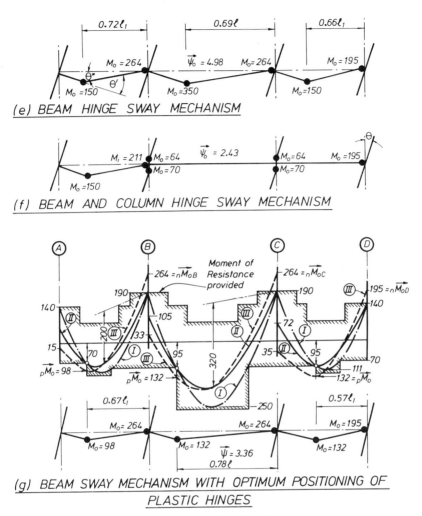

(e) BEAM HINGE SWAY MECHANISM

(f) BEAM AND COLUMN HINGE SWAY MECHANISM

(g) BEAM SWAY MECHANISM WITH OPTIMUM POSITIONING OF PLASTIC HINGES

Fig. 4.73 (*Continued*)

combination $(D + 1.3L_r + \vec{E})$, shown by the pointed curve (1) in Fig. 4.73(*d*), gives a maximum negative moment of $170 + 30 = 200$ units at the right-hand end of the center span. Only 5% moment reduction is required if the design moment derived from Fig. 4.73(*a*) is to be used. It is seen that for the load combination $(D + 1.3L_r + \vec{E})$, the negative moments due to gravity at the left-hand end of each span are reduced. Curve 1 of Fig. 4.73(*d*) also shows that moment reversals at supports will not occur as a result of the specified earthquake forces on the structure.

It was pointed out in Section 4.5.1(*c*) that to ensure adequate curvature ductility at a beam section, and in accordance with the requirements of several codes, the flexural compression reinforcement should be at least 50% of the tension reinforcement at the same section. Therefore, if a story mechanism is to form during a severe earthquake, with two plastic beam hinges at overstrength in each span, developed at the column faces, the required lateral force would need to be increased very considerably. This should result in the moment pattern shown by the full-line curves (2) in Fig. 4.73(d). In this frame the end moments at flexural overstrength are such that $_n\vec{M}_{oB} = 2_p\vec{M}_{oB}$ and $_n\vec{M}_{oD} = 2_p\vec{M}_{oA}$, where the subscripts n and p refer to negative and positive moments, respectively, and the arrow indicates the direction of earthquake attack that is being considered. For example, $_n\vec{M}_{oB} - 1.39 \times 190 = 264$ and $_p\vec{M}_{oB} = 0.5 \times 264 = 132$ units. It is also evident that for this hypothetical plastic hinge pattern the positive flexural reinforcement in each span would need to be increased greatly if plastic hinges within the span were to be avoided. The system flexural overstrength factor describing this situation would be

$$\psi_o = \frac{\Sigma_{Bn}^D\vec{M}_{oi} + \Sigma_{Ap}^C\vec{M}_{oi}}{\Sigma_A^D M_E} \qquad (4.84)$$

For the example structure, using Eq. (4.84), the following values are obtained from Fig. 4.73(d) curve (2) and Fig. 4.73(c):

$$\vec{\psi}_o = \frac{(264 + 264 + 195) + (98 + 132 + 132)}{40 + 70 + 70 + 40} = \frac{1085}{220} = 4.93$$

By considering beam overstrengths ($\lambda_o = 1.25$) [Section 3.2.4(e)], the lateral force resistance of the frame could be $4.93/1.39 = 3.55$ times that required by the code-specified lateral forces, E. However, with this very large increase of strength, strain hardening should not be expected.

If the steps of the capacity design for columns summarized in Section 4.6.8 were now to be followed, using $\omega_{min} = 1.3$, the strength of the columns would need to be at least 4.6 times that required by design earthquake forces. In certain situations this may be achieved without having to increase the column size or reinforcement above that required for gravity loading alone. However, the seismic shear requirements, which follow from the recommendations of Section 4.6.7, are likely to be excessive. Therefore, the excessive strength of the frame with respect to lateral forces should be reduced.

4.9.3   Deliberate Reduction of Lateral Force Resistance

(a)   Minimum Level of Lateral Force Resistance   The strength of acceptable mechanisms other than those preferred in ductile frames needs to be estab-

lished first. This level of resistance may be made larger than that required to absorb code-specified lateral forces because:

1. For a given overall displacement ductility demand, relocated plastic hinges in the span of beams will lead to increased rotational ductility demands in these hinges.
2. A more conservative approach might be warranted when a large number of plastic hinges in columns are necessary to complete the hinge mechanisms for the entire frame.

As Fig. 4.14(a) shows, the plastic hinges rotations in a beam can be approximated [Section 4.5.1(d)] in terms of the average slope of the inelastic frame, θ; thus

$$\theta' = \frac{l}{l^*}\theta \qquad (4.85)$$

Using the approximation that increased strength with respect to lateral forces will result in proportional reduction in displacement ductility demand [Section 2.3.4] during the design earthquake, the design forces should be magnified by the factor

$$m = l/l^*_{\text{average}} \geq 1 \qquad (4.86)$$

where the average values of the lengths ratios (Fig. 4.14) for all affected beams of the frame may be used. Further refinements would seldom be justified.

For the second alternative, whereby plastic hinges in interior columns, as shown in Fig. 4.72, are chosen to limit lateral force resistance, the designer should consider the following factors when choosing an appropriate value for m.

1. The importance of the building and the possible difficulties to be encountered with column repair
2. The relative magnitude of axial compression on the columns affected

Recent research [P8] has shown that adequately confined regions of columns [Section 3.6.1(a)] will ensure very ductile behavior. Hence the traditional conservatism with respect to the acceptances of plastic hinges in columns is no longer justified.

For convenience, the desirable minimum lateral force resistance of a fully plastic gravity-load-dominated subframe may be expressed in terms of the system overstrength factor thus:

$$\psi_{o,\min} \geq m\psi_{o,\text{ideal}} \qquad (4.87)$$

where, as discussed in Section 1.3.3(g), $\psi_{o,\text{ideal}} = \lambda_o/\phi$.

By taking into account the level of code-specified lateral forces in relation to the elastic response of the structure to the design earthquake, an upper-bound value of ψ_o should also be established. It would be pointless to increase the strength of the structure beyond the level associated with elastic response in an attempt to ascertain the desired hierarchy in the formation of plastic hinges. Hence it is suggested that the capacity design of gravity-load-dominated frames be limited to structures for which

$$\psi_o < R/\lambda_o \qquad (4.88)$$

where R is the force reduction factor studied in Section 2.4.3(b)(iv), and λ_o is the materials overstrength factor given in Section 3.2.4(e).

(b) Beam Sway Mechanisms This mechanism of a subframe in each story can be maintained only if plastic hinges in the spans of beams are permitted to form. For example, the moment pattern shown by the dashed lines (3) in Fig. 4.73(d) and relevant to beam overstrengths may be considered to govern the design of beams. The required positive moment strength of the short and long example beams in the midspan regions is found from Fig. 4.73(a) not to be less than 111 and 250 moment units, respectively. Hence when $\lambda_o/\phi \approx 1.4$, the flexural overstrength of plastic hinges in the span regions of approximately 150 and 350 units would be developed, as shown in Fig. 4.73(d). The beam mechanism so developed is shown in Fig. 4.73(e). The corresponding (clockwise turning) positive moments at column centerlines can be readily determined from considerations of equilibrium for each span. The system overstrength factor for the beam mechanism of Fig. 4.73(e) is found to be

$$\vec{\psi}_o = \frac{(87 + 264) + (230 + 264) + (56 + 195)}{220} = 4.98$$

Thus the relocation of plastic hinges from column faces, as shown in Fig. 4.73(e), did not reduce the lateral force resistance of this example subframe. Therefore, this mechanism would also require columns of unjustifiably excessive strength.

If this mechanism had proved to be feasible, the code-specified lateral force resistance of the frame should have been increased, in accordance with Eq. (4.86) and Fig. 4.14, with the factor

$$m = \left(\frac{1}{0.72} + \frac{1}{0.69} + \frac{1}{0.66}\right)\frac{1}{3} = 1.45$$

where the relevant values for l^*, together with the beam mechanism, are shown in Fig. 4.73(e). Thus the required overall overstrength factor for the example structure should not need to be more than

$$\vec{\psi}_{o,\,min} = 1.45 \times 1.39 = 2.02 < 4.98$$

(c) Introduction of Plastic Hinges in Columns   As shown in Fig. 4.72, a frame mechanism involving column hinges may be the other alternative to reduce lateral force resistance. For the example structure, the mechanism of Fig. 4.73(f) may be developed as follows:

1. The outer columns must not be permitted to yield. Hence plastic hinges will need to develop in adjacent beams. The moment patterns of Fig. 4.73(d) suggest that in beam A–B it will be more convenient to allow a positive beam hinge to form some distance away from column A, while a negative plastic hinge in beam A–B at the interior column B will also need to develop.

2. With the plastic hinge in the first span developing overstrength (150 units) and that at the interior column judged at its ideal strength ($264/1.25 = 211$ units), because, as Fig. 4.73(f) shows, of the relatively small hinge rotation θ^* there, the bending moment for the entire span is given. This is shown by curve (4) in Fig. 4.73(d). The positive moment developed at column A is 73 units at this stage. The beam plastic hinge at column D is assumed to develop its overstrength at 195 units.

3. The maximum feasible moment inputs into the exterior columns are thus known. By assuming a strength magnification $m = 1.75 > 1.45$ for this case, with the aim of reducing plastic rotational demands on the column plastic hinges, the system overstrength factor needs to be $\psi_{o,\,min} = 1.75 \times 1.39 = 2.43$. Therefore, by rearranging Eq. (4.84), the total minimum required moment input into the two interior columns is

$$\sum_{B}^{C} \vec{M}_{oi} = \sum_{A}^{D} \vec{M}_E - {}_p\vec{M}_{oA} - {}_n\vec{M}_{oD} = 2.43 \times 220 - 73 - 195 = 267 \text{ units}$$

Thus the two plastic hinges in each of the two identical interior columns must, in the presence of the axial compression to be carried, develop total flexural overstrength of $267/2 \approx 134$ units. In Fig. 4.73(f) this has been arbitrarily split into 64 and 70 units by considering the effect of the larger axial compression on the columns just below the beams.

4. From joint equilibrium at column B, it follows that the ideal flexural strength at the left-hand end of beam B–C must be at least $(211 - 267/2) = 77$ units. This involves tension in the top reinforcement. However, from Fig. 4.73(a) it is evident that the available ideal strength at this section will not be less than $190/0.9 = 211$ units. It is also evident that the beam moments at

column C are not critical and their accurate determination is not necessary. In Fig. 4.73(d), curve (4) shows that ideal beam strengths of 211 and 77 units have also been assumed at column C. The hinge mechanism associated with this moment pattern is shown in Fig. 4.73(f).

5. Column D may now be designed in accordance with the principles of Section 4.6 for a beam centerline moment of

$$M_{col} = \phi_o \omega M_E$$

where in this case $\phi_o = 195/40 = 4.88$ and for the top column $M_E = 25$ units, as shown in Fig. 4.73(c). In a low-rise one-way frame, the likely value of ω is 1.3. Therefore, the ideal flexural strength of column D should be on the order of

$$M_i = 4.88 \times 1.3 \times 25 = 159 \text{ units}$$

It is evident that for the same direction of earthquake attack the identically reinforced column A will be adequate to resist the much smaller moment input (73 units) at the other end of the frame.

6. The interior columns could be made quite small to resist, at ideal strength, a moment of the order of

$$M_i = 70/1.25 = 56 \text{ units}$$

as shown in Fig. 4.73(f).

The example showed that despite providing some 75% ($m = 1.75$) excess seismic resistance, no additional reinforcement of significance is likely to be required in any of the members on account of earthquake forces. This is because of gravity load dominance.

While the outer columns will require only a limited amount of transverse reinforcement, in accordance with the principles of Section 4.6.11, the end regions of the inner columns, above and below the beam, where plastic hinges are expected, will need to be fully confined, as explained in Section 4.6.11(e). Moreover, splices in the interior columns must be placed at midheight of the story.

(d) Optimum Location of Plastic Hinges in Beams   A careful location of beam plastic hinges may result in maximum possible reduction of the lateral force resistance of beam sway mechanisms. To illustrate the corresponding strategy that may be employed, another solution for the frame of Fig. 4.73 will be considered and compared with previous alternatives.

In this case flexural reinforcement in the beams is provided to satisfy the gravity load demands shown in Fig. 4.73(a). The curtailment of the reinforcement leads to a dependable moment of resistance, shown by the stepped shaded envelope in Fig. 4.73(g). The gravity-induced moment demands are shown by curves marked (I), and these are identical with those produced in Fig 4.73(a) after moment redistribution. Thus maximum dependable negative moments of 190 and 140 units occur at the interior and exterior columns, respectively. The corresponding negative moments at the development of flexural overstrength, due to earthquake attack $\vec{E}$, will be 264 and 195 units. These are the same as those used in Fig. 4.73(d).

The optimum location of the plastic hinges in each span will depend on utilization of the minimum amount of positive (bottom) flexural reinforcement. It was stated in Section 4.5.1(c) that to ensure adequate rotational ductility, the bottom beam reinforcement at column faces should not be less than one-half of the amount of top beam reinforcement at the same section. If this condition is satisfied in this example structure, the dependable positive moment of resistance of the beams at the interior and exterior columns will be approximately 95 and 70 units, respectively. The locations at which this amount of bottom reinforcement will be just adequate to meet the gravity load demands are readily found. The location in each span is shown by a small circle in Fig. 4.73(g). These points may also be chosen for plastic hinges for the load condition $U = D + 1.3L_r + E$. Excess bottom reinforcement should be provided beyond these points in each span to ensure that the central regions remain elastic.

With the location of two potential plastic hinges, to be developed during a large earthquake, the appropriate bending moment diagram for each span can be readily drawn. These are shown by the curves marked II in Fig. 4.73(g). At the development of flexural overstrengths, the positive plastic hinge moments $_pM_o$ will increase by 39%, to 132 and 98 units, respectively. By fitting the bending moment due to gravity loads ($D + 1.3L_r$) only, the beam moment introduced to the left-hand column in each span is finally found. The corresponding moments at the development of flexural overstrength are shown by the dashed curves marked III in Fig. 4.73(g). Thus the system flexural overstrength factor for this case will be

$$\vec{\psi}_o = \frac{(15 + 264) + (-33 + 264) + (35 + 195)}{220} = 3.36$$

From the hinge mechanisms and hinge positions shown in Fig. 4.73(g), it is found that $m = 1.51$.

It is seen that with the optimum location of beam plastic hinges, the lateral force resistance of this subframe could be reduced by 33% compared with that of the beam sway mechanism studied earlier and shown in Fig. 4.73(e). This resistance is, however, still 38% in excess of that obtained with the admission of plastic hinges in columns, as shown in Fig. 4.73(f) (i.e., $3.36/2.43 = 1.38$).

4.9.4   Design for Shear

Beam shear forces may readily be derived once the moment pattern, such as shown by curve 4 in Fig. 4.73(d), is established. The shear forces should be based on the development of flexural overstrength of the beam hinges if these occur in a span. In other spans shear strength will be controlled by gravity loading alone. Also, these shear forces should be used to establish the earthquake-induced column axial forces. Such column forces are not likely to be critical in gravity-load-dominated frames.

When column hinging is admitted, the design of the interior columns for shear resistance must be based on simultaneous hinging at overstrength at both ends of the column. Since the design shear force so derived is an upper-bound value, it may be matched by ideal rather than dependable shear strength, as outlined in Section 3.4.1. The design shear force across the exterior (nonhinging) columns of gravity-load-dominated frames may be determined as in earthquake-dominated frames, as set out in Section 4.6.7.

In the design of interior beam column joints, attention must be paid to the fact that the adjacent plastic hinges are to develop in the columns rather than in the beams. This will influence the selection of the sizes of column bars as outlined in Section 4.8.6(b).

4.10   EARTHQUAKE-DOMINATED TUBE FRAMES

4.10.1   Critical Design Qualities

In many situations it will be advantageous to assign the major part or entire lateral force resistance to peripheral frames. In these structures, as seen in Fig. 1.12(e), more closely spaced columns may be used, while widely spaced interior columns or walls will primarily carry gravity loads only. Deeper exterior spandrel beams may be used without having to increase story height.

As there are only two frames to resist the major part of the total lateral forces in each direction, the resulting seismic actions on members may be rather large. Because of the relatively shorter spans, the influence of gravity load, with the exceptions of the top stories, will be only secondary. As a result, the following design features may be expected:

1. Due to the large flexural capacity of peripheral spandrel beams developed over relatively short spans, large shear forces may be generated in them. This may lead to early deterioration in energy dissipation in plastic hinge regions because of sliding shear effects. It may necessitate the use of diagonal shear reinforcement across the potential plastic hinge zone, as discussed in Section 3.3.2(b).

2. In comparison with columns of interior frames, closely spaced peripheral columns will carry smaller gravity loads. The absence of significant axial compression on the columns, combined with the presence of

possibly large amounts of beam flexural reinforcement, may create more severe conditions at interior beam–column joints.

While conventional design approaches, outlined in this chapter for ductile frames, may result in a satisfactory structure, alternative solutions, which cope more efficiently with the critical features above, should be explored.

One of these solutions, using the concepts of coupling of structural walls, discussed in Section 5.6, is outlined next.

4.10.2 Diagonally Reinforced Spandrel Beams

Some reduction in the demand for shear reinforcement in beam–column joints will result if the joint can be assured to remain elastic during an earthquake attack. Relevant principles were discussed in Section 4.8.7(e). This necessitates the relocation of beam plastic hinges away from column faces, as shown in Figs. 4.16 and 4.17. However, if both plastic hinges in each relatively short span are moved toward the center of the span, the distance between them may become too short. As Fig. 4.14 shows, this may result in excessive curvature ductility demands. Moreover, to control sliding shear displacements, diagonal reinforcement may be required in both of the plastic hinge regions in each span.

Details of the reinforcement in Fig. 4.74 satisfy the foregoing constraints while assuring adequate ductility capacity. The principles of behavior are

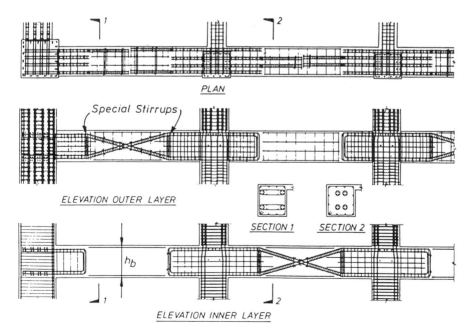

Fig. 4.74 Diagonally reinforced spandrel beams for tube frames.

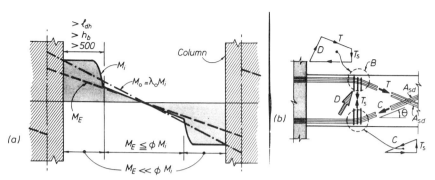

Fig. 4.75 Moment envelopes for and details of diagonally reinforced spandrel beams.

based on the moment patterns shown in Fig. 4.75(a). The moment to be resisted is that shown by the line M_E. Small gravity load effects are neglected in this example. Following the principles outlined in Section 5.4.5(b), inclined reinforcement is then provided in the central region, as shown in Fig. 4.74, to satisfy the normal requirement that $M_u = M_E \leq \phi M_i$.

The central diagonally reinforced portion of the beams, when subjected to reversed cyclic inelastic displacements, will behave like similar coupling beams examined in some detail in Section 5.4.5(b).

The remainder of the beam, as well as the joints and columns, are intended to remain elastic. To ensure this, the overstrength of the central region needs be considered. This is readily evaluated from the relationship $M_o = \lambda_o M_i = \phi_o M_E$. When additional flexural reinforcement in the end regions of spans is provided in such a fashion that the ideal strength of each beam section at column faces, M_i, is equal to or larger than the moment demand at the same section, when the flexural overstrength of the central plastic region, M_o, is developed, no yielding at the column face will occur. This is shown in Fig. 4.75(a). Hence the joint region will remain elastic. After the number of bars to be used in each region of a span has been chosen, the length of the central region over which the beam bars are bent over can readily be determined.

In the example of Fig. 4.74, there are exactly twice as many bars at the column face as there are in the central region. This also enables ready prefabrication and assembly of the beam reinforcing cages, as seen in Fig. 4.76.

4.10.3 Special Detailing Requirements

1. To ensure that the diagonal compression bars can sustain yield strength without buckling, adequate transverse reinforcement must be provided, as in coupling beams of structural walls. This is examined in Section 5.6.2 and Fig. 5.55.

Fig. 4.76 Construction of diagonally reinforced spandrel beams [B8]. (Courtesy of Holmes Consulting Group.)

2. The elastic end region of the beam must be made long enough to ensure that yield penetration from the inelastic center part does not reach the joint region. Suggested minimum lengths are shown in Figs. 4.71 and 4.75. This length must also be sufficient to allow the required hooked anchorage l_{dh} of the main bars [Eq. (3.67)] to be accommodated.

3. Adequate web reinforcement in the elastic end region of the spandrel should be provided to resist, together with the contribution of the concrete, V_c, the shear force associated with the overstrength of the central region.

4. Particular attention must be paid to transverse reinforcement, which is required around diagonal bars where they are bent from the horizontal, as shown in Fig. 4.75(b). It is suggested that the transverse bars be designed to resist in tension at least a force 1.2 times the outward thrust generated in the bend of the main beam bars when the same develop overstrength in compression. The designer should ensure by inspection on the site that the vertical special stirrups shown in Figs. 4.74 and 4.75(b) are accurately placed around the bends of bars in each vertical layer, or that equivalent force transfer by other arrangements is achieved.

5. Bearing stresses against the concrete, developing a large diagonal compression force D in the region B of Fig. 4.75(b), need to be checked.

The vertical component of the force D is the total beam shear $V_o = 2\lambda_o f_{yd} A_{sd} \sin \theta$. To relieve bearing stresses under the bent bars, short transverse bars of the type shown in Fig. 4.69(b) may be necessary.

6. The beam must be wide enough to accommodate offset diagonal bars, which must bypass each other at the center of the spandrel.

7. This arrangement also has been used successfully in the construction of precast elements for multistory tube-framed buildings. Cruciform or T-shaped beam–column elements were used, with provisions for site connections at the midspan of beams.

4.10.4 Observed Beam Performance

The response of one-half of a diagonally reinforced spandrel beam to simulated seismic loading revealed excellent performance [P16, P30]. This is seen in Fig. 4.77. The slight deterioration of the beam after extremely large imposed displacement ductility was due to yielding of the special transverse ties, marked X in the insert, and crushing of the concrete due to bearing stresses. The horizontal elastic end region was found to be adequate. Only at the end of the test were yield strains being approached in one layer of the beam bars near the face of the column. Yielding over the entire length of the diagonal bars was observed. The beam sustained 116% of its ideal strength when the test was terminated with the imposition of a displacement ductility factor of $\mu = 18$. The maximum reversed shear stress of $0.6\sqrt{f_c'}$ (MPa) ($7.2\sqrt{f_c'}$ (psi)) in a conventionally reinforced beam would have resulted in a sliding shear failure.

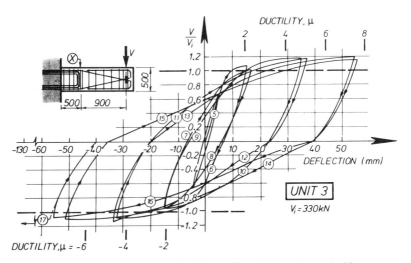

Fig. 4.77 Seismic response of a diagonally reinforced spandrel beam.

4.11 EXAMPLES IN THE DESIGN OF AN EIGHT-STORY FRAME

4.11.1 General Description of the Project

An eight-story reinforced concrete two-way frame for an office building is to be designed. In the following examples the sequence of the design and a few specific features are considered. Necessarily, certain simplifications had to be introduced for the sake of brevity. Wherever necessary, reference is made to sections, equations, or figures of the text, and some explanation, not normally part of a set of design calculations, is also offered.

The floor plan and the chosen framing system is shown in Fig. 4.78. A two-story penthouse placed eccentrically in one direction will require torsional seismic effects to be considered. Openings in the floor and nonstructural walls at and around an elevator and stair-well have been deliberately omitted from the plan to allow the utilization of double symmetry in this example design. In Fig. 4.78, preliminary member sizes, which can be used for the analysis, are also shown.

Emphasis is placed on those aspects that consider earthquake effects in the design. For this reason the design of the cast-in-place two-way floor slab, a routine operation, is not included. Where necessary a few design aids are reproduced. However, calculations largely follow first principles. Where appropriate, reference is made to design steps to determine column design actions, summarized in Section 4.6.8.

The design of an exterior and an interior beam spanning east–west and an exterior and an interior column is illustrated. Actions at or immediately below the level 3 only are considered. The base section of an interior column is also checked. Member sizes have been chosen that satisfy the stability requirements of Section 3.4.3.

4.11.2 Material Properties

Concrete strength $f'_c = 30$ MPa (4.35 ksi)

Yield strength of steel used in:

Deformed bars in beams (D) $f_y = 275$ MPa (40 ksi)

Plain bars for stirrups and ties (R) $f_y = 275$ MPa (40 ksi)

Deformed bars in columns (HD) $f_y = 380$ MPa (55 ksi)

4.11.3 Specified Loading and Design Forces

(a) Gravity Loads

(1) Floors: 120-mm (4.7-in.)-thick slab
at 24 kN/m^3 (153 lb/ft^3) $= 2.88$ kPa (60 lb/ft^2)

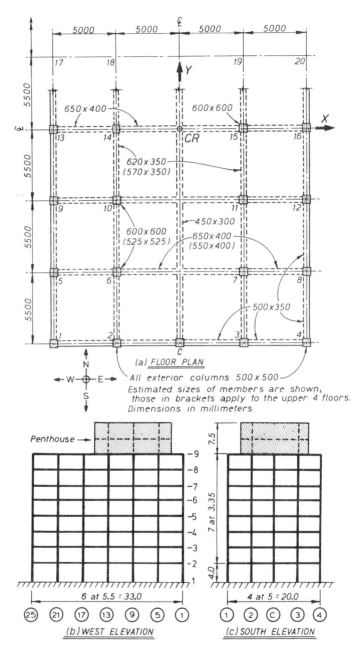

Fig. 4.78   Framing system of an eight-story example building. (1000 mm = 3.28 ft.)

Floor finish, ceiling, services,
   and movable partitions $= 1.20$ kPa (25 lb/ft^2)

Total dead load on slabs $D = \overline{4.08 \text{ kPa } (85 \text{ lb/ft}^2)}$

Live load on all floors and the roof $L = \overline{2.50 \text{ kPa } (52 \text{ lb/ft}^2)}$

(2) Curtain walls, glazing, etc., supported
    by periphery beams only, extending
    over floor height of 3.35 m (11.0 ft) $D = 0.50$ kPa (10 lb/ft^2)

(3) Preliminary analysis indicates that due to the penthouse, each of the six interior columns is subjected to a dead load of 300 kN (68 kips) and a live load of 100 kN (23 kips).

(b) Earthquake Forces   From the assumed first period of $T = 1.08$ of the building, applicable to both directions of seismic attack, the final base shear coefficient for the seismic zone was found to be 0.09. For the equivalent mass at each floor, 1.1 times the dead load is considered. Ten percent of the base shear is to be applied at roof level and the remaining 90% is to be distributed according to Eq. (2.32(b)). The mass of the penthouse is lumped with that of the roof.

4.11.4 Stiffness Properties of Members

In accordance with the suggestions made in Table 4.1, the following allowances will be made for the effects of cracking on the stiffness of various members:

$$\text{Beams:} \qquad I_e = 0.35 I_g$$

$$\text{Exterior columns:} \qquad I_e = 0.60 I_g$$

$$\text{Interior column:} \qquad I_e = 0.80 I_g$$

Only members of the lower stories are considered in this example. Dimensions are expressed in millimeters. Relative stiffness are used (i.e., $k = I_e/l$), as in Appendix A.

(a) Members of East–West Frames

(1) Column 1: $k_c = 0.6(500^4/12)/3350$ $= 0.93 \times 10^6$ mm^3 (57 in.^{3})
     Columns 2, C, and 5 are as column 1, $k_c = 0.93 \times 10^6$ mm^3 (57 in.^{3})
     Column 6: $k_c = 0.8(600^4/12)/3350$ $= 2.58 \times 10^6$ mm^3 (157 in.^{3})

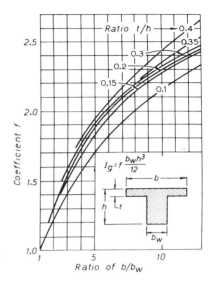

Fig. 4.79 Coefficient for moment of inertia of flanged sections.

(2) *Beams* 1–2, 2–C, *etc.*: From Fig. 4.3 the effective width for this L beam is estimated for stiffness purposes as follows:

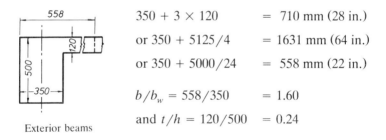

$350 + 3 \times 120$	$= 710$ mm (28 in.)
or $350 + 5125/4$	$= 1631$ mm (64 in.)
or $350 + 5000/24$	$= 558$ mm (22 in.)
$b/b_w = 558/350$	$= 1.60$
and $t/h = 120/500$	$= 0.24$

Exterior beams

Using the chart of Fig. 4.79, it is found that $f \approx 1.2$. Hence

$$k = 0.35(1.2 \times 350 \times 500^3/12)/5000 = 0.31 \times 10^6 \text{ mm}^3 \ (19 \text{ in.}^3)$$

(3) *Beams* 5–6, 7–8, *etc.*: The effective flange width from Fig. 4.3 is

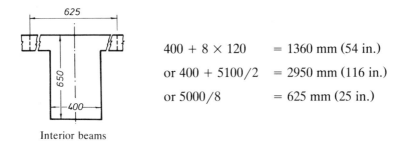

Interior beams

$$400 + 8 \times 120 \qquad = 1360 \text{ mm } (54 \text{ in.})$$

$$\text{or } 400 + 5100/2 \quad = 2950 \text{ mm } (116 \text{ in.})$$

$$\text{or } 5000/8 \qquad = 625 \text{ mm } (25 \text{ in.})$$

From Fig. 4.79 with $b/b_w = 625/400 = 1.56$, with $t/h = 0.18$ it is found that $f = 1.2$. Hence

$$k = 0.35(1.2 \times 400 \times 650^3/12)/5000 = 0.77 \times 10^6 \text{ mm}^3 \ (47 \text{ in.}^3)$$

(4) *Beams* 6–7, *etc.*: (Section as for beam 5–6.) Effective flange width is

$$400 + 8 \times 120 \qquad = 1360 \text{ mm } (54 \text{ in.})$$

$$\text{or } 400 + 5100/2 \qquad = 2950 \text{ mm } (116 \text{ in.})$$

$$\text{or } 10{,}000/4 \qquad = 2500 \text{ mm } (98 \text{ in.})$$

with

$$b/b_w = 1360/400 \qquad = 3.40$$

$$t/h = 120/650 \qquad = 0.18$$

$$f \qquad \approx 1.60$$

Hence

$$k = 0.35(1.6 \times 400 \times 650^3/12)/10{,}000$$

$$= 0.51 \times 10^6 \text{ mm}^3 \ (31 \text{ in.}^3)$$

(b) *Members of North–South Frames*

(1) *Columns* 1, 2, C, 5, *etc.*, *E–W*: $k_c = 0.93 \times 10^6 \text{ mm}^3 \ (57 \text{ in.}^3)$
    Column 6, *E–W*: $k_c = 2.58 \times 10^6 \text{ mm}^3 \ (157 \text{ in.}^3)$
(2) *Beams* 1–5, 5–9, *etc.*: These are similar to beam 1–2. Therefore,

$$\text{effective width} = 350 + 5500/24 = 579 \text{ mm}^2 \ (23 \text{ in.})$$

By similarity to beam 1–2,

$$k \approx 0.31(5/5.5)10^6 = 0.28 \times 10^6 \text{ mm}^3 \ (17 \text{ in.}^3)$$

(3) *Beams* 2–6, 6–10, *etc.*: Effective flange width

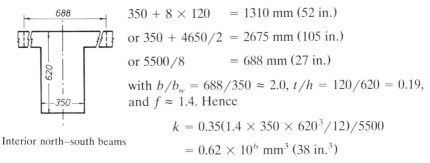

Interior north–south beams

$350 + 8 \times 120 \quad = 1310$ mm (52 in.)

or $350 + 4650/2 = 2675$ mm (105 in.)

or $5500/8 \qquad = 688$ mm (27 in.)

with $b/b_w = 688/350 \approx 2.0$, $t/h = 120/620 = 0.19$, and $f \approx 1.4$. Hence

$$k = 0.35(1.4 \times 350 \times 620^3/12)/5500$$
$$= 0.62 \times 10^6 \text{ mm}^3 \ (38 \text{ in.}^3)$$

(4) The *secondary beam* at column C may be assumed to have a span of 11 m (36 ft) for the purpose of the approximate lateral-force analysis. The effective width of the flange is approximately

Secondary beam

$11,000/8 \quad = 1375$ mm (54 in.)

with $b/b_w \ = 1375/300 = 4.6$

with $t/h \quad = 120/450 = 0.27$ and $f \approx 1.8$

$$k = 0.35(1.8 \times 300 \times 450^3/12)/1100 = 0.13 \times 10^6 \text{ mm}^3 \ (8 \text{ in.}^3)$$

4.11.5 Gravity Load Analysis of Subframes

The tributary floor areas for each beam are determined from first principles using the model of Fig. 4.4. Details of this are not given in subsequent calculations.

In recording the moments, the following abbreviations are used:

FEM: fixed-end moments, using C_1

SSM: midspan moments for simply supported spans, using C_2

The appropriate moment coefficients C_1 and C_2 are taken from Table 4.6.

TABLE 4.6   Moments Coefficient C

μ	C_1	C_2
0	9.60	6.00
0.1	9.63	6.02
0.2	9.67	6.09
0.3	9.75	6.21
0.4	9.91	6.36
0.5	10.08	6.55
0.6	10.34	6.76
0.7	10.61	7.01
0.8	11.00	7.30
0.9	11.42	7.64
1.0	12.00	8.00

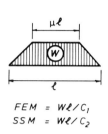

$FEM = W\ell/C_1$
$SSM = W\ell/C_2$

(1) *Beams 1–2, 2–C, etc.:* The tributary area is $A = 5^2/4 = 6.3$ m^2 (68 ft^2). As this is less than 20 m^2 (214 ft^2), in accordance with Section 1.3.1(b) no reduction for live load will be made (i.e., $r = 1.0$).

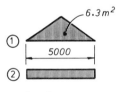

Loading pattern

(1) *D*: 6.3 × 4.08   = 25.7 kN (5.8 kips)

   L: 6.3 × 2.50   = 15.8 kN (3.6 kips)

(2) Weight of beam

   0.35(0.5 − 0.06)24   = 3.70 kN/m (0.25 kip/ft)

   Curtain wall

   0.5 × 3.35   = 1.68 kN/m (0.12 kip/ft)

   Total   = 5.38 kN/m (0.37 kip/ft)

   D: 5.38 × 5.00   = 26.90 kN (6.1 kips)

Load	FEM		SSM	
(1) *D*: 25.7 kN	×5/9.6	= 13.4 kNm	×5/6 = 21.4 kNm	
(2) *D*: 26.9 kN	×5/12	= 11.2 kNm	×5/8 = 16.8 kNm	
	Σ = 52.6 kN(11.8 kips)	= 24.6 kNm(18.2 kip-ft)	= 38.2 kNm (28.2 kip-ft)	
(1) *L*: 15.8 kN	×5/9.6	= 8.2 kNm	×5/6 = 13.2 kNm (9.7 kip-ft)	

(2) *Secondary Beam*

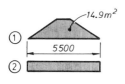

Loading pattern

$$A = 14.9 \text{ m}^2 < 20 \text{ m}^2; \quad \therefore r = 1.0$$

(1) *D*: $14.9 \times 4.08 = 60.8$ kN (13.7 kips)

   L: $14.9 \times 2.50 = 37.3$ kN (8.4 kips)

(2) Weight of beam

   D: $0.3(0.45 - 0.12)24 \times 5.5$

   $= 13.1$ kN (2.9 kips)

   $\mu = (5 - 5.5)/5.5 \approx 0.1$

   for Table 4.6.

Load	FEM	SSM

(1) *D*: 60.8 kN   ×5.5/9.63 = 34.7 kNm   ×5.5/6.02 = 55.5 kNm

(2) *D*: 13.1 kN   ×5.5/12   =   6.0 kNm   ×5.5/8 =   9.0 kNm

$\Sigma = 73.9$ kN (16.6 kips) = 40.7 kNm   (30.0 kip-ft) = 64.5 kNm (47.6 kip-ft)

(1) *L*: 37.3 kN   ×5.5/9.63 = 21.3 kNm   ×5.5/6 = 34.1 kNm (25.2 kip-ft)

(3) *Beam* 5–6

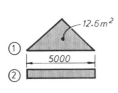

Loading pattern

$$A = 2 \times 6.3 = 12.6 \text{ m}^2 < 20 \text{ m}^2; \quad \therefore r = 1.0$$

(1) *D*: $12.6 \times 4.08 = 51.4$ kN (11.6 kips)

   L: $12.6 \times 2.50 = 31.5$ kN (7.1 kips)

(2) Weight of beam

   $0.4(0.65 - 0.12)24 \times 5 = 25.4$ kN (5.7 kips)

Load	FEM	SSM

(1) *D*: 51.4 kN   ×5/9.6   = 26.8 kNm   ×5/6 = 42.8 kNm

(2) *D*: 25.4 kN   ×5/12   = 10.6 kNm   ×5/8 = 15.9 kNm

$\Sigma = 76.8$ kN (17.3 kips) = 37.4 kNm (27.6 kip-ft) = 58.7 kNm (43.3 kip-ft)

(1) *L*: 31.5 kN   ×5/9.6   = 16.4 kNm   ×5/6 = 26.3 kNm (19.4 kip-ft)

(4) *Beam* 6–7

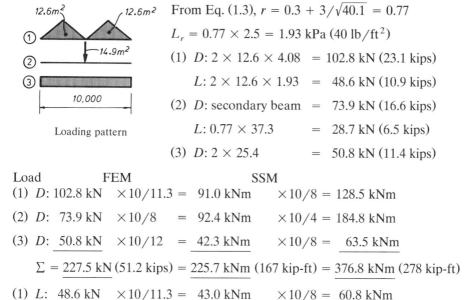

$A = 2 \times 12.6 + 14.9 = 40.1 > 20 \text{ m}^2$

From Eq. (1.3), $r = 0.3 + 3/\sqrt{40.1} = 0.77$

$L_r = 0.77 \times 2.5 = 1.93 \text{ kPa } (40 \text{ lb/ft}^2)$

(1) D: $2 \times 12.6 \times 4.08 \quad = 102.8 \text{ kN } (23.1 \text{ kips})$

$\quad\; L$: $2 \times 12.6 \times 1.93 \quad = \; 48.6 \text{ kN } (10.9 \text{ kips})$

(2) D: secondary beam $\quad = \; 73.9 \text{ kN } (16.6 \text{ kips})$

$\quad\; L$: $0.77 \times 37.3 \qquad\quad = \; 28.7 \text{ kN } (6.5 \text{ kips})$

(3) D: $2 \times 25.4 \qquad\qquad = \; 50.8 \text{ kN } (11.4 \text{ kips})$

Loading pattern

Load	FEM		SSM	
(1) D: 102.8 kN	$\times 10/11.3 =$	91.0 kNm	$\times 10/8 =$	128.5 kNm
(2) D: 73.9 kN	$\times 10/8 \quad =$	92.4 kNm	$\times 10/4 =$	184.8 kNm
(3) D: 50.8 kN	$\times 10/12 \quad =$	42.3 kNm	$\times 10/8 =$	63.5 kNm

$\Sigma = 227.5 \text{ kN } (51.2 \text{ kips}) = 225.7 \text{ kNm } (167 \text{ kip-ft}) = 376.8 \text{ kNm } (278 \text{ kip-ft})$

(1) L: 48.6 kN	$\times 10/11.3 =$	43.0 kNm	$\times 10/8 =$	60.8 kNm
(2) L: 28.7 kN	$\times 10/8 \quad =$	35.9 kNm	$\times 10/4 =$	71.8 kNm
(3) L: —	$=$	—	$=$	—

$\Sigma = \; 77.3 \text{ kN } (17.4 \text{ kips}) = \; 78.9 \text{ kNm } \; (58 \text{ kip-ft}) = 132.6 \text{ kNm } (98 \text{ kip-ft})$

(5) *Frame* 1–2–C–3–4: For gravity loading on all spans, assume that points of inflection are at midheights of columns. Because of symmetry the beam is fully fixed at C. Hence the absolute flexural stiffnesses for columns and beams are $6k_c$ and $4k_b$, respectively, where values of k were given in Section 4.11.4.

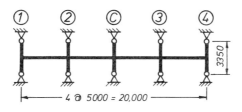

Model of frame

Distribution factors:

Col. 1: 6 × 0.93 = 5.58 → 0.45   Col. 2: 5.58 → 0.41

Beam: 4 × 0.31 = 1.24 → 0.10   Beam: 1.24 → 0.09

Col. 1:   = 5.58 → 0.45   2Σ = 13.60 → 1.00

Σ = 12.40 → 1.00

Fixed-end moments are derived for the load combination [Eq. (1.5a)] (1.4D + 1.7L) in the table below, using previous results. The combination (D + 1.3L) [Eq. (1.6a)] is obtained by proportions using the multiplier (D + 1.3L)/(1.4D + 1.7L) derived from fixed-end moments only as follows:

FEM (1.4D + 1.7L) = 1.4 × 24.6 + 1.7 × 8.2 = 48.4 kNm (35.7 kip-ft)

FEM (D + 1.3L) = 24.6 + 1.3 × 8.2 = 35.3 kNm (26.1 kip-ft)

Multiplier: 35.3/48.4 = 0.73 to be applied to final moments as an acceptable approximation.

Distribution of moments (kNm)

	Col. 1	1–2		2–1	Col. 2	2–C		C	
Load	0.45	0.10		0.09	0.41	0.09		—	
1.4D + 1.7L$_r$	—	− 48.4		48.4	—	− 48.4		48.4	
	21.8	5.0	→	2.5					
		− 0.1	←	− 0.2	− 1.0	− 0.2		− 0.1	
	21.8	− 43.5		50.7	− 1.0	− 48.6		48.3	× 0.73
D + 1.3L$_r$	15.9	− 31.8		37.0	− 0.8	− 0.8		35.3	

(1 kNm = 0.738 kip-ft)

(6) *Frame* 5–6–7–8: Because of symmetry, the beam stiffnesses for the outer and inner spans are 4k and 2k, respectively [see Section 4.11.4(a)].

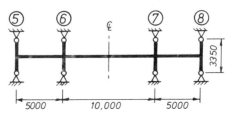

Model of frame

Distribution factors:

Col. 5: $6 \times 0.93 = 5.58 \rightarrow 0.392$ Beam 6–5: $4 \times 0.7\,7 = 3.08 \rightarrow 0.089$

Beam: $4 \times 0.77 = 3.08 \rightarrow 0.216$ Col. 6: $6 \times 2.58 = 15.48 \rightarrow 0.441$

Col. 5: $= 5.58 \rightarrow 0.392$ Beam 6–7: $2 \times 0.5\,1 = 1.02 \rightarrow 0.029$

$\Sigma = 14.24 \rightarrow 1.000$ Col. 6: $= 15.48 \rightarrow 0.441$

$\Sigma = 35.06 \rightarrow 1.000$

FEM $(1.4D + 1.7L_r)$: span 5–6: $1.4 \times 37.4 + 1.7 \times 16.4 = 80.2$ kNm (59.2 kip-ft)

span 6–7: $1.4 \times 225.7 + 1.7 \times 78.9 = 450.1$ kNm (332.2 kip-ft)

FEM $(D + 1.3L_r)$:   span 5–6: $37.4 + 1.3 \times 16.4$ $= 58.7$ kNm (43.3 kip-ft)

span 6–7: $225.7 + 1.3 \times 78.9$ $= 328.3$ kNm (242.3 kip-ft)

Multiplier: $328.3/450.1 \approx 0.73$

Distribution of moments (kNm)

	Col. 5	5–6		6–5	Col. 6	6–7	
Load	0.392	0.212		0.089	0.441	0.029	
$1.4D + 1.7L_r$	—	− 80.1		80.1	—	− 450.1	
		16.5	←	32.9	163.2	10.7	
	24.9	13.5	→	6.7			
		− 0.3	←	− 0.6	− 3.0	− 0.2	
	0.1	0.1		—			
	25.0	− 50.3		119.1	160.2	− 439.6	× 0.73
$D + 1.3L_r$	18.3	− 36.7		86.9	116.9	− 320.9	

(1 kNm = 0.738 kip-ft)

(7) *Gravity Load on Column 5*: By similarity to the example given in Fig. 4.4(a), the tributary area to column 5 at each floor is $5.5 \times 0.5 \times 5 = 13.8$ m^2 (150 ft^2). Therefore, the components of dead load at a typical lower floor are as follows:

D: Floor slab 13.8×4.08 $= 56.3$ kN (12.7 kips)

Beam 5–6 $0.5 \times 25.4(4.45/5)$ $= 11.2$ kN (2.5 kips)

Beam 1–5–9
+ (curtain wall) $= 5.38 \times 5.0$ $= 26.9$ kN (6.1 kips)

Column $\approx 0.5 \times 0.5 \times 24 \times 3.35 = 20.1$ kN (4.5 kips)

$\Sigma = 114.5$ kN (25.8 kips)

At the upper four levels

D: as at lower four levels   114.6 kN (25.8 kips)

Reduction for beam 5–6:

$$0.5(0.1 \times 0.4 \times 24 \times 4.49) \quad = - \quad 2.2 \text{ kN } (0.5 \text{ kips})$$

$$\Sigma = \quad 112.4 \text{ kN } (25.3 \text{ kips})$$

Top level

D: as at upper levels   112.4 kN (25.3 kips)

less 50% of column $0.5 \times 20.1 = -$ 10.0 kN (2.2 kips)

$$\Sigma = \quad 102.4 \text{ kN } (23.1 \text{ kips})$$

Total dead load below level 3

$$P_D = 102.4 + 3 \times 112.4 + 3 \times 114.5 = \quad 783.1 \text{ kN } (176.1 \text{ kips})$$

L: tributary area
$$A = 7 \times 13.8 = 96.6 \text{ m}^2 \ (1040 \text{ ft}^2) > \ 20 \text{ m}^2$$

$$\therefore \quad r = 0.3 + 3/96.6 = 0.605$$

$$P_{Lr} = 0.605 \times 2.5 \times 96.6 \qquad = -146.1 \text{ kN } (32.9 \text{ kips})$$

(8) *Gravity Load on Column* 6

Tributary area per level $= 5.5 \times 0.5(5 + 10) = 41.3 \text{ m}^2 \ (444 \text{ ft}^2)$

At lower levels

D: Floor slab $41.3 \times 4.08 = 168.5$ kN (37.9 kips)

Beam 5–6–7 $50.8(6.9/10) = \ 35.1$ kN (7.9 kips)

Beam 2–6–10 $(0.62 - 0.12)0.35 \times 24 \times \ 4.9 = \ 20.6$ kN (4.6 kips)

Secondary beam $0.5 \times 13.1 = \ 6.5$ kN (1.5 kips)

Column $0.6 \times 0.6 \times 24 \times 3.23 = \ 27.9$ kN (6.3 kips)

$$\Sigma = 258.6 \text{ kN } (58.2 \text{ kips})$$

At upper levels

D: As at lower floors		=	258.6 kN (58.2 kips)
Less for beam 5–6–7:	$(0.1 \times 0.4 \times 24)6.9 = -$		6.6 kN (1.5 kips)
Less for beam 2–6–10:	$(0.05 \times 0.35 \times 24)5 = -$		2.1 kN (0.5 kips)
Less for column	$0.084 \times 24 \times 3.23 = -$		6.5 kN (1.5 kips)
		$\Sigma =$	243.4 kN (54.8 kips)

Top level

D: As at upper levels		=	243.4 kN (54.8 kips)
Less 50% column	$0.5(27.9 - 6.5) = -$		10.7 kN (2.4 kips)
Penthouse		=	300.0 kN (67.5 kips)
		$\Sigma =$	532.7 kN (119.9 kips)

Total dead load below level 3

P_D $= 532.7 + 3 \times 243 + 3 \times 258.6$ $= 2039.0$ kN (458.8 kips)

Live load, L: $A = 2 \times 40 + 7 \times 41.3 = 369$ m^2 (3970 ft^2)

$$r = 0.3 + 3/\sqrt{369.1} = 0.456$$

P_{Lr} $= 0.456(100 + 7 \times 41.3 \times 2.5)$ $= 375.0$ kN (84.4 kips)

4.11.6 Lateral Force Analysis

(a) Total Base Shear

(1) *Concentrated Masses at Levels 2 to 5*

Floor, inclusive partitions	$20 \times 33 \times 4.08$	=	2693 kN (606 kips)
Peripheral beams and curtain walls			
$2(6 \times 5 \times 5.38) + 2(4 \times 4.5 \times 5.38)$		=	516 kN (116 kips)
Interior beams, east–west	$5(4 \times 25.4) =$		508 kN (114 kips)
Interior beams, north–south	$2(6 \times 20.6) =$		247 kN (56 kips)
Secondary beam (north–south)	$6 \times 13.1 =$		79 kN (18 kips)
Exterior columns	$(10 + 10) \times 20.1 =$		402 kN (90 kips)
Interior columns	$10 \times 27.9 =$		279 kN (63 kips)
		$\Sigma =$	4724 kN (1063 kips)

(2) *Concentrated Masses at Levels 6 to 8*

As at lower floors	$=$	4724 kN (1063 kips)
Less for east–west beams	$(2.2 + 6.6)10 = -$	88 kN (20 kips)
Less for north–south beams	$10 \times 2.1 = -$	21 kN (5 kips)
Less for columns	$10 \times 6.5 = -$	65 kN (15 kips)
	$\Sigma =$	4550 kN (1024 kips)

(3) *Concentrated Mass at Roof Level*

As at upper floors	$=$	4550 kN (1024 kips)
Less for 50% of columns $0.5(20 \times 20.1 + 10 \times 21.4) = -$		308 kN (69 kips)
From above roof (penthouse)	$6 \times 300 =$	1800 kN (450 kips)
	$\Sigma =$	6042 kN (1359 kips)

(4) *Equivalent Weights Assumed to be Subject to Horizontal Accelerations*

Roof level	$1.1 \times 6042 \approx 6600$ kN (1485 kips)
Levels 6 to 8	$1.1 \times 4550 \approx 5000$ kN (1125 kips)
Levels 2 to 5	$1.1 \times 4724 \approx 5200$ kN (1170 kips)
Total equivalent weight	

$$\Sigma W_x = 6600 + 3 \times 5000 + 4 \times 5200 = 42{,}400 \text{ kN (9540 kips)}$$

(5) *Total Base Shear*

$$0.09 \times 42{,}400 = 3816 \text{ kN (859 kips)}$$

(b)   Distribution of Lateral Forces over the Height of the Structure   Using Eqs. (2.32*a*) and (2.32*b*) and allocating 10% of the base shear, 382 kN (86 kips), to the roof level, the remaining 90% (3434 kN) (733 kips) is distributed as shown in the table of the determination of lateral design forces.

Determination of Lateral Design Forces

Level	W_x (MN)	h_x (m)	$h_x w_x$ (MNm)	$\dfrac{h_x w_x}{\Sigma h_x w_x}$	F_x (kN)	V_x (kN)
9	6.60	27.45	181.2	0.266	1295^a	
						1295
8	5.00	24.10	120.5	0.177	608	
						1903
7	5.00	20.75	103.8	0.153	526	
						2428
6	5.00	17.40	87.0	0.128	440	
						2868
5	5.20	14.05	73.1	0.107	367	
						3235
4	5.20	10.70	55.6	0.082	282	
						3517
3	5.20	7.35	38.2	0.056	192	
						3709
2	5.20	4.00	20.8	0.031	106	
						3816
1	—	0.00	—	0.000	—	
			680.2	1.000	3816	

$^a F_9 = (0.266 \times 0.9 + 0.1) \, 3816 = 1295$ kN.
The meaning of the floor forces, F_x, and story shear forces, V_x, is shown in Fig. 1.9.
1 kN = 0.225 kips.

(c) Torsional Effects and Irregularities

(1) *Static and Design Eccentricities*: The static eccentricity e_y of the sum of the floor forces F_x (i.e., the story shear V_x) is gradually reduced toward level 1. The lateral force corresponding to the weight of the penthouse acts with an eccentricity of 5.5 m (18 ft) and its magnitude is:

$$F_p = (1.1 \times 1800/6600)(1295 - 382) = 271 \text{ kN (61 kips)}$$

Thus the eccentricity of the total force acting above level 8 is

$$e_y = (271/1295)5.5 = 1.15 \text{ m (3.8 ft)}$$

For the design of the beams at level 3, the average of the story shear forces V_2 and V_3 will be considered:

$$V_{2,3} = 0.5(3709 + 3517) = 3613 \text{ kN (813 kips)}$$

and the corresponding eccentricity is

$$e_y = (271/3613)5.5 = 0.41 \text{ m } (1.34 \text{ ft})$$

Thus in accordance with Section 2.4.3(g) and Eq. (2.37), the design eccentricity is, with $B_b = 33.5$ m,

$$e_{dy} = 0.41 + 0.1 \times 33.5 \approx 3.8 \text{ m } (12.5 \text{ ft})$$

This will govern the design of all bends in the southern half of the plan (Fig. 4.78).

(2) *Check on Irregularity*: As Fig. 4.78 shows, structural irregularities in this system are negligible. However, to illustrate the application of Section 4.2.6, a check will be carried out. For this, information given in Section 4.11.6(d) is used.

(i) *Irregularity in Elevation*: To check the effect of stiffness variations in accordance with the limits suggested in Section 4.2.6(a), consider the shear stiffness of the interior column 6 at the upper levels.

Column 6: $k_c = 525^4/(12 \times 3350) = 1.89 \times 10^6 \text{ mm}^4$

Beam 5–6 with $b/b_w = 1.56$ and $t/h = 0.22$, $f = 1.2$

$k_{56} = 0.35 \times 1.2 \times 400 \times 550^3/(12 \times 5000) = 0.466 \times 10^6 \text{ mm}^4$

Beam 6–7 with $b/b_w = 3.40$, $t/h = 0.22$, $f = 1.62$

$k_{67} = 0.35 \times 1.62 \times 400 \times 550^3/(12 \times 10,000)$

$\qquad = 0.314 \times 10^6 \text{ mm}^4$

Hence from Eqs. (A7) and (A.11)

$$\bar{k} = 2(0.466 + 0.314)/(2 \times 1.89) = 0.0413, \qquad D_{6x} = 0.323$$

At lower levels, from Table 4.7, $D_{6x} = 0.513$ and

$$D_{6x}/D_{6,\text{avg}} = 0.323/0.418 = 0.77 > 0.6 \quad \text{satisfactory}$$

The differences in the stiffness of the perimeter columns are even smaller.

(ii) *Irregularity in Plan*: Significant eccentricity arises only at roof level, where it was shown that $e_y = 1.15$ m. The radius of gyration of story stiffness is, from Table 4.7,

$$r_{dy} = \sqrt{\frac{I_p}{\sum D_{xi}}} = \sqrt{\frac{1431 \times 10^{12}}{9.774 \times 10^6}} = 12.1 \times 10^3 \text{ mm } (39.7 \text{ ft})$$

Thus $e_y/r_{dy} = 1.15/12.1 = 0.095 < 0.15$. This is satisfactory, and hence consideration of torsional effects by this approximate "static" analysis is acceptable.

(d) Distribution of Lateral Forces Among All Columns of the Building   To carry out the approximate lateral force analysis given in Appendix A, the basic stiffness parameters $\bar{k}$ and D_i are to be determined for typical members from Eqs. (A.7) and (A.11). These are set out below. The relative stiffnesses k are those given in Section 4.11.4. D values are given in 10^6 mm^3 units.

(1) $\bar{k}$ and D_i Values: East–West Earthquake

$$\text{Col. 1: } \bar{k} = \frac{2 \times 0.31}{2 \times 0.93} = 0.333, \qquad D_{1x} = \frac{0.333}{2.333} \times 0.93 = 0.133$$

$$\text{Col. 2: } \bar{k} = \frac{4 \times 0.31}{2 \times 0.93} = 0.666, \qquad D_{2x} = \frac{0.66}{2.66} \times 0.93 = 0.232$$

$$\text{Col. 5: } \bar{k} = \frac{2 \times 0.77}{2 \times 0.93} = 0.828, \qquad D_{5x} = \frac{0.828}{2.828} \times 0.93 = 0.272$$

$$\text{Col. 6: } \bar{k} = \frac{2(0.77 + 0.51)}{2 \times 2.58} = 0.496, \qquad D_{6x} = \frac{0.496}{2.496} \times 2.58 = 0.513$$

(2) $\bar{k}$ and D_i Values: North–South Earthquake

$$\text{Col. 1: } \bar{k} = \frac{2 \times 0.28}{2 \times 0.93} = 0.301, \qquad D_{1y} = \frac{0.301}{2.301} \times 0.93 = 0.122$$

$$\text{Col. 2: } \bar{k} = \frac{2 \times 0.62}{2 \times 0.93} = 0.666, \qquad D_{2y} = \frac{0.666}{2.666} \times 0.93 = 0.232$$

$$\text{Col. C: } \bar{k} = \frac{2 \times 0.13}{2 \times 0.93} = 0.140, \qquad D_{cy} = \frac{0.140}{2.140} \times 0.93 = 0.061$$

$$\text{Col. 5: } \bar{k} = \frac{4 \times 0.28}{2 \times 0.93} = 0.602, \qquad D_{5y} = \frac{0.602}{2.602} \times 0.93 = 0.215$$

$$\text{Col. 6: } \bar{k} = \frac{4 \times 0.62}{2 \times 2.58} = 0.481, \qquad D_{6y} = \frac{0.481}{2.481} \times 2.58 = 0.500$$

(3) *Distribution of Unit Story Shear Force in East–West Direction*: Because of assumed double symmetry, the center of rigidity of the lateral-force-resisting system, defined in Section 4.2.5(*b*), is known to be at the center of the floor plan. This is the center of the coordinate system chosen, shown as CR in Fig. 4.5. The computation of the distribution of column shear forces in

TABLE 4.7   Column Shear Forces (E–W)[a]

(1)	(2)	(3)	(4)	(5)	(6)	(7)	(8)	(9)
Col.	n	y (m)	D_{ix}	yD_{ix}	$y^2 D_{ix}$	$\dfrac{D_{ix}}{\Sigma D_{ix}}$	$\dfrac{yD_{ix}e_{dy}}{I_p}$	$\dfrac{V_{ix}}{V_x}$
1	4	16.5	0.133	2.19	36.2	0.0136	0.0058	0.0194
2	4	16.5	0.232	3.83	63.2	0.0237	0.0102	0.0339
C	2	16.5	0.232	3.83	63.2	0.0237	0.0102	0.0339
5	4	11.0	0.272	2.99	32.9	0.0278	0.0080	0.0358
6	4	11.0	0.513	5.64	62.1	0.0525	0.0150	0.0675
9	4	5.5	0.272	1.50	8.2	0.0278	0.0040	0.0318
10	4	5.5	0.513	2.82	15.5	0.0525	0.0075	0.0600
13	2	—	0.272	—	—	0.0278	—	0.0278
14	2	—	0.513	—	—	0.0525	—	0.0525
			9.774		998.8	0.9996		

[a](1) Column identification; (2) number of columns of this type; (3) distance from center of rigidity; (4) D values from Sections 4.11.6(d) and (2) in 10^6-mm^3 units; (7) distribution factors for unit story shear resulting from story translation [Fig. 1.10(a)] only, given by the first term of Eq. (A.20); (8) column shear forces induced by torsion due only to unit story shear force [Fig. 1.10(c)]. The value of I_p given by Eq. (A.19) is, from Tables 4.7 and 4.8,

$$I_p = 998.8 + 432.0 = 1431 \quad \text{and} \quad e_{dy} = -3.8 \text{ m } (12.5 \text{ ft})$$

$$e_{dy}/I_p = 3.8/1431 = 2.66 \times 10^{-3}$$

(9) Total column shear due to story shear $V_x = 1.0$ kN, from Eq. (A.20). The values shown are applicable only to columns in the southern half of the plan shown in Fig. 4.78. In the northern half of the floor plan, the torsional terms are subtractive. For those columns a design eccentricity of $e_{dy} = 3.35 - 0.41 = 2.94$ m (9.6 ft) would need to be considered. For practical reasons, however, structural symmetry would be preserved.

accordance with the procedure given in Appendix A is presented in Tables 4.7 and 4.8.

(e)   Actions in Frame 5–6–7–8 Due to Lateral Forces (Step 1 of Column Design)   With the information obtained, the moments and forces in the frame members are now readily determined. The results are shown in Fig. 4.80. In the following only explanatory notes are provided to show the sequence of computing the lateral-force-induced actions, together with some intermediate computations [Section 4.6.8].

Because the stiffnesses in the upper stories are slightly less, strictly an evaluation of the column shear forces, using appropriate values for k, would need to be carried out. The differences in this case are very small. Hence it

TABLE 4.8   Column Shear Forces (N–S)

(1)	(2)	(3)	(4)	(5)	(6)	(7)	(8)	(9)
Col.	n	x (m)	D_{iy}	xD_{iy}	x^2D_{iy}	$\dfrac{D_{iy}}{\Sigma\,D_{iy}}$	$\dfrac{xD_{iy}e_{dx}}{I_p}$	$\dfrac{V_{iy}}{V_y}$
1	4	10.0	0.122	1.22	12.2	0.0141	0.0017	0.0158
2	4	5.0	0.232	1.16	5.8	0.0269	0.0017	0.0286
C	2	—	0.061	—	—	0.0070	—	0.0070
5	10	10.0	0.215	2.15	21.5	0.0249	0.0031	0.0280
6	10	5.0	0.500	2.50	12.5	0.0580	0.0036	0.0616
			8.627		432.0	1.000		

(1) to (7) These columns contain terms similar to those presented in Table 4.7. (8) In the north–south direction, $e_s = 0$. Therefore, the design eccentricity is $e_d = \pm 0.1 \times 20.5 = 2.05$ m (6.7 ft). Hence $e_{dx}/I_p = \pm 2.05/1431 = \pm 1.43 \times 10^{-3}$.

will be assumed that the values of $\bar{k}$ and D (Appendix A) are the same over the full height of each column.

(1) The assumed points of contraflexure along the columns are first found with the aid of Appendix A, according to Eq. (A.21).

For column 5 with $\bar{k} = 0.828$, as no corrections for changing stiffnesses or story heights need to be made, the locations of points of contraflexure are from Eq. (A.21) and Table A.1 as follows: $\eta = \eta_0 = 0.68, 0.5, 0.5, 0.5, 0.45, 0.45, 0.42$, and 0.35 for the eight stories in ascending order of the story numbers.

Similarly, for column 6 with $\bar{k} \approx 0.50$, we find that $\eta = 0.75, 0.55, 0.50, 0.45, 0.45, 0.45, 0.38$, and 0.25. The location of points of contraflexure (η times story height) above each floor is thus determined as seen in Fig. 4.80.

(2) Using the column shear coefficients from Table 4.7 and the total story shear forces given in the table of Section 4.11.6(b), the column shear forces are recorded at each floor. For example, for column 5 in the first story, $V_{5x} = 0.0358 \times 3816 = 137$ kN (31 kips), and for column 6 in the fourth story $V_{6x} = 0.0675 \times 3235 = 218$ kN (49 kips).

(3) With the known column shear forces and the assumed points of contraflexure, the terminal moments in each column can be computed. For example, for column 5 at level 1, $M_{5x} = 137 \times 2.72 = 373$ kNm (275 kip-ft).

(4) The sums of the column moments above and below a floor give the beam moments at the outer columns. At interior joints the sum of the column terminal moments can be distributed between the two beams in proportion of

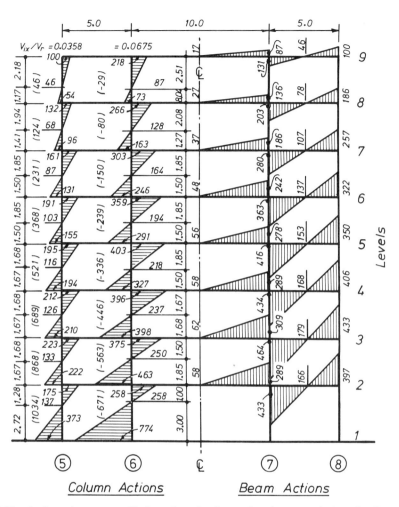

Fig. 4.80 Actions due to specified earthquake forces for the example interior frame (1 kN = 0.225 kips, 1 kNm = 0.738 kip-ft).

the stiffnesses; that is, for beam 6–5,

$$d_{65} = \frac{k_{65}}{k_{65} + k_{67}} = \frac{0.77}{0.77 + 0.51} = 0.60, \qquad d_{67} = 0.40$$

For example, the end moment for the long span at level 2 becomes $M = 0.40(258 + 463) = 289$ kNm (213 kip-ft).

(5) From the beam terminal moments, the beam shear forces are readily determined. For example, the lateral-force-induced shear force across the short span at level 3 is $V = (464 + 433)/5 = 179$ kN (40 kips).

(6) Finally, the column axial loads are derived from the summation of the beam shear forces. For example, in the seventh story, axial forces are for column $5 = 46 + 78 = 124$ kN (28 kips), and for column $6 = -46 + 17 - 78 + 27 = -80$ kN (19 kips).

(7) The critical design quantities for the beams at level 3 due to the lateral forces, as shown in Fig. 4.80, do not differ by more than 5% from those derived from a computer analysis for the entire building.

(f) Actions for Beam 1–2–C–3–4 Due to Lateral Forces The complete pattern of moments, shear, and axial forces may be obtained for this frame in the same fashion as shown for the interior frame in the preceding section. Because only the design of beam 1–2–C at level 3 will be carried out in this illustrative example, column actions are not required at any other level. Therefore, only the beam moments, extracted from the information given in Table of Section 4.11.6(*b*), and Table 4.7 are required here. The column shear forces in kN (kips) are as follows:

	Exterior Column	Interior Columns
Below level 3	$0.0194 \times 3710 = 72$ (16)	$0.0339 \times 3710 = 126$ (28)
Above level 3	$0.0194 \times 3518 = 68$ (15)	$0.0339 \times 3518 = 119$ (27)

The locations of points of contraflexure, η, for the two stories are obtained from Table A.1 as follows:

	Exterior Column	Interior Column
$\bar{k}$:	0.333	0.666
Below level 3	0.57	0.50
Above level 3	0.50	0.50

Hence the lateral-force-induced beam actions are for the half frame as shown in Fig. 4.81.

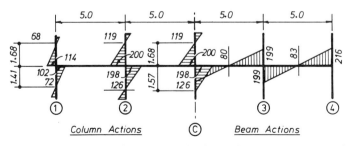

Fig. 4.81 Actions due to specified earthquake forces for the example exterior beam.

4.11.7 Design of Beams at Level 3

(a) Exterior Beams

(1) *Combined Moments*: Before the bending moments are drawn for the purpose of superposition, the beam terminal moments from gravity M_g and from earthquake forces $\overleftarrow{M}_E$ or $\overrightarrow{M}_E$, as discussed in Section 4.3, can be determined algebraically as follows:

Terminal Moments in Beam 1–2–C–3–4 in kNm (kip-ft)[a]

(1) End of Beam	(2) M_g $(D + 1.3L_r)$	(3) $\overleftrightarrow{M}_E$	(4) $M\hat{u} = M_g \pm M_E$
1–2	−32 (24)	±216 (159)	+184 (136)
			−248 (183)
2–1	37 (27)	±199 (147)	+238 (176)
			−160 (119)
2–C	−36 (27)	±199 (147)	+163 (120)
			−235 (173)
C–2	35 (26)	±199 (147)	+234 (173)
			−164 (121)
		$2\Sigma = 1624\ (1199)$	$\Sigma = 1624\ (1199)$

[a](1) The first number refers to the centerline of the column at which the moment is being considered. (2) M_g is obtained from Section 4.11.5(5). Moments applied to the end of a beam are taken positive here when rotating clockwise. (3) M_E is given in Fig. 4.81. (4) These moments are plotted by thin straight lines in Fig. 4.82 for each direction of earthquake attack, $\overrightarrow{E}$ and $\overleftarrow{E}$.

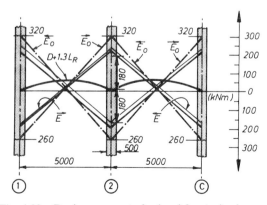

Fig. 4.82 Design moments for level 3 exterior beams.

The baseline of the moment diagram shown in Fig. 4.82 results from gravity only. From Section 4.11.5(1) the maximum ordinate of this at midspan is $38.2 + 1.3 \times 13.2 = 55$ kNm (41 kip-ft). It is seen that the beam design is dominated entirely by lateral forces.

By recalling the aims of moment redistribution set out in Section 4.3.2, the moments at each end of the beams could be made approximately the same. This would be $1624/8 = 203$ kNm (150 kip-ft) and involves a maximum moment redistribution of $(248 - 203)100/248 \approx 18\% < 30\%$ (i.e., less than the suggested limit). The moments so obtained are shown in Fig. 4.82 by heavy straight lines. The critical beam moments at column faces scale approximately to ± 180 kNm (133 kip-ft).

(2) *Flexural Reinforcement for Beam 1–2–C*: Assumed beam size is 500×350 mm (19.7×13.8 in.), as shown in Fig. 4.78.

$$\text{Cover to main beam bars} = 40 \text{ mm } (1.6 \text{ in.})$$

Maximum grade 275 (40 ksi) bar size that can be used at interior joints must be estimated [Section 4.8.6(b)]. The relevant parameters affecting bond quality are, from Eq. (4.57c), $\xi_m = 1.55$, or ignoring the beneficial effects of axial compression, $\xi_p = 1.0$. $\xi_t = 0.85$. $\xi_f = 1.0$. Hence, from Eq. (4.56),

$$d_b = \left[5.4 \times 1.0 \times 0.85 \times 1.0 \times \sqrt{30}\ /(1.55 \times 1.25 \times 275)\right]500 = 23.6 \text{ mm}$$

Hence D24 (0.94-in.-diameter) bars may be used. The effective width of the tension flange of these L beams is, in accordance with Section 4.5.1(b) and Fig. 4.12(b), $b_e = 5000/4 + 350/2 = 1425$ mm (4.7 ft). The effective overhang of the slab is thus $1424 - 350 = 1075$ mm.

Because no detailed design for the slab has been carried out in this example, it will be assumed that all parts of the slab adjacent to a beam face contain 0.25% reinforcement parallel to the beam. Therefore, the top reinforcement, contributing to the flexural strength of the beam, will be taken as

$$A_{s1} = 0.0025 \times 120 \times 1075 = 323 \text{ mm}^2 \ (0.5 \text{ in.}^2)$$

The distance to the centroid of tension and compression reinforcement from the edge of the section will be assumed as $d' = 53$ mm (2.1 in.). Therefore, with $\phi = 0.9$ and $jd \approx 500 - 2 \times 53 = 394$ mm (15.5 in.) from Eq. (4.10),

$$\pm A_s \approx 180 \times 10^6/(0.9 \times 275 \times 394) \approx 1900 \text{ mm}^2 \ (2.95 \text{ in.}^2)$$

Try four D24 bars in the top and in the bottom of the section (1809 mm²).

The total top steel area will then be

$$A_s = 1809 + 323 = 2132 \text{ mm}^2 \ (3.3 \text{ in.}^2)$$

and

$$A'_s = 1809 \text{ mm}^2 \ (2.8 \text{ in.}^2)$$

is the area of bottom bars.

It is evident that for practical reasons equal flexural reinforcement, to resist $M_u = \pm 180$ kNm at every section, could not be provided. However, a small amount of supplementary moment redistribution could readily be carried out to align moment demands more closely with the available resistance. It will be shown subsequently that the total resistance of the bent is adequate. The maximum reinforcement content is with $d = 500 - 53 = 443$ mm (17.6 in.^{2}).

$$\rho = 2132/(350 \times 447) = 0.0136 > 0.4\sqrt{30}\,/275 = 0.008 = \rho_{\min}$$

[Eq. (4.12)]. From Eq. (4.13),

$$\rho_{\max} = \left(\frac{275}{275}\right) \frac{1 + 0.17(30/7 - 3)}{100} \left(1 + \frac{1809}{2132}\right) = 0.023 > \rho$$

Hence the reinforcement content is satisfactory.

Now determine the flexural overstrength at the critical sections of potential plastic hinges, as outlined in Section 4.5.1(*e*). The centroid of tension reinforcement, arranged as shown, is, from extreme fiber, $d' = 40 + 12 \approx 53$ mm (2.1 in.), as assumed. For all positive moments with $\lambda_o = 1.25$, $M_o = 1.25 \times 275 \times 1809 \times 394 \times 10^{-6} = 245$ kNm (181 kip-ft) is the flexural overstrength. For negative moments with $A_s = 2132$ mm^2, $M_o = 289$ kNm (213 kip-ft).

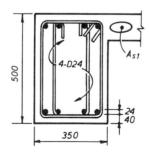

The bending moments at overstrength, with these values at column faces, are plotted in Fig. 4.82, and these are shown as $\vec{E}_o$ or $\overleftarrow{E}_o$. The resulting moments at the column centerlines are 260 and 320 kNm, respectively. The adequacy of the design for lateral forces can now be checked by forming the ratio of the scaled overstrength story moments and the corresponding moments from the lateral force analysis, given in Section 4.11.7(*a*) and Fig. 4.81, which is the system overstrength factor:

$$\psi_o \approx \frac{4 \times 260 + 4 \times 320}{1624} = 1.43 > 1.39 = 1.25/0.9$$

This indicates a very close design. Gravity load alone obviously does not control beam reinforcement.

(3) *Shear Strength*:   The design for shear would follow the procedure shown for the interior beams 5–6–7–8. Therefore, details are not given here. The maximum shear stress is on the order of only $0.04f_c'$ MPa. R10 (0.39-in.-diameter) stirrups on 110-mm (4.3-in.) centers would resist the entire shear at overstrength in the plastic hinge regions.

(b)   Interior Beams (Step 2 of Column Design)

(1) *Combined Moments for $D + 1.3L_r + E$*:   As for the exterior beam, the terminal moments from gravity loads M_g and earthquake forces M_E are superimposed in the table below.

Terminal Moments in Beam 5–6–7–8 in kNm (kip-ft)[a]

(1) End of Beam	(2) M_g $(D + 1.3L_r)$	(3) $\overset{\leftrightarrow}{M}_E$	(4) $M_u = M_g \pm M_E$
5–6	−37 (27)	±433 (320)	+396 (292) −470 (247)
6–5	87 (64)	±464 (342)	+551 (407) −377 (278)
6–7	−321 (237)	±309 (228)	−12 (9) −630 (465)
	$\Sigma M_g = 0$	$\Sigma M_E = 2412$ (1780)	$\Sigma M_u = 2412$ (1780)

[a]See footnotes for table in Section 4.11.7(a).

The moment diagram may now be constructed as follows:

   (i) The midspan moments for each span are established from Sections 4.11.5(3) and (4) as follows:

   Short-span SSM: $58.7 + 1.3 \times 26.3 \approx 93$ kNm (69 kip-ft)

   Long-span SSM: $376.8 + 1.3 \times 132.6 \approx 549$ kNm (405 kip-ft)

   (ii) The terminal moments $M_g + M_E$ from the table above are plotted by thin straight lines. Values are shown only in the right-hand half of the bending moment diagram in Fig. 4.83(a) by lines $\vec{E}$ and $\overset{\leftarrow}{E}$.

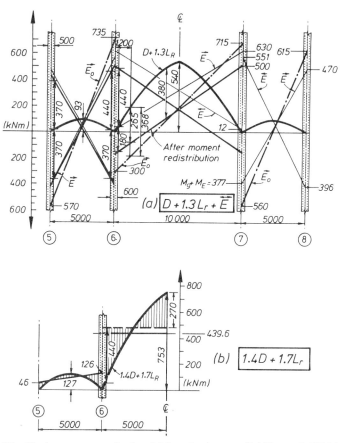

Fig. 4.83 Design moments for level 3 interior beams. (1 kNm = 0.738 kip-ft.)

(iii) Moment redistribution is carried out in accordance with the principles of Section 4.3. The maximum reduction is applied at the ends of the center span [i.e., $(630 - 500)/630 \approx 0.21 < 0.30$]. Thereby the critical negative moments at the interior column faces are made 440 kNm (325 kip-ft), and all other moments at the ends of the short spans are made approximately 370 kNm (273 kip-ft). These moments are shown in Fig. 4.83(a) by the heavy straight lines.

(iv) An approximate check with scaled terminal moments is carried out to ensure that no moments were lost in the process of redistribution. As the table shows, these scaled terminal moments must be equal to $\Sigma M_E = 2412$ kNm (1780 kip-ft).

(2) *Combined Moments for 1.4D + 1.7L$_r$:* For only one-half of the beam, the gravity moments are plotted in Fig. 4.83(b). It is seen that in the side spans, gravity-load-induced moments are insignificant. The center moment at midspan of span 6–7 is, from Section 4.11.5(4), SSM = 1.4 × 376.8 + 1.7 × 132.6 ≈ 753 kNm (556 kip-ft), and the support column centerline negative moment is, from moment distribution of Section 4.11.5(6), 440 kNm (325 kip-ft). It is seen that a small moment increase at the support will bring the critical moment at the column face there to 440 kNm (325 kip-ft), the same as for the previous load combination. The midspan moment of 270 kNm (199 kip-ft) is not critical.

(3) *Determination of the Flexural Reinforcement (Step 3 of Column Design):* As Fig. 4.78 shows, the assumed size of the beam is 650 × 400 mm (25.6 × 15.7 in.). When selecting suitable beam bar sizes the anchorage conditions at the exterior columns [Section 4.8.11(a) and Fig. 4.69(a)] and within the joint at interior columns [Section 4.8.6(b) and Eq. (4.56)] must be considered.

At column 5, D24 (0.94-in.-diameter) bars will be used. From Section 3.6.2(a)(ii) and Eq. (3.68), the basic development length for a hooked bar is $l_{hb} = 0.24 \times 24 \times 275/\sqrt{30} = 289$ mm (11.4 in.), and with allowance for confinement within the joint core, $l_{dh} = 0.7 \times 289 = 202$ mm (8 in.). From Fig. 4.69(a) the available development length is $l_{dh} = 500 - 10 \times 24 - 50 = 210$ mm (8.3 in.). Hence D24 is the maximum usable bar size. Less development length is available to beam bars in the second layer. Hence reinforcement there will be limited to D20 (0.79-in.-diameter) bars.

By considering conditions similar to those for the exterior beams 1–2–C, bars which may be used at interior joints with an allowance for two-way frames ($\xi_f = 0.9$) may have a diameter on the order of $d_b \approx (0.9 \times 600/500)23.6 = 25.5$ mm. To allow for more flexibility in curtailment of beam bars, D24 (0.94 in.) and D20 (0.79 in.) bars will in general be used.

(i) At column 5 the effective overhang of the tension flange is, from Section 4.5.1(b) and Fig. 4.12: $b_e = 2 \times 5000/4 - 400 = 2100$ mm (82.7 in.). Therefore, the assumed contribution of slab reinforcement is

$$A_{s1} = 0.0025 \times 120 \times 2100 = 630 \text{ mm}^2 \ (0.98 \text{ in.}^2)$$

Assume that $d' = 40 + 24 + 0.5 \times 24 = 76$, say 75 mm (3 in.)

$\therefore \quad d = 650 - 75 = 575$ and $d - d' = 500$ mm (19.7 in.)

and $(+)A_s = 370 \times 10^6/(0.9 \times 275 \times 500)$

$$\approx 3000 \text{ mm}^2 \ (4.65 \text{ in.}^2)$$

Try four D24 + four D20 = 3067 mm^2 in the bottom of the section.

$$M_i = 422 \text{ kNm} \ (311 \text{ kip-ft})$$

Provide four D24 and two D20 bars in the top of the beam so that

$$(-)A_s = 2439 + 630 = 3069 \text{ mm}^2$$
$$M_i = 422 > 370/0.9 = 411 \text{ kNm}$$

(ii) At column 6, the effective slab width being larger than 2100 mm, more slab reinforcement will participate. The effective tensile steel area in the flanges will be assumed to be $A_{s1} = 1000 \text{ mm}^2$ (1.55 in.²). Assume that $d - d' = 500$ mm.

$$(-)A_s = 440 \times 10^6/(0.9 \times 275 \times 500) \approx 3560 \text{ mm}^2 \text{ (5.5 in.}^2)$$

Use six D24 bars = 2714 mm² (4.2 in.²).

$$(-)A_{s,\text{provided}} = 2714 + 1000 = 3714 \text{ mm}^2 \text{ (5.8 in.}^2)$$
$$\therefore \quad M_i = 511 \text{ kNm (377 kip-ft)}$$
$$\rho = 3714/(400 \times 575) = 0.0161 \quad \text{satisfactory}$$

From Section 4.5.1(c) the minimum bottom steel to be used at this section is $A_s' = 0.5A_s = 1857 \text{ mm}^2$ (2.9 in.). The bottom beam reinforcement to the left of column 6 is the same as at column 5. Hence carry through four D24 bars from the short span into the long span, giving $A_s' = 1810 \approx 1857 \text{ mm}^2$, and terminate four D20 bars at the inner face of column 6 as shown in Fig. 4.84.

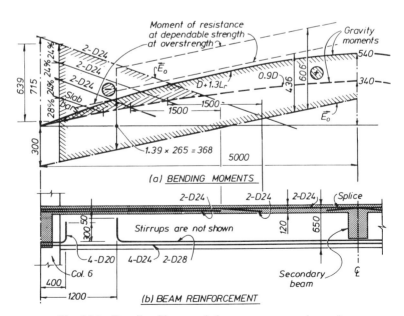

Fig. 4.84   Details of beam reinforcement near column 6.

This figure also shows how the top and bottom bars should be curtailed. The maximum extent of the negative moment will occur when the load $E_o + 0.9D$ is considered. Two D24 bars resist 24% of the maximum moment, respectively, at the face of column 6. According to Section 3.6.3, these bars need to be extended by 1500 mm (4.9 ft) past the points at which, according to the bending moment diagram, they are not required to resist any tension. Figure 4.84 shows how these points have been determined.

(iii) The positive moment of resistance of four D24 (0.94-in.-diameter) bars in the long span is as follows: The effective width of the compression flange in the span is, from Fig. 4.3:

$$b = 400 + 16 \times 120 = 2320 \text{ mm } (7.6 \text{ ft})$$
$$\therefore \quad a = 1810 \times 275/(0.85 \times 30 \times 2320) = 8.4 \text{ mm } (0.33 \text{ in.})$$
$$\text{Take } jd = d - 0.5a \approx (650 - 52) - 5 = 593 \text{ mm } (23.2 \text{ in.})$$
$$\therefore \quad M_i = 1810 \times 275 \times 593 \times 10^{-6} = 295 \text{ kNm } < 380/0.9$$
$$= 422 \text{ kNm } (311 \text{ kip-ft})$$

[See Fig. 4.83(a).]

In order to place the positive plastic hinge nearer to column 6, and to satisfy positive moment requirements at midspan, additional bars must be placed in the bottom of the beam. As Fig. 4.83(a) shows, at 1200 mm (3.9 ft) to the right of column 6, the ideal moment demand is $265/0.9 \approx 295$ kNm (218 kip-ft). Hence at this location a plastic hinge may be formed with four D24 (0.93-in.-diameter) bars in the bottom. Beyond this section provide two additional D28 (1.1-in.-diameter) bars, which will increase the flexural strength to approximately 436 kNm (322 kip-ft), so that yielding in the central region of the long span, where bottom bars will need to be spliced, can never occur. This is shown in Fig. 4.84.

(iv) The flexural overstrengths of the potential plastic hinge sections, as detailed, are as follows:

At 5–6:	$\overrightarrow{M}_o = 1.25 \times 422 = 528$ kNm (390 kip-ft)	
	$\overleftarrow{M}_o = 1.25 \times 422 = 528$ kNm (390 kip-ft)	
At 6–5:	$\overrightarrow{M}_o = 1.25 \times 511 = 639$ kNm (471 kip-ft)	
	$\overleftarrow{M}_o \qquad\qquad = 528$ kNm (390 kip-ft)	
At 6–7:	$\overleftarrow{M}_o \qquad\qquad = 639$ kNm (471 kip-ft)	
At 1200 mm from column 6:	$\overrightarrow{M}_o = 1.25 \times 295 = 368$ kNm (272 kip-ft)	

These moments are plotted in Fig. 4.83(a), and from these the moments at column centerlines are obtained. For the sake of clarity they are shown for the force direction $\vec{E}_o$ only. This completes step 4 of column design. When these moments are compared with those of Fig. 4.80, the overstrength factors, subsequently required for the design of the columns, are found as follows:

At col. 5: $\quad \vec{\phi}_o = 570/433 \qquad\qquad\qquad = 1.32, \; \overleftarrow{\phi}_o = 1.42$

At col. 6: $\quad \vec{\phi}_o = (735 + 300)/(464 + 309) = 1.34, \; \overleftarrow{\phi}_o = 1.65$

At col. 7: $\quad \vec{\phi}_o = (715 + 560)/(464 + 309) = 1.65, \; \overleftarrow{\phi}_o = 1.34$

At col. 8: $\quad \vec{\phi}_o = 615/433 \qquad\qquad\qquad = 1.42, \; \overleftarrow{\phi}_o = 1.32$

(This is step 5 of column design.)

From Section 1.3.3(g) the system overstrength factor, indicating the closeness of the design for seismic requirements, is

$$\psi_o = (570 + 735 + 300 + 715 + 560 + 615)/2412 = 1.45 > 1.39$$

(4) *Design for Shear Strength*

(i) *Span 5–6*: From Section 4.5.3 and Eq. (4.18), using the load combination given by Eq. (1.7), we find at column centerlines:

$$\vec{V}_5 = (615 + 560)/5 + (76.8 + 31.5)/2 = 235 + 54 = 289 \text{ kN (65 kips)}$$
$$\overleftarrow{V}_5 = (570 + 735)/5 - 0.9 \times 76.8/2 \qquad\qquad = 226 \text{ kN (51 kips)}$$

The variations of shear due to gravity load along the span are small; therefore, the shear forces above will be considered for the design of the entire span. At the right-hand support,

$$\vec{V}_6 = (735 + 570)/5 + (76.8 + 31.2)/2 = 315 \text{ kN (71 kips)} > \overleftarrow{V}_6$$

Therefore, at the development of beam overstrength,

$$v_i < 315{,}000/(400 \times 575) = 1.37 \text{ MPa (194 psi)} < 0.16 f_c' \; [\text{Eq. (3.31)}]$$

Check whether sliding shear is critical [Section 3.3.2(b)]. From Eq. (3.44),

$$r = -\frac{(615 + 560)/5 - 38.4}{(735 + 570)/5 + 38.4} = -0.66$$

and from Eq. (3.43),

$$v_i = 0.25(2 - 0.66)\sqrt{30} = 1.83 > 1.37 \text{ MPa}$$

Hence no diagonal shear reinforcement is required.
When $v_c = 0$, from Eq. (3.40)

$$A_v/s = v_i b_w/f_y = 1.37 \times 400/275$$
$$= 1.99 \text{ mm}^2/\text{mm} \left(0.078 \text{ in.}^2/\text{in.}\right)$$

The spacing limitations for stirrup ties in the plastic hinge regions are, from Section 3.6.4,

$$s \leq 6d_b = 6 \times 24 = 144 \text{ mm} (5.7 \text{ in.})$$

or

$$s \leq 150 \text{ mm} (5.9 \text{ in.})$$

or

$$s \leq d/4 = 143 \text{ mm} (5.6 \text{ in.})$$

Also consider lateral support for compression bars (Section 4.5.4):

$$\Sigma A_b = 452 + 314 = 766 \text{ mm}^2$$

to allow for two bars in the vertical plane. Hence, from Eq. (4.19b),

$$A_{te}/s = \Sigma A_b f_y/(1600 f_{yt}) = 766 \times 275/(1600 \times 275)$$
$$= 0.48 \text{ mm}^2/\text{mm for one leg} < 1.99/4$$

Thus shear requirements govern. Using four R10 (0.39-in.-diameter) stirrup legs with $A_v = 314 \text{ mm}^2$, $s = 314/1.99 = 158$ mm (6.2 in.). Hence use four R10 legs on $140 < 143$ mm centers.

In the central (elastic) region of the beam using Eq. (3.33) with $\rho_w \geq 1810/(400 \times 575) = 0.008$,

$$v_c \geq (0.07 + 10 \times 0.008)\sqrt{f_c'} = 0.82 \text{ MPa} (119 \text{ psi})$$
$$A_v/s < (1.37 - 0.82)400/275 = 0.80 \text{ mm}^2/\text{mm} (0.31 \text{ in.}^2/\text{in.})$$
$$s < d/2 = 287 \text{ mm} (11.3 \text{ in.})$$

Use two legs of R10 at 200-mm (7.9-in.) spacing, so that

$$A_v/s = 157/200 = 0.79 \approx 0.80 \text{ mm}^2/\text{mm}$$

(ii) *Span 6–7*

$$V_6 = (715 + 300)/10 + (227.5 + 77.3)/2 = 102 + 152$$
$$= 254 \text{ kN } (57 \text{ kips})$$

There is no shear reversal due to earthquake action. The maximum shear due to factored gravity load alone is of similar order. Provide stirrups in the end region as in span 5–6. The shear calculations for the remainder of the noncritical span are not given here.

4.11.8   Design of Columns

(a)   Exterior Column 5 at Level 3   Consider the section just below level 3. Subsequently, the section above this level would also need to be checked.

(1) The earthquake-induced axial force would be derived from Eq. (4.30). However, in this example the beam shear forces V_{oe} at levels above level 3 are not available. Therefore, it will be assumed that with an average value of $\phi_o = 1.50$ for the upper-level beams, the term ΣV_{oe} is, from Fig. 4.80, $1.5 \times 868 = 1302$ kN (293 kips).

With the period T_1 taken as 1.0 s, from Table 4.3, $\omega = 1.6$ for a two-way frame (step 6), and from Table 4.5, $R_v = 0.875$ for seven levels. Hence, from Eq. (4.30),

$$P_{eq} = 0.875 \times 1302 = 1139 \text{ kN } (256 \text{ kips})$$

(2) For maximum compression on the column (step 7), for a load combination $U = D + L_R + \overleftarrow{E}_o$,

$$P_u = 782.7 + 146.1 + 1139 = 2068 \text{ kN } (465 \text{ kips})$$

From Eq. (4.33), Fig. 4.80, and with $\phi_o = 1.42$, the design column shear force (step 8) is

$$V_u = 1.6 \times 1.42 \times 133 = 302 \text{ kN } (68 \text{ kips})$$

so that from Fig. 4.80 and with Eq. (4.28) and $R_m = 1$ (step 9),

$$M_u = 1(1.42 \times 1.6 \times 223 - 0.3 \times 0.65 \times 302) = 448 \text{ kNm } (330 \text{ kip-ft})$$

(3) For minimum compression or for tension load from $0.9D + \overrightarrow{E}_o$,

$$P_u = 0.9 \times 782.7 - 1139 = -435 \text{ kN } (99 \text{ kips}) \text{ (tension)}$$

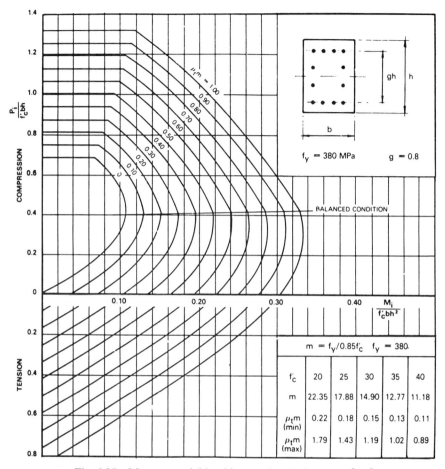

Fig. 4.85 Moment–axial load interaction design chart [N3].

With $\vec{\phi}_o = 1.32$, $V_{col} = 1.6 \times 1.32 \times 133 = 281$ kN (63 kips). With $P_u/f'_c A_g = -435,000/(30 \times 500^2) = -0.058$, from Table 4.4, $R_m = 0.60$, and hence from Eq. (4.28),

$$M_u = 0.60(1.32 \times 1.6 \times 223 - 0.3 \times 0.65 \times 281) = 250 \text{ kNm } (184 \text{ kip-ft})$$

(4) Using the chart of Fig. 4.85, giving the ideal strength of the column section, with $g = (500 - 2 \times 40 - 24)/500 \approx 0.8$, $m = 14.9$, $P_u/f'_c bh = -0.058$, and

$$M_u/f'_c bh^2 = 250 \times 10^6/(30 \times 500^3) = 0.067$$
$$\rho_t = 0.25/14.9 = 0.0168$$

When axial compression is considered with

$$P_u/f_c'bh = 2,068,000/(30 \times 500^2) = 0.276$$
$$M_u/f_c'bh^2 = 448 \times 10^6/(30 \times 500^3) = 0.119$$
$$\rho_t < 0.0168$$

Hence we require that

$$A_s = 0.0168 \times 500^2 = 4200 \text{ mm}^2 \ (6.5 \text{ in.}^2)$$

Provide four HD24 (0.94-in.-diameter) and eight HD20 (0.79-in.-diameter) bars = 4323 mm² (6.7 in.²). This is to be reviewed when the beam–column joint is designed.

(5) The design of the transverse reinforcement for this column is not given, as it is similar to that for column 6. When maximum compression of $P_u = 2068$ kN (465 kips) is acting, the required confining reinforcement in accordance with Section 4.6.11(e) is $A_{sh} = 2.17$ mm²/mm (0.085 in.²/in.).

(b)   Interior Column 6 at Level 3

(1) The estimated earthquake-induced axial load, with an assumed average value of $\phi_o = 1.64$ at the upper levels, is, by similarity to the previous approximation from Fig. 4.80 (step 7),

$$P_{eq} = 0.875 \times 1.64 \times 563 = 808 \text{ kN } (187 \text{ kips})$$

For maximum compression considering $D + L_R + E_o$ from Section 4.11.5(8) is

$$P_u = 2039 + 375 + 808 \approx 3222 \text{ kN } (725 \text{ kips})$$

With $\phi_o = 1.34$ [Section 4.11.7(b)(3)]

$$V_u = 1.6 \times 1.34 \times 250 = 536 \text{ kN } (121 \text{ kips}) \quad (\text{step 8})$$
$$M_u = 1(1.34 \times 1.6 \times 375 - 0.3 \times 0.65 \times 536)$$
$$= 699 \text{ kNm } (516 \text{ kip-ft}) \quad (\text{step 9})$$

For minimum compression considering $0.9D + E_o$,

$$P_u = 0.9 \times 2039 - 868 \approx 1027 \text{ kN } (231 \text{ kips})$$

Therefore, with $R_m = 1$ and from Section 4.11.7(b)(3), $\overleftarrow{\phi}_o = 1.65$ and

$$V_u = 1.6 \times 1.65 \times 250 = 660 \text{ kN } (149 \text{ kips})$$
$$M_u = 1.6 \times 1.65 \times 375 - 0.3 \times 0.65 \times 660 = 861 \text{ kNm } (636 \text{ kip-ft})$$

(2) The steel requirements are with $g \approx 0.8$ from the chart of Fig. 4.85 with $1{,}027{,}000/(30 \times 600^2) = 0.095$ and $861 \times 10^6/(30 \times 600^3) = 0.133$,

$$\rho_t = 0.28/14.9 = 0.019$$

It is evident that the load with $P_u = 3222$ kN (725 kips) is not critical. Hence we require that

$$A_{st} = 0.019 \times 600^2 = 6840 \text{ mm}^2 \ (10.6 \text{ in.}^2)$$

Sixteen HD24 (0.94-in.-diameter) bars $= 7238$ mm^2 (11.2 in.^{2}) could be provided.

The section above level 3 should also be checked because, as Fig. 4.80 shows, the design moment is somewhat larger while the minimum axial compression load on the column is smaller. It will be found that approximately 8% more steel area is required than at the section below level 3 [i.e., 7400 mm^2 (11.5 in.^{2})]. Provide four HD28 (1.1-in.-diameter) and 12 HD24 (0.94-in.-diameter) bars giving 7892 mm^2 (12.2 in.^{2}) [$\rho_t = 0.0219$].

(3) *Transverse reinforcement*

(i) The shear strength for $\overleftarrow{E}_o$, from the preceding section, is

$$V_u = 660 \text{ kN } (149 \text{ kips}) \text{ and with } d = 0.8h_c,$$

$$v_i = 660{,}000/(0.8 \times 600^2) = 2.29 \text{ MPa } (332 \text{ psi}) = 0.076 f_c'$$

From Eqs. (3.33) and (3.34) with $A_s \approx 0.3 A_{st}$,

$$\rho_w \approx 0.3 \times 7892/(0.8 \times 600^2) = 0.008$$

$$v_b = (0.07 + 10 \times 0.008)\sqrt{30} = 0.82 \text{ MPa } (119 \text{ psi})$$

As no plastic hinge is expected,

$$v_c = (1 + 3 \times 0.095)0.82 = 1.05 \text{ MPa } (152 \text{ psi})$$

from Eq. (3.40); therefore,

$$A_v/s = (2.29 - 1.05)600/275 = 2.71 \text{ mm}^2/\text{mm } (0.107 \text{ in.}^2/\text{in.})$$

The shear associated with the other direction of earthquake attack $\overrightarrow{E}_o$ ($V_u = 536$ kN and $P_u = 3222$ kN) is obviously not critical.

(ii) Confinement requirements from Eq. (3.62):

$$P_u/f_c' A_g = 3{,}222{,}000/(30 \times 600^2) = 0.30$$

$$h'' \approx 600 - 2 \times 40 + 10 = 530 \text{ mm } (20.9 \text{ in.})$$

Hence if a plastic hinge with significant curvature ductility would develop, the transverse reinforcement required, with $k = 0.35$ and $A_g/A_c = (600/530)^2 = 1.28$, is

$$A_{sh}/s_h = [0.35(30/275)1.28(0.30 - 0.08)]530$$
$$= 5.7 \text{ mm}^2/\text{mm} \ (0.225 \text{ in.}^2/\text{in.})$$

For columns above level 1, according to Section 4.6.11(e), however,

$$A_{sh}/s_h = 0.5 \times 5.7 = 2.85 > 2.71 \text{ mm}^2/\text{mm}$$

This requirement is similar to but slightly more severe than that for shear.

(iii) Spacing limitations from Sections 3.3.2(a) and 3.6.4: $s_h \leq 6d_b = 6 \times 24 = 144$ mm (5.7 in.) and $s_h \leq h_c/4 = 600/4 = 150$ mm (5.9 in.)

(iv) Stability requirements for the four HD28 (1.1-in.-diameter) corner bars [Eq. (4.19b)] require that

$$\frac{A_{te}}{s} = \frac{616 \times 380}{1600 \times 275} = 0.532 \text{ mm}^2/\text{mm} \ (0.021 \text{ in.}^2/\text{in.}) \text{ per leg}$$

(v) Splice requirements from Eq. (3.70) above level 3 indicate

$$\frac{A_{tr}}{s} = \frac{28 \times 380}{50 \times 275} = 0.774 \text{ mm}^2/\text{mm} \ (6.03 \text{ in.}^2/\text{in.})$$

per leg for HD28 (1.1-in.-diameter) and 0.663 mm²/mm for HD24 (0.94-in.-diameter) bars.

(vi) When using the arrangement shown in Fig. 4.31(b), the number of effective legs is 5.4. Hence

$$A_{sh}/s_h = 2.85/5.4 = 0.53 \text{ mm}^2/\text{mm} \ (0.021 \text{ in.}^2/\text{in.}) \text{ per leg}$$

As splice requirements will control the design in the end region, we require R12 (0.47-in.-diameter) peripheral ties [$A_{te} = 113$ mm² (0.18 in.²)] at $s = 113/0.774 = 146$ mm and intermediate R10 (0.39-in.-diameter) ties at $78.5/0.663 = 118$-mm spacing. Provide R10 and R12 ties on 120-mm (4.7-in.) centers.

(vii) The end region of the column is defined in Section 4.6.11(e) and Fig. 4.33.

(viii) In the middle portion of the column, shear requirements will govern. Hence $A_v/s = 2.71$ mm²/mm (0.107 in.²/in.). With

$A_v = 2 \times 113 + 3.4 \times 78.5 = 493$ mm^2 (0.76 in.^{2}), $s = 493/2.71 = 182$ mm. Use 180-mm (7.1-in.) spacing and this will satisfy all spacing requirements for this region.

(c) Interior Column 6 at Level 1   To illustrate the different requirements for transverse reinforcement in a potential plastic hinge region, some features of the base section design are presented.

(1) Estimated axial column loads are, from Section 4.11.5(8),

$$\text{Dead load: } 2039 + 259 \approx 2300 \text{ kN (516 kips)}$$
$$\text{Live load:} \qquad\qquad \approx 420 \text{ kN (95 kips)}$$

Earthquake-induced axial force from Fig. 4.80 with allowance for force reduction from Table 4.5 with $\omega < 1.3$ and hence $R_v = 0.88$ is

$$P_{eq} = 0.88(1.64 \times 671) = 968 \text{ kN (218 kips)}$$

(2) The base moment is from Sections 4.6.3(b) and 4.6.4(c) and Figs. 4.22 and 4.80,

$$M_u = \omega M_E/\phi = 1.1 \times 774/0.9 = 946 \text{ kNm (698 kip-ft)}$$

(3) For the minimum axial compression,

$$P_u = 0.9 \times 2300 - 968 = 1102 \text{ kN (248 kips)}$$

we find from Fig. 4.85 that $\rho_t = 0.023$. Provide 16 HD28 (1.1-in.-diameter) bars, $A_{st} = 9856(15.3 \text{ in.}^2)$, $\rho_t = 0.0274$. The moment with maximum axial compression

$$P_u = 2300 + 1.0 \times 420 + 968 = 3688 \text{ kN (830 kips)}$$

does not govern the design.

(4) Transverse reinforcement

(i) Because column shear in the first story is related to the flexural overstrength of the base section, this must be estimated. From the reinforcement provided ($\rho_t = 0.0274$) and Fig. 4.85, the ideal strengths are found to be

$$M_i = 1082 \text{ kNm (799 kip-ft) when } P_i = P_{min} = 1102 \text{ kN (248 kips)}$$

and

$$M_i = 1250 \text{ kNm (923 kip-ft) when } P_i = P_{max} = 3688 \text{ kN (830 kips)}$$

Based on Fig. 3.22 for low axial compression when 1.13 is replaced by $\lambda_o = 1.4$,

$$M_{max} = \lambda_o M_i = 1.4 \times 1082 = 1515 \text{ kNm } (1118 \text{ kip-ft})$$

and for the large axial compression from Eq. (3.28),

$$M_{max} = 1.40 + 2.35\left(\frac{3688 \times 10^3}{30 \times 600^2} - 0.1\right)1250$$
$$= 1921 \text{ kNm } (1418 \text{ kip-ft})$$

Hence by assuming that the relevant flexural overstrength factor for column 6 at level 2 is $\phi_o = 1035/721 \approx 1.50$, from Eq. (4.34) and Figs. 4.78 and 4.80

$$V_u = (1515 + 1.6 \times 1.5 \times 258)/4 = 534 \text{ kN } (120 \text{ kips}) \text{ with } P_{min}$$

and

$$V_u = (1921 + 1.6 \times 1.5 \times 258)/4 = 635 \text{ kN } (143 \text{ kips}) \text{ with } P_{max}$$

Thus shear stresses are, from Eq. (3.33),

$$v_b = (0.07 + 10 \times 0.0274/3)\sqrt{30} = 0.88 \text{ MPa } (128 \text{ psi})$$

Where $P_i = P_{min}$ is considered:

From Eq. (3.38): $v_c = 4 \times 0.88\sqrt{1102 \times 10^3/(30 \times 600^2)}$
$$= 1.12 \text{ MPa } (163 \text{ psi})$$
From Eq. (3.29): $v_i = 534,000/(0.8 \times 600^2)$
$$= 1.85 \text{ MPa } (269 \text{ psi})$$

Hence, from Eq. (3.40),

$$A_v/s = (1.85 - 1.12)600/275 = 1.59 \text{ mm}^2/\text{mm}$$

When $P_i = P_{max}$ is considered, from Eq. (3.38),

$$v_c = 4 \times 0.88\sqrt{(3688 \times 10^3)/(30 \times 600^2)} = 2.06 \text{ MPa } (298 \text{ psi})$$

and

$$v_i = 635,000/(0.8 \times 600^2) = 2.20 \text{ MPa } (320 \text{ psi})$$

As $(v_i - v_c)$ is rather small, this case is not critical.

(ii) For confinement of the plastic hinge region with $P_i = P_{max}$, from Eq. (3.62),

$$\frac{A_{sh}}{s_h} = 0.35 \times \frac{30}{275} \times 1.28 \left(\frac{3688 \times 10^3}{30 \times 600^2} - 0.08 \right) 530$$

$$= 6.77 \text{ mm}^2/\text{mm} \ (0.267 \text{ in.}^2/\text{in.})$$

(iii) Stability requirements, which are the same for the HD28 (1.1-in.-diameter) bars as at level 3, indicate that

$$A_{te}/s = 5 \times 0.532 = 2.66 \text{ mm}^2/\text{mm} \ (0.11 \text{ in.}^2/\text{in.})$$

(iv) The splicing of column bars at the base is not permitted.
(v) Thus confinement requirements govern and when R12 (0.47-in.-diameter) ties with 5.4 effective legs, as shown in Fig. 4.31(b), are used, the spacing of the sets will be

$$s_h = 5.4 \times 113/6.77 = 90 \text{ mm} \ (3.5 \text{ in.})$$

Alternatively, R16 (0.63-in.-diameter) ties in 150-mm (6-in.) centers may be used.

From Section 4.6.11(e)(i) and Fig. 4.33, the extent of the end region at the column base to be confined is 1200 mm, i.e.

$$l_0 = 1.5 \times 600 = 900 \text{ mm} < 0.3 \times 4000 = 1200 \text{ mm} \ (3.94 \text{ ft})$$

because $P_{max}/(f'_c A_g) > 0.3$.

4.11.9 Design of Beam–Column Joints at Level 3

(a) Interior Joint at Column 6

(1) From the evaluation of the overstrength factors in Section 4.11.7(b)(3)(iv) it appears that earthquake actions corresponding with $\overleftarrow{E}_o$ will be critical. For this case, from Section 4.11.8(b), the minimum axial compression force above level 3 results from

$$0.9P_D = 0.9(2039 - 259) \quad = 1602 \text{ kN} \ (360 \text{ kips})$$

$$P_{eq} = 0.89 \times 1.64 \times 446 \quad = -651 \text{ kN} \ (146 \text{ kips})$$

$$\therefore \quad P_u \quad\quad\quad\quad\quad\quad = 951 \text{ kN} \ (214 \text{ kips})$$

From Section 4.8.3(b), Eq. (4.48), or Fig. 4.83(a),

$$\overleftarrow{V}_{col} \approx (715 + 560)/3.35 = 381 \text{ kN } (87 \text{ kips})$$

From Sections 4.11.7(b)(3)(i) and (ii), the internal beam tension forces are

$$T = 1.25 \times 275 \times 3714 \times 10^{-3} = 1277 \text{ kN } (287 \text{ kips})$$

and

$$T' = 1.25 \times 275 \times 3069 \times 10^{-3} = 1055 \text{ kN } (237 \text{ kips})$$

and thus from Eq. (4.47b), the horizontal joint shear force is

$$V_{jh} = 1277 + 1055 - 381 = 1951 \text{ kN } (439 \text{ kips})$$

From Eq. (4.73),

$$v_{jh} = 1,951,000/600^2 = 5.42 \text{ MPa } (786 \text{ psi})$$

The joint shear stress should not exceed [Eq. (4.74b)]

$$v_{jh} = 0.2 f'_c = 6 \text{ MPa} \quad \text{satisfactory}$$

(2) To evaluate the necessary amount of joint shear reinforcement, the contribution of axial compression on the column, in accordance with Eq. (4.77), needs to be estimated. From Tables 4.7 and 4.8 it is seen that

$$C_j = \frac{V_E}{V_{Ex} + V_{Ey}} \approx \frac{0.0565}{0.0565 + 0.0580} \approx 0.49$$

Hence for joint design, $P_u = 0.49 \times 951 = 466$ kN (105 kips). Therefore, from Eq. (4.67b),

$$V_{sh} = \left[1.15 - 1.3 \times 466,000/(30 \times 600^2)\right]1277 = 1397 \text{ kN } (314 \text{ kips})$$

The required joint shear reinforcement is thus, from Eq. (4.70a) or (4.70b),

$$A_{jh} = 1397 \times 10^3/275 = 5080 \text{ mm}^2 \ (7.9 \text{ in.}^2)$$

A_{jh} represents a large amount of transverse reinforcement in the joint. Alternatives for its placement should be considered.

> (i) If ties are arranged in four sets on approximately $500/4 = 125$ mm (5 in.) centers, as in the column sections above and below the joint shown in Fig. 4.31(b), the area of one leg should be $A_b \approx$

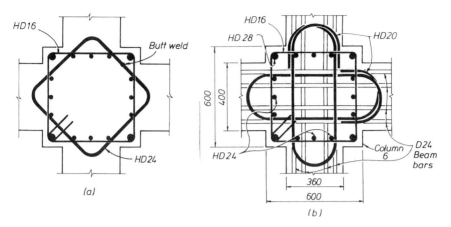

Fig. 4.86 Tie arrangements at the interior joint of the example frame.

$5080(4 \times 6.4) = 198$ mm^2 (0.31 in.^{2}), where the number of effective legs is $4 + 2 \times 0.5 + 2/\sqrt{2} = 6.4$. Thus R16 (0.63-in.-diameter) ties could be used with a total area of

$$A_{jh} = 4 \times 6.4 \times 201 = 5146 \text{ mm}^2 \ (8.0 \text{ in.}^2)$$

(ii) The same arrangement using high-strength steel with $f_y = 380$ MPa (55 ksi) would allow the use of only three sets with spacing of 160 mm (6.3 in.).

(iii) Alternatively, three sets of HD16 (0.63-in.-diameter) peripheral and HD20 (0.79-in.-diameter) or HD24 (0.94-in.-diameter) intermediate ties, as shown in Fig. 4.86, could be provided when steel with $f_y = 450$ MPa (65 ksi) is available. The contribution of this arrangement is, according to Fig. 4.86(b),

$$V_{sh} = 3(2 \times 201 + 2 \times 314)450 \times 10^{-3}$$
$$= 1390 \text{ kN} \approx 1397 \text{ kN} \ (314 \text{ kips})$$

and when, according to Fig. 4.86(a),

$$V_{sh} = 3(2 \times 201 + 1.4 \times 452)450 \times 10^{-3}$$
$$= 1397 = 1397 \text{ kN} \ (314 \text{ kips})$$

(iv) Another type of arrangement for horizontal joint reinforcement is that shown in Fig. 4.65(b). The use of HD20 (0.79-in.-diameter) hoops with overlapping hooks or butt welded splices within relatively small horizontal beam haunches can relieve congestion within the joint core while providing for improved anchorage of beam bars, as discussed in Section 4.8.10(c).

(3) Considerations of vertical joint shear reinforcement require, from Eq. (4.49), $V_{jv} = 1951 \times 650/600 = 2114$ kN (476 kips), and from Eq. (4.71b),

$$A_{jv} = (0.5 \times 2114 - 466)\frac{1000}{380} = 1555 \text{ mm}^2 \text{ (2.4 in.}^2\text{)}$$

$$A_{jv,\text{provided}} = 6 \text{ HD24 (0.94-in.-diameter) bars} = 2712 \text{ mm}^2 \text{ (4.2 in.}^2\text{)}$$

(b) Exterior Joint at Column 5

(1) Critical conditions are likely to result for the $0.9D + \vec{E}_o$ combination of actions for which:

From Section 4.11.8(a)(3): axial tension $= P_u = -435$ kN (-99 kips)

From Section 4.11.7(b)(3):
$$T = 3067 \times 1.25 \times 275 \times 10^{-3} \qquad = 1054 \text{ kN (237 kips)}$$

From Fig. 4.83:
$$V_{\text{col}} = 570/3.35 \qquad\qquad = 170 \text{ kN (38 kips)}$$

From Eq. (4.79):
$$V_{jh} = 1054 - 170 \qquad\qquad = 884 \text{ kN (199 kips)}$$

From Eq. (4.73):
$$v_{jh} = 884,000/500^2 \qquad\qquad = 3.54 \text{ MPa (513 psi)}$$

which is not critical.

(2) Determine the required horizontal joint shear reinforcement with $\beta = A_s/A_s' = 1.0$ from Eq. (4.82)

$$V_{sh} = (1/1.25)(0.7 + 435,000/(30 \times 500^2))1054$$
$$= 639 \text{ kN (146 kips)}$$
$$\therefore \quad A_{jh} = 639,000/275 = 2324 \text{ mm}^2 \text{ (3.60 in.}^2\text{)}$$

Provide four sets of R16 (0.63-in.-diameter) and R12 (0.47-in.-diameter) ties with 628 mm^2 (0.97 in.^{2})/set $= 2512$ mm^2 (3.89 in.^{2}).

(3) The vertical joint shear reinforcement is, from Eq. (4.72b) with $V_{jv} \approx$ (650/500) 884 $= 1149$ kN (259 kips),

$$A_{jv} = (0.5 \times 1149 + 435)1000/380 = 2657 \text{ mm}^2 \text{ (4.12 m}^2\text{)}$$

$$A_{jv,\text{provided}} \approx 4 \times 314 = 1256 \text{ mm}^2 \text{ (1.95 in.}^2\text{)}$$

By providing 12 HD24 bars in this column ($\rho_t = 0.0217$), the vertical steel area is increased by 1107 mm^2 (1.72 in.^{2}).

$$A_{jv,\,\text{provided}} \approx 2362 \text{ mm}^2 = 0.89 A_{jv,\,\text{required}}$$

(4) An alternative design could consider placing two D20 (0.79-in.-diameter) bottom bars to the top of the beam section at column 5. Thereby the positive and negative moment capacities would change by approximately 20%, with no change occurring in the total frame strength. As a consequence, the need for horizontal joint shear reinforcement would also be reduced to

$$A_{jh} \approx 0.8 \times 2360 = 1890 \text{ mm}^2 \ (2.92 \text{ in.}^2)$$

allowing only three sets of ties to be used. The area of the required vertical joint shear reinforcement would thus be reduced to

$$A_{jv} \approx (0.5 \times 0.8 \times 1149 + 435)1000/380 = 2354 \text{ mm}^2 \ (3.65 \text{ m}^2)$$

which can be provided in the modified column.

When the increased top reinforcement is in tension, joint shear forces will correspondingly increase by approximately 20%. However, in this case a minimum axial compression of $P_u = 0.9 \times 782.7 + 1139 = 1843$ kN (415 kips) would need to be considered. Significantly less horizontal joint shear reinforcement [$A_{jh} \approx 704$ mm^2 (1.09 in.^{2})] and no vertical joint shear reinforcement would be required for this case.

The example shows that axial tension on a column leads to severe joint shear requirements.

5 Structural Walls

5.1 INTRODUCTION

The usefulness of structural walls in the framing of buildings has long been recognized. When walls are situated in advantageous positions in a building, they can form an efficient lateral-force-resisting system, while simultaneously fulfilling other functional requirements. For buildings up to 20 stories the use of structural walls is often a matter of choice. For buildings over 30 stories, structural walls may become imperative from the point of view of economy and control of lateral deflection [B4].

Because a large fraction of, if not the entire, lateral force on the building and the horizontal shear force resulting from it is often assigned to such structural elements, they have been called shear walls. The name is unfortunate, for it implies that shear might control their behavior. This need not be so. It was postulated in previous chapters that with few exceptions, an attempt should be made to inhibit inelastic shear modes of deformations in reinforced concrete structures subjected to seismic forces. It is shown in subsequent sections how this can also be achieved readily in walled structures. To avoid this unjustified connotation of shear, the term *structural walls* will be used in preference to *shear walls* in this book.

The basic criteria that the designer will aim to satisfy are those discussed in Chapter 1 (i.e., stiffness, strength, and ductility). Structural walls provide a nearly optimum means of achieving these objectives. Buildings braced by structural walls are invariably stiffer than framed structures, reducing the possibility of excessive deformations under small earthquakes. It will thus often be unnecessary to separate the nonstructural components from the lateral-force-resisting structural system. The necessary strength to avoid structural damage under moderate earthquakes can be achieved by properly detailed longitudinal and transverse reinforcement, and provided that special detailing measures are adopted, dependable ductile response can be achieved under major earthquakes.

The view that structural walls are inherently brittle is still held in many countries as a consequence of shear failure in poorly detailed walls. For this reason some codes require buildings with structural walls to be designed for lower ductility factors than frames. A major aim of this chapter is to show that the principles of the inelastic seismic behavior of reinforced concrete components developed for frames are generally also applicable to structural

362

walls and that it is relatively easy to dissipate seismic energy in a stable manner [P39, P42]. Naturally, because of the significant differences in geometric configurations in structural walls, both in elevation and sections, some modifications in the detailing of the reinforcement will be required.

In studying various features of inelastic response of structural walls and subsequently in developing a rational procedure for their design, a number of fundamental assumptions are made:

1. In all cases studied in this chapter, structural walls will be assumed to possess adequate foundations that can transmit actions from the superstructure into the ground without allowing the walls to rock. Elastic and inelastic deformations that may occur in the foundation structure or the supporting ground will not be considered in this chapter. Some seismic features of foundations, including rocking, are, however, reviewed in Chapters 6 and 9.

2. The foundation of one of several interacting structural walls does not affect its own stiffness relative to the other walls.

3. Inertia forces at each floor are introduced to structural walls by diaphragm action of the floor system and by adequate connections to the diaphragm. In terms of in-plane forces, floor systems (diaphragms) are assumed to remain elastic at all times.

4. The entire lateral force is resisted by structural walls. The interaction of frames with structural walls is, however, considered in Chapter 6.

5. Walls considered here are generally deemed to offer resistance independently with respect to the two major axes of the section only. It is to be recognized, however, that under skew earthquake attack, wall sections with flanges will be subjected to biaxial bending. Suitable analysis programs to evaluate the strength of articulated wall sections subjected to biaxial bending and axial force, are available. They should be employed whenever parts of articulated wall sections under biaxial seismic attack may be subjected to significantly larger compression strains than during independent orthogonal actions.

5.2 STRUCTURAL WALL SYSTEM

To facilitate the separation of various problems that arise with the design of structural walls, it is convenient to establish a classification in terms of geometric configurations.

5.2.1 Strategies in the Location of Structural Walls

Individual walls may be subjected to axial, translational, and torsional displacements. The extent to which a wall will contribute to the resistance of overturning moments, story shear forces, and story torsion depends on its geometric configuration, orientation, and location within the plane of the building. The positions of the structural walls within a building are usually dictated by functional requirements. These may or may not suit structural planning. The purpose of a building and the consequent allocation of floor space may dictate arrangements of walls that can often be readily utilized for lateral force resistance. Building sites, architectural interests, or clients' desires may lead, on the other hand, to positions of walls that are undesirable from a structural point of view. In this context it should be appreciated that while it is relatively easy to accommodate any kind of wall arrangement to resist wind forces, it is much more difficult to ensure satisfactory overall building response to large earthquakes when wall locations deviate considerably from those dictated by seismic considerations. The difference in concern arises from the fact that in the case of wind, a fully elastic response is expected, while during large earthquake demands, inelastic deformations will arise.

In collaborating with architects, however, structural designers will often be in the position to advise as to the most desirable locations for structural walls, in order to optimize seismic resistance. The major structural considerations for individual structural walls will be aspects of symmetry in stiffness, torsional stability, and available overturning capacity of the foundations. The key in the strategy of planning for structural walls is the desire that inelastic deformations be distributed reasonably uniformly over the whole plan of the building rather than being allowed to concentrate in only a few walls. The latter case leads to the underutilization of some walls, while others might be subjected to excessive ductility demands.

When a permanent and identical or similar subdivision of floor areas in all stories is required, as in the case of hotel construction or apartment buildings, numerous structural walls can be utilized not only for lateral force resistance but also to carry gravity loads. Typical arrangements of such walls are shown in Fig. 5.1. In the north–south direction the lateral force per wall will be small as a result of a large number of walls. Often, code-specified minimum levels of reinforcement in the walls will be adequate to ensure elastic response even to large earthquakes. Behavior in the east–west direction of the structure in Fig. 5.1(a) will be more critical, because of reduced wall area and the large number of doors to be provided.

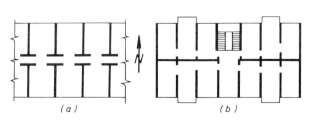

(a)   *(b)*

Fig. 5.1   Typical wall arrangements in hotels and apartment buildings.

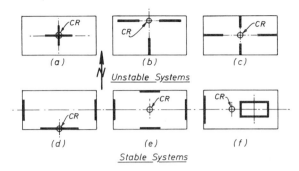

Fig. 5.2   Examples for the torsional stability of wall systems.

Numerous walls with small length, because of door openings, shown in Fig. 5.1(*b*), will supplement the large strength of the end walls during seismic attack in the north–south direction. Lateral forces in the east–west direction will be resisted by the two central walls which are connected to the end walls to form a T section [Fig. 5.1(*b*)]. The dominance of earthquake effects on walls can be conveniently expressed by the ratio of the sum of the sectional areas of all walls effective in one of the principal directions to the total floor area.

Apart from the large number of walls, the suitability of the systems shown in Fig. 5.1 stems from the positions of the centers of mass and rigidity being close together or coincident [Section 1.2.3(*b*)]. This results in small static eccentricity. In assessing the torsional stability of wall systems, the arrangement of the walls, as well as the flexural and torsional stiffness of individual walls, needs to be considered. It is evident that while the stiffness of the interior walls shown in Fig. 5.1(*b*) is considerable for north–south seismic action, they are extremely flexible with respect to forces in the east–west direction. For this reason their contribution to the resistance of forces acting in the east–west direction can be neglected.

The torsional stability of wall systems can be examined with the aid of Fig. 5.2. Many structural walls are open thin-walled sections with small torsional rigidities. Hence in seismic design it is customary to neglect the torsional resistance of individual walls. Tubular sections are exceptions. It is seen that torsional resistance of the wall arrangements of Fig. 5.2(*a*) (*b*) and (*c*) could only be achieved if the lateral force resistance of each wall with respect to its weak axis was significant. As this is not the case, these examples represent torsionally unstable systems. In the case of the arrangement in Fig. 5.2(*a*) and (*c*), computations may show no eccentricity of inertia forces. However, these systems will not accommodate torsion, due to other causes described in Section 1.2.3(*b*) and quantified in Section 2.4.3(*g*), collectively referred to as accidental torsion.

Figure 5.2(*d*) to (*f*) show torsionally stable configurations. Even in the case of the arrangement in Fig. 5.2(*d*), where significant eccentricity is

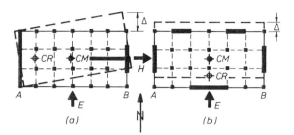

Fig. 5.3   Torsional stability of inelastic wall systems.

present under east–west lateral force, torsional resistance can be efficiently provided by the actions induced in the plane of the short walls. However, eccentric systems, such as represented by Fig. 5.2(d) and (f), are particular examples that should not be favored in ductile earthquake-resisting buildings unless additional lateral-force-resisting systems, such as ductile frames, are also present.

To illustrate the torsional stability of inelastic wall systems, the arrangements shown in Fig. 5.3 may be examined. The horizontal force, H, in the long direction can be resisted efficiently in both systems. In the case of Fig. 5.3(a) the eccentricity, if any, will be small, and the elements in the short direction can provide the torsional resistance even though the flange of the T section may well be subject to inelastic strains due to the seismic shear H.

Under earthquake attack E in the short direction, the structure in Fig. 5.3(a) is apparently stable, despite the significant eccentricity between the center of mass (CM) and center of rigidity (CR), defined in Section 1.2.3(b) and shown in Fig. 1.12. However, no matter how carefully the strengths of the two walls parallel to E are computed, it will be virtually impossible to ensure that both walls reach yield simultaneously, because of inevitable uncertainties of mass and stiffness distributions. If one wall, say that at B, reaches yield first, its incremental stiffness will reduce to zero, causing excessive floor rotations as shown. There are no walls in the direction transverse to E (i.e., the long direction) to offer resistance against this rotation, and hence the structure is torsionally unstable.

In contrast, if one of the two walls parallel to E in Fig. 5.3(b) yields first, as is again probable, the walls in the long direction, which remain elastic under action E, stabilize the tendency for uncontrolled rotation by developing in-plane shears, and the structure is hence torsionally stable.

Elevator shafts and stair wells lend themselves to the formation of a reinforced concrete core. Traditionally, these have been used to provide the major component of lateral force resistance in multistory office buildings. Additional resistance may be derived, if necessary, from perimeter frames as shown in Fig. 5.4(a). Such a centrally positioned large core may also provide sufficient torsional resistance.

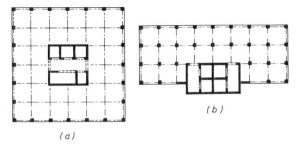

Fig. 5.4 Lateral force resistance provided by reinforced concrete cores.

When building sites are small, it is often necessary to accommodate the core close to one of the boundaries. However, eccentrically placed service cores, such as seen in Fig. 5.4(*b*) lead to gross torsional imbalance. It would be preferable to provide torsional balance with additional walls along the other three sides of the building. Note that providing one wall only on the long side opposite the core for torsional balance is inadequate, for reasons discussed in relation to the wall arrangement in Fig. 5.3(*a*). If it is not possible to provide such torsional balance, it might be more prudent to eliminate concrete structural walls either functionally or physically and to rely for lateral force resistance on a torsionally balanced ductile framing system. In such cases the service shaft can be constructed with nonstructural materials, carefully separated from the frame so as to protect it against damage during inelastic response of the frame.

For better allocation of space or for visual effects, walls may be arranged in nonrectilinear, circular, elliptic, star-shaped, radiating, or curvilinear patterns [S14]. While the allocation of lateral forces to elements of such a complex system of structural walls may require special treatment, the underlying principles of the seismic design strategy, particularly those relevant to torsional balance, remain the same as those outlined above for the simple rectilinear example wall systems.

In choosing suitable locations for lateral-force-resisting structural walls, three additional aspects should be considered:

1. For the best torsional resistance, as many of the walls as possible should be located at the periphery of the building. Such an example is shown in Fig. 5.50(*a*). The walls on each side may be individual cantilevers or they may be coupled to each other.

2. The more gravity load can be routed to the foundations via a structural wall, the less will be the demand for flexural reinforcement in that wall and the more readily can foundations be provided to absorb the overturning moments generated in that wall.

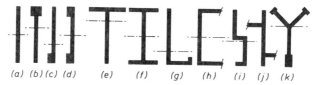

(a) (b) (c) (d)   (e)   (f)   (g)   (h)   (i)   (j)   (k)

Fig. 5.5   Common sections of structural walls.

3. In multistory buildings situated in high-seismic-risk areas, a concentration of the total lateral force resistance in only one or two structural walls is likely to introduce very large forces to the foundation structure, so that special enlarged foundations may be required.

5.2.2   Sectional Shapes

Individual structural walls of a group may have different sections. Some typical shapes are shown in Fig. 5.5. The thickness of such walls is often determined by code requirements for minima to ensure workability of wet concrete or to satisfy fire ratings. When earthquake forces are significant, shear strength and stability requirements, to be examined subsequently in detail, may necessitate an increase in thickness.

Boundary elements, such as shown in Fig. 5.5(b) to (d), are often present to allow effective anchorage of transverse beams. Even without beams, they are often provided to accommodate the principal flexural reinforcement, to provide stability against lateral buckling of a thin-walled section and, if necessary, to enable more effective confinement of the compressed concrete in potential plastic hinges.

Walls meeting each other at right angles will give rise to flanged sections. Such walls are normally required to resist earthquake forces in both principal directions of the building. They often possess great potential strength. It will be shown subsequently that when flanges are in compression, walls can exhibit large ductility, but that T- and L-section walls, such as shown in Fig. 5.5(e) and (g), may have only limited ductility when the flange is in tension.

Some flanges consist of long transverse walls, such as shown in Fig. 5.5(h) and (j). The designer must then decide how much of the width of such wide flanges should be considered to be effective. Code provisions [A1] for the effective width of compression flanges of T and L beams may be considered to be relevant for the determination of dependable strength, with the span of the equivalent beam being taken as twice the height of the cantilever wall.

As in the case of beams of ductile multistory frames, it will also be necessary to determine the flexural overstrength of the critical section of ductile structural walls. In flanged walls this overstrength will be governed primarily by the amount of tension reinforcement that will be mobilized

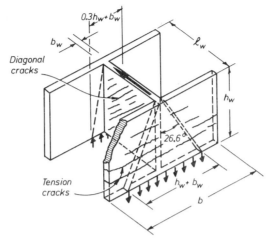

Fig. 5.6   Estimation of effective flange widths in structural walls.

during a large inelastic seismic displacement. Thus some judgment is required in evaluating the effective width of the tension flange. The width assumed for the compression flange will have negligible effect on the estimate of flexural overstrength.

A suggested approximation for the effective width in wide-flanged structural walls is shown in Fig. 5.6. This is based on the assumption that vertical forces due to shear stresses introduced by the web of the wall into the tension flange, spread out at a slope of $1:2$ (i.e., $26.6°$). Accordingly, with the notation of Fig. 5.6, the effective width of the tension flange is

$$b_{\text{eff}} = h_w + b_w \le b \qquad (5.1a)$$

For the purpose of the estimation of flexural overstrength, the assumption above is still likely to be unconservative. Tests on T-section masonry walls [P29] showed that tension bars within a spread of as much as $45°$ were mobilized.

As stated earlier, the flexural strength of wall sections with the flange in compression is insensitive with respect to the assumed effective width. It should be noted, however, that after significant tension yield excursion in the flange, the compression contact area becomes rather small after load reversal, with outer bars toward the tips of the flange still in tensile strain. It may be assumed that the effective width in compression is

$$b_{\text{eff}} = 0.3h_w + b_w \le b \qquad (5.1b)$$

The approximations above represent a compromise, for it is not possible to determine uniquely the effective width of wide flanges in the inelastic state. The larger the rotations in the plastic hinge region of the flanged wall, the larger the width that will be mobilized to develop tension. The foundation system must be examined to ensure that the flange forces assumed can, in fact, be transmitted at the wall base.

5.2.3   Variations in Elevation

In medium-sized buildings, particularly apartment blocks, the cross section of a wall, such as shown in Fig. 5.5, will not change with height. This will be the case of simple prismatic walls. The strength demand due to lateral forces reduces in upper stories of tall buildings, however. Hence wall sizes, particularly wall thickness, may then be correspondingly reduced.

More often than not, walls will have openings either in the web or the flange part of the section. Some judgment is required to assess whether such openings are small, so that they can be neglected in design computations, or large enough to affect either shear or flexural strength. In the latter case due allowance must be made in both strength evaluation and detailing of the reinforcement. It is convenient to examine separately solid cantilever structural walls and those that are pierced with openings in some pattern.

(a)   Cantilever Walls Without Openings   Most cantilever walls, such as shown in Fig. 5.7(a), can be treated as ordinary reinforced concrete beam–columns. Lateral forces are introduced by means of a series of point loads through the floors acting as diaphragms. The floor slab will also stabilize the wall against lateral buckling, and this allows relatively thin wall sections, such as shown in

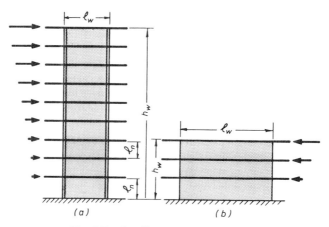

Fig. 5.7   Cantilever structural walls.

Fig. 5.5, to be used. In such walls it is relatively easy to ensure that when required, a plastic hinge at the base can develop with adequate plastic rotational capacity.

In low-rise buildings or in the lower stories of medium- to high-rise buildings, walls of the type shown in Fig. 5.7(b) may be used. These are characterized by a small height-to-length ratio, h_w/l_w. The potential flexural strength of such walls may be very large in comparison with the lateral forces, even when code-specified minimum amounts of vertical reinforcement are used. Because of the small height, relatively large shearing forces must be generated to develop the flexural strength at the base. Therefore, the inelastic behavior of such walls is often strongly affected by effects of shear. In Section 5.7 it will be shown that it is possible to ensure inelastic flexural response. Energy dissipation, however, may be diminished by effects of shear. Therefore, it is advisable to design such squat walls for larger lateral force resistance in order to reduce ductility demands.

To allow for the effects of squatness, it has been suggested [X3] that the lateral design force specified for ordinary structural walls be increased by the factor Z_1, where

$$1.0 < Z_1 = 2.5 - 0.5h_w/l_w < 2.0 \qquad (5.2)$$

It is seen that this is applicable when the ratio $h_w/l_w < 3$. In most situations it is found that this requirement does not represent a penalty because of the great inherent flexural strength of such walls.

While the length of wall section and the width of the flanges are typically constant over the height of the building, the thickness of the wall [Fig. 5.8(a)], including sometimes both the web and the flanges [Fig. 5.8(f)], may be reduced in the upper stories. The reduction of stiffness needs to be taken into account when the interaction of several such walls, to be discussed in Section 5.3.2(a), is being evaluated. More drastic changes in stiffness occur when the length of cantilever walls is changed, either stepwise or gradually, as seen in Fig. 5.8(b) to (e). Tapered walls, such as shown in Fig. 5.8(d), are

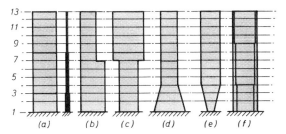

Fig. 5.8   Nonprismatic cantilever walls.

structurally efficient. However, care must be taken in identifying the locations and lengths of potential plastic hinge regions, as these will critically affect the nature of detailing that has to be provided. The inefficiency of the tapered wall of Fig. 5.8(e), sometimes favored as an architectural expression of form, is obvious. If a plastic hinge would need to be developed at the base of this wall, its length would be critically restricted. Therefore, for a given displacement ductility demand, excessive curvature ductility would develop (Section 3.5.4). Such walls may be used in combination with ductile frames, in which case it may be advantageous to develop the wall base into a real hinge.

(b) Structural Walls with Openings In many structural walls a regular pattern of openings will be required to accommodate windows or doors or both. When arranging openings, it is essential to ensure that a rational structure results, the behavior of which can be predicted by bare inspection [P1]. The designer must ensure that the integrity of the structure in terms of flexural strength is not jeopardized by gross reduction of wall area near the extreme fibers of the section. Similarly, the shear strength of the wall, in both the horizontal and vertical directions, should remain feasible and adequate to ensure that its flexural strength can be fully developed.

Windows in stairwells are sometimes arranged in such a way that an extremely weak shear fiber results where inner edges of the openings line up, as shown in Fig. 5.9(a). It is difficult to make such connections sufficiently ductile and to avoid early damage in earthquakes, and hence it is preferable to avoid this arrangement. A larger space between the staggered openings would, however, allow an effective diagonal compression and tension field to develop after the formation of diagonal cracks [Fig. 5.9(b)]. When suitably reinforced, perhaps using diagonal reinforcement, distress of regions between openings due to shear can be prevented, and a ductile cantilever response due to flexural yielding at the base only can be readily enforced.

Overall planning may sometimes require that cantilever walls be discontinued at level 2 to allow a large uninterrupted space to be utilized between levels 1 and 2. A structure based on irrational concepts, as seen in Fig. 5.10(a), may result, in which the most critical region is deliberately weakened. Shear transfer from the wall to foundation level will involve a soft-story

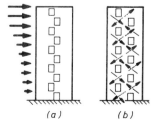

Fig. 5.9 Shear strength of walls as affected by openings.

(a)   (b)

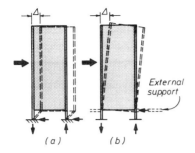

Fig. 5.10 Structural walls supported on columns.

sway mechanism with a high probability of excessive ductility demands on the columns. The overturning moment is likely to impose simultaneously very large axial forces on one of the supporting columns. This system must be avoided! However, often it is possible to transfer the total seismic shear above the opening by means of a rigid diaphragm connection, for example, to other structural walls, thus preventing swaying of the columns (props). This is shown in Fig. 5.10(*b*). Because of the potential for lateral buckling of props under the action of reversed cyclic axial forces, involving yielding over their full length, they should preferably be designed to remain elastic.

Extremely efficient structural systems, particularly suited for ductile response with very good energy-dissipation characteristics, can be conceived when openings are arranged in a regular and rational pattern. Examples are shown in Fig. 5.11, where a number of walls are interconnected or coupled to each other by beams. For this reason they are generally referred to as coupled structural walls. The implication of this terminology is that the connecting beams, which may be relatively short and deep, are substantially weaker than the walls. The walls, which behave predominantly as cantilevers, can then impose sufficient rotations on these connecting beams to make them yield. If suitably detailed, the beams are capable of dissipating energy over the entire height of the structure. Two identical walls [Fig. 5.11(*a*)] or two walls of differing stiffnesses [Fig. 5.11(*b*)] may be coupled by a single line of

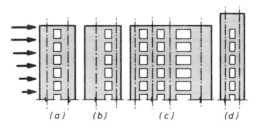

Fig. 5.11 Types of coupled structural walls.

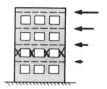

Fig. 5.12   Undesirable pierced walls for earthquake resistance.

beams. In other cases a series of walls may be interconnected by lines of beams between them, as seen in Fig. 5.11(*c*). The coupling beams may be identical at all floors or they may have different depths or widths. In service cores, coupled walls may extend above the roof level, where lift machine rooms or space for other services are to be provided. In such cases walls may be considered to be interconnected by an infinitely rigid diaphragm at the top, as shown in Fig. 5.11(*d*). Because of their importance in earthquake resistance, a detailed examination of the analysis and design of coupled structural walls is given in Section 5.3.2(*c*).

From the point of view of seismic resistance, an undesirable structural system may occur in medium to high-rise buildings when openings are arranged in such a way that the connecting beams are stronger than the walls. As shown in Fig. 5.12, a story mechanism is likely to develop in such a system because a series of piers in a particular story may be overloaded, while none of the deep beams would become inelastic. Because of the squatness of such conventionally reinforced piers, shear failure with re-stricted ductility and poor energy dissipation will characterize the response to large earthquakes. Even if a capacity design approach ensures that the shear strength of the piers exceeded their flexural strength, a soft-story sway mechanism would result, with excessive ductility demands on the hinging piers. A more detailed examination of this issue is given in Section 7.2.1(*b*). Such a wall system should be avoided, or if it is to be used, much larger lateral design force should be used to ensure that only reduced ductility demand, if any, will arise.

Designers sometimes face the dilemma, particularly when considering shear strength, as to whether they should treat coupled walls, such as shown in Fig. 5.11(*a*) or (*d*), as two walls interconnected or as one wall with a series of openings. The issue may be resolved if one considers the behavior and mechanisms of resistance of a cantilever wall and compares these with those of coupled walls. These aspects are shown qualitatively in Fig. 5.13, which compares the mode of flexural resistance of coupled walls with different strength coupling beams with that of a simple cantilever wall.

It is seen that the total overturning moment, M_{ot}, is resisted at the base of the cantilever [Fig. 5.13(*a*)] in the traditional form by flexural stresses, while in the coupled walls axial forces as well as moments are being resisted. These

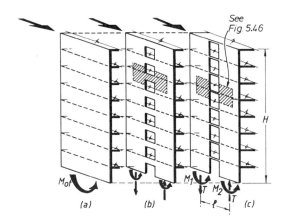

Fig. 5.13   Comparison of flexural resisting mechanisms in structural walls.

satisfy the following simple equilibrium statement:

$$M_{ot} = M_1 + M_2 + lT \tag{5.3}$$

for the coupled walls to carry the same moment as the cantilever wall.

The magnitude of the axial force, being the sum of the shear forces of all the coupling beams at upper levels, will depend on the stiffness and strength of those beams. The derivation of axial forces on walls follows the same principles which apply to columns of multistory frames, presented in Section 4.6.6. For example, in a structure with strong coupling beams, shown in Fig. 5.13(b), the contribution of the axial force to the total flexural resistance, as expressed by the parameter

$$A = Tl/M_{ot} \tag{5.4}$$

will be significant. Hence this structure might behave in much the same way as the cantilever of Fig. 5.13(a) would. Therefore, one could treat the entire structure as one wall.

When the coupling is relatively weak, as is often the case in apartment buildings, where, because of headroom limitation, coupling by slabs only is possible, as shown in Fig. 5.13(c), the major portion of the moment resistance is by moment components M_1 and M_2. In this case the value of A [Eq. (5.4)] is small. One should then consider each wall in isolation with a relatively small axial load induced by earthquake actions. An example of the interplay of actions in coupled walls is given in Section 5.3.2(b) and Fig. 5.22.

In recognition of the significant contribution of appropriately detailed beams to energy dissipation at each floor in walls with strong coupling [Fig.

5.13(b)], it has been suggested [X8] that they be treated as ductile concrete frames. Accordingly, the force reduction factor, R, for ductility, given in Table 2.4, may be taken as an intermediate value between the limits recommended for slender cantilever walls and ductile frames, depending on the efficiency of coupling defined by Eq. (5.4) thus:

$$5 \leq R = 3A + 4 \leq 6 \tag{5.5}$$

when

$$1/3 \leq A = Tl/M_{ot} \leq 2/3 \tag{5.6}$$

Various aspects of coupling by beams are examined in Section 5.3.2(b).

A rational approach to the design of walls with significant irregular openings is discussed in Section 5.7.8(d).

5.3 ANALYSIS PROCEDURES

5.3.1 Modeling Assumptions

(a) Member Stiffness To obtain reasonable estimates of fundamental period, displacements and distribution of lateral forces between walls, the stiffness properties of all elements of reinforced concrete wall structures should include an allowance for the effects of cracking. Aspects of stiffness estimates were reviewed in Section 1.1.2(a) and shown in Fig. 1.8. Displacement ($0.75\Delta_y$) and lateral force resistance ($0.75S_i$), relevant to wall stiffness estimate in Fig. 1.8, are close to those that develop at first yield of the distributed longitudinal reinforcement.

1. The stiffness of cantilever walls subjected predominantly to flexural deformations may be based on the equivalent moment of inertia I_e of the cross section at first yield in the extreme fiber, which may be related to the moment of inertia I_g of the uncracked gross concrete section by the following expression [P26]:

$$I_e = \left(\frac{100}{f_y} + \frac{P_u}{f_c' A_g} \right) I_g \tag{5.7}$$

where P_u is the axial load considered to act on the wall during an earthquake taken positive when causing compression and f_y is in MPa. (The coefficient 100 becomes 14.5 when f_y is in ksi.)

In ductile earthquake-resisting wall systems, significant inelastic deformations are expected. Consequently, the allocation of internal design actions in accordance with an elastic analysis should be considered only as one of several acceptable solutions that satisfy the unviolable requirements of internal and external equilibrium. As will be seen subsequently, redistribution of design actions from the elastic solutions are not only possible but may also be desirable.

Deformations of the foundation structure and the supporting ground, such as tilting or sliding, will not be considered in this study, as these produce only rigid-body displacement of cantilever walls. Such deformations should, however, be taken into account when the period of the structure is being evaluated or when the deformation of a structural wall is related to that of adjacent frames or walls which are supported on independent foundations. Elastic structural walls are very sensitive to foundation deformations [P37].

2. For the estimation of the stiffness of diagonally reinforced coupling beams [P21] with depth h and clear span l_n (Section 5.4.5),

$$I_e = 0.4 I_g \Big/ \Big[1 + 3(h/l_n)^2 \Big] \qquad (5.8a)$$

For conventionally reinforced coupling beams [P22, P23] or coupling slabs,

$$I_e = 0.2 I_g \Big/ \Big[1 + 3(h/l_n)^2 \Big] \qquad (5.8b)$$

In the expressions above, the subscripts e and g refer to the equivalent and gross properties, respectively.

3. For the estimation of the stiffness of slabs connecting adjacent structural walls, as shown in Fig. 5.13(c), the equivalent width of slab to compute I_g may be taken as the width of the wall b_w plus the width of the opening between the walls or eight times the thickness of the slab, whichever is less. The value is supported by tests with reinforced concrete slabs, subjected to cyclic loading [P24]. When flanged walls such as shown in Fig. 5.6 are used, the width of the wall b_w should be replaced by the width of the flange b.

4. Shear deformations in cantilever walls with aspect ratios, h_w/l_w, larger than 4 may be neglected. When a combination of "slender" and "squat" structural walls provide the seismic resistance, the latter may be allocated an excessive proportion of the total lateral force if shear distortions are not accounted for. For such cases (i.e., when $h_w/l_w < 4$) it may be assumed that

$$I_w = \frac{I_e}{1.2 + F} \qquad (5.9a)$$

where

$$F = 30 I_e / h_w^2 b_w l_w \qquad (5.9b)$$

In Eq. (5.9) some allowance has also been made for deflections due to anchorage (pull-out) deformations at the base of a wall.

Deflections due to code-specified lateral static forces may be determined with the use of the equivalent sectional properties above. However, for consideration of separation of nonstructural components and the checking of drift limitations, the appropriate amplification factors that make allowance for additional inelastic drift, given in codes, must be used.

(b) Geometric Modeling   For cantilever walls it will be sufficient to assume that the sectional properties are concentrated in the vertical centerline of the wall (Fig. 5.11). This should be taken to pass through the centroidal axis of the wall section, consisting of the gross concrete area only. When cantilever walls are interconnected at each floor by a slab, it is normally sufficient to assume that the floor will act as a rigid diaphragm. Effects of horizontal diaphragm flexibility are discussed briefly in Section 6.5.3. By neglecting wall shear deformations and those due to torsion and the effects of restrained warping of an open wall section on stiffness, the lateral force analysis can be reduced to that of a set of cantilevers in which flexural distortions only will control the compatibility of deformations. Such analysis, based on first principles, can allow for the approximate contribution of each wall when it is subjected to deformations due to floor translations and torsion, as shown in Section 5.3.2(a). It is to be remembered that such an elastic analysis, however approximate it might be, will satisfy the requirements of static equilibrium, and hence it should lead to a satisfactory distribution also of internal actions among the walls of an inelastic structure.

When two or more walls in the same plane are interconnected by beams, as is the case in coupled walls shown in Figs. 5.11 and 5.12, in the estimation of stiffnesses, it will be necessary to account for more rigid end zones where beams frame into walls. Such structures are usually modeled as shown in Fig. 5.14. Standard programs written for frame analyses may then be used. It is emphasized again that the accuracy of geometric stiffness modeling may vary considerably. This is particularly true for deep membered structures, such as shown in Fig. 5.14. In coupled walls, for example, axial deformations may be significant, and these affect the efficiency of shear transfer across the coupling system. It is difficult to model accurately axial deformations in deep members after the onset of flexural cracking.

Figure 5.15 illustrates the difficulties that arise. Structural properties are conventionally concentrated at the reference axis of the wall, and hence

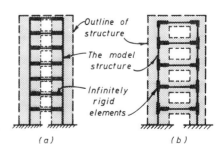

Fig. 5.14 Modeling of deep-membered wall frames.

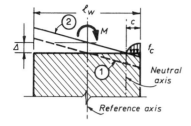

Fig. 5.15 Effect of curvature on uncracked and cracked wall sections.

under the action of flexure only, rotation about the centroid of the gross concrete section is predicted, as shown in Fig. 5.15, by line 1. After flexural cracking, the same rotation may occur about the neutral axis of the cracked section, as shown by line 2, and this will result in elongation Δ, measured at the reference axis. This deformation may affect accuracy, particularly when the dynamic response of the structure is evaluated. However, its significance in terms of inelastic response is likely to be small. It is evident that if one were to attempt a more accurate modeling by using the neutral axis of the cracked section as a reference axis for the model (Fig. 5.14), additional complications would arise. The position of this axis would have to change with the height of the frame due to moment variations, as well as with the direction of lateral forces, which in turn might control the sense of the axial force on the walls. These difficulties may be overcome by employing finite element [P25] analysis techniques. However, in design for earthquake resistance involving inelastic response, this computational effort would seldom be justified.

(c) Analysis of Wall Sections   The computation of deformations, stresses, or strength of a wall section may be based on the traditional concepts of equilibrium and strain compatibility, consistent with the plane section hypothesis. Because of the variability of wall section shapes, design aids, such as standard axial load–moment interaction charts for rectangular column sections, cannot often be used. Frequently, the designer will have to resort to the working out of the required flexural reinforcement from first principles [Section 3.3.1(c)]. Programs to carry out the section analysis can readily be developed for minicomputers. Alternatively, hand analyses involving successive approximations for trial sections may be used such as shown in design examples at the end of this chapter. With a little experience, convergence can be fast.

The increased computational effort that arises in the section analysis for flexural strength, with or without axial load, stems from the multilayered arrangement of reinforcement and the frequent complexity of section shape. A very simple example of such a wall section is shown in Fig. 5.16. It represents one wall of a typical coupled wall structure, such as shown in Fig.

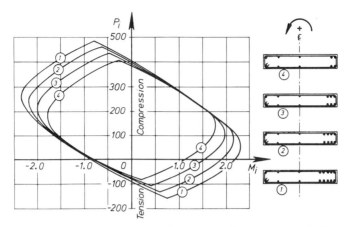

Fig. 5.16 Axial load–moment interaction curves for unsymmetrically reinforced rectangular wall section.

5.11. The four sections are intended to resist the design actions at four different critical levels of the structure. When the bending moment (assumed to be positive) causes tension at the more heavily reinforced right-hand edge of the section, net axial tension is expected to act on the wall. On the other hand, when flexural tension is induced at the left-hand edge of the section by (negative) moments, axial compression is induced in that wall. Example calculations are given in Sections 5.5.2 and 5.6.2.

The moments are expressed as the product of the axial load and the eccentricity, measured from the reference axis of the section, which, as stated earlier, is conveniently taken through the centroid of the gross concrete area rather than through that of the composite or cracked transformed section. It is expedient to use the same reference axis also for the analysis of the cross section. It is evident that the plastic centroids in tension or compression do not coincide with the axis of the wall section. Consequently, the maximum tension or compression strength of the section, involving uniform strain across the entire wall section, will result in axial forces that act eccentrically with respect to the reference axis of the wall. These points are shown in Fig. 5.16 by the peak values at the top and bottom meeting points of the four sets of curves. This representation enables the direct use of moments and forces, which have been derived from the analysis of the structural system, because in both analyses the same reference axis has been used.

Similar axial load–moment interaction relationships can be constructed for different shapes of wall cross sections. An example for a channel-shaped section is shown in Fig. 5.17. It is convenient to record in the analysis the neutral-axis positions for various combinations of moments and axial forces, because these give direct indication of the curvature ductilities involved in developing the appropriate strengths, an aspect examined in Section 5.4.3(*b*).

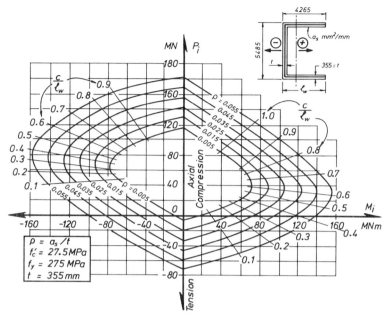

Fig. 5.17 Axial load–moment interaction relationships for a channel-shaped wall section.

Because axial load P_i will vary between much smaller limits than shown in Fig. 5.17, in design office practice only a small part of the relationships shown need be produced. For walls subjected to small axial compression or axial tension, linear interpolations will often suffice.

5.3.2 Analysis for Equivalent Lateral Static Forces

The choice of lateral design force level, based on site seismicity, structural configurations and materials, and building functions has been considered in detail in Chapter 2. Using the appropriate model, described in preceding sections, the analysis to determine all internal design actions may then be carried out. The outline of analysis for two typical structural wall systems is given in the following sections.

(a) Interacting Cantilever Walls The approximate elastic analysis for a series of interacting prismatic cantilever walls, such as shown in Fig. 5.18, is based on the assumption that the walls are linked at each floor by an infinitely rigid diaphragm, which, however, has no flexural stiffness. Therefore, the three walls shown and so linked are assumed to be displaced by identical amounts at each floor. Each wall will thus share in the resistance of

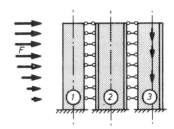

Fig. 5.18 Model of interacting cantilever walls.

a story force, F, or story shear, V, or overturning moment, M, in proportion to its own stiffness thus:

$$F_i = \frac{I_i}{\Sigma I_i} F \quad \text{or} \quad V_i = \frac{I_i}{\Sigma I_i} V \quad \text{or} \quad M_i = \frac{I_i}{\Sigma I_i} M \qquad (5.10)$$

where the stiffness of the walls is proportional to the equivalent moment of inertia of the wall section as discussed in Section 5.3.1(a).

The stiffness of rectangular walls with respect to their weak axis, relative to those of other walls, is so small that in general it may be ignored. It may thus be assumed that as for wall 1 in Fig. 5.19, no lateral forces are introduced to such walls in the relevant direction. A typical arrangement of walls within the total floor plan is shown in Fig. 5.19. The analysis of this wall system is based on the concepts summarized in Appendix A. The shear force, V, applied in any story and assumed to act at the point labeled CV in Fig. 5.19, may be resolved for convenience into components V_x and V_y. Uniform deflection of all the walls would occur only if these component story shear forces acted at the center of rigidity (CR), the chosen center of the coordinate system for which, by analogy to the derivations of Eq. (A.20), the following conditions are satisfied:

$$\Sigma x_i I_{ix} = \Sigma y_i I_{iy} = 0 \qquad (5.11)$$

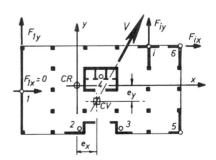

Fig. 5.19 Plan layout of interacting cantilever walls.

where I_{ix}, I_{iy} = equivalent moment of inertia of wall section about the x and y axis of that section, respectively

x_i, y_i = coordinates of the wall with respect to the shear centers of the wall sections labeled $1, 2, \ldots, i$ and measured from the center of rigidity (CR)

Hence for the general case, shown in Fig. 5.19, the shear force for each wall at a given story can be found from

$$V_{ix} = \frac{I_{iy}}{\Sigma I_{iy}} V_x + \frac{(V_x e_y - V_y e_x) y_i I_{iy}}{\Sigma (x_i^2 I_{ix} + y_i^2 I_{iy})} \tag{5.12a}$$

$$V_{iy} = \frac{I_{ix}}{\Sigma I_{ix}} V_y + \frac{(V_x e_y - V_y e_x) x_x I_{ix}}{\Sigma (x_i^2 I_{ix} + y_i^2 I_{iy})} \tag{5.12b}$$

where $(V_x e_y - V_y e_x)$ is the torsional moment of V about CR, $\Sigma (x_i^2 I_{ix} + y_i^2 I_{iy})$ is the rotational stiffness of the wall system, and e_x and e_y are eccentricities measured from the center of rigidity (CR) to the center of story shear (CV). Note that the value of e_y in Fig. 5.19 is negative. With substitution of I_i for D_i, the meaning of the expressions above are identical to those given in Sections (f) and (g) of Appendix A.

The approximations above are also applicable to walls with variable thickness provided that all wall thicknesses reduce in the same proportions at the same level, so that the wall stiffnesses relative to each other do not change. When radical changes in stiffnesses (I_{ix} and I_{iy}) occur in some walls, the foregoing approach may lead to gross errors, and some engineering judgment will be required to compensate for this. Alternatively, a more accurate analysis may be carried out using established computer techniques.

The validity of the assumption that the interaction of lateral force resisting components is controlled by an infinitely rigid connection of the floor diaphragm is less certain in the case of structural walls than in the case of structures braced by interacting ductile frames. The in-plane stiffness of walls and floor slabs, especially in buildings with fewer than five stories, may be comparable. Thus diaphragm deformations in the process of horizontal force transfer can be significant, particularly when precast floor systems are used. For such buildings, variations of the order of 20 to 40%, depending on slab flexibility, have been predicted in the distribution of horizontal forces among elastically responding walls [U3]. This aspect is of importance when assessing the adequacy of the connection of precast floor panels to the structural walls. Diaphragm flexibility is examined further in Section 6.5.3.

The adequacy of the connections between the floor system, expected to function as a diaphragm, and structural walls must be studied at an early stage of the design process. Large openings for services are often required immediately adjacent to structural walls. These may reduce the effective

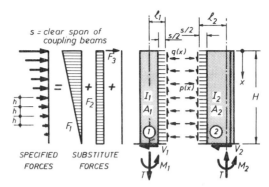

Fig. 5.20   Modeling of the lateral forces and the structure for the laminar analysis coupled walls.

stiffness and strength of the diaphragm and hence also the effectiveness of the poorly connected wall. In buildings with irregular plans, such as an L shape, reentrant corners in the floor system may invite early cracking and consequent loss of stiffness [P37] [Section 1.2.3(a)].

The larger the expected inelastic response of the cantilever wall system, the less sensitive it becomes to approximations in the elastic analysis. For this reason the designer may utilize the concepts of inelastic force redistribution to produce more advantageous solutions, and this is discussed in Section 5.3.2(c).

(b) Coupled Walls   Some of the advantages that coupled wall structures offer in seismic design were discussed in Section 5.2.3(b). Analysis of such structures may be carried out using frame models, such as shown in Fig. 5.14, or that of a continuous connecting medium. The latter, also referred to as laminar analysis, reduces the problems of a highly statically indeterminate structure to the solution of a single differential equation. Figure 5.20 shows the technique used in the modeling with which discrete lateral forces or member properties are replaced by equivalent continuous quantities. This analysis, rather popular some 25 years ago, has been covered extensively in the technical literature [B1, C6, P1, R2, R3] and is not studied further here. It has been employed to derive the quantities given in Fig. 5.22, which are used here to illustrate trends in the behavior of coupled walls.

To study the effect of relative stiffnesses on the elastic behavior of coupled walls, the results of a parametric study of an example service core structure, with constant sectional dimensions as shown in Fig. 5.21, will be examined. The depth for the two rectangular, 300-mm-thick coupling beams at each floor will be varied between 1500 and 250 mm. Beam stiffnesses are based on Eq. (5.8a) when the depth is 400 mm or more, and on Eq. (5.8b) when the

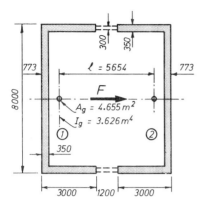

Fig. 5.21 Dimensions of an example service core structure.

depth is less than 400 mm. Also, coupling by a 150-mm-thick slab with 1200-, 600-, and 350-mm effective widths is considered. Figure 5.22 summarizes the response of this structure to lateral forces which, in terms of patterns in Fig. 5.20, were of the following magnitudes: $F_1 = 2000$ kN, $F_2 = 700$ kN, and $F_3 = 300$ kN.

Figure 5.22(a) compares the bending moments in the two walls which would have been developed with both walls remaining uncracked, with those obtained if some allowance for cracking in the tension wall 1 only is made. For both cases, beams with 1000 mm depth were considered. It is seen that the redistribution of moments due to cracking of wall 1 is significant.

Figure 5.22(c) compares the variation of laminar shear forces, q (force per unit height), over the full height of the structure as the depth of the beams is varied. It is seen that with deep beams the shear forces are large in the lower third of the structure and reduce rather rapidly toward the top. This is due to the fact that under the increased axial load on the walls, axial deformations in the top stories become more significant and these relieve the load on the coupling beams. On the other hand, the shear, q, in shallower beams is largely controlled by the general slope of the wall; hence a more uniform distribution of its intensity over the height results. The outermost curve shows the distribution of vertical shear that would result in a cantilever wall (i.e., with infinitely rigid coupling). This curve is proportional to the shear force diagram that would be obtained from the combined forces F_1, F_2, and F_3 on the structure.

Within a wide range of beam depths, the shear force and hence the axial force T, being the sum of the beam shear forces, does not change significantly in the walls, particularly in the upper stories. This is seen in Fig. 5.22(d). There is a threshold of beam stiffness (depth), however, below which the axial force intensity begins to reduce rapidly. This is an important feature of behavior, and designers may make use of it when deciding how efficient coupled walls should be.

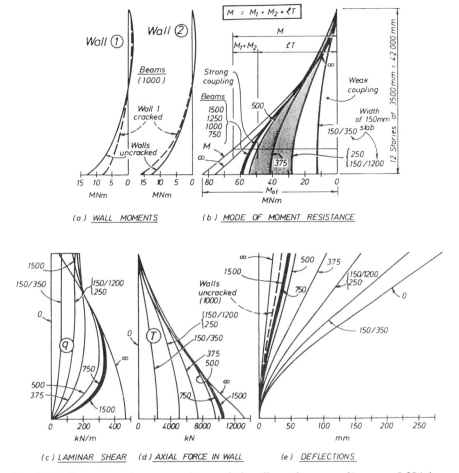

(a) WALL MOMENTS   (b) MODE OF MOMENT RESISTANCE

(c) LAMINAR SHEAR   (d) AXIAL FORCE IN WALL   (e) DEFLECTIONS

Fig. 5.22 Response of an example coupled wall service core. (1 mm = 0.394 in., 1 MNm = 735 kip-ft, 1 kN/m = 0.0686 kip/ft, 1 kN = 0.225 kip).

The interplay of internal wall forces at any level is expressed by Eq. (5.3):

$$M = M_1 + M_2 + lT$$

and this is shown in Fig. 5.22(b). It is seen again that in this example with beams 500 mm or deeper, very efficient coupling is obtained because the lT component of moment resistance is large. Significant increase of beam stiffness does not result in corresponding increase of this component. However, when beams shallower than 250 mm and particularly when 150-mm slabs, irrespective of effective widths, are used, the moment demand on the walls $(M_1 + M_2)$ increases rapidly. With the degeneration of the coupling system, the structure reverts to two cantilevers. The shaded range of the lT component of moment resistance shows the limits that should be considered in terms of potential energy dissipation in accordance with Eq. (5.6). Finally, Fig. 5.22(e) compares the deflected shapes of the elastic structure. It shows the dramatic effects of efficient coupling and one of the significant benefits in terms of seismic design, namely drift control, which is obtained.

(c) Lateral Force Redistribution Between Walls   The principles of the redistribution of design actions in ductile frames, estimated by elastic analyses, were discussed in Section 1.4.4(iv) and in considerable detail in Section 4.3. Those principles are equally applicable to wall structures studied in this chapter because with specific detailing they will possess ample ductility capacity [S2].

In Section 5.3.2(a) the elastic analysis of interconnected cantilever walls, such as shown in Fig. 5.18, was presented. During a large earthquake plastic hinges at the base of each of the three walls in Fig. 5.18 are to be expected. However, the base moments developed need not be proportional to those of the elastic analysis. Bending moments, and correspondingly lateral forces, may be redistributed during the design from one wall to another when the process leads to a more advantageous solution. For example, wall 3 in Fig. 5.18 might carry considerably larger gravity loads than the other two walls. Therefore, larger design moments may be assigned to this wall without having to provide proportionally increased flexural tension reinforcement. Moreover, it will be easier to transmit larger base moments to the foundation of wall 3 than to those of the other two walls.

It is therefore suggested [X3] that if desirable in ductile cantilever wall systems, the design lateral force on any wall may be reduced by up to 30%. This force must then be redistributed to other walls of the system, there being no limit to the amount by which the force on any one wall could be increased.

The advantages of the redistribution of design action can be utilized to an even greater degree in coupled walls, such as shown in Figs. 5.21 and 5.23(d). The desired full energy-dissipating mechanism in coupled walls will be similar to that in multistory frames with strong columns and weak beams, as shown in Fig. 1.19(a). This involves the plastification of all the coupling beams and the development of a plastic hinge at the base of each of the walls, as seen in Fig. 5.23(d) with no inelastic deformation anywhere else along the height of the walls. This is because the walls are usually very much stronger than the connecting beams.

The elastic analysis for such a structure (Figs. 5.21 and 5.22) may have resulted in bending moments M_1 and M_2 for the tension and compression walls, respectively, as shown by full lines in Fig. 5.23(a) and (b). In this analysis it is assumed that the elastic redistribution of moments due to the effects of cracking, as outlined in Section 5.3.1(a), has already been considered. In spite of M_1 being smaller than M_2, more tension reinforcement is

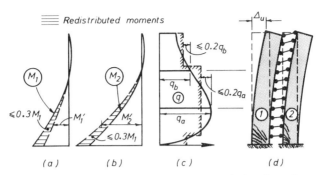

Fig. 5.23 Ductile response of an example coupled wall service core.

likely to be required in wall 1 because it will be subjected to large lateral force induced axial tension [Fig. 5.22(d)]. The flexural strength of wall 2, on the other hand, will be enhanced by the increased axial compression. It is therefore suggested that if desirable and practical, the moments in the tension wall be reduced by up to 30% and that these moments be redistributed to the compression wall. This range of maximum redistributable moments is shown in Fig. 5.23(a) and (b). The limit of 30% is considered to be a prudent measure to protect walls against excessive cracking during moderate earthquakes. Moment redistribution from one wall to another also implies redistribution of wall shear forces of approximately the same order.

Similar considerations lead to the intentional redistribution of vertical shear forces in the coupling system. It has been shown [P1] that considerable ductility capacity can be provided in the coupling beams. Hence they will need to be designed and detailed for very large plastic deformations. This is considered in Section 5.4.5. A typical elastic distribution of shear forces in coupling beams, in terms of laminar shear, q, is shown in Fig. 5.23(c). Coupling beam reinforcement should not be varied continuously with the height, but changed in as small a number of levels as possible. Shear and hence moment redistribution vertically among coupling beams can be utilized, and the application of this is shown by the stepped shaded lines in Fig. 5.23(c). It is suggested that the reduction of design shear in any coupling beam should not exceed 20% of the shear predicted for this beam by the elastic analysis. It is seen that with this technique a large number of beams can be made identical over the height of the building.

When shear is redistributed in the coupling system, it is important to ensure that no shear is "lost." That is, the total axial load introduced in the walls, supplying the lT component of the moment resistance, as seen in Fig. 5.22(b), should not be reduced. Therefore, the area under the stepped and shaded lines of Fig. 5.23(c) should not be allowed to be less than the area under the curve giving the theoretical elastic laminar shear, q. Neither should the strength of the coupling system significantly *exceed* the demand, shown by the continuous curve, because this may unnecessarily increase the overturning capacity of the structure, thus overloading foundations. It will be shown in the complete example design of a coupled wall structure in Section 5.6 how this can be readily checked.

While satisfying the moment equilibrium requirements of Eq. (5.3), it is also possible to redistribute moments between the $(M_1 + M_2)$ and lT components, involving a change in the axial force, T, and hence in shear forces in the coupling system. However, this is hardly warranted because with the two procedures above only, as illustrated in Fig. 5.23, usually a practical and economical allocation of strength throughout the coupled structural wall system can readily be achieved.

5.4 DESIGN OF WALL ELEMENTS FOR STRENGTH AND DUCTILITY

5.4.1 Failure Modes in Structural Walls

A prerequisite in the design of ductile structural walls is that flexural yielding in clearly defined plastic hinge zones should control the strength, inelastic deformation, and hence energy dissipation in the entire structural system [B14]. As a corollary to this fundamental requirement, brittle failure mechanisms or even those with limited ductility should not be permitted to occur. As stated earlier, this is achieved by establishing a desirable hierarchy in the failure mechanics using capacity design procedures and by appropriate detailing of the potential plastic regions.

The principal source of energy dissipation in a laterally loaded cantilever wall (Fig. 5.24) must be the yielding of the flexural reinforcement in the plastic hinge regions, normally at the base of the wall, as shown in Fig. 5.24(b) and (e). Failure modes to be prevented are those due to diagonal tension [Fig. 5.24(c)] or diagonal compression caused by shear, instability of thin walled sections or of the principal compression reinforcement, sliding shear along construction joints, shown in Fig. 5.24(d), and shear or bond

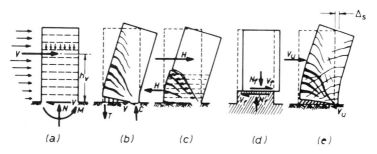

Fig. 5.24   Failure modes in cantilever walls.

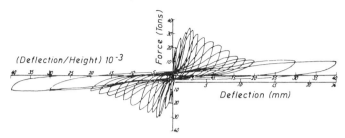

Fig. 5.25 Hysteretic response of a structural wall controlled by shear strength.

failure along lapped splices or anchorages. An example of the undesirable shear-dominated response of a structural wall to reversed cyclic loading is shown in Fig. 5.25. Particularly severe is the steady reduction of strength and ability to dissipate energy.

In contrast, carefully detailed walls designed for flexural ductility and protected against a shear failure by capacity design principles exhibit greatly improved response, as seen in Fig. 5.26, which shows a one-third full size cantilever structural wall with rectangular cross section. The test unit simu-

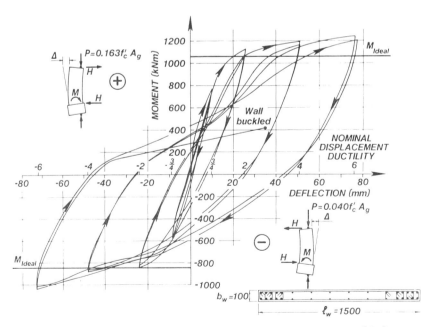

Fig. 5.26 Stable hysteretic response of a ductile wall structure [G1].

lates one wall of a coupled wall structure that was subjected to variable axial compression between the limits shown. It is seen that a displacement ductility of approximately 4 has been attained in a very stable manner [G1, P44]. Failure due to inelastic instability, to be examined subsequently, occurred only after two cycles to a displacement ductility of 6, when the lateral deflection was 3.0% of the height of the model wall [P44].

The hysteretic response shown in Fig. 5.26 also demonstrates that the flexural overstrength developed depends on the imposed ductility. The ideal flexural strengths shown were based on measured yield strength of the vertical bars which was 18% in excess of the specified yield strength that would have been used in the design. Thus it can be seen that when a displacement ductility of 4 was imposed in the positive direction, the strength of the wall was approximately 32% in excess of that based on specified yield strength.

The observed hysteretic behavior of well-detailed structural walls is similar to that of beams. Plastic rotational capacity may, however, be affected by axial load and shear effects, and these will be examined subsequently. Also, shear deformations in the plastic hinge region of a cantilever wall may be significantly larger than in other, predominantly elastic, regions [O2].

5.4.2 Flexural Strength

(a) Design for Flexural Strength   It was shown in Section 3.3.1(*b*) and Fig. 3.21 that because of the multilayered arrangement of vertical reinforcement in wall sections, the analysis for flexural strength is a little more complex than that for beam sections such as seen in Fig. 3.20. Therefore, in design, a successive approximation technique is generally used. This involves initial assumptions for section properties, such as dimensions, reinforcement content, and subsequent checking (i.e., analysis) for the adequacy of flexural strength. This first assumption may often be based on estimates which in fact can lead close to the required solution, and this is illustrated here with the example wall section shown in Fig. 5.27.

Wall dimensions are generally given and subsequently may require only minor adjustments. Moment M and axial load P combinations with respect to the centroidal axis of the wall section are also known. Thus the first

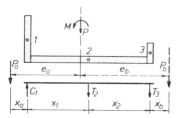

Fig. 5.27 Example wall section.

estimate aims at finding the approximate quantity of vertical reinforcement in the constituent wall segments, such as 1, 2, and 3 in Fig. 5.27. The amount of reinforcement in segment 2 is usually nominated and it often corresponds to the minimum recommended by codes. However, this assumption need not be made because any reinforcement in area 2 in excess of the minimum is equally effective and hence will correspondingly reduce the amounts required in the flange segments of the wall. By assuming that all bars in segment 2 will develop yield strength, the total tension force T_2 is found. Next we may assume that when $M_a = e_a P_a$, the center of compression for both concrete and steel forces C_1 is in the center of segment 1. Hence the tension force required in segment 3 can be estimated from

$$T_3 \approx \frac{x_a P_a - x_1 T_2}{x_1 + x_2}$$

and thus the area of reinforcement in this segment can be found. Practical arrangement of bars can now be decided on. Similarly, the tension force in segment area 1 is estimated when $M_b = e_b P_b$ from

$$T_1 \approx \frac{x_b P_b - x_2 T_2}{x_1 + x_2}$$

Further improvement with the estimates above may be made, if desired, by checking the intensity of compression forces. For example, when P_a is considered, we find that

$$C_1 = P_a + T_2 + T_3$$

and hence with the knowledge of the amount of reinforcement in segment 1, to provide the tension force T_1, which may now function as compression reinforcement, the depth of concrete compression can be estimated. It is evident that in the example of Fig. 5.27, very little change in the distance x_1 and hence in the magnitude of T_3 is likely to occur.

With these approximations the final arrangement of vertical bars in the entire wall sections can be made. Subsequently, the ideal flexural strength M_i and the flexural overstrength ($M_{o,w}$), based on f_y and $\lambda_o f_y$, respectively, of the chosen section can be determined using the procedure given in Section 3.3.1(c). This will also provide the depth of the compression zone of the wall section, c, an important quantity, which, as Section 5.4.3 will show, indicates the ductility capacity and the need, if any, for confining parts of compressed regions of the wall section.

(b)   Limitations on Longitudinal Wall Reinforcement   For practical reasons the ratio of longitudinal (i.e., vertical) wall reinforcement to the gross concrete area, ρ_l, as given by Eq. (5.21), over any part of the wall should not be less than $0.7/f_y$ (MPa) $(0.1/f_y$ (ksi)) nor more than $16/f_y$ (MPa) $(2.3/f_y$ (ksi)). The upper limit, which controls the magnitude of the maximum steel force, is likely to cause congestion when lapped splices are to be provided. It has also been recommended [X5] that in boundary elements of walls, such as the edge regions of sections shown in Figs. 5.26, 5.39, 5.51, and 5.56, concentrated vertical reinforcement not less than $0.002b_w l_w$ should be provided.

The lower limit stems from traditional recommendations of codes [A1, X3], where the primary concerns were shrinkage and temperature effects. In some countries this practice is considered to be excessive, and nominal reinforcement content in walls as low as 0.1% are used. Obviously, designers must ensure that requirements for wind force resistance, which may exceed those due to design earthquakes, are also satisfied. Of concern is the fact that when too little reinforcement is used in walls, cracks, few in numbers, when they form, can become unacceptably wide. This is because the reinforcement provided is insufficient to replace the tensile strength of the surrounding concrete, significantly increased during the high strain rate imposed by an earthquake, and bars will instantaneously yield with crack formation. Thus response to a moderate earthquake may result in structural damage requiring costly repair. Moreover, during more intense shaking comparable to the design earthquake, excessively large tensile strains may be imposed on bars. Because of the extremely large range of strain variation in bars crossing these widely spaced cracks and partial buckling when in compression, fracture of bars may set in after only a few cycles of displacement reversals. This has been observed in the 1985 Chile earthquake [W6].

In reinforced concrete walls that are thicker than 200 mm, preferably two layers of reinforcement, one near each face of the wall, should be used. In regions where the wall section is to be confined, the horizontal spacing of vertical bars should not exceed 200 mm (8 in.), and in other (i.e., elastic) regions, 450 mm (18 in.) or three times the thickness of the wall. The diameter of bars used in any part of a wall should not exceed one-eighth of the thickness of the wall. Several of these recommendations are based on engineering judgment and traditional practice rather than on specific studies.

(c) Curtailment of Flexural Reinforcement Typical bending moment diagrams under the specified lateral forces are shown for coupled structural walls in Figs. 5.23(a) and (b). If the flexural reinforcement were to be curtailed exactly in accordance with the moment so indicated, plastic hinges could form with equal probability anywhere along the height of such walls during a strong earthquake. This would be undesirable from a design point of view because potential plastic hinges require special and necessarily more expensive detailing, as discussed in Section 5.4.3. Also, when plastic hinges form at some height above the base of the wall, the curvature ductility, to attain a required displacement ductility, is greatly increased. Moreover, as in the case of beams, the shear strength of reinforced concrete walls will

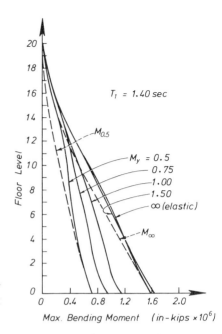

Fig. 5.28 Dynamic moment envelopes for a 20-story cantilever wall with different base yield moment strengths [F2].

diminish in regions where yielding of the flexural reinforcement occurs. This would then necessitate additional horizontal shear reinforcement at all levels. It is more rational to ensure that plastic hinges can develop only in predetermined locations, logically at the bases of walls, by providing flexural strength over the remainder of the wall, which is in excess of the likely maximum demands.

Bending moment envelopes, covering moment demands that arise during the dynamic response, are different from bending moment diagrams resulting from code-specified equivalent lateral forces. This may be readily shown with modal superposition techniques [B11]. Similar results are obtained from time-history analyses of inelastic wall structures using a variety of earthquake records [B11, F2, I1]. Typical bending moment envelopes obtained analytically for 20-story cantilever walls with different base yield moment strengths and subjected to particular ground excitations are shown in Fig. 5.28 [F2]. It is seen that there is an approximate linear variation of moment demands during both elastic and inelastic dynamic response of the walls to ground shaking. For the sake of comparison, bending moments due to static forces, corresponding to 10% of the base shear being applied at the top and 90% in the form of an inverted triangularly distributed force, are shown in Fig. 5.28 by dashed lines for two cases.

As a consequence, it is recommended that the flexural reinforcement in cantilever walls be curtailed so as to give not less than a linear variation of moment of resistance with height. The interpretation of this suggestion is

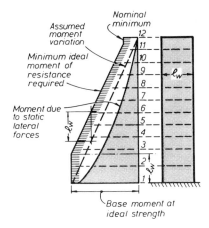

Fig. 5.29 Recommended design moment envelope for cantilever walls.

shown in Fig. 5.29. Once the critical wall section at the base has been designed and the exact size and number, as well as the positions of flexural bars, have been established, the ideal flexural strength of this section, to be developed in the presence of the appropriate axial load on the wall, can be evaluated. The shaded bending moment diagram in Fig. 5.29 shows moments that would result from the application of the lateral static force pattern with this ideal strength developed at the base. The straight dashed line represents the minimum ideal flexural strength that should be provided in terms of the recommendation above. When curtailing vertical bars the effect of diagonal tension on the internal flexural tension forces must be considered, in accordance with the recommendations of Section 3.6.3. Accordingly, the tension shift is assumed to be equal to the length of the wall l_w. Hence bars to be curtailed should extend by a distance not less than the development length l_d beyond the level at which according to the shaded bi-linear envelope they are required to develop yield strength.

The demand for flexural reinforcement in a cantilever wall is not proportional to the bending moment demand, as indicated, for example, by an envelope, as in the case of prismatic beams, because axial compression is also present. If the flexural steel content is maintained constant with height, the flexural resistance of the section will reduce with height because the axial compression due to gravity and/or earthquake effects becomes smaller. This will be evident from Fig. 5.17. Cantilever walls are normally subjected to axial compression well below the level associated with *balanced strains*, and the $M-P$ interaction relationship clearly shows that in that range the wall section is rather sensitive to the intensity of axial compression. As examples in Section 5.6.2 will show, this issue is seldom critical, but conservatism with curtailment is justified.

The recommended procedure for curtailment is compared in Fig. 5.30, with the analytically predicted moment demand resulting from the inelastic

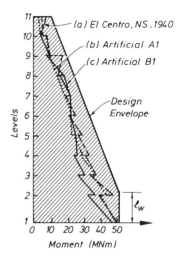

Fig. 5.30 Comparison of dynamic moment demands in a coupled wall.

dynamic response of a coupled wall to three different earthquake records [T3].

(d) **Flexural Overstrength at the Wall Base** As in the case of ductile frames, several subsequent aspects of the design of wall systems depend on the maximum flexural strength that could be developed in the walls. In accordance with the definitions in Section 1.3.3(*d*) and (*f*), this is conveniently quantified by the wall flexural overstrength factor, defined as

$$\phi_{o,w} = \frac{\text{flexural overstrength}}{\text{moment resulting from code forces}} = \frac{M_{o,w}}{M_E} \qquad (5.13)$$

where both moments refer to the base section of the wall. To ensure ductile wall response by means of a plastic hinge at the wall base only, it will be necessary to amplify all subsequent actions, such as shear forces acting on the foundations, by this factor, and to proportion other components so as to remain essentially elastic under these actions.

An important and convenient role of the factor $\phi_{o,w}$ is to measure the extent of any over- or underdesign by choice, necessity, or as a result of an error made [Section 1.3.3(*f*)]. Whenever the factor $\phi_{o,w}$ exceeds the optimal value of λ_o/ϕ [Eq. (1.12)] the wall possesses reserve strength. As higher resistance will be offered by the structure than anticipated when design forces were established, it is expected that corresponding reduction in ductility demand in the design earthquake will result. Often, benefits may be derived when, as a consequence, design criteria primarily affected by ductility capacity may be met for the reduced ($\mu_{\Delta r}$) rather than the anticipated (μ_Δ) ductility. Hence in the following sections reductions in expected ductility demands μ_Δ will be made so that the reduced ductility becomes

$$\mu_{\Delta r} = \frac{\lambda_o/\phi}{\phi_{o,w}}\mu_\Delta \tag{5.14}$$

Values of λ_o [Section 3.2.4(e)] are known and $\phi = 0.9$. This ratio will be incorporated in several equations that follow which serve the purpose of checking ductile performance.

Example 5.1   A wall has been designed for an overturning moment of $M_E = 10$ MNm (7380 kip-ft) corresponding with $\mu_\Delta = 5$. Reinforcement with $f_y = 400$ MPa and overstrength factor $\lambda_o = 1.4$ has been provided, considering also the practicality of the placement of bars. As a result, the moment capacity of the wall at overstrength is found to be $M_{o,w} = 18.7$ MNm (13800 kip-ft), so that $\phi_{o,w} = 18.7/10 = 1.87$. Hence the required ductility capacity can be expected to be reduced from $\mu_\Delta = 5$ to

$$\mu_{\Delta r} = \left(\frac{1.4/0.9}{1.87}\right)5 = 4.2$$

5.4.3   Ductility and Instability

(a)   Flexural Response   The ability of a particular section to sustain plastic rotations, as measured by the curvature ductility, follows from the same simple principles used to evaluate flexural strength, as reviewed in Section 3.3.1. It was shown (Fig. 3.26) that at the development of flexural strength the ratio of the concrete compression strain in the extreme fiber ϵ_{cm} to the neutral axis depth c_u quantifies the associated curvature ϕ_m. With the definition of yield curvature ϕ_y, the curvature ductility μ_ϕ is then readily determined (Section 3.5.2).

Strain profiles 1 and 2 in Fig. 5.31 show that for the same extreme concrete strain ϵ_c, compression on a rectangular wall section, requiring a

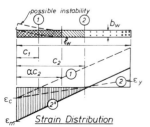
Strain Distribution

Fig. 5.31   Strain patterns for rectangular wall sections.

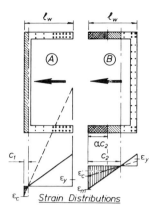

Fig. 5.32 Strain profiles showing ductility capacity in channel-shaped walls.

large neutral axis depth c_2, will result in much smaller curvature than in the case of no or small axial compression load.

If two walls are part of an interconnected wall system, such as shown in Fig. 5.18, similar curvatures will be required in both walls. This means that one wall with a large compression load will have to develop the strain profile shown by line $2'$ in Fig. 5.31 to attain the same curvature as the less heavily loaded wall, given by line 1. This would imply concrete compression strains in the extreme fiber, ϵ_m, considerably in excess of the critical value, ϵ_c. Clearly, such curvature could not be sustained unless the concrete subjected to excessive compressive strains is confined. This is examined in Section 5.4.3(e).

The effect of sectional configuration on the ductility potential of a wall section can be studied with the example of a channel-shaped wall in Fig. 5.32. In the case of wall A, subjected to earthquake forces in the direction shown, the potential width of the compression zone is considerable. Consequently, only a small compression zone depth c_1 is needed to balance the wall axial compression load and the internal tension forces resulting from the longitudinal reinforcement placed in the webs. Hence the ensuing strain gradient, corresponding to ϵ_c and shown by the dashed line, is extremely large. It is probable that such a large curvature would not need to be developed under even an extreme earthquake, and possibly the one shown by the full line would be adequate. Thus concrete compression strains might remain subcritical at all times. In cases such as this, even at moderate ductility demands, moments well in excess of the ideal flexural strength may be developed because of strain hardening of the steel located in regions of large tensile strains.

Wall B of Fig. 5.32, on the other hand, requires a large neutral axis depth, c_2, to develop a compression zone large enough to balance the tension forces generated in the flange part of the section and the axial force on the wall. Even specified minimum wall reinforcement placed in a long wall, as in Fig. 5.32, can develop a significant tension force when yielding. As the dashed-line strain profile indicates, the curvature developed with the ideal strength will in this case not be sufficient if the same displacement as in wall A, shown by the full-line strain profiles, is imposed. The excessive concrete compression strain in the vicinity of the tip of the stems of wall B, will require confinement to be provided if a brittle failure is to be avoided.

It is thus seen that the depth of the neutral axis, c, relative to the length of the wall, l_w, is a critical quantity. If a certain curvature ductility is to be attained, the ratio c/l_w may need to be limited unless critically compressed regions of the wall section are confined. This is examined in Section 5.4.3(e).

As in the case of beams and columns, the ideal flexural strength of wall sections is based on specified material properties f'_c and f_y. During large inelastic displacement pulses, particularly when large curvature ductility demands arise, such as shown for wall A in Fig. 5.32, much larger moments may be developed at the critical wall section. In accordance with the principles of capacity design this strength enhancement needs to be taken into account. It can be quantified with the flexural overstrength factor ϕ_o [Section 1.3.3(f)], which in the case of cantilever walls is the ratio of the overstrength moment of resistance to the moment resulting from the code-specified lateral forces expressed by Eq. (5.13), where both moments refer to the base section of the wall.

(b) Ductility Relationships in Walls   The displacement ductility capacity of $\mu_\Delta = \Delta_u/\Delta_y$ of walls studied so far depends on the rotational capacity of the plastic hinge at the base. It is most conveniently expressed in terms of curvature ductility capacity, which, when necessary, can readily be evaluated when the wall section is designed for strength [Section 3.3.1(c)]. Definitions of types of ductility and their relationships to each other, reviewed in Section 3.5, are applicable. A detailed study of the parameters affecting the ductility capacity of reinforced masonry walls is presented in Section 7.2.4, and the conclusions there are also relevant to reinforced concrete walls.

One of the major parameters affecting curvature ductilities in walls is the length of the plastic hinge l_p (Fig. 3.27), which cannot be defined with great precision. Its magnitude is affected primarily by the length of the wall l_w, the moment gradient at the base (i.e., shear), and axial load intensity. Plastic tensile strains at one edge of a wall section will invariably extend over a greater height of the wall than inelastic compression strains, if any, at the opposite edge. Therefore, it is not possible to define a unique section above the wall base which separates elastic and inelastic regions.

Typical values of the plastic hinge length are such that $0.3 < l_p/l_w < 0.8$. By expanding Eq. (3.59) and using two different suggestions [P28] for the estimation of the plastic hinge length, general trends in the relationship between displacement and curvature ductility factors for cantilever walls can be established, and this is shown in Fig. 5.33. An important feature, often overlooked, is that for a given displacement ductility μ_Δ, the curvature ductility demand μ_ϕ increases with increased aspect ratio A_r of walls.

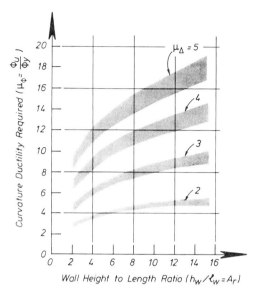

Fig. 5.33   Variation of the curvature ductility ratio at the base of cantilever walls with the aspect ratio and the imposed displacement ductility demand [P28].

(c)   Wall Stability   When parts of a thin-wall section are subjected to compression strains, the danger of instability due to out-of-plane buckling arises. The concern for this lead to recommendations [X3], largely based on traditional concepts of Eulerian buckling of struts and on engineering judgment, to limit the thickness of walls in the potential plastic hinge regions, typically to about one-tenth of the height of a wall in the first story. The use of flanges or enlarged boundary elements to stabilize walls is therefore encouraged [X3].

More recent studies [G1, G2] revealed, however, that potential for out-of-place buckling of thin sections of ductile walls depends more on the magnitude of the inelastic tensile strains imposed on that region of the wall, which on subsequent moment reversal is subjected to compression. The perceived major factors affecting instability, and the means by which these may be approximated, are described in this section.

At large curvature ductilities μ_ϕ considerable tensile strains may be imposed on vertical bars placed at the extreme tension edge of a section, such as shown in Fig. 5.26. At this stage uniformly spaced wide, nearly horizontal cracks across the width of the section develop over the extent of plasticity. Typical crack patterns in the plastic hinge region may be seen in Fig. 5.37(a). In a somewhat idealized form, these are shown in Fig. 5.34(a) and (c) for walls with thickness b and single and double layers of vertical reinforcement, respectively. During the subsequent reversal of wall displace-

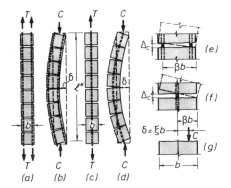

Fig. 5.34 Deformations leading to out-of-plane buckling.

ments, and hence unloading, the tensile stresses in these bars reduce to zero while the width of cracks remains large. A reversal of lateral force, and in the case of coupled walls an increase of axial compression on the wall, will eventually produce compression stresses in the bars. Until the cracks close, the internal compression force within the wall section must be resisted by the vertical reinforcement only.

At this stage the flexural compression force C within the thickness b of the wall may not coincide with the centroid of the vertical reinforcement, as shown in Fig. 5.34(b) and (d). The eccentricity may result in rotation of blocks of concrete bound by adjacent horizontal cracks [Fig. 5.34(e) and (f)]. Thus significant out-of-plane curvature may develop at the stage when contact between crack boundaries at one face of the wall occurs, as illustrated in Fig. 5.34(b) and (d). The bending moment $M = \delta C$ at the center of a wall strip, shown in Fig. 5.34(b) and (d), with buckling length l^* may then cause an out-of-plane buckling failure of the wall well before cracks would fully close and before the flexural strength of the wall section could be developed.

It must be appreciated that the phenomenon is more complex than what idealizations in Fig. 5.34 suggest. Small dislocated concrete particles and misfit of crack faces, caused by sliding shear displacements, may also influence the eneven closure of wide cracks. Bending of the wall about the weak axis due to large inelastic deformations of the lateral force resisting system in the relevant direction may aggrevate the situation. However, the initiation of out-of-plane displacement δ will depend primarily on the crack width Δ_c and the arrangement of the vertical reinforcement within the thickness of the wall, as suggested in Fig. 5.34(e) and (f). Crack width, in turn depends on the maximum tensile strain ϵ_{sm} imposed on the vertical bars in the preceding displacement cycle. Using first principles and limited experimental evidence [G1], some guidance with respect to the critical wall thickness b_c is presented in the following paragraphs.

With the aid of the models in Fig. 5.34, it may be shown from first principles that instability will occur when $b \leq b_c = l^*\sqrt{\epsilon_{sm}/8\xi\beta}$, where from Fig. 5.34($g$), ξ defines the critical eccentricity in terms of the wall thickness b, and β quantifies the angular rotations for a given crack width, as shown in Fig. 5.34(e) and (f). It may be assumed that steel strains exceeding that at yield will extend over a height above the base equal to the length of the wall l_w [Fig. 5.37(a)], and that the estimated maximum steel strains ϵ_{sm}, corresponding with the curvature ductility to be expected, will develop only in the lower half of that region [Fig. 5.37(b)]. With this assumption the buckling length is $l^* = 0.5l_w$. With a conservative estimate for the extrapolated yield curvature in accordance with Fig. 3.26(a), of $\phi_y = 0.0032/l_w$, the maximum steel strains ϵ_{sm} can be predicted as a function of the curvature ductility (Section 3.5.2) demand μ_ϕ. Experimental studies [G2] have shown that when out-of-plane displacements δ are relatively small, they reduce or disappear upon complete closure of cracks. However, with increased curvature ductility, increased displacement δ [Fig. 5.34(b) and (d)] do not recover completely, and with repeated load cycles, out-of-plane displacements increase progressively. The threshold of critical displacement was found to be on the order of $\delta = b/3$. Hence by taking $\xi = \frac{1}{3}$ it can be shown that the critical wall thickness in the compressed end of a wall section in the plastic hinge region can be estimated by

$$b_c = 0.017l_w\sqrt{\mu_\phi} \qquad \text{when } \beta = 0.8 \qquad (5.15a)$$

and

$$b_c = 0.022l_w\sqrt{\mu_\phi} \qquad \text{when } \beta = 0.5 \qquad (5.15b)$$

where β is defined in Fig. 5.34(e) and (f) and is taken conservatively as 0.8 when two layers of vertical bars are used in a wall.

From fundamental relationships between curvature and displacement ductilities, examined in Section 3.5.4 and given for walls in Fig. 5.33, the critical wall thickness b_c can be related to the displacement ductility capacity μ_Δ on which the design of the wall is based. Such relationships, based on a plastic hinge length $l_p = 0.2l_w + 0.044h_w$, are given in Fig. 5.35.

By assuming that the buckling length will not exceed 80% of the height of the first story h_1, in evaluating the minimum wall thickness from Fig. 5.35, l_w need not be taken larger than $1.6h_1$. It is also recommended that b should not be less than $h_1/16$.

When the critical thickness b_c is larger than the web thickness b_w, a boundary element with area A_{wb} should be provided so that

$$b_c^2 \leq A_{wb} \geq b_c l_w/10 \qquad (5.16)$$

The interpretation Eq. (5.16) is illustrated in Fig. 5.36, where limitations on the dimensions of boundary elements are summarized.

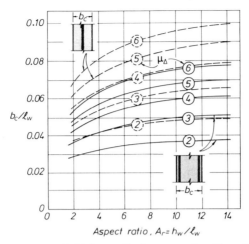

Fig. 5.35 Critical wall thickness displacement ductility relationship.

Another equally important purpose of the recommended dimensional limitations of wall sections and the use of boundary elements, such as shown in Fig. 5.36, is to preserve the flexural strength of the wall section after the complete closure of wide cracks, shown in Fig. 5.34. Once out-of-plane displacements occur, the distribution of concrete compression strains and stresses across the wall thickness b will not be uniform [Fig. 5.34(e) to (g)]. To develop the necessary total compression force in the wall section, shown as C_c in Fig. 3.21(b), the neutral-axis depth c will need to increase. A softening of the concrete near the compressed edge of the wall section, as a result of cyclic variations of the out-of-plane displacements [Fig. 5.34(b) and (d)], may eventually lead to a significant increase in the compression zone of the wall with some reduction in flexural strength. Of prime concern is, however, the increase in concrete compression strains in regions of the wall

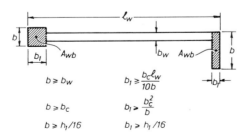

Fig. 5.36 Minimum dimensions of boundary elements of wall sections in the plastic hinge region.

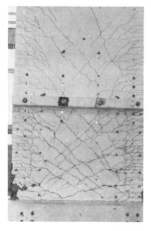  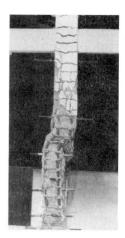

Fig. 5.37 Diagonal cracking and buckling in the plastic hinge region of a structural wall [G1].

where, because of the anticipation of noncritical compression strains, no confinement [Section 5.4.3(*e*)] has been provided. Thus crushing of the concrete outside the confined end region may result in a brittle failure. Figure 5.37(*c*) shows such a failure initiated by crushing of unconfined concrete some distance away from the edge seen.

Example 5.2 A reinforced concrete cantilever wall, such as wall 5 in Fig. 5.50, is 6.5 m (21.3 ft) long and 20 m (65.6 ft) high and is to have a displacement ductility capacity of $\mu_\Delta = 5$. The height of the first story is $l_1 = 4$ m (13.1 ft). With $l_w = 6.5 > 1.6 \times 4 = 6.4$ m and $A_r = 20/6.5 = 3.08$ from Fig. 5.35, we find that

$$b_c = 0.054 \times 6400 = 346 \text{ mm } (13.6 \text{ in.}) > 4000/16 = 250 \text{ mm } (9.8 \text{ in.})$$

Example 5.3 One 18-story reinforced concrete coupled wall with $l_w = 5.5$ m (18 ft), $l_1 = 4$ m (13.1 ft), and $A_r = 10$ with a displacement ductility capacity of $\mu_\Delta = 6$ will require a thickness in the first story of

$$b_c = 0.075 \times 5500 = 413 \text{ mm } (16.4 \text{ in.}) > 4000/16$$

From Eq. (5.16) the area of boundary element should not be less than $A_{wb} = 413^2 \approx 170{,}000$ mm^2 (264 in.^{2}) or $A_{wb} = 413 \times 5500/10 = 227{,}000$ mm^2 (352 in.^{2}) A 500-mm (20-in.)-square boundary element may be provided. Alternatively, a flange, as at the right-hand edge of the wall section in Fig. 5.36, can be arranged, with minimum thickness $b_1 > 4000/16 = 250$ mm (10 in.) and a flange length of $b \geq 227{,}000/250 = 908 \approx 1000$ mm (40 in.).

When boundary elements are provided, the wall thickness b_w will be governed by requirements for shear strength (Section 5.4.4(b)).

Example 5.4 A reinforced masonry block cantilever wall with the following properties is to be used: $\mu_\Delta = 3.5$, $l_w = 3$ m (9.8 ft), $h_1 = 3$ m (9.8 ft), $h_w = 12$ m (39.3 ft), $A_r = 12/3 = 4$. A single layer of reinforcement will be used, and hence, from Fig. 5.35,

$$b_c = 0.06 \times 3000 = 180 \text{ mm } (7 \text{ in.}) < 3000/16 = 188 \text{ mm}$$

(d) Limitations on Curvature Ductility It has been shown that the ultimate curvature of a wall section is inversely proportional to the depth c of the compression zone (Fig. 5.31). It is thus apparent that given a limiting strain in the extreme compression fiber, adequate curvature ductility can be assured by limiting the compression zone depth. It has also been found that the relationship between curvature ductility and displacement ductility depends on the aspect ratio of a wall, as shown in Fig. 5.33.

Because relatively small axial load due to gravity needs to be carried and the flexural reinforcement content is generally small, in the great majority of wall sections, the depth of compression is also small. This is generally the case for rectangular and symmetrical flanged wall sections, and therefore the curvature ductility capacity of such sections is in excess of the probable ductility demand during a major earthquake. For this type of wall section a crude but rather conservative simple check may be made to estimate the maximum depth of the compression zone c, which would allow the desired curvature to develop.

As shown in Section 3.5.2(a), the yield curvature of the wall section may be approximated by

$$\phi_y = (\epsilon_y + \epsilon_{ce})/l_w \tag{5.17}$$

where ϵ_y is the yield strain of the steel assumed at the extreme wall fiber and ϵ_{ce} is the elastic concrete compression strain developed simultaneously at the opposite edge of the wall. If desired, the value of ϵ_{ce} may be determined from a routine elastic analysis of the section. However, for the purpose of an approximation that will generally overestimate the yield curvature, it may be assumed that $\epsilon_y = 0.002$ and $\epsilon_{ce} = 0.0005$. The latter value would necessitate a rather large quantity of uniformly distributed vertical reinforcement in a rectangular wall, in excess of 1%. With this estimate the extrapolated yield curvature shown in Fig. 3.26(a) [Eq. (5.17)] becomes

$$\phi_y = 1.33(0.002 + 0.0005)/l_w \approx 0.0033/l_w$$

By relating the curvature ductility μ_ϕ to the associated displacement ductility μ_Δ demand, which was assumed when selecting the appropriate force reduction factor R in Section 2.4.3, the information provided in Fig. 5.33 may be utilized. Again by making limiting assumptions for cantilever walls, such as $A_r = h_w/l_w \leq 6$ and $\mu_\Delta \leq 5$, quantities that are larger than those encountered in the great majority of practical cases, we find from Fig. 5.33 that $\mu_\phi \approx 13$. Hence by setting the limiting concrete compression strain, associated with the development of the desired ultimate curvature of $\phi_u = 13 \times 0.0033/l_w = 0.043/l_w$, at $\epsilon_c = 0.004$, we find from

$$\phi_u = \frac{0.004}{c_c} = \frac{0.043}{l_w}$$

that the maximum depth of compression is

$$c_c \approx l_w/10 \qquad\qquad (5.18a)$$

In tests at the University of California, Berkeley, average curvatures ranging from $0.045/l_w$ to $0.076/l_w$ were attained in walls with $l_w = 2388$ mm (94 in.) while displacement ductility ratios were on the order of $\mu_\Delta = 9$ (VI).

When this order of curvature ductility (i.e., $\mu_\phi \approx 13$) is developed, and $\epsilon_c = 0.004$, maximum tensile strains will approach 4%, and hence significant strain hardening of the steel will occur. Hence at this stage the flexural overstrength of the wall section $M_{o,w}$ [Section 1.3.3(f)] will be mobilized. In accordance with capacity design procedures, the flexural overstrength of the wall base section, as detailed, will need to be computed [Section 5.4.2(d)]. Therefore, it is more convenient to relate c_c to the flexural overstrength rather than the required flexural strength M_E [Section 1.3.3(b)], which corresponds to the selected reduction factor R (i.e., displacement ductility factor μ_Δ), as described in Section 2.4.3. Thus, by making allowances in proportions of excess or deficiency of flexural strength and ductility demands, Eq. (5.18a) can be modified to

$$c_c = \frac{M_{o,w}}{(\lambda_o M_E/\phi)}\frac{5}{\mu_\Delta}\frac{l_w}{10} = \frac{M_{o,w}}{2.2\lambda_o\mu_\Delta M_E}l_w \qquad (5.18b)$$

It is emphasised that Eq. (5.18b) serves the purpose of a conservative check. If it is found that c, computed at flexural overstrength, is less than c_c, no further attention need be given to the compressed concrete, as maximum strains are expected to remain below $\epsilon_c = 0.004$.

A more detailed estimate of the critical value of the depth of compression, taking into account variations in aspect ratio A_r and the yield strength of the tension reinforcement, may be made by expansion of the relationships given in Fig. 5.33 in this form

$$c_c = \frac{k_c M_{o,w}}{(\mu_\Delta - 0.7)(17 + A_r)\lambda_o f_y M_E} l_w \tag{5.18c}$$

where $k_c = 3400$ MPa or 500 ksi. Unless the aspect ratio A_r exceeds 6, Eq. (5.18c) will always predict larger values of c_c than Eq. (5.18b). If it is found that c is larger than that given in Eq. (5.18c), extreme concrete strains in excess of $\epsilon_c = 0.004$ must be expected and, accordingly, to sustain the intend ductility, the compressed concrete needs to be confined. This is considered in the next section.

When T- or L-shaped wall sections or those with significantly more reinforcement at one edge than at the other are used, yield curvature ϕ_y should be checked from first principles. In this case the critical depth of compression in the section can be estimated from

$$c = \frac{0.004}{\mu_\phi \phi_y} l_w \tag{5.18d}$$

(e) Confinement of Structural Walls From the examination of curvature relationships in the simple terms of c/l_w ratio, it is seen that in cases when the computed neutral-axis depth is larger than the critical value c_c, at least a portion of the compression region of the wall section needs to be confined. The provision of confining reinforcement in the compression region of the potential plastic hinge zone of a structural wall must address the two interrelated issues of the concrete area to be confined and the quantity of hoops to be used. The confinement of longitudinal reinforcement to avoid buckling is another issue.

(i) Region of Compression Zone to Be Confined: The definition of the area of confinement may be approached with the precept that unconfined concrete should not be assumed to be capable of sustaining strain in excess of 0.004. The strain profile (1) in Fig. 5.38 indicates the ultimate curvature, ϕ_u, that might be necessary to enable the estimated displacement ductility, μ_Δ, for a particular structural wall to be sustained when the theoretical concrete strain in the extreme compression fiber reaches 0.004. The value of the associated neutral axis depth, c_c, may be estimated by Eq. (5.18). To achieve the same ultimate curvature in the wall when the neutral-axis depth c is larger, as shown by strain profile (2) in Fig. 5.38, the length of section subjected to compression strains larger than 0.004 becomes $\alpha' c_c$. It is this length that should be confined. From the geometry shown in Fig. 5.38, $\alpha' = 1 - c_c/c$. However, some conservatism in the interpretation of the simple curvature and ductility relationships, shown in Figs. 5.31 and 5.32 and Eq. (5.18), should be adopted. This is because during reversed cyclic loading

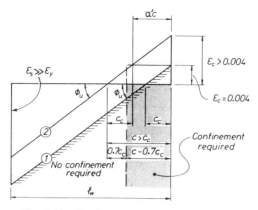

Fig. 5.38   Strain patterns for wall sections.

the neutral-axis depth tends to increase, due to the gradual reduction of the contribution of the cover concrete as well as that of the confined core to compression strength, or the out-of-plane bending of the compression zone of thin sections, discussed in Section 5.4.3(c).

It is therefore suggested that the length of wall section to be confined should not be less than αc, where

$$\alpha = (1 - 0.7c_c/c) \geq 0.5 \tag{5.19}$$

whenever $c_c/c < 1$. When c is only a little larger than c_c, a very small and impractical value of α is obtained. The lower limit (i.e., 0.5) is suggested for this case. The application of this approach is shown in Section 5.6.2.

(ii) Quantity of Confining Reinforcement: The principles of concrete confinement to be used are those relevant to column sections, examined in 3.6.1(a), with the exceptions that very rarely will the need arise to confine the entire section of a wall. Accordingly, using the nomenclature of Section 3.6.1, it is recommended that rectangular or polygonal hoops and supplementary ties surrounding the longitudinal bars in the region to be confined should be used so that

$$A_{sh} = 0.3s_h h'' \left(\frac{A_g^*}{A_c^*} - 1 \right) \frac{f_c'}{f_{yh}} \left(0.5 + 0.9\frac{c}{l_w} \right) \tag{5.20a}$$

$$A_{sh} = 0.12s_h h'' \frac{f_c'}{f_{yh}} \left(0.5 + 0.9\frac{c}{l_w} \right) \tag{5.20b}$$

whichever is greater. In practice, the ratio c/l_w will seldom exceed 0.3.

In the equations above:

A_g^* = gross area of the wall section that is to be confined in accordance with Eq. (5.19)

A_c^* = area of concrete core within the area A_g^*, measured to outside of peripheral hoop legs

The area to be confined is thus extending to αc_2 from the compressed edge as shown by crosshatching in the examples of Figs. 5.31 and 5.32.

For the confinement to be effective, the vertical spacing of hoops or supplementary ties, s_h, should not exceed one-half of the thickness of the confined part of the wall or 150 mm (6 in.), whichever is least [X3]. When confinement is required, walls with a single layer should not be used for obvious reasons.

(iii)   Vertical Extent of Regions to Be Confined: Confining transverse rein-forcement should extend vertically over the probable range of plasticity for the wall, which for this purpose should be assumed to be equal to the length of the wall, l_w [X3], or one-sixth of the height, h_w, whichever is greater, but need not exceed $2l_w$.

An application of this procedure is given in Section 5.6.2.

(iv)   Stabilizing of Longitudinal Bars: A secondary purpose of confinement is to prevent the buckling of the principal vertical wall reinforcement, where it may be subjected to yielding in compression. The approach to the stability of bars in compression in beams and columns was studied in Sections 4.5.4 and 4.6.11(*d*). The same requirements are also relevant to vertical bars in walls.

It is considered that in regions of potential yielding of the longitudinal bars within a wall with two layers of reinforcement, only those bars need be supported laterally, which contribute substantially to compression strength. Typically, affected bars will occur in the edge regions or boundary elements of wall sections. Accordingly, transverse hoops or ties with cross-sectional area A_{te}, given by Eq. (4.19), and with vertical spacing s_h not exceeding six times the diameter of the vertical bar to be confined should be provided where the longitudinal wall reinforcement ratio ρ_l, computed from Eq. (5.21), exceeds $2/f_y$ (MPa) $(0.29/f_y$ (ksi)).

The vertical reinforcement ratio that determines the need for transverse ties should be computed from

$$\rho_l = \Sigma\, A_b/bs_v \tag{5.21}$$

where the terms of the equation, together with the interpretation of the requirements above, are shown in Fig. 5.39. The interpretation of Eq. (5.21)

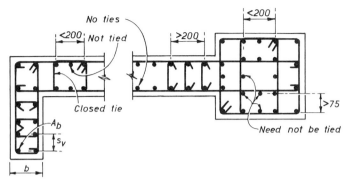

Fig. 5.39   Transverse reinforcement in potential yield zones of wall section.

with reference to the wall return at the left-hand end of Fig. 5.39 is as follows: $\rho_l = 2A_b/bs_v$.

The distance from the compression edges of walls over which vertical bars should be tied, when $\rho_l > 2/f_y$ (MPa) $(0.29/f_y$ (ksi)), should be not less than $c - 0.3c_c$ or $0.5c$. Over this distance of the compression zone the yielding of the vertical reinforcement must be expected. However, with reversed cyclic loading, compression yielding of vertical bars may occur over a much larger distance from the extreme compressed edge of the wall, because bars that have yielded extensively in tension must yield in compression before concrete compression can be mobilized. It is unlikely, however, that at large distances from the compression edge of the wall section, the compression reinforcement ratio ρ_l will exceed $2/f_y$ (MPa) $(0.29/f_y$ (ksi)).

Vertical bars arranged in a circular array and the core so confined by spiral or circular hoop reinforcement in a rectangular boundary element, such as at the right-hand side of the wall section in Fig. 5.39, have been found to be very effective [V1] even though a larger amount of concrete is lost after spalling of the cover.

In areas of the wall in upper stories, where $\rho_l > 2/f_y$ (MPa) $(0.29/f_y$ (ksi)) and where no compression yielding is expected, the lateral reinforcement around such bars should satisfy the requirements applicable to the noncritical central region of columns in ductile frames [A1].

(v) Summary of Requirements for the Confinement of Walls: The requirements of transverse reinforcement in the potential yield region of a wall are summarized for an example wall section in Fig. 5.40.

1. When for the direction of applied lateral forces (north) the computed neutral-axis depth exceeds the critical value, c_c, given by Eqs. (5.18*b*) and (5.18*c*), reinforcement confining the concrete over the outer αc length of the

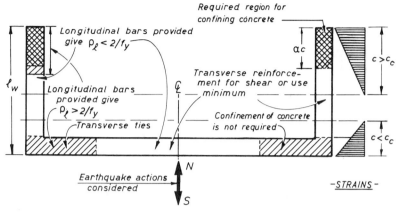

Fig. 5.40 Regions of a wall section where transverse reinforcement is required for different purposes (f_y in MPa).

compression zone, shown by crosshatching, should be provided in accordance with Section 5.4.3(e) (ii).

2. In the single shaded flange part of the channel-shaped wall, over a distance $c - 0.3c_c$, antibuckling ties around vertical bars of the type shown in Fig. 5.39 should be provided in accordance with Eq. (4.19) when $\rho_l > 2/f_y$ (MPa) ($0.29/f_y$ (ksi)).

3. In the web portion of the channel-shaped wall, vertical bars need to be confined, using antibuckling ties in accordance with Section 5.4.3(e)(4) because $\rho_l > 2/f_y$ (MPa) ($0.29/f_y$ (ksi)). The affected areas are shaded.

4. In all other areas, which are unshaded, the transverse (horizontal) reinforcement need only satisfy requirements for shear.

5. Some judgment is necessary to decide whether confinement of the compressed concrete is necessary at other locations, for example at the corners where flanges join the web part of the section in Fig. 5.40 when the wall is subjected to skew (bidirectional) earthquake attack.

5.4.4 Control of Shear

(a) Determination of Shear Force To ensure that shear will not inhibit the desired ductile behavior of wall systems and that shear effects will not significantly reduce energy dissipation during hysteretic response, it must not be allowed to control strength. Therefore, an estimate must be made for the maximum shear force that might need to be sustained by a structural wall during extreme seismic response to ensure that energy dissipation can be confined primarily to flexural yielding.

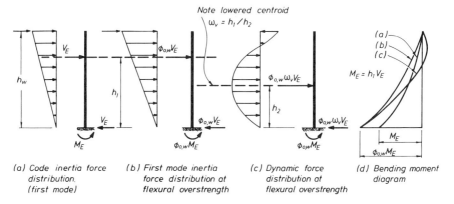

(a) Code inertia force   (b) First mode inertia   (c) Dynamic force   (d) Bending moment
    distribution.   force distribution at   distribution at   diagram
    (first mode)   flexural overstrength   flexural overstrength

Fig. 5.41   Comparison of code-specified and dynamic lateral forces.

The approach that may be used stems from the capacity design philosophy, and its application is similar to that developed for ductile frames in Chapter 4. Allowance needs to be made for flexural overstrength of the wall $M_{o,w}$ and for the influence of higher mode response distorting the distribution of seismic lateral forces assumed by codes.

In comparison with beams, there is a somewhat larger uncertainty involved in walls with respect to the influence of material properties. Wall sections with a small neutral-axis depth, such as shown in Fig. 5.31, will exhibit greater strength enhancement due to early strain hardening. The flexural strength of compression dominated wall sections will increase significantly if at the time of the earthquake the strength of the concrete is considerably in excess of the specified strength, f'_c. However, this type of wall is rare.

Increase of shear demand may result from dynamic effects. During a predominantly first-mode response of the structure, the distribution of inertial story forces will be similar to that shown in Fig. 5.41(a) and (b).

The force pattern is similar to that of standard code-specified static forces. The center of inertial forces is typically located at approximately $h_1 \simeq 0.7h_w$ above the base. At some instants of response, displacement and accelerations may be strongly influenced by the second and third modes of vibration, resulting in story force distributions as seen in Fig. 5.41(c), with the resultant force being located much lower than in the previous case [M17]. Shapes in the second and third modes of vibration of elastic cantilevers (Fig. 2.24(b)) with fixed or hinged bases are very similar. This suggests that the formation of a plastic hinge at the wall base may not significantly affect response in the second and third modes. While a base plastic hinge will greatly reduce wall actions associated with first mode response, it can be expected that those resulting from higher mode responses of an inelastic cantilever will be comparable with elastic response actions [K8]. A plastic hinge may still form at the wall base under the distribution of forces shown in Fig. 5.41(c) because wall flexural strength is substantially lower than "elastic" response strength. As a consequence, the induced shear near the base corresponding to flexural hinging with second- and third-mode response is larger than that in the first mode. However, the probability of a base hinge developing overstrength in a higher-mode response is not high, because of reduced plastic rotations. Bending moments associated with force patterns shown in Fig. 5.41(a) to (c) are shown in Fig. 5.41(d).

The contribution of the higher modes to shear will increase as the fundamental period of the structure increases, implying that shear magnification will increase with the number of stories. From a specific study of this problem [B11], the following recommendation [X3] has been deduced for estimation of the total design shear.

$$V_u = V_{\text{wall}} = \omega_v \phi_{o,w} V_E \qquad (5.22)$$

where V_E is the horizontal shear demand derived from code-specified lateral static forces, $\phi_{o,w}$ is as defined by Eq. (5.13), and $\omega_v = h_1/h_2$, as shown in Fig. 5.41(c), is the dynamic shear magnification factor, to be taken as

$$\omega_v = 0.9 + n/10 \qquad (5.23a)$$

for buildings up to six stories, and

$$\omega_v = 1.3 + n/30 \qquad (5.23b)$$

for buildings over six stories, where n is the number of stories, which in Eq. (5.23b) need not be taken larger than 15, so that $\omega_v \leq 1.8$.

Theoretical consideration [A11] and parametric analytical studies [M17] indicate that dynamic shear magnification is likely also to be a function of expected ground accelerations. The inclusion of more accurate predictions of shear stiffness, fundamental period instead of the number of stories, leading to modal limit forces, suggests that improved analytical predictions for the value of the dynamic magnification factor will be available [K8]. Predictions by Eq. (5.22) compare with results obtained from these studies for rather large accelerations.

As subsequent examples will show, the design shear force at the base of a structural wall, derived from Eq. (5.22), can become a critical quantity and may control the thickness of the wall. Although Eq. (5.22) was derived for shear at the base of the wall, it may be used to amplify the code-level shear at heights above the base. However, this is an approximation, and imposed shear force envelopes based on inelastic dynamic analyses have been suggested [I1]. Because the magnitude of the design shear at greater heights above the base is much less than that near the base of the wall, and because at these heights inelastic flexural response is suppressed, the design and prediction for shear strength in the upper stories will seldom be critical.

Some walls, particularly those of low- or medium-rise buildings, may have inherent flexural strength well in excess of that required, even with minimum reinforcement content. In such a wall little or no flexural ductility demand will arise, and it will respond essentially within the elastic domain. Provided that it resists all or the major fraction of the total shear for the building, it is therefore unnecessary to design such a wall for shear which would be in excess of the elastic response demand. Hence the design shear force for such a wall may be limited to

$$V_{\text{wall}} \leq \mu_\Delta V_E / \phi \qquad (5.24a)$$

When the overstrength of one wall in an interconnected wall system, such as shown in Fig. 5.19, is disproportionately excessive; that is, when $\phi_{o,w} \gg \psi_{o,w}$ [Sections 1.3.3(f) and (g)], the required shear strength of the affected wall need not exceed its share of the total shear of an elastically responding system. By similarity to Eq. (5.24a) and using Eq. (5.10), this can be quantified as

$$V_{\text{wall}} \leq \frac{I_i}{\Sigma I_i} \mu_\Delta V_E \Big/ \phi \qquad (5.24b)$$

The shear demand so determined, V_{wall}, should be equal to or less than the ideal shear strength of the wall, V_i. However, as explained in Table 3.1, for this upper-bound estimate of earthquake resistance a strength reduction factor of $\phi = 1.0$ is appropriate.

(b)   Control of Diagonal Tension and Compression

(i)   *Inelastic Regions*: As in the case of beams, it must be recognized that shear strength will be reduced as a consequence of reversed cyclic loading involving flexural ductility. However, the uniform distribution of both horizontal and vertical reinforcement in the web portion of wall sections is considered to preserve better the integrity of concrete shear-resisting mechanisms [C12], expressed by the quantity v_c, given by Eq. (3.39).

Because of the additional and unwarranted computational effort involved in evaluating the effective depth, d, in structural wall sections, it is customary [A1, X3], as in the case of column sections, to assume that $d = 0.8l_w$, and hence the average shear stress [Eq. (3.29)] at ideal strength, v_i, is

$$v_i = V_i / 0.8 b_w l_w \qquad (5.25)$$

Web reinforcement, consisting of horizontal bars, fully anchored at the extremities of the wall section, must then be provided in accordance with Eq. (3.40). The vertical spacing of bars, s_v, should not exceed 2.5 times the thickness of the wall, or 450 mm [Section 3.3.2(a)(vii)].

Experiments have shown that other conditions being equal, the hysteretic response of structural walls improves when the web reinforcement consists of

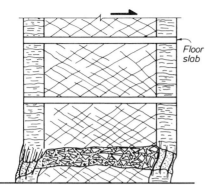

Fig. 5.42 Web crushing in a wall after several load cycles with large ductility demands [V1].

smaller-diameter bars placed with smaller spacing [I2]. The provisions above should ensure that diagonal tension failure across plastic hinges does not occur during a very large earthquake.

Diagonal compression failure may occur in walls with high web shear stresses, even when excess shear reinforcement is provided. As a consequence, codes [A1, X3, X5] set an upper limit on the value of v_i [Section 3.3.2(a)(ii)]. Because the web of the wall may be heavily cracked diagonally in both directions, as seen in Fig. 5.37, the diagonal compression strength of the concrete required to sustain the truss mechanism may be reduced dramatically. Therefore, it is recommended [Section 3.3.2(a)(ii)] that in this region the total shear stress be limited to 80% of that in elastic regions [Eq. (3.30)] (i.e., $v_{i\,max} \leq 0.16 f_c'$).

Tests conducted by the Portland Cement Association [O1] and the University of California at Berkeley [B12, V1] have demonstrated, however, that, despite the limitation on maximum shear stress above, web crushing in the plastic hinge zone may occur after a few cycles of reversed loading involving displacement ductility ratios of 4 or more. When the imposed ductilities were only 3 or less, a shear stress equal to or in excess of $0.16 f_c'$ could be attained. Web crushing, which eventually spreads over the entire length of the wall, can be seen in Fig. 5.42. When boundary elements with a well-confined group of vertical bars were provided, significant shear after the failure of the panel (web) could be carried because the boundary elements acted as short columns or dowels. However, it is advisable to rely more on shear resistance of the panel, by preventing diagonal compression failure, rather than on the second line of defense of the boundary elements. To ensure this, either the ductility demand on a wall with high shear stresses must be reduced, or, if this is not done, the shear stress, used as a measure of diagonal compression, should be limited as follows:

$$v_{i,\,max} \leq \left(\frac{0.22 \phi_{o,w}}{\mu_\Delta} + 0.03 \right) f_c' < 0.16 f_c \leq 6\,\text{MPa (870 psi)} \quad (5.26)$$

For example, in coupled walls with typical values of the overstrength factor $\phi_{o,w} = 1.4$ and $\mu_{\Delta} = 5$, $v_{i,\max} = 0.092 f'_c$. In a wall with restricted ductility, corresponding values of $\phi_{o,w} = 1.4$ and $\mu_{\Delta} = 2.5$ would give $v_{i,\max} = 0.153 f'_c$, close to the maximum suggested. The expression also recognizes that when the designer provides excess flexural strength, giving a larger value of $\phi_{o,w}$, a reduction in ductility demand is expected, and hence Eq. (5.26) will indicate an increased value for the maximum admissible shear stress.

(ii) Elastic Regions: Since ductility demand will not arise in the upper stories of walls, if designed in accordance with capacity design principles and the moment envelope of Fig. 5.29, shear strength will not be reduced. Several of the restrictions applicable to inelastic regions are then unnecessary, and the general requirements of Section 3.3.2 need be satisfied only.

(c) Sliding Shear in Walls   Well-distributed reinforcement in walls provides better control of sliding than in beams where sliding, resulting from high-intensity reversed shear loading, can significantly affect the hysteretic response. This is because more uniformly distributed and embedded vertical bars in the web of the wall provide better crack control and across the potential sliding plane better dowel shear resistance. Another reason for improved performance is that most walls carry some axial compression due to gravity, and this assists in closing cracks across which the tension steel yielded in the previous load cycle. In beams several small cracks across the flexural reinforcement may merge into one or two large cracks across the web, thereby forming a potential plane of sliding, as seen in Fig. 3.24. Because of the better crack control and the shear stress limitation imposed by Eq. (5.26), it does not appear to be necessary, except in some cases of low-rise ductile walls, examined in Section 5.7.4, to provide diagonal reinforcement across the potential sliding planes of the plastic hinge zone, as has been suggested in Section 3.3.2(*b*) for some beams. The spacing of vertical bars in walls crossing potential horizontal sliding planes, such as construction joints should not exceed $2\frac{1}{2}$ times the wall thickness or 450 mm (18 in.). Across sliding planes in the plastic hinge region, a much closer spacing, typically equal to the wall thickness, is preferable.

Construction joints represent potential planes of weakness where excessive sliding displacements may occur [P1]. Therefore, special attention should be given to careful and thorough roughening of the surface of the hardened concrete. The principles of shear friction concepts may then be applied. Accordingly, vertical reinforcement crossing the construction joints should be determined from Eq. (3.42). Commonly, the designer simply checks that the total vertical reinforcement provided is in excess of that required as shear friction reinforcement. In assessing the effective reinforcement that can provide the necessary clamping action, all the vertical bars placed in wall sections, such as shown in Fig. 5.5(*a*) to (*d*) may be considered. Because shear transfer occurs primarily in the web, vertical bars placed in wide flanges, such as seen in Fig. 5.5(e) to (h), should not be relied upon.

In coupled walls with significant coupling [i.e., when according to Eq. (5.4), $A > 0.33$], the structure may be considered as one cantilever, both walls may be considered to transfer the entire shear, and the earthquake-induced axial load need not then be considered on individual component elements. However, construction joints of walls of weakly coupled structures, when $A < 0.33$, should preferably be considered as independent units with gravity and earthquake-induced axial load acting across such joints.

5.4.5   Strength of Coupling Beams

(a)   Failure Mechanisms and Behavior   The primary purpose of beams between coupled walls (Fig. 5.11) during earthquake actions is the transfer of shear from one wall to the other, as shown in Fig. 5.22(c). In considering the behavior of coupling beams it should be appreciated that during an earthquake significantly larger inelastic excursions can occur in such beams than in the walls that are coupled. Moreover, during one half-cycle of wall displacement, several moment reversals can occur in coupling girders, which are rather sensitive to changes in wall curvature. This is caused mainly by the response of the structure in the second and third modes of vibration. Thus during one earthquake significantly larger numbers of shear reversals can be expected in the beams than in the walls [M9].

Many coupling beams have been designed as conventional flexural members with stirrups and with some shear resistance allocated to the concrete. Such beams will inevitably fail in diagonal tension, as shown in Fig. 5.43(a). This was experienced, for example, in the 1964 Alaska earthquake [U2] in the city of Anchorage (see Fig. 1.5). It is evident that the principal diagonal failure crack will divide a relatively short beam into two triangular parts. Unless the shear force associated with flexural overstrength of the beam at the wall faces can be transmitted by vertical stirrups only, a diagonal tension failure will result. In such beams it is difficult to develop full flexural strength even under monotonic loading [P22], and therefore such conventional beams are quite unsuitable [P1] for energy dissipation implied in Fig. 5.23(d).

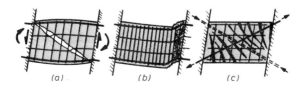

Fig. 5.43 Mechanisms of shear resistance in coupling beams.

When conventional shear reinforcement is based on capacity design principles, some limited ductility can be achieved. However, after only a few load reversals, flexural cracks at the boundaries will interconnect and a sudden sliding shear failure, such as shown in Fig. 5.43(*b*), will occur [P1]. This has been verified with individual beam tests [P23] as well as with reinforced concrete coupled wall models [P1, P31]. Under reversed cyclic loading it is difficult to maintain the high bond stresses along the horizontal flexural reinforcement, necessary to sustain the high rate of changes of moment along the short span. Such horizontal bars, shown in Fig. 5.43(*a*) and (*b*), tend to develop tension over the entire span, so that shear is transferred primarily by a single diagonal concrete strut across the beam.

This consideration leads to the use of a bracing mechanism that utilizes diagonal reinforcement in coupling beams as shown in Figs. 5.43(*c*) and 5.45. While on first loading the necessary diagonal compression force is transmitted primarily by the concrete, this force is gradually transferred fully to the diagonal reinforcement, which is shown by the parallel dashed lines in Fig. 5.43(*c*). This is because these bars would have been subjected to large inelastic tensile strains in the preceding response cycle, as are those bars that are shown by the full line in Fig. 5.43(*c*). Note that the diagonal bars are either in tension or in compression over the full length, and hence bond problems within the coupling beam do not arise. This transfer of diagonal tension and compression to the reinforcement results in a very ductile behavior with excellent energy-dissipating properties [P32–P34]. Beams so reinforced can then sustain the large deformations imposed on them during the inelastic response of coupled walls as illustrated in Fig. 5.23(*d*) [P1, P31].

(b) Design of Beam Reinforcement The design of diagonally reinforced beams for coupled wall structures follows from first principles [P1]. Once the dimensions of the beam are known, the design shear force at midspan (point of zero moment) is simply resolved into appropriate diagonal components. This is shown in Fig. 5.55 for an example beam. From the diagonal tension force the area of the diagonal bars is then readily found.

During the inelastic response of coupled walls the concrete in the beams will become gradually ineffective in resisting diagonal compression, and the diagonal bars must be capable of carrying the full compression components of the shear force. Hence adequate transverse ties or rectangular spirals must be provided to prevent premature buckling of the main diagonal bars. The amount of ties required should be based on the principles outlined in Section 3.6.4 using Eq. (4.19), and it is recommended that the spacing of ties or pitch of spiral should not exceed 100 mm, irrespective of the size of the diagonal bars [P1].

The mechanism of diagonally reinforced coupling beams, as in Figs. 5.43(*c*) and 5.55, is based solely on consideration of equilibrium, and therefore it is independent of the slenderness of the beam (i.e., the inclination of the diagonal bars). Hence the principles are applicable in all situations as long as shear forces due to transverse gravity loading over the span are negligible. When coupling beams are as slender as normal beams, which are used in ductile frames, distinct plastic hinges may form at the ends, and these can be detailed as in beams. The danger of sliding shear failure and reduction in energy dissipation increases with increased depth-to-span ratio, h/l_n, and with increased shear stresses. Therefore, it is recommended that in coupling beams of structural walls, the entire seismic design shear and moment should be resisted by diagonal reinforcement in both directions unless the earthquake-induced shear stress is less than

$$v_i = 0.1(l_n/h)\sqrt{f_c'} \quad \text{(MPa)}; \qquad v_i = 1.2(l_n/h)\sqrt{f_c'} \quad \text{(psi)} \quad (5.27)$$

It should be noted that this severe limitation is recommended because coupling beams can be subjected to much larger plastic rotational demands than spandrel beams of similar dimensions in frames. Plastic rotations at the ends of beams in ductile frames are approximately proportional to the rotation of adjacent joints. In coupled walls or frames of the type in Fig. 5.14, however, beam distortions are further amplified by the relative vertical displacements of the edges of adjacent walls, as seen in Fig. 5.23(d). There is no limitation on the inclination of the diagonal bars.

Shear stress, v_i, as a measure of diagonal compression, is meaningless in diagonally reinforced beams. Because the diagonal compression force can be fully resisted by reinforcement, no limitations on maximum shear stress need be imposed [Section 3.3.2(a)(ii)].

The diagonal bars are normally formed in a group of four or more, as shown for the example beam in Fig. 5.55, and attention must be paid to detailing of these bars so that clashes do not occur within the beam between bars that cross each other or with the wall reinforcement. Also, as Fig. 3.32 shows, there is a concentration of anchorage forces in the adjacent coupled walls. It is therefore recommended that the development length for a group of diagonal bars l_d^* be taken as 1.5 times the standard development length l_d for individual bars [Section 3.6.2(c)]. Nominal secondary (basketing) reinforcement, as shown in Fig. 5.55, should be provided to hold the cracked concrete of the beam in position.

There may be some additional horizontal reinforcement, placed in a connecting floor slab, which might interact with the coupling beam. As Fig. 5.44(a) shows, rotating rigid bodies, attached to a homogeneous isotropic coupling beam, will introduce diagonal compression and tensile stresses, but the total length of horizontal fibers does not change. In diagonally cracked beams, however, the length of the tension diagonal will increase by an amount significantly larger than the shortening of the compression diagonal. As a result, as Fig. 5.44(b) shows, all horizontal reinforcement, irrespective of its level within the depth of the coupling beam, will be subject to tension.

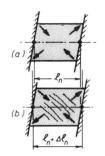

Fig. 5.44 Lengthening of inelastic coupling beams.

Thus horizontal reinforcement, parallel to the beam in connected slabs, will increase the resistance of the beam.

Figure 5.45 shows a particular example in which a slab is attached to the top of the beam. The tributary area of the slab reinforcement A_{ss}, as in the case of beams studied in Section 4.5.1(b), cannot be determined accurately. From first principles and using the notation given in Fig. 5.45, it is found that the moment capacity at the right-hand side of the coupling beams may increase to

$$M_r = M_l + T_h z_b = (A_{sd} \cos \alpha + A_{ss})f_y z_b \qquad (5.28)$$

and hence the ideal shear strength of the beam becomes

$$Q_i = \frac{M_l + M_r}{l_n} = (2A_{sd} \cos \alpha + A_{ss})\frac{z_b}{l_n}f_y \qquad (5.29)$$

As a result of the moment and shear strength increment, a diagonal compression force $C_c = T_h/\cos \alpha$, carried by a concrete strut, shown shaded

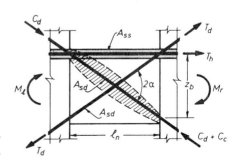

Fig. 5.45 Contribution of slab reinforcement to strength of a coupling beam.

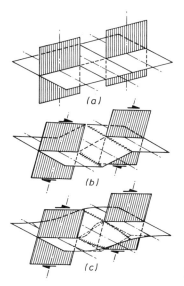

(a)

(b)

(c)

Fig. 5.46 Coupling of walls solely by slabs.

in Fig. 5.45, will develop. Depending on the position of the horizontal steel with area A_{ss}, the increase of flexural strength $\Delta M = z_b T_h$ will affect the wall on the right- or left-hand side, or both walls equally with $0.5 z_b T_h$ if the slab, as in cases of spandrel beams at the exterior of a building, is attached at middepth of the coupling beams. The horizontal D10 bars in the beam of Fig. 5.55 have deliberately been provided with short development length to prevent significant contribution to coupling beam strength.

(c) Slab Coupling of Walls   Although slabs provide relatively weak coupling of walls, as shown for an example structure in Fig. 5.22, their role should be considered [B9, C9]. The region of slab coupling shown shaded in Fig. 5.13(c) at various stages of its response is reproduced in Fig. 5.46. When sufficiently large rotations occur in the walls during an earthquake, slab yield-line moments develop as shown in Fig. 5.46(b), and significant shear transfer across the opening may result. However, parts of the slab distant from the wall may not be as efficient because transverse bending and hence torsional distortions will reduce curvature near the edges, as shown in Fig. 5.46(c). Shear transmission from slab into the wall will occur mainly around the inner toe of the wall section, where curvature ductility in the slab will be maximum. It is therefore to be expected that local shear failure of the slab, due mainly to punching, may occur in this area. Torsional cracking of the slab and shear distortions around the toe are responsible for the rather poor hysteretic response of this system [P24]. It has therefore been suggested that slab

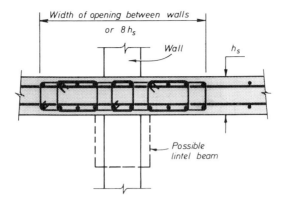

Fig. 5.47 Concentrated reinforcement in slab coupling.

coupling should not be relied on as a significant source of energy dissipation in ductile coupled wall systems [P24].

The concentration of well-confined slab reinforcement in a relatively narrow band across the slab, as shown in Fig. 5.47, will somewhat improve behavior. The small-diameter stirrup ties provide some additional shear strength and when sufficiently closely spaced will delay buckling of compression bars in the plastified regions. However, their contribution to the control of punching shear is insignificant.

The efficiency of such a beam strip will be improved if the shear resisting mechanism around the wall toes is strengthened. This may be achieved by

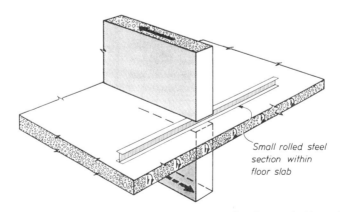

Fig. 5.48 Control of punching shear at the toes of walls coupled by a slab.

Fig. 5.49   Shear failure of a shallow lintel beam spanning between coupled walls.

placing a short rolled steel section across the toe, between the upper and lower layers of slab reinforcement, as shown in Fig. 5.48. The concept employed is similar to that governing the design of shear head reinforcement in flat slab construction.

The contribution of such slab reinforcement to the flexural strength of a small lintel beam across the opening, as shown by the dashed outline in Fig. 5.47, might be excessive even if the concentrated slab reinforcement shown in this figure is not adopted. This means that when large inelastic deformations are imposed by the walls, a shear failure of such a lintel beam might be inevitable. Such a situation, encountered during a test [P24], is shown in Fig. 5.49.

5.5   CAPACITY DESIGN OF CANTILEVER WALL SYSTEMS

5.5.1   Summary

Having examined the features of behavior, analyses and detailing of cantilever walls relevant to ductile seismic performance, in this section the main conclusions are summarized while, step by step, the application of the capacity design philosophy is reviewed. A numerical example, illustrating the execution of each design step, is presented in Section 5.5.2.

Step 1: Review of the Layout of Cantilever Wall Systems. The positioning of individual walls to satisfy architectural requirements is to be examined from a structural engineering point of view. In this respect the following aspects are of particular importance:

(a) Regularity and preferably, symmetry in the positioning of walls within the building to reduce adverse torsional effects [Sections 1.2.3(b) and 5.2.1 and Figs. 1.12, 5.2, and 5.3].

(b) Efficiency of force transfer from diaphragms to walls where large openings exist in the floors (Fig. 1.11).

(c) Checking of the configuration of walls in elevation (Figs. 5.8 to 5.10) to ensure that feasible shear resistance and flexural strength with adequate ductility capacity can readily be achieved.

(d) A review of foundation conditions to ensure that overturning moments, particularly where significant gravity loads cannot be routed to a cantilever wall, can be transmitted to the soil. Implications for the foundation structure (Section 9.4) and the possible rocking of walls (Section 9.4.3) should be studied.

Step 2: Derivation of Gravity Loads and Equivalent Masses. After estimating the likely sizes of all structural components and the contribution of the total building content, as well as code-specified live loads:

(a) Design dead and live loads (Section 1.3.1) and their combinations (Section 1.3.2) are derived for each wall of the cantilever system.

(b) From the total gravity loads over the entire plan of the building, the participating weights W_i (masses) at all floors (Section 2.4.3) are quantified.

Step 3: Estimation of Earthquake Design Forces. Using the procedure described in Section 2.4.3, estimate the total design base shear V_b [Eq. (2.26)] and evaluate the component forces F_r [Section 2.4.3(c)] to be assigned to each level. This will require, among other variables enumerated in Section 2.4.3, an estimation of the period T of the structure and the ductility capacity μ_Δ (i.e., the force reduction factor R). With some design experience this can be made readily, but values must be subsequently confirmed.

Step 4: Analysis of the Structural System. With the evaluation of section properties for all walls, the actions due to lateral forces can be distributed [Section 5.3.2(a)] in proportion to wall stiffnesses. In this the design eccentricities of story shear forces [Section 2.4.3(g)] are to be considered. To determine the critical condition for each wall in each direction of seismic attack, different limits for torsion effects must be examined. If cantilever walls interact with frames, the principles outlined in Chapter 6 are applicable.

Step 5: Determination of Design Actions. For each wall the appropriate combination of gravity load and lateral force effects are determined, using appropriate load factors (Section 1.3.2), and critical design quantities with respect to each possible direction of earthquake attack are found. Spot checks may be made to ascertain that the chosen wall dimensions will be adequate. A redistribution of design actions should also be considered [Section 5.3.2(*c*)].

Step 6: Design for Flexural Strength. For each wall this involves:

(a) The determination of the amount and arrangement of vertical flexural reinforcement at the base, using approximations [Section 5.4.2(*a*)] or otherwise.

(b) Checking that limits of reinforcement content, bar sizes, and spacing are not exceeded [Section 5.4.2(*b*)] and that chosen wall dimensions satisfy stability criteria [Section 5.4.3(*c*)].

(c) Using enhanced yield strength $\lambda_o f_y$, the determination of flexural overstrength of the base section $M_{o,w}$. As bar arrangements are finalized, the analysis of the section, including the finding of the depth of compression c, can be carried out [Section 3.3.1(*c*)].

(d) The checking of ductility capacity ($c \leq c_c$) and the need for confining a part of the compressed regions of the wall section over the height of the plastic hinge [Section 5.4.3(*e*)].

(e) Consideration of sections at higher levels and curtailment of the vertical reinforcement [Section 5.4.2(*c*) and Fig. 5.29].

Step 7: Design for Shear Strength. Magnified design wall shear forces V_{wall} [Eq. (5.22)] and corresponding shear stresses [Eq. (5.25)] are determined. The latter are compared with maximum allowable values in both the potential plastic hinge [Eq. (5.26)] and the elastic [Eq. (3.30)] regions. For each of these two regions the contribution of the concrete to shear strength v_c is found [Eqs. (3.39) and (3.36)], and hence the necessary amount of horizontal shear reinforcement is evaluated [Eq. (3.40)]. Sliding shear resistance is checked [Section 5.4.4(*c*) and Eq. (3.42)].

Step 8: Detailing of Transverse Reinforcement. The final stage of wall design involves:

(a) The determination of transverse hoop or tie reinforcement in the end regions of wall sections to satisfy confining requirements for compressed vertical bars [Section 5.4.3(*e*)(3) and Eq. (4.19)] or possibly for the compressed concrete [Eqs. (5.19) and (5.20)] with due regard to limitations on tie spacing [Section 5.4.3(*e*)]

(b) Determination of the spacing and anchorage of horizontal wall shear reinforcement [Section 3.3.2(*a*)(vii)].

5.5.2 Design Example of a Cantilever Wall System

(a) General Description of Example Figure 5.50(*a*) shows the floor plan of a regular six-story, 20-m (65.6-ft)-high building for a department store, set out on a 6-m (19.7-ft) grid system. The height of the first story is 4 m (13.1 ft). The floor is of waffle slab construction. The entire lateral force resistance is assigned to nine cantilever structural walls. The floor slab is assumed to be very flexible, so that no flexural coupling is assumed to exist between any of

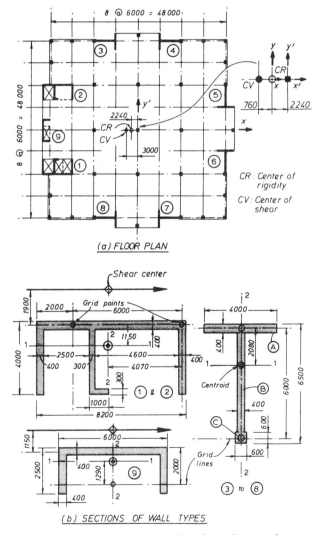

Fig. 5.50 Floor plan and principal wall sections for a six-story department store.

TABLE 5.1   Wall Stiffnesses

i	axis	I_g (m^4)^a	F^b	I_w (m^4)^c
1	1-1	12.4	0.14	5.55
	2-2	63.3	0.87	18.35
6^d	1-1	19.7	0.34	7.68
	2-2	2.2	0.06	1.05
9	1-1	2.2	0.05	1.06
	2-2	20.4	0.38	7.75

^{a}In this analysis only relative stiffnesses are required. Hence no allowance in accordance with Eq. (5.7) has been made for effects of cracking.
^{b}F is obtained from Eq. (5.9b).
$^c I_w$ is obtained from Eq. (5.9a).
^{d}The gross area of wall 6 is $A_g = 4.16$ m^2.

the walls. The purpose of this design example is to show how the actions for the most critical T-shaped wall (Nos. 5 and 6) may be found and then to design the reinforcement for only that wall at its base.

The following data are used:
Design ductility factor $\qquad\qquad\qquad\qquad\qquad \mu_\Delta = 4.0$
Material properties $\qquad\qquad\qquad\qquad\qquad f'_c = 35$ MPa (5000 psi)
$\qquad\qquad\qquad\qquad\qquad\qquad\qquad\qquad f_y = 400$ MPa (58 ksi)
Total design base shear for the entire structure $\quad$ 13 MN (2900 kips)
Total seismic overturning design moment
$\quad$ for the building $\qquad\qquad\qquad\qquad\qquad$ 240 MNm (176,000 kip-ft)
Total axial compression on one wall (No. 5) $\qquad P_D = 8.0$ MN (1800 kips)
$\quad$ due to dead and reduced live load, respectively. $\;P_{Lr} = 1.3$ MN (290 kips)

Sectional properties have been derived from the dimensions shown in Fig. 5.50(b), and these are given in Table 5.1 for each typical wall. The appropriate coordinates of the centroids or shear centers of wall sections are obtained from Fig. 5.50(b). The center of shear (CV), that is the center of mass of the building in the first story is 3 m to the left of the central column.

(b) *Design Steps*   Design computations in this section are set out following the steps listed in Section 5.5.1. Where necessary, additional explanatory notes have been added and references are given throughout to relevant sections and equations in the preceding chapters.

Step 1: Review of Layout. As can be seen in Fig. 5.50, the positioning of walls in plan is near symmetrical, and hence the system provides optimum torsional resistance. Some attention will need to be given to the detailing of floor reinforcement to ensure that in the vicinity of floor openings, inertia forces at each level are effectively introduced to the shear areas of walls 1, 2, and 9. Openings in walls, if any, are assumed in this example to be small enough to be neglected. Foundations are assumed to be adequate.

Step 2: Gravity Loads. Only the base section of walls 5 and 6 in Fig. 5.50 is considered in this example. Results of gravity load analysis are given in Section 5.5.2(a). It is assumed that moments introduced to the walls by gravity loads are negligibly small.

Step 3: Earthquake Design Forces. Because it is not the purpose of this example to illustrate aspects of seismic force derivation, only the end results, in terms of the overturning moment and shear for the entire wall system at the base, are given in Section 5.5.2(a). With allowance for torsional effects the total required strength of the systems S_u [Section 1.3.3(a)] must meet these criteria.

Step 4: Analysis of System. Using the assumption and approximations of Section 5.3.1(a), the stiffness [i.e., the equivalent moment of inertia with respect to the principal axes 1–1 and 2–2 identified in Fig. 5.50(b)] of each wall section is determined. Details of these routine calculations are not given here, but the results are assembled in Table 5.1.

TABLE 5.2 Distribution of Unit Seismic Base Shear in the Y Direction

i	I_{ix} (m^4)	S'_{iy}	$x_i'^a$ (m)	$x_i' S'_{iy}$ (m)	x_i^a (m)	$x_i I_{ix}$ (m^5)	$x_i^2 I_{ix}$ (m^6)	S''_{iy}	S_{iy}
1	5.55	0.145	−22.07	−3.200	−19.83	−110.1	2,183	−0.010	0.135
2	5.55	0.145	−22.07	−3.200	−19.83	−110.1	2,183	−0.010	0.135
3	1.05	0.027	−8.08	−0.218	−5.84	−6.1	36	−0.001	0.026
4	1.05	0.027	8.08	0.218	10.32	10.8	111	0.001	0.028
5	7.68	0.200	24.00	4.800	26.24	201.5	5,287	0.018	**0.218**
6	7.68	0.200	24.00	4.800	−26.24	201.5	5,287	0.018	**0.218**
7	1.05	0.027	8.08	0.218	10.32	10.8	111	0.001	0.028
8	1.05	0.027	−8.08	−0.218	−5.84	−6.1	36	−0.001	0.026
9	7.75	0.202	−26.95	−5.444	−24.71	−191.5	4,732	−0.017	0.185
	38.41	1.000		−2.244^c		0.7^b	19,966	−0.001^b	0.999^b

^{a}Coordinates for walls 1, 2, and 9 are measured to the approximate shear centers of these setions, as shown in Fig. 5.50(b).
^{b}Small error due to rounding off.
^{c}The center of rigidity (CR) is $x \approx -2.24$ m from the central column [Fig. 5.50(a)].

TABLE 5.3 Distribution of Unit Seismic Base Shear in the X Direction

i	I_{iy} (m^4)	S'_{ix}	y_i (m)	$y_i^2 I_{iy}$ (m^6)	S''_{ix}	S_{ix}
1	18.35	0.260	-13.90	3545	-0.028	0.232
2	18.35	0.260	13.90	3545	0.028	0.288
3	7.68	0.109	24.00	4424	0.020	0.129
4	7.68	0.109	24.00	4424	0.020	0.129
5	1.05	0.015	8.08	69	0.001	**0.016**
6	1.05	0.015	-8.08	69	-0.001	0.014
7	7.68	0.109	-24.00	4424	-0.020	0.089
8	7.68	0.109	-24.00	4424	-0.020	0.089
9	1.06	0.015	0.00	—	—	0.015
	70.58	1.001^a		24,924	0.000	1.001^a

^{a}Small error due to rounding off.

From the procedure described in Section 5.3.2(a), the share of each wall in the total resistance to lateral forces is determined in Tables 5.2 and 5.3. For convenience, a unit shear force is first distributed among the nine walls. Equations (5.12a) and (5.12b) can then be simplified as follows:

$$S_{ix} = S'_{ix} + S''_{ix}$$

and

$$S_{iy} = S'_{iy} + S''_{iy}$$

where S'_{ix} and S'_{iy} are the shear contributions due to story translations and S''_{ix} and S''_{iy} are the shear contributions due to story twist, as a result of the eccentrically applied unit story shear.

Therefore, when considering a unit shear force acting in the y direction with eccentricity e_x, the terms in Eq. (5.12) reduce to

$$S'_{iy} = I_{ix}/\Sigma I_{ix} \quad \text{and} \quad S''_{iy} = e_x x_i I_{ix}/\Sigma(x_i^2 I_{ix} + y_i^2 I_{iy})$$

These values are recorded in the tables.

From Table 5.2 it is seen that the center of rigidity (CR) is 2.24 m to the left of the central column. Hence the static eccentricity is $e_{sx} = 2.24 - 3.00 = -0.76$ m. Considering also an accidental eccentricity [Eq. (2.37)] of 10% of the lateral dimension of the building, commonly required by building codes (i.e., $\alpha B_b \approx 0.1 \times 48.4 = 4.84$ m), the total design eccentricity $e_d = e_x$

affecting most critically walls (5) and (6) will be

$$e_x = e_{sx} + \alpha B_b = -0.76 + 4.84 = 4.08 \text{ m } (13.4 \text{ ft})$$

With this value, the torsional terms with respect to unit shear in the y direction

$$S''_{iy} = 4.08 x_i I_{iy}/(19{,}966 + 24{,}924) = x_i I_{iy}/11{,}002$$

and the corresponding value of $S_{iy} = S'_{iy} + S''_{iy}$ for each wall is recorded in the last column of Table 5.2.

For unit shear action along the x axis, the design eccentricity is taken $e_y = 4.84$ m and the corresponding torsional terms S''_{ix} are entered in Table 5.3. It is seen that due to lateral forces in the y direction, the most severely loaded T-shaped walls are Nos. 5 and 6. Thus analyses assign 21.8% and 1.6% of the total base shear and overturning moment to these walls in the y and x directions, respectively.

Step 5: Design Actions for Wall 5. The design base shear and moment for wall 5, acting in the y direction, are from the previous analysis:

$$V_u = V_E = 0.218 \times 13 = 2.834 \text{ MN } (635 \text{ kips})$$

and $$M_u = M_E = 0.218 \times 240 = 52.32 \text{ MNm } (38{,}500 \text{ kip-ft})$$

The axial compression due to gravity, to be considered simultaneously, is

$$P_u = P_D + 1.3 P_{Lr} = 8.0 + 1.3 \times 1.3 = 9.69 \text{ MN } (2170 \text{ kips})$$

or $$P_u = 0.9 P_D = 0.9 \times 8.0 \qquad\qquad = \underline{7.20 \text{ MN } (1610 \text{ kips})}.$$

As the average axial compression is rather small, the wall is tension dominated. The maximum demand for tension reinforcement will occur with minimum compression, when $P_u/f'_c A_g = 7.20 \times 10^6/(35 \times 4.16 \times 10^6) \approx 0.050$.

Figures 5.50(b) and 5.51 show details of wall 5. Preliminary analysis indicates that more tension reinforcement is required in area element C than in element A. For a better distribution of vertical bars, approximately 13% lateral force redistribution from wall 5 to wall 6 will be attempted in accordance with the suggestions of Section 5.3.2(c), so that the design

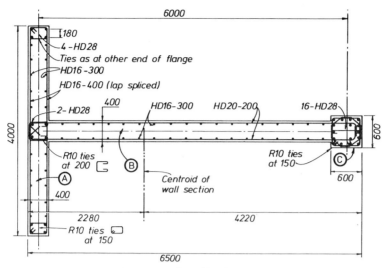

Fig. 5.51 Details of wall 5.

moment causing compression in area element *A* will be reduced to

$$M_E = 0.87 \times 52.32 = 45.52 \text{ MNm (33,500 kip-ft)}$$

while the moment in wall 6, causing compression in area element *C* of that wall, will be increased to

$$M_E = 1.13 \times 52.32 = 59.12 \text{ MNm (43,500 kip-ft)}$$

The appropriate value of ϕ, for walls designed according to capacity design principles, is 0.9 (Table 3.1). Hence the ideal strength requirements are when flexural compression force acts as follows:
At wall element *A*:

$$M_i = 45.52/0.9 = 50.6 \text{ MNm (37,200 kip-ft)}$$
$$P_i = 7.2/0.9 = 8.0 \text{ MN (1790 kips)}$$

$e = 6.32$ m (20.7 ft) from the centroid.
At wall element *C*:

$$M_i = 59.12/0.9 = 65.7 \text{ MNm (48,300 kip-ft)}$$
$$P_i = 8.0 \text{ MN (1790 kips)}$$

$e = 8.21$ m (26.9 ft) from the centroid.

Step 6: Design of Wall 5 for Flexure

(a) Flexural Reinforcement: There are three cases to be considered, each being dependent on the region of the wall base section, shown in Fig. 5.50(*b*), that is subjected to flexural compression.

Case (i) considers the flange, element *A*, in compression due to bending about the 1–1 axis.

Provide two layers of HD16 (0.63-in.-diameter) bars on 300-mm (11.8-in.) centers in the 400-mm (15.7-in.)-thick wall in all regions where no larger bars are present (Fig. 5.51). This corresponds to

$$a_s = 2 \times 201/0.3 = 1340 \text{ mm}^2/\text{m} \left(0.63 \text{ in.}^2/\text{ft}\right)$$

reinforcement, giving $\rho = 1340/(1000 \times 400) = 0.00335$, that is, somewhat more than normally considered to be a minimum (i.e., $\rho = 0.7/f_y = 0.00175$) [Section 5.4.2(*b*)].

Assume the center of compression to be 100 mm from the edge of the flange, and at least in the preliminary analysis, neglect the contribution to strength of all bars in the flange. Use approximations similar to those outlined in Section 5.4.2(*a*).

The tension force in wall element *B* (i.e., the web) is

$$T_b = (6.5 - 0.4 - 0.6)1340 \times 400 \times 10^{-3} = 5.5 \times 536$$

$$= 2948 \text{ kN } (660 \text{ kips})$$

The moment contributions of the force components about the assumed center of compression are then from Fig. 5.50(*b*) or Fig. 5.51:

due to T_b: $(0.5 \times 5.5 + 0.3)2948$ $=$ 8,991 kNm (6600 kip-ft)

due to P_i: $-(6.32 - 2.28 + 0.1)8000$ $= -33,120$ kNm (24,300 kip-ft)

∴ due to T_c: (the tension force in element *C*) $= -24,129$ kNm (17,700 kip-ft)

Hence

$$T_c \approx 24,129/(6.5 - 0.3 - 0.1) \qquad = \quad 3,956 \text{ kN } (886 \text{ kips})$$

Therefore, the approximate area of reinforcement required in element *C* is

$$A_{s, \text{required}} \geq 3956 \times 10^3/400 \approx 9890 \text{ mm}^2 \left(15.3 \text{ in.}^2\right)$$

Try 16 HD28 (1.1-in.-diameter) bars in area element C with $A_s = 9852$ mm^2 (15.3 in.^{2}). Therefore,

$$T_c = 9852 \times 400 \times 10^{-3} = 3941 \text{ kN (883 kips)}$$

Now check the center of compression, neglecting all reinforcement in the flange. The total compression to be resisted in the flange (i.e., element A), is from simple equilibrium requirement for vertical forces

$$C_a = T_b + T_c + P_i = 2948 + 3941 + 8000 = 14{,}889 \text{ kN (3335 kips)}$$

Therefore,

$$a = C_a/(0.85f_c'b) = 14{,}889 \times 10^3/(0.85 \times 35 \times 4000) = 125 \text{ mm (4.9 in.)}$$

Hence the center of compression is approximately $125/2 = 63 < 100$ mm from the edge of the flange.

The inclusion of the contribution of reinforcement close to the neutral axis ($c = 125/0.81 = 154$ mm), approximately in the middle of the 400-mm-thick flange, will make only negligible changes. Hence this approximation remains acceptable.

The moment of resistance with respect to the centroidal axis of the wall in the presence of $P_i = 8000$ kN (1790 kips) compression will be therefore

$$M_i = x_a C_a = 14{,}889(2.28 - 0.5 \times 0.125) = 33{,}016 \text{ kNm (24,270 kip-ft)}$$

$$x_b T_b = 2{,}948(4.22 - 2.75 - 0.6) \qquad\quad = 2{,}565 \text{ kNm (1,880 kip-ft)}$$

$$x_c T_c = 3{,}941(4.22 - 0.3) \qquad\qquad\qquad\underline{= 15{,}449 \text{ kNm (11,360 kip-ft)}}$$

$$M_{i,\text{required}} = 50{,}600 < M_{i,\text{provided}} \qquad\quad = 51{,}030 \text{ kNm (37,510 kip-ft)}$$

This agreement is acceptable and a new trial is not warranted.

With strength enhancement of the steel by 40% (i.e., $\lambda_o = 1.4$), the flexural overstrength factor [Eq. (5.13)] for this wall becomes

$$\phi_{o,w} = \frac{M_{o,w}}{M_E} = \frac{1.4 \times 51{,}030}{52{,}320} = 1.365$$

Case (ii) considers now the effect of moment reversal when area element C is in compression. Assume a neutral-axis depth from the edge of 600-mm (23.6-in.)-square element C (Fig. 5.51) $c = 880$ mm (34.6 in.). Hence the depth of the concrete compression block is $a = 0.81 \times 880 = 713$ mm (28 in.). From the corresponding strain profile, the average compression stress in the 16 HD28 (1.1-in.-diameter) bars in element C is found to be $f_s = 298$ MPa (43.2 ksi) if $\epsilon_c = 0.003$ is assumed.

The internal forces are then as follows:

In element C, $C_c = 0.85 \times 35 \times 600^2 \times 10^{-3}$ = 10,710 kN (2400 kips)

$\quad C_s = 9,852 \times 298 \times 10^{-3}$ = 2,936 kN (658 kips)

In element B, C_s (neglect steel compression)   = —

$\quad C_c = 0.85 \times 35 \times (713 - 600)400 \times 10^{-3}$ = 1,345 kN (301 kips)

Total internal compression   = 14,991 kN (3359 kips)

External compression   = 8,000 kN (1790 kips)

Required internal tension   = 6,991 kN (1569 kips)

In element B, $T_b \approx (5.5 - 0.5)536$ = 2,680 kN (600 kips)

Hence the tension force required in element

$\quad A$ is T_a = 4,311 kN (969 kips)

The area of reinforcement required in the flange is thus:

$\quad A_{sa} = 4,311 \times 10^3/400$ = 10,778 mm^2 (16.7 in.^{2})

$\quad$ Try 22 HD16 (0.63-in.-diameter) bars   = 4,422 mm^2 (6.9 in.^{2})

and 10 HD28 (1.1-in.-diameter) bars   = 6,158 mm^2 (9.5 in.^{2})

$\quad A_{sa}$ provided   = 10,580 mm^2 (16.4 in.^{2})

$\quad\quad\quad\quad\quad\quad\quad\quad\quad\quad (T_a$ = 4,232 kN) (945 kips)

Check now the moment of resistance of internal forces with $P_i = 8000$ kN (1790 kips) using coordinates from the centroid of wall section. The component moments are:

Element C: $x_c(C_c + C_s) = (4.22 - 0.3)(10,710 + 2936) = 53,492$ kNm (39,300 kip-ft)

Element B: $x_b C_c = (4.22 - 0.6 - 0.5 \times 0.113)1345$ = 4,793 kNm (3,520 kip-ft)

$\quad\quad\quad\quad x_b T_b = (0.5 \times 5 - 2.28 + 0.4)2680$ = 1,662 kNm (1,220 kip-ft)

Element A: $x_a T_a = (2.28 - 0.2)4232$ = 8,803 kNm (6,470 kip-ft)

$\quad\quad M_{i,\text{required}} = 65,700 < M_{i,\text{provided}}$ $\approx$ 68,750 kNm (50,530 kip-ft)

This is close enough! A new trial for the neutral-axis depth, c, is not required.

The sum of the resistances of walls 5 and 6 will then be

$$M_5 + M_6 = 51.03 + 68.75 \approx 119.8 \text{ MNm } (88{,}000 \text{ kip-ft}) > 2 \times 52.32/0.9$$

$$= 116.3 \text{ MNm } (85{,}500 \text{ kip-ft})$$

Hence for case (ii) for wall 5, $\phi_{o,w} = 1.4 \times 68.75/52.32 = 1.84$. However, for the combined strength of the two walls,

$$\phi_{o,w} = 1.4 \times 119.8/(2 \times 52.32) = 1.60$$

a value very close to the optimum given by Eq. (1.12) (i.e., $\lambda_o/\phi = 1.4/0.9 = 1.56$), showing that with the choice of reinforcement the strength is only 3% in excess of that required.

Case (iii) considers earthquake action for wall 5 in the x direction. Lateral forces assigned to wall 5 will cause bending about axis 2–2 of the section shown in Fig. 5.50(b). The moment and shear demands for this case are, from Table 5.3 and Section 5.5.2(a),

$$M_E = 0.016 \times 240 = 3840 \text{ kNm } (2822 \text{ kip-ft})$$

and 　　　　　$V_E = 0.016 \times 13 = 208 \text{ kN } (47 \text{ kips})$

while axial compression in the range of 7200 kN (1610 kips) $< P_u <$ 9690 kN (2170 kips) need be considered as in the previous two cases.

A rough check will indicate that the flexural resistance of the section, as shown in Fig. 5.51, is ample. For example, if one considers the resistance of the four HD28 (1.1-in.-diameter) bars only at each end of the flange (element A), it is found that

$$M_i \approx 4 \times 616 \times 400 \times 10^{-3}(4.00 - 2 \times 0.13) \approx 3686 \text{ kNm } (2710 \text{ kip-ft})$$

neglecting the significant contributions of the axial compression and all other bars.

Hence instead of designing the section all that needs to be done is to estimate its ideal and hence its overstrength as detailed in Fig. 5.51. The axial load present will be assumed $P_u = 0.9P_D = 7200$ kN and $P_i = 8000$ kN (1790 kips).

Assume that $c = 1400$ mm; therefore, $a = 1134$ mm (45 in.). From the corresponding strain distribution along the section, with $\epsilon_c = 0.003$, the steel stress at the axis 2–2 will be $(2000 - 1400)600/1400 = 257$ MPa (37 ksi).

Hence the internal tensile forces are in:

Element C: $T_c = 9852 \times 257 \times 10^{-3}$ $\qquad$ $=$ $\quad$ 2,532 kN (567 kips)

Element B: $T_b = 5.5 \times 1340 \times 257 \times 10^{-3}$ $\qquad$ $=$ $\quad$ 1,895 kN (424 kips)

Element A: $(4\,\text{HD28})T_a = 2464 \times 400 \times 10^{-3}$ $\quad$ $=$ $\quad$ 986 kN (220 kips)

$\qquad$ $(12\,\text{HD16})T_a'' \approx 2412 \times 400 \times 10^{-3}$ $\quad$ $=$ $\quad$ 965 kN (216 kips)

$\qquad$ $(2\,\text{HD28})T_a'' = 1232 \times 257 \times 10^{-3}$ $\quad$ $=$ $\quad$ 317 kN (71 kips)

External compression P_i $\qquad\qquad\qquad$ $=$ $\quad$ 8,000 kN (1790 kips)

Internal compression required is then $\qquad$ 14,695 kN (3288 kips)

Internal compression provided in:

Element A: $C_c = 0.85 \times 35 \times 1134 \times 400 \times 10^{-3} =$ $\;$ 13,495 kN (3022 kips)

$\qquad$ $C_s = T_a'$ $\qquad\qquad\qquad\qquad$ $=$ $\quad$ 986 kN (220 kips)

$\quad$ HD16 bars approximately $\qquad\qquad$ $\approx$ $\quad$ 300 kN (67 kips)

$\qquad\qquad\qquad\qquad$ 14,695 $\quad$ $\approx$ $\;$ 14,781 kN (3310 kips)

This is a satisfactory approximation. The moment about axis 2–2 is thus due to:

T_c, T_b, and T_a'' $\qquad\qquad\qquad$ $=$ $\qquad$ 0 kNm

T_a': $986(2.0 - 0.13)$ $\qquad$ $=$ $\;$ 1,843 kNm (1354 kip-ft)

T_a'': 965×1.0 $\qquad\qquad$ $=$ $\quad$ 965 kNm (709 kip-ft)

C_c: $13,495(2.0 - 0.5 \times 1.134)$ $\quad$ $=$ 19,338 kNm (14,213 kip-ft)

C_s: $986(2.0 - 0.13)$ $\qquad\qquad$ $=$ $\;$ 1,843 kNm (1354 kip-ft)

HD16 bars: 300×1.5 $\qquad\qquad$ $=$ $\quad$ 450 kNm (331 kip-ft)

$M_{i,\text{required}} = 3840/0.9 = 4267(3136)$ $\quad M_i = 24{,}439$ kNm (17,963 kip-ft)

Hence $\phi_{o,w} = 1.4 \times 24{,}439/3840 = 8.91$. This indicates a very large reserve flexural strength.

(b) Code-Specified Limits

(1) The reinforcement content of element C is $\rho_l = 9852/600^2 = 0.0274 < 16/f_y = 0.04$, and thus with $\rho_l = 0.00335$ in the 400-mm width of the wall, the limits in Section 5.4.2(b) are well satisfied.

(2) The recommendations for stability criteria of the wall section indicate (Fig. 5.36) that the flange thickness should be at least $4000/16 = 250$ mm (10 in.) < 400 mm. The critical thickness of the opposite end of the wall section is from Fig. 5.35 with $\mu_\Delta = 4$, $A_r = 20/6.5 = 3.1$.

$$l_w = 6500 > 1.6 \times h_1 = 1.6 \times 4000 = 6400 \text{ mm (21 ft)}$$

$$b_c = 0.047 \times 6400 = 301 \text{ mm (11.9 in.)} < 600 \text{ mm}$$

Hence all stability criteria are satisfied.

(c) Flexural Overstrength: The approximate analysis given in (a) of this step shows that no further refined analysis is required and that the maximum value of the overstrength factor, to be used for shear design in the y direction in step 7, is $\phi_{o,w} = 1.84$.

(d) Ductility Capacity: It was seen that the largest neutral axis depth was $c = 880$ mm (34.6 in.). The critical value is, however, from Eq. (5.18b),

$$c_c = \frac{1.4 \times 68.75}{2.2 \times 1.4 \times 4 \times 52.32} \times 6500 = 970 \text{ mm (38.2 in.)} > 880 \text{ mm}$$

When the larger compression load of $P_u = 9.69$ MN is used, the neutral-axis depth will increase. Estimate this value of c using approximate values of internal forces involved in the previous section analysis.

Compression forces:	T_a = as before	=	4,232 kN (948 kips)
	T_b = estimate	=	2,300 kN (515 kips)
	$P_i = 9.69 \times 10^3/0.9$	=	10,767 kN (2412 kips)
Required total internal compression		=	17,299 kN (3875 kips)
C_c = in element C as before		=	10,710 kN (2399 kips)
C_s = assuming $f_s = 380$ MPa		=	3,744 kN (839 kips)
∴ C_c = in element B will have to be		=	2,845 kN (637 kips)

Depth of compression in the web required is thus

$$a' = 2845 \times 10^3/(0.85 \times 35 \times 400) = 239 \text{ mm (9.4 in.)}$$
$$\text{Total depth: } a = 600 + 239 \qquad = 839 \text{ mm (33.0 in.)}$$

Therefore,

$$c = 839/0.81 = 1036 \text{ mm} > 970 \text{ mm} \ (38.2 \text{ in.}) = c_c$$

Hence confinement of the concrete may be required. The more accurate Eq. (5.18c) would give

$$c_c = \frac{3400 \times 1.4 \times 68.75}{(4 - 0.7)(17 \times 20/65)1.4 \times 400 \times 52.32} \times 6500$$

$$= 1096 > 1036 \text{ mm} \ (46.8 \text{ in.})$$

Hence reinforcement for confining the concrete is not required.

The neutral-axis depth for forces in the other direction with the flange-in compression was shown to be very much smaller [i.e., $c \approx 154$ mm (6.1 in.)].

(e)   Wall Sections at Higher Levels: The curtailment of flexural reinforcement, in accordance with Figs. 5.29, is not considered in this example.

Step 7: Shear Strength. From Eq. (5.23a) or Eq. (5.23b) for a six-story wall the dynamic shear magnification is

$$\omega = 0.9 + 6/10 = 1.3 + 6/30 = 1.5$$

Hence from Eq. (5.22) the critical shear force on the wall, when the boundary (element C) is in tension, is

$$V_u = V_{\text{wall}} = 1.5 \times 1.84 \times 2834 = 7822 \text{ kN} \ (1752 \text{ kips})$$

From Eq. (5.24a),

$$V_u = V_{\text{wall}} < \mu_\Delta V_E = 4 \times 2834 = 11,336 \text{ kN} \ (8332 \text{ kips})$$

Because capacity design procedure is relevant, the value of $\phi = 1.0$, and hence from Eq. (5.25),

$$v_i = 7822 \times 10^3/(0.8 \times 6500 \times 400) = 3.76 \text{ MPa} \ (545 \text{ psi})$$

The maximum acceptable shear stress is, from Eq. (5.26),

$$V_{i,\text{max}} = \left(\frac{0.22 \times 1.84}{4} + 0.03 \right)35 = 4.59 \text{ MPa} \ (666 \text{ psi}) > 3.76 \text{ MPa} \ (545 \text{ psi})$$

and from Eq. (3.31),

$$V_{i,\text{max}} = 0.16 \times 35 = 5.6 \text{ MPa} \ (812 \text{ psi}) > 3.76 \text{ MPa}$$

and thus the web thickness is satisfactory. With a minimum axial compression of P_u = 7.2 MN (1613 kips) on the wall, from Eq. (3.39),

$$v_c = 0.6\sqrt{7.2/4.16} = 0.79 \text{ MPa (115 psi)}$$

and hence by rearranging Eq. (3.40),

$$\frac{A_v}{s} = \frac{(3.76 - 0.79)400}{400} = 2.97 \text{ mm}^2/\text{mm } (0.117 \text{ in.}^2/\text{in.})$$

Using HD20 (0.79-in.-diameter) horizontal bars in each face, the spacing becomes

$$s = 2 \times 314/2.97 = 211 \approx 200 \text{ mm (7.9 in.)}$$

A significant reduction of horizontal shear reinforcement will occur above the end region of the wall ($l_p \approx l_w$ = 6.5 m) at approximately the level of the second floor, where the thickness of the wall may also be reduced. It may readily be shown that sliding shear requirements [Section 5.4.4(c)] are satisfied.

Finally, the wall shear strength with respect to earthquake attack in the X direction must be examined. Case (iii) in step 6 showed that $\phi_{o,w}$ = 8.91. Accordingly, from Eq. (5.22) the design shear force would become

$$V_u = V_{wall} = 1.5 \times 8.91 \times 208 = 2780 \text{ kN (623 kips)}$$

corresponding to

$$v_i = 2780 \times 10^3/(0.8 \times 400 \times 4000) = 2.17 \text{ MPa (315 psi)}$$

Because walls 5 and 6 together represent only 3% of the total required lateral force resistance (Table 5.3) in the X direction, a nearly ninefold increase of flexural resistance at overstrength represents only about $(8.91/1.56)3\% = 17\%$ increase in the overstrength of the entire wall system. This could be developed during a major earthquake with a relatively small decrease in overall ductility demand. However, the structural overstrength at the base of Wall 5, $M_{o,w}$ = 24.44 MNm (17,960 kip-ft), represents an eccentricity of $e \approx 24.44/8.0 = 3.06$ m of the gravity load with respect to the 2–2 axis of the wall. It is not likely that a moment of this magnitude could be absorbed by the foundations for walls 5 and 6 before the onset of rocking.

It is decided to provide HD16 bars in elements A (i.e., the flange) at 400 mm spacing. This gives $\rho = 2 \times 200/(400 \times 400) = 0.0025$, a quantity only a little more than the minimum generally recommended for shear reinforcement [Section 3.3.2(a)(vi)]. Using Eq. (3.40), this steel content corresponds to a shear stress of

$$v_i - v_c = \frac{A_v f_y}{b_w s} = \frac{400 \times 400}{400 \times 400} = 1.0 \text{ MPa (145 psi)}$$

which, with $v_c = 0.79$ MPa (115 psi), provides a total shear strength of

$$V_i = (1 + 0.79)(0.8 \times 400 \times 4000)10^{-3} = 2291 \text{ kN (513 kips)} = 0.82V_u$$

Alternatively, Eq. (5.24*b*) may be used, according to which the ideal shear strength of wall 5 in this direction need not exceed

$$V_i = V_{\text{wall}}/\phi = S_x \mu_\Delta V_{\text{base}}/\phi = 0.016 \times 4 \times 1300/0.85 = 980 \text{ (220 kips)}$$
$$< 2291 \text{ kN}$$

Step 8: Transverse Reinforcement. Details of the transverse hoop reinforcement, shown in Fig. 5.51, are not given here because of the detailed study of a similar example in Section 5.6.2.

5.6   CAPACITY DESIGN OF DUCTILE COUPLED WALL STRUCTURES

5.6.1   Summary

By similarity to Section 5.5.1, the main conclusions of the capacity design of coupled structural walls are summarized here, step by step, to aid the designer in its application. A detailed design example follows in Section 5.6.2.

Step 1: Geometric Review. Before the static analysis procedure begins, the geometry of the structure should be reviewed to ensure that in the critical zones compact sections, suitable for energy dissipation, will result [Sections 5.4.3(*c*) and 3.4.3]. All aspects listed in step 1 (Section 5.5.1) for cantilever wall systems are also applicable.

Step 2: Gravity Loads and Equivalent Masses. The derivation of these quantities is as for cantilever systems listed in step 2 of Section 5.5.1.

Step 3: Earthquake Design Forces. In the estimate of horizontal design forces, again the principles outlined for cantilevers in step 3 of Section 5.5.2 may be followed. The ductility capacity of the system can be approximated only as the efficiency of the coupling beams [Eqs. (5.5) and (5.6)] is not yet known. As adjustments can be made readily in step 5. It is best at this stage to assume that, for example, $\mu_\Delta = 5$.

Step 4: Analyses of Structural System. With the evaluation of the lateral static forces, the complete analysis for the resulting internal structural actions, such as moments, forces, and so on, can be carried out. In this the modeling assumptions of Section 5.3.1 should be observed. Usually, a frame analysis, using the model of Fig. 5.14, will be used. Typical patterns of actions are shown in Fig. 5.22. If the coupled walls are part of a cantilever wall system, the procedure summarized in step 4 of Section 5.5.2 may be followed. If the walls interact with frames, the principles of Chapter 6 are relevant.

Step 5: Confirmation of the Appropriate Ductility Factor and Design Forces. Having obtained the moments and axial forces at the base of the coupled wall structure, the moment parameter $A = Tl/M_{ot}$ [Eq. (5.6)], discussed in Section 5.2.3(b), can be determined. With the use of Eq. (5.5) the required value of the reduction factor R, and hence the appropriate ductility factor μ_Δ, can be found. If this differs from that assumed earlier (i.e., $\mu_\Delta = 5$), simply all quantities of the elastic analysis may be proportionally adjusted.

Step 6: Checking Demands on Foundations. To avoid unnecessary changes in the design later, at this stage it should be checked whether the foundation structure for the coupled walls would be capable of transmitting at least 1.5 times the overturning moment, M_{ot} [Eq. (5.3)], to the foundation material (soil). It is to be remembered that in a carefully designed superstructure in which no excess strength of any kind has been allowed to develop, at least λ_o/ϕ times the overturning moment M_{ot} resulting from code forces may be mobilized during large inelastic displacements [see Section 1.3.3(f)]. The foundation system must have a potential strength in excess of the overstrength of the superstructure ($\phi_{o,w}M_{ot}$); otherwise, the intended energy dissipation in the coupled walls may never develop. This issue is examined in greater detail in Chapter 9.

Step 7: Design of Coupling Beams. Taking flexure and shear into account, the coupling beams at each floor can be designed. Normally, diagonal bars in cages should be used. A strength reduction factor of $\phi = 0.9$ is appropriate. Particular attention should be given to the anchorage of caged groups of bars [Section 3.6.2(c) and Fig. 3.32] and to ties which should prevent inelastic buckling of individual diagonal bars [Eq. (4.19)]. The beam reinforcement should match the shear demand as closely as possible. Excessive coupling beam strength may lead to subsequent difficulties in the design of walls and foundations. To achieve this, shear redistribution vertically among several or all beams [Fig. 5.23(c)] may be used in accordance with Section 5.3.2(c). If necessary, the contribution of slab reinforcement, shown in Fig. 5.45, to coupling beam strength should be estimated [Eq. (5.29)].

Step 8: Overstrength of Coupling Beams. To ensure that the shear strength of the coupled wall structure will not be exceeded and that the maximum load demand on the foundation is properly assessed, the overstrength of the potential plastic regions must be estimated. Accordingly, the shear over-strength, Q_{io} of each coupling beam, as detailed and based on the overstrength $\lambda_o f_y$ of the diagonal and if relevant, slab reinforcement, is determined.

Step 9: Determination of Actions on the Walls. To find the necessary vertical reinforcement in each of the coupled walls (Fig. 5.22) at the critical base section, the following loading cases should be considered in accordance with Section 1.3.2.

(i) $P_u = 0.9 P_D - P_E$ (axial tension or minimum compression) and M_{u1} [Eq. (1.6b)].

(ii) $P_u = P_E + P_D + 1.3 P_{LR}$ or $P_u = P_E + 0.9 P_D$ axial compression and M_{u2} [Eq. (1.6a)], where

$\quad P_u$ = axial design load on a wall corresponding with required strength

$\quad P_E$ = axial tension or compression induced in the wall by the lateral static forces only, shown as T in Figs. 5.20 and 5.22

$\quad P_D$ = axial compression due to dead load

$\quad P_{Lr}$ = axial compression due to reduced live load rL [Eq. (1.3)].

$\quad M_{u1}$ = moment at the base developed concurrently with earthquake-induced axial tension force [Fig. 5.22(a)]

$\quad M_{u2}$ = moment at the base developed concurrently with earthquake-induced axial compression force [Fig. 5.22(a)]

(iii) If case (i) above is found to result in large demand for tension reinforcement, or for other reasons, a redistribution of the design moments from the tension wall to the compression wall may be carried out in accordance with Section 5.3.2(c), within the following limits:

$$\text{(a)} \quad M'_{u1} = M_{u1} - \Delta M \geq 0.7 M_{u1}$$

$$\text{(b)} \quad M'_{u2} = M_{u2} + \Delta M \leq M_{u2} + 0.3 M_{u1}$$

where M'_{u1} and M'_{u2} are the new design moments for the tension and compression walls, respectively, after moment and lateral force redistribution has been carried out, as shown in Fig. 5.23(a) and (b).

Step 10: Design of Wall Base Sections. With the information above and using $\phi = 0.9$, the dimensions and reinforcement at the base of each wall can be determined. The procedure suggested in Sections 5.4.2(a) and (b) may be used, considering each wall for both directions of the applied lateral forces. It is emphasized that, as a general rule, a refinement of the approximate flexural design (Section 5.4.2) at this stage is not warranted because flexural capacities of the walls, as detailed, will be checked in step 12.

Step 11: Earthquake-Induced Axial Forces at Overstrength. The maximum feasible earthquake-induced axial force induced in one of the coupled walls would be obtained from the summation of all the coupling beam shear forces at overstrength Q_{io}, applied to the wall above the section that is considered. For structures with several stories, this is considered to be an unnecessarily conservative estimate, and accordingly, it is recommended that the wall axial force at overstrength be estimated with

$$P_{Eo} = \left(1 - \frac{n}{80}\right) \sum_i^n Q_{io} \qquad (5.30)$$

where n = number of beams above level i, and its value should not be taken larger than 20.

Step 12: Overstrength of the Entire Structure. To estimate the maximum likely overturning moment that could be developed in the fully plastic mechanism of the coupled wall structure, it is necessary to assume gravity loads that are realistic and consistent with such a seismic event. Accordingly, for this purpose only, the axial forces to be sustained by the walls at overstrength may be estimated by neglecting the presence of live load, as follows:

For tension of minimum compression: $P_{1o} = P_{Eo} - P_D$ (5.31a)

For compression: $P_{2o} = P_{Eo} + P_D$ (5.31b)

It is now possible to estimate the flexural overstrength of each wall section at the base, as detailed, which may be developed concurrently with the axial forces above. The flexural overstrengths, based on material strengths defined by λ_o in Section 3.2.4(e), so derived for the tension and compression walls, respectively, are M_{1o} and M_{2o}. By similarity to Eq. (5.13), the overstrength factor for the entire coupled wall structure is obtained from

$$\phi_{o,w} = \frac{M_{1o} + M_{2o} + lP_{Eo}}{M_E} \qquad (5.32)$$

where M_E is the total overturning moment at the base due to code-specified horizontal forces, shown as M_{ot} in Fig. 5.22(b). If the value of $\phi_{o,w}$ obtained from Eq. (5.32) is less than λ_o/ϕ, the design should be checked further for possible errors. On the other hand, if the value of $\phi_{o,w}$ is much in excess, the design should be reviewed to detect the source of excess strength. Care must be taken with the interpretation of $\phi_{o,w}$ when different grades of steel with different values of λ_o are used in the coupling beams (λ_o'') and the walls (λ_o') when axial force reduction in accordance with Eq. (5.30) is used.

In such cases, to satisfy conditions for the required strength of the structure as specified, it should be shown that

$$\phi_{o,w} \geq \frac{\lambda'_o(M_{1E} + M_{2E}) + \lambda''_o(1 - n/80)lP_E}{\phi(M_{1E} + M_{2E} + (1 - n/80)lP_E)} \qquad (5.33)$$

where it is assumed that a reduction by $n/80$ of the contribution of wall axial force to overturning moment capacity at overstrength is acceptable.

Step 13: Wall Design Shear Forces. Using the concepts of inelastic force redistribution outlined in Section 5.3.2(c), the maximum shear force for one wall of a coupled wall structure may be estimated at any level from

$$V_u = V_{i,\text{wall}} = \omega_v \phi_{o,w}\left(\frac{M_{io}}{M_{1o} + M_{2o}}\right)V_E; \qquad i = 1,2 \qquad (5.34)$$

where ω_v = dynamic shear magnification factor in accordance with Eq. (5.23)
$\quad\quad V_E$ = shear force on the entire coupled wall structure at the corresponding level, derived by the initial elastic analysis for code-specified forces with the appropriate μ_Δ factor as obtained in step 5
$\omega_v \phi_{o,w} \leq \mu_\Delta/\phi$ in accordance with Eq. (5.24a)

The term in parentheses Eq. (5.34) makes an approximate allowance for the distribution of shear forces between the two walls, which, at the development of overstrength, is likely to be different from that established with the initial elastic analysis.

Step 14: Design of Shear Reinforcement at the Base of the Walls. The required horizontal shear reinforcement may be determined now. In assessing the contribution of the concrete to shear-resisting mechanism, in accordance with Section 3.3.2(a) (iv), the effects of the axial forces P_{1o} and P_{2o}, as appropriate, should be taken into account.

Step 15: Check of Sliding Shear Capacity of Walls. In the plastic hinge region, and especially at construction joints, the adequacy of the vertical reinforcement to provide the necessary clamping force in accordance with Section 3.3.2(b) is to be checked. In this respect two coupled walls may be treated as one cantilever wall.

Step 16: Confinement of Wall Sections. From the force combinations considered above, the positions of the neutral axes relative to the compressed edges of the wall sections are readily obtained. For the regions of the wall section over which, in accordance with Section 5.4.3(e), transverse hoops and ties are required to stabilize vertical bars [Eq. (4.19)] or to confine compressed concrete [Eqs. (5.19) and (5.20)] is now to be determined.

Step 17: Consideration of the Required Strength of Walls at Higher Levels. For the purpose of establishing the curtailment of the principal vertical wall reinforcement, a linear bending moment envelope along the height of each wall should be assumed, as shown in Fig. 5.29. This is intended to ensure that the likelihood of flexural yielding due to higher mode dynamic responses along the height of the wall is minimized. No special requirements exist for the provision of shear reinforcement, as plastic hinges are not expected at higher levels of the building.

Step 18: Foundation Design. The actions at the development of the over-strength of the superstructure, P_{1o}, P_{2o}, M_{1o}, and M_{2o} and wall shear force, $V_{wall} = V_{1,wall} + V_{2,wall}$, should be used as design forces acting on the foundations. For ductile coupled walls, the foundation structure should be capable of absorbing these actions within its ideal strength capacity as discussed in Chapter 9.

5.6.2 Design Example of Coupled Walls

(a) Design Requirements and Assumptions   Lateral forces on a 10-story symmetrical building are resisted in one direction by coupled walls, positioned at each end of the building. In the other direction over six 7.35-m (24.1-ft)-long bays, reinforced concrete frames provide the necessary seismic resistance.

The overall dimensions of the coupled wall structure, together with assumed member sizes, are shown in Fig. 5.52, where the specific lateral static forces are also recorded. A displacement ductility capacity of $\mu_\Delta = 5$ was initially assumed. The estimated period of the structure was $T = 0.87$ s. The appropriate zone and importance factors (Section 2.4.3) were 0.217 and 1.0, respectively. Hence from Fig. 2.25 the corresponding $C_{T,S}$ coefficient was taken as 1.66. The total weight of the building was estimated at 67,200 kN (15,050 kips), so that with 10% allowance for accidental torsion, from Eq. (2.26) the design base shear was assessed as

$$V_{base} = 0.5 \times 1.1(0.217 \times 1.66 \times 1.0 \times 67200/5) = 2660 \text{ kN (596 kips)}$$

for the coupled walls at each end of the building. The total gravity load, including the weight of the walls and beams, and live loads introduced at each floor to the coupled wall structure at each end of the building, is shown in Table 5.4.

The gravity load may be assumed to cause uniform compression in each of the two walls. The floor area tributary to the wall at each level is 45 m². The

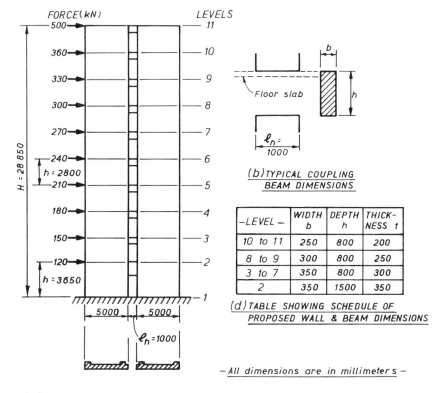

(b) TYPICAL COUPLING BEAM DIMENSIONS

-LEVEL -	WIDTH b	DEPTH h	THICK-NESS t
10 to 11	250	800	200
8 to 9	300	800	250
3 to 7	350	800	300
2	350	1500	350

(d) TABLE SHOWING SCHEDULE OF PROPOSED WALL & BEAM DIMENSIONS

− All dimensions are in millimeters −

(a) DIMENSIONS OF SHEAR WALL & LOADING

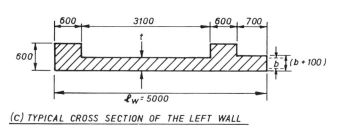

(c) TYPICAL CROSS SECTION OF THE LEFT WALL

Fig. 5.52 Overall dimensions of example coupled wall.

following strength properties are intended to be used:

Beam reinforcement	$f_y = 275$ MPa (40 ksi),	$\lambda_o = 1.25$
Wall reinforcement	$f_y = 380$ MPa (55 ksi),	$\lambda_o = 1.40$
Hoop or tie reinforcement	$f_y = 275$ MPa (40 ksi)	
Concrete	$f_c' = 30$ MPa (4350 psi)	
(above level 3)	$f_c' = 25$ MPa (3625 psi)	

TABLE 5.4   Gravity Loads for the Structure Shown in Fig. 5.52

At Level:	Dead Load		Live Load	
	kN	kips	kN	kips
10	320	72	80	18
9	350	78	100	22
7 and 8	400	90	100	22
6	430	96	100	22
2 to 5	480	108	100	22
1	500	112	120	27
Total at level 1	4320	970	1000	224

Two beams, spanning over six bays in the long direction of the building, frame into each of the walls at the 600-mm (23.6-in.)-wide projections. Analysis for earthquake forces in the long direction indicated that 0.8% longitudinal reinforcement, placed in each of the 600×600 mm equivalent column sections, is adequate for this purpose.

The coupled walls are assumed to be fully fixed at level 1 by a 30-m (98-ft)-long and 7.5-m (24.6-ft)-deep foundation wall.

Steps of the design process follow those summarized in Section 5.6.1.

(b) Design Steps

Step 1: Geometric Review

(i)   Minimum wall dimension [Section 5.4.3(c)]

$$\text{From Fig. 5.36: } b_w \geq 3650/16 \quad = 228 \text{ mm (9 in.)}$$
$$\text{From Fig. 5.35: } b_w \geq 0.069 \times 5000 = 345 \text{ mm (13.6 in.)}$$

assuming that $\mu_\Delta = 6$ and $A_r = 28.85/5 \approx 6$.

(ii)   For the stability of coupling beams from Eq. (3.48):

$$l_n/b_w \quad = 1000/200 \quad = 5 < 25$$
$$\text{or } l_n h/b_w^2 = 1000 \times 800/200^2 \quad = 20 < 100$$
$$\text{or} \quad = 1000 \times 1500/350^2 = 12 < 100$$

These are compact sections.

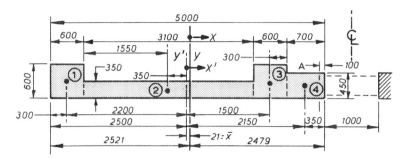

Fig. 5.53   Wall dimensions for computing properties (1000 mm = 39.37 in.).

Step 2: Gravity Loads and Equivalent Masses. The gravity loads are given in Table 5.4. Allowance for the equivalent floor masses has been made when deriving the lateral forces given in Fig. 5.52(a).

Step 3: Earthquake Design Forces. The total estimated base shear $V_{\text{base}} = 2660$ kN has been distributed over the height of the wall [Eq. (2.32)] as shown in Fig. 5.52(a). After the analysis of the structure, adjustments will be made in step 5.

Step 4: Analysis of the Structure

(a) Member Properties: The assumed wall dimensions, applicable from level 1 to level 7, are shown in Fig. 5.53. For the purpose of computing the properties of the wall section, an arbitrary axis Y', located at midway, has been used and the resulting routine calculations for sectional properties are set out in Table 5.5. For convenience four subareas, numbered 1 to 4, are used, to which frequent reference is made.

TABLE 5.5   Wall Properties

i	x' (10^3 mm)	A_i (10^3 mm^2)	$x'A_i$ (10^6 mm^3)	x [a] (10^3 mm)	xA_i (10^6 mm^3)	x^2A_i [b] (10^9 mm^4)	I_y [b] (10^9 mm^4)
1	−2.20	360	−792	−2.221	−800	1777	10
2	−0.35	1085	−380	−0.371	−403	149	869
3	1.50	360	540	1.479	532	787	11
4	2.15	315	677	2.129	671	1429	13
		2120	45		0	4142	904

[a] $\bar{x} = 45 \times 10^6 / (2120 \times 10^3) = 21$ mm
[b] $I_{yy} = (4142 + 904)10^9 = 5.046 \text{ m}^4$

To allow for effects of cracking [Section 5.3.1(a)], assume the following equivalent values:

$$\text{Tension wall:} \quad I_e = 0.5 I_g = 0.5 \times 5.046 = 2.52 \text{ m}^4$$
$$A_e = 0.5 A_g = 0.5 \times 2.12 = 1.06 \text{ m}^2$$
$$\text{Compression wall:} \quad I_e = 0.8 I_g = 4.04 \text{ m}^4$$
$$A_e = A_g = 2.12 \text{ m}^2$$

When the order of lateral-force-induced axial forces on the walls are known, the assumptions above will be revised [Eq. (5.7)] and adjustments in design actions made if necessary. For the purpose of the elastic analysis for lateral static forces, the second moment of area of the walls at upper levels has been reduced, corresponding with the reduced wall thicknesses, without, however, recalculating the new centroidal axis positions. This approximation is acceptable, particularly in the light of the crude allowance for the reduction of stiffness due to cracking.

To allow for cracking and significant shear distortions in the diagonally reinforced coupling beams, Eq. (5.8a) is used:

$$I_e = \frac{0.4 I_b}{1 + 3\left(h^2/l_n\right)}$$

where from Fig. 5.52, $l_n = 1000$ mm. Values of the effective second moment of area I_e are given in Table 5.6.

(b) *Actions Due to Earthquake Forces*: Using a conventional frame analysis, the beam shear forces, wall bending moments, axial forces, and shear were derived. These are shown in Fig. 5.54.

Step 5: Confirmation of μ_Δ and Design Forces. The total overturning moment on the structure, as derived from the lateral forces shown in Fig. 5.52(a), is

TABLE 5.6 Beam Properties

At Level:	h (m)	b (m)	I_b (m⁴)	h/l_n	I_e (m⁴)
10–11	0.80	0.25	0.0107	0.80	0.00147
8–9	0.80	0.30	0.0128	0.80	0.00175
3–7	0.80	0.35	0.0149	0.80	0.00204
2	1.50	0.35	0.0984	1.50	0.00507

(1 m = 39.27 in., 1 m⁴ = 2.4 × 10⁶ in.⁴)

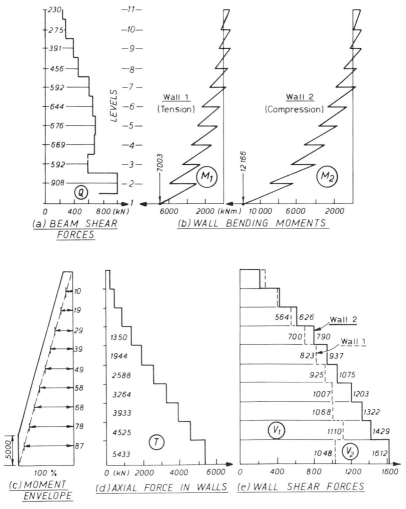

Fig. 5.54 Actions induced in the example coupled wall structure by the lateral static forces.

51,540 kNm. This may also be verified from the expression for internal equilibrium:

$$M_{ot} = M_1 + M_2 + lT = 7003 + 12,166 + 5.958 \times 5433 = 51,538$$
$$\approx 51,540 \text{ kNm } (37,880 \text{ kip-ft})$$

where the magnitudes of M_1, M_2, and T, applicable to the base of the structure, are given in Fig. 5.54(b) and (d). The distance between the centroids of the walls (see Fig. 5.53) is $(2 \times 2479 + 1000) = 5958$ mm (19.53 ft). Therefore, the ratio of the moment resisted by the axial forces (i.e., coupling beams) in the walls and the total overturning moment is, from Eq. (5.4),

$$A = \frac{lT}{M_{ot}} = \frac{5.958 \times 5433}{51,540} = 0.628$$

Therefore, instead of $\mu_\Delta = 5$ as assumed in Section 5.6.2(a), the design ductility factor in accordance with Eq. (5.5), may be increased to

$$R = \mu_\Delta = 3 \times 0.628 + 4 = 5.88 > 5.0$$

Also, analysis indicated that the period of the structure was slightly larger than assumed and this allows the C_{TS} coefficient to be reduced to 1.60. Hence the lateral design forces shown in Fig. 5.52(a) and their effects given in Fig. 5.54 may be adjusted by the factor $(5/5.88)(1.60/1.66) = 0.82$.

Step 6: Checking Demands on the Foundations. It is assumed that the average value of $\lambda_o = 1.35$ is relevant [Section 3.2.4(e)], and thus a carefully designed structure at overstrength can be expected to develop a total overturning moment at the base:

$$\lambda_o M_{ot}/\phi = 1.35(0.82 \times 51.540)/0.9 = 63,400 \text{ kNm } (46,600 \text{ kip-ft})$$

The foundation structure is assumed to be capable of absorbing this moment.

Step 7: Design of Coupling Beams.

(a) *Beams at Levels 3 to 9*: Gravity load effects on these beams are neglected. From Eq. (5.27),

$$v_i = 0.1 l_n \sqrt{f_c'} / h = 0.1 \times 1000\sqrt{30} / 800 = 0.68 \text{ MPa } (100 \text{ psi})$$

whereas the minimum shear stress induced in the coupling beam at roof level [see Fig. 5.54(a)], with $\phi = 0.85$, is on the order of

$$v_i = Q_u/(\phi b_w d) = 0.82 \times 230,000/(0.85 \times 250 \times 0.8 \times 800)$$
$$= 1.39 \text{ MPa } (202 \text{ psi}) > 0.68 \text{ MPa}$$

Therefore, diagonal reinforcement should be used in all coupling beams [Section 5.4.5(b)] to resist the entire earthquake-induced shear force. The contribution of slab reinforcement (Fig. 5.45) is assumed to be negligible.

The approximate position of the diagonal reinforcement and the corresponding centerline dimensions for typical coupling beams at upper levels are

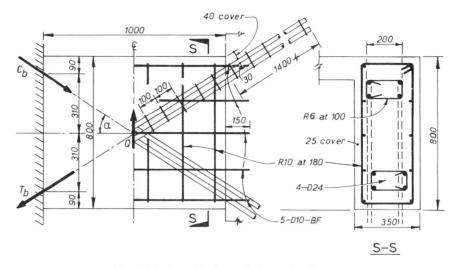

Fig. 5.55 Details of a typical coupling beam.

shown in Fig. 5.55. From this it is seen that the diagonal forces are

$$C_b = T_b = Q/(2\sin\alpha)$$

where $\tan\alpha = 310/500 = 0.62$, (i.e., $\alpha = 31.8°$), so that, in general, the area of diagonal steel required is, with $\phi = 0.9$,

$$A_{sd} = T_b/(\phi f_y) = Q/(2\phi f_y\sin\alpha)$$
$$= Q/(2\times0.9\times275\times0.527) = Q/261 \text{ (MPa)}$$

To resist the maximum shear force at level 5, where $Q_u = 0.82\times676 = 554$ kN, we find that $A_{sd} = 554,000/261 = 2123$ mm^2(3.29 in.^{2}). We can use four D28 (1.1-in.-diameter) bars with $A_{sd} = 2463$ mm^2 (3.82 in.^{2}) or consider four D24 (0.94-in.-diameter) bars with $A_{sd} = 1810$ mm^2 (2.81 in.^{2}).

Make use of redistribution of shear forces vertically among several coupling beams, noting that in accordance with Section 5.3.2(c), up to 20% moment redistribution, and hence shear redistribution between coupling beams, is considered acceptable.

Hence try to use four D24 (0.94-in.-diameter) bars in all beams from the third to ninth levels. The dependable shear strength of one such beam is $Q_u = (1810/2140)554 = 472$ kN (106 kips). The total shear force to be resisted over these seven levels is, from Fig. 5.55(a),

$$\Sigma Q_u = 0.82(592 + 669 + 676 + 644 + 592 + 456 + 391)$$
$$= 3295 \text{ kN (738 kips)}$$

The total dependable vertical shear strength provided with four D24 (0.94-in.-diameter) bars over the same seven levels is

$$Q_u = 7 \times 472 = 3304 > 3295 \text{ kN}(738 \text{ kips})$$

This is satisfactory!

The maximum reduction involved in this shear force redistribution is

$$(554 - 472)100/554 = 14.8\% < 20\%$$

Transverse reinforcement required around the D24 bars ($A_b = 452 \text{ mm}^2$) to prevent buckling in accordance with Section 5.4.5(b) and Eq. (4.19) is

$$A_{te} = \frac{\Sigma A_b f_y}{16 f_{yt}} \frac{s}{100} = \frac{452 \times 275 \times s}{16 \times 275 \times 100} = 0.283s \ (\text{mm}^2)$$

where $s \leq 100$ mm (4 in.) or $s \leq 6d_b = 6 \times 24 = 144$ mm (5.67 in). Hence use $A_{te} = 0.283 \times 100 = 28.3 \text{ mm}^2$ (0.044 in.²), that is, R6 (0.24-in.-diameter) ties on 100-mm (4-in.) centers.

(i) The required development length for these D24 (0.94-in.-diameter) bars would normally be [Eq. (3.64)]

$$l_{db} = \frac{1.38 A_b f_y}{c\sqrt{f_c'}} = \frac{1.38 \times 452 \times 275}{27\sqrt{30}} = 1160 \text{ mm } (3.8 \text{ ft})$$

where from Fig. 5.55 the center-to-center distance between bars in the vertical plane is $2c_s = 24 + 30 = 54$ mm (2.13 in.). The development length of this group of four bars is, however, to be increased by 50% in accordance with Section 5.4.5(b). Therefore,

$$l_d = 1.5 \times 1160 \simeq 1750 \text{ mm } (5.75 \text{ ft})$$

(ii) Alternatively, when transverse ties are also used within the wall, the development length may be reduced. With R6 (0.24-in.-diameter) ties at 100-mm (4-in.) spacing, in accordance with Section 3.6.2(a) and Eq. (3.66),

$$k_{tr} = A_{tr} f_{yt}/10s = 28.3 \times 275/(10 \times 100) = 7.8 \text{ mm } (0.31 \text{ in.})$$

and hence the reduction factor is, from Eq. (3.65),

$$\frac{c}{c + k_{tr}} = \frac{27}{27 + 7.8} = 0.776$$

and thus $l_d = 0.776 \times 1.5 \times 1160 \simeq 1350 < 1400$ mm (4.6 ft), which has been provided.

Details of these beams are shown in Fig. 5.55.

Also, provide some nominal (basketing) reinforcement in these coupling beams to control cracking. Say, use 10 R10 (0.39-in.-diameter) horizontal bars, giving

$$\rho_h = 10 \times 78.5/(350 \times 800) = 0.0028 > 0.0025 = 0.7/f_y(0.1/f_y)$$

and vertical R10 (0.39-in.-diameter) ties on 180-mm (7-in.) centers, giving

$$\rho_v = 2 \times 78.5/(180 \times 350) = 0.0025$$

(b) *Beams at Levels 10 and 11*: Provide four D16 (0.63-in.-diameter) bars in each diagonal direction [$A_{sd} = 804$ mm^2 (1.25 in.^{2})]. This provides a dependable shear resistance of

$$Q_u = 2\phi A_s f_y \sin \alpha = 2 \times 0.9 \times 804 \times 275 \times 0.527 \times 10^{-3}$$

$$= 210 \text{ kN } (47 \text{ kips})$$

From Fig. 5.54(*a*) the average shear force required is

$$Q_{u,\text{required}} = 0.82(275 + 230)/2 = 208 < 210 \text{ kN}$$

For transverse reinforcement with a maximum spacing of $6 \times 16 = 96 \simeq$ 100 mm (4 in.), R6 (0.24-in.-diameter) ties may be used as in beams at lower floors.

Development lengths may be determined as for the D24 bars. That calculation is omitted here.

(c) *Beam at Level 2*: As this beam is 1500 mm (59 in.) deep, it is found that $\alpha \simeq 50°$ and $\sin \alpha = 0.766$. Hence

$$A_{sd} = 0.82 \times 908,000/(2 \times 0.9 \times 275 \times 0.766) = 1976 \text{ mm}^2(3.06 \text{ in.}^2)$$

Use four D28 (1.1-in.-diameter) bars [$A_{sd} = 2463$ mm^2 (3.82 in.^{2})] in each direction, with similar details as given for the beams at the upper levels. However, use R10 (0.39-in.-diameter) ties on 100-mm (4-in.) centers along the diagonal D28 (1.1-in.-diameter) bars.

Step 8: Overstrength of Coupling Beams. The overstrength in shear, Q_{oi}, of these 800-mm (31.5-in.)-deep beams, considering the contribution of diagonal bars only, may be derived from first principles and from Fig. 5.55 thus:

$$Q_{oi} = (\lambda_o A_s f_y) 2 \sin \alpha = 1.25 \times 275 \times 2 \times 0.527 \times 10^{-3} A_s = 0.362 A_s \text{ (kN)}$$

where Q_{oi} is in kN and A_s is in mm². Therefore,

$$Q_{oi} = 0.362 \times 1810 = 655 \text{ kN (147 kips)}$$

from the third to ninth levels. The overstrength in shear of the two beams at levels 10 and 11 is

$$Q_{io} = \lambda_o Q_u / \phi = 1.25 \times 210 / 0.9 = 291 \text{ kN (65 kips)}$$

The overstrength of the level 1 beam is

$$Q_o = 1.25 \times 275 \times 2463 \times 2 \times 0.766 \times 10^{-3} = 1297 \text{ kN (291 kips)}$$

Step 9: Determination of Actions on the Walls

(a) *Tension Wall at the Base*: From Fig. 5.54 the following quantities are found:

$$M_{u1} = 0.82 \times 7003 = 5740 \text{ kNm (4220 kip-ft)}$$

and from Table 5.4,

$$\text{Dead load: } P_D = 0.5 \times 4320 = 2160 \text{ kN (484 kips)}$$
$$\text{Live load: } P_{Lr} = rP_L = 0.5 \times 1000r$$

where from Eq. (1.3) with a total tributary area

$$A = 10 \times 45 = 450 \text{ m}^2; \qquad r = 0.3 + 3/\sqrt{450} = 0.44$$

Hence

$$P_{Lr} = 500 \times 0.44 = 220 \text{ kN (49 kips)}$$

The tension force is

$$P_E = -0.82 \times 5433 \text{ (tension)} = -4482 \text{ kN (1004 kips)}$$

Thus the combined axial forces to be considered are

$$P_u = 0.9P_D + P_E = 0.9 \times 2160 - 4482 \text{ (tension)} = -2538 \text{ kN (569 kips)}$$

The eccentricity is

$$e = M_{u1}/P_u = 5740/2538 = 2.26 \text{ m } (7.41 \text{ ft})$$

The ideal strength of the base section should sustain $P_i = 2538/0.9 = 2820$ kN (632 kips) tension with $e = 2.26$ m (7.41 ft).

(b)   Compression Wall at the Base: As moment redistribution from wall 1 to wall 2 will be necessary, action on wall 2 will be determined after the design of wall 1 in step 10.

Step 10: Design of Wall Base Sections

(a)   Consideration of Minimum Reinforcement: Minimum vertical reinforcement in 350-mm (13.8-in.)-thick wall, from Section 5.4.2(*b*), is

$$\rho_{l, \min} = 0.7/f_y = 0.7/380 = 0.0018$$

Using HD12 (0.47-in.-diameter) bars at 300 mm (11.8 in.) in both faces will give

$$\rho_l = 226/(350 \times 300) = 0.00215 > 0.0018$$

Minimum reinforcement in 600-mm (23.6-in.)-square area components (elements 1 and 3 in Fig. 5.53) is, according to the requirement for frames [Section 5.6.2(*a*)],

$$A_s = 0.008 \times 600^2 = 2800 \text{ mm}^2 \ (4.46 \text{ in.}^2)$$

To meet requirements for bar arrangements in such a column, provide at least 12 bars. Assume that 12 HD20 (0.79-in.-diameter) bars $= 3768$ mm^2 (5.84 in.^{2}) will be provided.

(b)   Flexural Reinforcement in the Tension Wall: Using the principles of Section 5.4.2(*a*), estimate the tension reinforcement required in area element 1, shown in Fig. 5.53, when axial tension acts on the wall. Because of net axial tension, assume initially that the center of internal compression forces is only 100 mm from the outer edge of element 4. Then the tension force [with $f_y = 380$ MPa (55 ksi)] is, from Fig. 5.53, in:

Element 1: $A_{s1}f_y = 380 A_{s1} \times 10^{-3}$ $=$ $0.38 A_{s1}$ kN

Element 2: $0.00215 \times 3100 \times 350 \times 380 \times 10^{-3} =$ 886 kN (198 kips)

Element 3: $3768 \times 380 \times 10^{-3}$ $= 1,432$ kN (321 kips)

Element 4: neglect presently $= -$

Moment contributions with reference to the center of compression, shown as line A within element 4 in Fig. 5.53, can then be computed. The subscripts used in the following refer to the identically labeled area elements. In step 9 it was found that $e = 2.26$ m (7.41 ft):

$$M_2 = 886(0.5 \times 3.1 + 0.6 + 0.7 - 0.1) \qquad = \qquad \underline{2,437 \text{ kNm (1791 kip-ft)}}$$

$$M_3 = 1432(0.3 + 0.7 - 0.1) \qquad = \qquad \underline{1,289 \text{ kNm (947 kip-ft)}}$$

$$P_i e_A = -2820(2.280 + 2.479 - 0.1) \qquad = -\underline{13,138 \text{ kNm (9656 kip-ft)}}$$

$$M_1 = \text{required to be } -(M_2 + M_3 - P_i e_A) = \qquad \underline{9,412 \text{ kNm (6918 kip-ft)}}$$

Therefore,

$$A_{s1} = 9412/[(5.0 - 0.3 - 0.1)380 \times 10^{-3}] = \qquad 5,381 \text{ mm}^2 \text{ (8.34 in.}^2)$$

We could try to use 12 HD24 (0.94-in.-diameter) bars $= 5428$ mm^2 (8.41 in.^{2}) in element 1. However, this tension reinforcement could be reduced by moment redistribution from wall 1 to wall 2. Wall 2 will have a much larger capacity to resist moments because of the large axial compression force present. Hence try to utilize the minimum reinforcement in this area element [i.e., 12 HD20 (0.79-in.-diameter) bars $= 3768$ mm^2 (5.84 in.^{2})]. However, this is, even after moment redistribution, likely to be short of that required. Try to use 14 HD20 (0.79-in.-diameter) bars, giving a tension force of $T_1 = 380 \times 4396 \times 10^{-3} = 1670$ kN (374 kips).

(c) Ideal Flexural Strength of the Tension Wall at the Base: Evaluate the moment of resistance of the wall section with the arrangement of bars suggested. Because of the close vicinity of the neutral axis, neglect the contribution of reinforcement in tension and compression in element 4, yet to be determined.

Total internal tension $= 1670 + 886 + 1432 \quad = \underline{3,988 \text{ kN (893 kips)}}$

External tension $\qquad\qquad\qquad\qquad P_i = \underline{2,820 \text{ kN (632 kips)}}$

$\therefore$ the internal compression force is $\qquad C = 1,168$ kN (261 kips)

Therefore, the depth of the concrete compression block is

$$a = 1,168,000/(0.85 \times 30 \times 450) = 102 \text{ mm (4.01 in.)}$$

that is, the neutral-axis depth is theoretically only $102/0.85 = 120$ mm (4.72 in.) from the compression edge at element 4.

The ideal moment of resistance with reference to the centroidal axis of section (i.e., axis Y in Fig. 5.53), using the distances x given in Table 5.5, is, therefore:

Element 1: $+1670 \times 2.221$	$=$	3,709 kNm (2726 kip-ft)
Element 2: $+886 \times 0.371$	$=$	329 kNm (242 kip-ft)
Element 3: -1432×1.479	$=$	$-2,118$ kNm (1557 kip-ft)
Element 4: $+1168(2.479 - 0.051)$	$=$	2,836 kNm (3496 kip-ft)

$$P_i = 1670 + 886 + 1432 - 1168 = 2820 \text{ kN}, \quad M_i = \quad 4{,}756 \text{ kNm (3496 kip-ft)}$$

Check against the required dependable flexural strength

$$\phi M_i / M_u = 0.9 \times 4756/5740 = 0.746 > 0.7$$

that is, 30% limitation on moment redistribution suggested in Section 5.3.2(c) is not exceeded if 14 HD20 (0.79-in.-diameter) bars are used in area element 1.

(d) Design of the Compression Wall for Flexure

(i) *Actions at the Wall Base.* Moment from Fig. 5.54(b) and as a result of 25.4% moment redistribution from wall 1.

$$M_2 = 0.82[12{,}166 + (1 - 0.746)7003] = 11{,}430 \text{ kNm } (8401 \text{ kip-ft})$$

Axial load from Fig. 5.54(d) and step 9:

$$P_u = P_E + P_D + 1.3P_L = 0.82 \times 5433 + 2160 + 1.3 \times 220$$
$$= 6899 \text{ kN } (1545 \text{ kips})$$

or

$$P_u = P_E + 0.9P_D = 0.82 \times 5433 + 0.9 \times 2160$$
$$= 6397 \text{ kN (critical) } (1433 \text{ kips})$$
$$e = 11{,}430/6397 = 1.787 \text{ m } (5.86 \text{ ft}) < 2.521 \text{ m (see Fig. 5.53)}$$

(ii) *Flexural Strength.* As the axial compression acts within the wall section, the existing tension reinforcement provided in area elements 2 and 3 will be adequate. Provide only nominal reinforcement in element 4. Use, say, six HD16 bars, giving

$$A_s = 1206 \text{ mm}^2 \ (1.87 \text{ in.}^2) > \rho_{l,\min} = 0.0018 \times 700 \times 450$$
$$= 567 \text{ mm}^2 \ (0.88 \text{ in.}^2)$$

The vertical reinforcement is shown in Fig. 5.56. As the flexural strength of the section appears to be ample, there is no need to work out its value. One may proceed immediately to the evaluation of the flexural overstrength when the adequacy of the section may be checked.

Step 11: Axial Load Induced in Walls by Coupling Beams at Overstrength. From the beam calculation in step 7, the sum of the shear forces expected to be developed with the overstrength of all beams is

$$\Sigma Q_{oi} = 1297 + 7 \times 655 + 2 \times 291 = 6464 \text{ kN (1448 kips)}$$

$$\text{Check: } 6464/(0.82 \times 5433) = 1.44 > \lambda_o/\phi = 1.25/0.9 = 1.39$$

The estimated maximum earthquake-induced axial force at the base of the walls may be obtained from Eq. (5.30).

$$P_{Eo} = \left(1 - \frac{n}{80}\right)\sum_1^{10} Q_{io} = \left(1 - \frac{10}{80}\right)6464 = 5656 \text{ kN (1267 kips)}$$

Similarly, between levels 3 and 4 with $n = 7$:

$$P_{Eo} = \left(1 - \frac{7}{80}\right)(6464 - 1297 - 2 \times 655) = 3520 \text{ kN (788 kips)}$$

Step 12: Overstrength of the Entire Structure. This requires the evaluation of Eq. (5.32).

(a) Flexural Overstrength of the Tension Wall at the Base: The axial load to be considered with the development of the flexural overstrength is, according to Eq. (5.31a),

$$P_{1,o} = P_{Eo} - P_D = 5656 - 2160 = 3496 \text{ kN (783 kips) (tension)}$$

where P_{Eo} was obtained from step 11.

By assuming that $\lambda_o = 1.4$ for the grade of steel of the vertical wall tension reinforcement, the total internal tension force in the section increases to $1.4 \times 3988 = 5583$ kN (1251 kips) [section (c) of step 10]. Therefore, the internal compression force becomes $5583 - 3496 = 2087$ kN (467 kips). Hence with say 25% increase in concrete strength

$$a = 2,087,000/(0.85 \times 1.25 \times 30 \times 450) = 145 \text{ mm (5.71 in.)}$$

Thus by computing moment contributions similar to those in the preceding section, the flexural overstrength of the tension wall becomes

$$M_{1o} \simeq 1.40(3709 + 329 - 2118) + 2087 \times 2.407 = 7711 \text{ kNm } (5668 \text{ kip-ft})$$

(b) Flexural Overstrength of the Compression Wall: From step 11 and Eq. (5.31b): $P_{2,o} = 5656 + 2160 = 7816$ kN (1751 kips) (compression). Estimate neutral-axis depth at $600/0.85 = 706$ mm (27.8 in.) from the face of element 1. Then, with reference to Fig. 5.54, the forces at the area elements are as follows:

Element 4: tension $1206 \times 1.40 \times 380/10^3 \ = \qquad 642$ kN (144 kips)

Element 3: tension $1.40 \times 1432 \qquad\qquad = \quad 2{,}005$ kN (449 kips)

Element 2: tension $1.15 \times 886 \qquad\qquad\quad = \quad \underline{1{,}019 \text{ kN } (228 \text{ kips})}$

(assume no strain hardening in element 2)

Total internal tension $\qquad\qquad\qquad\qquad \underline{3{,}666 \text{ kN } (821 \text{ kips})}$

Element 1: the total compression force is

$P_{2,o}$ + sum of tension forces = $7816 + 3666 = 11{,}482$ kN (2572 kips)

Assume that the outer half

of the compression bars yield $0.5 \times 1670 \ = \qquad 835$ kN (187 kips)

Assume that the inner half of compression

bars are stressed to $0.5f_y \ \ 0.5 \times 0.5 \times 1670 \ = \quad \underline{418 \text{ kN } (94 \text{ kips})}$

Required compression force in concrete $\quad = 10{,}229$ kN (2291 kips)

Again assuming that $f'_c = 1.25 \times 30 = 37.5$ MPa (6550 psi):

$$a = 10{,}229 \times 10^3 / (0.85 \times 37.5 \times 600) = 535 \text{ mm } (21 \text{ in.})$$
$$c = 535/0.85 = 629 < 706 \text{ mm } (27.8 \text{ in.}) \text{ assumed}$$

Evaluate the moment contributions of element forces about the centroidal axis, using distances of Fig. 5.53 or Table 5.5 thus:

Element 4: $642 \times 2.129 \qquad\qquad\qquad = + \ 1{,}367$ kNm (1005 kip-ft)

Element 3: $2005 \times 1.479 \qquad\qquad\qquad = + \ 2{,}965$ kNm (2179 kip-ft)

Element 2: $-1019 \times 0.371 \qquad\qquad\quad = - \qquad 378$ kNm (278 kip-ft)

Element 1: $10{,}230(2.521 - 0.5 \times 0.535) \quad = +23{,}053$ kNm (16,944 kip-ft)

$835(2.521 - 0.25 \times 0.6) \qquad\qquad = + \ 1{,}980$ kNm (1455 kip-ft)

$418(2.521 - 0.75 \times 0.6) \qquad\qquad = + \quad \underline{866 \text{ kNm } (637 \text{ kip-ft})}$

$$M_{2,o} = \underline{\quad 29{,}853 \text{ kNm } (21{,}942 \text{ kip-ft})}$$

Thus the overstrength of the compression wall is very large. However, this cannot be reduced significantly.

(c) Flexural Overstrength of Entire Structure: The total overturning moment, M_{ot}, was derived in step 5. Hence from Eq. (5.32) and with $P_{Eo} = 5656$ kN (1267 kips), from step 11,

$$\phi_{o,w} = \frac{M_{1,o} + M_{2,o} + P_{Eo}l}{M_{ot}}$$

$$= \frac{7711 + 29,853 + 5656 \times 5.958}{0.82 \times 51540} = 1.686$$

If every section would have been designed to meet "exactly" the strength requirements, the overstrength factor could be obtained from Eq. (5.33) and Fig. 5.54 by

$$\phi_{o,w} = \frac{1.4(7003 + 12,166) + 1.25 \times 5.958(1 - 0.2)5433}{0.9(7003 + 12,166 + 5.958(1 - 0.2)5433} = 1.46$$

Thus the structure as designed possesses approximately

$$\frac{1.686 - 1.46}{1.46}100 \simeq 15\%$$

excess strength, despite the reduction of lateral-force-induced design axial load at overstrength in accordance with Eq. (5.30). As very close to minimum vertical reinforcement has been used in all parts of the wall sections, this excess strength cannot be reduced significantly.

Step 13: Wall Design Shear Forces. The total required ideal shear strength associated with the development of the flexural overstrength of the structure is, in accordance with Eq. (5.22),

$$V_{\text{wall}} = \omega_v \phi_{o,w} V_E$$

where, from Eq. (5.23*b*), $\omega_v \simeq 1.6$, and from step 12, $\phi_{o,w} = 1.686$, while

$$V_E = 0.82 \times 2660 = 2180 \text{ kN } (488 \text{ kips})$$

Hence

$$V_{\text{wall}} = 1.6 \times 1.686 \times 2180 = 5880 \text{ kN } (1317 \text{ kips})$$

This may be distributed between the two walls in proportion of the flexural overstrengths at the base [Eq. (5.34)]:

$$V_u = V_{1,\text{wall}} = \frac{M_{1o}}{M_{1,o} + M_{2,o}} V_{\text{wall}} = \frac{7711}{7711 + 29{,}853} \times 5880$$

$$= 1207 \text{ kN } (270 \text{ kips})(21\%)$$

$$V_{2,\text{wall}} = \frac{29{,}853}{7711 + 29{,}853} \times 5880 = 4673 \text{ kN } (1047 \text{ kips})(79\%)$$

Because of the relatively large flexural overstrength of wall 2, 79% of the design shear has been assigned to it. According to the initial elastic analysis and Fig. 5.54(e), only 1612 kN (361 kips) (i.e., 60.6% of the total shear) was assigned to wall 2.

Step 14: Design of Shear Reinforcement at the Base of the Walls. Using the information from Section 5.4.4(b), we find that for the compression wall,

$$V_{2,\text{wall}} = V_{\text{max}} = 4673 \text{ kN } (1047 \text{ kips})$$

$\phi = 1.0$, and $d = 0.8l_w$. From Eq. (5.25),

$$v_i = 4{,}673{,}000/(0.8 \times 5000 \times 350) = 3.34 \text{ MPa } (484 \text{ psi})$$

According to Eq. (5.26), the shear stress in the plastic hinge region should, however, not exceed

$$v_{i,\text{max}} \leq \left(\frac{0.22\phi_{o,w}}{\mu_\Delta} + 0.03\right)f'_c = \left(\frac{0.22 \times 1.686}{5.88} + 0.03\right)30$$

$$= 2.79(405) < 0.16f'_c = 4.8(696) < 6.0 \text{ MPa } (870 \text{ psi})$$

To overcome this limitation, $v_i = 3.34$ MPa (484 psi) $> v_{i,\text{max}} = 2.79$ MPa (405 psi) on shear stress in the plastic hinge region of ductile walls, increase the wall thickness in the lower two stories from 350 (13.8 in.) to 380 mm (15 in.) and use concrete with $f'_c = 35$ MPa (5075 psi), Then, from Eq. (5.26),

$$v_{i,\text{max}} = 2.79 \times 35/30 = 3.26 \text{ MPa } (473 \text{ psi})$$

and

$$v_i = 3.34 \times 350/380 = 3.08 \text{ (447 psi)} < 3.26 \text{ MPa } (473 \text{ psi})$$

The gross area of the wall increases to

$$A_g = 2{,}120{,}000 + 3100 \times 30 = 2{,}213{,}000 \text{ mm}^2 \text{ (3430 in.}^2)$$

Hence using Eq. (3.39) and $P_{2,o} = 7816$ kN (1751 kips) from step 12(b),

$$v_c = 0.6\sqrt{\frac{P_e}{A_g}} = 0.6\sqrt{\frac{7,817,000}{2,213,000}} = 1.13 \text{ MPa (164 psi)}$$

and from Eq. (3.40),

$$\frac{A_v}{s} = \frac{v_i - v_c}{f_y} b = \frac{3.08 - 1.13}{380} \times 380 = 1.95 \text{ mm}^2/\text{mm } (0.077 \text{ in.}^2/\text{in.})$$

Using HD16 (0.63-in.-diameter) bars at both faces the spacing, s becomes $s = 2 \times 201/1.95 = 206$ mm. Use $s = 200$ mm (7.87 in.).

The shear strength of wall 1 when subjected to tension must be at least

$$V_{1,\text{wall}} = 1207 \text{ kN (270 kips)}$$

Thus

$$v_i = 1,207,000/(0.8 \times 5000 \times 380) = 0.80 \text{ MPa (116 psi)}$$

In this case $v_c = 0$ and $(v_i - v_c) = 0.80$ MPa (116 psi). As $0.80 < (3.08 - 1.13) = 1.95$ MPa (283 psi), the stirrups provided for wall 2 are ample.

Note that by increasing the wall thickness from 350 mm (13.8 in.) to 380 mm (15 in.) the vertical reinforcement provided in the web need not be changed because with HD12 (0.47-in.-diameter) bars provided at 300-mm (11.8-in.) spacing, we have

$$\rho = 2 \times 113/(300 \times 380) = 0.00198 > 0.0018 = \rho_{\min}$$

Therefore, previous computations for flexure are not affected.

These shear computations apply to the potential plastic hinge region, which according to Section 5.4.3(e) extends above level 1 by $l_w = 5000$ mm (16.4 ft) or by $h_w/6 = 28,850/6 = 4808$ mm (15.8 ft) < 5000 mm. For practical reasons use the arrangement of reinforcement shown in Fig. 5.56 up to level 3.

Step 15: Check of the Sliding Shear Capacity of the Walls. It is evident that when a plastic hinge develops in the tension wall, flexural cracks could theoretically penetrate to within 120 mm of the compression edge. Consequently, the resistance against sliding shear failure must rely more on dowel action of the vertical wall reinforcement than on shear friction along the length of the wall. However, sliding shear, if it does occur, must develop in both walls simultaneously before a failure can occur. Thus redistribution of shear from the tension wall to the compression wall must occur, much the same way as resistance against sliding is redistributed from the flexural

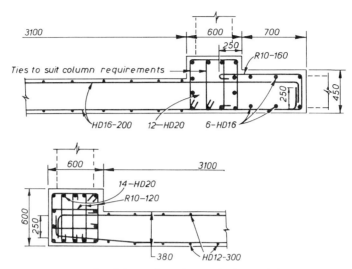

Fig. 5.56   Details of the wall section at the base of the structure.

tension zone to the compression zone in a single cantilever wall. Therefore, for the purpose of sliding, two walls coupled by beams and floor slabs may be treated as one wall, as suggested in Section 5.4.4(c).

Therefore, when using Eq. (3.42) in conjunction with capacity design,

$$A_{vf} = \frac{V_u - \phi \mu P_u}{\phi \mu f_y}$$

the total shear at the development of overstrength is, from step 13,

$$V_u = V_{wall} = 5880 \text{ kN } (1317 \text{ kips}) = 2.69 V_E$$

and the minimum axial compression [i.e., $P_D = 4320$ kN (968 kips)] only should be considered. Therefore, from Eq. (3.42) with $\phi = 1.0$ and a friction factor $\mu = 1.4$,

$$A_{vf} = (5,880,000 - 1 \times 1.4 \times 4,320,000)/(1 \times 1.4 \times 380) < 0$$

However, if the surface roughness is inadequate, $\mu = 1.0$ should be taken and hence $A_{vf} = 4015$ mm^2 (6.22 in.^{2}).

From Fig. 5.56 and previous calculations, it is seen that the total vertical reinforcement provided in four groups in one wall is exactly

$$A_{st} = 4396 + 2486 + 3768 + 1206 = 11,856 \text{ mm}^2 \gg 0.5 \times 4121 \text{ mm}^2$$

Alternatively, the compression wall with $V_{wall} = 4673$ kN (1047 kips) and axial compression of $P_u = P_{2,o} = 7816$ kN (1751 kips) could be examined. In this case no vertical reinforcement would be required for shear friction because

$$\phi \mu P_u > P_u = 7816 > V_u = 4673 \text{ kN}$$

If the tension wall is examined in isolation, it will be found, as expected, that sliding shear becomes very critical, Again using Eq. (3.42) with $V_u = V_{1,wall} = 1207$ kN (270 kips) and from step 12, $P_u = P_{1,o} = -3496$ kN (783 kips), we find that when

$$\mu = 1.4, \qquad A_{vf} = (1{,}207{,}000 + 1.4 \times 3{,}496{,}000)/(1.4 \times 380)$$

$$= 11{,}470 \text{ mm}^2 \ (17.8 \text{ in.}^2)$$

$$\mu = 1.0, \qquad A_{vf} = (1{,}207{,}000 + 1.0 \times 3{,}496{,}000)/(1.0 \times 380)$$

$$= 12{,}376 \text{ mm}^2 \ (19.2 \text{ in.}^2)$$

and these quantities are comparable with the total steel content of the wall [i.e., $A_{st} = 11{,}856$ mm^2 (18.4 in.^{2})].

As pointed out, sliding shear failure of wall 1 cannot, however, occur in isolation. The quantities above imply that when these extreme loads occur more shear may be redistributed from wall 1 onto wall 2, which has some reserve capacity.

Step 16: Confinement of Wall Sections

(a) *Confinement of Compressed Concrete*: With respect to the need for confinement, the critical position of the neutral axis from the compression edge of the section is to be established. From Eq. (5.18b),

$$c_c = \frac{M_{o,w}}{2.2 \lambda_o \mu_\Delta M_E} l_w = \frac{29{,}853 \times 5000}{2.2 \times 1.4 \times 5.88 \times 0.82 \times 12166} = 826 \text{ mm } (32.5 \text{ in.})$$

From step 12(b) the estimated neutral-axis depth of 629 mm (24.8 in.) based on $f'_c = 30$ MPa (4350 psi) is less than this. Because of shear requirements it was decided to use $f'_c = 35$ MPa (5075 psi). Hence $c \approx 629 \times 30/35 = 540$ mm (21.3 in.). Therefore, confinement of the concrete in the compression zone is not necessary.

A larger compression zone would have been required without the (600 × 600) boundary element. For the purpose of illustrating the application of the relevant requirements, it will be shown how the compression zone of the wall could have been reinforced if the enlargement of the boundary element 1 in Fig. 5.56 was to be omitted. This is shown in Fig. 5.57.

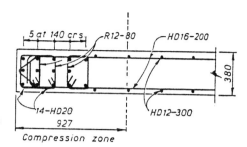

Fig. 5.57 Alternative detailing of the confined region of the wall section.

With a 380-mm (15-in.)-wide section, using concrete upgraded to $f'_c = 35$ MPa (5075 psi), the neutral-axis depth would have been, from step $12(b)$, approximately

$$c = (600/380)629(35/37.5) = 927(36.5 \text{ in.}) > 826 \text{ mm } (32.5 \text{ in.})$$

If one were to use the more refined expression for the critical neutral-axis depth [i.e., Eq. (5.18c)],

$$
\begin{aligned}
c_c &= \frac{3400 M_{o,w}}{(\mu_\Delta - 0.7)(17 + A_r)\lambda_o f_y M_E} l_w \\
&= \frac{3400 \times 29{,}853 \times 5000}{(5.88 - 0.7)(17 + 28{,}850/5000)1.4 \times 380 \times 0.82 \times 12{,}166} \\
&= 811 < 927 \text{ mm}
\end{aligned}
$$

Therefore, the use of confinement of the region is necessary.

The 14 HD20 (0.79-in.-diameter) bars in the boundary (column) element of Fig. 5.56 could be rearranged as shown in Fig. 5.57.

In accordance with Section 5.4.3(e) and Eq. (5.19),

$$\alpha = 1 - 0.7 \times 811/927 = 0.389 < 0.5$$

and hence confine the outer half of the 927-mm (36.5-in.)-long compression zone thus:

The gross area of the region to be confined is

$$A_g^* = 380 \times 0.5 \times 927 = 176{,}100 \text{ mm}^2 \ (273 \text{ in.}^2)$$

The corresponding core area is, assuming that R12 (0.47-in.-diameter) ties with 28-mm (1.1-in.)-cover concrete will be used,

$$A_c^* = (380 - 2 \times 34)(0.5 \times 927 - 34) = 134{,}000 \text{ mm}^2 \ (208 \text{ in.}^2)$$

Because [see Eqs. (5.20a) and (5.20b)]

$$0.3\left(\frac{A_g^*}{A_c^*} - 1\right) = 0.3\left(\frac{176,100}{134,000} - 1\right) = 0.094 < 0.12$$

Eq. (5.20b) is applicable. This is

$$A_{sh} = 0.12 s_h h'' \frac{f_c'}{f_{yh}}\left(0.5 + 0.9\frac{c}{l_w}\right)$$

where

$$s_h \le 6d_b = 6 \times 20 = 120 \text{ mm } (4.7 \text{ in.})$$

or

$$s_h < 0.5(380 - 2 \times 34) = 156 \text{ mm } (6.1 \text{ in.})$$
$$h'' = 140 \text{ mm } (5.5 \text{ in.}) \text{ (for one tie leg from Fig. 5.58)}$$
$$c = 927 \text{ mm } (36.5 \text{ in.}), \qquad f_c' = 35 \text{ MPa } (5075 \text{ psi}),$$
$$f_{yh} = 275 \text{ MPa } (40 \text{ ksi})$$

Hence

$$A_{sh} = 0.12 \times 120 \times 140(35/275)(0.5 + 0.9 \times 927/5000)$$
$$= 171 \text{ mm}^2 \text{ } (0.27 \text{ in.}^2)$$

Use R12 (0.47-in.-diameter) ties at $s_h = (113/171)120 = 79 \approx 80$ mm (3.1 in.).

It is seen in Fig. 5.57 that for practical reasons the confined length of 560 mm (22 in.) is in excess of that required [i.e., $0.5 \times 927 = 464$ mm (18.2 in.)]. If the wall section shown in Fig. 5.57 is to be used, a stability check will be required. With the aspect ratio $A_r = 28.85/5 \approx 5.8$ and a design ductility of $\Delta_\mu = 5.9$, from Fig. 5.35 it is found that $b_c/l_w = 0.069$ and hence $b_{c,\text{min}} = 0.069 \times 5000 = 345$ mm (13.6 in.) < 380 mm (15 in.). This is satisfactory.

(b) *Confinement of Longitudinal Compression Reinforcement*: The HD20 (0.79-in.-diameter) bars in area element 1 of Fig. 5.53 will need to be confined over the entire length of the potential plastic hinge (i.e., up to level 3).

From the requirements of Section 5.4.3(e)(iv),

$$\rho_l = \frac{\Sigma A_b}{bs_v} = \frac{4396}{600^2} = 0.0122 > \frac{2}{f_y} = 0.0053 \quad \text{and}$$

$$s_h \le 6 \times 20 = 120 \text{ mm } (4.7 \text{ in.})$$

and from Eq. (4.19), the area of a confining tie around each HD20 (0.79-in.-diameter) bar is

$$A_{te} = \frac{\Sigma A_b f_y}{16 f_{yt}} \times \frac{s_h}{100} = \frac{314 \times 380}{16 \times 275} \times \frac{120}{100}$$

$$= 33 \text{ mm}^2 < 78.5 \text{ mm}^2 \ (0.12 \text{ in.}^2)$$

Consideration of earthquake forces transverse to the plane of the walls may require more transverse reinforcement within the 600×600 mm columns.

The arrangement of ties, together with all other reinforcement, is shown in Fig. 5.56. At the inner edges of the walls, in area element 4,

$$\rho_l = 6 \times 201/(450 \times 700) = 0.0038 < 0.0053$$

and thus these HD16 (0.63-in.-diameter) bars need not be confined. Nominal U-shaped R10 (0.39-in.-diameter) ties may be provided on 160-mm (6.3-in.) centers.

Step 17: Consideration of the Required Strength of Walls at Higher Levels. The recommended linear design bending moment envelope (Fig. 5.29), in terms of the ideal moment strength of the base section, is reproduced in Fig. 5.54(c). The flexural reinforcement required at any one level to resist the moment, which reduces linearly up the wall, must be extended so as to be fully effective in tension at distance l_w above the level that is being considered. Therefore, such reinforcement must extend by at least the development length, l_d, beyond the level obtained from the envelope in Fig. 5.54(c). This is in recognition of the tension shift implied in the general requirements for the development of flexural reinforcement (Section 3.6.3). The envelope in Fig. 5.29, labeled as "minimum ideal moment of resistance required," does not take into account the effect of axial load on the flexural strength of the wall.

(a)   *Tension Wall at Level 3*: The wall section just above level 3, where according to Fig. 5.54(c) approximately 78% of the base moment capacity should be sustained, will be checked. In the light of the approximations made, such as the simple moment envelope, overly accurate analysis is not justified.

Design actions for this section may thus be derived as follows:

   (i)   From calculations given in step 10(c), the ideal flexural strength at the base is 4756 kNm (3496 kip-ft), while the concurrently acting axial tension was 2820 kN (632 kips). Hence at level 2 we may assume that we require

$$M_i = 0.78 \times 4756 = 3710 \text{ kNm (2727 kip-ft)}$$

(ii)   From Fig. 5.54(d) at above, level 3

$P_E = 0.82 \times 3933 = 3225$ kN (722 kips) (tension)

$P_D = 2160 - 0.5(500 + 480) = 1670$ kN (374 kips)

$P_u = P_E - 0.9P_D = 3225 - 0.9 \times 1670 = 1722$ kN (386 kips) (tension)

$P_i = 1722/0.9 = 1913$ kN (428 kips)

(iii) $e = 3710/1913 = 1.94$ m (6.4 ft). It is evident that significant reduction of vertical reinforcement may be achieved only in area element 1 (see Figs. 5.53 and 5.56). The vertical reinforcement in the stem of the wall, which at this level is 300 mm (11.8 in.) thick, may be reduced to HD12 (0.47 in. diameter) at 350 (13.8 in.), giving

$$\rho = 2 \times 113/(300 \times 350) = 0.00215 > 0.0018 = \rho_{min}$$

(iv)   Hence the steel forces to be considered in the various wall area elements, shown in Fig. 5.53 and step 10(c), are as follows:

Element 1: $= 0.38 A_{s1}$ kN

Element 2: $0.00215 \times 3100 \times 300 \times 380 \times 10^{-3} = 760$ kN (70 kips)

Element 3: unchanged (minimum) $= 1432$ kN (321 kips)

Element 4: neglect $= -$

The moment contributions computed with reference to the assumed center of compression, at, say, 50 mm from compression edge (close to line A in Fig. 5.53), are:

$M_2 = 760(0.5 \times 3.1 + 0.6 + 0.7 - 0.05)$ $=$ 2128 kNm (1564 kip-ft)

$M_3 = 1432(0.3 + 0.7 - 0.05)$ $=$ 1360 kNm (1000 kip-ft)

$-Pe'_A = 1912(1.940 + 2.479 - 0.05)$ $= -$ 8354 kNm (6140 kip-ft)

$M_1 =$ required $=$ 4866 kNm (3577 kip-ft)

Hence

$$A_{s1} = 4866 \times 10^3/[(5.0 - 0.3 - 0.05)380] = 2754 \text{ mm}^2 \ (4.27 \text{ in.}^2)$$

Six HD20 (0.79-in.-diameter) and six HD16 (0.63-in.-diameter) bars, giving $A_{st} = 3090$ mm^2 (4.79 in.^{2}) could be used in element 1. These bars would have to extend above the level by 5000 mm (16.4 ft) plus the development length. However, the reinforcement content in the element as a column is now reduced to $\rho_t = 3090/600^2 = 0.0086$, close to the absolute minimum

recommended (Section 3.4.2), and this may not satisfy the moment requirements for this column in the other direction of the building.

It is evident also that only minor reduction of reinforcement up the
building in area element 1 is possible. This shows that, as is generally the
case, it is not difficult to meet the apparently conservative requirements of
Fig. 5.29.

(v) The shear force may be assessed from the proportional reduction of
the total shear at this level. From Fig. 5.54(e) and step 13, according to which
wall 1 with axial tension resists 21% of the total shear:

$$V_i = 0.21 \times 0.82(1068 + 1322)\omega_v\phi_{o,w}$$

and $$= 412 \times 1.6 \times 1.686 = 1110 \text{ kN } (249 \text{ kips})$$

$$v_i = 1{,}110{,}000/(0.8 \times 5000 \times 300) = 0.93 \text{ Mpa } (135 \text{ psi})$$

This is rather small and the shear force with axial compression is likely to
govern the design.

(b) Compression Wall at Level 3: It was seen that at the base, only nominal
tension reinforcement was required in area element 4. This reinforcement
cannot be reduced significantly with height. Any reduction of compression
reinforcement in element 1 has negligible effect on the flexural strength of
the wall section. Hence the section immediately above level 3 should be
adequate and should require no further check for flexural strength. However,
to illustrate the procedure, the ideal flexural strength of the section will be
evaluated.

(i) The design forces that should be considered are as follows: From
evaluation of the flexural overstrength of the section in step 12(b), the ideal
moment of resistance at the base is $29{,}853/(0.90 \times 1.686) = 19{,}670$ kNm
(14,460 kip-ft), where $\phi = 0.90$ and from step 12(c), $\phi_{o,w} = 1.656$. Therefore
from Fig. 5.54(c), the ideal moment demand at the second floor is estimated
as

$$M_i = 0.78 \times 19{,}670 = 15{,}300 \text{ kNm } (11{,}250 \text{ kip-ft})$$

The live load to be considered is with a reduction factor of

$$r = 0.3 + 3/\sqrt{8 \times 45} = 0.46$$

$$P_{Lr} = 0.46 \times 0.5(1000 - 220) = 179 \text{ kN } (44 \text{ kips})$$

Therefore,

$$P_u = P_E + P_D + 1.3P_{Lr} = 1670 + 1.3 \times 179 + 3224 = 5127 \text{ kN } (1148 \text{ kips})$$

Therefore,

$$P_i = 5127/0.9 \approx 5700 \text{ kN } (1277 \text{ kips})$$

(ii)   By assuming that the center of internal compression of the wall section is at the center of area element 1, the moment contributions about this center are as follows:

Element 4: $(1206 \times 0.38)(5.000 - 0.35 - 0.30)$ = 1994 kNm (1466 kip-ft)

Element 3: $1432(5.000 - 1.000 - 0.30)$ = 5298 kNm (3894 kip-ft)

Element 2: $760(0.5 \times 3.100 + 0.30)$ = 1406 kNm (1033 kip-ft)

Element 1: neglect = —

= 8698 kNm (6393 kip-ft)

Therefore, $e_1 = 8698/5700 = 1.526$ m (5 ft) to the left of the center of element 1. Hence $e = (2.521 - 0.30) + 1.526) = 3.747$ m (12.3 ft) from the y axis of Fig. 5.53. Therefore,

$$M_i = 3.747 \times 5700 = 21,360 > 15,300 \text{ kNm } (11,250 \text{ kft})$$

which is satisfactory. With this magnitude of strength reserve, no further check is warranted. With less margin, the assumed center of compression at the center of area element 1 may be checked thus:

Total tension = $1206 \times 0.38 + 1432 + 760$ = 2650 kN (594 kips)

External compression load = 5700 kn (1277 kips)

Internal compression must be = 8350 kN (1870 kips)

Compressed steel in element 1 (14 HD20) = 1670 kN (374 kips)

Concrete at assumed center must resist = 6680 kN (1496 kips)

Therefore, the area of compression is

$$6,680,000/(0.85 \times 25) = 314,400 \text{ mm}^2 \ (487 \text{ in.}^2)$$

The assumed area in compression is $600^2 = 360,000$ mm^2 (558 in.^{2}). This is close enough.

(iii)   The shear force, being 79% of the total magnified shear, is from Fig. 5.54(e) and step 13 or from a comparison with the shear in wall 1 at the same level:

$$V_i = \frac{0.79}{0.21} \times 1110 \simeq 4176 \text{ kN (935 kips)}$$

$$v_i = 4176 \times 10^3 / (0.8 \times 5000 \times 300) = 3.48 \text{ MPa (505 psi)} < 0.2 f_c'$$

$$= 5 \text{ MPa (725 psi)}$$

[Eq. (3.30)]. The determination of the contribution of the concrete to shear strength, v_c, in the elastic wall may be based on Eq. (3.36) conservatively with the beneficial effect of axial load neglected [i.e., $v_c = 0.27 \sqrt{f_c'} = 1.35$ MPa (196 psi)]. Hence

$$\frac{A_v}{s_h} = \frac{(3.48 - 1.35)300}{380} = 1.68 \text{ mm}^2/\text{mm } (0.066 \text{ in.}^2/\text{in.})$$

Using HD12 (0.47-in.-diameter) bars in each face of the 300-mm-thick wall, the spacing will be

$$s_h = 2 \times 113/1.68 = 134 \text{ say } 150 \text{ mm (5.9 in.)}$$

or HD 16 (0.63-in.-diameter) bars on 250-mm (9.8-in.) centers may be used. It is seen that the shear reinforcement at this level is still considerably in excess of the minimum wall reinforcement:

$$\rho_h = 2 \times 113/(300 \times 150) = 0.0050 > 0.0018 = \rho_{\min}$$

where the vertical wall reinforcement cannot be reduced any further. It may readily be shown that sliding shear along well-prepared construction joints is not critical at upper floors.

Step 18: Foundation Design. The design of the foundation structure is not considered here. However, an example of the design process for a similar structure is given in Section 9.5.

A Revision of Stiffness. In the absence of the knowledge of axial forces on the walls, approximations with respect to the effects of cracking on flexural stiffnesses were made in step 4. Subsequently, it was found that for

Wall 1: $P_{1u} = P_D - P_E = 2160 - 0.82 \times 5433 = -2295$ kN (514 kips)

and

Wall 2: $P_{2u} = P_D + P_E = 2160 + 0.82 \times 5433 = 6615$ kN (1482 kips) and hence from Eq. (5.7) for

Wall 1: $I_e = \left(\dfrac{100}{f_y} - \dfrac{P_{1u}}{f'_c A_g} \right) I_g = \left(\dfrac{100}{380} - \dfrac{2295 \times 10^3}{30 \times 2120 \times 10^3} \right) I_g = 0.23 I_g$

Wall 2: $I_e = \left(\dfrac{100}{f_y} + \dfrac{P_{2u}}{f'_c A_g} \right) I_g = \left(\dfrac{100}{380} + \dfrac{6615 \times 10^3}{30 \times 2120 \times 10^3} \right) I_g = 0.37 I_g$

The analysis could have been repeated with these reduced stiffnesses and more realistic values for deflections would be obtained. However, the bending moment intensities will not be affected significantly because they depend primarily on the ratio I_1/I_2, which was assumed to be $0.5/0.8 = 0.625$ and was found with the revised stiffnesses to be $0.23/0.37 = 0.621$. Much more radical changes in moments were introduced subsequently with in step $10(c)$ 25% of the moment for the tension wall being redistributed to the compression wall.

5.7 SQUAT STRUCTURAL WALLS

5.7.1 Role of Squat Walls

Squat structural walls with a ratio of height, h_w, to length, l_w, of less than 2 or 3 find wide application in seismic force resistance of low-rise buildings. They are also used in high-rise structures, where they may make a major contribution to lateral force resistance when extending only over the first few stories above foundation level. On the basis of their response characteristics, squat walls may be divided into three categories.

1. *Elastic Walls.* In low-rise buildings the potential strength of squat walls may be so large that they would respond in the fully elastic domain during the largest expected earthquake in the locality. The majority of squat walls belong to this group.

2. *Rocking Walls.* In many cases squat walls may provide primary lateral force resistance while supporting comparatively small vertical force. In such cases the wall's lateral force-resisting capacity may be limited by simple statics to overturning capacity unless tension piles are provided or substantial foundation members link the wall to adjacent structural elements. A feasible, though largely untested design approach, considered in Chapter 9, allows such walls to rock on specially designed foundations, while possessing greater flexural and shear strength than that corresponding with overturning or rocking. This ensures elastic action in the wall.

3. *Ductile Walls.* In many cases, squat walls, with foundations of adequate strength to prevent overturning, may not practically be designed to respond elastically to design-level ground shaking. In such cases considerable ductility

may need to be developed. This third category of squat walls needs closer examination. These walls usually occur in low-rise buildings, where a few walls must resist total horizontal inertia force on a building without a rocking mechanism. In multistory framed buildings the major portion of seismic shear force may need to be transferred from frames to structural walls which extend only a few stories above foundation level.

There are cases when the flexural strength of squat walls is so large that it is difficult to match it with corresponding shear resistance. Such walls could eventually fail in shear. It must be recognized that such failure could be accepted provided that the wall response is associated with dutility demands which are much less than those envisaged for the more slender walls considered in previous sections. Such squat walls may be classified as structures with restricted ductility.

5.7.2   Flexural Response and Reinforcement Distribution

Even though the plane-sections hypothesis may be extensively violated in squat walls, the significance at full flexural strength, when most reinforcement is at yield and therefore independent of errors in strain calculations, will not be great. Consequently, the standard approach used for predicting flexural strength is likely to be satisfactory for the design of squat walls. This was reviewed in Section 3.3.1(*c*).

Evenly distributed vertical reinforcement theoretically results in lower curvature ductilities at the ultimate state, but this arrangement is preferable because it results in an increased flexural compression zone and in improved conditions for shear friction and dowel action, features that are significant in assessing sliding shear resistance, discussed in the next section. In fact, with typically low axial force levels, which are common for squat walls, the reduction in curvature ductility with distributed reinforcement is not significant [P1]. Potential ductility factors associated with a concrete strain of 0.004 in the extreme fiber are greatly in excess of ductility demands expected under seismic response.

At maximum seismic response, extreme fiber compression strains are likely to be less than 0.003, resulting in only moderate concrete compression stresses. This is beneficial because, as a result of the transmission of large shear forces, concrete in the flexural compression zone is subjected to severe stress conditions. These are examined in subsequent sections.

5.7.3   Mechanisms of Shear Resistance

Because of relative dimensions, boundary conditions, and the way transverse forces (shear) are introduced to squat walls, mechanisms of shear resistance appropriate to reinforced concrete beams are not wholly applicable. In particular, apart from the contribution of horizontal shear reinforcement, a

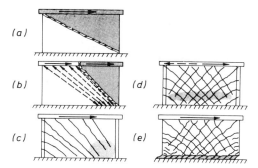

Fig. 5.58 Shear failure modes in squat walls.

significant portion of shear introduced at the top of a squat cantilever wall is transmitted by diagonal compression directly to the foundations [B13].

(a) Diagonal Tension Failure. When horizontal shear reinforcement is insufficient, a corner-to-corner diagonal tension failure plane [Fig. 5.58(*a*)] may develop. As the diagonal tension strength of squat walls is significantly affected by the way by which force is introduced at the top edge, care is necessary in assessing this mode in various design situations.

Diagonal tension failure may also develop along a steeper failure plane [Fig. 5.58(*b*)]. If a path is available to transfer the shear force to the rest of the wall, such a diagonal crack need not result in failure. The use of a tie beam at the top of the wall is an example of this method. It is evident that there are ways to redistribute shear force along the top edge to minimize diagonal tension and enhance force transfer more efficiency to the foundation by diagonal compression.

(b) Diagonal Compression Failure. When the average shear stress in the wall section is high and adequate horizontal shear reinforcement has been provided, concrete may crush under diagonal compression. This is not uncommon in walls with flanged sections [Fig. 5.58(*c*)], which may have very large flexural strengths. When reversed cyclic forces are applied so that two sets of diagonal shear cracks develop, diagonal compression failure may occur at a much lower shear force. Transverse tensile strains and intersecting diagonal cracks, which cyclically open and close, considerably reduce the concrete's compressive strength. Often the crushing of the concrete rapidly spreads [B13] over the entire length of the wall [Fig. 5.58(*d*)]. Diagonal compression failure results in dramatic and irrecoverable loss of strength and must be avoided when designing ductile walls. Limitations on maximum shear stress at the wall's flexural strength are intended to ensure that shear compression failure will not curtail ductile response.

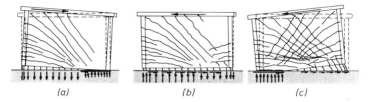

Fig. 5.59   Development of the sliding shear mechanism.

(c)   Phenomenon of Sliding Shear. By limiting nominal shear stress and providing adequate horizontal shear reinforcement, shear failures by diagonal compression or tension can be avoided, as outlined above. Inelastic deformations required for energy dissipation would then be expected to originate mainly from postyielding strains generated in the vertical flexural reinforcement. However, after a few cycles of displacement reversals causing significant yielding in the flexural reinforcement, sliding displacement can occur at the base or along flexural cracks which interconnect and form a continuous, approximately horizontal shear path [Fig. 5.58(*e*)]. Such sliding displacements are responsible for a significant reduction of stiffness, particularly at low force intensities at the beginning of a displacement excursion. As a consequence energy dissipation is reduced [P35].

The development of this mechanism is illustrated in greater detail in Fig. 5.59. In the first cycle, involving large flexural yielding, the major part of the shear force at the cantilever wall base must be transmitted across the flexural compression zone [Fig. 5.59(*a*)]. Because the concrete in the comparatively small flexural compression zone has not yet cracked, horizontal shear displacements along the base section are insignificant.

Cracks will also develop across the previous flexural compression zone after force reversals, while bars that have yielded considerably in tension will be subject to compressive stresses. Until the base moment reaches a level sufficient to yield these bars in compression, a wide, continuous crack will develop along the wall's base [Fig. 5.59(*b*)]. Along this crack the shear force will be transferred primarily by dowel action of the vertical reinforcement. Because of the flexible nature of this mechanism, relatively large horizontal shear displacements must occur at this stage of the response. These sliding shear displacements will be arrested only after yielding of the compression steel occurs, closing the crack at the compression end of the wall and allowing flexural compression stresses to be transmitted by the concrete [Fig. 5.59(*c*)]. Due to sliding shear displacements that occurred during this displacement reversal, the compression in the flexural compression zone is transmitted by uneven bearing across crack surfaces. This, in turn, leads to a reduction of both the strength and stiffness of the aggregate interlock (shear friction) mechanism.

Fig. 5.60   Failure of a squat wall by sliding shear [P35].

With subsequent inelastic displacement reversals, further deterioration of shear friction mechanisms along the plane of potential sliding is expected. Due to a deterioration of bond transfer along the vertical bars and the Bauschinger effect, the stiffness of the dowel shear mechanism [P1] will also be drastically reduced. Eventually, the principal mode of shear transfer along the base will be by kinking of the vertical bars, as shown in Fig. 5.58(*e*). The failure by sliding of a squat rectangular test wall is seen in Fig. 5.60.

5.7.4   Control of Sliding Shear

The mechanisms of shear friction, consisting of shear transfer by dowel action of reinforcement crossing the plane of potential sliding and shear transfer by aggregate interlock, are well established [P1, M7, M8]. In these studies the effect of precracking as well as the type of preparation of a construction joint surface [P1] has been examined, using specimens in which the potential failure plane was subjected to shear only. However, at the base of a squat wall, where continuous cracking is likely to be initiated along a construction joint, bending moments also need to be transferred. Consequent shear transfer along the critical sliding plane will then be restricted to the vertical wall reinforcement and flexural compression zone, where cyclic opening and closing of cracks (Fig. 5.59) will take place.

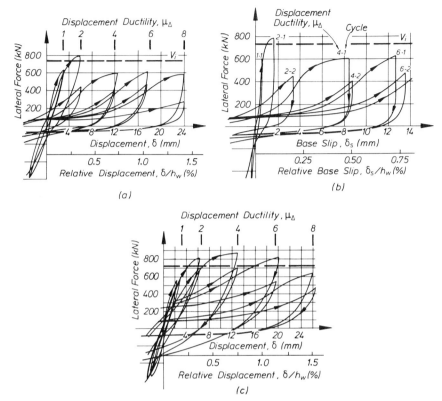

Fig. 5.61 Hysteretic response of squat flanged walls controlled by sliding at the base [P35].

Tests of walls [P35] have shown the detrimental effects of excessive sliding shear displacements as well as the marked improvement of response when some diagonal reinforcement crossing the sliding plane was used to reduce displacements and increase the forces resisting sliding shear. Figure 5.61 compares the response of two squat test walls with small flanges ($h_w/l_w = 1700/3000 = 0.57$) [P35]. Only a part of the approximately symmetrical hysteresis loops are shown. The dramatic reduction of strength, particularly in the second cycles of loading to the same level of ductility, is evident in Fig. 5.61(a). This wall failed in sliding shear, with the web eventually slicing through the flange, as illustrated in Fig. 5.62(b). Figure 5.61(b) shows the observed sliding responses only. By comparing Fig. 5.61(a) and (b) it is seen that at a ductility of $\mu_\Delta = 6$, approximately 75% of the total displacement was due to sliding along the base. After attaining a μ_Δ value of 4, the stiffness of the wall [Fig. 5.61(a)] was negligible upon load reversal.

Figure 5.61(*c*) shows the improved response of an identical wall in which diagonal reinforcement in the form shown in Fig. 5.63(*a*) was used to resist only 30% of the computed shear V_i. It may be noted that a ductility of $\mu_\Delta = 4$, the probable upper limit such a wall would be designed for, corresponded to a deflection equal to only 0.7% of the story height. With cumulative ductility the test unit with diagonal reinforcement dissipated approximately 70% more energy than its counterpart without diagonal bars.

Current understanding of the phenomenon of sliding shear allows only tentative recommendations to be made as to when and how much diagonal reinforcement should be used in squat walls. Identification in the following sections of parameters critical to sliding shear is likely to assist the designer when a relevant decision is to be made.

(*a*) *Ductility Demand.* Tests have shown [P1] that as long as cracks remain small, consistent with the elastic behavior of reinforcement, the strength of shear transfer, primarily by aggregate interlock action, is in excess of the diagonal tension or compression force-carrying capacity of a member. Therefore, sliding shear is not a controlling factor in the design of elastically responding structural walls. However, when flexural yielding begins during an earthquake, shear transfer is restricted mainly to the alternating flexural compression zones of the wall section. This shear transfer mechanism becomes much more flexible. Because of the drastic reduction in contact area between crack faces during flexural rotations, interface shear stresses will increase rapidly. This, and crack misfit, in turn, lead to grinding of the concrete and a consequent reduction of the limiting friction factor. The ensuing deterioration of the response will increase with the number of inelastic displacement cycles imposed and particularly with the magnitude of yielding imposed in any one cycle. The need to control sliding will thus increase with increased ductility demand.

By relating the flexural strength of a squat wall as constructed, in terms of overstrength [Eq. (5.13)] to the strength required to ensure elastic response, the ratio of strength deterioration

$$R_d = 1.6 - 2.2\phi_{o,w}/\mu_\Delta \leq 1.0 \tag{5.35}$$

may be derived [P35], which quantifies the effect of ductility demand only on the need for remedial measures to boost the effectiveness of sliding shear mechanisms in order to improve the energy—dissipating potential. For example, for a wall with only 40% overstrength, designed for a displacement ductility demand of $\mu_\Delta = 3.5$, Eq. (5.35) gives $R_D = 1.6 - 2.2 \times 1.4/3.5 = 0.72$, suggesting that significant improvements in sliding shear transfer at the base need be made. On the other hand, for a wall with significant reserve strength, so that $\phi_{o,w} = 1.8$, and designed to resist a lateral force corresponding to $\mu_\Delta = 2.5$, we find that $R_D = 0.02$. This indicates adequate performance of a conventionally reinforced squat wall.

(b) Sliding Shear Resistance of Vertical Wall Reinforcement. In conventionally reinforced squat walls, a particularly critical situation arises every time the force reverses after an excursion into the inelastic range of response. At this stage a crack could be open over the full length of the wall as shown in Fig. 5.59(*b*). Until contact between faces of the crack at the compression side of the wall is reestablished, the entire flexural and shear force at this section must be transferred by the vertical reinforcement. Since extensive yielding has occurred in the flexural tension zone of the wall, vertical bars in this region must be subjected to significant compression yielding before the previously formed crack can close. To achieve this, a moment close to that required to develop the plastic moment of resistance of the steel section, consisting of vertical bars only, will need to be applied unless significant axial compression is also acting on the wall. This moment may be at least one-half the flexural strength of the reinforced concrete wall section. Thus a large shear force may need to be transferred by the vertical bars only.

Dowel action of vertical bars [P1] is associated with significant shear displacements. Most of the vertical bars will yield before the crack can close. Hence only some of the bars in the elastic core of the section could contribute to sliding shear resistance by dowel action. As seen in fig. 5.60, the more significant contribution of the vertical reinforcement to sliding shear resistance by kinking of the bars is mobilized only after a slip of several millimeters has occurred.

If sliding is to be controlled before the closure of the critical crack in the flexural compression zone, diagonal reinforcement capable of resisting at least 50% of the full shear force would need to be provided. Moreover, in general, diagonal reinforcement may contribute to the flexural strength of the wall and hence will also increase the shear load.

By extrapolating from test results [P1] for dowel action along construction joints, it is estimated that the dowel shear resistance is on the order of

$$V_{do} = 0.25 A_{sw} f_y \qquad (5.36)$$

where A_{sw} is the total area of the vertical reinforcement in the web of a squat wall. This contribution of dowel action to sliding shear may then be assumed to be sustained when the diagonal reinforcement shown in Fig. 5.63 yields and thus allows the sliding displacement necessary to mobilize this action to occur.

(c) Relative Size of Compression Zone. It is emphasized that strength with respect to sliding across a yielding wall section is derived principally from shear friction in the flexural compression zone. Shear friction tests [M7, M8] show that there is an upper limit on interface shear stress beyond which

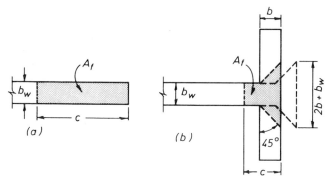

Fig. 5.62 Theoretical effective area for interface shear transfer in the flexural compression zone.

clamping forces, supplied by reinforcement and/or external compressions, do not increase shear strength. This limiting (sliding) shear strength attained with monotonic loading is on the order of $0.35f'_c$ [M7]. With cyclic displacements it is unlikely that a shear stress in excess of $0.25f'_c$ could be attained. It was found in specific tests [M8] that approximately 20% reduction of strength occurred with reversed cyclic loading when the slip did not exceed 2 mm. Slip in excess of this magnitude was observed in tests with squat walls [P35].

It is therefore recommended that the sliding shear strength of the flexural compression zone be assessed as

$$V_f = 0.25f'_c A_f \tag{5.37}$$

where the effective area of shear friction A_f is shown in Fig. 5.62.

The significant role of the depth of the compression zone c should be noted. With constant shear applied to the wall, the base moment, and hence c, will increase with height, h_w. As a corollary, with constant height and applied shear, c will decrease with increasing wall length, l_w. Thus the contribution of the flexural compression zone to sliding shear strength will increase with the height-to-length ratio h_w/l_w. The neutral-axis depth is thus a measure of this important parameter; it also accounts for the contribution of any axial compression force that may be present. The beneficial effect of axial compression load due to gravity is likely to be rather small in low-rise buildings. Squat walls near the base of medium- to high-rise buildings, however, may carry significant compression load, and this should significantly boost the sliding shear resistance of rectangular walls.

Tests have shown that a squat wall with flanges [Fig. 5.5(f)] showed distress due to the stem punching through the flange, as indicated in Fig. 5.62(b), much earlier than a rectangular wall [Fig. 5.5(a)], the two walls being identical in every other respect [P35].

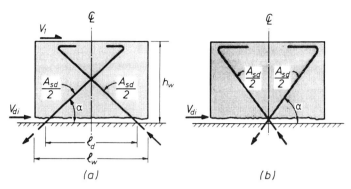

Fig. 5.63   Diagonal reinforcement in squat walls.

In many cases it will be found that the interface shear strength, V_f, of the effective flexural compression area, A_f, is less than the maximum shear force V_o to be expected. In such cases supplementary strength to resist sliding should be provided.

(d) Effectiveness of Diagonal Shear Reinforcement. When diagonal reinforcement is used to control sliding shear, it is necessary to consider also its contribution to flexural strength. For the commonly used arrangement shown in Fig. 5.63(a), it may be shown that when bars across the sliding plane yield in tension and compression, as appropriate, they can resist a moment of

$$M_d = 0.5 l_d A_{sd} f_{yd} \sin \alpha = h_w V_1 \qquad (5.38)$$

where A_{sd} is total area of diagonal reinforcement used, f_{yd} the yield strength of the diagonally placed steel, l_d the horizontal moment arm for the diagonal reinforcement only, and α the inclination of the symmetrically arranged diagonal bars.

The sum of the horizontal components of these diagonal steel forces at the base is, however, larger than the shear V_1 necessary to develop M_d. Therefore, the diagonal bars can provide resistance to sliding shear originating from other mechanisms, such as flexural resistance of the vertically reinforced concrete wall. This additional resistance against sliding shear is, from Fig. 5.63(a)

$$V_{di} = A_{sd} f_{yd} \cos \alpha - V_1 = A_{sd} f_{yd} [\cos \alpha - (l_d/2h_w) \sin \alpha] \quad (5.39)$$

It is seen that the effectiveness of diagonal reinforcement in resisting sliding shear increases with smaller bar inclination, α, and the decrease of distance, l_d. If so detailed the resultants of the diagonal steel forces may intersect at the base, so that $l_d = 0$, as shown in Fig. 5.63(b). Thus when the shear displacement is large enough to cause diagonal bars in Fig. 5.63(b) to yield in tension and compression, their contribution to flexural resistance, Eq. (5.38), will diminish, and the entire strength may then be used to resist sliding shear associated with flexural overstrength derived from only the vertical wall reinforcement. If shear displacement does not occur, the arrangement of diagonal bars in Fig. 5.63(b) will increase the flexural strength of the wall by the same order as that resulting from the diagonal bars shown in Fig. 5.63(a).

To ensure that sliding shear failure will not occur at another horizontal wall section, effective diagonal bars should cross all sections within a distance $0.5l_w$ or $0.5h_w$, whichever is less, above the critical base section.

(e) Combined Effects. The contributions of dowel shear force, V_{do}, derived from Eq. (5.36), and the effective flexural compression zone to resistance of sliding shear, V_f [Eq. (5.37)] can then be considered together with the allowance for the expected ductility demand, with the aid of the factor R_d defined by Eq. (5.35). Accordingly, it is suggested that diagonal reinforcement should be provided to resist a shear force

$$V_{di} = R_d \frac{V_{Eo} - V_{do} - V_f}{V_{Eo}}(V_{Eo} - V_1) \qquad (5.40)$$

where $V_{Eo} = \phi_{o,w}V_E$ is the shear force developed with the flexural overstrength of the base section. The examples in Section 5.7.8 given an appreciation of the significance of each term.

5.7.5 Control of Diagonal Tension

Conservatively, it may be assumed that shear force V_{Eo} at the wall's top is introduced by uniform shear flow. Hence the shaded element in Fig. 5.64, bound by a potential diagonal failure plane at $45°$, should receive a share of

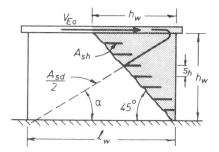

Fig. 5.64 Model for the control of shear failure by diagonal tension.

the shear force $h_w V_{Eo}/l_w$ when $h_w/l_w \leq 1$. This force may then be assumed to be transferred by the horizontal shear reinforcement, with area A_{sh} and spaced vertically at s_h intervals, and by the horizontal component of the force developed in the diagonal tension bars with area $A_{sd}/2$, if they have been provided. All relevant quantities may be obtained from first principles.

To satisfy the requirement that

$$(h_w/l_w)V_{Eo} \leq V_c + V_s + V_{dh}$$

with the usual assumption that $V_c = 0$, the horizontal shear reinforcement A_{sh} must resist a shear force

$$V_s = (h_w/l_w)V_{Eo} - V_{dh} \qquad (5.41a)$$

where

$$V_s = h_w A_{sh} f_{yh}/s_v \qquad (5.41b)$$

and the contribution of the diagonal bar in tension is

$$V_{dh} = (A_{sd}/2)f_{yd} \cos \alpha \qquad (5.41c)$$

5.7.6 Framed Squat Walls

Structural walls in Japan are traditionally provided with substantial boundary members appearing as both beams and columns. The behavior of such squat walls, of the configuration shown in Fig. 5.65(a), has been studied extensively over the last 25 years, particularly at Kyushu University [T4, T5]. In the design of these walls, emphasis has been placed on strength rather than on ductility. The postulated mechanism of shear transfer is similar to that encountered in infilled frames. Frame action of the boundary members combines with the shear strength of the web, which provides primarily a diagonal compression field, as suggested in Fig. 5.65(a). After the breakdown of the web due to diagonal tension, the column members are expected to provide shear resistance. The vertical boundary members, reinforced in the same fashion as columns, are intended to prevent a sliding shear failure, such as shown in Fig. 5.60. Due to the overturning moment the reinforcement in column members is not expected to yield. The hysteretic response of such walls is rather poor. Figure 5.25 is an example.

The study and use of such squat walls has also strongly influenced the modelling in Japan of the behavior of multistory cantilever walls. Wall

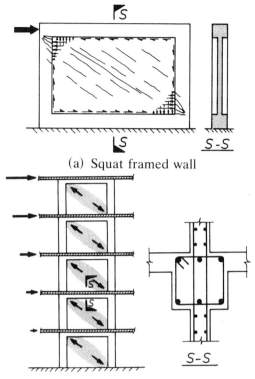

(a) Squat framed wall

(b) Multistory wall with framed panels

Fig. 5.65 Framed structural walls.

elements in each story of the structure shown in Fig. 5.65(*b*) are expected to behave much like the squat unit shown in Fig. 5.65(*a*). The horizontal beam element at the slab wall junction is assumed to act as a tension member of a truss while the web portion provides the corner-to-corner diagonal compression strut, as suggested in Fig. 5.65(*b*). However, tests do not support such mechanisms. Diagonal cracks form at steeper angles with a pattern similar to that seen in Fig. 5.37. The mechanism suggested in Fig. 5.65(*b*) would imply significant diagonal compression stress concentrations at corners and also inefficient utilization of the horizontal reinforcement in the web of the wall. It is unlikely that the additional concrete and reinforcement in the beam element, shown in the section in Fig. 5.65(*b*), would improve the strength or the behavior of such walls [I2]. Such beam elements would only be justified to provide anchorage for flexural reinforcement in beams, which frame into such a wall.

5.7.7 Squat Walls with Openings

Walls with small aspect ratios in many low-rise buildings may contain openings for doors and windows. Improvements in the design of these walls in seismic regions has received attention only relatively recently. It was customary in design to neglect the effects of openings on the strength and behavior of such walls unless a relatively simple frame model could be conceived. There is ample evidence that apart from providing nominal diagonal bars at corners of openings to arrest the widening of cracks, which usually develop there first, very little was generally done to aid satisfactory seismic performance.

The seismic design of walls with significant openings, such as shown in Fig. 5.66 may readily be undertaken with the use of "strut and tie" models [S13, Y3]. These enables an admissible load path for the horizontal earthquake forces from floor levels down to the foundations to be established. Critical magnitudes of tension forces can be derived from statics. Compression forces to be transmitted in struts are seldom critical. With a rational design strategy, accompanied by careful detailing, significant ductility capacity can be built into these structures, contrary to common belief.

Figure 5.66(a) and (b) show examples for the choice of two models for an example squat wall with openings. Each model is suitable for the seismic response corresponding with lateral forces in a given direction to be considered. If ductile response is to be assured, the designer should choose particular tension chords in which yielding can best be accommodated. For example, members $F-E$ and $G-H$ in Fig. 5.66(a) represent a good choice for this purpose. After the evaluation of the overturning moment capacity of the

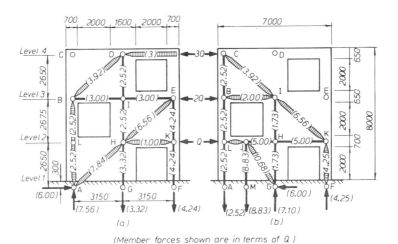

(Member forces shown are in terms of Q)

Fig. 5.66 Strut and tie models for a squat wall with openings.

entire structure at overstrength, development of tensile forces in members *F–K* and *G–H* are readily evaluated. Corresponding forces in all other members can be estimated and hence reinforcement may be provided so as to ensure that no yielding in other ties, such as *B–E*, can occur. This is simply the application of capacity design philosophy to a truss-type structure. A similar exercise will establish forces [Fig. 5.66(*b*)] generated in the other model during reversal of the earthquake forces, *Q*. Correspondingly, the amount of tension reinforcement in members *A–L* and *M–J* is found.

It is now necessary to study the possible effects of one inelastic system [Fig. 5.66(*a*)] on the ductile response of the other [Fig. 5.66(*b*)]. Such a study will confirm the appropriateness for the choice of "ductile links" (Fig. 1.18) in the "chain of resistance" and will assist in the identification of areas where special attention of the designer is required to aid energy dissipation. The following conclusions may be drawn from a study of the example structure in Fig. 5.66:

1. Compression strains in concrete struts are likely to remain small. They should be kept small because in general these members are not suitable for energy dissipation.

2. The magnitude of the earthquake force V_{Eo} at overstrength [Fig. 5.66(*b*)] is determined by the amount of reinforcement provided for ties *A–L*, *M–J*, and *G–H*. These members are relatively long, and hence large inelastic elongations can be achieved with moderate inelastic tensile steel strains.

3. Similarly, the magnitude of V_{Eo} may be determined by the capacity at overstrength of ties *G–H* and *F–K* [Fig. 5.66(*a*)].

4. A significant fraction of the inelastic tensile strains that might be imposed on members *A–L* and *F–K* are recoverable because upon force reversal both members become struts.

5. Using capacity design principles, reinforcement for all other ties, (i.e., *B–E* and *J–K*) should be determined to ensure that yielding cannot occur. As these members carry only tension, yielding with cyclic displacements may lead to unacceptable cumulative elongations. Such elongations would impose significant relative secondary displacement on the small piers adjacent to openings, particularly those at *F–K* and *L–B*. The resulting bending moment and shear forces, although secondary, may eventually reduce the capacity of these vital struts.

6. Careful detailing of the reinforcement is necessary as follows:
   (a) All node points where full anchorage, preferably with effective hooks, must be provided for bars of the ties to enable three coplanar forces to equilibrate each other.
   (b) Because of cyclic force reversals in ductile members such as *A–B* and *F–E*, reinforcing bars in cages must be stabilized against

buckling by closely spaced ties, as in plastic hinge regions of beams or columns and the end regions of walls.

(c) The need for the use of diagonal bars, crossing the wall base at $A–G$, may arise to control shear sliding.

(d) Note that in this example it has been assumed that lateral forces at different levels are introduced at the tension edge of the wall. If these forces are applied at the compression edge of the wall, the horizontal tie forces will correspondingly increase at each level.

Strut and tie models, reviewed here briefly, are powerful tools in the hands of the imaginative designer, to enable satisfactory and predictable ductile behavior of walls with irregular openings to be assured. A design example is given in Section 5.7.8(d).

5.7.8 Design Examples for Squat Walls

(a) Squat Wall Subjected to a Large Earthquake Force

(i) *Design Requirements and Properties*: A one-story cantilever wall with dimensions as shown in Fig. 5.67(a) is to carry a lateral force of $V_u = 1600$ kN (358 kips) assumed to be uniformly distributed along its top edge. This force was based on a displacement ductility capacity of $\mu_\Delta = 2.5$. For the purpose of determining the necessary reinforcement, gravity load effects, being rather small, are neglected. The material properties to be used are as follows: $f_c' = 25$ MPa (3625 psi) and $f_y = 275$ MPa (40 ksi).

(ii) *Preliminary Estimates*: If the wall is 250 mm (10 in.) thick, the ideal shear stress at flexural overstrength would be with $\phi_o = 1.4$ on the order of

$$v_i = \phi_o V_E / (0.8 b_w l_w) = 1.4 \times 1,600,000 / (250 \times 0.8 \times 7000)$$
$$= 1.6 \text{ MPa } (232 \text{ psi})$$

and is thus well within permissible values.

As the vertical reinforcement in the stem consists of D12 (0.47-in.-diameter) bars on 300-mm (11.8-in.) centers in each face of the wall, Section 5.4.2(b) is satisfied as

$$\rho_l = 2 \times 113 / (250 \times 300) = 0.00301 > \rho_{l, \min} = 0.7 / f_y = 0.00255$$

This reinforcement can resist a base moment of approximately

$$M_1 = (\rho_l b_w l_w) f_y (0.5 - 0.05) l_w$$
$$= (0.00301 \times 250 \times 7000) 275 \times 0.45 \times 7000 / 10^6$$
$$= 4563 \text{ kNm } (3354 \text{ kip-ft})$$

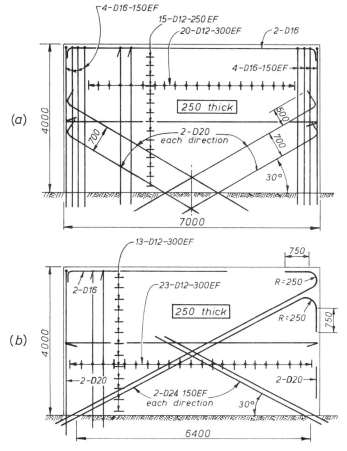

Fig. 5.67 Details of example squat walls.

The total required ideal flexural strength at the base is, however,

$$M_i = h_w V_E/\phi = 4 \times 1600/0.9 = 7111 \text{ kNm (5227 kip-ft)}$$

Therefore, reinforcement in the end region must be provided to resist

$$M_2 = M_i - M_1 = 7111 - 4563 = 2548 \text{ kNm (1873 kip-ft)}$$

The area of reinforcement required to resist this is approximately

$$A_{s2} = 2548 \times 10^6/[275(7000 - 2 \times 275)] = 1437 \text{ mm}^2 \text{ (2.23 in.}^2)$$

Try to use eight D16 (0.63-in.-diameter) bars [1608 mm^2 (2.49 in.^{2})] at each end of the wall.

(iii) Flexural Strength of Wall: With the vertical reinforcement arranged as in Fig. 5.67(a), the ideal flexural strength is determined more accurately. As the neutral-axis depth will be rather small, assume that only 50% of the eight D16 bars in the end region will yield in compression. With this the depth of the compression block can be estimated as

$$a = \frac{(40 \times 113 + 4 \times 201)275}{0.85 \times 25 \times 250} = 276 \text{ mm } (10.9 \text{ in.})$$

The neutral-axis depth is $c = 276/0.85 \approx 325$ mm (12.8 in.). The ideal flexural strength, M_i, is then approximately:

$(40 \times 113 \times 275)(0.5 \times 7000 - 0.5 \times 276)10^{-6} = 4{,}179$ kNm (3072 kip-ft)

$(8 \times 201 \times 275)(7000 - 275 - 0.5 \times 276)10^{-6} = \underline{2{,}913 \text{ kNm}}$ (2141 kip-ft)

acceptable as $M_i = 7111$ (5227 kip-ft) $\approx 7{,}092$ kNm (5213 kip-ft)

The flexural overstrength at the base will be approximately $\lambda_o M_i = 1.25 \times 7092 = 8870$ kNm (6519 kip-ft), and hence $\phi_{o,w} = 8870/(4 \times 1600) = 1.39$ and $V_{Eo} \approx 2224$ kN (498 kips). Note that concentration of vertical reinforcement at wall edges has detrimental effects on sliding shear resistance similar to those of flanges [Fig. 5.62(b)].

(iv) Requirements for Diagonal Reinforcement: These are based on Eq. (5.40), which is evaluated as follows: From Eq. (5.35),

$$R_d = 1.6 - 2.2\phi_{o,w}/\mu_\Delta = 1.6 - 2.2 \times 1.39/2.5 = 0.38$$

The dowel shear resistance of the vertical reinforcement with total area, conservatively taken as $A_{sw} = 0.003 \times 250 \times 7000 = 5250$ mm^2 (8.14 in.^{2}), is, from Eq. (5.36),

$$V_{do} = 0.25 A_{sw} f_y = 0.25 \times 5250 \times 275 \times 10^{-3} = 361 \text{ kN } (81 \text{ kips})$$

The sliding shear resistance of the concrete in the flexural compression zone with $c = 325$ mm is, from Eq. (5.37) and Fig. 5.62(a),

$$V_f = 0.25 f'_c c b_w = 0.25 \times 25 \times 325 \times 250 \times 10^{-3} = 509 \text{ kN } (114 \text{ kips})$$

Hence from Eq. (5.40) with $V_1 = 0$, the shear to be resisted by diagonal reinforcement is

$$V_{di} = R_d(V_{Eo} - V_{do} - V_f) = 0.38(2224 - 361 - 509) = 515 \text{ kN (115 kips)}$$

This is 23% of the design shear force V_{Eo}. Therefore, from Eq. (5.39), in which $V_1 = 0$ and $l_d = 0$, or from first principles with $\alpha = 30°$,

$$A_{sd} = V_{di}/(f_{yd} \cos \alpha) = 515 \times 10^3/(275 \cos 30°) = 2162 \text{ mm}^2 \text{ (3.35 in.}^2)$$

Use four D20 (0.79-in-diameter) bars in each direction with $A_{sd} = 2512$ mm² (3.89 in.²), as shown in Fig. 5.67(a).

(v) *Control of Diagonal Tension*: In accordance with the concepts of Fig. 5.64, the area of horizontal shear reinforcement is obtained from Eq. (5.41c):

$$V_{dh} = 0.5A_{sd}f_{yd} \cos \alpha = 0.5 \times 2512 \times 275 \times 0.866 \times 10^{-3}$$
$$= 299 \text{ kN (67 kips)}$$

From Eq. (5.41a),

$$V_s = (h_w/l_w)V_{Eo} - V_{dh} = (4/7)2224 - 299 = 972 \text{ kN (218 kips)}$$

Hence from Eq. (5.41b),

$$\frac{A_{sh}}{s_h} = \frac{V_s}{h_w f_{yh}} = \frac{972,000}{4000 \times 275} = 0.88 \text{ mm}^2/\text{mm (0.035 in.}^2/\text{in.)}$$

If D12 (0.47.-diameter) bars are used, the required spacing will be

$$s_h = 2 \times 113/0.88 = 257 \simeq 250 \text{ mm (10 in.)}$$

as shown in Fig. 5.67(a).

(b) Alternative Solution for a Squat Wall Subjected to a Large Earthquake Force

(*i*) *Design Requirements*: Consider the squat wall of the previous example but attempt to utilize the diagonal reinforcement for flexural resistance also. All other properties are as given in Section 5.7.8(a).

(ii) *Preliminary Estimates*: With the arrangement shown in Fig. 5.67(b), taking $\alpha = 30°$ and $l_d = 6400$ mm (21 ft), assume that to control sliding, the shear force to be resisted in accordance with Eq. (5.40) is approximately 20% of the shear at flexural overstrength:

$$V_{di} \approx 0.2 \times 1.4V_E = 0.2 \times 1.4 \times 1600 = 448 \text{ kN } (100 \text{ kips})$$

Therefore, from Eq. (5.39),

$$A_{sd} \simeq V_{di} \Bigg/ \left[f_{yd} \left(\cos \alpha - \frac{l_d}{2h_w} \sin \alpha \right) \right]$$

$$= 448,000 \Bigg/ \left[275 \left(0.866 - \frac{6.4}{2 \times 4} 0.5 \right) \right] = 3496 \text{ mm}^2 \ (5.42 \text{ in.}^2)$$

Try to use four D24 (0.94-in.-diameter) bars in each direction with $A_{sd} = 8 \times 452 = 3616$ mm^2 (5.6 in.^{2}) at a slope of 30°, as shown in Fig. 5.67(b).
  The moment resisted by the diagonal bars is, from Eq. (5.38),

$$M_d = 0.5 \times 6400 \times 3616 \times 275 \times \sin 30° /10^6 = 1591 \text{ kNm } (1169 \text{ kip-ft})$$

The moment resisted by the vertical D12 bars in the stem may be assumed to be, as in Section 5.7.8(a)(ii),

$$M_1 = 4{,}563 \ \text{kNm } (3{,}354 \text{ kip-ft})$$

M_i required from the given force $= 7{,}111 \ \text{kNm } (5{,}227 \text{ kip-ft})$

Therefore, $M_2 = \ 957 \ \text{kNm } (703 \text{ kip-ft})$

Hence provide reinforcement in the end regions:

$$A_{s2} \simeq 957 \times 10^6/[275(7000 - 2 \times 50)] = 504 \text{ mm}^2 \ (0.78 \text{ in.}^2)$$

Provide two D20 (0.79-in.-diameter) bars [628 mm^2 (0.97 in.^{2})] at each end.

(iii) *Flexural Strength of Wall*: With the arrangement of the vertical reinforcement as shown in Fig. 5.67(b), check the flexural strength. Assume that only 42 D12 (0.47-in.-diameter) bars in tension need to be balanced by concrete compression stresses. Hence

$$a \simeq 42 \times 113 \times 275/(0.85 \times 25 \times 250) = 246 \text{ mm } (9.7 \text{ in.})$$

and

$$c \simeq 246/0.85 = 289 \text{ mm } (11.4 \text{ in.})$$

Thus

$$M_1 = 42 \times 113 \times 275(3500 - 0.5 \times 246)/10^6 = 4{,}407 \text{ kNm } (3{,}239 \text{ kip-ft})$$

$$M_2 = 2 \times 314 \times 275(7000 - 100)/10^6 \qquad = 1{,}192 \text{ kNm } (876 \text{ kip-ft})$$

from Eq. (5.38) as shown above $\qquad M_d = \underline{1{,}591 \text{ kNm } (1{,}169 \text{ kip-ft})}$

satisfactory $\qquad M_i = 7111 < 7{,}190 \text{ kNm } (5{,}285 \text{ kip-ft})$

(iv) *Requirements for Diagonal Reinforcement*: According to Eq. (4.39), the four D24 (0.94-in.-diameter) bars placed in each direction provide shear resistance against sliding:

$$V_{di} = 3616 \times 275\left(\cos 30° - \frac{6.4}{2 \times 4} \sin 30°\right)\bigg/10^3 = 463 \text{ kN } (104 \text{ kips})$$

From Eq. (5.38),

$$V_1 = M_d/h_w = 1591/4 = 398 \text{ kN } (89 \text{ kips})$$

For this wall $V_{Eo} \simeq 1.25 \times 7190/4 = 2247$ kN; therefore, $\phi_{o,w} = 2247/1600 = 1.40$ and $v_i = 2247 \times 10^3/(0.8 \times 250 \times 7000) = 1.61$ MPa (233 psi). Therefore, from Eq. (5.40), the required shear resistance for sliding is:

From Eq. (5.35): $R_d = 1.6 - 2.2 \times 1.4/2.5 \qquad = 0.37$

From Eq. (5.36): $V_{do} =$ as in the previous case $\qquad = 361 \text{ kN } (81 \text{ kips})$

From Eq. (5.37): $V_f = 0.25 \times 25 \times 289 \times 250 \times 10^{-3} = 452 \text{ kN } (101 \text{ kips})$

$$V_{di} = R_d \frac{V_{Eo} - V_{do} - V_f}{V_{Eo}}(V_{Eo} - V_1)$$

$$= 0.37\left(\frac{2247 - 361 - 452}{2247}\right)(2247 - 398) = 437 \text{ kN } (98 \text{ kips})$$

The V_{di} value provided is $(3616/3496)\,448 = 463$ kN (104 kips). Therefore, this arrangement is satisfactory.

(v) *Control of Diagonal Tension*: Again we obtain:

From Eq. (5.41c): $V_{dh} = 0.5 \times 3616 \times 275 \times \cos 30° = 431 \text{ kN } (97 \text{ kips})$

From Eq. (5.41a): $V_s = (4/7)2247 - 431 \qquad = 853 \text{ kN } (191 \text{ kips})$

From Eq. (5.41b): $A_{sh}/s_h = 853{,}000/(4000 \times 275) \quad = 0.775 \text{ mm}^2/\text{mm}$

Place D12 (0.47-in.) bars in each face, as shown in Fig. 5.67(*b*), at

$$s_h = 2 \times 113/0.775 = 292 \text{ mm} \simeq 300 \text{ mm } (11.8 \text{ in.})$$

(c) Squat Wall Subjected to a Small Earthquake Force. A 200-mm (7.9 in.)-thick wall with the same dimensions and properties as shown in Fig. 5.67 is to resist a shear force of $V_E = 700$ kN (157 kips) only.

Minimum reinforcement in the wall is likely to be adequate. Hence provide in the stem D10 (0.39-in.-diameter) bars in each face on 300-mm (11.8-in.) spacing.

$$\rho_l = 2 \times 78.5/(200 \times 300) = 0.00262 > 0.00255 = \rho_{l,\,min}$$

Also provide two D16 (0.63-in.-diameter) bars at each of the vertical edges.

Assume that the depth of the flexural compression stress block will be on the order of

$$a = \frac{7000 \times 200 \times 0.00262 \times 275}{0.85 \times 25 \times 200} \simeq 237 \text{ mm (9.33 in.)}$$

Therefore, with some approximation for the amount and location of the tension bars (D10) with an internal lever arm of $(7000 - 0.5 \times 6450 - 113) = 3656$ mm (12 ft):

$$M_1 \simeq (6450 \times 200 \times 0.00262 \times 275)(3656)/10^6 = 3{,}398 \text{ kNm (2498 kip-ft)}$$

$$M_2 \simeq 2 \times 201 \times 275(7000 - 100)/10^6 \qquad = \underline{\quad 763 \text{ kNm (561 kip-ft)}}$$

M_i required $= 700 \times 4/0.9$

$$= 3111 < M_i, \text{ available} \qquad = 4{,}161 \text{ kNm (3058 kip-ft)}$$

At overstrength $V_{Eo} \approx 1.25 \times 4161/4 \qquad = 1{,}300 \text{ kN (291 kips)}$

and $\phi_{o,w} = 1300/700 = 1.86$. The first term of Eq. (5.40) is in this case $R_d = (1.6 - 2.2 \times 1.86/2.5) = 0.037$ from Eq. (5.35), and hence the required resistance against sliding is negligible. No diagonal reinforcement need be provided.

The shear stress is

$$v_i = 1{,}300{,}000/(0.8 \times 200 \times 7000) = 1.16 \text{ MPa (168 psi)} < 0.16 f_c'$$
$$= 4.0 \text{ MPa (580 psi)}$$

with $V_s = (4/7)1300 = 743$ kN (166 kips):

$$\frac{A_{sh}}{s_h} = \frac{743{,}000}{4000 \times 275} = 0.675 \text{ mm}^2/\text{mm} \left(0.027 \text{ in.}^2/\text{in.}\right)$$

Using D10 (0.39-in.-diameter) horizontal bars at each face, the vertical spacing is

$$s_h = 2 \times 78.5/0.675 = 233 \approx 225 \text{ mm (8.9 in.)}$$

(d)   Squat Wall with Openings. A three-story ductile squat wall with openings and dimensions as shown in Fig. 5.68 is to be designed to resist lateral forces together with gravity loads. For this purpose a strut-and-tie model is chosen, as in Fig. 5.66. The design lateral forces, corresponding with an assumed displacement ductility capacity of $\mu_\Delta = 3$, are such that $Q_u = 80$ kN (17.9 kips). Hence the member forces may readily be determined from Fig. 5.66. The relatively small gravity loads due to dead load alone are approximated by a number of forces at node points given in Fig. 5.68(b).

To accommodate two layers of reinforcement in each of the horizontal and vertical directions, a wall thickness of 200 mm (7.9 in.) was chosen with 20-mm (0.79-in.) cover to the horizontal D10 (0.39-in.-diameter) bars seen in Figs. 5.68(a) and (c). The material strength properties are $f_y = 300$ MPa (43.5 ksi) ($\lambda_o = 1.25$) and $f'_c = 25$ MPa (3625 psi) and $\phi = 0.9$.

(i)   Required Capacity of the Foundations: From the given lateral forces the total overturning moment at the wall base is $M_{ot} = 37.23Q_u = 2978$ kNm (2189 kip-ft). At the development of a ductility of $\mu_\Delta = 3$, with some overstrength, the overturning moment can be expected to increase to at least $\phi_{o,w}M_{ot} \approx 1.5 \times 2978 \approx 4500$ kNm (3308 kip-ft). With a total dead load on the wall of $W_D = 372$ kN (83 kips), this moment corresponds to an eccentricity of $e \approx 12$ m (39.4 ft). This indicates that a very substantial foundation structure is required to enable the ductile wall to be sustained.

(ii)   Determination of Internal Forces and the Required Tension Reinforcement: With a member force F_i the area of tension steel required is $A_{si} = F_i/(\phi f_y) = F_i/270$ (mm^2).

(1)   Consider earthquake action $\overleftarrow{Q}$ as in Fig. 5.66(a).

$$F_{fe} = (4.24 \times 80 - 0.9 \times 62) \qquad = 283 \text{ kN (63 kips)}$$

$$A_s = 1050 \text{ mm}^2 \text{ (1.63 in.}^2)$$

Try four D16 (0.63-in.-diameter) and

four D10 (0.39-in.-diameter) $= 1118$ mm (1.73 in.^{2})

$$(\rho_t = 0.80\%)$$

$$F_{gh} = (3.32 \times 80 - 0.9 \times 186) \qquad = 98 \text{ kN (22 kips)}$$

$$A_s = 363 \text{ mm}^2 \text{ (0.56 in.}^2)$$

or $F_{hi} = (2.52 \times 80 - 0.9 \times 124)$ $= 90 < 98$ kN

Try four D12 (0.47-in.-diameter) and
eight D10 (0.39-in.-diameter) bars $= 1080$ mm^2 (1.67 in.^{2})

$$(\rho_t = 0.34\%)$$

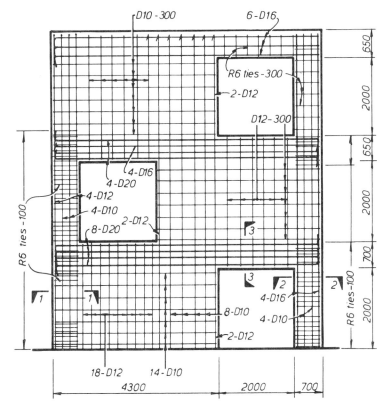

(a) Elevation

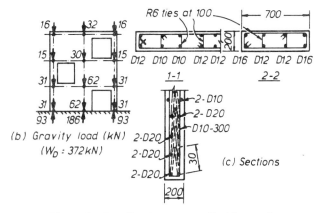

(b) Gravity load (kN)
(W_D : 372kN)

(c) Sections

Fig. 5.68 Details of a squat wall with openings.

Flexural overstrengths developed in tension with $\lambda_o f_y = 375$ MPa (54 ksi):

$$F_{fk,o} = 375 \times 1118 \times 10^{-3} = 419 \text{ kN (94 kips)}$$
$$F_{gh,o} = 375 \times 1080 \times 10^{-3} = 405 \text{ kN (91 kips)}$$

Compression at A, including total dead load:

$$F_{a,o} = 419 + 405 + 372 = 1196 \text{ kN (268 kips)}$$

Theoretical depth of neutral axis:

$$c = 1,196,000/[(0.85 \times 25 \times 200)0.85] = 331 \text{ mm (13 in.)} \approx 0.047 l_w$$

(i.e., very small). Moment capacity at the base at overstrength:

$$\phi_{o,w} M_{ot} \approx (419 + 1196)3.15 = 5087 \text{ kNm (3739 kip-ft)}$$

of which $(372 \times 3.15)100/5084 = 23\%$ is due to gravity loads. Hence

$$\phi_{o,w} = 5087/2978 = 1.71 > \lambda_o/\phi = 1.39$$

Check the shear capacity of the strut A–H by estimating the approximate magnitudes of all vertical forces acting at node A thus

$$\overleftarrow{V}_{Eo} \approx (1196 - 93 - 1.71 \times 2.52 \times 80)/(3150/2650) = 901 \text{kN (202 kips)}$$

As this is larger than $1.71 \times 6 \times 80 = 821$ kN (184 kips) the inclination of strut A–H would be somewhat steeper than that assumed in Fig. 5.66(a). It is seen that the corresponding average shear stress is rather small:

$$v_i \approx 821,000/(0.8 \times 4300 \times 200) = 1.19 \text{ MPa (173 psi)} < 0.05 f_c'$$

Member B–I–E is shown in Fig. 5.66(a) to carry a tensile force of $3Q$. However, if the lateral forces would be applied at the opposite edge of the wall, this force would increase to $5Q$. Design reinforcement to resist $4.5Q$ with $\phi = 1.0$.

$$F_{be,o} = 1.71 \times 4.5 \times 80 \qquad\qquad = 616 \text{ kN (138 kips)}$$
$$A_s = 2053 \text{ mm}^2 (3.18 \text{ in.}^2)$$

Provide four D20 (0.79-in.-diameter) and

four D16 (0.63-in.-diameter) bars $= 2060 \text{ mm}^2 (3.19 \text{ in.}^2)$

As curtailment of the reinforcement is impractical, other vertical tension members in Fig. 5.66(a) are not critical.

(2) Now consider earthquake actions $\vec{Q}$ in the other direction, as shown in Fig. 5.66(b).

$$F_{a1} < F_{1b} = (2.52 \times 80 - 0.9 \times 62) \qquad\qquad = 146 \text{ kN (33 kips)}$$

$$A_s = 541 \text{ mm}^2 \text{ (0.84 in.}^2\text{)}$$

Provide four D12 (0.47-in.-diameter) and

four D10 (0.39-in.-diameter) bars $\qquad = 766 \text{ mm}^2 \text{ (1.19 in.}^2\text{)}$

$$(\rho = 0.55\%)$$

$$F_{mj} < 8.83 \times 80 - 0 \qquad\qquad = 706 \text{ kN (158 kips)}$$

$$A_s = 2614 \text{ mm}^2 \text{ (4.05 in.}^2\text{)}$$

Provide only 18 D12 (0.47-in.-diameter) bars $\quad = 2034 \text{ mm}^2 \text{ (3.15 in.}^2\text{)}$

$$(\rho = 0.51\%)$$

Forces at flexural overstrength:

$$F_{a1,o} = 375 \times 766 \quad = 287 \text{ kN (64 kips)}$$
$$F_{mj,o} = 375 \times 2034 \quad = 763 \text{ kN (171 kips)}$$
$$F_{gh,o} = \text{as before} \quad = 405 \text{ kN (91 kips)}$$

Compression force at F, including dead load:

$$F_{f,o} = 287 + (763 - 763) + 405 + 372 = 1064 \text{ kN (238 kips)}$$
$$f_c \approx 1{,}064{,}000/(700 \times 200) = 7.6 \text{ MPa (1100 psi) (small)}$$

Moment at overstrength about center of structure:

$$\phi_{o,w} M_{ot} = (287 + 1064)3.15 + 763 \times 1.8 = 5629 \text{ kNm (4137 kip-ft)}$$
$$\phi_{o,w} = 5629/2978 = 1.89 > 1.39$$

The shear capacity at overstrength from the model in Fig. 5.66(b) is at G from a consideration of the tensile capacity of member $M-J$:

$$V_{Eo} = (6/8.83)763 = 518 < 1.89 \times 6 \times 80 = 907 \text{ kN (230 kips)}$$

However, the inclination of the strut $G-J$ (56°) has been assumed overly conservatively. A more realistic inclination would be 47° and this would increase the horizontal component of the strut $G-J$ to $V_{Eo} \approx 763/\tan 47° = 711 \text{ kN (159 kips)}$. Hence the overstrength factor based on this shear capacity and controlled by the capacity in tension of member M–J is only

$$\phi_{o,w} = 711/(6 \times 80) = 1.48 > 1.39$$

Now the tie L–K may be (conservatively) designed for a tension force of 711 kN; $A_s = 2370$ mm^2 (3.67 in.^{2}). Provide eight D20 (0.79-in.-diameter) bars = 2512 mm^2 (3.89 in.^{2}). Other tension members in the model of Fig. 5.66(b) above level 2 are not critical.

(iii)   Detailing of the Reinforcement: The arrangement of reinforcement corresponding with the calculations above is shown in Fig. 5.68. The larger-diameter horizontal bars are provided with standard hooks to enable the horizontal forces to be transmitted in the immediate vicinity of the nodes of the strut and tie model. In noncritical areas minimum reinforcement ($\rho = 0.29\%$) has been provided. Because restricted ductility demand has been assumed, R6 ties around bars in the boundary elements are provided [Eq. (8.4)] at 100-mm spacing. From Eq. (4.19) the area of a tie leg should be

$$A_{te} = \frac{201 \times 100}{1600} = 12.6 \text{ mm}^2 < A_{te, \text{provided}} = 28.3 \text{ mm}^2 \ (0.044 \text{ in.}^2)$$

As Fig. 5.68(a) shows, the spacing of these ties should be extended beyond the ends of the 700×200 mm column elements to control possible vertical splitting of the concrete in the anchorage zones of the horizontal DH20 (0.79-in.-diameter) and DH16 (0.63-in.-diameter) bars.

To illustrate the criticality under seismic actions of certain members of this strut and tie model, particularly severe simulating earthquake forces were chosen. As a consequence the quantity of reinforcement in this 200 mm (7.9 in.) thick wall with openings is rather large. In most situations much smaller reinforcement content will be found to be sufficient.

6 Dual Systems

6.1 INTRODUCTION

In Chapters 4 and 5 design procedures and detailing requirements for reinforced concrete ductile frames and ductile structural walls were examined. In many buildings, however, these two types of structural forms appear together. When lateral force resistance is provided by the combined contribution of frames and structural walls, it is customary to refer to them as a *dual system* or a *hybrid structure*.

Dual systems may combine the advantages of their constituent elements [B14]. Ductile frames, interacting with walls, can provide a significant amount of energy dissipation, when required, particularly in the upper stories of a building. On the other hand, as a result of the large stiffness of walls, good story drift control during an earthquake can be achieved, and the development of story mechanisms involving column hinges (i.e., soft stories), as shown in Fig. 1.14(e) and (f), can readily be avoided.

Despite the attractiveness and prevalence of dual systems, it is only recently that research effort has been directed toward developing relevant seismic design methodologies [B17, B18]. This research, involving analytical studies of existing building [A9, C13] and experimental work, using static [P27] and shake table tests, has indicated a potential for excellent inelastic seismic response [A5, B15]. Therefore, in this chapter we concentrate on the behavior of dual systems, with strong emphasis on inelastic response, the interaction between frames and walls, and overall response.

Under the action of lateral forces, a frame will deform primarily in a shear mode, whereas a wall will behave like a vertical cantilever with primary flexural deformations, as shown in Fig. 6.1(b) and (c). Compatibility of deformations requires that frames and walls sustain at each level essentially identical lateral displacements [Fig. 6.1(d)]. Because the preferred displacement mode of the two elements shown in Fig. 6.1(b) and (c) is modified, it is found that the walls and frames share in the resistance of story shear forces in the lower stories, but tend to oppose each other at higher levels. The mode of sharing the resistance to lateral forces between walls and frames of a dual system is also strongly influenced by the dynamic response characteristics and development of plastic hinges during a major seismic event, and it may be quite different from that predicted by an elastic analysis. Consequently, in the case of dual systems, simplified elastic analyses are likely to be mislead-

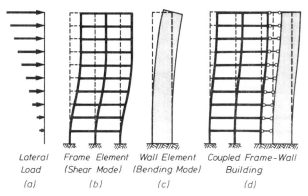

Lateral   Frame Element   Wall Element   Coupled Frame-Wall
Load   (Shear Mode)   (Bending Mode)   Building

(a)   (b)   (c)   (d)

Fig. 6.1 Deformation patterns due to lateral forces of a frame, a wall element, and a dual system.

ing. In particular, the common practice of allocating a portion of the lateral forces to the frames and the remainder to the walls, each of which are then independently analyzed, is entirely inappropriate. Interaction based on compatibility of deformations of the two elements *must* be considered.

Although several variants affecting the interaction of frames and walls are discussed, it is not possible here to review all possible combinations. However, the approach presented is capable of being extended to cater for unconventional solutions. Necessarily, when doing so, the use of some engineering judgment will be required.

6.2 CATEGORIES, MODELING, AND BEHAVIOR OF ELASTIC DUAL SYSTEMS

In the following, some different categories of interacting frames and structural walls are described, and appropriate analytical modeling techniques for the assessment of their elastic response are reviewed briefly. The results of such analyses may be used as the basis for allocating member strength. However, in some cases significant adjustment will need to be made. Suggestions are made for choices of suitable energy-dissipating systems in dual systems.

6.2.1 Interacting Frames and Cantilever Walls

Figure 6.2(*a*) shows in plan the somewhat idealized disposition of frames and walls in a 12-story symmetrical example structure. The properties of all walls and frames may be conveniently lumped into a single frame and a single cantilever wall, as shown in Fig. 6.2(*b*). Although cantilever walls are shown

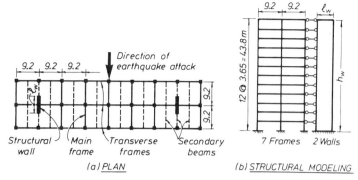

Fig. 6.2 Modeling of a typical wall frame system.

in Fig. 6.2, tubular cores [Figs. 5.4(*a*) and 5.21] or coupled structural walls [Fig. 5.13(*b*) and (*c*)], interacting with frames, are also frequently used.

As outlined in Section 1.2.3(*a*), it is customary to assume that floor slabs at all levels have infinite in-plane rigidity. Such diaphragms will then enable story displacements of all frames and walls to be established from a simple linear relationship, as shown in Fig. 1.10. However, when diaphragms are relatively slender and when large concentrated lateral story forces need to be introduced to relatively stiff walls, particularly when these walls are spaced far apart, the flexibility of floor diaphragms may need to be taken into account. This issue is reviewed briefly in Section 6.5.3.

The extensionally infinitely rigid horizontal connection between lumped frames and walls at each floor, shown by links in Fig. 6.2(*b*), enables the analysis of such lateral structures subjected to lateral forces to be carried out speedily. Initially, it will be assumed that full rotational fixity is provided by the foundation structure at the bases of both walls and columns. The influence of foundation rotation can, however, be significant, and this is considered in Section 6.2.3.

Typical results of such analyses are shown for three elastic example structures in Fig. 6.3. The buildings chosen are in plan, as shown in Fig. 6.2(*a*). They consist of seven two-bay frames and two cantilever walls. To illustrate the effects of wall stiffness on load sharing between these component structures, the length of the walls, l_w, considered was 4, 6, and 8 m, respectively (13, 19.7, and 26 ft). These represent relative wall stiffnesses of approximately 0.13, 0.42, and 1.00. Each 12-story building was subjected to identical horizontal forces derived with the use of equations in Section 2.4.3(*c*), resulting in identical total overturning moments at each level, as shown in Fig. 6.3(*a*). As expected, with increased wall stiffness (i.e., wall length) the contribution of the walls to the resistance of the base moment increases. However, at upper levels all walls become less effective and their

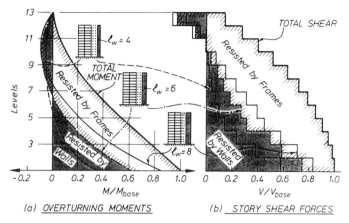

Fig. 6.3 Wall and frame contributions to the resistance of overturning moments and story shear forces in three elastic example structures.

contribution to moment resistance at midheight of the building becomes negligible. The differences between the total moment at any level and the share of the walls is then resisted by the seven frames. This sharing of the resistance of overturning moment is emphasized for the 6-m (19.7-ft) wall in Fig. 6.3(a) by shading. Because of the gross incompatibility of deformations of independent components in the upper stories, shown in Fig. 6.1(b) and (c), the frames are required to resist overturning moments at those levels that are larger than the total produced by the external lateral forces.

Figure 6.3(b) shows the sharing of horizontal story shear forces between the walls and the frames. It is seen that the more flexible the walls, the more rapidly does their contribution to shear resistance diminish with height. For example, using a 4-m (13-ft) wall, more than 80% of story shear forces above the third floor have to be resisted by the columns of the frames. Figure 6.3 emphasizes the fact that cantilever walls of hybrid systems may make significant contributions to lateral force resistance, but only in the lower stories. The results of dynamic inelastic response analyses will be compared in Section 6.3 with those of elastic analyses for static forces.

As a matter of convenience the relative contribution at the base of all cantilever walls to the shear resistance of the total lateral static forces on the building may be expressed by the ratio of the sum of the horizontal shear forces assigned to the walls and the total shear to be resisted, both values taken at the base of the structure. This wall shear ratio, η_v, [Eq. (6.11)] will be used subsequently to estimate the maximum likely wall shear demands during dynamic response. The ratio is not applicable to the moment contribution of the walls. For the three example structures chosen, the relevant values for increasing wall lengths are $\eta_v = 0.59, 0.75$, and 0.83, as Fig. 6.3(b) shows.

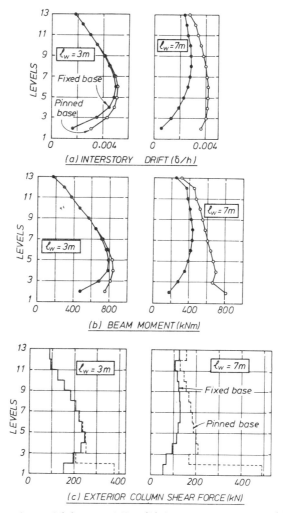

Fig. 6.4 Comparison of (*a*) story drifts, (*b*) beam moments, and (*c*) column shear forces.

The contribution of columns and beams to the total lateral force resistance of a structure, similar to that seen in Fig. 6.2, with interacting flexible and relatively stiff cantilevers, is shown by the full-line curves in Fig. 6.4(*b*) and (*c*). As expected, the patterns in the distribution with height of beam moments and column shear forces are very similar. The stiffening effect of the larger wall, in terms of the control of interstory drift, is seen when the distributions in Fig. 6.4(*a*) are compared.

As the flexural response of walls is intended to control deflections in dual systems, the danger of developing soft stories should not arise. The designer

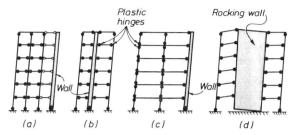

Fig. 6.5 Energy-dissipating mechanisms associated with different dual systems.

may therefore freely choose those members or localities in the frames where energy dissipation should take place. A preferable and practical mechanism for the type of frame of Fig. 6.2 is shown in Fig. 6.5(a). In this frame, plastic hinges are made to develop in all the beams and at the base of all vertical elements. At roof level, plastic hinges may form in either the beams or the columns. The main advantage of this mechanism is in the detailing of the potential plastic hinges. Generally, it is easier to detail beam rather than column ends for plastic rotation. Moreover, the avoidance of plastic hinges in columns allows lapped splices to be constructed at the bottom end rather than at midheight of columns in each upper story. The procedure is examined in detail in Section 6.4.

When long-span beams are used, and in particular when gravity loads rather than earthquake forces govern the strength of beams, it may be preferable to allow the development of plastic hinges at both ends of all columns over the full height of the structure, as shown in Fig. 6.5(c).

6.2.2 Ductile Frames and Walls Coupled by Beams

Instead of being isolated free-standing cantilevers, as shown in Fig. 6.1, structural walls may be connected by continuous beams in their plane to adjacent frames. The model of such a system is shown in Fig. 6.6(a). Beams with span lengths l_1 and l_2 are rigidly connected to the walls. These

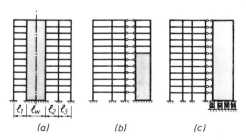

Fig. 6.6 Modeling of different types of dual systems.

structures can be modeled as frames in which beams connected to the wall are extended by infinitely rigid arms attached to the centerline of the wall, as shown in Fig. 5.14(b).

This type of system could also be utilized in the building shown in Fig. 6.2(a) if the walls were to be connected to the adjacent columns by primary lateral load-resisting beams. In that case the entire structural system would consist of seven ductile frames, shown in Fig. 6.2(b) and two coupled frame walls of the type given in Fig. 6.6(a).

Before the design of individual members can be finalized, it is necessary to identify clearly the locations in beams and columns, at which plastic hinges are intended, to enable the capacity design procedure to be applied. A possible mechanism that can be utilized in this type of system is shown in Fig. 6.5(b). Beam hinges at or close to the wall edges must develop. However, at columns, the designer may decide to allow plastic hinges to form in either the beams or the columns, above and below each level, as shown in Fig. 6.5(c).

6.2.3 Dual Systems with Walls on Deformable Foundations

It is customary to assume that cantilever walls are fully restrained against rotations at the base. It is recognized, however, that full base fixity for such large structural elements is very difficult, if not impossible, to achieve. Foundation compliance may result from soil deformations below footings and/or from deformations occurring within the foundation structure, such as piles. Base rotation is a vital component of wall deformations. Therefore, it may significantly affect the stiffness of cantilever walls and hence possibly their share in the lateral force resistance within elastic dual systems. The reluctance by designers to address the problem may be attributed to our limitations in being able to estimate reliably stiffness properties of soils (Section 9.4.6).

To illustrate the sensitivity of dual systems, some features of the elastic response [G2] of the previous example structures to lateral forces are examined briefly in the following. The 12-story example structure is shown in Fig. 6.2. Walls of length $l_w = 3$ m (9.8 ft) and 7 m (23 ft) are considered, representing common extremes in wall stiffness. The shear ratios η_v at the base, discussed in Section 6.2.1, for these two types of walls with fully fixed base are 0.44 and 0.80, respectively. Typical modeling of interacting wall frames, with walls on deformable foundations, is illustrated in Fig. 6.6(c). The springs at the wall base in Fig. 6.6(c) may conveniently be substituted with a deformable foundation beam, such as shown in Fig. 6.7. A range of base restraints have been studied [G2], but for the purpose of this discussion, only the extreme limits of the stiffness, K, shown in Fig. 6.7 are considered. Thus when $K = 0$, a wall with pinned base is assumed.

Some important and revealing results of elastic analyses are shown in Figs. 6.4 and 6.8, which compare the behavior of walls with both fixed and hinged bases. It is seen that the elastic response to lateral static forces of the

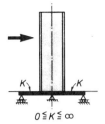

$0 \leqq K \leqq \infty$

Fig. 6.7   Modeling of partial base restraint of cantilever walls.

structure with flexible walls is not significantly affected by the degree of base restraint, but with the stiffer wall ($l_w = 7$ m) the influence of foundation compliance is considerable [Fig. 6.8(a)]. The differences in response due to extremes in base restraint become smaller at higher levels. However, as expected, drift control particularly in the lower stories, is strongly affected by the degree of wall base restraint when stiff walls are used, as seen in Fig. 6.4(a). An important change due to loss of base fixity is in the increase of shear force across columns of the first story, as seen in Fig. 6.4(c).

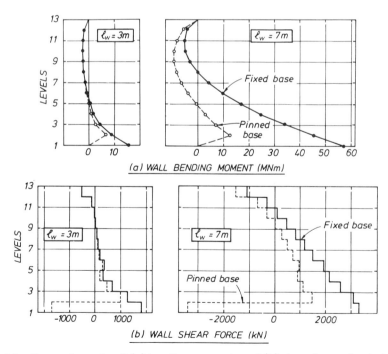

Fig. 6.8   Comparison of wall (a) bending moments and (b) shear forces due to lateral static forces for 12-story buildings with fixed- and pinned-base walls.

The effects on walls of the introduction of a hinge at the base of cantilever walls are shown in Fig. 6.8. Moments induced in flexible cantilever walls above the first floor are hardly affected by changes in base restraint [Fig. 6.8(a)]. However, profound changes occur in stiff walls, which, when fully fixed at the base, resist a very significant portion of the total overturning moment on the building. Because of the reduction on wall moments from that developed at level 2 to zero at a pinned base, a reversal of wall shear forces occurs in the first story. As Fig. 6.8(b) shows that this shear reversal is significant regardless of whether flexible or stiff walls are used. As a result of this the combined shear resistance of all the columns in the first story needs to be in excess of the total base shear applied to the dual system. This accounts for the dramatic increase of column base shear demand shown in Fig. 6.4(c) and points to the need of studying during the design the transfer of shear forces from walls to columns via the first floor, acting as a diaphragm.

It is emphasized that Figs. 6.4 and 6.8 illustrated extreme cases of base restraint which would rarely occur in real structures. The results of these simple analyses suggest, however, that when walls with moderate stiffnesses are used, so that $\eta_v < 0.5$, the effects of foundation compliance are likely to be negligible with respect to deformations, moments, and shear forces, above level 2, that would result from given lateral static forces.

6.2.4   Rocking Walls and Three-Dimensional Effects

Loss of wall base restraint will also occur if parts of footings under walls can uplift during response. In the extreme, rocking of a wall about a point close to the compression edge at the base could occur, as illustrated in Figs. 6.5(d) and 6.9. This may have profound effects on the behavior of dual systems.

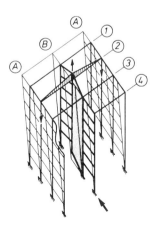

Fig. 6.9 Activation of transverse frames by rocking walls [B15].

Because of the rigid-body displacement of such walls, rotations along the height of the wall, of the same order as that at the foundation, will be introduced at every level [Fig. 6.5(*d*)], increasing ductility demand in beams of the frame at the tension side of the wall.

Similar distortion of walls at upper levels may also result, however, in walls with fixed base after plastic hinges with significant plastic rotations develop at the base. As Fig. 5.24(*b*) suggests, the deformations of the compressed concrete over the plastic hinge length are rather small in comparison with the tensile deformations at the opposite edge of the wall. The dissimilar plastic deformations at the edges of a wall result in rigid-body rotations at higher levels [Fig. 5.24(*b*)], much the same as in the case of rocking walls [B15]. As Fig. 6.9 shows, this introduces twisting into the adjacent floor system, and in particular will cause uplift of those transverse beams that frame into the tension edge of the wall.

The more important effects of wall rocking in dual systems that should be considered in the design, are:

1. The imposed deformations on beams transverse to the wall, shown in Fig. 6.9, if they have not been assigned a primary role in earthquake resistance, should be considered, particularly in the detailing of the reinforcement [B15] and the assessment of shear strength.

2. When transverse beams have substantial flexural capacity, they may introduce at every floor significant eccentric compression force at or near the tension edge of a wall. At the upward-moving parts of the walls, the top reinforcement in the transverse beams will be in tension. At these locations the effective width of the floor slab, acting as a tension flange [Section 4.5.1(*b*)], could be particularly large [Y2]. Moreover, if the wall is on an external face of a building, the eccentricity of these beam shear forces may induce significant lateral bending into the wall. Thus the flexural resistance of the wall may be significantly increased by the reactions from the transverse beams. This increase in flexural strength will in turn result in increased horizontal shear forces, which must be taken into account if, in accordance with capacity design principles, a premature shear failure of the wall is to be prevented [M12].

3. Increased axial load on walls may also necessitate a review of the required transverse confining reinforcement in the wall sections within the plastic hinge length [Section 5.4.3(*e*)].

4. Tilting walls of the type shown in Fig. 6.9 may mobilize additional energy-dissipating mechanisms: for example, in transverse beams. Plastic hinges in all beams attached to the tension edge in the plane of the wall may be subjected to increased plastic rotations, as shown in Fig. 6.5(*d*).

5. Shear forces introduced to uplifted transverse beams will cause axial tension in columns at the other end of these beams. If this axial tension is not

taken into account, plastic hinges may develop in these columns, which may not have been detailed accordingly [Y2].

6. When attention is paid to details of the three-dimensional effects above, the contribution of tilting walls to overall seismic response is beneficial [M12].

6.2.5 Frames Interacting with Walls of Partial Height

Although in most buildings structural walls extend over the full height, there are cases when for architectural or other reasons, walls are terminated below the level of the top floor. A model of such a structure is shown in Fig. 6.6(b).

Because of the abrupt discontinuity in total stiffnesses at the level where walls terminate, the seismic response of these structures is viewed with some concern. Gross discontinuities are expected to result in possibly critical features of dynamic response which are not predicted by routine elastic analyses for static forces. It is suspected that the regions of discontinuity may suffer premature damage and that local ductility demands during the largest expected earthquakes might exceed the ability of affected components to deform in the plastic range without significant loss of resistance.

On the other hand, elastic analyses for lateral static forces show that structural walls in the upper stories may serve no useful structural purpose. Figure 6.3 suggests that the termination of walls below the top floor may in fact reduce the force demand on the frames in the upper stories. In the following the response to static forces of some example structures, similar to those used previously, are compared to enable some general conclusions to be drawn with respect to effects of partial height walls.

The geometric and stiffness properties of these 12-story example structures are assumed to be the same as those of their parent structures, considered in the preceding section. Both fully fixed- and pinned-base walls with lengths of 3 and 7 m (9.8 and 23 ft) are considered and the same lateral forces are used for each structure, to allow a meaningful comparison to be made. Wall heights of 0, 3, 6, 9, and 12 stories are considered. Wall elements are assumed to extend to the chosen heights continuously from level 1. The effects of the termination of the lower end of walls above level 1 were not considered in this study. As discussed in Section 1.2.3(c), these types of structures should be viewed as being unsatisfactory in seismic areas. For this reason they are not examined here.

Decreasing the wall height has little effect on drifts experienced by the example dual systems with 3-m (9.8-ft)-long fixed-base walls. This is because of the relatively large flexibility of these walls. Figure 6.10(a) shows the band of envelopes for story drifts encountered for the five different types of dual systems with 3-m (9.8-ft)-long pinned- or fixed-base walls. In the case of structures with 7-m (23-ft) fixed base walls, the reduction in wall height alters the drift pattern more dramatically, as Fig. 6.10(c) shows. Drift at levels

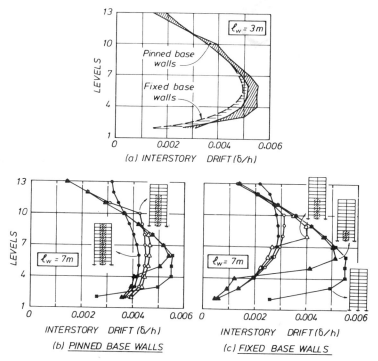

Fig. 6.10 Interstory drift distributions for dual systems with walls of different heights and base restraints.

above the shortened walls tend to converge rapidly to those occurring in the pure frame structure. Relaxation of wall base fixity, in this case the introduction of a pin at the base of 7-m (23-ft) walls, increases wall flexibility and hence lessens the influence of wall height. This is shown in Fig. 6.10(*b*). An interesting but not unexpected feature of the drift distributions, shown in Fig. 6.10(*b*) and (*c*), is the fact that the absence of a wall element in the top stories of the buildings actually leads to reduced drifts in the upper two or three stories.

Patterns of story drift distributions for structures with different wall heights and wall base conditions provide a good indication also of the response of the frames. Column shear forces, in particular moments at beam ends, are approximately proportional to story drifts. Hence it is not surprising that any variation in the boundary conditions of the 3-m (9.8-ft) walls does not affect beam moments significantly at any level. Appreciable changes in column shear forces occur only in the lower two stories. The more dramatic changes resulting from the contribution of the 7-m (23-ft) walls with a fixed base are shown in Fig. 6.11(*a*). As may be seen, these changes are similar to

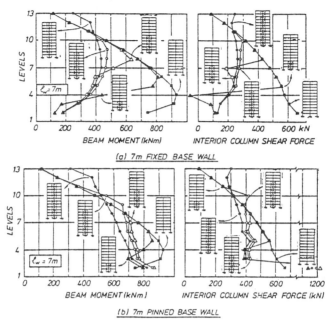

Fig. 6.11 Beam moments and column shear forces as affected by the height of 7 m (23 ft) walls with different base restraints.

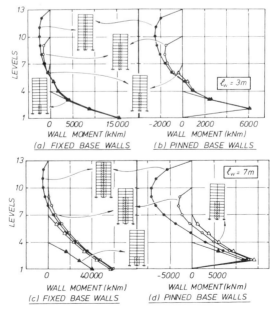

Fig. 6.12 Moments for walls of different heights, stiffnesses, and base restraints.

512

those experienced with story drifts, shown in Fig. 6.10(c). This similarity also exists for pinned-base walls, as Figs. 6.11(b) and 6.10(b) show.

Bending moments generated in the walls of these example dual systems are compared in Fig. 6.12. It is seen that while the magnitudes of moments at the lower levels are sensitive to wall stiffness, the influence of variable wall heights on moments in the critical regions are negligible. In conclusion, it may be said that for identical wall base restraints, the contribution of partial-height walls to both flexural and shear resistance of lateral static forces in the critical lower half of the structure is not affected significantly by the height of the walls.

6.3 DYNAMIC RESPONSE OF DUAL SYSTEMS

The dual systems described in the preceding section were designed in accordance with the procedure described in Section 6.4, and subsequently subjected to dynamic inelastic time-history analyses under the El Centro 1940 NS accelerogram to investigate possible differences between the dynamic response and the static response predicted in the preceding section [G2]. The most important results of these analyses are listed below and summarized in Figs. 6.13 and 6.14.

1. The natural period was lengthened by pinning the wall base. This was particularly relevant for the 7-m (23-ft) wall. This could affect seismic design forces.

2. Roof-level maximum displacements were not influenced significantly by the degree of base fixity.

3. The degree of base fixity influenced the vertical distribution of inter-story drifts, but not the peak magnitude. For fixed-base walls, maximum interstory drifts occurred in the upper floors, while for pinned-base walls, maximum interstory drifts occurred in the lower third of the building.

4. Wall bending moments envelopes [Fig 6.13(a)] bear little resemblance to those resulting from static analysis [Fig 6.8(a)]. Of particular importance is the lack in the dynamic response of a region of low moment in the midheight region in the vicinity of the point of contraflexure predicted by elastic analysis. The discrepancy results from the importance of higher mode effects on wall moments in the upper levels.

5. Wall shear forces, shown in Fig. 6.13(b), do not indicate the high base shear force predicted for pinned-base walls by the elastic analyses and shown in Fig. 6.8(b). Shear forces at upper levels of the walls exceed considerably elastic analysis predictions, again as a result of high mode effects. An explanation of this phenomenon in cantilever walls was given in Section 5.4.4(a). It should be noted, however, that wall shear forces at these levels

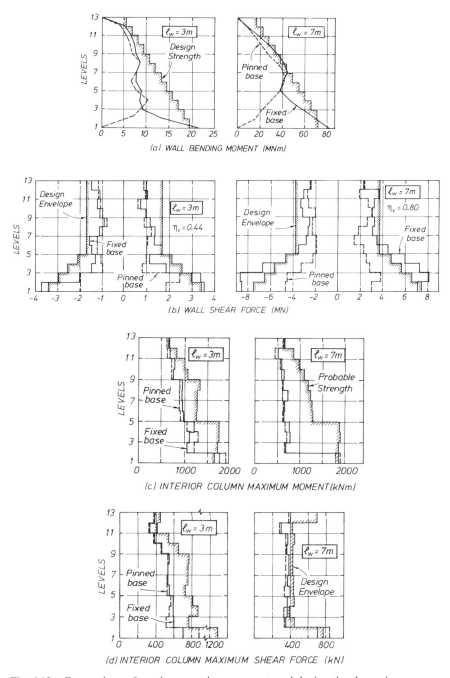

Fig. 6.13 Comparison of maximum actions encountered during the dynamic response of the example dual systems with recommended design values.

predicted by elastic analyses are very small, and hence a very large increase during dynamic response does not necessarily lead to critical design values.

6. Interior column moments and shears [Fig. 6.13(c) and (d), respectively] are less than predicted by the elastic analyses. Although column shear forces in the first story were consistently larger when pinned bases were used, the increases did not approach the magnitudes suggested by the elastic analysis shear patterns. At higher levels, the differences between fixed- and pinned-base conditions for the wall had a negligible influence on the column forces. It is emphasized that the results presented in Fig. 6.13 relate to one earthquake record and one structure. Different trends may result in other cases [G2]. In particular, period shift with shorter and stiffer structures may be accompanied by increased intensity of response, as discussed in Section 2.3.

To evaluate the relevance to design for moments, shears, and so on, derived from elastic analyses for static forces, a number of dual systems with partial height walls were analyzed using the El Centro 1940 NS record. The example buildings with 0-, 3-, 6-, 9-, and 12-story walls with both fully fixed and pinned bases were designed in the manner described in Section 6.4. The "frame only" system was designed for only 80% of the lateral forces applied to the dual systems. Maximum interstory drift indices for the structures are shown in Fig. 6.14. These generally confirm trends displayed by the elastic analyses (Fig. 6.10). Although some increase in drift occurs above the top of partial-height walls, as expected, the increases are moderate. Pinned-wall buildings indicate drift indices that are more constant over the height of the structure, with lesser discontinuities in the envelopes at points of wall termination.

These analyses [G1, G2] showed that good seismic response may also be expected from well-detailed frame-wall structures in which the wall compo-

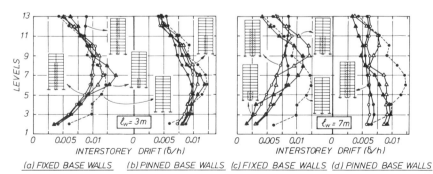

Fig. 6.14 Interstory drift envelopes for 12-story buildings with partial-height walls (El Centro, 1940).

nents extend from ground level to only partway up the building. Unduly high ductility demands in frame members near wall termination points were not observed. Also, loss of wall base restraint, investigated by considering the behavior of the extreme cases of pinned-base wall structures, did not suggest seriously impaired performance, as was also found for the full-height walls in dual systems discussed earlier in this section. Trends presented above have been confirmed by other studies using different earthquake records [S8].

6.4   CAPACITY DESIGN PROCEDURE FOR DUAL SYSTEMS

As explained earlier, the dominant feature of the capacity design strategy is the a priori establishment of a rational hierarchy in strength between components of the structural system. Accordingly, the approach to the design of each primary lateral-force-resisting component of a dual system, which is to be protected against yielding or a brittle failure, such as due to shear, can be described with the simple general expression for the required ideal strength, S_i:

$$S_i \geq \omega \phi_o S_E \tag{6.1}$$

where S_E is the required strength of the member selected for energy dissipation, as determined by elastic analysis techniques for the appropriate code-specified lateral forces; ϕ_o is the overstrength factor, defined by Eqs. (1.11) and (1.12); and ω is a dynamic magnification factor that quantifies deviations in strength demands on the member to be protected during inelastic earthquake response from the demand indicated by the elastic analysis. The design steps that follow are similar to those listed for ductile frames in Section 4.6.8 and for structural walls in Sections 5.5 and 5.6.

Step 1. Derive the bending moments and shear forces for all members of the dual system subjected to the code-specified lateral forces only. These actions are subscripted "E." The lateral static forces are obtained with the principles outlined in Section 2.4.3. In the analysis of the elastically responding structure, due allowance should be made for the effects of cracking on the stiffness of both frame members (Section 4.1.3) and walls [Section 5.3.1(a)]. Both frames and walls are generally assumed to be fully restrained against rotations at their base. However, when warranted, provisions should be made for partial elastic restraint at wall bases (Fig. 6.7).

Step 2. Superimpose the beam bending moments obtained in step 1 upon corresponding beam moments derived for appropriately factored gravity loading on the structure. Load factors chosen must be those specified in the relevant design code. Details of the superposition of actions due to gravity and lateral forces, with the use of a particular set of load factors [X8], are given in Section 4.3.

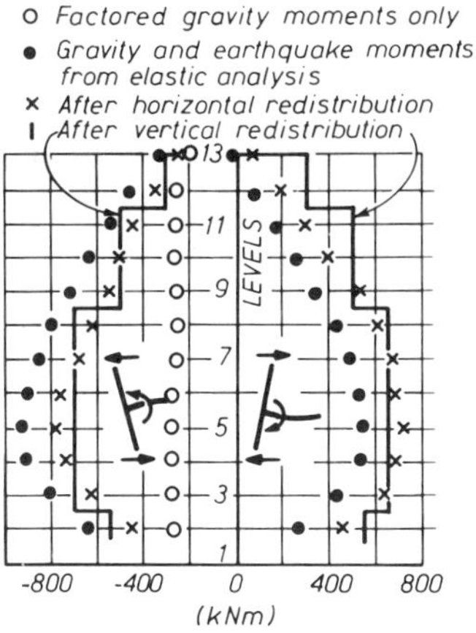

○ Factored gravity moments only
● Gravity and earthquake moments
   from elastic analysis
× After horizontal redistribution
I After vertical redistribution

Fig. 6.15 Redistribution of design moments among beams framing into an exterior column of a dual system.

Step 3. If advantageous, redistribute design moments obtained in step 2 horizontally at a level between any or all beams in each bent, and rationalize beam moments in the same span at different levels. The concepts, aims, and techniques of moment redistribution along continuous beams in frames have been examined in Section 4.3. It was pointed out that most design aims can readily be achieved if the reduction of peak moments in any beam span resulting from redistribution does not exceed 30%, provided that equilibrium criteria are not violated.

One of the advantages that may result from moment redistribution along beams is the reduction of a peak negative beam moment at an exterior column, which is, for example, associated with the moment combination $U = D + 1.3L_r + \overleftarrow{E}$. The reduction is achieved at the expense of increasing the (usually noncritical positive) moment at the same section associated with the moment combination $U = D + 1.3L_r + \overrightarrow{E}$. In the latter case the gravity and earthquake moments, superimposed in step 2, oppose each other. An example in Fig. 6.15 shows the magnitudes of beam design moments at each level at an exterior column of the frame in Fig. 6.2 at various stages of the analysis. The gravity moments (always negative), shown by circles, are changed by the addition of earthquake moments $\overleftarrow{E}$ or $\overrightarrow{E}$, to values shown by solid circles, where both sets of moments were derived from elastic analyses.

If moment redistribution is carried out between beams at each floor, for the example in the structure of Fig. 6.2, the beam moments at an exterior column may be changed to those shown by crosses in fig. 6.15. It is seen that the negative and positive moment demands are now comparable in magnitude. It may be recalled that this redistribution of beam moments also involves the redistribution of shear forces between columns without, however, changing the total shear to be resisted by all columns of the bent. To optimize practicality of design, whereby beams of identical strength are preferred over the largest possible number of adjacent levels, some rationalization in the vertical distribution of beam design moments should also be considered. In the example of Fig. 6.15 the design moments shown by crosses may be adjusted so as to result in magnitudes shown by the continuous stepped lines. It is seen that beams of the same flexural strength could be used over several floors. The stepped line has been chosen in such a way that the area enclosed by it is approximately the same as that within the curve formed by the crosses.

Vertical rationalization of beam design moments implies some changes also in the column moments. Hence the total shear assigned to columns of a particular story may decrease (the fifth story in Fig. 6.15), while in other stories (the second story in Fig. 6.15) it will increase. To ensure that there is no decrease in the total story shear resistance intended by the code-specified lateral forces, there must be a horizontal redistribution of shear forces between vertical elements of the structure (i.e., columns and walls). It will be shown subsequently that the upper regions of walls will be provided with sufficient shear and flexural strength to readily accommodate additional shear forces shed by upper-story columns. The principles involved here are similar to those used in the design of coupling beams of coupled structural walls and shown in Fig. 5.23(c).

To safeguard against premature yielding in beams during small earthquakes, the reduction of beam moments resulting from moment redistribution and rationalization should not exceed 30%. Other load combinations, such as gravity load alone, must also be checked to ensure that beam strength is based on the most critical load combination.

Step 4. Design all critical beam sections so as to provide required flexural strengths, and detail the reinforcement for all beams in all frames. These routine steps require determination of the size and number of reinforcing bars to be used to resist moments along all beams in accordance with the demands of moment envelopes obtained after moment redistribution. It is important at this stage to locate the two potential plastic hinges in each span [Fig. 6.5(a)] for each of the two directions of earthquake attack. In locating plastic hinges that require the bottom (positive) flexural reinforcement to yield in tension, moment combinations $U = D + 1.3L_r + E$ and $U = 0.9D + E$ should both be considered, as each combination may indicate a different hinge position. Detailing of the beams should then be carried out in accordance of the principles of Section 4.5.

Step 5. In each beam determine the flexural overstrength of each of the two potential plastic hinges corresponding with each of the two directions of earthquake attack. The procedure, incorporating allowance for strain hardening of the steel and the possible participation in flexural resistance of all reinforcement present in the structure as built, is the same as that used in the design of beams of ductile frames (Section 4.5.1). The primary aim is to estimate the maximum possible moment input from beams to adjacent columns.

Step 6. Determine the lateral displacement-induced vertical shear force. V_{Eo}, associated with the development of flexural overstrength of the two plastic hinges in each beam span, for each of the two directions of earthquake attack. These shear forces are readily obtained [Section 4.5.3(a)] from the flexural overstrengths of potential plastic hinges, determined in step 5. When combined with gravity-induced shear forces, the design shear envelope for each beam span is obtained, and the required shear reinforcement can then be determined [Section 4.5.3(b)]. The horizontal displacement-induced maximum beam shear forces, V_{Eo}, are used subsequently to determine the maximum lateral displacement-induced axial column load input at each floor.

Step 7. Determine the beam flexural overstrength factor ϕ_o, at the centerline of each column at each level for both directions of earthquake attack. The evaluation of the factor ϕ_o at each node point of a frame, subsequently used to estimate the maximum moments that could be introduced to columns by fully plastified beams, is described in detail in Section 4.6.3. The beam moments at column centerlines can readily be obtained graphically from the design bending moment envelopes, after the flexural overstrength moments at the exact locations of the two plastic hinges along the beam have been plotted (Fig. 4.18).

Step 8. Evaluate the column design shear forces in each story from

$$V_u = V_{\text{col}} = \omega_c \phi_o V_E \tag{6.2}$$

where the column dynamic shear magnification factor, ω_c, is 2.5, 2.0, and 1.3 for the bottom, top, and intermediate stories, respectively. The design shear force in the bottom-story columns should not be less than

$$V_u = V_{\text{col}} = \frac{M_{o,\text{col}} + 1.3\phi_o M_{E,\text{top}}}{l_n + 0.5h_b} \tag{6.3}$$

where $M_{o,\text{col}}$ = flexural overstrength of the column base section consistent with the axial force and moment associated with the direction of earthquake attack

$M_{E,\text{top}}$ = value of M_E for the column at the centerline of the beams at level 2

l_n = clear height of the column

h_b = depth of the beam at level 2

The procedure for the evaluation of column design shear forces is very similar to that used in the capacity design of ductile frames in Section 4.6.7.

It reflects a higher degree of conservatism because of the intent to avoid a column shear failure in any event. Case studies [G2, G3] show that despite the apparent severity of Eqs. (6.2) and (6.3), shear requirements for columns are very seldom critical because of the very low value of V_E predicted by elastic analysis (step 1).

Step 9. Estimate in each story the maximum probable lateral displacement-induced axial load on each column from

$$P_{Eo} = R_v \Sigma V_{Eo} \tag{6.4}$$

where

$$R_v = (1 - n/67) \geq 0.7 \tag{6.5}$$

is a reduction factor that takes into account the number of floors, n, above the story under consideration. The magnitudes of the maximum lateral displacement-induced beam shear forces V_{Eo} at each level were obtained in step 6. The probability of all beams above a particular level simultaneously developing plastic hinges at flexural overstrength diminishes with the number of floors above that level. The reduction factor R_v makes an approximate allowance for this. Equation (6.5) gives the same values as in Table 4.5 when the dynamic moment magnification, ω, is 1.3 or less.

Step 10. Determine the total design axial force on each column for each of the two directions of earthquake attack from

$$P_{u,\text{max}} = P_D + P_{Lr} + P_{Eo} \tag{6.6}$$

and

$$P_{u,\text{min}} = 0.9P_D - P_{Eo} \tag{6.7}$$

where P_D and P_{Lr} are axial forces due to dead and reduced live loads, respectively. The procedure in this step is the same as that used for columns of ductile frames, summarized in Section 4.6.8. The magnitude of P_{Eo} was obtained in step 9. Axial force combinations are in accordance with Eq. (1.7).

Step 11. Obtain the design moments for columns above and below each level from

$$M_u = R_m(\omega \phi_o M_E - 0.3h_b V_u) \tag{6.8}$$

where ω = dynamic moment magnification factor, the value of which is given in Fig. 6.16(a) when full height walls are used

ϕ_o = beam overstrength factor applicable at the level and corresponding to the direction of lateral forces under consideration

h_b = depth of the beam that frames into the column

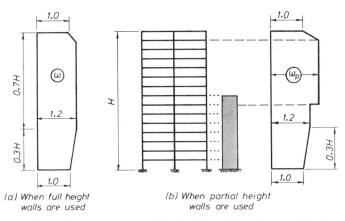

(a) When full height walls are used

(b) When partial height walls are used

Fig. 6.16 Dynamic moment magnification factor for columns of dual systems.

and
$$0.75 \leq R_m = 1 + 0.5(\omega - 1)\left(10\frac{P_u}{f_c' A_g} - 1\right) \leq 1 \qquad (6.9)$$

is a design moment reduction factor applicable where the range of axial forces is such that $-0.15 \leq P_u/(f_c' A_g) \leq 0.10$, where P_u is to be taken negative when causing axial tension.

These requirements are very similar to those recommended for columns of ductile frames in Section 4.6.5(c), shown in Fig. 4.22, and summarized in Section 4.6.8. It was outlined in Section 4.6.4 that the main purpose of the dynamic moment magnification factor, ω, was to allow for increases of column moment demands either above or below a beam due to higher mode participations of the frame during dynamic response. Modal shapes in dual systems are largely controlled by deformations of the walls. For this reason, full-height walls will protect columns to a large extent with respect to large local moment demands due to higher mode effects. Dynamic analyses have shown [G1, G2] that in the case of dual systems a value of $\omega = 1.2$ is sufficient to protect columns at upper levels against the development of plastic hinges in all cases when full-height walls are present. The probable flexural strength of columns designed in accordance with these recommendations is compared in Fig. 6.13(c) with moments predicted by dynamic analyses.

Because the value of the dynamic moment magnification factor for columns in dual systems is relatively small (i.e., $\omega \leq 1.2$), the reduction of column design moments due to small axial compression or axial tension will seldom exceed 20%. To simplify computations, the designer may use $R_m = 1.0$ [Eq. (6.9)]. When the reduction factor, R_m, is used in determining the amount of column reinforcement, the design shear V_u, obtained in step 8, may also be reduced proportionally.

Having obtained the critical design quantities for each column (i.e., M_u from step 11 and V_u from step 8), the required flexural and shear reinforcement at each critical section can be found. Because the design quantities have been derived from beam overstrengths input (Section 1.3.4), the appropriate strength reduction factor for these columns is $\phi = 1.0$. End regions of columns need further checking to ensure that the transverse reinforcement provided satisfies the requirements for confinement, stability of vertical reinforcing bars, and lapped splices, as outlined in Sections 3.6.1 and 4.6.11.

The design of columns at the base, where the development of a plastic hinge in each column must be expected, is the same as for columns of ductile frames. Partial-height walls, such as shown in Fig. 6.6(b), also provide the same degree of protection against column hinge formation as do full-height walls, but only up to one story below the top of the wall. As in all other cases of dual systems, hinge development in columns must also be expected at the base of these structures.

Columns of the dual system extending above the level of the top ends of partial-height walls enjoy less protection against hinging. Therefore, in these structures greater column flexural strength is desirable if the postulated hierarchy in column–beam strengths is also to be maintained at levels where walls are absent. However, in comparison with frames without walls, some protection for columns in stories, above the level at which partial-height walls terminate, does exist. Therefore, it is suggested that the maximum value of the dynamic magnification factor for columns of dual systems with partial-height walls, ω_p, be linearly interpolated between values relevant to pure frames, ω, given in Table 4.3, and 1.2, given in Fig. 6.16(a):

$$\omega_p = \omega - (h_w/H)(\omega - 1.2) \tag{6.10}$$

where h_w is the height of the wall and H is the total height of the structure. Equation (6.10) applies to columns from one story below the top of the wall to one story below roof level. Dynamic moment magnification factors for dual systems with partial-height walls ω_p are shown in Fig. 6.16(b).

Step 12. Determine the appropriate gravity- and earthquake-induced axial forces on walls. In the example structure (Fig. 6.2), it was implicitly assumed that lateral forces on the building do not introduce axial forces to the cantilever walls. For this situation the design axial forces due to gravity loads on the walls are typically $P_u = P_D + 1.3P_{Lr}$ or $P_u = 0.9P_D$. If the walls are connected to columns via rigidly connected beams, as shown, for example, in fig. 6.6(a), the lateral force induced axial forces on the walls are obtained from the initial elastic analysis of the structure (step 1). Similarly, this applies when, instead of cantilever walls, coupled structural walls share with frames in lateral force resistance.

Step 13. Determine the maximum bending moment at the base of each wall and design the necessary flexural reinforcement, taking into account the most

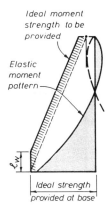

Fig. 6.17 Design moment envelopes for walls of dual systems.

adverse combination with axial forces on the wall. This simply implies that the requirements of strength design be satisfied. The appropriate combination of actions are $M_u = M_E$ and P_u from step 12. The exact arrangement of bars within each wall section at the base, as built, must be determined to allow the flexural overstrength [Section 5.4.2(d)] of the section to be estimated.

Step 14. When curtailing the vertical reinforcement at upper levels of walls, provide flexural resistance not less than that given by the moment envelope in Fig. 6.17. The envelope shown is similar to but not the same as that recommended for cantilever walls (Fig. 5.29). It specifies slightly larger flexural resistance in the top stories by relating the wall top moment to the maximum reversed moment predicted by the elastic analysis in step 1 [see Fig. 6.3(a)]. It is important to note that the envelope is related to the ideal flexural strength of a wall at its base, as detailed, rather than the moment required at that section by the analysis for lateral forces. The envelope refers to effective ideal flexural strength. Hence vertical bars in the wall must extend by at least the full development length beyond levels indicated by the envelope. The aims in choosing this design envelope were discussed in Section 5.4.2(c).

Figure 6.13(a) compares wall moment demands encountered during the analysis for the El Centro 1940 earthquake records with the moment envelopes derived from Fig. 6.17. Although moment demands may approach the suggested design envelope, curvature ductility demands have been found at upper levels to be negligibly small, and the formation of a plastic hinge at those levels is considered to be unlikely.

Where flexibility of the wall base (foundations) is considered in the analysis in step 1, patterns of wall moments will result which are between the limits found for full and zero base fixity, as seen in Fig. 6.8(a). The designer may then choose to allow the formation of a wall plastic hinge at level 2 if it is found that the moment predicted by the elastic analysis for that level is higher than at level 1. Using the principles relevant to Fig. 6.17, a moment envelope can readily be constructed to ensure that the formation of plastic hinges above level 2 is precluded. As Fig. 6.13(a) suggests, wall moments at higher levels of dual systems are not sensitive to the degree of base fixity.

Step 15. Determine the magnitude of the flexural overstrength factor $\phi_{o,w}$ for each wall. The meaning and purpose of this factor, $\phi_{o,w} = M_{Eo}/M_E$ was outlined for cantilever walls in Section 5.4.2(d). Strictly, for walls there are two limiting values of flexural overstrength, M_{Eo}, which could be considered. These are the moments developed in the presence of two different axial force intensities (i.e., $P_{u,\max}$ and $P_{u,\min}$) found in step 10. However, it is considered to be sufficient for the intended purpose to evaluate flexural overstrength developed with axial compression on cantilever walls due to dead load alone.

Step 16. Compute the wall shear ratio η_v. For convenience the relative contribution of all walls to the required total lateral force resistance at the base V_E is expressed by the shear ratio

$$\eta_v = \left(\sum_{i=1}^{n} V_{i,\text{wall},E} / V_{E,\text{total}} \right)_{\text{base}} \tag{6.11}$$

As outlined in Section 6.2.1, it applies strictly to the base of the structure [Fig. 6.3(b)].

Step 17. Evaluate for each wall the design shear force, V_u, at the base from

$$V_u = V_{\text{wall,base}} = \omega_v^* \phi_{o,w} V_{\text{wall},E} \tag{6.12}$$

and

$$\omega_v^* = 1 + (\omega_v - 1)\eta_v \tag{6.13}$$

where ω_v is the dynamic shear magnification factor relevant to cantilever walls, obtained from

$$\omega_v = 0.9 + n/10 \quad \text{when } n \le 6 \tag{5.23a}$$

or $\qquad \omega_v = 1.3 + n/30 \le 1.8 \quad \text{when } n > 6 \tag{5.23b}$

where n is the number of stories above the wall base.

The approach developed for the shear design of walls in dual systems is an extension of the two-stage methodology used for cantilever walls and developed in Section 5.4.4. In the first stage, the design shear force is increased from the initial (step 1) value, to that corresponding with the development of a plastic hinge at flexural overstrength at the base of the wall. This is achieved with the introduction of the flexural overstrength factor, $\phi_{o,w}$, obtained in step 15. In the next stage, allowance is made for amplification of the base shear force during the inelastic dynamic response of the structure. While a plastic hinge develops at the base of a wall, due to the contribution of higher modes of vibration, the centroid of inertia forces over the height of the building may be in a significantly lower position than that predicted by the conventional analysis for lateral forces (Fig. 5.41). The larger the number of stories used as an approximate measure of the fundamental period of vibration T, the more important is the participation of higher modes. The dynamic shear magnification factor for cantilever walls, ω_v, given in Eq. (5.23), makes allowance for this phenomenon.

It has also been found [G2] that for a given earthquake record, the dynamically induced base shear forces in walls of dual systems increased with an increased participation of such walls in the resistance of the total base shear for the entire structure. Wall participation is quantified by the shear ratio, η_v, obtained in step 16. The effect of the shear ratio upon the magnification of the maximum wall shear force is estimated by Eq. (6.13). It is seen that when $\eta_v \leq 1$, $1 \leq \omega_v^* \leq \omega_v$.

Design criteria for shear strength will often be found to be critical. At the base the thickness of walls may need to be increased on account of Eq. (6.12) in conjunction with the maximum shear stress limitations given by Eqs. (3.31) and (5.26). Typically, when using grade 400 (58 ksi) vertical wall reinforcement in a 12-story hybrid structure, where the walls have been assigned 60% of the total base shear resistance, it will be found that with $\phi_{o,w} \simeq 1.6$, $\phi = 1.0$, $\omega_v = 1.7$, $\eta_v = 0.6$, and $\omega_v^* = 1.42$, the ideal shear strength will need to be $V_{\text{wall}} = 1.42 \times 1.6 \times V_E = 2.27 V_E$. Thus Eq. (6.12) implies very large apparent reserve strength in shear. However, dynamic analyses of dual systems [G2] consistently predicted shear forces similar to, or exceeding, those required by Eq. (6.12).

Step 18. In each story of each wall, provide shear resistance not less than that given by the shear design envelope of Fig. 6.18. As Fig. 6.3(b) shows, shear demands predicted by analyses for static load may be quite small in the upper halves of walls. As can be expected, during the response of the building to vigorous seismic excitations, much larger shear forces may be generated at these upper levels. A linear scaling up of the shear force diagram drawn for static lateral force, such as shown in Fig. 6.3(b), in accordance with Eq. (6.12), would give an erroneous prediction of shear demands in the upper stories. The shear design envelope shown in Fig. 6.18 was developed from dynamic analyses of dual systems. It is seen that the envelope gives the required shear strength in terms of the base shear for the wall, which was obtained in step 17. It must be appreciated that lateral static forces prescribed by codes give poor prediction of shear demands on walls during earthquakes [B15].

Wall shear forces encountered during the response of the previously quoted example dual system to the El Centro 1940 NS record are compared

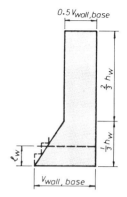

Fig. 6.18 Envelope for design shear forces for walls of dual systems.

with the recommended design shear envelope in Fig. 6.13(*b*). With the aid of the design shear envelope, the required amount of horizontal (shear) wall reinforcement at any level may readily be found. In this, attention must be paid to the different approaches used to estimate the contribution of the concrete to shear strength, v_c, in the potential plastic hinge and the elastic regions of a wall [Eqs. (3.39) and (3.36)].

Step 19. In the potential plastic hinge region of the walls, provide adequate transverse reinforcement to supply the required confinement to parts of the flexural compression zone and to prevent premature buckling of vertical bars. These detailing requirements for ductility are the same as those recommended for cantilever and coupled structural walls, presented in detail in Section 5.4.

6.5 ISSUES OF MODELING AND DESIGN REQUIRING ENGINEERING JUDGMENT

The proposed capacity design procedure and the accompanying discussion of the behavior of dual systems, presented in the previous sections, are by necessity restricted to simple and regular structural systems. The variety of ways in which walls and frames may be combined may present problems to which a satisfactory solution will require, as in many other structures, the application of engineering judgment. This may necessitate some rational adjustments in the outlined 19-step procedure. In the following, a few situations are mentioned where judgment in the application of the proposed design methodology will be necessary. Some directions for promising approaches are also suggested.

6.5.1   Gross Irregularities in the Lateral-Force-Resisting System

It is generally recognized that the larger the departure from symmetry and regularity in the arrangement of lateral-force-resisting elements within a building, the less confidence the designer should have in predicting likely seismic response. Examples of irregularity were given in Section 1.2.3.

6.5.2   Torsional Effects

Codes make simple provisions for torsional effects. The severity of torsion is commonly quantified by the distance between the center of rigidity of the lateral-force-resisting structural system and the center of mass discussed in Section 1.2.3(b). In reasonably regular and symmetrical buildings, this distance (horizontal eccentricity) does not change significantly from story to story. Errors due to inevitable variations of eccentricity over building height and effects during torsional dynamic compliance are thought to be compensated for by code-specified amplifications of the computed (static) eccentricities [Section 2.4.3(g)]. The corresponding assignment of additional lateral force to resisting elements, particularly those situated at greater distances from the center of rigidity (shear center), are intended to compensate for torsional effects. Because minimum and maximum eccentricities, at least with respect to the two principal (orthogonal) directions of earthquake attack, need to be considered, the structural system, as designed, will possess increased translational resistance when compared with an identical system in which torsion effects were ignored.

It was emphasized that the contributions of walls to lateral force resistance in dual systems usually change dramatically over the height of the building. Examples were shown in Fig. 6.3. For this reason in unsymmetrical systems the position of the center of rigidity may also change significantly from floor to floor. For the purpose of illustrating variation of eccentricity with height, consider the example structure shown in Fig. 6.2, but slightly modified. Assume that instead of the two symmetrically positioned walls shown in Fig. 6.2(a), two 6-m (19.7-ft)-long walls are placed in the vertical plane at 9.2 m (30.2 ft) from the left-hand end of the building, as shown in Fig. 6.19, and that the right-hand wall is replaced by a standard frame. The two walls, when displaced laterally by the same amount as the frames, would in this example structure resist 75% of the total shear in the first story [Fig. 6.3(b)], and the center of rigidity would be 19.03 m (62.4 ft) from the center of the building. In the eighth story the two walls become rather ineffective, as they resist only about 12% of the story shear (i.e., approximately as much as one frame). At this level the eccentricity becomes negligible. At higher levels, the frames and the walls work against each other (Fig. 6.3) in resisting lateral forces, hence inducing torsion in the opposite direction. As Fig. 6.19 shows, the computed static eccentricities would vary considerably in this example

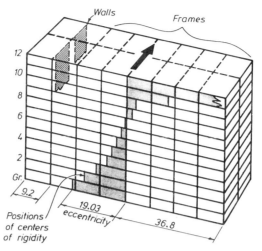

Fig. 6.19 Variation of computed torsional eccentricities in an asymmetrical 12-story dual system.

building between opposite limits at the bottom and top stories. Torsional effects on individual columns and walls will depend on the total torsional resistance of the system, including the periphery frames along the long sides of the building. Because of the complexity of response resulting from this phenomenon, it is recommended that member forces be determined from three-dimensional elastic analyses for such structures.

6.5.3   Diaphragm Flexibility

For most buildings, floor deformations associated with diaphragm actions are negligible. However, when structural walls resist a major fraction of the seismically induced inertia forces in long and narrow buildings, the effects of in-plane floor deformations on the distribution of resistance to frames and walls may need to be examined. Figure 6.20 shows plans of a building with three different positions of identical walls. The building is similar to that

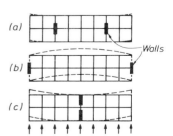

Fig. 6.20   Diaphragm flexibility.

shown in Fig. 6.2(*a*). The contribution of the two walls to total lateral force resistance is assumed to be the same in each of the three cases. Diaphragm deformations associated with each case are shown approximately to scale by the dashed lines. Diaphragm deformations in the case of Fig. 6.20(*a*) would be negligibly small in comparison with those of the other two cases. In deciding whether such deformations are significant, the following aspects should be considered:

1. If elastic response is considered, the assignment of lateral forces to some frames [Fig. 6.20(*b*) and (*c*)] would be clearly underestimated if diaphragms were to be assumed to be infinitely rigid. In-plane deformations of floors under distributed floor lateral forces, which may be based on simple approximations, should be compared with interstory drifts predicted by standard elastic analyses. Such a comparison will then indicate the relative importance of diaphragm flexibility. A diaphragm should be considered flexible when the maximum lateral deformation of the diaphragm is more than twice the average story drift in the associated story [X4].

2. In ductile structures subjected to strong earthquakes, significant inelastic story drifts are to be expected. The larger the inelastic deformations, the less important are differential elastic displacements between frames and walls that would result from diaphragm deformations.

3. As Fig. 6.3(*b*) illustrated, the contribution of walls to lateral force resistance in dual systems diminishes with the distance measured from the base. Therefore, according to the elastic analysis results, at upper levels lateral forces are more evenly distributed among frames and walls. This suggests greatly reduced diaphragm in-plane shear and flexural actions. However, dynamic analysis showed [Fig. 6.13(*b*)] that contrary to the predictions of elastic analysis, significant wall shear forces, a good measure of diaphragm action, can also be generated at upper levels.

These observations emphasize the need to pay attention [X4] to the vital role of floor systems, acting as diaphragms, details of which have been presented in Section 1.2.3(*a*). In particular it is important that adequate continuous horizontal reinforcement be provided at the edges of reinforced concrete diaphragms to ensure that, as Fig. 6.20 indicates, they can act as beams with ample flexural resistance.

6.5.4 Prediction of Shear Demand in Walls

A number of case studies of the dynamic response of structures of the type shown in Fig. 6.2, typically with walls 3 to 7 m (9.8 and 23 ft) long, have indicated that the capacity design procedure set out in Section 6.4 led to structures in which:

1. Inelastic deformations during the El Centro event remained within limits currently envisaged in most building codes. Typically story drifts did not exceed 1% of story heights.

2. Plastic hinges in the columns above level 1 were not predicted by dynamic analyses even when extremely severe earthquake records were used [G1].

3. Derived column design shear forces implied sufficient protection against shear failure without the use of excessive shear reinforcement.

4. Curvature ductility demands at the base of both columns and walls of the example dual systems, designed with an assumed displacement ductility capacity of $\mu_\Delta = 4$, remained well within the limits readily attained in appropriately detailed laboratory specimens.

5. Predicted shear demands at upper levels of walls were satisfactorily catered for by the envelope shown in Fig. 6.18. However, detailed analytical studies examining the duration of high moment and shear demand in walls of dual systems [P49] lead to the conclusion that the concern stemming from the less-than-satisfactory correlation between recommended design shear force levels for walls, with maxima obtained from analytical predictions, as seen in Fig. 6.13(*b*), could be dismissed because:

   (a) Predicted peak shear forces were of very short duration. While there was no experimental evidence to prove it, it was felt that shear failure during real earthquakes could not occur within a few hundredths of a second.

   (b) Analytical shear predictions are sensitive to the modeling techniques adopted. These in turn have certain limitations, particularly when diagonal cracking in both directions and the effect of flexural yielding on diagonal crack behavior is to be taken into account.

   (c) The probable shear strength of a wall is in excess of the ideal strength used in design.

   (d) Some increase in shear resistance during short pulses could be expected due to increased strain rates that would prevail.

   (e) Some inelastic shear deformation during the very few events of peak shear should be acceptable.

   (f) Walls and columns were found *not* to be subjected simultaneously to peak shear demands. Therefore, the danger of shear failure at the base, for the building as a whole, should not arise.

   (g) The simultaneous occurrence during an earthquake record of predicted peak shear and peak flexural demands in the walls was found to be about the same as the occurrence of peak shear demands. This means that when maximum shear demand occurred, it generally coincided with maximum flexural demands. The recommendations for the shear strength of walls in the plastic hinge region in Section 5.4.4 [Eq. (5.22)] were based on this precept.

6.5.5 Variations in the Contributions of Walls To Earthquake Resistance

The study of the seismic response of dual systems has shown, as was to be expected, that the presence of walls significantly reduced the dynamic moment demands on columns. This is because in the presence of walls, which control the deflected shape of the system, frames become relatively insensitive to higher modes of vibrations. This was recognized by the introduction of a smaller dynamic column moment magnification factor, $\omega = 1.2$, at levels above the base, as shown in Fig. 6.16.

The contribution of all walls to lateral force resistance was expressed by the wall shear ratio, η_v, introduced in design step 16. The minimum value η_v used in the example structure with two 3-m (9.8-ft) walls was 0.44. The question arises as to the minimum value of the wall shear ratio, η_v, relevant to a dual system, for the design of which the proposed procedure in Section 6.4 is still applicable. As the value of η_v diminishes, indicating that lateral force resistance must be assigned primarily to frames, parameters of the design procedure must approach values applicable to the capacity-designed framed buildings described in Section 4.6. At a sufficiently low value of this ratio, say $\eta_v \leq 0.1$, a designer may decide to ignore the contribution of walls. Walls could then be treated as secondary elements which would need to follow, without distress, displacements dictated by the response of frames. The minimum value of η_v for which the procedure in Section 6.4 is applicable has not been established. It is felt that $\eta_v = 0.33$ might be an appropriate limit. For dual systems for which $0.1 < \eta_v < 0.33$, a linear interpolation of the relevant parameters, applicable to ductile frames and ductile dual systems, seems appropriate. These parameters are ω_c, ω_v^*, and R_m.

7 Masonry Structures

7.1 INTRODUCTION

A fundamental issue to be resolved is whether elastic or strength design approaches should be used. Masonry has lagged behind other materials in the adoption of a strength, or limit-states design approach, and is still generally designed by traditional methods to specified stress levels under service loads. The reason that codes have not moved to strength design appears to be in the dubious belief that behavior of masonry structures can be predicted with greater precision at service load levels than at ultimate levels. There is, in fact, little evidence to support this belief. At service load levels, the influence of shrinkage (or swelling), creep, and settlement will often mean that actual stress levels are significantly different from values predicted by elastic theory. Further, the "plane-sections-remain-plane" hypothesis may be invalid in many cases, particularly for squat masonry walls under in-plane loading. Ultimate strength behavior is, however, rather insensitive to these aspects, so moments and shears at ultimate strains can be predicted with comparative accuracy. There is now adequate test information, particularly for reinforced masonry [M18, P55, P56, P57], to support the application of strength methods developed for concrete, to masonry structures, whether brickwork or blockwork.

Elastic theory has drawbacks. In specifying elastic theory, many codes persist in treating the combination of axial load and bending moment on masonry compression members by requiring that stresses satisfy

$$f_a/F_a + f_b/F_b \leq 1.0 \tag{7.1}$$

where f_a and f_b are the computed stresses under the axial and bending moments, calculated independently, and F_A and F_B are the permitted stress levels for pure axial load and pure bending. As is well known, this approach, implying that direct superposition of stresses is applicable, is invalid when analysis of cracked sections is used for flexure. It results in extremely conservative designs.

Two examples are presented below to examine further inconsistencies inherent in elastic design.

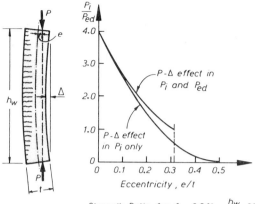

Fig. 7.1 Elastic and ultimate loads compared for an eccentrically loaded unreinforced slender wall.

1. *Eccentrically Loaded Unreinforced Slender Wall.* Figure 7.1 examines the behavior of a slender unreinforced wall subjected to vertical load with end eccentricity = e. Moments at midspan are increased by the lateral displacement Δ. Sahlin [S11] presents an exact solution for this case, but a simpler, approximate solution for load capacity based on elastic buckling may be found with the use of notation in Fig. 7.1.

$$P = 0.75 f_c t \left[y + \sqrt{y^2 - \frac{f_c}{6E_m} \left(\frac{h_w}{t} \right)^2} \right] \qquad (7.2)$$

where $y = \frac{1}{2} - e/t$ is the dimensionless distance from the extreme compression fiber to the line of action of the load at the top and bottom of the wall, h_w/t is the slenderness ratio, and f_c is the stress at the extreme compression fiber at midheight. Equation (7.2) applies for $e > t/6$, and is always accurate to within 4% of the exact solution. It may be solved for a specified maximum allowable stress (elastic design), or solved for maximum P, either by using trial values of f_c or by differentiating Eq. (7.2) to find the value of f_c for maximum P (i.e., ultimate strength). Figure 7.1 compares results for a wall of slenderness ratio $h_w/t = 25$, $E_m = 600 f'_m$, maximum allowable stress for elastic design = $0.2 f'_m$, and different levels of end eccentricity. Two curves are shown, one where the maximum elastic design load P_{ed} is based on solution of Eq. (7.2) for $f_c = 0.2 f'_m$, and the other where additional moments due to the P–Δ effect are ignored. The ultimate value for zero eccentricity was based on the Euler buckling load. It will be seen that where P–Δ effects are included in the estimate of both elastic and ultimate load, the

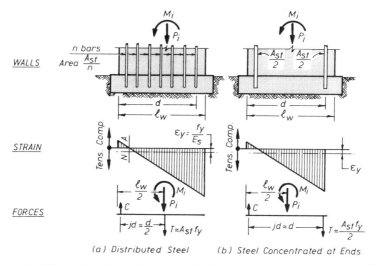

Fig. 7.2 Effect of steel distribution on the flexural capacity of masonry walls.

curve ends at $e/t = 0.313$, since Eq. (7.2) indicates instability for maximum stress less than $f_c = 0.2f'_m$ at higher eccentricities. It is apparent that elastic theory provides inconsistent protection against failure, with elastic design becoming progressively unsafe as end eccentricity increases.

2. *Distribution of Flexural Reinforcement in Masonry Walls.* As a second example of unsound results obtained from elastic theory, the behavior of reinforced masonry walls is examined. Figure 7.2 shows two walls of identical dimensions and axial load level, reinforced with the same total quantity of flexural reinforcement, A_{st}. In Fig. 7.2(*a*), this reinforcement is uniformly distributed along the wall length, while in Fig. 7.2(*b*), the reinforcement is concentrated in two bundles of $A_{st}/2$, one at each end of the wall. Elastic theory indicates that the arrangement shown in Fig. 7.2(*b*) is more efficient, typically resulting in an allowable moment about 33% higher than for the distributed reinforcement of Fig. 7.2(*a*). However, for the typically low steel percentages and low axial loads common in masonry buildings, the flexural capacity is insensitive to the steel distribution. For uniformly distributed reinforcement [Fig. 7.2(*a*)] the small neutral-axis depth will ensure tensile yield of virtually all vertical reinforcement. With the notation given in Fig. 7.2, this results in an ideal flexural strength of

$$M_i \approx A_{st} f_y \frac{d}{2} + P_i \frac{l_w}{2} \qquad (7.3)$$

For reinforcement concentrated near the ends of the wall, the tension force, at $\frac{1}{2}A_{st}f_y$ is approximately half that for the distributed case but at roughly twice the lever arm, so the flexural capacity remains effectively unaltered.

For typical levels of axial load and flexural reinforcement content, the difference in ultimate moment capacity of the two alternatives of Fig. 7.2 will be less than 5%. As will be shown later, there are good reasons for adopting an even distribution of flexural reinforcement rather than concentrating bars at the wall ends. This option would be difficult to choose if the design was to allowable working stress levels.

The considerations discussed above indicate that strength design is more likely than an elastic design to produce consistently safe masonry buildings. When seismic forces are considered, the case for strength becomes overwhelming. As discussed in Section 2.4, most codes specify seismic lateral force coefficients that are reduced from elastic response levels, typically by a factor of about 4, implying considerable ductility demand. A masonry building designed to allowable stress levels at the code level of lateral force may still attain its ideal strength under the design earthquake, but with a reduction in the required structure ductility level. It is thus clear that elastic design does not protect against inelastic action. A more realistic approach is to accept that the ideal strength of a masonry structure will be attained and to design accordingly by ensuring that the materials and structural systems adopted are capable of sustaining the required ductility without excessive strength or stiffness degradation.

7.2 MASONRY WALLS

7.2.1 Categories of Walls for Seismic Resistance

(a) Cantilever Walls Figure 7.3(a) illustrates the masonry structural system preferred for ductile seismic response. Seismic forces are carried by simple cantilever walls. Where two or more such walls occur in the same plane, linkage between them is by flexible floor slabs rather than by stiff coupling beams. This is to minimize moment transfer between walls. Columns, acting as props, may be used in conjunction with the walls to assist with gravity load resistance. Openings within the wall elevation should be kept small enough to ensure that basic vertical cantilever action is not affected. Energy dissipation should occur only in carefully detailed plastic hinges at the base of each wall.

The displacement ductility capacity of the walls will be dictated by the plastic rotation capacity θ_p of the plastic hinges [see Fig. 7.3(b)], and as was

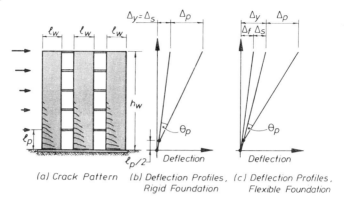

(a) Crack Pattern   (b) Deflection Profiles,   (c) Deflection Profiles,
                        Rigid Foundation   Flexible Foundation

Fig. 7.3   Cantilever walls linked by flexible floor slabs.

shown in Section 3.5.4(*b*), can be expressed as

$$\mu_\Delta = 1 + \frac{\theta_p}{\Delta_y}\left(h_w - \frac{l_p}{2}\right) \tag{7.4}$$

where Δ_y is the yield displacement at height h_w, and h_w is the wall height. Factors affecting μ_Δ, including the influence of foundation flexibility, are discussed in detail in Section 7.2.4.

(b)   Coupled Walls with Pier Hinging   Traditionally, masonry construction has generally consisted of peripheral masonry walls pierced by window and door openings, as idealized in Figs. 7.4(*a*) and 7.5(*a*). Under lateral forces hinging may initiate in the piers [Fig. 7.4(*a*)] or in the spandrels [Fig. 7.5(*a*)].

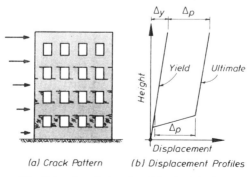

(a) Crack Pattern   (b) Displacement Profiles

Fig. 7.4   Coupled walls with pier hinging.

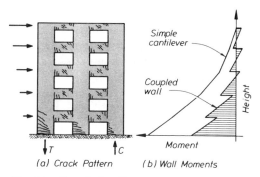

(a) Crack Pattern   (b) Wall Moments

Fig. 7.5 Coupled walls with spandrel hinging.

In the former and more common case, the piers will be required to exhibit substantial ductility unless designed to resist elastically the displacements resulting from the design earthquake. Plastic displacement due to flexure or shear will inevitably be concentrated in the piers of one story, generally the lowest, with consequential extremely high ductility demand at that level.

Consider the deflection profiles at first yield and ultimate, illustrated in Fig. 7.4(b). If design is on the basis of a specified or implied ductility μ_Δ at roof level, plastic displacement

$$\Delta_p = (\mu_\Delta - 1)\Delta_y \qquad (7.5)$$

is required. Under a typical triangular distribution of lateral seismic forces the yield load will be attained when the flexural or shear strength of the bottom-level piers is reached. Other elements of the structure will not have yielded at this stage. Subsequent deformation of the yielding piers will occur at constant or decreasing lateral force, thus ensuring that the other structural elements are protected from inelastic action. The plastic deformation Δ_p is thus concentrated in the yielding piers. If the structure has n stories and pier height is half the story height, the elastic displacement over the height of the piers, Δ'_e, at yield will be

$$\Delta'_e = \Delta_y/2n \qquad (7.6)$$

assuming a linear yield displacement profile. From Eqs. (7.5) and (7.6) the displacement ductility factor $\mu_{\Delta p}$ required of the pier will then be

$$\mu_{\Delta p} = 1 + \Delta_p/\Delta'_e = 2n(\mu_\Delta - 1) + 1 \qquad (7.7)$$

Thus for a 10-story masonry wall, designed for a displacement ductility factor $\mu_\Delta = 4$, the ductility required of the piers would be $\mu_{\Delta p} = 61$. Extensive experimental research on masonry pier units at the University of California, Berkeley [M18] has indicated extreme difficulty in obtaining reliable ductility levels an order of magnitude less than this value. It is concluded that the structural system of Fig. 7.4(a), with ductile piers, is only suitable if very low displacement ductilities are required.

Equation (7.7) can be inverted to yield the safe displacement ductility factor μ_Δ for a coupled wall with pier hinges, based on a specified maximum displacement ductility factor, $\mu_{\Delta p}$, in the piers:

$$\mu_\Delta = 1 + 0.5(\mu_{\Delta p} - 1)/n \qquad (7.8)$$

Thus for the 10-story example structure, discussed above, a safe pier ductility factor of $\mu_{\Delta p} = 4$ would result in a structure displacement ductility of only $\mu_\Delta = 1.15$. It is apparent that the structure would effectively have to be designed for lateral forces corresponding to full elastic response.

(c) *Coupled Walls with Spandrel Hinging*   Occasionally, openings in masonry walls will be of such proportions that spandrels will be relatively weaker than piers, and behavior will approximate coupled walls with crack patterns as illustrated in Fig. 7.5(a). As discussed in detail in Section 5.6, well-designed coupled walls in reinforced concrete constitute an efficient structural system for seismic resistance. However, high ductility demand is generated in the coupling beams, particularly at upper levels of the building. Although these ductility levels can be accommodated in reinforced concrete, the low ultimate compression strain of masonry make such a system generally unsuitable for masonry structures. Rapid strength and stiffness degradation of the masonry spandrels could be expected, resulting in an increase in wall moments from that which is characteristic of coupled walls to those appropriate to simple linked cantilevers, as seen in Fig. 7.5(b). If the wall moment capacities have been proportioned on the basis of ductile coupled wall action, the moment increase implied at the wall base will not be possible, and the consequence will be excessive ductility demand of the wall base plastic hinges.

Design options are thus to design for a reduced overall displacement ductility, or to separate the spandrels from the walls by flexible joints to avoid damage in the spandrels due to wall rotation. As the latter alternative is structurally equivalent to the simple vertical cantilever design of Fig. 7.3, the best alternative is to avoid the use of spandrels entirely.

(d) *Selection of Primary and Secondary Lateral-Force-Resisting Systems*   Some masonry wall structures do not lend themselves to rational analysis under lateral forces, as a consequence of the number, orientation, and complexity of shape of the load-bearing walls. In such cases it is rational to consider the walls to consist of a primary system, which carries gravity loads and the entire seismic lateral force, and a secondary system, which is designed to support

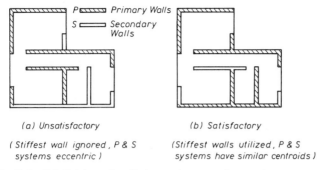

Fig. 7.6 Subdivision of walls into primary and secondary systems.

gravity loads and face loads only. This allows simplification of the lateral force analysis in cases where the extent of wall area exceeds that necessary to carry the code seismic forces. However, although it is assumed in the analysis that the secondary walls do not carry any in-plane forces, it is clear that they will carry an albeit indeterminate proportion of the lateral force. Consequently, they must be detailed to sustain deformations to which they may be subjected, by specifying similar standards as for structural walls, although code-minimum requirements will normally be adopted. To ensure satisfactory behavior results, the natural period should be based on a conservatively assessed stiffness of the composite primary/secondary system.

No secondary wall should have a stiffness greater than one-fourth that of the stiffest wall of the primary system. This is to ensure that the probability of significant inelastic deformation developing in secondary walls is minimized and the integrity of secondary walls for the role of gravity load support is maintained. Long, stiff secondary walls may be divided into a series of more flexible walls by the incorporation of vertical control joints at regular centers. A further requirement in selecting the primary and secondary systems of walls, discussed in Section 7.2.1(*d*), is that the centers of rigidity of the two systems should be as close as possible, to minimize torsional effects. Figure 7.6 shows acceptable and unacceptable division of a complex system of walls into primary and secondary systems.

(e) *Face-Loaded Walls* Masonry walls may also be required to support seismic forces by face-load or out-of-plane bending. Examples are the loading of masonry retaining walls by seismic earth pressures, and the inertial response of walls to transverse seismic accelerations. Where the walls also provide primary seismic resistance of a building by in-plane action, the effects of simultaneous response in the in-plane and out-of-plane directions must be considered.

7.2.2 Analysis Procedure

Analysis procedures required to obtain reasonable estimates of fundamental period, displacement, and distributions of lateral forces between walls have been considered in some detail in Section 5.3.1 for reinforced concrete structural walls. These procedures, with the obvious substitution of f'_m for f'_c where appropriate, are equally applicable to masonry walls.

T-section walls are common in masonry construction, and when the direction of loading parallel to the web is to be considered, different values of stiffness will apply depending on whether the flange or the web is in compression. When the flange is in compression, the stiffness may be based on the web section alone. When the web is in compression, the web section alone may again be used, but an additional axial force, equal to the tensile yield capacity of the flange reinforcement, should be applied to the wall when using Eq. (5.7). This will result in significantly increased stiffness.

7.2.3 Design for Flexure

(a)   Out-of-Plane Loading and Interaction with In-Plane Loading   Because of the multiaxial nature of ground shaking under seismic forces, walls will be subjected to simultaneous vertical, in-plane, and out-of-plane (face-load) response. While the in-plane response will primarily be a result of the resistance of the wall to inertia forces from other parts of the structure, such as floor masses, the out-of-plane response will be due to the inertia mass of the walls themselves responding to the floor-level excitation. As discussed in more detail in Section 7.8.2(*d*), the out-of-plane response of multistory walls is complex, as it involves the modification of the ground motion by the in-plane response of the seismic structural system perpendicular to the wall being considered. Further modification may be due to floor flexibility. The input motion that the face-loaded wall "sees" will be very different from the earthquake ground motion and will have strong harmonic components, representing the natural frequencies of the structure as a whole.

Consider the typical four-story wall shown in Fig. 7.7(*a*). The envelope of floor-level accelerations under in-plane response is shown in Fig. 7.7(*b*). If the structure is designed for low or moderate ductility, the maximum absolute accelerations at the upper floors will exceed those at lower floors. At ground level, the peak acceleration will, of course, be equal to peak ground acceleration; that is, $a_1 = a_g$. The design envelope of in-plane moments, shown in Fig. 7.7(*c*), exceeds those resulting from the code distribution of forces at levels above the base due to higher-mode effects, as discussed in Section 5.4.2(*c*).

Assuming that the wall in Fig. 7.7 is part of an essentially symmetrical structure and that the floors act as a rigid diaphragm, Fig. 7.7(*b*) can be taken to represent the peak out-of-plane accelerations of the wall at floor levels. These will induce out-of-plane moments, shown in Fig. 7.7(*d*), whose magnitude will be larger in the upper than for the lower stories.

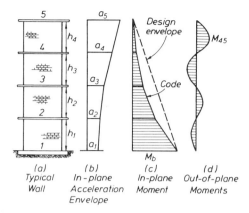

Fig. 7.7   Response of a masonry wall to biaxial excitation.

Figure 7.7(*d*) assumes that the first out-of-plane mode of the wall is excited, with wall displacements at alternate stories being out of phase, implying contraflexure points at the floor levels. This is more conservative than assuming excitation of any higher mode, which would result in lower maximum bending moments. This is a reasonable assumption, because the lowest mode will generally be the easiest to excite. Also, due to in-plane action, tension cracking will reduce the ability of the walls to provide restraining moments at floor levels, even if the different levels of wall all responded in phase rather than out of phase, as assumed. The result will be actions that approximate simple support at floor levels. However, because of the somewhat conservative nature of the assumption, it is realistic to assume that there is no interaction between the in-plane and out-of-plane bending moments. In this context it should be noted that maximum in-plane moments occur at the wall base, while maximum out-of-plane moments occur in the top story, where in-plane moments are low.

For the wall of Fig. 7.7, using the notation of Fig. 7.7, the maximum out-of-plane moment M_{45} will be approximately

$$M_{45} = 0.1 m a_{45} h_4^2 \tag{7.9}$$

for unit horizontal length of wall, where m is the wall mass per unit area, and a_{45} is the maximum response acceleration at midheight of the wall in the fourth story. This may be much greater than either a_4 or a_5, the floor accelerations, due to the possibility of resonant response as a result of near coincidence of natural periods of out-of-plane response of the wall, T_w, and transverse in-plane response of the structure as a whole, T_s.

Figure 7.8 shows the extent of amplification of wall response that can be obtained from steady-state elastic behavior. The amplification depends on

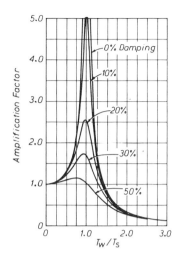

Fig. 7.8 Amplification in walls of floor response by wall flexibility, for steady-state sinusoidal excitation.

the period ratio T_w/T_s and the level of damping. In the real situation of earthquake response, the amplification will be less because the driving force (floor displacements) will be transient instead of steady. Even in the event of exact coincidence of periods, unexpectedly large out-of-plane response of the wall would result in inelastic response, increasing the effective out-of-plane wall period T_w, thus reducing response. It is thus not necessary to design for the highest level of amplification implied by Fig. 7.7. Conversely, it would be unsafe to adopt a low-amplification factor on the basis of a large T_w/T_s ratio, such as $T_w/T_s = 2$. Further, it must be realized that out-of-plane response of the wall may be excited by higher-mode transverse response of the structure, which can result in high local accelerations, and that there are a number of possible out-of-plane modes with closely spaced frequencies. Figure 7.9 shows the first four out-of-plane mode shapes and periods for a three-story masonry wall, as an example. The modes include the effect of the transverse structural stiffness, and hence are not symmetrical.

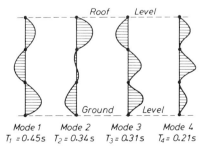

Fig. 7.9 Wall out-of-plane mode shapes and periods for a three-story masonry building.

Mode 1   Mode 2   Mode 3   Mode 4
$T_1 = 0.45$s $T_2 = 0.34$s $T_3 = 0.31$s $T_4 = 0.21$s

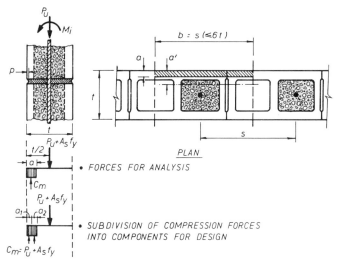

Fig. 7.10   Analysis and design of wall section subjected to out-of-plane moment.

It is thus extremely likely that some degree of resonance will occur between out-of-plane and transverse structure response. However, because of the many factors involved, it would be impractical to attempt to quantify the level of amplification accurately without resort to dynamic inelastic time-history analysis. This will not be feasible, nor warranted for routine designs, but typical analyses of this kind have resulted in amplification factors of 1.5 to 2.5.

It is thus recommended that out-of-plane response be based on a conservative amplification factor of 2.5 times the maximum feasible floor acceleration, and that a design ductility factor of 1 be used in assessing the maximum out-of-plane moment. The limitation on ductility is to ensure an adequate reserve for the case of unexpectedly high amplification, and is further advisable since the mechanism by which simultaneous ductility in-plane and out-of-plane could occur is hard to visualize, particularly at the edge of the wall in tension under in-plane loading. As shown in Example 7.1, these recommendations will not be onerous in typical design situations.

(b) *Section Analysis for Out-of-Plane Flexure*   Figure 7.10 illustrates the procedure for analysis and design of a masonry wall for out-of-plane moments. The effective width b of wall acting in compression to balance the tension force of each bar must be calculated. In Fig. 7.10 the effective width is taken equal to the bar spacings. When $s > 6t$, it should be assumed that $b = 6t$.

For analysis the procedure follows normal strength theory for flexure. That is, assuming a tension failure:

Depth of compression block: $\quad a = \dfrac{P_u + A_s f_y}{0.85 f'_m b}$ $\qquad$ (7.10)

Ideal moment capacity: $\qquad M_i = (P_u + A_s f_y)(d - a/2)$ $\quad$ (7.11)

where

$$d = t/2 - p \qquad (7.12)$$

and p is the depth of pointing of the mortar bed. Except for deep-raked joints ($p > 3$mm), it is normal to set $p = 0$ in Eq. (7.12).

In most cases depth a will lie wholly within the thickness of the face shell. For partially grouted walls with high reinforcement contents or high axial load it is possible that the depth a could exceed the face shell thickness ($a = a'$ in Fig. 7.10), and Eq. (7.10) will not apply. In this case normal T-beam flexural theory should be used to find a'.

Equation (7.10) assumes unconfined masonry. If confining plates are used, different parameters describe the shape of the stress block, as shown in Section 3.2.3(g), and these must be used in the analysis. As with concrete design, a check should be made to ensure that a tension failure occurs as assumed.

Example 7.1   A fully grouted 190-mm (7.5-in.)-thick multistory concrete wall of unit weight 20 kN/m^3 (127 lbs/ft^3) is reinforced vertically by D12 bars (0.47 in.) [$f_y = 275$ MPa (40 ksi)] on 600-mm (23.6-in.) centers. The shell thickness is 40 mm (1.58 in.). The story heights are 3 m (9.7 ft) and maximum floor accelerations of $0.4g$ are expected at the upper levels. Assuming that $f'_m = 8$ MPa (1160 psi), and conservatively ignoring any axial load in the upper stories, check whether the reinforcement is adequate.

SOLUTION: Assuming an amplification factor of 2.5, the maximum response acceleration of the walls in the upper stories will be $1.0g$.

$\qquad$ Wall weight/unit area $= 0.19 \times 20 = 3.80$ kN/m^2 (79.4 lb/ft^2)

From Eq. (7.12),

$\qquad M_{as} = 0.1mah^2$

$\qquad\qquad = 0.1 \times 3.80 \times 1.0 \times 3^2 = 3.42$ kNm/m (0.77 kip-ft/ft)

D12 (0.47-in.-diameter) bars on 600-mm (23.6-in.) centers correspond to

$$A_s = 113/0.6$$
$$= 188 \text{ mm}^2/\text{m} \ (0.089 \text{ in.}^2/\text{ft})$$

From Eq. (7.10)

$$a = \frac{188 \times 275}{0.85 \times 8 \times 1000} = 7.6 \text{ mm} \ (0.30 \text{ in.})$$

Clearly, a tension failure would occur. Therefore,

$$M_i = (A_s f_y)\left(\frac{t}{2} - \frac{a}{2}\right)$$
$$= 188 \times 275(95 - 3.8)10^{-6} = 4.72 \text{ kNm/m} \ (1.06 \text{ kip-ft/ft})$$

Thus $M_i/M_{as} = 4.72/3.42 = 1.38$, and the reinforcement is adequate. Based on full wall thickness rather than effective depth, the reinforcement provided represents a steel ratio $A_s/bt = 0.001$. As mentioned in Section 7.2.8(a), this is about the minimum level permitted by most codes. Since this is sufficient to provide the moment capacity for a 1.0g lateral response, it will be appreciated that seismic out-of-plane loading will rarely be critical.

(c) *Design for Out-of-Plane Bending* The typical out-of-plane design situation will involve establishing the appropriate amount of reinforcement to provide a required ultimate moment capacity M_u, with a given axial load level of P_u. The required ideal design actions are thus

$$M_i \geq M_u/\phi; \qquad P_i = P_u$$

where ϕ is the flexural strength reduction factor (Section 3.4.1). Note that the axial load has not been factored up, since in the typical seismic design situation P_u is wholly or largely due to gravity loads, and M_u is wholly or largely due to seismic forces. For the typical wall section, which will be at less than balanced load, dividing P_u by ϕ will result in an increase in moment capacity, and is hence nonconservative.

It is convenient to consider the ideal flexural strength M_i to consist of two components: a moment M_p sustained by axial load and a moment M_s sustained by reinforcement. Hence

$$M_i = M_p + M_s \tag{7.13}$$

Therefore, in Fig. 7.10, the depth of the compression zone a is divided into a_1 and a_2, resulting from P_u and A_s, respectively. Thus the moment for a unit length of wall sustained by the axial load is, from $a_1 = P_u/0.85f'_m$,

$$M_p = P_u\left(\frac{t}{2} - \frac{a_1}{2}\right)$$ (7.14)

The moment to be sustained by reinforcement with A_s is

$$M_s = M_i - M_p$$

Assuming that a is small compared with $t/2$,

$$a_2 \approx a_1\frac{M_s}{M_p}$$ (7.15)

Hence

$$M_s = A_sf_y\left(\frac{t}{2} - a_1 - \frac{a_2}{2}\right)$$

and thus

$$A_s = \frac{M_s}{f_y[(t/2) - a_1 - (a_2/2)]}$$ (7.16)

Because of the limited options for bar spacing [normally, multiples of 200 mm (8 in.) for blockwork], greater accuracy in design is not warranted. However, having chosen a suitable bar size and spacing to satisfy Eq. (7.16), the moment capacity should be checked using Eqs. (7.10) to (7.12).

Example 7.2 A 190-mm (7.5-in.)-thick fully grouted concrete block wall is subjected to an axial load due to dead weight of 27.7 kN/m (1.90 kips/ft) and is required to resist a moment of $M_u = 15$ kNm/m (3.37 kip-ft/ft). Design the flexural reinforcement, given $\phi = 0.8$ and $f'_m = 12$ MPa (1740 psi).

SOLUTION: Using factored loads according to $U = 0.9D + E$,

$$M_i \geq 15/0.8 = 17.75 \text{ kNm/m (3.99 kip-ft/ft)}$$
$$P_u = 0.9P_D = 25.0 \text{ kN/m (1.71 kips/ft)}$$

The moment sustained by the axial load is from Eq. (7.14) with

$$a_1 = \frac{25 \times 10^3}{0.85 \times 12 \times 10^6} = 2.45 \text{ mm (0.096 in.)}$$

$$M_p = 25(95 - 1.2)10^{-3} = 2.35 \text{ kNm/m (0.528 Kft/ft)}$$

Hence, moment to be sustained by reinforcement

$$M_s = 17.75 - 2.35 = 16.4 \text{ kNm/m } (3.69 \text{ Kft/ft})$$

From Eq. (7.15),

$$a_2 \approx 2.45 \times \frac{16.4}{2.35} = 17.1 \text{ mm } (0.673 \text{ in.})$$

∴

From Eq. (7.16)

$$A_s f_y \geq \frac{16.4 \times 10^3}{(95 - 2.5 - 0.5 \times 17.1)10^3} = 195 \text{ kN/m } (13.4 \text{ K/ft})$$

Thus, for

$$f_y = 275 \text{ MPa } (40 \text{ ksi}), \qquad A_s \geq 710 \text{ mm}^2/\text{m } (0.336 \text{ in}^2/\text{ft})$$

for

$$f_y = 380 \text{ MPa } (55 \text{ ksi}), \qquad A_s \geq 513 \text{ mm}^2/\text{m } (0.242 \text{ in}^2/\text{ft})$$

The design options are for

$$f_y = 275 \text{ MPa } - \text{D20@400 } (0.79 \text{ in@15.7 in}) \text{ crs} = 785 \text{ mm}^2/\text{m}$$
$$f_y = 380 \text{ MPa } - \text{DH12@200 } (0.47 \text{ in@7.9 in}) \text{ crs} = 565 \text{ mm}^2/\text{m}$$
$$\text{or} \quad - \text{DH20@600 } (0.79 \text{ in@23.6 in}) \text{ crs} = 523 \text{ mm}^2/\text{m}$$

All three options satisfy the required ideal flexural strength of 17.75 kNm/m.

(d)   Analysis for In-Plane Bending   In Section 7.1 it was established that the in-plane flexural strength of a wall was effectively independent of whether the flexural reinforcement was concentrated at the wall ends, or uniformly distributed along the wall. Uniform distribution of reinforcement is, however, to be preferred for a number of reasons. Reinforcement concentrated at wall ends causes bond and anchorage problems because of the limited grout space. Moreover it increases the tendency for splitting of the masonry compression zone as a result of compression bar buckling, particularly under cyclic inelastic response to seismic forces. If this occurs, strength and stiffness degrade rapidly, and if the flexural reinforcement is lapped with starter bars at the base, total collapse can result. Distributed reinforcement is not subject to these faults to the same extent, and has the added advantage of enhancing shear performance. The distributed reinforcement increases the magnitude of the flexural masonry compression force, thus improving compression–shear

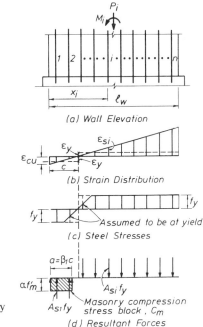

Fig. 7.11 Analysis of rectangular masonry
wall section subjected to in-plane flexure.

transfer. It also provides a clamping force along the wall base and this helps
to limit slip of the wall along the base. For walls with low axial load, base slip
has been shown to be the most significant cause of degradation of hysteresis
loops in well designed walls [P56]. On the basis of these arguments, it has
been recommended [X13] that flexural reinforcement be essentially uniformly
distributed along the wall length.

Figure 7.11 illustrates the procedure developed in Section 3.3.1(c) for
deriving the ideal flexural strength of a rectangular wall with distributed
reinforcement. The wall contains n bars of individual areas $A_{s1} \cdots A_{si} \cdots$
A_{sn}. Because of generally low axial load P_i and reinforcement ratio $\rho_t =$
$\Sigma A_{si}/(l_w t)$, the neutral axis depth c will be small compared with the wall
length l_w. Since the ultimate compression strain of the masonry, ϵ_{cu}, will
normally exceed the reinforcement yield strain ϵ_y even for unconfined
masonry (unless very high strength reinforcement is used), virtually all
reinforcement will be at yield in either tension or compression, as shown in
Fig. 7.11(b). It is a justifiable approximation to assume all flexural reinforcement is at yield, since the influence on the total moment capacity of
overestimating the stress of bars close to the neutral axis will be small.

Analysis requires an iterative procedure to find the neutral axis position.
The following sequence is suggested, using the notation of Fig. 7.11 and
general expressions $f_{\text{average}} = \alpha f'_m$ and $a = \beta c$ to define the equivalent
rectangular stress block for the masonry.

1. Assume a value for the depth of

$$a = \frac{P_i + 0.5\Sigma_{i=1}^n A_{si} f_y}{\alpha f_m' t} \tag{7.17}$$

This initial estimate assumes that 75% of the steel is in tension and 25% in compression.

2. Calculate $c = a/\beta$.
3. Calculate masonry compression force $C_m = \alpha f_m' t a$.
4. On the basis of the current value for c, bars 1 to j are in compression. Thus the steel compression force is

$$C_s = \sum_{i=1}^{j} A_{si} f_y$$

5. Thus bars $(j + 1)$ to n are in tension. Hence the steel tension force is

$$T = \sum_{i=j+1}^{n} A_{si} f_y$$

6. Check whether equilibrium is satisfied.

$$C_m + C_s - T \gtrless P_i \tag{7.18}$$

If $C_m + C_s - T > P_i$, reduce a in proportion.
If $C_m + C_s - T < P_i$, increase a in proportion.

7. Repeat steps 2 to 6 until the agreement in Eq. (7.17) is within a tolerance of 2 to 5%, which is adequate. This normally takes only two or three cycles, and with some experience less.

8. Take moments of all forces, approximately in equilibrium, about any convenient point. The neutral axis is generally particularly suitable. Hence

$$M_i = C_m\left(c - \frac{a}{2}\right) + \sum_{i=1}^{n} \left| f_y A_{si} (c - x) \right| + P_i\left(\frac{l_w}{2} - c\right) \tag{7.19}$$

All terms inside the summation are positive.

Example 7.3 The unconfined 190-mm (7.5-in.)-thick concrete masonry wall in Fig. 7.12 is vertically reinforced with six DH20 bars [0.79 in. $f_y = 380$ MPa (55 ksi)] and supports an axial load of 200 kN (45 kips). Given that $f_m' = 8$ MPa (1160 psi), calculate its ideal in-plane flexural strength.

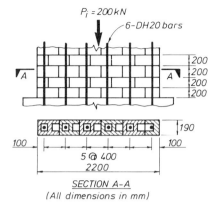

Fig. 7.12   Example masonry wall. (1 kN = 0.225 kip; 1 mm = 0.03937 in.)

SECTION A-A
(All dimensions in mm)

SOLUTION: Area of one DH20 bar = 314 mm² (0.487 in.²).

1. Estimate a from Eq. (7.17). As the wall is unconfined, $\alpha = 0.85$.

$$a = \frac{200 \times 10^3 + 3 \times 314 \times 380}{0.85 \times 8 \times 190} = 432 \text{ mm} \ (17 \text{ in.})$$

2. $\beta = 0.85$; hence $c = 432/0.85 = 508$ mm (20 in.).

3. Masonry compression force:

$$C_m = 0.85 \times 8 \times 190 \times 432 \times 10^{-3} = 558 \text{ kN} \ (125 \text{ kips})$$

4. With $c = 508$ mm, the neutral axis is almost at bar 2. Assume that this bar has zero stress. Thus one bar is in compression and four are in tension.

$$C_s = 1 \times 314 \times 380 \times 10^{-3} = 119 \text{ kN} \ (26.8 \text{ kips})$$

5. $T = 4 \times 314 \times 380 \times 10^{-3} = 477$ kN (107 kips)

6. Check equilibrium from Eq. (7.18):

$$C_m + C_s - T = 558 + 119 - 477 = 200 \text{ kN} = P_i$$

In this case the initial estimate for a was correct and no iteration is necessary. Go to step 8.

8. Moments about neutral axis: Eq. (7.19) can be used thus:

$$M_i = C_m\left(c - \frac{a}{2}\right) + A_{si}f_y\Sigma|c - x| + P_i\left(\frac{l_w}{2} - c\right)$$

$$= [558(508 - 216) + 119(408 + 392 + 792 + 1192 + 1592)$$

$$+ 200(1100 - 508)]10^{-3}$$

$$= 802 \text{ kNm} \ (592 \text{ kip-ft})$$

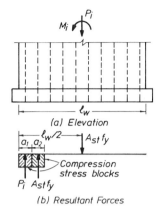

(a) Elevation

(b) Resultant Forces

Fig. 7.13 Assumed forces for the flexural design of a rectangular masonry wall section.

(e) Design for In-Plane Bending In Section 7.1 it was established that the in-plane ideal flexural strength of a wall was little affected by the extent to which the reinforcement was concentrated at the ends, or evenly distributed along the length of the wall, provided that the total quantity was the same. It thus follows that a wall with the same total quantity of reinforcement concentrated at the center of the wall would have a very similar flexural strength. Although this is not suggested as a practical arrangement, of course, it can be used conceptually to estimate the quantity of reinforcement required for a given wall.

The method is identical to that used for face-loaded walls. It resolves the total moment into components sustained separately by the axial load and by the reinforcement. The procedure is summarized with reference to Fig. 7.13, where the depth a of the equivalent rectangular compression stress block is again divided into a_1 and a_2, as in Fig. 7.10.

M_u is the required dependable moment capacity; hence $M_i \geq M_u/\phi$. The moment sustained by axial load P_i with

$$a_1 = P_i/\alpha f'_m t$$

$$\text{is } M_p = P_i\left(\frac{l_w}{2} - \frac{a_1}{2}\right) \tag{7.20}$$

The moment to be sustained by reinforcement is thus $M_s = M_i - M_p$ and, from Eq. (7.15), $a_2 \approx a_1(M_s/M_p)$. Therefore,

$$M_s = A_s f_y\left(\frac{l_w}{2} - a_1 - \frac{a_2}{2}\right)$$

and thus

$$A_s \geq \frac{M_s}{f_y[(l_w/2) - a_1 - (a_2/2)]} \tag{7.21}$$

A suitable choice of bar size and spacing may now be made to satisfy Eq. (7.21). If necessary, the flexural strength may be checked using Eqs. (7.17) to (7.19). Note that in this case the value of $a = a_1 + a_2$ should be used as the initial estimate for a.

(f) *Design of a Confined Rectangular Masonry Wall* A rectangular concrete masonry wall 3.2 m long and 0.19 m wide (10.5 ft × 7.5 in.) is confined by 3-mm (1/8-in.)-thick galvanized steel plates within the plastic hinge region. It supports an axial dead load of 520 kN (117 kips). Given $f'_m = 10$ MPa (1450 psi), and $K = 1.2$, design the flexural reinforcement such that the wall can dependably support a seismic bending moment of $M_u = 1200$ kNm (886 kip-ft).

Assuming a tension-dominated wall, $P_u = 0.9 \times 520 = 468$ kN (105.2 kips):

$$\text{Axial load ratio: } \frac{P_u}{f'_m A_g} = \frac{468 \times 10^3}{10 \times 3200 \times 190} = 0.077$$

Therefore, from Eq. (3.47), $\phi = 0.696$ and hence

$$M_i \geq 1200/0.696 = 1724 \text{ kNm (1272 kip-ft)}$$

Since the wall is confined, the moment sustained by axial load is, from Section 3.2.3(g), with $\alpha = 0.9K = 0.9 \times 1.2 = 1.08$ and

$$a_1 = \frac{468 \times 10^3}{1.08 \times 10 \times 190} = 228 \text{ mm (8.98 in.)}$$

from Eq. (7.20),

$$M_p = 468(0.5 \times 3200 - 0.5 \times 228)10^{-3} = 695 \text{ kNm (512 kip-ft)}$$

Therefore, the moment to be sustained by the reinforcement is

$$M_s = M_i - M_p = 1724 - 695 = 1029 \text{ kNm (759 kip-ft)}$$

From Eq. (7.15),

$$a_2 \approx 228 \times 1029/695 = 338 \text{ mm (13.3 in.)}$$

and from Eq. (7.21)

$$A_s = \frac{1029 \times 10^6}{275 \times (1600 - 228 - 0.5 \times 338)} = 3110 \text{ mm}^2 \ (4.82 \text{ in.}^2)$$

The best practical solution is to use D16 bars on 200-mm centers (0.63 in. at 7.9 in.), giving

$$A_s = 16 \times 201 = 3216 \text{ mm}^2 \ (4.99 \text{ in.}^2)$$

This is about 3% more than that required. It is informative to check the capacity, to see how adequate the idealization of all reinforcement at the wall center was.

Calculate the neutral-axis position. Figure 3.17 shows that $\beta = 0.96$ for confined masonry. Hence with $a = 228 + 338 = 566$ mm (22.3 in.),

$$c = a/\beta = 566/0.96 = 590 \text{ mm} \ (23.2 \text{ in.})$$

Since bars will be on 100-, 300-, 500-, 700-mm (etc.) centers from the compression end, 3 bars are in compression and 13 are in tension.

Masonry compression: $C_m = \alpha f'_m a t$

$$= 1.08 \times 10 \times 566 \times 190 \times 10^{-3} = 1161 \text{ kN (261 kips)}$$

Steel compression: $C_s = 3 \times 201 \times 275 \times 10^{-3}$ $= 166$ kN (37.3 kips)

Steel tension: $T = 13 \times 201 \times 275 \times 10^{-3}$ $= 719$ kN (162 kips)

From Eq. (7.18), check the equilibrium:

$$C_m + C_s - T = 1161 + 166 - 719 = 608 > 468 \text{ kN}$$

Thus $C_m + C_s$ are too high. Reduce C_m by 140 kN, to 1021 kN, to give balance. This will give

$$c = 590 \times 1021/1161 \approx 520 \text{ mm} \ (20.5 \text{ in.})$$

and

$$a = 520 \times 0.96 \quad = 499 \text{ mm} \ (19.6 \text{ in.})$$

Since $c > 500$, 3 bars are still in compression and 13 in tension, so C_s and T remain unchanged, and the forces are in equilibrium.

Finally, by taking moments about the neutral axis and using Eq. (7.19), we obtain

$$M_i = \{1021(520 - 250) + 201 \times 275[3 \times (520 - 300) + 13(1900 - 520)]$$

$$+ 468(1600 - 520)\}10^{-3}$$

$$= 1809 \text{ kNm} \ (1335 \text{ kip-ft})$$

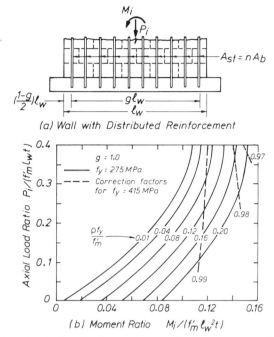

(a) Wall with Distributed Reinforcement

(b) Moment Ratio $M_i/(f'_m \ell_w^2 t)$

Fig. 7.14 Design charts for the flexural strength of masonry walls with uniformly distributed reinforcement. (1 MPa = 145 psi.)

This exceeds the required strength of 1724 kNm by 5%, and hence the design is satisfactory.

The foregoing methods for analysis and design of rectangular masonry walls have been developed at some length because the principles are basic to the design and analysis of any section shape, as discussed briefly for flanged walls in Section 7.2.3(g). It is, however, possible to develop design charts for flexural strength of rectangular masonry walls in a nondimensional form, suitable for general use. Figure 7.14 shows such a design aid to determine the flexural strength of masonry walls with uniformly distributed reinforcement. The axial load and ideal bending moment are expressed in the dimensionless forms $P_i/f'_m l_w t$ and $M_i/f'_m l_w^2 t$ in terms of the dimensionless parameter $\rho f_y/f'_m$. The range of this parameter (0.01 to 0.20) includes all practical designs.

Although the curves in Fig. 7.14(b) are plotted for $f_y = 275$ Mpa (40 ksi), the results are only weakly dependent on f_y (except insofar as f_y influences the parameter $\rho f_y/f'_m$). Plotted on the curves are correction factors, shown as dashed lines, for $f_y = 415$ MPa (60 ksi). For most cases the required correction is less than 1%.

The design aids of Fig. 7.14 are given only for $g = 1$, where g is the ratio of distance between extreme vertical bars, to wall length [see Fig. 7.14(a)]. This is because the end bar will normally be 100 mm (4 in.) from the wall

TABLE 7.1   Moment Coefficients m_i for Rectangular Masonry Walls with Uniformly Distributed Reinforcement[a]: f_y = 275 MPa (40 ksi), g = 1.0

$\dfrac{\rho f_y}{f_m'}$	Axial Load Ratio $P_u/(f_m' l_w t)$								
	0	0.05	0.10	0.15	0.20	0.25	0.30	0.35	0.40
0.0100	0.0052	0.0279	0.0480	0.0652	0.0795	0.0910	0.0995	0.1052	0.1080
0.0200	0.0101	0.0322	0.0519	0.0687	0.0826	0.0938	0.1021	0.1076	0.1102
0.0400	0.0194	0.0406	0.0593	0.0754	0.0887	0.0993	0.1072	0.1123	0.1147
0.0600	0.0284	0.0487	0.0666	0.0819	0.0946	0.1047	0.1122	0.1170	0.1193
0.0800	0.0370	0.0565	0.0737	0.0883	0.1005	0.1101	0.1172	0.1218	0.1238
0.1000	0.0454	0.0641	0.0805	0.0946	0.1062	0.1154	0.1221	0.1265	0.1284
0.1200	0.0535	0.0714	0.0873	0.1007	0.1119	0.1207	0.1271	0.1312	0.1329
0.1400	0.0613	0.0786	0.0938	0.1068	0.1175	0.1259	0.1320	0.1359	0.1375
0.1600	0.0690	0.0856	0.1003	0.1127	0.1230	0.1311	0.1369	0.1406	0.1421
0.1800	0.0764	0.0925	0.1066	0.1186	0.1285	0.1362	0.1418	0.1453	0.1466
0.2000	0.0837	0.0992	0.1128	0.1244	0.1339	0.1413	0.1467	0.1500	0.1512

[a]$M_i = m_i(f_m' l_w^2 t)$.

end, with a spacing of at least 200 mm (8 in.) between bars. Distributing the area of bar over a width equal to the spacing to get a fully distributed lamina, as assumed in the analyses, thus implies an effective value of $g \geq 1$. Adopting $g = 1.0$ is thus always conservative. The results are not significantly affected by the value of g unless the axial load or reinforcement content are very high. For convenience, design values are also presented in Table 7.1.

The curves of Fig. 7.14(b) and Table 7.1 have been prepared without making the assumption that all reinforcement is at yield. Thus they are more accurate than hand analyses using the methods described previously, particularly when reinforcement ratios or axial loads are high.

(g) Flanged Walls   The extension of the analysis methods described in the preceding section to flanged walls is straightforward. For design purposes the axial load and reinforcement content for the wall may be divided into lumped loads, and steel areas acting at the center of each element (flange, web, etc.) of the flanged section in similar fashion to that for rectangular walls. Effective flange widths for flexural strength calculations may be taken to be the same as for reinforced concrete walls, given in Section 5.2.2.

7.2.4   Ductility Considerations

For reinforced concrete cantilever walls it is reasonably straightforward to ensure adequate curvature ductility capacity even without provision of confining reinforcement within the plastic hinge regions. Although a measure of confinement can be provided to masonry walls by confining plates in critical mortar beds within the plastic hinge region, most reinforced masonry walls will be unconfined. Since the ultimate compression strain is lower than for reinforced concrete, and since the degradation of the compression zone resulting from the characteristic vertical splitting failure mode of masonry is very rapid, the available ductility capacity needs to be checked carefully to ensure that it satisfies the capacity assumed in setting the level of lateral seismic forces. Even for confined masonry walls, the degree of confinement afforded by the confining plates will be less than for reinforced concrete walls, and ductility should be limited.

(a) Walls with Rectangular Section Consideration of deflection profiles at yield and ultimate for the linked cantilever wall system preferred for seismic design (Fig. 7.3) are identical to those presented in Section 3.5.3. Making the approximation suggested in Section 3.5.4(c), that the plastic hinge length l_p may be taken as half the wall length l_w for many cases, Eq. (3.58) may be simplified to

$$\mu_\Delta = 1 + \frac{3}{2A_r}(\mu_\phi - 1)\left(1 - \frac{1}{4A_r}\right) \tag{7.22}$$

where $A_r = h_w/l_w$ is the wall aspect ratio.

For a given wall length l_w and axial load P_u, the curvature ductility factor $\mu_\phi = \phi_u/\phi_y$ will be constant. It is thus apparent from Eq. (7.22) that the available displacement ductility decreases as the aspect ratio A_r increases. Figure 5.33, derived for reinforced concrete walls, also shows this trend.

When cantilever walls are constructed on a flexible foundation [see Fig. 7.3(c)], foundation compliance will increase the yield displacement by an amount proportional to the foundation rotation but will have no influence on plastic displacement, which originates entirely within the wall plastic hinges. Equation (7.22) can be adjusted to incorporate foundation compliance in the form

$$\mu_{\Delta f} = 1 + \frac{3}{2fA_r}(\mu_\phi - 1)\left(1 - \frac{1}{4A_r}\right) \tag{7.23a}$$

or

$$\mu_{\Delta f} = 1 + \frac{\mu_\Delta - 1}{f} \tag{7.23b}$$

where f is a constant expressing the increase in flexibility due to foundation deformation ($f = 1$ for a rigid foundation).

This reduction in available displacement ductility is particularly important for very stiff structures ($T < 0.35$), where the increase in elastic flexibility may be associated with an increased spectral response. In Section 3.5.5 we discuss

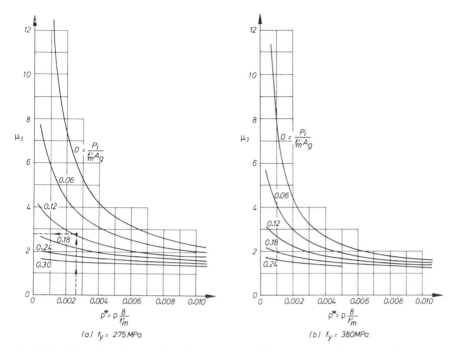

Fig. 7.15 Ductility of unconfined masonry walls for aspect ratio $A_r = 3$. (1 MPa = 145 psi.)

other aspects, including shear deformation, which may reduce available displacement ductility.

From Eqs. (7.22) and (7.23), the factor, other than aspect ratio A_r and foundation flexibility coefficient f, affecting displacement ductility capacity will be the curvature ductility factor ϕ_u/ϕ_y. This will be a function of the material properties f'_m and f_y, the axial load ratio $P_u/(f'_m l_w t)$, and the total longitudinal reinforcement ratio $\rho = A_{st}/l_w t$. Figures 7.15 and 7.16 contain ductility charts for unconfined masonry walls and for masonry walls confined by 3-mm confining plates which provide a confinement ratio of $\rho_s = 0.00766$ and $f_{yh} = 315$ MPa (45.6 ksi) for two flexural reinforcement yield strengths, $f_y = 275$ MPa (40 ksi) and $f_y = 380$ MPa (55 ksi) [P59].

The charts are in dimensionless form, for different levels of axial load ratio, and are plotted against a modified reinforcement ratio of:

For unconfined walls: $\rho^* = 8\rho/f'_m$

For confined walls: $\rho^* = 10.42\rho/Kf'_m$

where $K = 1 + \rho_s(f_{yh}/f'_m)$ is the strength enhancement ratio resulting from

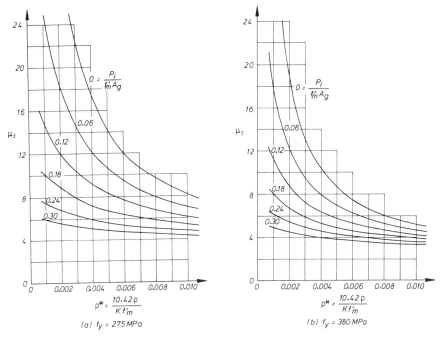

Fig. 7.16 Ductility of masonry walls of aspect ratio $A_r = 3$ confined with 3-mm plates ($\rho_s = 0.00766$, $f_{yh} = 315$ MPa). (1 MPa = 145 psi.)

confinement [see Section 3.2.3(f)], and 10.42 is the value of Kf'_m for $f'_m = 8$ MPa (1160 psi), $\rho_s = 0.00766$, and $f_{yh} = 315$ MPa (45.7 ksi). Using these modified reinforcement ratios enables the influence of f'_m to be included in one chart rather than presenting a series of charts for different f'_m values.

The charts have been prepared for rigid-based walls of aspect ratio $h_w/l_w = 3$. For walls of other aspect ratios, and for flexible foundation conditions, the true displacement ductility value μ_A is given by

$$\mu_A = 1 + \frac{3.3(\mu_3 - 1)(1 - 0.25/A_r)}{fA_r} \qquad (7.24)$$

where μ_3 is the ductility given by Fig. 7.15 or 7.16 for the appropriate condition and $A_r = 3$.

The design charts of Figs. 7.15 and 7.16 together with Eq. (7.24) indicate that the ductility of rectangular masonry walls decreases as axial load, reinforcement ratio, reinforcement yield stress, or aspect ratio increase, but increases as the masonry compression strength increases. The substantial

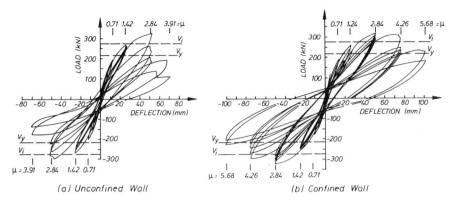

(a) Unconfined Wall

(b) Confined Wall

Fig. 7.17   Load–deflection behavior of concrete masonry test walls with high aspect ratio.

theoretical increase in ductility capacity resulting from confinement is apparent by comparing Figs. 7.15 and 7.16.

Figure 7.17 shows load–displacement hysteresis loops for two 6-m (19.7-ft)-high walls of effective aspect ratio A_r = 3.75, with ρ = 0.0072, $P_i/(f'_m l_w t)$ = 0.072, f'_m = 25 MPa (3620 psi), and f_y = 430 MPa (62.4 ksi). One of the walls was unconfined while the other had 600-mm (24 in.)-long confining plates in the bottom seven mortar courses at each end of the wall [P55]. Both walls were subjected to the same moderate level of axial load and were loaded laterally with a single force at the top of the wall. The loops of the confined wall [Fig. 7.17(b)] indicate markedly improved behavior compared to those for the unconfined wall [Fig. 7.17(a)]. As seen in Fig. 7.18, this improvement is also apparent from comparison of the condition of the two walls at moderate ductility levels.

Nevertheless, the improvement from confinement in these walls was not as marked as expected, and the apparent available ductility factor from the tests, of about μ_Δ = 5, was less than the predicted value of μ_Δ = 7. The reason for this lies in the influence of the lapped flexural starter bars at the wall base DH16 (0.63 in.). Starter bars from the foundation beam extended approximately 1000 mm (39.4 in.) into the wall. In the central region of the lap, the increased reinforcement content resulted in a local increase in flexural strength and a marked reduction in curvature. The effect of this was to reduce the effective plastic hinge length to approximately l_p = 0.2l_w, forcing much higher curvatures, and hence higher compression strains, at the wall base. This effect is clearly apparent in Fig. 7.19.

This behavior is undesirable, and where possible, lapping of flexural reinforcement should not occur within the potential plastic hinge region. The use of long starter bars can cause construction difficulties, particularly when it is necessary to "thread" blocks over the starters. However, if the base of

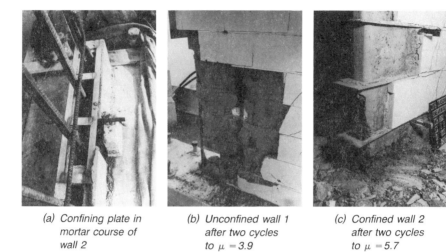

(a) Confining plate in
    mortar course of
    wall 2

(b) Unconfined wall 1
    after two cycles
    to $\mu = 3.9$

(c) Confined wall 2
    after two cycles
    to $\mu = 5.7$

Fig. 7.18   Details of concrete masonry test walls of high aspect ratio.

the lap can be lifted at least $0.5l_w$ above the wall base, substantially improved behavior can be expected. If this is not practical, the ductility capacity should be checked using $l_p = 0.2l_w$, rather than $0.5l_w$ as has been assumed in the charts of Figs. 7.15 and 7.16. It should be noted that use of open-end blocks will mean that the blocks can be moved laterally into position, and thus the construction difficulties associated with long starter bars will be minimized.

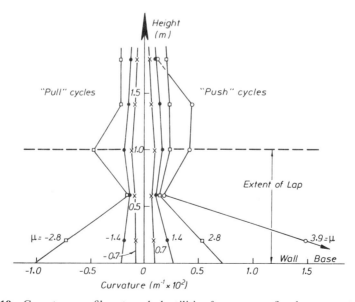

Fig. 7.19   Curvature profiles at peak ductilities for an unconfined masonry test wall.

Example 7.4 A rectangular unconfined masonry wall of height 16 m (52.5 ft) and length 4 m (13.1 ft) is reinforced with D16 bars [f_y = 275 MPa (40 ksi)] on 400-mm (15.7-in.) centers. If f'_m = 8 MPa (1160 psi), calculate the available displacement ductility factor. The appropriate axial load is 720 kN (162 kips). The reinforcement ratio is ρ = 201/(400 × 190) = 0.00264.

SOLUTION: Since f'_m = 8 MPa, ρ^* = ρ.

$$\text{Axial load ratio: } \frac{P_i}{f'_m l_w t} = \frac{720 \times 10^3}{8 \times 4000 \times 190} = 0.118$$

Thus from Fig. 7.15(a), μ_3 = 2.77 (follow the dashed line). From Eq. (7.24), assuming rigid base conditions (i.e., f = 1.0) for a wall with aspect ratio A_r = 16/4 = 4.0,

$$\mu_4 = 1 + \frac{3.33(2.77 - 1)(1 - 0.25/4)}{4} = 2.37$$

In many cases this level of available ductility capacity would be insufficient for the assumed level of lateral force. Assuming that $\mu_\Delta \geq 4$ was required, some redesign would be needed. Options available would be:

1. Use confining plates at the wall base. Figure 7.16(a) and Eq. (7.24) would indicate an available ductility of μ_4 = 7.9, more than twice the required level.
2. Changing section dimensions. Increasing the wall thickness t would reduce both axial load and reinforcement ratios, hence would increase ductility, but at the expense of increased structural weight and cost.
3. Adopt a higher design value for f'_m. For example, if the compression strength is doubled to f'_m = 16 MPa (2320 psi), the axial load ratio becomes 0.5 × 0.118 = 0.059 and the effective reinforcement ratio ρ^* = 0.5 × 0.00264 = 0.00132. Thus Fig. 7.15(a) indicates that μ_3 = 5.5. Substituting into Eq. (7.24) results in μ_4 = 4.6.

Obviously, the option of increasing f'_m will be the most satisfactory if high compression strengths can be dependably obtained. Increasing f'_m will only be necessary for the plastic hinge region of the wall.

(b) Walls with Nonrectangular Section The design charts of Figs. 7.15 and 7.16 cannot be used directly for other than rectangular-section walls. For flanged walls, an analytical approach based on calculation of the actual curvature ductility factor ϕ_u/ϕ_y, and substitution into Eq. (7.22) may be necessary. However, as has been noted in Section 5.4.3(a), when the flange of a wall is in compression, the ultimate curvature ϕ_u will be very large, because

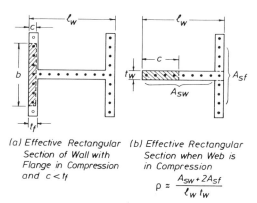

(a) Effective Rectangular
Section of Wall with
Flange in Compression
and $c < t_f$

(b) Effective Rectangular
Section when Web is
in Compression

$$\rho \approx \frac{A_{sw} + 2A_{sf}}{l_w t_w}$$

Fig. 7.20   Equivalent sections for ductility calculations of flanged masonry walls.

of the small compression depth c at ultimate moment capacity resulting from the great width of the compression zone. For such walls it will rarely be necessary to check ductility. However, if the compression zone is entirely within the flange thickness (i.e., $c \leq t_f$), the ductility capacity may be found from Fig. 7.15 and Eq. (7.24), using an "equivalent" reinforcement ratio $\rho = A_s/l_w b$ and an equivalent axial load ratio $P_i/f_m l_w b$, where b is the effective width of the flange, as shown in Fig. 7.20(a).

For T sections where the web is in compression, the available ductility can be quite low because of the increased depth of the compression zone, shown in Figs. 7.20(b) and 7.21, resulting from the effect of the flange tension reinforcement. A conservative estimate of ductility capacity may be obtained from Fig. 7.15 or 7.16 and Eq. (7.24) for an equivalent rectangular wall with dimensions equal to the web dimensions, carrying the entire T-section axial load, and with an equivalent reinforcement ratio

$$\rho_w = \frac{A_{sw} + 2A_{sf}}{l_w t_w} \tag{7.25}$$

where A_{sw} is the total web steel area, A_{sf} is the total flange steel area, within the effective flange width, and t_w and l_w are web thickness and length, respectively, as shown in Fig. 7.20(b). Effective widths of flanges may be made using Eqs. (5.1a) and (5.1b) and Fig. 5.6.

In comparing ductility capacity from Figs. 7.15 and 7.16 with required design levels, any flexural overstrength resulting from excess reinforcement provided should be considered when establishing the expected ductility demand.

7.2.5 Design for Shear

(a) Design Shear Force Considerations for design of wall structures to avoid inelastic shear response are presented in Section 5.4.4 and are equally applicable to masonry structures. Shear failure of masonry elements must be avoided by a capacity design process that takes into account potential material strengths, strain hardening, and dynamic amplification of shear force. Thus the design shear force V_u will be related to the shear V_E corresponding to code-specified lateral forces by Eq. (5.22), reproduced here for convenience:

$$V_u = V_{\text{wall}} = \omega_v \phi_{o,w} V_E \qquad (5.22)$$

where ω_v is the dynamic shear amplification factor, defined by Eq. (5.23).

In Eq. (5.22), the overstrength factor $\phi_{o,w}$ relates the maximum feasible flexural strength of the wall to the required strength corresponding to the code distribution of lateral force [Eq. (5.13)]. The value of $\phi_{o,w}$ will typically exceed that appropriate for concrete wall structures because of lower flexural strength reduction factors adopted in masonry design [Eq. (3.47)], and because of the increased probability of excess flexural reinforcement being provided, as a result of limited options for bar spacing.

It should also be recognized that for walls, the design load combination for assessing required flexural strength will not generally be the appropriate combination for assessing required shear strength. For example, the flanged wall in Fig. 7.21(a) will have flexural reinforcement for the sense of lateral

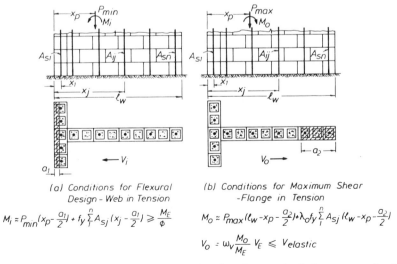

(a) Conditions for Flexural
Design - Web in Tension

(b) Conditions for Maximum Shear
-Flange in Tension

$$M_i = P_{min}\left(x_p - \frac{a_1}{2}\right) + f_y \sum_1^n A_{sj}\left(x_j - \frac{a_1}{2}\right) \geq \frac{M_E}{\phi}$$

$$M_o = P_{max}\left(\ell_w - x_p - \frac{a_2}{2}\right) + \lambda_o f_y \sum_1^n A_{sj}\left(\ell_w - x_p - \frac{a_2}{2}\right)$$

$$V_o = \omega_v \frac{M_o}{M_E} V_E \leq V_{elastic}$$

Fig. 7.21 Capacity design principles for estimating the required shear strength of a flanged masonry wall.

forces shown generally dictated by minimum axial load, with the moment inducing compression in the flange and tension in the web.

Much higher flexural strength will result on reversal of the direction of lateral loading, putting the flange reinforcement in tension, in combination with maximum axial load [Fig. 7.21(b)]. This condition, with reinforcement at the appropriate level of overstrength, should be used for assessing the required shear strength of the web of the element. Both expressions in Fig. 7.21 for moment capacity are developed by taking moments about the center of the compression block.

A similar condition to that represented by Fig. 7.21 occurs with masonry pilasters built integrally with long walls. The pilasters may be constructed out of special pilaster units, and intended to act as a column, or may be constructed as a stub wall perpendicular to the long wall. In each case, the interaction between the long wall and the pilaster must be considered carefully. Reinforcement in the long wall will significantly increase the flexural strength of the pilaster or stub wall, resulting in high seismic shear forces. To assess an upper bound on the shear force, an approach similar to that illustrated in Fig. 7.21 must be adopted, with the length of the long wall contributing to flexure being based on capacity design principles (Section 5.2.2). Alternatively, the shear in the pilaster or stub wall should be conservatively based on forces corresponding to elastic response to the design-level earthquake.

(b)   ***Shear Strength of Masonry Walls Unreinforced for Shear***   There have been numerous experimental studies of the shear strength of unreinforced masonry. Figure 7.22 illustrates some of the test methods that have been used to develop shear stresses in masonry elements. None of the methods in Fig.

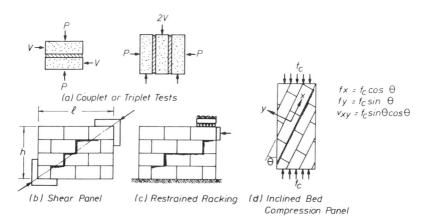

$$f_x = f_c \cos \theta$$
$$f_y = f_c \sin \theta$$
$$v_{xy} = f_c \sin\theta\cos\theta$$

(a) Couplet or Triplet Tests

(b) Shear Panel   *(c) Restrained Racking*   *(d) Inclined Bed Compression Panel*

Fig. 7.22   Methods of testing shear strength in masonry construction.

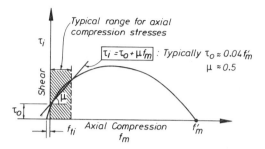

Fig. 7.23   Shear–compression interaction.

7.22 give a good representation of the actual behavior under seismic forces, where the cyclic reversal of force direction coupled with the influence of crack propagation along mortar beds by flexural action may cause a reduction in the true shear strength compared with the values measured in simple monotonic tests where flexural cracking is inhibited.

Although the different test methods tend to give different shear strengths, all indicate a strong dependence of shear stress τ_i on transverse compression stress (f_m). The general form of the shear strength equation used is

$$\tau_i = \tau_o + \mu f_m \tag{7.26}$$

also shown in Fig. 7.23.

In fact, this is a simplification of the real interaction between shear and compression, which is illustrated for the full range of stresses in Fig. 7.23. It will be seen that the linear equation (7.26) provides an adequate representation of shear strength over the typical range of axial compression for uncracked masonry. However, after cracking, local compression stresses can increase to values close to the compression strength. In such cases, Eq. (7.26) is clearly invalid. Values for the constants τ_o and μ vary with test method and type of masonry. Typical ranges of values are $0.1 \leq \tau_o \leq 1.5$ MPa $(15 \leq \tau_o \leq 218$ psi) and $0.3 \leq \mu \leq 1.2$.

(c) Design Recommendations for Shear Strength   The principles for determining shear strength developed in Sections 3.3.2 and 5.4.4 may be applied to design of masonry walls, with minor modifications as follows. The ideal shear strength of a masonry wall reinforced for shear, by analogy to Eq. (3.32), is

$$V_i = V_m + V_s \tag{7.27}$$

where $V_m = v_m b_w d$ is the contribution of the masonry to shear strength, and V_s is the contribution of shear reinforcement.

(i) *Contribution of the Masonry,* v_m: Most current reinforced masonry design codes specify values for v_m that do not vary with axial load level. Some put $v_m = 0$ when the total shear stress $V_i/b_w d$ exceeds some specified level requiring the total shear force to be carried by transverse reinforcement. In view of the discussion above, this seems unrealistically conservative for well-designed masonry. Recent extensive experimental research in Japan [M19] and the United States [S17] support the following equations.

1. In all regions except potential plastic hinges,

$$v_m = 0.17\sqrt{f'_m} + 0.3(P_u/A_g) \quad \text{(MPa)}; \qquad 2.0\sqrt{f'_m} + 0.3(P_u/A_g) \quad \text{(psi)}$$
$$(7.28a)$$

but not greater than

$$v_m = 0.75 + 0.3(P_u/A_g) \quad \text{(MPa)}; \qquad 110 + 0.3(P_u/A_g) \quad \text{(psi)} \quad (7.28b)$$

nor greater than

$$v_m = 1.3 \text{ MPa} \quad (190 \text{ psi}) \qquad (7.28c)$$

The experimental data support an increased shear strength for walls with aspect ratio less than $h_w/l_w = 1$. However, more research is needed to quantify the increase.

2. In regions of plastic hinges there is comparatively little experimental evidence on which to base design expressions for strength of masonry shear-resisting mechanisms in potential plastic hinge regions. Because of this, at least two codes [X10, X13] that specifically address the problem require that $v_m = 0$ in such regions, regardless of axial load level. From examination of recent rest data [M19, S17] the following conservative recommendations are drawn:

$$v_m = 0.05\sqrt{f'_m} + 0.2(P_u/A_g) \quad \text{(MPa)}; \qquad 0.6\sqrt{f'_m} + 0.2(P_u/A_g) \quad \text{(psi)}$$
$$(7.29a)$$

but not greater than

$$v_m = 0.25 + 0.2(P_u/A_g) \quad \text{(MPa)}; \qquad 36 + 0.2(P_u/A_g) \quad \text{(psi)} \quad (7.29b)$$

nor greater than

$$v_m = 0.65 \text{ MPa} \quad (94 \text{ psi}) \qquad (7.29c)$$

(ii) *Contribution of Shear Reinforcement*: The design of horizontal shear reinforcement for masonry walls follows the principles developed in Section 3.3.2(*a*) (v), and widely accepted for reinforced concrete. Thus from Eqs.

(3.41) and (3.32), the required horizontal steel area A_{sh} at vertical spacing s_h is

$$A_{sh} = \frac{(V_i - V_m)s_h}{f_{yh}d} \qquad (7.30)$$

where d is the effective depth, normally taken as $0.8l_w$ for rectangular wall sections with distributed reinforcement. A larger value of d is appropriate for flanged walls when the flange is in tension.

When the aspect ratio is less than unity, the considerations of Section 5.7 apply. However, since it will rarely be feasible to place diagonal wall reinforcement, as shown for example in Fig. 5.63, the required horizontal reinforcement, from Eq. (5.41b) will be

$$A_{sh} = \frac{(V_i - V_m)s_h}{f_{yh}h_w} \qquad (7.31)$$

It will be seen that this differs from Eq. (7.30) only by substitution of h_w for d. For uniformity of design approach, and because of potentially greater degradation of V_m for squat walls because of sliding, it is recommended that Eq. (7.30) be adopted regardless of wall aspect ratio.

(d)   Effective Shear Area   In the discussion above, the effective shear area $b_w d$ was introduced. Although these terms are familiar from reinforced concrete design [Section 3.3.2(a)], they are less obviously calculated for masonry, because of the typically large spacing of flexural reinforcement, and because of the influence of ungrouted vertical flues, if these exist. Figure 7.24 shows the appropriate definitions of b_w and d for a number of different conditions [X13].

(i)   In-Plane Shear Force: For fully grouted walls, $d = 0.8l_w$ and $b_w = t$ [Fig. 7.24(a)]. For partial grouting, $d = 0.8l_w$ and $b_w = t - b_f$, where $b_f = $ maximum width of ungrouted flue [Fig. 7.24(b)]. Thus for partially grouted walls, the effective section width for shear will be the net thickness of the face shells. This limitation is necessary to satisfy requirements of continuity of shear flow, and to avoid the possibility of vertical shear failure up a continuous ungrouted flue. Typically, for partially grouted concrete hollow cell masonry, as shown in Fig. 7.24, $b_w = 60$ mm (2.36 in.).

(ii)   Face-Loaded Shear Force: For fully grouted walls with centrally located reinforcement, $d = t/2$ and $b_w = s_v$, but $b_w \leq 4t$ for running bond, and $b_w < 3t$ for stack bond [Fig. 7.24(c)]. For partially grouted walls with centrally located reinforcement, $d = t/2$, and $b_w = s_v$ less the length of any ungrouted flues, where s_v is the spacing of vertical reinforcement, not to be taken larger than $4t$ for walls constructed in running bond, nor larger than $3t$ for walls constructed in stack bond [Fig. 7.24(d)]. In both face-loaded cases,

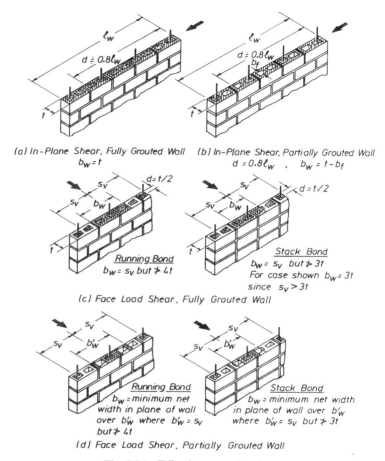

(a) In-Plane Shear, Fully Grouted Wall
$b_w = t$

(b) In-Plane Shear, Partially Grouted Wall
$d = 0.8\ell_w$, $b_w = t - b_f$

Running Bond
$b_w = s_v$ but $\not> 4t$

Stack Bond
$b_w = s_v$ but $\not> 3t$
For case shown $b_w = 3t$
since $s_v > 3t$

(c) Face Load Shear, Fully Grouted Wall

Running Bond
$b_w = $ minimum net width in plane of wall over b'_w where $b'_w = s_v$ but $\not> 4t$

Stack Bond
$b_w = $ minimum net width in plane of wall over b'_w where $b'_w = s_v$ but $\not> 3t$

(d) Face Load Shear, Partially Grouted Wall

Fig. 7.24 Effective areas for shear.

b_w is the effective width of masonry corresponding to each vertical bar. The total effective width is found by summation of the individual b_w values.

(e) Maximum Total Shear Stress Although ductile flexural behavior has been obtained from masonry walls with maximum total shear stresses as high as 2.8 MPa (400 psi), the data base is not extensive, and further research is needed to adequately identify upper limits to shear strength. The following provisions [X13] are considered to be suitably conservative until more data are available. Limit shear stress in potential plastic hinge regions to

$$v_i = \frac{V}{b_w d} \le 0.15 f'_m \quad \text{but} \ \le 1.8 \text{ MPa } (260 \text{ psi}) \qquad (7.32a)$$

For other regions,

$$v_i \leq 0.2f'_m \qquad \text{but} \ \leq 2.4 \text{ MPa (350 psi)} \qquad (7.32b)$$

7.2.6 Bond and Anchorage

Reliable data for anchorage requirements for reinforcement in masonry walls are not available. Behavior is complicated by the difficulty of assessing grout strength in the wall, the propensity for vertical splitting of masonry in the compression zone, the presence of planes of weakness parallel to reinforcement caused by horizontal or vertical mortar joints, and by the action of cyclic tension/compression stress reversals occurring under seismic response.

In regions where inelastic flexural action cannot develop (i.e., other than in potential plastic hinge regions), the development length l_d should be not less than

$$l_d = 0.12f_y d_b \ \text{ (MPa)}; \qquad l_d = 0.83f_y d_b \ \text{ (ksi)} \qquad (7.33a)$$

where d_b is the diameter of the bar. Equation $(7.33a)$ results in a development length of approximately $50d_b$ when $f_y = 414$ MPa (60 ksi).

Reinforcement should not be lapped in potential hinge regions. Where lapping cannot be avoided, the length given by Eq. $(7.33a)$ should be increased by approximately 50%, so that

$$l_d = 0.18f_y d_b \ \text{ (MPa)}; \qquad l_d = 1.24f_y d_b \ \text{ (ksi)} \qquad (7.33b)$$

It is common practice for vertical starter bars to be cast into the foundation beam for masonry walls and lapped with the main wall reinforcement, which is placed after laying up perhaps a story height of wall. This avoids the problem of having to thread hollow cell units over reinforcement that may extend 3 or 4 m above the wall base. However, the resultant lapping within the potential hinge region is likely to result in poor performance during seismic response.

The load–displacement hysteresis loop shown in Fig. 7.17 is for masonry walls with lapped starter bars at the wall base, with the lap length 25% longer than required to satisfy Eq. $(7.33a)$. The unconfined wall [Fig. 7.17(a)] suffered rapid degradation of performance after vertical splitting of the end regions at the wall base resulted in bond failures of the lapped bars at the ends of the wall [Fig. 7.18(a)].

Another similar wall subjected to reduced axial load and using extended lap length, satisfying Eq. $(7.33b)$, behaved better than the earlier unconfined wall. However, it, too, eventually suffered a bond failure at the wall ends. A wall with confining plates behaved much better, due to the action of the plates in inhibiting vertical splitting. Bond failure did not occur in that wall.

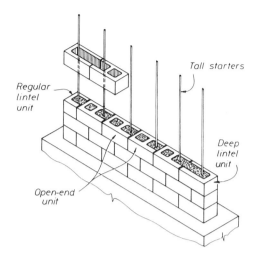

Fig. 7.25 Laying of open-end masonry blocks to avoid threading of blocks over long starter bars.

As noted in Section 7.2.4(*a*) and shown in Fig. 7.19, lapping of the vertical reinforcement in the potential hinge region has the further undesirable effect of concentrating the plasticity into a shorter hinge region, thus increasing peak compression strains and inducing vertical splitting at the wall ends earlier than would have been expected for walls without lapped starter bars.

Clearly, efforts must be made to avoid the lapping of vertical reinforcement in potential hinge regions. Figure 7.25 shows how this can be achieved using open-end units, with lintel [typically, 200 mm (8 in.)] or deep lintel units at the wall ends and a bar spacing equal to the unit module [i.e., 400 mm (16 in.)]. The top layer shown completed has been constructed from right to left, with the units moved laterally into position. In the next layer, the open-end blocks will be reversed (open end to the left) and the wall constructed from left to right. In this way the need to thread bars over high starters is avoided. Alternatively, when this approach is not feasible, due to tighter reinforcement spacing or other constraints, the base of the lap should be moved up as far as possible from the wall base, but not less than a height equal to one-fourth of the wall length.

To be fully effective on either side of any potential inclined crack, it is essential that shear reinforcement be anchored adequately at both ends. This generally requires a hook or bend at the end of the reinforcement, as shown in Fig. 7.26. Hooking the bar round the end vertical reinforcement [Fig. 7.26(*a*)] is the best solution for anchorage. However, in some cases it may induce excessive congestion at end flues and may result in incomplete grouting of the flue. In such cases bending the reinforcement up or down into

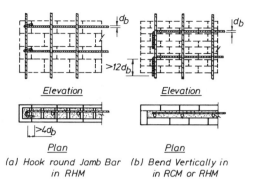

Fig. 7.26   Anchorage of horizontal shear reinforcement.

the end vertical flue, as shown in Fig. 7.26(b), may be considered. This detail should not be used for short walls. Tests on walls with this detail of anchorage [P55, P56] have indicated satisfactory performance. Hook dimensions should be the same as specified for reinforcement in concrete construction [A1, X3, X11]. Horizontal shear reinforcement should not be lap spliced within the plastic hinge region.

7.2.7   Limitations on Wall Thickness

Many codes [X10, X13] limit the ratio of unsupported wall height (i.e., story height) to wall thickness to reduce the probability of lateral instability of compression zones. However, as explained in Section 5.4.3(c), buckling potential is more properly related to wall length, and expected ductility required when the potential buckling area is subjected to inelastic tension in the opposite sense of loading direction. The considerations developed in Section 5.4.3(c) also apply for masonry walls. However, as a result of the conservative nature of masonry flexural strength design and difficulties in satisfying Eq. (5.15b) for long walls, it is recommended that the displacement ductility factor μ_Δ used in Fig. 5.35 to determine minimum wall thickness be based on the ideal rather than the dependable strength.

Where the wall height is less than three stories, a greater slenderness should be acceptable. In such cases, or where inelastic flexural deformations cannot develop, the wall thickness should satisfy

$$t \leq 0.05 l_n \tag{7.34}$$

7.2.8   Limitations on Reinforcement

(a) Minimum Reinforcement   Where structural considerations indicate that very little reinforcement is required, a minimum amount must be provided to ensure adequate control of shrinkage, swelling, and thermal effects. Most

codes specify a minimum of 0.07% of the gross cross-sectional area of wall, taken perpendicular to the reinforcement considered, both vertically and horizontally, with the sum of the vertical and horizontal reinforcement ratio being not less than 0.2%:

$$\rho_v \geq 0.0007; \qquad \rho_h \geq 0.0007; \qquad \rho_v + \rho_h \geq 0.002$$

(b)   Maximum Reinforcement   It is necessary to set practical upper limits for reinforcement in flues and cavities to ensure that a construction method is achieved that will allow an adequate vibrator space, and also to limit bond stresses to achievable limits. It is recommended that wherever possible there should be only one bar in each flue, except at laps, for walls of thickness $t \leq 190$ mm (7.5 in.). The maximum area of reinforcement, A_{sf}, in a flue or cavity should not exceed

$$A_{sf} = (8/f_y)A_f \quad (\text{MPa}); \qquad A_{sf} = (1.16/f_y)A_f \quad (\text{ksi}) \quad (7.35a)$$

except at laps, when the total area may be increased to

$$A_s = (13/f_y)A_f \quad (\text{MPa}); \qquad A_s = (1.89/f_y)A_f \quad (\text{ksi}) \quad (7.35b)$$

where A_f is the area of flue or tributary area of cavity (see Fig. 7.27).

For 190-mm (7.5-in.) concrete blockwork with typical flue dimensions of about 120 mm × 150 mm (18,000 mm^2) (28 in.^{2}), Eq. (7.35) would permit a maximum area of 524 mm^2 (0.81 in.^{2}) for grade 275 (40 ksi) reinforcement, or 360 mm^2 (0.56 in.^{2}) for grade 400 (60 ksi) reinforcement; for example, one D24 (0.94-in.) bar, two D16 (0.63-in.) bars, or one DH20 (0.79-in.) bar per flue. At laps the corresponding areas of 851 mm^2 (1.32 in.^{2}) for grade 275 and 585 mm^2 (0.91 in.^{2}) for grade 400 would permit lapping one D20 or two D16 bars in one flue. Lapping one DH20 would marginally exceed the area limitations.

When considering the maximum area of reinforcement in a horizontal flue of a bond beam, Eq. (7.35) should be used. However, it is permissible to use the full height of the block, rather than the height of cavity above a depressed web, when assessing the flue area A_f in a wall constructed of hollow unit masonry.

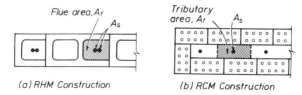

Flue area, A_f

A_s

Tributary area, A_f

A_s

(a) RHM Construction   (b) RCM Construction

Fig. 7.27   Flue and tributary area (A_f) definition for maximum reinforcement ratio.

(c) *Maximum Bar Diameter* A limitation on maximum bar diameter is needed to ensure against the undesirable situation of having very large diameter bars at large spacings. Although Eq. (7.35) will provide some protection against this in hollow unit masonry, it will not provide the necessary protection for grouted cavity masonry. In all cases the diameter of bars used in walls should not exceed:

1. One-fourth of the least dimension of the flue or cavity containing the reinforcement, or
2. One-eighth of the gross wall thickness.

This requirement limits maximum bar diameter in 140-mm (5.6-in.) block-work to 16 mm (0.63 in.) and in 190-mm blockwork to 20 mm (0.79 in.). Due to different available bar sizes in the United States, a No. 7 bar (22.2 mm) would be acceptable in nominal 8-in. masonry construction.

(d) *Bar Spacing Limitations* Within plastic hinge regions it is important that the vertical and horizontal reinforcement be sufficiently closely spaced to provide a "basketing" action under intersecting diagonal tension cracks. There should be at least four vertical bars, spaced at not greater than 400 mm (16 in.). Horizontal shear reinforcement should be spaced not greater than 400 mm (16 in.) nor $l_w/4$ vertically.

When the wall is designed to respond elastically to the design level earthquake, and where the corresponding shear stress is less than half of that given by Eq. (7.28), shear reinforcement is not required. However, a minimum area of 0.07% should still be used in the horizontal direction. Provided that the structure is small, say two stories or less, it is permissible to concentrate this reinforcement in story-height bands. This approach should not, however, be adopted for walls constructed in stack bond, nor in very seismically active areas because of the reduction in structural integrity resulting from the one-way reinforcement.

(e) *Confining Plates* Where confining plates are used to enhance the ductility capacity of potential plastic hinge regions of masonry walls, the following requirements should be met:

1. The plate material should be stainless or galvanized steel.
2. The minimum effective area of confining plate in each horizontal direction cut by a vertical section of area $s_h h''$ should be

$$A_p = 0.004 s_h h'' \qquad (7.36)$$

where s_h is the vertical spacing of the confining plates and h'' is the lateral dimension of the confined core shown in Fig. 7.28.

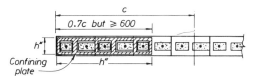

Fig. 7.28   Confining plates in potential plastic hinge regions. (1 mm-0.0394 in.)

3. Confinement should be provided from the critical section (e.g., the wall base) for a height not less than the plastic hinge length [see Section 5.4.3(e)(iii)] and for a horizontal distance not less than 0.7c nor 600 mm (24 in.) from the extreme compression fiber. The latter requirement is illustrated in Fig. 7.28. Covering the most highly strained 70% of the compression zone ensures that no unconfined fibers are subjected to strains in excess of the unconfined compression strain of 0.003 [see Section 3.2.3(g)]. The minimum length of 600 mm (24 in.) ensures that the end two vertical flues and any reinforcement they contain are tied back into the body of the wall, minimizing the influence of the weak header joints close to the extreme compression fiber. The confining plates must be continuous over masonry header joints for the full required length. The required plate area transverse to the wall should be obtained by cross-links above each masonry unit web, and should not inhibit grout placing, or vibrating.

7.3   MASONRY MOMENT-RESISTING WALL FRAMES

As discussed in Section 7.2.1, the preferred masonry structural system for resisting lateral inertia forces resulting from seismic response of buildings is the simple cantilever wall. An alternative reinforced masonry structural system that is suitable for ductile response is the moment-resisting wall frame, whose proportions in terms of ratio of bay length to story height is more or less typical of reinforced concrete frames rather than of coupled structural walls. Such an example is shown in Fig. 7.29. Although the system could be constructed from either reinforced grouted brick masonry or hollow unit masonry, the illustrations in this section will relate to the latter. The column units shown in Fig. 7.29 are constructed using standard units rather than special pilaster units, which effectively act only as permanent formwork. Figure 7.30(a) shows the joint region of a planar wall frame such as that discussed in this section. The alternative "frame" structure using pilaster unit columns, shown in Fig. 7.30(b), will only be suitable for light one- or two-story structures.

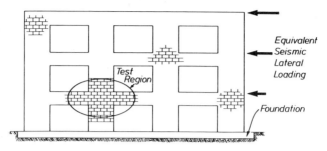

Fig. 7.29 Ductile wall frame in reinforced masonry.

7.3.1 Capacity Design Approach

Capacity design principles developed for reinforced concrete frames in Chapter 4 may be applied in simplified form in the design of masonry wall frames. Thus plastic hinges should be forced to form at beam ends, with plastic hinging of the wall-like column members avoided by use of the weak beam/strong column approach. Other undesirable mechanisms, such as beam or wall shear failures, or joint failures, must be avoided since these are brittle and do not possess the fundamental requisite of ductile response, namely the ability to deform inelastically during repeated cyclic displacement response without significant strength or stiffness degradation.

The flexural reinforcement of the potential plastic hinge region is detailed to ensure a dependable flexural strength no less than that corresponding to the code distribution of lateral seismic forces, but all other parts of the structure are detailed to ensure that their strength exceeds that correspond-

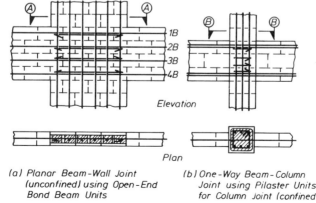

Fig. 7.30 Interior beam–wall and beam–column joints in masonry.

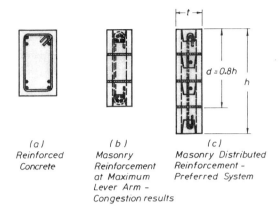

(a)
Reinforced
Concrete

(b)
Masonry
Reinforcement
at Maximum
Lever Arm –
Congestion results

(c)
Masonry Distributed
Reinforcement –
Preferred System

Fig. 7.31 Distribution of reinforcement in concrete and masonry beams.

ing to a maximum feasible strength of the potential plastic hinge regions. This maximum feasible strength is, of course, very much higher than the required dependable strength of the beam hinges, since the latter will include a strength reduction factor and be based on conservative estimates of material strengths (masonry compression strength and reinforcement yield strength) rather than actual strengths. This procedure recognizes that it is the actual flexural strength rather than the dependable strength that will be developed at the plastic hinges.

7.3.2 Beam Flexure

It is conventional practice in reinforced concrete beam flexural design for seismic resistance to concentrate the reinforcement in two layers, near the top and bottom faces of the beam as shown in Fig. 7.31(*a*), thus obtaining the maximum possible lever arm. Where gravity moments are small compared with seismic moments, as will generally be the case for normally proportioned frames in regions of high seismicity, the quantity of reinforcement in each of the two layers will essentially be equal. An attempt to adopt the same practice for masonry, as in Fig. 7.31(*b*), will result in extreme congestion because of the limited beam width, and will make placing of shear reinforcement and grouting of the beam very difficult. Consequently as shown in Fig. 7.31(*c*), the reinforcement should be uniformly distributed down the beam depth.

As is the case for walls, there is little penalty in uniformly distributing the reinforcement, as comparative analyses show that the flexural strength of a beam with reinforcement uniformly distributed through the depth is only a few percent less than that of a beam with the same total quantity of reinforcement concentrated in two equal layers adjacent to the top and bottom faces. Moreover, as with wall design, there are advantages to the uniform distribution. In addition to easing grouting and steel placement difficulties, the uniformly distributed reinforcement is better restrained against compression buckling, provides better resistance to sliding shear by dowel action, and results in a somewhat greater depth of the flexural compression zone, enhancing interface shear transfer.

Beam depths will need to be greater than for reinforced concrete beams designed to carry the same moment, because of reduced beam widths and limitations on reinforcement quantities. To facilitate placement of the longitudinal flexural reinforcement, depressed-web bond-beam units should be adopted throughout when construction using hollow-cell units. To enhance structural integrity the units should have at least one open end. In Fig. 7.31(c) the bottom block is laid inverted to allow placement of the lower flexural reinforcement at maximum effective depth and to facilitate cleanout of mortar droppings prior to grouting.

Analysis for flexural strength of beam members with distributed flexural reinforcement can be carried out using the methods developed in Section 7.2.3(e) for rectangular-section walls. Alternatively, the design charts of Fig. 7.14 may be used to estimate flexural strength, with the axial load set to zero. A strength reduction factor of $\phi = 0.85$ is recommended for beam flexure (Section 3.4.1).

7.3.3 Beam Shear

Within potential plastic hinge regions, wide flexural cracks are to be expected, propagating from top and bottom faces of the beam as the inertial forces change direction. As a consequence, masonry shear-resisting mechanisms are likely to be greatly weakened. It is thus advisable to carry all shear force on shear reinforcement within a region equal to the beam depth from the wall face.

Using conventional reinforced concrete strength theory and capacity design principles, the required area, A_v, of shear reinforcement is determined using Eqs. (7.27) and (7.30), where d, the effective depth, may be taken as $0.8h$ [see Fig. 7.31(c)] and the shear force must correspond to the maximum feasible flexural strength. Evaluating the flexural overstrength, as outlined in Section 4.5.1(e), and incorporating the effects of the flexural strength reduction factor of $\phi = 0.85$, the design shear force will need to be between 1.47 and 1.65 times the shear force corresponding to development of dependable flexural strength, to avoid the possibility of beam shear failure.

7.3.4 Column Flexure and Shear

To ensure against column plastic hinges or shear failures, design column actions in masonry construction should be based on the procedure developed in Section 4.6 for reinforced concrete columns.

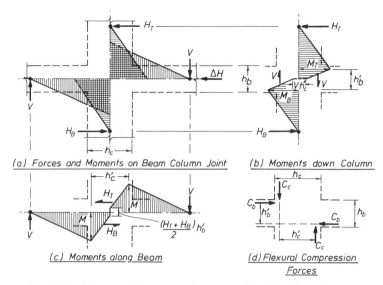

Fig. 7.32   Forces and moments for computing joint shear forces.

7.3.5   Joint Design

The design of the joint region between beams and columns requires special consideration. Two aspects are of particular importance:

1. The width of the joint (parallel to the beam axis) must be sufficient to allow the necessary change in beam reinforcement stress through the joint to be developed by bond.
2. The quantities and reinforcement in the joint must be adequate to carry the shear forces developed in the joint by the moment gradient across the joint.

Beam–column joint design has been considered in some depth in Section 4.8. There are, however, significant differences in construction between reinforced concrete and planar masonry frames, which influence the design requirements for the latter. In particular, the distributed nature of beam reinforcement and the lack of confinement of the core region by rectangular hoops in masonry joints needs consideration.

(a)   *Limitations on Bar Sizes*   Figure 7.32(a) shows the seismic forces the moments acting on a typical interior joint of a masonry frame, such as the circled region of Fig. 7.29. Because of the moment reversal across the joint, beam reinforcement may be yielding in compression at one side of the joint, and yielding in tension on the other side [Fig. 7.32(c)]. Consequently, the

joint width, which is equal to the column width, h_c, must be at least equal to the sum of tension and compression development lengths for the reinforcement. Thus for a given column width, the diameter of the beam flexural reinforcement d_b must be limited. This limitation must be more stringent than those relevant to reinforced concrete joints. Testing of wall–beam joints [P50, P60] supports the following relationship:

$$d_b \le 5.0 h_c/f_y \quad (\text{MPa}); \qquad d_b \le 0.725 h_c/f_y \quad (\text{ksi}) \qquad (7.37)$$

Similar considerations limit the diameter of column flexural reinforcement due to the moment gradient vertically through the joint [Fig. 7.32(b)]. However, since the capacity design approach will reduce column reinforcement stresses, less stringent requirements are appropriate thus:

$$d_b \le 8.0 h_b/f_y \quad (\text{MPa}); \qquad d_b \le 1.16 h_b/f_y \quad (\text{ksi}) \qquad (7.38)$$

where h_b is the beam height.

The limitations of Eqs. (7.37) and (7.38) are very much more severe than recommended in Section 4.8.6 for reinforced concrete beam–column joints. The very high bond stresses which correspond to the low h_c/d_b ratios, recommended for concrete joints, can develop only as a result of lateral confinement of the joint core region by closed joint stirrups, which also control splitting. Since reinforcement in a masonry beam–column joint will all be in the same plane, lateral confining pressures to control splitting cannot develop. A masonry wall-frame test where beam bar diameter exceeded by 25% that permitted by Eq. (7.37) performed well up to displacement ductility factors of $\mu_\Delta \le 3$, but then degraded as a result of bar slipping through the joint [P50]. In view of the gradual nature of the resulting degradation, the increased conservatism of Eq. (7.37) should be adequate.

(b) Joint Shear Forces   The joint shear forces may be evaluated by consideration of the moment gradient horizontally and vertically through the joint. This procedure, which differs slightly from the approach developed for concrete joints in Section 4.8, is more convenient for beams with distributed reinforcement but produces identical results. The relevant forces and moments for computation of horizontal joint shear force V_{jh} are shown in Fig. 7.32(b). Note that the moment Vh'_c from the beam shear forces (assumed equal in this example) assist in effecting the change in column moment in Fig. 7.32 from M_T at the top of the joint to M_B at the bottom of the joint, and hence the horizontal joint shear will be

$$V_{jh} = \frac{M_T + M_B - Vh'_c}{h'_b} \qquad (7.39)$$

with a corresponding joint shear stress of

$$v_{jh} = V_{jh}/h_c t \tag{7.40}$$

where t is the thickness of the joint.

Similarly, the vertical joint shear force, V_{jv}, across the joint will be given by

$$V_{jv} \approx \frac{1}{h'_c}\left(2M - \frac{H_T + H_B}{2}h'_b\right) \tag{7.41}$$

assuming that the beams have equal moment capacity, M in both loading directions. The moments and shear forces used in Eqs. (7.39) and (7.41) should correspond to development of maximum feasible beam moment capacity, to satisfy capacity design principles. Note that in Eqs. (7.39) and (7.41), the effective joint dimensions h'_b and h'_c are less than the external joint dimensions (h_b and h_c) since the change in moment gradient or joint faces will not occur over an infinitely short distance, as idealized in Fig. 7.32. The appropriate values of h'_b and h'_c, as seen in Fig. 7.32, will be the distances between centers of flexural compression forces, C_b or C_c, in the beams and columns at opposite sides of the joint, as shown in Fig. 7.32(d).

Conventional reinforced concrete joint design theory [X3] would require all of V_{jh} to be carried by horizontal joint reinforcement in the form of stirrups, unless axial load levels are high. However, the distributed nature of the beam flexural reinforcement will mean that maximum moment will be provided by a combination of a masonry compression force and steel tensile forces, as shown in Fig. 7.33, rather than by a steel couple, as will be more common for reinforced concrete joints with flexural reinforcement concentrated near the top and bottom faces. Consequently, as shown in Fig. 7.33, a diagonal

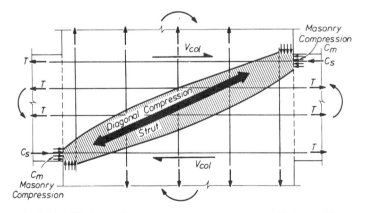

Fig. 7.33 Horizontal forces acting on a masonry wall–beam joint.

compression strut can develop at all times, carrying part of the joint shear, in much the same way as in a joint in reinforced concrete, shown in Fig. 4.44a. Test results [P50] indicate that a simplified form of Eq. (4.66) can be adopted for the amount of shear carried by masonry, namely

$$V_{mh} = 0.4V_{jh} \qquad (7.42)$$

By a similar argument, it is permissible to carry part of the vertical joint shear force V_{jh} by the diagonal strut, simplifying in the procedure of Section 4.8.7 to

$$V_{mv} = 0.6V_{jv} \qquad (7.43)$$

Equations (7.42) and (7.43) do not permit additional shear to be carried when axial loads on the column are high. Although this is acknowledged to be conservative, it is felt to be advisable in the absence of relevant test data.

(c) Maximum Joint Shear Stress   A limitation of the total joint shear stress, derived from Eq. (7.40), is necessary to ensure against degradation under cyclic loading. Again, since specific relevant test data are sparse, conservative recommendations must be made. Accordingly, shear limitations for regions outside plastic hinge regions, and given in Section 7.2.5(e), are suggested:

$$v_{jh} = v_i \leq 0.2f'_m \quad \text{but} \quad \leq 2.4 \text{ MPa} \quad (350 \text{ psi}) \qquad (7.32b)$$

Because of the large joint dimensions necessary to satisfy development lengths for beam flexural reinforcement, the limitation implied by Eq. (7.32b) will rarely govern.

7.3.6 Ductility

A displacement ductility factor of $\mu_\Delta = 4$ can be assumed [P50] provided that the beam longitudinal reinforcement ratio is limited to

$$\rho_t = A_{st}/h_b t \leq 0.15f'_m/f_y \qquad (7.44)$$

Where Eq. (7.44) cannot be satisfied, a more detailed check on ductility capacity is appropriate, using the methods developed in Section 3.5 and illustrated in Fig. 7.34. In particular, it must be recognized that although ductility is confined to the beam plastic hinges by a capacity design approach, the system ductility capacity will be less than the corresponding beam ductility capacity, as a result of the contribution of column elastic displacements Δ_{ct} and Δ_{cb} [see Fig. 7.34(b)] joint shear rotation θ_{je}, and also due to the reduced ratio of column plastic rotation θ_{cp} to beam plastic rotation θ_p resulting from the distance from the beam plastic hinge center to the column

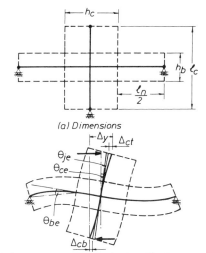

(a) Dimensions

(b) Elastic Components of Yield Displacement

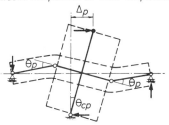

(c) Plastic Displacement from Beam Hinging

Fig. 7.34 Elastic and plastic displacement in a masonry beam–column unit.

center [see Fig. 7.34(c)]. Using an approach similar to that developed in Section 3.5.5(c), it may be shown that the system ductility capacity $\mu_{\Delta s}$ may be related to the beam ductility capacity $\mu_{\Delta b}$ by the relationship

$$\mu_{\Delta s} = 1 + \frac{\mu_{\Delta b} - 1}{C} \qquad (7.45)$$

where C is the ratio of yield displacement, including column and joint flexibility, to displacement resulting from beam rotation alone. C is typically about 1.5 for typical masonry wall frames.

In assessing elastic stiffness for period calculations, the effective length of beam and wall members should include an amount for strain penetration into the joint. Test results indicate that increasing the member length by one-third of the member depth results in good agreement between predicted and measured yield displacements [P50].

7.3.7 Dimensional Limitations

Limitations on beam and column dimensions should reflect the greater flexibility of masonry members compared with concrete members. Consequently, the following dimensional limits should be met.

Member length/thickness ratio: $l_n/b_w \leq 24$ (7.46a)

Beam length/depth ratio: $l_n/h_b \geq 3$ (7.46b)

Wall members need not satisfy the requirements of Eq. (7.46b). In addition, the beam should not be less than four masonry units in depth.

The length of potential plastic hinge regions may be taken as equal to the depth h of the hinging member. Equation (7.44) provides an upper limit to reinforcement in beams. To ensure that an adequate spread of plasticity occurs, the minimum steel provided in beams should satisfy

$$\rho_t = A_{st}/h_b t \geq 0.7/f_y \quad \text{(MPa)}; \qquad \rho_t = A_{st}/h_b t \geq 0.10/f_y \quad \text{(ksi)} \quad (7.47)$$

Reinforcement and other dimensional limitations for columns of the sort envisaged in this section (i.e., wall-like columns of the same width as the beams) should conform to the requirements of masonry walls given in Sections 7.2.7 and 7.2.8.

7.3.8 Behavior of a Masonry Wall–Beam Test Unit

The concept of a ductile masonry wall frame is relatively new and cannot be accepted without some experimental support. Although the extent of relevant testing is minimal, results from testing ductile masonry frames designed to recommendations similar to those presented above have been very satisfactory [P50, P60]. Figure 7.35 shows the hysteresis loops obtained for a full-sized joint unit [P60] constructed of 190-mm (7.5-in.)-thick concrete masonry blockwork, representing the circled region of a typical frame, shown in Fig. 7.29 with a story height of 3 m (9.85 ft) and a bay length of 5 m (16.4 ft). Beam depth h_b was 800 mm (31.5 in.) (four blocks) with four D20 (0.79-in.) longitudinal bars, giving a flexural reinforcement ratio of 0.0083. This is slightly in excess of the maximum recommended by Eq. (7.44). The column depth h_c of 1800 mm exceeded the bond requirements of Eq. (7.37). Reinforcement for shear in columns and joints was based on the principles outlined in the preceding section. It will be seen from Fig. 7.35 that stable hysteresis loops at, or close to, the theoretical lateral force H_i, corresponding to development of beam flexural plastic hinges on both sides of the joint, up to displacement ductility factors of $\mu = 4$, were obtained. At higher ductilities, spalling of face shells within the plastic hinge region and buckling of compression reinforcement resulted in gradual degradation of performance.

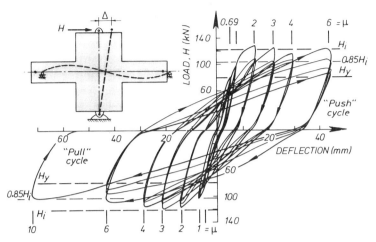

Fig. 7.35 Hysteretic response of a masonry wall-beam test unit.

The behavior compared favorably with a predicted ductility capacity of $\mu_f = 3$, based on Section 7.3.6.

The test unit depicted in Fig. 7.35 was designed in accordance with more conservative recommendations than listed in previous sections, but included those in a design code [X13]. A subsequent test on a wall-frame unit designed to violate the maximum beam reinforcement ratio suggested herein, and with inadequate column width by 20% to satisfy Eq. (7.37), also performed well [P50], indicating the conservative nature of these design recommendations.

7.4 MASONRY-INFILLED FRAMES

7.4.1 Influence of Masonry Infill on Seismic Behavior of Frames

It is a common misconception that masonry infill in structural steel or reinforced concrete frames can only increase the overall lateral load capacity, and therefore must always be beneficial to seismic performance. In fact, there are numerous examples of earthquake damage, some of which are shown in Chapter 1, that can be traced to structural modification of the basic frame by so-called nonstructural masonry partitions and infill panels. Even if they are relatively weak, masonry infill can drastically alter the intended structural response, attracting forces to parts of the structure that have not been designed to resist them. Two examples are illustrated below to examine this behavior.

Consider the floor plan of a symmetrical multistory reinforced concrete frame building with masonry-infill panels on two boundary walls, as shown in

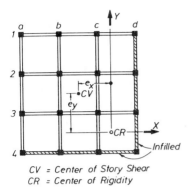

CV = Center of Story Shear
CR = Center of Rigidity

Fig. 7.36 Floor plan of a multistory reinforced concrete frame building with infill of two boundary frames.

Fig. 7.36. This may be compared with Fig. 1.10(d). If the masonry infill is ignored in the design phase, it may be assumed that each frame in each direction (i.e., frames 1, 2, 3, and 4 in the x direction, and frames a, b, c, and d in the y direction) is subjected to very similar seismic lateral forces, because of the structural symmetry. The true influence of the infill on frames 4 and d will be to stiffen these frames relative to the other frames. The consequence will be that the natural period of the structure will decrease, and seismic forces will correspondingly increase. Further, the proportion of the total seismic shear transmitted by the infilled frames will increase because of the increased stiffness of these frames relative to the other frames. The structure will also be subjected to seismic torsional response because of the shift in the center of rigidity [see Section 1.2.2(b)]. Thus for seismic response along the x and y axes, respectively, the torsional moments will be proportional to $M_{tx} = V_j e_y$ and $M_{ty} = V_j e_x$, respectively, where V_j is the total horizontal story shear and e_x and e_y are the eccentricities shown in Fig. 7.36.

The high shear forces generated in the infilled frames are transmitted primarily by shear stresses in the panels. Shear failure commonly results, with shedding of masonry into streets below, or into stairwells, with great hazard to life.

A second example is illustrated in Fig. 7.37, which shows masonry infill that extends for only part of the story height, to allow for windows. Again the infill will stiffen the frame, reducing the natural period and increasing seismic forces. If the frame is designed for ductile response to the design-level earthquake, without consideration of the effect of the infill, plastic hinges might be expected at the top and bottom of columns, or, preferably, in beams at the column faces. These hinges could develop at a fraction of the full design-level earthquake. The influence of the infill will be to inhibit beam hinges and stiffen the center and right column (for the direction of lateral load shown), causing plastic hinges to form at top of the column and top of the infill, as shown in Fig. 7.37. The consequence will be a dramatic increase

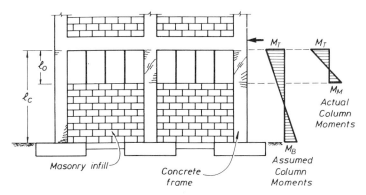

Fig. 7.37 Partial masonry infill in concrete frame.

in column shears. The design level of shear force in the column will be

$$V_D = \frac{M_T + M_B}{l_c} \qquad (7.48)$$

where l_c is the clear story height, and M_T and M_B are moments at the top and bottom of the first-story columns. These moments should be based on capacity design principles with the base moment M_B at flexural overstrength and the top moment M_T corresponding to flexural overstrength of the beam plastic hinges, with dynamic amplification effects taken into account as recommended in Section 4.6.4. However, in reality, a structure incorporating partial infill, such as that shown in Fig. 7.37, is unlikely to have been subjected to the sophistication of capacity design, and it is more probable that M_T and M_B will be moments directly derived from elastic analysis under the code distribution of lateral forces.

Regardless of the design philosophy adopted, Eq. (7.48) will underestimate the likely shear force, which, with the notation in Fig. 7.37, will be

$$V_D^* = \frac{M_T + M_M}{l_o} \qquad (7.49)$$

where l_o is the height of the window opening. Equation (7.49) corresponds to development of plastic hinges at the top of the column and at the top of the infill. If the column is not designed for the higher shear force of Eq. (7.49), shear failure can be expected. It should be noted that this higher shear force, corresponding to formation of plastic hinges, as shown, can develop because the original design was based on large ductility capacity. Hence the higher shear force will be developed, but at lower ductility demands.

When masonry infill of the type implied in Fig. 7.36 is to be used, there are two design alternatives. The designer may effectively isolate the panel from frame deformations by providing a flexible strip between the frame and panel, filled with a highly deformable material such as polystyrene. Alternatively, the designer may allow the panel and frame to be in full contact, and design both for the seismic forces to which they may be subjected. The first option, of isolation, is not very effective, as it is neither possible nor desirable to provide flexibility at the base of the panel. Moreover, it is difficult to provide support against out-of-plane seismic forces.

Isolated panels must be fully reinforced to carry the out-of-plane forces, because compression membrane action, which can assist in resisting in-plane loads, as will subsequently be established, is eliminated by the flexibility of the strip between the frame and the panel. Shear connection between frame and panel through the flexible layer will need to be designed for flexibility in the plane of the infill panel, while remaining stiff and strong enough to carry the out-of-plane reactions from inertial response back into the frame.

The most effective way of providing this behavior is to lay up the infill panel before the upper beam is poured, separating the top of the panel from the beam with a layer of flexible material. Shear connection to the beam can be provided by extending the panel vertical reinforcement into the beam and taping layers of flexible material (e.g., polystyrene) to the sides of the reinforcement in the in-plane direction, up to the beam midheight. After the beam concrete is placed, the flexible material will allow relative in-plane movement of panel and frame, while restricting out-of-plane relative movements.

The foregoing considerations will often mitigate against the use of isolated panels, and the subsequent discussion will be limited to interacting structural infill, where the role of the infill in influencing stiffness and strength is fully considered in the design process.

7.4.2   Design of Infilled Frames

(a)   In-Plane Stiffness   At low levels of in-plane lateral force, the frame and infill panel will act in a fully composite fashion, as a structural wall with boundary elements. As lateral deformations increase, the behavior becomes more complex as a result of the frame attempting to deform in a flexural mode while the panel attempts to deform in a shear mode, as shown in Fig. 7.38(a). The result is separation between frame and panel at the corners on the tension diagonal, and the development of a diagonal compression strut on the compression diagonal. Contact between frame and panel occurs for a length z, shown in Fig. 7.38(a).

The separation may occur at 50 to 70% of the ideal lateral shear capacity of the infill for reinforced concrete frames, and at very much lower loads for steel frames. After separation, the effective width of the diagonal strut, w, shown in Fig. 7.38(a), is less than that of the full panel.

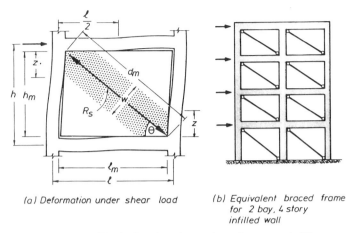

(a) Deformation under shear load

(b) Equivalent braced frame for 2 bay, 4 story infilled wall

Fig. 7.38   Equivalent bracing action of masonry infill.

Natural-period calculations should be based on the structural stiffness after separation occurs. This may be found by considering the structure as an equivalent diagonally braced frame, where the diagonal compression strut is connected by pins to the frame corners. Figure 7.38(b) shows the equivalent system for a two-bay, four-story frame. Analytical expressions have been developed [S12] based on a beam-on-elastic-foundation analogy modified by experimental results which show that the effective width w of the diagonal strut depends on the relative stiffnesses of frame and panel, the stress–strain curves of the materials, and the load level. However, since a high value of w will result in a stiffer structure, and therefore potentially higher seismic response, it is reasonable to take a conservatively high value of

$$w = 0.25d_m \tag{7.50}$$

where d_m is the diagonal length. This agrees reasonably well with published charts [S12], assuming typical masonry-infill properties and a lateral force level of 50% of the ultimate capacity of the infilled frame.

(b)   In-Plane Strength   There are several different possible failure modes for masonry infilled frames, including:

1. Tension failure of the tension column resulting from applied overturning moments.
2. Sliding shear failure of the masonry along horizontal mortar courses generally at or close to midheight of the panel.

3. Diagonal tensile cracking of the panel. This does not generally constitute a failure condition, as higher lateral forces can be supported by the following failure modes.
4. Compression failure of the diagonal strut.
5. Flexural or shear failure of the columns.

In many cases the failure may be a sequential combination of some of the failure modes above. For example, flexural or shear failure of the columns will generally follow a sliding shear failure, or diagonal compression failure, of the masonry. For a particular infilled frame, the strength associated with the various possible failure modes should be evaluated and the lowest value used as the basis for design.

(i) Tension Failure Mode: For infilled frames of high aspect ratio, the critical failure mode may be flexural, involving tensile yield of the steel in the tension column, acting as a flange of the composite wall, and of any vertical steel in the tension zone of the infill panel. Under these conditions the frame is acting as a cantilever wall, and a reasonably ductile failure mode can be expected. Design can then be in accordance with the recommendations for walls given in Chapter 5 and Section 7.2.

(ii) Sliding Shear Failure: If sliding shear failure of the masonry infill occurs, the equivalent structural mechanism changes from the diagonally braced pin-jointed frame of Fig. 7.38(*b*) to the knee-braced frame shown in Fig. 7.39. The support provided by the masonry panel forces column hinges to form at approximately midheight and top or bottom of the columns or may result in column shear failure. Initially, all the shear will be carried by the infill panel, but as the sliding shear failure develops, the increased displacements will induce moments and shears in the columns.

The equivalent diagonal strut compression force R_s to initiate horizontal shear sliding depends on the shear friction stress τ_f and the aspect ratio of

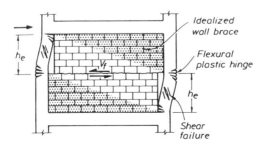

Fig. 7.39   Knee-braced frame model for sliding shear failure of masonry infill.

the panel, expressed by the angle θ in Fig. 7.38(a). It should be assumed that the panel carries no vertical load due to gravity effects, because of difficulties in constructing infill with a tight connection with the overlying beam of the frame, and also because vertical extension of the tension column will tend to separate the frame and panel along the top edge. Consequently, the clamping force across the potential sliding surface will be due only to the vertical component of the diagonal compression force R_s. The maximum shear force V_f that can be resisted by the panel is thus

$$V_f = \tau_o l_m t + \mu R_s \sin \theta$$

but from Fig. 7.38(a),

$$V_f = R_s \cos \theta = (l_m/d_m) R_s$$

and with $h_m/l_m \approx h/l$,

$$R_s = \frac{\tau_o}{1 - \mu(h/l)} d_m t \qquad (7.51)$$

Substituting the recommended values of $\tau_o = 0.03 f_m$ [Section 7.2.5(b)] and $\mu = 0.3$, into Eq. (7.51) yields the diagonal force to initiate sliding as

$$R_s = \frac{0.03 f'_m}{1 - 0.3(h/l)} d_m t$$

For a multibay frame of n bays, the base shear force to initiate sliding will thus be

$$V_b = \sum_{i=1}^{n} (R_{si} \cos \theta_i) \qquad (7.52a)$$

If the n bays all have the same length, Eq. (7.52a) reduces to

$$V_b = \frac{n 0.03 f'_m}{1 - 0.3(h/l)} l_m t \qquad (7.52b)$$

After sliding initiates, the columns and the panel share in the resistance of the shear force. The failure shear force for panels, such as shown in Fig. 7.39, may thus be estimated as

$$V_i = \sum_{i=1}^{n+1} \frac{2}{h_e} (M_{ct} + M_{cc})_i + V_b \qquad (7.53)$$

where M_{ct} and M_{cc} are the ideal flexural strength of the tension and

compression columns, respectively, including the effects of axial forces resulting from gravity loads and overturning moments. Equation (7.53) is based on
the assumption that column shear strength is greater than that corresponding
to the flexural hinging pattern at midheight and top or bottom of the column,
shown in Figs. 7.37 and 7.39.

The shear friction force V_b will degrade rapidly with cycling and should
conservatively be ignored in calculating the ductile shear capacity of this
failure mode. The effective column height h_e between column hinges (Fig.
7.39) is approximately half the story height h, both for exterior columns and
for columns between adjacent infill panels, as in the latter case hinges tend to
form close to the quarter points. Thus for a knee-braced frame n bays wide
with $n + 1$ columns, the ultimate shear capacity is

$$V_i = \frac{4}{h} \sum_{i=1}^{n+1} M_{ci} \tag{7.54}$$

where M_{ci} is the ideal flexural strength of the ith column, including axial
force effects. Column shear reinforcement should be based on a capacity
design approach using overstrength column moments to avoid column shear
failure.

Equation (7.52) should be used to predict the force required to initiate this
failure mode, for comparison with the flexural failure moment [Section
7.4.2(b)(ii)] or the diagonal crushing force [Section 7.4.2(b)(iii)]. If a ductile
response based on this mode of inelastic response is attempted, the value of
V_i given by Eq. (7.54) should exceed the shear given by Eq. (7.52).

(iii) Compression Failure of Diagonal Strut: For typical masonry-infill panels, diagonal tensile splitting will precede diagonal crushing [S12]. However,
the final panel failure force will be dictated by the compression strength,
which may thus be used as the ultimate capacity.

The following form of the diagonal compression failure force has given a
conservative agreement with test results [L2, T7]

$$R_c = \tfrac{2}{3} z t f'_m \sec \theta \tag{7.55}$$

where z defines the vertical contact length between panel and column, as
shown in Fig. 7.38(a), given by

$$z = \frac{\pi}{2} \left(\frac{4E_c I_g h_m}{E_m t \sin 2\theta} \right)^{1/4} \tag{7.56}$$

where E_c and I_g are the modulus of elasticity and moment of interia of the
concrete columns, E_m and h_m are the modulus of elasticity and height of the
infill, and θ is the angle between the diagonal strut and the horizontal, as
shown in Fig. 7.38(a). On inelastic cycling the capacity of the diagonal strut

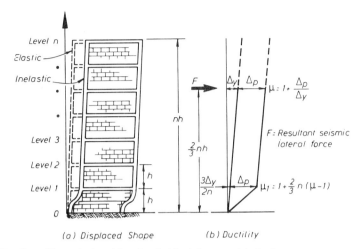

Fig. 7.40 Ductility relationships for infilled frame with inelastic soft-story displacements.

will degrade, and the behavior will approximate the knee-braced frame of Fig. 7.39.

*(c) **Ductility***   The flexural mode of inelastic response involving tensile column yield possesses adequate ductility, but is uncommon. The compression column must be well confined to take the high flange compression loads that will typically develop for an infilled frame (composite structural wall) slender enough to develop this mode. The full confinement required by Section 3.6.1 for the appropriate axial load level should be provided over the full height of the potential plastic hinge (i.e., the length l_w of the composite wall).

Work on carefully designed infilled frames with closely spaced vertical and horizontal reinforcement in the infill, spliced to dowel starters in the columns and beams, has shown that high ductilities can be obtained from panels deforming in an inelastic shear mode [K5]. However, this mode, and the sliding shear or diagonal crushing modes of inelastic displacement, resulting in the knee-braced frame behavior represented in Fig. 7.39, imply a soft-story displacement mode [Fig. 1.14(f)]. In this case the level of displacement ductility required of the hinging soft story will be very large for frames more than a few stories high.

Figure 7.40 shows the inelastic deformed shape and "yield" and maximum displacement profiles for an infilled frame with n stories. All inelastic displacement is expected in the bottom story. The structure displacement ductility, related to displacements at the height of the resultant of the seismic lateral loads (typically, about two-thirds of the building height for four or more stories), is $\mu = 1 + \Delta_p/\Delta_y$. Assuming a linear displacement profile at yield, the yield displacement over the height of the first (hinging) story will be

$\Delta_{y1} \approx 3\Delta_y/2n$. Since all the plastic displacement Δ_p occurs in this story, the displacement ductility capacity μ_1 required of the first story will be

$$\mu_1 = \frac{\Delta_{y1} + \Delta_p}{\Delta_{y1}} = \frac{\Delta_y + 2n\Delta_p/3}{\Delta_y} = 1 + \frac{2}{3}n(\mu - 1) \qquad (7.57)$$

Thus for a 10-story building designed for a structure displacement ductility of $\mu = 4$, the ductility required at the first story would be approximately $\mu_1 = 21$.

It is thus recommended that infilled frames with sliding shear failure or diagonal compression failure should be designed to resist the full lateral forces corresponding to elastic response to the design level earthquake, unless it can be shown that the infilled frame will rock on its foundations at a lower level of seismic response. In this case a capacity design approach (Section 9.4.3) may be used to set a safe minimum lateral seismic base shear.

(d) Out-of-Plane Strength   If the infill panel is reinforced and adequately connected to the surrounding frame, the response to out-of-plane inertia forces can be assessed treating the panel as a laterally loaded slab with the appropriate boundary conditions. Flexural strength can be assessed using techniques presented in Sections 7.2.3(b) and (c).

Masonry panels that are unreinforced in their plane may still be able to resist the out-of-plane inertial response without failure. Dynamic shaking of large unreinforced masonry panels confined by stiff frame members has indicated that the panels can resist very large out-of-plane accelerations with no apparent signs of distress.

The reason for this perhaps unexpectedly good performance is the resistance provided by compression membrane action. Figure 7.41(a) shows the lateral deformation of an infill panel as it cracks under lateral inertial accelerations. Because the diagonal length of the two half-height panels, separated by the midheight crack, exceeds the half clear story height $h_m/2$, diagonal compression struts, C, between the compression zones at top, midheight, and base of the panel are formed. The behavior is similar to compression membrane forces in slabs with stiff boundary beams. It has been shown [P2] that for infinitely rigid beams and perfect initial fit between the strip of panel and beams, shown in Fig. 7.41(a), the compression membrane strength envelope can be calculated from the following set of equations:

Compression zone depth: $c = t/2 - \Delta/4$ (7.58)

Compression strut force/unit width: $C = 0.72 f'_m c$ (7.59)

Moment capacity at hinges/unit width: $m_i = (C/2)(t - 0.85c)$ (7.60)

Equivalent response acceleration: $a = (8/mh_m^2)(2m_i - C\Delta)$

$\qquad\qquad\qquad\qquad\qquad\qquad\qquad\qquad\qquad\qquad\qquad\qquad\qquad (7.61)$

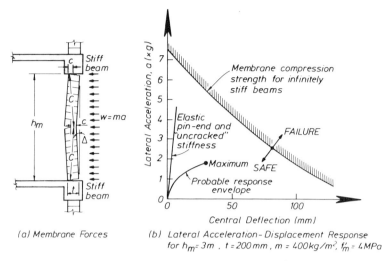

(a) Membrane Forces

(b) Lateral Acceleration-Displacement Response
for $h_m = 3m$, $t = 200\,mm$, $m = 400kg/m^2$, $f'_m = 4MPa$

Fig. 7.41   Compression membrane action in infill panels. (1 m = 3.28 ft; 1 MPa = 145 psi; 1 kg = 2.205 lb.)

where the symbols are defined in Fig. 7.41(a). Equations (7.58) to (7.61) do not include the effects of elastic flexural deformations, and thus do not apply at small values of Δ.

In Fig. 7.41(b), the relationship between displacement Δ and acceleration a implied by Eqs. (7.58) to (7.61) is plotted for typical panel properties in terms of multiples of the acceleration due to gravity, g. Also shown is the relationship based on elastic uncracked behavior of the panel, assuming it to be pin-ended, and the general shape of the probable response envelope, with allowance for cracking and membrane actions. It is conservatively assumed that the membrane compression forces develop only in the vertical direction, between beams, and that there is no frictional restraint between columns and the panel on the sides. In fact, horizontal compression struts may also be set up, requiring the panel to develop more complex crack patterns, analogous to those predicted by yield-line theories for two-way slabs [P2], resulting in still higher failure loads.

It is apparent from Fig. 7.41(b) that under ideal conditions, compression membrane action can provide a high factor of safety against failure under inertial loading. However, the potential strength resulting from compression membrane action is greatly diminished as the stiffness of the boundary members decreases [P2]. Moreover, it is very sensitive to initial gaps between the panel and the boundary members. These gaps could be a result of panel shrinkage, or difficulty in providing a sound connection between top of the panel and the beam above, or due to separation between panel and frame under simultaneous in-plane response, as shown, for example, in Fig. 7.38(a). Where the infill has openings for windows, the potential for membrane action is, of course, further diminished. Where in-plane shear stresses are sufficient to cause distress to the infill panel, it is extremely unlikely that effective compression membrane action could develop for in-plane response.

These considerations lead to the conclusion that unreinforced infill should not be considered as a satisfactory structural material except, perhaps, in low-rise (one- or two-story) buildings with very stiff boundary elements.

7.5 MINOR MASONRY BUILDINGS

7.5.1 Low-Rise Walls with Openings

The principle of designing with simple structural masonry elements for seismic resistance should apply to minor low-rise buildings as well as to more major buildings. Whenever possible, simple cantilever wall elements should be used, to avoid inevitable problems arising from large openings in walls. Where openings must be provided, the location and size should be chosen to optimize performance under lateral loads. Rational and irrational opening layouts were examined in Section 5.2.3.

Since the designer will often be faced with situations where, for architectural or other reasons, the structure cannot be simplified into cantilever elements, an approximate design technique will be described in the subsequent sections suitable for walls with large openings.

7.5.2 Stiffness of Walls with Openings

Figure 7.42 shows a two-story masonry wall with large window openings. Under a lateral shear force V applied at roof slab level, the total displacement at the same level can be considered to be comprised of a displacement

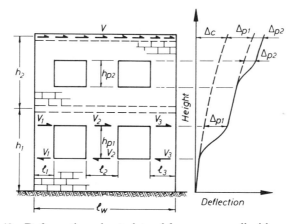

Fig. 7.42   Deformations due to lateral forces on a wall with openings.

Δ_c, of the gross wall acting as a cantilever, plus additional deflections Δ_{p1} and Δ_{p2} resulting from flexibility of the piers at levels 1 and 2.

The cantilever displacement Δ_c consists of flexural and shear components. An approximate assessment is

$$\Delta_c = \frac{V(h_1 + h_2)^3}{3 E_m I_e} + \frac{1.2 V(h_1 + h_2 - h_{p1} - h_{p2})}{G_m A_e} \tag{7.62}$$

where I_e and A_e are the effective stiffness and area based on cracked sections [Section 5.3.1(a)]. The shear modulus, G_m, may be taken as $0.4 E_m$. In Eq. (7.62), the shear deflection considered is only that in portions of the wall above and below openings, since the shear deflection of the piers will be considered separately. However, the full wall height must be used to provide a best estimate of the flexural deformation. The underestimate of Δ_c resulting from the increased axial stress in the piers under the overturning moment will generally be negligible.

The effective moment of inertia may be found using Eq. (5.7). Since diagonal shear cracking is likely to be limited to the piers, the reduction in shear stiffness for the main body of the wall is not likely to be extensive. It is recommended that a simple approximation of $A_e = 0.5 A_{\text{gross}}$ be taken in Eq. (7.62).

Pier displacements Δ_{p1} and Δ_{p2} over the height of the openings, shown in Fig. 7.42, must be the same. The total shear force V is distributed to the piers in proportion to their stiffness. Thus, in the first story,

$$\Delta_{p1} = V_1\left(\frac{h_{p1}^3}{12 E_m I_{e1}} + \frac{1.2 h_{p1}}{G_m A_{e1}}\right) = V_2\left(\frac{h_{p1}^3}{12 E_m I_{e2}} + \frac{1.2 h_{p1}}{G_m A_{e2}}\right)$$

$$= V_3\left(\frac{h_{p1}^3}{12 E_m I_{e3}} + \frac{1.2 h_{p1}}{G_m A_{e3}}\right) \tag{7.63}$$

where I_{e1} and A_{e1} are the effective moments of inertia and shear area of pier 1, and so on. Setting $G_m = 0.4 E_m$, the relative stiffnesses of the piers k_i are thus, for $i = 1$ to 3,

$$k_i = \frac{V_i}{\Delta_{p1}} = \frac{12 E_m I_{ei}}{h_{p1}^3(1 + F)} \tag{7.64}$$

where

$$F = \frac{36 I_{ei}}{h_{p1} A_{ei}} \tag{7.65}$$

The shear forces taken by individual piers are then

$$V_i = \frac{k_i}{\sum_1^3 k_i} V \qquad (7.66)$$

and the displacement Δ_{p1} is hence

$$\Delta_{p1} = \frac{V_i}{k_i} = \frac{V}{\sum_1^3 k_i} \qquad (7.67)$$

The displacement Δ_{p2} at the second level can be found in similar fashion, based on the dimensions of the piers at level 2.

The overall wall stiffness is then

$$K = \frac{V}{\Delta_c + \Delta_{p1} + \Delta_{p2}} \qquad (7.68)$$

The value of stiffness given by Eq. (7.68) can be used for period calculations and to determine the distribution of total seismic lateral forces between the various walls. Although the analysis, based on the details given in Fig. 7.42, assumed only one lateral force at roof level, the stiffness so derived in Eq. (7.68) may be used for the distribution of lateral shear forces in each story, the magnitude of which will normally be different.

7.5.3 Design Level of Lateral Force

Design levels of lateral force will depend not only on the wall stiffness but on the level of ductility adopted for the building as a whole. Because of the undesirable modes of inelastic displacement that must develop in walls with large openings, discussed in Section 7.2.1, the walls should be designed for the full seismic force corresponding to elastic response, or for a limited ductility level (i.e., $\mu_\Delta \leq 2$). In some cases response under the high lateral forces corresponding to these low ductility values will be dictated by the overturning capacity of the foundation elements (Section 9.4.3). In such cases the wall elements should be designed for lateral forces of 1.2 times those corresponding to overturning. It is emphasized that there is no risk of actual collapse due to overturning, as the level of energy input required to cause instability by overturning will be very much greater than that required to initiate uplift. Foundation rocking, examined in Section 9.4.3, is a useful method of base isolation, limiting seismic forces in the supported walls.

7.5.4 Design for Flexure

(a) *Piers* Figure 7.43 shows the complete set of forces to which the piers at level 1 are subjected under the action of lateral inertia forces F_1 and F_2. Shear forces V_1, V_2, and V_3 are given by Eq. (7.66), where $V = F_1 + F_2$. The

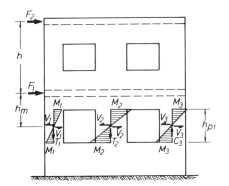

Fig. 7.43 Seismic forces in a lower story.

moments M_1, M_2, and M_3 are found assuming the piers to be fully fixed at top and bottom. Hence, for $i = 1$ to 3, $M_i = V_i h_{p1}/2$.

In addition to the moments and shear forces, the piers are subjected to axial forces, shown in Fig. 7.43 as T_1, T_2, and C_3, induced by the overturning moments. These must be calculated before designing the reinforcement for the piers. Since there are no flexural moments in the piers at midheight, the total overturning moment at pier midheight must be carried by these axial forces.

The pier seismic axial forces are most conveniently calculated from the shear forces induced in the spandrel beams. These in turn are found from the spandrel moments, which are based on moment equilibrium of the intersection points 1, 2, and 3 between pier and spandrel centerlines, as shown in Fig. 7.44. The pier moment diagrams, designated as M_p, for level 1 are extrapolated up to the joints 1, 2, and 3 to give moments M_{pb1}, M_{pb2}, and M_{pb3}, respectively. Similarly, pier moment diagrams for level 2 are extrapolated down to the same joints to give M_{pt1}, M_{pt2}, and M_{pt3}. Spandrel

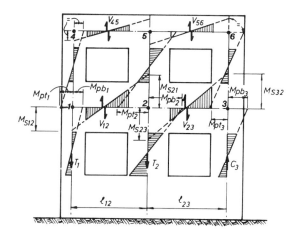

Fig. 7.44 Equilibrium conditions at joint centers of a wall with openings.

moments at the joints, designated in Fig. 7.44 as M_s, must be in equilibrium with the extrapolated pier moments. Hence:

At joint 1:
$$M_{s12} = M_{pb1} + M_{pt1}$$
(7.69a)

At joint 3:
$$M_{s32} = M_{pb3} + M_{pt3}$$
(7.69b)

At joint 2 it is reasonable to assume that

$$M_{s21} \approx (l_{23}/l_{21}) M_{s23}$$
(7.69c)

Hence

$$M_{s21} = \frac{l_{23}}{l_{12} + l_{23}} (M_{pb2} + M_{pt2})$$
(7.69d)

and M_{s23} is obtained from Eq. (7.69c).

The spandrel shear forces are found from the slope of the moment diagrams:

$$V_{12} = \frac{M_{s12} + M_{s21}}{l_{12}}$$
(7.70a)

$$V_{23} = \frac{M_{s23} + M_{s32}}{l_{23}}$$
(7.70b)

Spandrel shears at the upper level, V_{45} and V_{56}, may be found in similar fashion. Finally, the axial forces in the lower piers are found from vertical equilibrium of the seismic forces:

$$T_1 = V_{12} + V_{45}$$
(7.71a)

$$T_2 = V_{23} + V_{56} - V_{12} - V_{45}$$
(7.71b)

$$C_3 = V_{23} + V_{56}$$
(7.71c)

There are other more direct methods for calculating the seismic axial forces in the piers, but since the spandrel moments and shears will be required for design of the spandrel reinforcement, the method described above will result in less total effort. The method is similar to that used for the approximate analysis of frames in Appendix A. Total axial load on the piers will be due to the combined effects of reduced gravity load (e.g., $0.9P_D$) plus the seismic axial loads.

If the design level of lateral forces is based on an assumption of pier ductility, the piers should have at least four vertical bars distributed along the length to resist the seismic moments. This approach is also desirable when the design forces are based on elastic response, although some relaxation of requirements is justified.

(b) Spandrels Spandrel moments were calculated from joint equilibrium considerations in the preceding section, as shown in Fig. 7.44. Flexural reinforcement for the spandrels should be based on the maximum moment at the pier faces, shown shaded in Fig. 7.44.

7.5.5 Design for Shear

Shear forces in spandrels and piers have been estimated and are given by Eqs. (7.66) and (7.70), respectively. Where the lateral design forces are based on assumption of ductility, with $\mu_\Delta = 2$, the shear forces should be amplified in accordance with capacity design principles given in Section 5.4.4. However, the minor nature of the buildings considered in this section, together with the approximate nature of the analysis and the low design level of ductility, may not justify the detailed analyses of Section 5.4. If such analyses are carried out, the combined effects of flexural overstrength and dynamic shear amplification will result in an increase of the shear force by a factor of at least 2. It is, however, unnecessary to design for a shear force greater than that corresponding to elastic response [Eq. 5.24(a)]. Consequently, it is recommended that design levels of shear force in spandrels and piers be based on elastic response ($\mu_\Delta = 1$) regardless of whether a level of $\mu_\Delta = 1$ or $\mu_\Delta = 2$ is adopted for flexural design.

However, where $\mu_\Delta = 2$ is used for flexural design, a shear strength reduction factor for shear of $\phi = 1.0$ may be used, since the design shear force, as above, has been amplified in capacity design procedure, albeit of a simplified nature. Where $\mu_\Delta = 1$ is used for both flexure and shear design (i.e., no ductility), $\phi = 0.85$ should be used for shear.

There will be a difference in the shear to be allocated to masonry shear-resisting mechanisms, dependent on whether an elastic or ductile approach is used to establish design moments. If pier moments are based on $\mu_\Delta = 2$, considerable ductility may be required of the piers, since pier displacement ductility exceeds structure displacement ductility for soft-story mechanisms. Consequently, shear carried by masonry should be limited to that for ductile plastic hinge zones of walls, given by Eq. (7.29). If pier moments are based on $\mu_\Delta = 1$, ductile response is not expected, and the less onerous provisions of Eq. (7.28) may be followed to determine v_m. The excess shear force in piers or spandrels above that which can be carried by masonry shear-resisting mechanisms must be carried by shear reinforcement in accordance with Section 7.2.5(c) (ii).

7.5.6 Ductility

As discussed in Section 7.5.3 it is inappropriate to design these structures for full ductility. Because of the low design ductility levels and the typically low reinforcement and axial load levels, checks for ductility capacity are unnecessary.

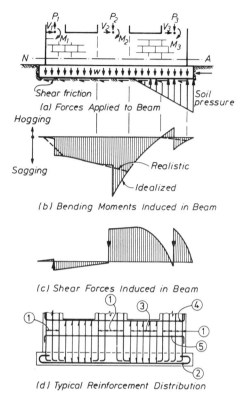

(a) Forces Applied to Beam

(b) Bending Moments Induced in Beam

(c) Shear Forces Induced in Beam

(d) Typical Reinforcement Distribution

Fig. 7.45   Design of wall base and foundation.

7.5.7   Design of the Wall Base and Foundation

The part of the wall below the lower openings in Figs. 7.42 to 7.44 requires special attention. Because of its proportions, this part of the wall will act in composite fashion with the foundation beam. This must be so considered in design. Figure 7.45(a) shows the wall base and foundation beam (which is considered as a concrete strip footing in this example) subjected to input forces and moments from the three piers, calculated in accordance with the preceding sections. Considerations of equilibrium result in a vertical soil pressure distribution as shown, implying uplift over part of the base. Note that a triangular distribution of soil pressure has been assumed, but if the maximum pressure exceeds the soil bearing capacity, a suitably modified distribution, with a maximum pressure equal to the bearing capacity, should be adopted. The passive lateral reaction at the end of the footing beam, and the shear friction along the soil–footing interface are also found from equilibrium considerations. Conservative results will be obtained if all the lateral reaction is assumed to be provided by shear friction at the base using data given in Table 9.1.

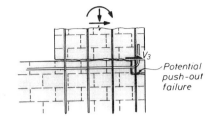

Fig. 7.46 Horizontal reinforcement to prevent push-out failure due to pier shear.

Bending moment [Fig. 7.45(b)] and shear force [Fig. 7.45(c)] diagrams may be established from statics. The diagrams imply sudden changes of moment and shear at the pier centerlines. However, actual variations will be more gradual, as shown, for example, by the dashed lines in Fig. 7.45(b). Design may be based on these more realistic distributions.

Figure 7.45(d) shows typical reinforcement requirements for the wall base/footing. The types referred to subsequently are identified by circled numbers. Flexural reinforcement from the piers (1) is brought down and anchored at the base of the footing. Flexural reinforcement for the footing (2) is based on the inverted T-section comprising the footing and wall base, using the moment diagram of Fig. 7.45(b). Large shear exists between piers 2 and 3, and vertical shear reinforcement (3) in the form of single-leg stirrups hooked over top and bottom horizontal reinforcement must be provided. Between piers 1 and 2 the shear is low, and only nominal reinforcement would be required. However, under seismic lateral forces acting in the opposite direction, the shear will again be large, requiring shear reinforcement similar to that between piers 2 and 3. For the example considered, with an unsymmetrical pier configuration, separate analyses would be required for the two directions of lateral force. The horizontal reinforcing bar (4) at the top of the wall base should have sufficient strength to transfer the shear force V_3 back into the body of the wall. Note that without this, there is a danger of the shear transfer within the flexural compression zone pushing out the wall end, as shown in Fig. 7.46. This bar should be anchored by bending up into the pier as shown. Finally, horizontal reinforcement (5) is placed to satisfy minimum reinforcement requirements.

7.5.8 Ductile Single-Story Columns

Masonry columns are frequently used to support minor single-story structures, such as factory buildings, awnings, and so on. These columns do not fit into the category considered in the preceding section, which was concerned with wall-like elements. Examples of typical designs commonly utilized are shown in Fig. 7.47.

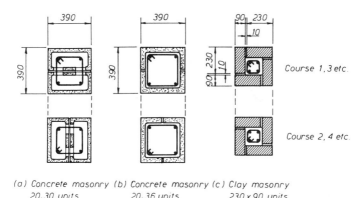

(a) Concrete masonry (b) Concrete masonry (c) Clay masonry
20.30 units   20.36 units   230 x 90 units

Fig. 7.47   Typical construction of masonry columns showing alternating courses.

Column sway mechanisms under lateral seismic load using masonry columns are permissible. However, since no test data are available for the ductile performance of masonry columns, it is recommended that the maximum design ductility factor be $\mu_\Delta = 2$, even if ductility calculations based on an ultimate compression strain of 0.003, using the approach developed for masonry walls in Section 7.2.3, indicates that high values are feasible.

Performance is likely to be significantly improved by the use of confining plates within the plastic hinge region. Confining plates should be in the form of closed rectangular hoops of galvanized steel. It should be noted that using a 3-mm ($\frac{1}{8}$-in.)-thick confining plate for the two examples of Fig. 7.47(a) and (b), and a width of 30 mm (1.2 in.) (corresponding approximately to the face-shell thickness), the volumetric ratio of confining reinforcement would be approximately $\rho_s = 0.005$. This is less than the value of $\rho_s = 0.0077$ used in confined masonry walls, on which the ultimate compression strain of 0.008 for confined masonry was based. Consequently, for confined masonry of this form of construction, either a 4-mm (0.157-in.) plate should be used, or a reduced ultimate strain of 0.006 adopted.

Conversely, with the type of construction using clay brick masonry in Fig. 7.47(c), quite high confining ratios are possible, and an ultimate compression strain of 0.008 is likely to be conservative. Where confined masonry columns are adopted, the maximum design ductility factor should be $\mu_\Delta = 2$, or the value calculated corresponding to the appropriate ultimate compression strain.

As an alternative design procedure, the masonry face shells may be ignored in analysis, with strength and ductility based on the concrete core alone. This approach would be particularly relevant for the form of construction that uses concrete channel section (pilaster) units, as shown in Fig.

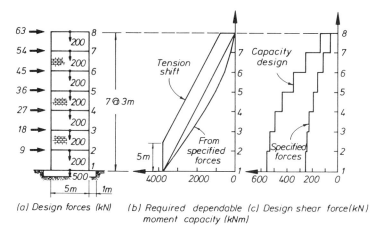

(a) Design forces (kN)   (b) Required dependable (c) Design shear force(kN)
                              moment capacity (kNm)

Fig. 7.48 Seven-story concrete masonry wall—design example. (1 m = 3.28 ft; 1 kN = 225 lb.)

7.47(b). In such cases, the core concrete and reinforcement must comply with the requirements for ductile concrete columns, given in Section 3.6.1(a).

7.6   DESIGN EXAMPLE OF A SLENDER MASONRY CANTILEVER WALL

The seven-story concrete masonry wall of Fig. 7.48 is to be designed for the seismic lateral forces shown, based on an assumed ductility of $\mu_\Delta = 4$. Minimum design gravity loads of 200 kN (45 kips), including wall self-weight, act on the wall at each floor and at roof level. The weight of the ground floor and footing are sufficient to provide overall stability under the base overturning moments. Determine the flexural and shear reinforcement for the wall. The wall width should be 190 mm (7.5 in.); f'_m must not exceed 24 MPa (3480 psi), and a lower design value is desirable. Assume that there is no base rotation due to foundation flexibility.

7.6.1   Design of Base Section for Flexure and Axial Load

From the lateral forces of Fig. 7.48(a) the wall base moment is

$$M_b = 3780 \text{ kNm } (33{,}450 \text{ kip-in.})$$

The distribution of bending moments corresponding to the code lateral forces, and the design envelope for bar curtailment [Section 5.4.2(c)], are shown in Fig. 7.48(b). The axial compression load at the base is

$$P_u = 0.9P_D = 0.9 \times 7 \times 200 \text{ kN} = 1260 \text{ kN (283 kips)}$$

Taking $f'_m = 12 \text{MPa}$ (1740 psi) initially, the strength reduction factor for flexure is, with

$$\frac{P_e}{f'_m A_g} = \frac{12{,}600 \times 10^3}{12 \times 5000 \times 190} = 0.11$$

and from Eq. (3.47), $\phi = 0.65$. The ideal moment at the base is $M_i = M_b/\phi = 5815 \text{ kNm}$ (51,460 kip-in.) The dimensionless parameters defined in Section 7.2.3(f) are

$$M_i/(f'_m l_w^2 t) = 5815 \times 10^6/(12 \times 5000^2 \times 190) = 0.102$$

and

$$P_u/(f'_m l_w t) = 0.110$$

From Fig. 7.14, $\rho f_y/f'_m = 0.155$. Hence with $f_y = 275$ MPa (40 ksi),

$$\rho = 0.155 \times 12/275 = 0.0068$$

7.6.2   Check of Ductility Capacity

With $\rho^* = \rho 8/f'_m = 0.0045$ and $P_u/f'_m A_g = 0.11$, from Fig. 7.15(a), $\mu_3 = 2.4$. The aspect ratio of the wall is $A_r = 7 \times 3/5 = 4.2$ and hence, from Eq. (7.24), with $f = 1.0$,

$$\mu_A = \mu_{4.2} = 1 + \frac{3.3(2.4 - 1)(1 - 0.25/4.2)}{1 \times 4.2} = 2.0$$

Thus the wall does not have adequate ductility!

7.6.3   Redesign for Flexure with $f'_m = 24$ MPa

The relevant parameters change to $P_u/f'_m A_g = 0.055$ and from Eq. (3.47), $\phi = 0.74$. Hence

$$M_i = 3780/0.74 = 5108 \text{ kN}_m \text{ (45,200 kip-in.)}$$

Therefore, from Fig. 7.14, with $M_i/(f'_m l_w^2 t) = 0.045$ and thus $\rho f_y/f'_m = 0.049$, the required reinforcement ratio is, with $f_y = 275$ MPa (40 ksi), significantly less than in the previous case; that is,

$$\rho = 0.049 \times 24/275 = 0.0043$$

7.6.4 Recheck of Ductility Capacity

From Fig. 7.15(a), with $\rho^* = \rho 8/f'_m = 0.0014$ and $P_u/f'_m A_g = 0.06$, $\mu_3 = 5.2$. Hence with $A_r = 4.2$ from Eq. (7.24),

$$\mu_{4.2} = 1 + 3.3(5.2 - 1)(1 - 0.25/4.2)/4.2 = 4.1 > 4.0$$

Thus the wall can just be designed for adequate ductility for $f'_m = 24$ MPa. If this strength could not be assured, the wall base would need to be confined.

7.6.5 Flexural Reinforcement

At the wall base,

$$A_{st} = \rho l_w t = 0.0043 \times 5000 \times 190 = 4085 \text{ mm}^2 \ (6.33 \text{ in.}^2)$$

This can be provided by 13 D20 (0.787-in.) bars (4082 mm²), that is, D20 bars on 400-mm (15.7-in.) centers along the wall base. These bars are required up to a height equal to l_w [i.e., 5 m (16.4 ft)] [see Fig. 7.48(b)]. Because of the marginal ductility capacity, it is important that the flexural reinforcement not consist of short starters lapped at the wall base. Since a spacing of 400 mm has been obtained, open-end blocks can be moved laterally into position, as shown in Fig. 7.25, and tall starters used. Lapping of bars will be required within the 5-m-high plastic hinge region, but no more than one-third of the bars should be lapped at any given level. From Eq. (7.33b), the required lap length within the plastic hinge region is

$$l_d = 0.18 \times f_y d_b = 0.18 \times 275 \times 20 = 990 \text{ mm} \ (39 \text{ in.})$$

The moment diagram of Fig. 7.48(b) and the moment capacity charts of Fig. 7.14 or Table 7.1 may be used to determine the reduction in flexural reinforcement with height. In fact, reductions are not great. At levels 3 to 7, the required reinforcement ratios are 0.0042, 0.0034, 0.0026, 0.0023, and 0.0014, respectively, compared with 0.0043 at the base. The only practical reduction is from D20 on 400-mm centers from the base to level 5, to D16 (0.63 in.) on 400-mm ($\rho = 0.0026$) centers from level 5 to the roof.

7.6.6 Wall Instability

Check the wall instability potential in accordance with Section 5.4.3(c). Since $\phi = 0.74$ and no excess strength has been provided, ductility corresponding to the required strength is on the order of $(\phi/0.9)\mu_\Delta \approx 0.8 \times 4.0 \approx 3.2$. Hence with an aspect ratio of $A_r = 4.2$, reading off from Fig. 5.35 it is found that

$$b_c = 0.055 \times 5000 = 275 > 192 \text{ mm}$$

or from Fig. 5.33, the curvature ductility is $\mu_\phi \approx 6.4$, and hence from Eq. (5.15b),

$$b_c \approx 0.022 \times 5000 \times \sqrt{6.4} = 278 > 192 \text{ mm}$$

The lateral stability of the wall is thus the governing influence on wall thickness, and nominal 300-mm (12-in.) masonry units will be required over the lowest story (levels 1 to 2). With the reinforcement as designed previously, the flexural strength of the wall will increase slightly, due to the reduction in depth of the flexural compression zone resulting from increased wall width. However, shear strength remains adequate and flexural ductility capacity is increased.

7.6.7 Design for Shear Strength

(a) Determination of Design Shear Force   As no excess reinforcement for flexure has been provided, the flexural overstrength factor for the wall is, from Eq. (5.13),

$$M_o/M_E = \lambda_o M_E/\phi M_E = \lambda_o/\phi = 1.25/0.74 = 1.69$$

Hence the design shear force at the wall base is, from Eq. (5.22),

$$V_{\text{wall}} = \omega_v \phi_{o,w} V_E = 1.53 \times 1.69 \times V_E = 2.59 V_E$$

where, from Eq. (5.23b),

$$\omega_v = 1.3 + n/30 = 1.3 + 7/30 = 1.53$$

From Fig. 7.52(c) the base shear is $V_E = 250$ kN (56.2 kips). Hence design the wall for

$$V_{\text{wall}} = 2.59 \times 250 = 648 \text{ kN } (145.7 \text{ kips})$$

This is less than the shear corresponding to elastic response (i.e., $4V_E$). With slender walls, response at flexural overstrength may be limited by the overturning capacity of the wall on its foundations. In this case, with dimensions and forces shown in Fig. 7.48(a) and an overstrength factor of 1.69, the wall is at incipient overturning. If the overturning capacity limits the overstrength base moment, a lower overstrength factor could be justified in shear calculations, provided that the consequencies of rocking on the foundation are appreciated by the designer (Section 9.4.3).

(b) Shear Stresses

$$V_i = V_{wall}/(b_w d) = 648 \times 10^3/(190 \times 0.8 \times 5000) = 0.85 \text{ MPa (123 psi)}$$

This is very small compared with the maximum permitted by Eqs. (7.32a) $v_i \leq 0.15 \times 24 = 3.6$ but not more than 1.8 MPa (261 psi).

(c) Shear Reinforcement

The contribution of masonry to shear strength at the base is estimated from Eq. (7.29):

$$v_m = 0.05\sqrt{f'_m} + 0.2(P_u/A_g) \leq 0.25 + 0.2(P_u/A_g) \leq 0.65 \text{ MPa}$$

$$= 0.05\sqrt{24} + 0.2 \times 1.26/0.95 = 0.51 \text{ MPa (74 psi)}$$

Hence provide shear reinforcement to carry

$$v_s = 0.85 - 0.51 = 0.34 \text{ MPa (50 psi)}$$

Thus from Eq. (3.41) or (7.30),

$$A_v/s = 0.34 \times 190/275 = 0.235 \text{ mm}^2/\text{mm } (0.0093 \text{ in.}^2/\text{in.})$$

Chose D12 bars at 400 mm (0.47 in. at 15.7 in.):

$$A_v/s = 113/400 = 0.283 \text{ mm}^2/\text{mm}$$

which is satisfactory. This is required over the assumed extent of the plastic region. That is, $l_p = l_w = 5000$ mm (16.4 ft) or $h_w/6 = 3500$ mm.

Outside the plastic region the shear requirements from Section 7.2.5(c)(ii) are as follows: Immediately above level 3, $V_{wall} = 2.59 \times 225 = 583$ kN (131 kips) and $P_u = 900$ kN (202 kips). From Section 7.2.5(c)(i) v_m is the lesser of the following: from Eq. (7.28a),

$$v_m = 0.17\sqrt{f'_m} + 0.3P_u/A_g = 0.17\sqrt{24} + 0.3 \times 900 \times 10^3/(0.95 \times 10^6)$$

$$= 1.12 \text{ MPa (162 psi)}$$

From Eq. (7.28b),

$$v_m = 0.75 + 0.3 \times 900 \times 10^3/(0.95 \times 10^6) = 1.03 \text{ MPa (149 psi)}$$

and from Eq. (7.28c),

$$v_m = 1.30 \text{ MPa (188 psi)}$$

However, at level 3,

$$v_i = 58,300/(190 \times 0.8 \times 5000) = 0.77 \text{ MPa } (112 \text{ psi})$$

Clearly, nominal horizontal reinforcement from levels 3 to 8 will be adequate. Code minimum is D12 bars on 800-mm centers ($\rho_h = 0.00074$). It is recommended, however, that D12 bars on 600-mm centers (0.472 in. at 23.6 in.) be used to avoid the wide spacing corresponding to the code minimum. Note that D12 bars at 600 mm provide a truss shear capacity [Eq. (3.41)] of

$$V_s = A_v f_y d/s = (113 \times 275 \times 0.8 \times 5000/600)10^{-3} = 207 \text{ kN } (46.5 \text{ kips})$$

This is 36% of the total shear at level 3.

7.7 DESIGN EXAMPLE OF A THREE-STORY MASONRY WALL WITH OPENINGS

The three-story concrete masonry wall of Fig. 7.49 is to be designed for the seismic lateral forces shown, which have been based on an assumed ductility of $\mu = 1$. That is, the wall is designed for lateral forces corresponding to

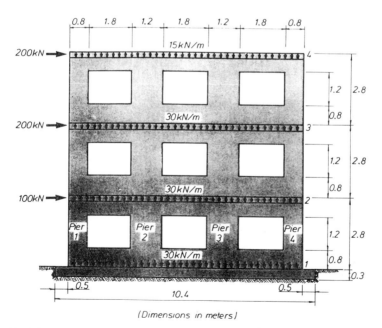

Fig. 7.49 Three-story masonry wall with openings. (1 kN = 0.225 kips; 1 m = 3.28 ft.)

elastic response to the design earthquake. Minimum gravity loads are 15 kN/m (1.03 kips/ft) at the roof and 30 kN/m (2.06 kips/ft) at levels 1, 2, and 3, including wall self-weight. At ground floor, the gravity load of 30 kN/m includes the footing weight. Design all reinforcement for the wall, using f'_m = 12 MPa (1740 psi) and f_y = 275 MPa (40 ksi). The wall thickness is 190 mm (7.63 in.), and the footing width is 1 m (39.4 in.). The maximum soil bearing capacity is 400 kPa (8.35 kips/ft^2).

7.7.1 Determination of Member Forces

(a) Pier Stiffnesses   Since the seismic lateral floor forces are already given, the overall wall stiffness is not required. However, the relative stiffness of the piers is required in order to apportion the total shear between the piers. For this purpose, an estimate of the axial forces in each pier is required.

1. For the purpose of gravity loading assume that axial stress is the same in all columns. Then

$$P/A_g = 75{,}000 \times 9.4/(4000 \times 190) = 0.93 \text{ MPa (135 psi)}$$

2. To determine the seismic axial forces, assume the point of contraflexure to be at midheight of the columns. At these levels all the seismic overturning moment is carried by axial forces. Hence the overturning moment at midheight of the lower level of piers is

$$M_{ot} = 200(7.0 + 4.2) + 100 \times 1.4 = 2380 \text{ kNm (21,060 kip-in.)}$$

The second moment of pier areas about the center line of the structure, shown in the sketch, is

$$I = 2(0.19 \times 0.8 \times 4.3^2 + 0.19 \times 1.2 \times 1.5^2) = 6.65 \text{ m}^4 \text{ (770 ft}^4\text{)}$$

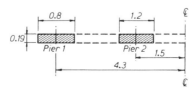

Plan view through lower-story piers

TABLE 7.2   Stiffness Properties of Masonry Piers

Parameter[a]	Units	Pier 1	Pier 2	Pier 3	Pier 4	Σ
(1) $0.9P_D$	kN	127	191	191	127	636
(2) P_E	kN	−234	−122	122	234	0
(3) P_u	kN	−107	69	313	361	636
(4) A_g	m^2	0.152	0.228	0.228	0.152	—
(5) I_g	m^4	0.0081	0.0274	0.0274	0.0081	—
(6) $P_u/f'_m A_g$	—	−0.059	0.025	0.114	0.198	—
(7) I_e/I_g	—	0.308	0.38	0.47	0.56	—
(8) I_e	10^{-3} m^4	2.43	10.4	12.9	4.54	—
(9) F	—	1.33	3.00	3.00	1.33	—
(10) k_i	10^{-3} m^4	0.96	2.48	3.07	1.79	8.30
(11) $k_i/\Sigma k_i$	—	0.116	0.298	0.370	0.216	1.000

[a](1) Axial load due to gravity (compression taken positive); (2) axial force due to specified lateral earthquake force; (3) total axial load; (4) gross sectional area of pier; (5) moment of inertia of pier based on gross (uncracked) concrete area; (7) based on Eq. (5.7); (9) based on Eq. (5.9a), which in this case reduces to $F = 3(l_w/h_p)^2$; (10) from Eq. (5.9a), $k = I_e/(1.2 + F)$; (11) stiffness ratio.

Thus the axial forces are:

On pier 1: $P_{E1} = A_g(My/I) = 0.152 \times 2380 \times (-4.3)/6.65$

$$= -234 \text{ kN } (-52.6 \text{ kips})$$

On pier 2: $P_{E2} = 0.228 \times 2380 \times (-1.5)/6.65 = -122 \text{ kN } (-27.4 \text{ kips})$

Corresponding compression forces are developed in piers 3 and 4.

3. The calculations of relative pier stiffnesses based on Sections 5.3.1 and 7.5.2 are summarized in Table 7.2.

(b)   Shear Forces and Moments for Members   The shear forces across the piers are found from Eq. (7.66) [i.e., $V_{Ei} = (k_i/\Sigma k_i)V_E$], and the pier moments at top and bottom of piers are $M_{Ei} = h_{pi}V_i/2 = 0.6V_{Ei}$. These pier forces are summarized in Table 7.3, which also includes pier moments extrapolated to the intersection points of spandrels and piers (Fig. 7.50). Thus, for first-story piers at level 2, for example,

$$M_{cl} = V_{Ei}(0.6 + 0.8) = 1.4V_{Ei}$$

Spandrel moments and shears are found from joint equilibrium using Eqs. (7.69) and (7.70). The resulting values for moment at the pier face and for shears are included in Fig. 7.50. It will be noted that the axial forces

TABLE 7.3   Pier Shear Forces and Moments
(1 kN = 0.225 kip, 1 kNm = 8.85 kip-in.)

Parameter[a]	Units	Pier 1	Pier 2	Pier 3	Pier 4	Σ
			First Story			
(1) V_{Ei}	kN	58.0	149.0	185.0	108.0	500.0
(2) M_{Ei}	kNm	34.8	89.4	111.0	64.8	
(3) M_{cl}	kNm	81.2	208.6	259.0	151.2	
			Second Story			
(1) V_{Ei}	kN	46.4	119.2	148.0	86.4	400
(2) M_{Ei}	kNm	27.8	71.5	88.8	51.8	
(3) M_{cl}	kNm	65.0	166.9	207.0	121.0	
			Third Story			
(1) V_{Ei}	kN	23.2	59.6	74.0	43.2	200
(2) M_{Ei}	kNm	13.9	35.8	44.4	25.9	
(3) $M_{cl,bot}$	kNm	32.5	83.4	103.6	60.5	
(4) $M_{cl,top}$	kNm	23.2	59.6	74.0	43.2	

[a](1) Shear force across pier; (2) moments of critical pier sections; (3) moments of spandrel center lines. The subscript i refers to the pier number.

in the piers in the first story in Fig. 7.50, which have been found with the use of Eq. (7.71) from equilibrium considerations of beam shears, differ significantly from those assumed at the start of this example for the purpose of calculating relative pier stiffnesses. A further iteration using the revised total pier axial loads could be carried out, but is of doubtful need considering the approximate nature of the analysis. Revised total axial loads ($P_u = 0.9P_D + P_{eq}$) are listed below.

	Total Axial Load P_u (kN) in Story:		
Pier	1	2	3
1	− 90.7	− 22.2	6.5
2	152.6	97.2	34.8
3	117.9	43.1	31.9
4	456.0	187.9	54.0

7.7.2   Design of First-Story Piers

(a)   Flexural Strength

1. The outer piers are designed for the worst pier 1 and pier 4 loadings. For pier 1:

$$P_u = -90.7 \text{ kN } (20.4 \text{ kips}), \qquad M_u = 34.8 \text{ kNm } (308 \text{ kip-in.})$$

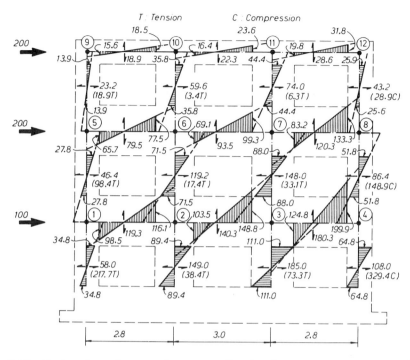

Fig. 7.50 Design shear forces and moments for the three-story masonry example structure.

Hence $P_u/(f'_m A_g) = -0.0497$; thus $\phi = 0.85$ and $M_i = 40.94$ kNm (362 kip-in.). With

$$M_i/(f'_m l_w^2 t) = 40.94 \times 10^6/(12 \times 800^2 \times 190) = 0.02806$$

from Table 7.1, $\rho f_y/f'_m = 0.103$.
    For pier 4:

$$P_u = 456.1 \text{ kN (102.5 kips)}; \qquad M_u = 64.8 \text{ kNm (573 kip-in.)}$$

Because $P_u/(f'_m A_g) = 0.250 > 0.1$, $\phi = 0.65$ and $M_i = 99.7$ kNm (882 kip-in.) With

$$M_i/(f'_m l_w^2 t) = 99.7 \times 10^6/(12 \times 800^2 \times 190) = 0.0683$$

from Table 7.1, $\rho f_y/f'_m < 0.01$; hence pier 1 governs. Thus we require that

$$\rho = 0.103 \times 12/275 = 0.00495$$

and

$$A_{st} = \rho l_w t = 0.00495 \times 800 \times 190 = 752 \text{ mm}^2 \ (1.17 \text{ in.}^2)$$

Provide four D16 (0.63 in.) bars in each pier: $A_{st} = 804 \text{ mm}^2$ (1.25 in.^{2}). This could be reduced somewhat if some shear was redistributed from pier 1 to pier 4 in accordance with Section 5.3.2(b).

2. Similarly for the inner piers, we find:

> Pier 2: $P_u = 152.6$ kN and $M_u = 89.4$ kNm
>
> Pier 3: $P_u = 117.9$ kN and $M_u = 111.0$ kNm

As pier 3 governs, from Eq. (3.47), with $P_u/f'_m A_g = 0.0431$, $\phi = 0.764$ and $M_i = 145.3$ kNm (1286 kip-in.). With $M_i/(f'_m l_w^2 t) = 0.0443$, we find $\rho f_y/f'_m = 0.056$, from Table 7.1. Hence $\rho = 0.00244$, requiring that

$$A_{st} = 0.00244 \times 1200 \times 190 = 556 \text{ mm}^2 \ (0.862 \text{ in.}^2)$$

Provide four D16 (0.63 in.) bars in each pier, giving $A_{st} = 804 \text{ mm}^2$ (1.25 in.^{2}).

This is more than required, and a solution using two D16 and two D12 bars would be satisfactory. However, since the structure is not being designed for ductility, moment overstrength has no consequential penalty in the form of increased design shear forces, and so on. To facilitate construction it is convenient to retain the one reinforcement size for all piers. The suggested layout is shown in Fig. 7.51.

(b)   *Shear Strength*   For pier 1 from Table 7.3,

$$V_i = V_u/\phi = 58.0/0.80 = 72.5 \text{ kN } (16.3 \text{ kips})$$

Hence shear stress

$$v_i = V_i/(0.8tl_w) = 72.5 \times 10^3/(190 \times 640) = 0.596 \text{ MPa } (86.4 \text{ psi})$$

The shear carried by masonry v_m is, from Eq. (7.28),

$$v_m = 0.17\sqrt{f'_m} + 0.3(P_u/A_g)$$
$$= 0.17\sqrt{12} + 0.3(-90.7 \times 10^3)/(0.152 \times 10^6) = 0.410 \text{ MPa } (59.5 \text{ psi})$$

Hence

$$v_s = v_i - v_m = 0.596 - 0.410 = 0.186 \text{ MPa } (27.0 \text{ psi})$$

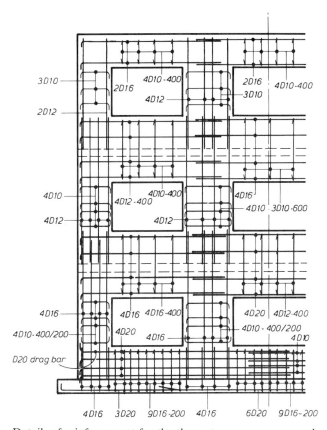

Fig. 7.51 Details of reinforcement for the three-story masonry example structure.

For pier 4 the steps above lead to

$$V_i = 108.0/0.8 = 135.0 \text{ kN (30.4 kips)}$$

$$v_i = 135 \times 10^3/(190 \times 640) = 1.11 \text{ MPa (161 psi)}$$

From Eq. (7.28a),

$$v_m = 0.17\sqrt{12} + 0.3 \times 456 \times 10^3/(0.152 \times 10^6) = 1.49 \text{ MPa (216 psi)}$$

Since this exceeds the requirement for pier 4, design of pier 1 governs, and

$$A_v/s = 0.186 \times 190/275 = 0.129 \text{ mm}^2/\text{mm} \left(0.0051 \text{ in.}^2/\text{in.}\right)$$

Use D10 (0.39-in.) bars at 400 mm (15.7 in.), giving

$$A_v/s = 78.5/400 = 0.196 \text{ mm}^2/\text{mm} \ (0.0077 \text{ in.}^2/\text{in.})$$

The spacing of 400 mm represents $l_w/2$, the maximum permissible for nonductile regions. Because of this, and also because the pier height of six blocks is not uniformly divisible into 400-mm centers, the layout in Fig. 7.51 with an extra bar at midheight of piers 1 and 4 is suggested.

For pier 2: $P_u = 152.6 \text{ kN} \ (34.3 \text{ kips})$

$V_i = 149.0/0.8 = 186.3 \text{ kN} \ (41.9 \text{ kips})$

For pier 3: $P_u = 117.9 \text{ kN} \ (26.5 \text{ kips})$

$V_i = 185/0.8 = 231.3 \text{ kN} \ (50.0 \text{ kips})$

Clearly, pier 3 governs. Following the steps above, from Eq. (7.28a),

$$v_m = 0.59 + 0.3 \times 117.9 \times 10^3/(0.228 \times 10^6) = 1.10 \text{ MPa} \ (160 \text{ psi})$$

$$v_i = 231.3 \times 10^3/(190 \times 0.8 \times 1200) \qquad = 1.27 \text{ MPa} \ (184 \text{ psi})$$

$$v_s = 1.27 - 1.10 \qquad\qquad\qquad\qquad\qquad = 0.17 \text{ MPa} \ (25 \text{ psi})$$

$$A_v/s = 0.17 \times 190/275 \qquad\qquad\qquad\quad = 0.117 \text{ mm}^2/\text{mm} \ (0.0046 \text{ in.}^2/\text{in.})$$

Again use D10 bars on 400-mm centers, giving

$$A_v/s = 78.5/400 = 0.196 \text{ mm}^2/\text{mm} \ (0.0077 \text{ in.}^2/\text{in.})$$

The layout can be the same as for piers 1 and 4.

The procedure for flexural and shear design of piers in the upper stories is the same as in the first story and is thus not included. Results in the form of required reinforcement are included in Fig. 7.51.

7.7.3   Design of Spandrels at Level 2

(a) Flexural Strength   Gravity load moments in the spandrels are only about 4% of the seismic moments. For simplicity they are ignored in this example. Considering spandrels 1–2 and 3–4, design for the maximum moment of 199.9 kNm (1770 kip-in.) adjacent to joint 4. Beam depth = 1.6 m.

$$P_u = 0, \qquad M_i = 199.9/0.85 = 235 \text{ kNm} (2080 \text{ kip-in.})$$

$$M_i/f_m' l_w^2 t = 235 \times 10^6/(12 \times 1600^2 \times 190) = 0.0403$$

From Table 7.1, $\rho f_y/f_m' = 0.088$ and with $f_y = 275$ MPa (40 ksi),

$\rho = 0.088 \times 12/275 = 0.00384$. Hence

$$A_{st} = 0.00384 \times 1600 \times 190 = 1167 \text{ mm}^2$$

Use four D20 (0.79 in.) bars, which gives $A_{st} = 1256 \text{ mm}^2$, which is sufficient.

These bars are placed in the top, third, sixth, and eighth (bottom) course, as shown in Fig. 7.51. Note that reinforcement placed in the floor slab, which will act in composite fashion with the spandrel masonry, forming the fourth course from the top, will also add to beam strength and could be used to reduce beam reinforcement. It would be possible to reduce the reinforcement for the moment adjacent to joint 3, but it is preferable to carry the bars through into the joint 3 region.

For spandrel 2–3, $M_u = 148.8$ kNm and $M_i = 175.0$ kNm (1549 kip-in.). Hence from $M_i/f'_m l_w^2 t = 0.0300$ and from Table 7.1, $\rho f_y/f'_m = 0.064$, so that with $f_y = 275$ Mpa, $\rho = 0.00279$. Hence $A_{st} = 849 \text{ mm}^2$. Providing four D16 (0.63 in.) bars results in a shortfall of 46 mm^2, which can be supplied by slab reinforcement.

*(b) **Shear Strength***   For spandrel 3–4 at level 2, $V_u = 180.3$ kN (40.5 kips). Hence $v_i = 180.3 \times 10^3/(0.8 \times 190 \times 0.8 \times 1600) = 0.93$ MPa (134 psi). Since $P_u = 0$, $v_m = 0.17\sqrt{f'_m} = 0.59$ MPa (85 psi) and thus $v_S = 0.34$ MPa (49 psi). With $f_y = 275$ MPa,

$$A_v/s = 0.34 \times 190/275 = 0.235 \text{ mm}^2/\text{mm} \ (0.0093 \text{ in.}^2/\text{in.})$$

Use D12 bars on 400-mm centers (0.47 in. at 15.7 in.):

$$A_v/s = 113/400 = 0.282 \text{ mm}^2/\text{mm} (0.0111 \text{ in.}^2/\text{in.})$$

For spandrel 2–3 at level 2, $V_u = 140.3$ kN (31.5 kips) and it will be found that D10 bars on 400-mm centers (0.39 in. at 15.7 in.) are sufficient.

The procedure is identical for spandrels at other levels and hence it is omitted. Details of the results are, however, shown in Fig. 7.51, where it should be noted that D10 bars on 400-mm centers provide a reinforcement ratio of 0.001, the minimum practical ratio.

7.7.4 Design of Wall Base and Foundation

*(a) **Load Effects***   Figure 7.52(a) shows dimensions of the foundation–wall base and the loads to which it is subjected. For vertical equilibrium, using data given in Table 7.3, the soil reaction force R is given by

$$R = 456.4 + 117.7 + 152.6 - 90.7 + 27 \times 9.4 = 890 \text{ kN (200 kips)}$$

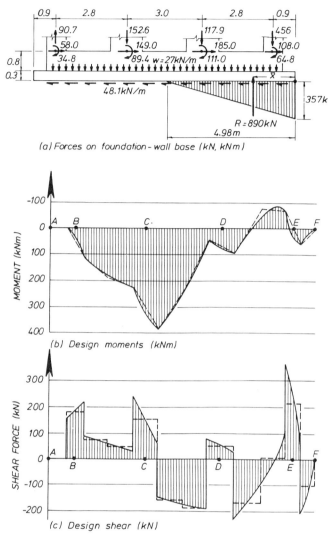

(a) Forces on foundation-wall base (kN, kNm)

(b) Design moments (kNm)

(c) Design shear (kN)

Fig. 7.52 Details of the wall base and foundations. (1 kN = 0.225 kips; 1 kNm = 0.738 kip-ft.)

Taking moments about the right end of the wall base:

$$R\bar{x} = (27 \times 9.4 \times 5.2) + (456.1 \times 0.9) + (117.9 \times 3.7) + (152.6 \times 6.7)$$
$$- (90.7 \times 9.5) - (500 \times 1.1) - (34.8 + 89.4 + 110.0 + 64.8)$$
$$= 1477 \text{ kNm } (245 \text{ kip-ft})$$

Hence the resultant soil pressure R is at $\bar{x} = 1477/890 = 1.66$ m (5.44 ft) from the right end. Assuming a linear soil pressure distribution results in the pressure block shown, with a maximum pressure of 357 kPa (7.46 kips/ft^2).

It will be assumed that the base shear of 500 kN is uniformly distributed along the base at the rate of $500/10.4 = 48.1$ kN/m (3.30 kips/ft). Figure 7.52(b) and (c) show the distributions of bending moment and shear force along the unit. Bending moments have been calculated at the level of the top of the concrete foundation slab. The variation of calculated moments and shear forces have been smoothed over the width of the piers, as suggested in Section 7.5.7.

(b) Flexural Strength   From Fig. 7.52(b) the maximum positive and negative moments are $+380$ kNm and -80 kNm, respectively. For the positive moment of $M_i = 380/0.85 = 447$ kNm (3956 kip-in.), from the sketch and assuming that $(d - a/2) = 900$ mm (35.4 in.):

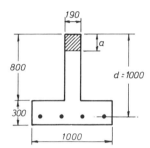

Section for +ve Moment

$$A_s f_y = 447 \times 10^3/0.9 = 497.0 \text{ kN } (111.8 \text{ kips})$$

Hence

$$a = A_s f_y/(0.85 f'_m t)$$

$$= 497 \times 10^3/(0.85 \times 12 \times 190) = 256 \text{ mm } (10.0 \text{ in.})$$

The revised tension force is then

$$A_s f_y = 447 \times 10^3/(1000 - 0.5 \times 256) = 512.6 \text{ kN } (115.2 \text{ kips})$$

requiring that

$$A_s = 512.6 \times 10^3/275 = 1862 \text{ mm}^2$$

Provide six D20 (0.79 in.) bars $= 1884$ mm^2

Tension shift (Section 3.6.3) (together with reverse direction loading) requires that this bottom reinforcement be continuous over the central region within 1.5 m from each end. Reduce the bars to three D20 for the remainder of the length. For the negative moment $M_i = 80/0.85 = 94.1$ kNm

(833 kip-in.) and with an internal lever arm (see the sketch) of $(d - a/2) \approx$ $400 + 280 = 680$ mm (26.8 in.),

$$A_s = 94.1 \times 10^6/(680 \times 275) = 503 \text{ mm}^2 \ (0.780 \text{ in.}^2)$$

A considerably larger amount is provided.

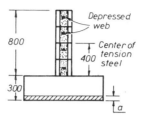

To aid shear transfer from piers 1 and 4 it is desirable to use a longer bar for the top of the foundation, and to bend it up into the pier as pointed out in Section 7.5.7 (Fig. 7.46). This enables the top D20 bar to transfer 86 kN of shear.

(c) **Shear Strength** A maximum shear force of 216 kN occurs in the outer spans. With $V_i = 216/0.8 = 270$ kN (60.8 kips),

$$v_i = 270 \times 10^3/(190 \times 0.8 \times 1100) = 1.61 \text{ MPa} \ (234 \text{ psi})$$

As $v_m = 0.59$ MPa, stirrups should resist $v_s = 1.02$ MPa (148 psi) and hence

$$A_v/s = 1.02 \times 190/275 = 0.70 \text{ mm}^2/\text{mm} \ (0.0277 \text{ in.}^2/\text{in.})$$

D20 bars on 400-mm centers (0.79 mm^2/mm) or D16 bars on 200-mm centers (1.00 mm^2/mm) may be provided. However, it will be difficult to hook the 20-mm-diameter bar at the top of the wall base, so use D16 bars on 200-mm centers (0.63 in. at 7.85 in.). The center span will require similar shear reinforcement.

(d) **Transverse Bending of Footing Strip** Maximum bearing pressure is $p_{max} = 357$ kPa (7.45 kips/ft^2). Allow a factor of safety of 2 against failure, since soil pressures are sensitive to lateral force level. Therefore, $p_u = 714$ kPa (14.9 kips/ft^2). Thus

$$M_{max} = 714 \times 0.405^2/2 = 58.6 \text{ kNm} \ (158 \text{ kip-in./ft})$$

Since this is a concrete structure, $\phi = 0.9$ and thus $M_i = 58.6/0.9 = 65.1$ kNm/m (176 kip-in./ft). With $(d - a/2) = 240$ mm,

$$A_s = 65.1 \times 10^6/(240 \times 275) = 986 \text{ mm}^2/\text{m} \ (0.466 \text{ in.}^2/\text{ft})$$

Use D16 bars on 200-mm centers (0.63 in. at 7.85 in.):

$$201/0.2 = 1005 \text{ mm}^2/\text{m } (0.475 \text{ in.}^2/\text{ft})$$

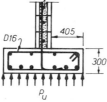

Because the pressure reduces rapidly from the end, provide D16 bars on 400-mm (15.7-in.) centers 2.5 m (8.2 ft) away from each end. In the sketch above, the D16 bars at 200 mm are shown as closed stirrups. Since the slab is more than 200 mm (7.85 in.) thick, nominal longitudinal shrinkage reinforcement is required in the top layer. Four D10 (4 × 0.39 in.) bars are adequate.

7.7.5 Lapped Splices in Masonry

From Eq. (7.33a), $l_d = 0.12 f_y d_b$. Hence in this design example the lap lengths are for D12, D16, and D20 bars on centers of 396, 528, and 660 mm (15.6, 20.8, and 26.0 in.), respectively. Figure 7.51 shows the recommended lap positions without dimensioning the lap lengths.

7.8 ASSESSMENT OF UNREINFORCED MASONRY STRUCTURES

7.8.1 Strength Design for Unreinforced Masonry

It is not recommended that new structures intended for seismic resistance be constructed of unreinforced masonry. However, it is still appropriate to discuss the seismic performance of unreinforced masonry structures because of the importance of assessment of the seismic risk associated with existing buildings. Traditionally, this has been done using elastic analysis techniques under simplified and generally unrealistically low lateral forces, and comparing results with specified acceptable stress limits. In Section 7.1 it was demonstrated that elastic analysis techniques were inappropriate for estimating the safety of eccentrically loaded unreinforced slender walls. In fact, the conclusion reached in that section, that strength considerations were more appropriate, applies generally to the assessment and design of unreinforced masonry structures. This may seem surprising, since it is not commonly considered that strength design can be applied to unreinforced structures. However, there are no conceptual difficulties involved, and the result is inevitably more realistic and intellectually satisfying.

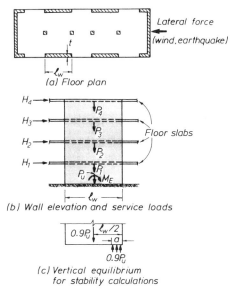

(a) Floor plan

(b) Wall elevation and service loads

(c) Vertical equilibrium
for stability calculations

Fig. 7.53 Unreinforced masonry wall under wind forces.

As an example of the application of strength design to unreinforced masonry, consider the general case of a four-story reinforced masonry wall subjected to in-plane lateral forces induced by wind or earthquake. In Fig. 7.53(a) and (b), a typical structural wall is subjected to floor loads P1 to P4 and lateral forces H1 to H4, resulting in a total axial force P_u and moment M_E at the wall base. Typically, axial compression stresses under P_u will be light and the maximum moment, M_E, permitted by elastic design will depend on the maximum allowable tension stress f_t for masonry:

$$M_E \leq \left(\frac{P_u}{l_w t} + f_t \right) \frac{l_w^2 t}{6} \qquad (7.72)$$

Strength design would require stability under an ultimate limit state which for wind forces might be defined as

$$U = 0.9D + 1.3W \qquad (7.73)$$

Note that reduced gravity load is adopted in Eq. (7.73). At the ultimate limit state, the forces involved in vertical equilibrium are as shown in Fig. 7.53(c), where the length of the equivalent rectangular compression block, a, is given by

$$a = 0.9P_u/0.85f_m' t \qquad (7.74)$$

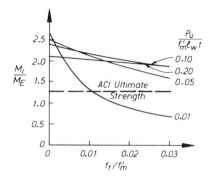

Fig. 7.54 Comparison between elastic and ultimate moment capacities for unreinforced masonry wall.

Equation (7.74) adopts the typical rectangular stress block for strength design of reinforced masonry. The ideal moment capacity will thus be

$$M_i = 0.9P_u \frac{l_w - a}{2} \qquad (7.75)$$

Figure 7.54 compares the ratio of ideal moment [Eq. (7.75)] to design elastic moment [Eq. (7.72)] for a range of axial load levels and allowable tension stress ratios f_t/f'_m. It will be seen that the level of protection against overturning afforded by elastic theory is inconsistent but is generally very high compared with typical strength design requirements, except when axial load levels are low and allowable tension stresses are high.

The maximum lateral wind forces that the wall can sustain are limited by the ultimate moment capacity given by Eq. (7.75). Any attempt to subject the building to higher wind forces would result in collapse by overturning. However, for seismic forces, the development of ultimate moment capacity and incipient rocking about a wall toe does not represent failure. The seismic lateral forces are related to ground acceleration and wall stiffness. Once the wall starts to rock, its incremental stiffness becomes zero, and any increase in ground acceleration will not increase forces on the wall. Failure can occur by overturning only if the acceleration pulse inducing rocking continues with the same sign for sufficient length of time to induce collapse. It is thus clear that collapse will be related to the seismic energy input. If the ground acceleration changes direction soon after rocking begins, the wall stabilizes and rocking ceases or reverses direction. Instability under out-of-plane seismic acceleration is, however, more problematical and is discussed in some detail in the next section.

7.8.2 Unreinforced Walls Subjected to Out-of-Plane Excitation

(a) Response Accelerations The response of unreinforced masonry walls to out-of-plane seismic excitation is one of the most complex and ill-understood areas of seismic analysis. In Section 7.2.3(*a*) the interaction between in-plane

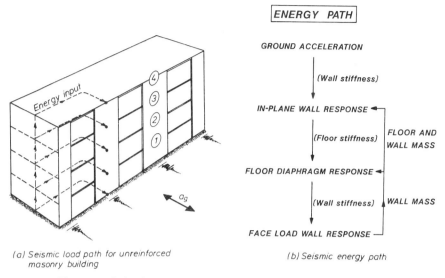

(a) Seismic load path for unreinforced
    masonry building

(b) Seismic energy path

Fig. 7.55 Seismic response of an unreinforced masonry building.

and out-of-plane response of reinforced masonry walls was investigated in some detail, and it was shown that low reinforcement ratios would be adequate to ensure elastic behavior under response accelerations amplified above ground excitation levels by in-plane wall and floor diaphragm response.

The interaction between in-plane and out-of-plane response is examined in more detail in Fig. 7.55 for an idealized unreinforced masonry building. The energy path is shown dashed in Fig. 7.55(a). The end walls, acting as in-plane structural walls, respond to the ground acceleration a_g with response accelerations that depend on height, wall stiffness, and contributory masses from the floor and face-loaded walls. The wall response accelerations at a given height act as input accelerations to the floor diaphragms. If these are rigid, the displacements and accelerations at all points along the floor will be equal to the end-wall displacements and accelerations. However, if the floor is flexible, as will often be the case for existing masonry buildings, response displacements and accelerations may well be modified from the end-wall values. The floor diaphragm response in turn becomes the input acceleration for the face-loaded wall. The ground acceleration has thus been modified by two actions: that of the end structural walls and that of the floor diaphragms before acting as an input acceleration to the face-loaded wall. The interactions implied by this behavior are described schematically in Fig. 7.55(b). The consequence of this complex interaction is that input accelerations to the face-loading wall at different floor levels will be of

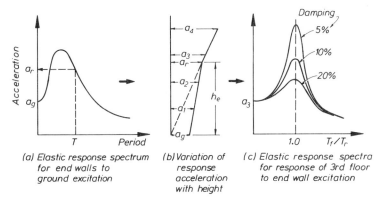

(a) Elastic response spectrum
for end walls to
ground excitation

(b) Variation of
response
acceleration
with height

(c) Elastic response spectra
for response of 3rd floor
to end wall excitation

Fig. 7.56 Flexible floor response to ground excitation.

different magnitude, and may be out of phase or have significantly different frequency composition.

Figure 7.56 describes the response in terms of response spectra. Figure 7.56(a) shows the elastic response spectrum for the in-plane response of the end walls to the ground excitation (a_g). For the fundamental period of transverse response, T, the response acceleration a_r can be calculated. It should be noted that the elastic response spectrum forms an upper bound to response, and a lower response acceleration will be appropriate if the end wall rocks on its base at less than the elastic response acceleration.

The response acceleration a_r refers to the acceleration at the effective center of seismic force, h_e. On the assumption of a linear first mode shape, the peak response accelerations at the various levels can be estimated by linear extrapolation. However, it must be realized that these accelerations are accelerations relative to ground acceleration. Thus, although the mode shape indicates zero acceleration at ground level, it is clear that the maximum absolute acceleration at this level is of course a_g, the peak ground acceleration.

At higher levels the peak absolute acceleration is less easy to define unless a full dynamic time-history computer analysis is carried out. It would be unrealistically conservative to add the peak ground acceleration to the peak response accelerations, since the two accelerations will not commonly occur simultaneously. In fact, in a resonant situation, the response and ground accelerations will be out of phase, and hence will subtract.

A conservative solution for estimating peak floor-level accelerations from the response spectrum is illustrated in Fig. 7.56(b). At heights above the center of seismic force, h_e, the peak accelerations are given by the mode shape from the response accelerations. That is, ground accelerations, which are as likely to decrease as to increase the absolute acceleration, are ignored. At heights less than h_e, the increasing significance of the ground acceleration is acknowledged by use of a linear design acceleration envelope from a_g at ground level to a_r at h_e.

The floor accelerations in Fig. 7.56(b) now become the input acceleration for floor response. Since the end-wall response will largely be comprised of energy at the natural period T of the transverse response, the response of the floor to the end-wall excitation will depend on the ratio of the natural floor period T_f to the wall period T, and the equivalent viscous damping, as shown in Fig. 7.56(c). For very stiff floors ($T_f = 0$), the response acceleration will be equal to the excitation acceleration. For very flexible floors, the response accelerations will be small, but for values of T_f/T_r close to 1.0, resonant response could occur, with high amplification of end-wall response.

As illustrated in Fig. 7.56(b), the level of response acceleration increases with height. Consequently, the roof level is subjected to the maximum accelerations. Since this is combined with the lowest gravity load [and hence lowest stability moment capacity; see Eq. (7.75)], failure of the wall is expected initially at the upper level. This agrees with many cases of earthquake damage to unreinforced masonry buildings, where the walls at upper levels have collapsed but are still standing at lower levels.

The final stage of the energy path for face-load excitation is represented by Fig. 7.57 for one of the stories of the multistory unreinforced masonry building. Inertial response of the wall in face loading is excited by the floor accelerations a_i and a_{i+1} below and above the wall. Although the response acceleration a_{ir} will vary with height up the wall and will be a maximum at midheight and minimum at the floor levels, it is not excessively conservative to assume a_{ir} to be constant over the floor height, as indicated in Fig. 7.57(a). The magnitude of a_{ir} depends again on the ratio of natural frequency of wall response to floor excitation frequency, indicated by the period ratio T_{ir}/T_f in Fig. 7.57(b). If the wall responds elastically without cracking,

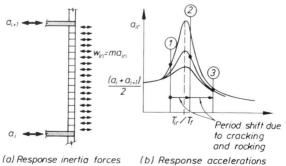

(a) Response inertia forces   (b) Response accelerations

Fig. 7.57   Inertia loads from out-of-plane response.

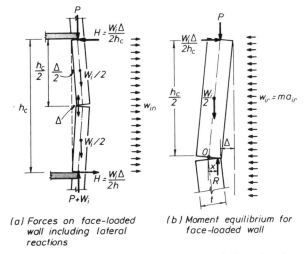

(a) Forces on face-loaded
wall including lateral
reactions

(b) Moment equilibrium for
face-loaded wall

Fig. 7.58   Out-of-plane response of unreinforced wall.

the response acceleration is comparatively easy to calculate. However, as the wall cracks and begins to rock (as discussed subsequently), the natural period will lengthen, changing the response amplification of input acceleration. Figure 7.57(b) shows two possibilities. With a moderate period shift from 1 to 2 in Fig. 7.56(b), coupled with light damping, the face-load response of the wall will be affected by resonance and will be substantially higher than the input acceleration. For larger displacements, the equivalent period may shift past the resonant range to point 3 in Fig. 7.57(b), resulting in lower response accelerations than input accelerations. However, for the large displacements necessary to cause structural collapse, the response period will be quite long, and equivalent viscous damping quite high. It thus seems reasonable to assume that the response acceleration is the average of the input accelerations, a_i and a_{i+1}.

(b)   Conditions at Failure and Equivalent Elastic Response   Figure 7.58 illustrates the condition representing failure for a face-loaded wall element, as discussed in the preceding section. The formation of cracks does not constitute wall failure, even in an unreinforced wall. Failure can occur only when the resultant compression force R in the compression zone of the central crack is displaced outside the line of action of the applied gravity loads at the top and bottom of the wall.

In developing the equations to predict conditions for wall failure, some simplifying assumptions are necessary. First, as mentioned above, it will be assumed that the response acceleration a_{ir} is constant up the floor height, h_c. Hence the lateral inertia force per unit area will be

$$w_{ir} = ma_{ir} \qquad (7.76)$$

where m is the wall mass per unit area of wall surface. The second assumption concerns the degree of end fixity for the wall at floor levels. It is conservatively assumed that the ends are simply supported (i.e., no end moments are applied). This would be appropriate if the walls at alternate story heights were displacing out of phase by 180°, which is a real possibility (see Fig. 7.7).

Figure 7.58(a) shows the forces acting on the wall. As well as the inertia load w_{ir}, there is the applied gravity load P, transmitted by walls and floors above, and the self-weight W_i of the wall, divided into two equal parts $W_i/2$, centered above and below the central crack as shown. The resulting gravity load R acting on the upper half of the wall has the magnitude

$$R = P + 0.5W_i \tag{7.77}$$

Horizontal reactions H are required for stability of the displaced wall. Taking moments about the base reaction, it is clear that

$$H = W_i \Delta/2h_c \tag{7.78}$$

where Δ is the central lateral displacement.

Moment equilibrium at the center of the wall [Fig. 7.58(b)] about point 0 requires that

$$Rx = \frac{w_{ir}h_c^2}{8} + \frac{W_i}{2}\frac{\Delta}{2} + P\Delta + \frac{W_i\Delta}{2h_c}\frac{h_c}{2} \tag{7.79}$$

Consequently, the response acceleration required to develop a displacement Δ is given by

$$a_{ir} = w_{ir}/m = (8/mh_c^2)R(x - \Delta) \tag{7.80}$$

The maximum possible value of the distance x occurs when a stress block at ultimate occurs under the resultant force R at the edge of the wall. Hence

$$x_{max} = t/2 - a/2 \quad \text{where} \quad a = R/0.85f'_m t \tag{7.81}$$

where t is the thickness of the wall considered. Instability will occur when $\Delta = x_{max}$.

(c) Load Deflection Relation for Wall

To assess the energy requirements at failure, it is necessary to develop the load–deflection relationship for the unreinforced masonry wall during out-of-plane response. Prior to cracking, the response is linear elastic. Figure 7.59(a) shows the stress conditions at the central section when cracking is about to occur, assuming zero tension

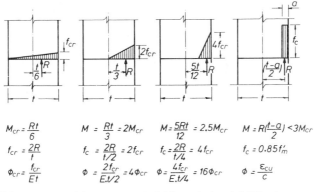

$$M_{cr} = \frac{Rt}{6}$$ $$M = \frac{Rt}{3} = 2M_{cr}$$ $$M = \frac{5Rt}{12} = 2.5M_{cr}$$ $$M = R\frac{(t-a)}{2} < 3M_{cr}$$

$$f_{cr} = \frac{2R}{t}$$ $$f_c = \frac{2R}{t/2} = 2f_{cr}$$ $$f_c = \frac{2R}{t/4} = 4f_{cr}$$ $$f_c = 0.85f_m'$$

$$\Phi_{cr} = \frac{f_{cr}}{Et}$$ $$\Phi = \frac{2f_{cr}}{E.t/2} = 4\Phi_{cr}$$ $$\Phi = \frac{4f_{cr}}{E.t/4} = 16\Phi_{cr}$$ $$\Phi = \frac{\varepsilon_{cu}}{c}$$

(a) At cracking (b) Half cracked (c) ¾ cracked (d) Ultimate

Fig. 7.59   Moments and curvatures at center of face-loaded wall.

strength. Thus, taking moments of the resultant force about the wall,

$$M_{cr} = Rt/6 \tag{7.82}$$

and

$$f_{cr} = 2R/t \tag{7.83}$$

where f_{cr} is the maximum compression stress, as shown in Fig. 7.59(a). The distributed lateral forces required to cause M_{cr} will be given by

$$M_{cr} = w_{ir}h_c^2/8$$

Hence

$$w_{ir} = 8M_{cr}/h_c^2 \tag{7.84}$$

Since $w_{ir} = ma_{ir}$, the acceleration required to cause cracking can readily be calculated. The displacement at the center of the elastic wall is given by

$$\Delta_{cr} = 5w_{ir}h_c^4/384EI \tag{7.85}$$

Figure 7.59(b) shows the stress distribution when the crack has propagated to the wall centroid. The resisting moment is now

$$M = Rt/3 = 2M_{cr} \tag{7.86}$$

At cracking, the curvature at the central section [Fig. 7.59(a)] was

$$\phi_{cr} = f_{cr}/Et \qquad (7.87)$$

For conditions represented by Fig. 7.59(b), the curvature will be

$$\phi = 2f_{cr}/(Et/2) = 4\phi_{cr} \qquad (7.88)$$

It may conservatively be assumed that the displacement Δ increases in proportion with the central curvature. Thus, for the stress conditions in Fig. 7.59(b),

$$\Delta = 4\Delta_{cr}$$

A more accurate estimate for Δ can be obtained by integrating the curvature distribution, but is probably not warranted when other approximations made in the analysis are considered. In fact, the errors are typically not large until very large displacements are obtained.

Figure 7.59(c) shows the appropriate calculations when the crack has propagated to three-fourths of the section depth, and Fig. 7.59(d) shows conditions at ultimate. Using the procedure above, the moment–curvature relationship developed may be converted to an equivalent moment–displacement relationship. Equation (7.80) then allows an acceleration versus central displacement plot to be drawn, since $M = Rx$.

It should be noted that calculations will normally indicate that instability occurs before the ultimate stress conditions, represented by Fig. 7.59(d), are reached. It should also be noted that the ultimate moment in Fig. 7.59(d) has a magnitude $M_i \leq 3M_{cr}$, where M_{cr} is the cracking moment, with the upper limit being approached only for very small axial loads R, when the compression block depth, a, is close to zero.

The form of the acceleration–displacement curve is indicated by the curved line in Fig. 7.60. This curve is elastic nonlinear. That is, the wall will "unload" down the same curve. It is suggested that an estimate of the

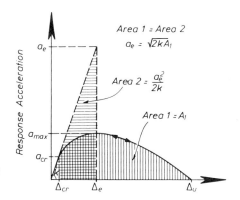

Fig. 7.60 Equal-energy principle for equivalent elastic stiffness.

equivalent linear elastic response acceleration a_e can be found by the equal-energy approach, equating an area under a linear acceleration–displacement line with the same initial stiffness k as that of the true wall acceleration–displacement curve, as shown in Fig. 7.60. If A_1 is the area under the true curve, then

$$A_2 = a_e^2/2k = A_1$$

Hence

$$a_e = \sqrt{2kA_1} \qquad (7.89)$$

In calculating the response using the methodology outlined above, some account of vertical acceleration should be taken, since this reduces the equivalent acceleration necessary to induce failure. Conservatively, it is suggested that a value equal to two-thirds of the peak lateral ground acceleration should be adopted.

(d)   Example of Unreinforced Masonry Building Response   The five-story unreinforced masonry building in Fig. 7.61 has perimeter walls 220 mm (8.7 in.) thick. The 20-m (65.6-ft)-long end walls support masses of 40 metric tons (88.2 kips) per floor (including self-weight) and 20 metric tons (44.1 kips) at roof level. Use the design elastic response spectrum of Fig. 7.61(b) to estimate the natural in-plane period of the end walls, and hence its response accelerations. Assuming that flexible floors amplify the end-wall accelerations by a factor of 2.0, calculate at what proportion of the design forces, failure of a longitudinal face-loaded wall in Fig. 7.61(a) would occur in stories 5, 3, and 1. Assume that the gravity load applied to the wall by the roof is 10 kN/m (0.685 kips/ft) and that at each lower floor it is 14 kN/m (0.959 kps/ft). The self-weight of the wall must be added.

Data:

Elastic modulus: $E = 1.0$ GPa (145 ksi).
Shear modulus: $G = 0.4$ GPa (58 ksi).
Brick density: 1900 kg/m^3 (119 lb/ft^3).
Masonry compression strength: $f_m' = 5$ MPa (725 psi).
Vertical accelerations: Assume $0.2g$ in conjunction with horizontal acceleration [i.e., two-thirds of peak ground horizontal acceleration in Fig. 7.61(b)]. The rotational stiffness of the wall on the foundation material is $k_\theta = 8500$ MNm/rad.

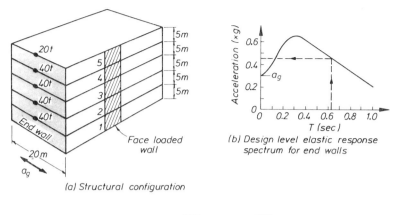

(a) Structural configuration

(b) Design level elastic response spectrum for end walls

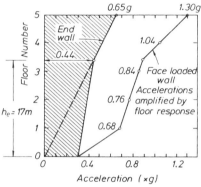

(c) Response accelerations for example building

Fig. 7.61 Design example for five-story masonry building.

(i) *Solution*: A simple simulation of the in-plane response of the end walls can be provided by an equivalent single-degree-of-freedom model of mass:

$$m_e = \sum_1^5 m_i = 4 \times 40 + 20 = 180 \text{ metric tons (397 kips)}$$

at an equivalent height

$$h_e = \frac{\sum m_i h_i^2}{\sum m_i h_i}$$

$$= \frac{20 \times 25^2 + 40(20^2 + 15^2 + 10^2 + 5^2)}{20 \times 25 + 40(20 + 15 + 10 + 5)} = 17.0 \text{ m (55.8 ft)}$$

(ii) *Wall Stiffness*: The gross wall sectional area is $A_g = 0.22 \times 20 = 4.4 \ m^2$. The wall moment of inertia

$$I_g = \frac{0.22 \times 20^3}{12} = 146.7 \ m^4 \ (16{,}980 \ ft^4)$$

The wall stiffness depends on the flexural, shear, and foundation flexibility, and using the equivalent height h_e shown in Fig. 7.61(c) can be expressed as

$$k_e = \frac{1}{\dfrac{h_e^3}{3EI_g} + \dfrac{1.2h_e}{A_g G} + \dfrac{h_e^2}{k_\theta}}$$

$$= \frac{1}{\dfrac{17^3}{3 \times 10^9 \times 146.7} + \dfrac{1.2 \times 17}{4.4 \times 0.4 \times 10^9} + \dfrac{17^2}{8.5 \times 10^9}}$$

$$= 17.6 \times 10^3 \ kN/m \ (1206 \ kips/ft)$$

(iii) *Natural Period*: For the equivalent one-degree-of-freedom model,

$$T = 2\pi \sqrt{\frac{m_e}{k_e}} = 2\pi \sqrt{\frac{180 \times 10^3}{17.6 \times 10^6}} \ second = 0.635s$$

From Fig. 7.61(b) the response acceleration at the center of the mass is 0.44 g.

(iv) *Moment to Cause In-Plane Rocking*: Check stability under $0.8D + E$. The compression block depth

$$a = \frac{0.8P_u}{0.85f_m' t} = \frac{0.8 \times 180 \times 9.8}{0.85 \times 5000 \times 0.22} = 1.51 \ m \ (59.4 \ in.)$$

Hence the restoring moment is

$$M_r = 0.8 \times 180 \times 9.8 \frac{l_w - a}{2} = 1.41 \left(\frac{20 - 1.51}{2} \right) = 13.1 \ MNm \ (9668 \ kip\text{-}ft)$$

The overturning moment at 0.44 response is

$$M_o = 180 \times 0.44 \times 9.8 \times 17 = 13.2 \ MNm \ (9742 \ kip\text{-}ft)$$

Thus the wall starts to rock at $(13.1/13.2)100\% = 99\%$ of the design earthquake.

Figure 7.61(c) shows the response accelerations for the end walls, using the recommendations of Section 7.8.2(a), with the floor accelerations amplified by the factor of 2 noted in the problem statement. Note that the ground-floor accelerations have not been amplified, on the assumption that this floor is rigidly connected to the ground.

(v) *Moment–Curvature Relationship of Face-Loaded Wall*: Consider a 1-m (39.4-in.)-wide strip of the wall at the fifth story. Referring to Fig. 7.58(a) with $P = 0.8 \times 10$ kN/m and $W = 0.8 \times 1.9 \times 9.8 \times 0.22 \times 5 = 16.4$ kN/m, we find that

$$R = P + W/2 = 16.2 \text{ kN/m (1.11 kips/ft)}$$

Note that a factor of 0.8 has been included for a vertical acceleration of $0.2\,g$.

Conditions at cracking (see Fig. 7.59):

$$f_{cr} = 2 \times 16.2/0.22 = 0.147 \text{ MPa (21.2 psi)}$$

Cracking moment [Eq. (7.82)]:

$$M_{cr} = 16.2 \times 0.22/6 = 0.595 \text{ kNm/m (0.134 kip-in./in.)}$$

Equivalent lateral force [Eq. (7.84)]:

$$w_{ir} = 8 \times 0.595/5^2 = 0.19 \text{ kN/m}^2 \left(4 \text{ lb/ft}^2\right)$$

Central displacement [Eq. (7.85)]:

$$\Delta_{cr} = \frac{5 \times 190 \times 5^4}{384 \times 10^9 \times 0.22^3/12} = 1.74 \text{ mm (0.0685 in.)}$$

Curvature [Eq. (7.87)]:

$$\phi_{cr} = 0.147/(10^3 \times 0.22) = 0.668 \times 10^{-3}/\text{m} \left(0.17 \times 10^{-4}/\text{in.}\right)$$

The unit weight of the wall is $1.9 \times 9.8 \times 2.2 = 4.10 \text{ kN/m}^2$ (85.7 lb/ft^2). Hence, from Eq. (7.80), the acceleration at cracking for a displacement of 1.74 mm will be

$$a_{ir} = \frac{8}{4.10 \times 25} \times 16.2\left(\frac{0.22}{6} - 0.00174\right) = 0.0442\,g$$

Thus the top story of the wall will crack at the very low response acceleration

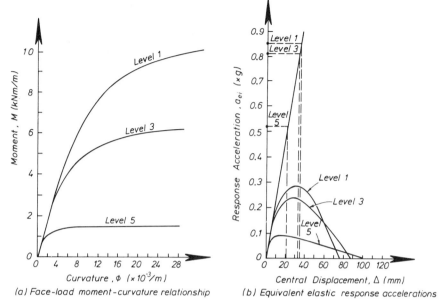

Fig. 7.62   Response of unreinforced masonry building example. (1 m = 39.4 in.)

of $0.0442\,g$. For a moment of $M = 2M_{cr}$ [see Fig. 7.59(b)],

$$\phi = 4\phi_{cr} = 2.67 \times 10^{-3}/\text{m} \ (0.68 \times 10^{-4}/\text{in.});$$

$$\Delta = 4 \times 1.74 \ \text{mm} = 6.96 \ \text{mm} \ (0.274 \ \text{in.})$$

The corresponding acceleration will be

$$a_{ir} = \frac{8}{4.10 \times 25} \times \left(\frac{0.22}{3} - 0.00696\right) = 0.0839g$$

Similar calculations for successively higher curvatures enable the moment–curvature and acceleration–displacement curves of Fig. 7.62 to be plotted. These figures also include curves for levels 3 and 1, based on similar calculations, but with axial load levels of:

For level 3: $R = 0.8 \times 89.3 = 71.4 \ \text{kN/m} \ (4.89 \ \text{kips/ft})$

For level 1: $R = 0.8 \times 158.3 = 126.6 \ \text{kN/m} \ (8.68 \ \text{kips/ft})$

The increased gravity load greatly improves the moment–curvature behavior, but because of the greater P–Δ effect on central moments, the acceleration–displacement curves are not enhanced to the same degree.

The areas under the three acceleration–displacement curves in Fig. 7.62(b) can be measured to give

$$\text{Level 5:} \quad A_5 = 5.27 \ (\text{mm} \times g \ \text{units})$$

$$\text{Level 3:} \quad A_3 = 12.8 \ (\text{mm} \times g \ \text{units})$$

$$\text{Level 1:} \quad A_1 = 14.1 \ (\text{mm} \times g \ \text{units})$$

Thus the equivalent elastic response accelerations to induce failure can be calculated from Eq. (7.89) for level 5 as

$$a_{e5} = \sqrt{2kA_5} \ g = \sqrt{2 \times 0.0254 \times 5.27} \ g = 0.52g$$

where $k = 0.0442g/1.74 \ \text{mm} = 0.0254g/\text{mm}$. Similarly, for levels 3 and 1 the accelerations are $a_{e3} = 0.81g$ and $a_{e1} = 0.85g$, respectively. These levels of equivalent response acceleration are shown in Fig. 7.62(b) and can be compared with the response accelerations corresponding to the design-level earthquake given in Fig. 7.61(c). For level 5, the average design acceleration is

$$a = \frac{1.04 + 1.30}{2} = 1.17g$$

Hence failure at level 5 is expected at $(0.52/1.17) \times 100\% = 45\%$ of the design-level earthquake.

7.8.3 Unreinforced Walls Subjected to In-Plane Excitation

The methodology developed in previous sections for estimating the level of earthquake excitation necessary to induce failure of a face-loaded wall can also be used for walls under in-plane loading. However, for large walls without openings, it will generally be found that no real instability will occur, and that the walls will simply rock on their bases. If the uplift displacements are too large, failure may occur gradually by shedding bricks from the tension end of the wall. Many unreinforced concrete structures subjected to seismic loading have shown signs of relative displacements at one or more levels. This can be attributed to rocking response with simultaneous accelerations in the face-loaded direction.

The behavior of walls with openings subject to in-plane forces requires deeper consideration, however. Figure 7.63(a) illustrates a typical example of a wall divided into four piers by openings. Figure 7.63(b) and (c) represent maximum shear forces that can be transmitted by a typical pier in a rocking mode and shear failure mode, respectively. For a rocking mode, stability is provided by the axial load P. Taking moments about the toe reaction P, and

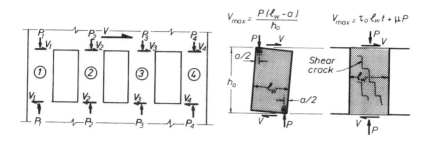

(a) Wall with openings   (b) Rocking pier   (c) Shear failure of pier

Fig. 7.63   Failure of unreinforced walls with piers.

considering the pier self-weight as insignificant,

$$P(l_w - a) = Vh_o$$

that is,

$$V = \frac{P(l_w - a)}{h_o} \qquad (7.90)$$

where, as before, $a = P/(0.85f'_m t)$ is the compression contact length at ultimate.

Very large displacements ($\Delta = l_w - a$) will be necessary to induce instability failure under the rocking mode. The analysis developed for face-loaded walls can be used to estimate maximum equivalent elastic response.

The shear force indicated by Eq. (7.90) can develop only if shear failure of the pier does not occur at a lower shear force. From Fig. 7.63(c) and Eq. (7.26), the shear force associated with shear failure will be

$$V = \tau_o l_w t + \mu P \qquad (7.91)$$

Equation (7.91) assumes that when incipient shear cracking develops, the effective shear stress is uniformly rather than parabolically distributed across the section. In such circumstances, and considering the repeated load reversals expected under seismic loading, it is advisable to set $\tau_o = 0$. Hence

$$V = \mu P \qquad (7.92)$$

Equations (7.90), (7.91), and (7.92) can be combined to find the critical aspect ratio for piers, to ensure that shear failure does not occur. Thus

$$\mu P > \frac{P}{h_o}\left(l_w - \frac{P}{0.85f'_m t}\right)$$

The average axial compression stress on the pier is $f_m = P/(l_w t)$. Then rearranging yields

$$\frac{h_o}{l_w} > \frac{1}{\mu}\left(1 - \frac{f_m}{0.85f'_m}\right) \tag{7.93}$$

In Section 7.2.5(b) a conservatively low value of $\mu = 0.3$ was recommended for design. Thus the requirement of the pier aspect ratio to avoid shear failure is

$$\frac{h_o}{l_w} > 3.3\left(1 - \frac{f_m}{0.85f'_m}\right) \tag{7.94}$$

Typically, f_m will be small compared with $0.85f'_m$, and the criterion $h_o/l_w \geq 3$ can be used as a useful initial check for sensitivity to shear failure. Piers more slender than this should generally rock before suffering shear failure. More squat piers may be considered to be at high risk of shear failure, leading to potential collapse.

It is recommended that the criterion of acceptable performance for piers be that shear failure cannot occur. Provided that rocking limits the shear capacity of all piers at a given level, simple addition of the pier shear forces gives the total story shear.

It should be noted that for face loading, the influence of wall openings is to increase the axial load on the piers, thus making them more stable. This aspect can easily be incorporated in the methodology developed above.

8 Reinforced Concrete Buildings with Restricted Ductility

8.1 INTRODUCTION

In many situations earthquakes are unlikely to impose significant ductility demands on certain types of structures (Fig. 8.1). This is also recognized in common seismic classification of structural systems, as discussed in Section 1.2.4(*b*). Usually, it will be advantageous to relax the strict requirements for the detailing of the potential plastic regions in structures of restricted ductility. There are a number of reasons why a building may be considered to be exposed to restricted ductility demands.

1. Inherently, many structures possess strength considerably in excess of the strength required to accommodate earthquake effects predicted for fully ductile systems [Fig. 8.1(*b*)]. This means that for a given earthquake attack, the ductility demand in such "strong" structures will be less. A typical example is a frame in which the proportioning of structural members is governed by actions due to gravity loads rather than code-specified seismic forces [Fig. 8.1(*c*)]. Some of the relevant issues for gravity-load-dominated frames were examined in Section 4.9. When the reduction of ductility demand is adequately quantified, an appropriate relaxation of seismic detailing is justified.

2. In situations where the detailing for full ductility is found to be difficult or considered to be too costly, the adoption of larger seismic design forces, to reduce ductility demands during earthquakes, may be a more attractive proposition. Greater economic benefits may well be derived from consequent simplifications in the detailing for reduced ductility and perhaps adopting a less than optimal philosophy for hinge location, as implied by the column hinge mechanism which would develop in the structure of Fig. 8.1(*a*).

3. There are moderate-sized structures, the configuration of which do not readily permit a clear classification in terms of structural types [Fig. 8.1(*b*)] to be made, as implied by Chapters 4 to 6. The precise modeling of such structures may often be difficult. Consequently, the prediction of their inelastic seismic response is likely to be rather crude. However, without incurring economic penalty, a more conservative design approach, relying on increased lateral force resistance and consequent reduction in ductility de-

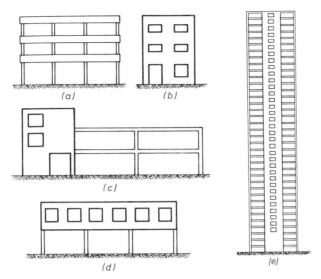

Fig. 8.1   Examples of structures with restricted ductility.

mands, may be more promising. Such a decision may also compensate for disturbing uncertainties about structural performance inherent in such structures.

4. In certain regions, particularly for tall buildings [Fig. 8.1(e)], the critical lateral design force intensity, necessitating fully elastic response, will originate from wind forces. Even though ample ductility may readily be provided, its full utilization in an earthquake is not expected. Consequently, detailing for reduced ductility in potential plastic regions, to meet demands of the largest expected earthquake, should be the rational solution.

The behavior during seismic excitations of such structures will be within limits exemplified by fully ductile and those of elastically responding structures. Although only limited research has been conducted to identify relevant features of the behavior of and the reduced detailing requirements for reinforced concrete structures of restricted ductility capacity, there is evidence that such structures, where properly constructed, satisfactorily met performance criteria during major earthquakes.

The aim of this chapter is therefore to suggest procedures which, while reasonably conservative, may be considered to be suitable to use in the design of concrete structures of restricted ductility without leading to economic sacrifices. The principles presented may also be used to estimate the ductility capacity, and hence the available seismic resistance, of existing structures, which may not have been designed or detailed for fully ductile seismic response.

It is quite feasible to adopt for these structures slightly modified strength design procedures, such as those used for nonseismic situations. For example, to ensure a reasonable hierarchy within a complete energy-dissipation mechanism, artifically low values of strength reduction factors for elements to be protected against hinging or shear failure may be adopted [A6]. However, once the simple principles of the capacity design philosophy for fully ductile structures are grasped by the designer, it is more rational, dependable, and convenient to adopt the same strategy for structures of restricted ductility. The presentation of material in these sections follows this decision.

8.2 DESIGN STRATEGY

The proposed design strategy for structures of restricted ductility is based on the following precepts:

1. Estimate the ductility capacity of a given structure. This may be based on preliminary calculations, the study of potentially critical regions within the structure, or on more sophisticated dynamic response analyses. However, sophistication will seldom be warranted in the design of these structures. Therefore, the designer is more likely to rely on experience, engineering judgment, code recommendations [X5, X7, X10], and general guides, such as given in Section 1.2.4(*b*) and Fig. 1.17.

2. Using the information provided in Chapter 2, the appropriate intensity of the equivalent lateral static forces, corresponding with a reduced ductility capacity μ_Δ, can readily be determined.

3. The determination of the critical design actions, which originate from appropriate combinations of factored gravity loads and earthquake forces, can be completed with the usual techniques of elastic analysis. This may require skillful conceptions for the appropriate mathematical modeling of possibly irregular structural systems. Because of restricted ductility potential, a redistribution of design actions associated with inelastic response should be used with moderation. As a guide (Section 4.3.4) the reduction of "design actions," such as maximum moments in beams at any section, should not exceed

$$\Delta M / M (\%) = 7\mu_\Delta - 5 < 30\% \qquad (8.1)$$

4. Unless fully elastic seismic response ($\mu_\Delta \leq 1$) is assured, the development of plastic regions within the structure should be expected, as in fully ductile systems. These regions should be clearly located. This is essential to enable the necessary detailing to be provided. From the study of possible

collapse mechanisms (i.e., complete kinematically admissible mechanisms) for the structure, locations at which excessively large member ductility demands may arise should be identified. For example, the high ductility demands associated with a column sway mechanism may be related to the structure ductility as indicated in Fig. 7.40. In such cases either a different energy-dissipating mechanism should be chosen, or alternatively, the overall ductility demand μ_Δ for the structure should be further reduced to ensure that the locally imposed ductility can be sustained.

5. The principles of capacity design, as outlined for fully ductile structures, should still be employed to ensure that parts of the structure intended to remain elastic are sufficiently protected against possible overload from adjacent plastic regions, and hence against brittle failure. However, significant simplifications may be made in the evaluation of the design actions to be used for the analysis of the potentially brittle elements. Simplifications may often be based on judgment, or be made as outlined in the following sections.

6. Some structures with large irregularities will necessitate gross approximations in analysis. It should be appreciated, for example, that there is a gradual transition from a deep-membered frame to a wall with openings. The size of openings will indicate whether dominant frame or wall behavior is to be expected. No attempt is made here to introduce comprehensive guidelines for such transient systems. The designer must use judgment. The strategy for restricted ductility implies, however, that such structures are relatively insensitive to the accuracy involved in overall analysis.

Particular difficulties with analyses may arise in irregular dual systems (Chapter 6). In such cases it may be convenient to reduce the structure to a primary lateral-force-resisting system, consisting of potentially effective members only. By conceptually excluding in the design certain elements from participation in lateral force resistance, a complex structure may be reduced to a simple system or subsystems of restricted ductility. Elements excluded from the primary system must then be considered as secondary elements, capable of carrying appropriate gravity loads. While lateral force effects on such elements may be ignored, critical regions should be detailed for limited ductility to enable the secondary system to maintain its role of supporting gravity loads when subjected to restricted lateral displacements, which will be controlled by the chosen primary system. Figure 8.1(c) shows an example structure for which this approach would be applicable, and Section 7.2.1(d) provides some advice for rational choices.

7. Finally, as a result of the expected reduction in ductility demands, a relaxation of the detailing requirements for potential plastic regions is justified. In subsequent sections a number of corresponding recommendations are made.

Because of the paucity of factual information, these suggestions are by necessity based on rational judgment, and on interpolations taken from the observed response of fully ductile and elastic structural components, rather than on experimental verifications of the performance of reinforced concrete structures with "adequate restricted ductility." Performance criteria for structures of restricted ductility should be the same as for fully ductile systems, as defined in Section 3.5.6.

8.3 FRAMES OF RESTRICTED DUCTILITY

In this section the design of components of frames with restricted ductility is described. Frequent references to Chapter 4 are made to point out similarities with or modification to approaches described there in greater detail for components of fully ductile frames. The design of columns for two different types of inelastic frame behavior are studied in separate sections. First, columns of multistory frames in which a strong column/weak beam mechanism is to be enforced are reviewed (Section 8.3.2). Typically, these are columns of multistory frames where strength with respect to lateral forces is governed by wind rather than earthquake forces. Subsequently, a fundamentally different approach to the design of columns, which are assigned to be part of a plastic story mechanism (soft story), is examined in Section 8.3.3. Such mechanisms may be utilized in low-rise frames, where a weak column/strong beam hierarchy is difficult or impossible to avoid. The displacement ductility capacity μ_Δ of such frames is not expected to exceed 3.5 (Fig. 1.17).

8.3.1 Design of Beams

(a) Ductile Beams As in fully ductile frames, the locations of potential plastic hinges, as shown in Fig. 4.15, may be established from the bending moment envelopes constructed for the appropriate combination of factored loads and forces. Because it is relatively easy to provide ample ductility in beams, plastic hinges at some distance away from columns, as seen in Fig. 4.14(*a*), resulting in increased curvature ductilities, may be adopted. This choice will often prove to be practical, because frames of restricted ductility are often dominated by gravity loads (Section 4.9).

While the requirements for minimum tension reinforcement [Eqs. (4.11) and (4.12)] must be retained, the maximum reinforcement content [Eq. (4.13)] may be increased to

$$\rho_{\max} = 7/f_y \text{ (MPa)}; = 1/f_y \text{ (ksi)} \qquad (8.2)$$

To ensure adequate curvature ductility and to provide for unexpected moment reversals, the requirement of Eq. (4.14), may be relaxed. Hence compression reinforcement corresponding to

$$\rho' = \frac{\mu_\Delta}{8}\rho \geq \frac{\rho}{4} \tag{8.3}$$

should also be provided in critical beam sections.

The presence in the plastic hinge region of transverse reinforcement, to prevent premature buckling of compression bars, in accordance with the principles described in Section 4.5.4, is important. However, because of expected reductions of inelastic steel strains, the spacing limitations could be relaxed so that a compression bar with diameter d_b, shown in Fig. 4.20, situated no farther than 200 mm (8 in.) from an adjacent beam bar, is held in position against lateral movement by a tie leg so that

$$6 \leq s/d_b = 16 - 2\mu_\Delta \leq 12 \tag{8.4}$$

The size of the ties should be as required for shear resistance, but generally not less than 10 mm ($\frac{3}{8}$ in.) in diameter.

Although imposed curvature ductility demands may not be large enough to cause strain hardening in the tension reinforcement, the flexural overstrength of the section at both plastic hinges in the span of a beam can be derived, as for ductile frames, with Eq. (4.15). To achieve improved accuracy, the use of a more realistic value of λ_o [Section 3.2.4(e)], allowing for reduced strength enhancement of the steel, will seldom be warranted. The corresponding lateral-displacement-induced shear forces can then be derived from first principles, as given by Eq. (4.18), and shear reinforcement provided in accordance with Section 3.3.2(a)(v). Because of smaller ductility demands and consequent reduced damage in the potential hinge zone, the contribution of the concrete to shear resistance [Section 3.3.2(a)(iv)] may be estimated with

$$v_c = (4 - \mu_\Delta)(v_b/3) \geq 0 \tag{8.5}$$

where v_b is given by Eq. (3.33). Even with complete shear reversal, diagonal shear reinforcement should not be required in plastic hinges of beams with restricted ductility.

(b) Elastic Beams   When frames of restricted ductility are chosen such that columns rather than beams are assigned to dissipate seismic energy [Fig. 8.1(a)], capacity design principles will again lead to a convenient and practical solution. The flexural overstrength of potential plastic hinges in such columns, immediately above and below a beam–column joint, can readily be estimated. The sum of these moments, without any magnification for dynamic effects, may then be used as moment input at the ends of the adjacent beams. As gravity loads will govern the strength of beams, as shown in Section 4.9, it

will be found that usually these earthquake-induced moments in the columns can be absorbed by the elastic beams without requiring additional flexural reinforcement. As ductility demand cannot arise in such beams, no special requirements for detailing arise. The approach to the estimation of shear forces in such beams is as described in Section 4.9.4.

8.3.2 Design of Columns Relying on Beam Mechanisms

(a) Derivation of Design Actions For the determination of bending moments and axial and shear forces required for the proportioning of column sections, the procedure outlined in Section 4.6 is recommended. However, significant simplifications are warranted. It should be remembered that with some exceptions, to be examined in Section 8.3.3, the aim is still to sustain a weak beam/strong column plastic frame system. For this purpose each of the design steps relevant to columns of ductile frames, summarized in Section 4.6.8, is examined and modified where appropriate.

Steps 1 to 4: These are relevant to frame analysis and to beam design and thus remain applicable as for fully ductile frames.

Step 5: When determining the beam flexural overstrength factor ϕ_o, as outlined in Sections 4.5.1(f) and 4.6.3, it will often be found the values of $\phi_{o,i}$ are considerably larger in frames of restricted ductility than those encountered in fully ductile frames. This is because gravity load effects commonly dominate those due to earthquakes, as demonstrated in Section 4.9.3.

Clearly, there is no need to design a column to resist forces larger than those corresponding to elastic response of the structure. Hence the maximum value of the overstrength factor ϕ_o to be used subsequently to determine column design moment may be limited to

$$\phi_{o,\max} = \mu_\Delta/\phi \tag{8.6}$$

Step 6: Because the overstrength factor for beams in structures with restricted ductility is likely to overestimate moment input from beams to columns, less conservative allowances for dynamic effects on column moments are warranted. Accordingly, it is suggested that the dynamic magnification factor, ω, should be:

(a) At ground floor and at roof level $\omega = 1.0$ or $\omega = 1.1$ for one- and two-way frames, respectively

(b) At all intermediate floors $\omega = 1.1$ and 1.3 for one- and two-way frames, respectively

Step 7: It was postulated that the earthquake-induced axial force on a column should be the sum of the earthquake-induced shear forces V_{Eo}, introduced to the column by all beams at overstrength above the level considered (i.e., $P_{Eo} = \Sigma V_{Eo}$). In determining the beam shear forces, V_{Eo}, the value ϕ_o need not be larger than given by Eq. (8.6). As earthquake-induced axial forces on columns are not likely to be critical design quantities, a reduction of these axial forces in accordance with Eq. (4.30) and Table 4.5, will seldom be warranted (i.e., conservatively, $R_v = 1.0$ may be assumed). Determine the maximum and minimum axial forces on the column for load combinations $(D + L_R + E_o)$ and $(0.9D + E_o)$ as for ductile frames.

Step 8: The procedure to determine column shear forces is the same as given in Section 4.6.8.

Step 9: If desired, the moment reduction factor R_m for columns, subjected to small axial compression, given in Table 4.4 may be used. However, this again will seldom be justified, and for most cases, conservatively $R_m = 1$ may be used. The critical design moments at the top and bottom ends of columns above level 1 are then found from

$$M_u = R_m(\phi_o \omega M_E - 0.3 h_b V_u) \tag{4.35}$$

where the column shear force may be assumed to be

$$V_u = 1.1 \phi_o V_E \text{ in one-way frames}$$

and

$$V_u = 1.3 \phi_o V_E \text{ in two-way frames}$$

noting that $\phi_o \leq \mu_\Delta / \phi$.

It is emphasized again that earthquake-induced actions based on the flexural overstrength of beam plastic hinges, as recommended above, are likely to be overconservative. Nevertheless, in terms of column dimensions and reinforcement content, they will seldom prove to be critical. The advantage of the procedure is its simplicity, whereby the more complex derivation of strength requirements, associated with moderate ductility demands, is avoided.

(b) Detailing Requirements for Columns   When arranging the vertical and transverse reinforcement in columns, the general principles outlined in Section 4.6.9 to 4.6.11 should be followed. However, some relaxation in detailing requirements may be made as follows:

1. While precautions have been taken to ensure that plastic hinges above level 1 will not develop in columns, the occurrence of high reversed stresses in the vertical column reinforcement should be recognized. For this reason

the arrangement and detailing of lapped splices of column bars should be in accordance with Sections 3.6.2(*b*) and 4.6.11(*f*).

2. Provided that $\mu_\Delta \leq 3$, lapped splices, if desired, may be used at the lower end of columns in the first story, where a plastic hinge is expected. In this case the transverse reinforcement around spliced bars, as shown in Fig. 3.31, should not be less than 1.3 times that given by Eq. (3.70).

3. Shear reinforcement in potential plastic hinges at the base of columns may be based on the assumption that the contribution of the concrete to shear strength v_c is as given for nonseismic situations by Eqs. (3.34) and (3.35) but reduced by Δv_c, where

$$\Delta v_c = \frac{\mu_\Delta - 1}{3} v_b \qquad (8.7)$$

However, v_c need not be taken less than given by Eq. (3.38).

Columns above level 1, designed in accordance with Section 8.3.2 and the elastic portion of first-story columns, should be designed for shear according to Eqs. (3.33), (3.34), and (3.40). The spacing limitation for horizontal shear reinforcement in columns should be as given in Sections 3.3.2(*a*)(vii) and 3.6.4.

4. Lateral reinforcement to prevent the buckling of compression bars in potential plastic hinge regions should be as required for beams in Section 8.3.1(*a*) by Eq. (8.4).

5. Confining reinforcement in the potential plastic hinge regions of columns should be in accordance with Section 3.6.1(*a*). These requirements [Eq. (3.62)] already incorporate allowances for curvature ductility demands and hence are also applicable to columns with restricted ductility. It may be assumed that the curvature ductility demand μ_ϕ is not more than 10 for columns designed in accordance with Section 8.3.2 when $\mu_\Delta \leq 3.5$.

8.3.3 Columns of Soft-Story Mechanisms

If it can be shown that the rotational and hence curvature ductility demand in potential plastic hinges of columns is not excessive, the development of plastic hinges can be admitted in any story. This means that soft-story mechanisms, such as shown in Fig. 8.2, could be accepted in frames with restricted ductility. The structures shown in Fig. 8.1(*a*) and (*d*) are typical examples. Often this mechanism would offer considerable advantages in low-rise frames and whenever gravity loads are dominant. It was shown in Section 4.9 that in such cases it may be difficult to develop the mechanism shown in Fig. 8.2(*a*), where columns are made stronger than beams. If column hinge mechanisms, such as shown in Fig. 8.2(*b*) to (*d*), control the inelastic response of the frame, it is relatively easy to make the beams

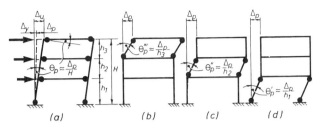

Fig. 8.2 Sway mechanisms in frames of restricted ductility.

stronger. The latter can then be assured to remain elastic, and the detailing requirements of Section 8.3.1(b) apply.

As a recommendation for an approximate but rational design procedure, consider the example frame shown in Fig. 8.2. In this it will be assumed that all frames shown are identical and subjected to the same lateral design forces. Therefore, the maximum elastic displacement for each frame at the top floor will be Δ_y. This then defines the displacement ductility factor $\mu_\Delta = \Delta_u / \Delta_y$ to be considered. The plastic hinge rotations in the beams and first-story columns will be on the order of $\theta_p = (\Delta_u - \Delta_y)/H$ if a weak beam/strong column system, as shown in Fig. 8.2(a), is chosen. It is also evident that unless significant differences in story stiffnesses exist, the member ductility demand in each of these eight plastic hinges will be approximately the same as the overall ductility demand. Recommendations for the design of such a frame of restricted ductility were given in the preceding two sections of this chapter.

In this section we discuss cases in which columns provide the weak links in the chain of resistance, as shown in Fig. 8.2(b) to (d). For the same overall ductility demand μ_Δ, column hinges will now be subjected to larger plastic hinge rotations (θ_p' or θ_p'' or θ_p''') and hence larger member ductility demands. These will be approximately

$$\mu_c = 1 + (H/h_i)(\mu_\Delta - 1) \qquad (8.8a)$$

where μ_c is the member displacement ductility demand in the column in the ith story with story height h_i.

Therefore, if the ductility demand μ_c for such a column is to be limited to a realistic and attainable magnitude, the ductility demand on the entire frame must be reduced to

$$\mu_\Delta = 1 + (h_i/H)(\mu_c - 1) \qquad (8.8b)$$

For example, if we wish to limit column ductility to $\mu_c \leq 4$, assuming that all stories in the frame of Fig. 8.2 have the same height, we find from Eq. (8.8b)

that the displacement ductility demand, which controls the intensity of the earthquake design force, is to be limited to $\mu_\Delta = 2$. If the frame is to have a displacement ductility capacity of $\mu_\Delta = 3$, Eq. (8.8a) will indicate that the column plastic hinges will need to develop a member displacement ductility of $\mu_c = 7$.

It is thus seen that plastic hinge rotations in columns of "soft stories" of multistory frames are extremely sensitive to overall displacement ductility demands. In frames of this type, with three or more stories, designed for restricted ductility ($\mu_\Delta < 3.5$), the curvature ductility demands in plastic hinges may exceed those encountered in weak beam/strong column systems designed for full displacement ductility capacity ($\mu_\Delta = 6$). Plastic hinge regions of such columns must therefore be detailed as described for fully ductile frames in Section 4.6.11. Moreover, lapped splices in columns with potential plastic hinges must be placed at midstory height.

For one- or two-story frames, soft-story mechanisms will not impose excessive ductility demands on plastic hinges of columns. Therefore, these may be detailed the same way as column hinges examined in Section 8.3.2(b), with the exception that in all relevant equations for detailing, the value of μ_Δ will need to be replaced with that of μ_c. For example, if a two-story frame with $h_1 = 4$ m (13.1 ft) and $h_2 = 3$ m (9.8 ft) is to be designed for earthquake forces corresponding with an overall ductility of $\mu_\Delta = 2$, then from Eq. (8.8a) the first-story column hinges should be designed and detailed using

$$\mu_c = 1 + \tfrac{7}{4}(2 - 1) = 2.75$$

8.3.4 Design of Joints

The behavior and design of joints subjected to earthquake forces has been presented in considerable detail in Section 4.8. Therefore, only those issues will be examined here for which some simplifications could be made as a consequence of reduced ductility demand on the structure. As a result of smaller inelastic steel strains, a lesser degree of deterioration within joint cores can be expected. This should result in improved contribution of the "concrete" in the joint core to shear resistance and also in better conditions for the anchorage of bars within the joint.

(a) Derivation of Internal Forces The forces that are applied to the joint were shown for interior joints in Section 4.8.5(b) using Fig. 4.48 and for exterior joints in Section 4.8.11(a) based on Fig. 4.66. The principles outlined apply equally to frames with restricted ductility.

If plastic hinges are to be expected at both faces of an interior column, the joint forces should be evaluated accordingly, as implied by Fig. 4.48(b), using Eq. (4.47). However, in beams of frames of restricted ductility, the bottom flexural reinforcement may not yield at the joint because a plastic hinge with tension in the bottom fibers may, if at all, develop at some distance away from the column face. Such examples are shown in Fig. 4.73(d) to (g). Once the flexural overstrengths of the two plastic hinges in each beam span have been determined, the corresponding moments at column faces and center lines are readily found.

To illustrate a case likely to be encountered in frames with restricted ductility, the frame shown in Fig. 4.73 may be studied. For the plastic beam hinge positions shown in Fig. 4.73(g), it is seen that a moment of $M_{oB} = 264$ units can be expected at the left of the interior column B and that this involves overstrength. However, because of the positions of the positive plastic hinges in span B–C, a moment of only 33 units can develop at the right of column B. Moreover, this moment produces compression and not tension in the bottom reinforcement. The total moment input into column B, which develops the joint shear forces, is therefore only $\Sigma M_{B,\,col} = 264 - 33 = 231$ units. This is considerably less than are similar moment inputs implied in Fig. 4.48 for fully ductile earthquake-dominated frames.

As the bending moment diagrams for each beam span, with two plastic hinges at overstrength, are available, the beam moments to be transmitted to the columns and the corresponding column shear force are readily found. Subsequently, the corresponding internal forces, such as shown in Fig. 4.48(b), can be derived. Using Eq. (4.47), the horizontal joint shear force V_{jh} is then found. Having obtained the joint shear forces V_{jh} and V_{jv}, the procedure described in Section 4.8.7 for fully ductile structures may be used to determine the joint shear reinforcement. However, this may be considered unduly conservative. Therefore, a joint design procedure, more appropriate when restricted ductility demand is expected, is presented in the following sections.

When column hinge mechanisms, such as shown in Figs. 4.72 and 8.2(b) to (d), are used, the process is simply reversed by determining first the maximum feasible column moments below and above the joint. The joint shear forces are then found from either the internal vertical or horizontal forces, the two sets being related to each other by the laws of joint equilibrium.

(b) Joint Shear Stresses Joint shear forces and hence joint stresses are not likely to be critical in this class of structures. However, it is preferable not to exceed the limits set by Eqs. (4.74a) and (4.74b).

(c) Usable Bar Diameters at Interior Joints In Section 4.8.6 various factors affecting the bond performance of bars passing through a joint were examined and Eq. (4.56) quantified these factors. In most situations it will be found that these requirements, catering for conditions more severe than those expected in frames with restricted ductility, can readily be complied with. However, if it is desired to use larger-diameter bars, the value of the critical parameter ξ_m may be evaluated from Eq. (4.57a). For example when the structure in Fig. 4.73(g) is considered again, Eq. (4.57a) will indicate that $\xi_m \equiv 1.13$ will be applicable to the critical top bars in the beam at column C.

This is because the tension stress in the bottom bars at the section to the right of column C will be only on the order of $(35/132)\lambda_o f_y$ and it may be assumed that the relative amount of bottom beam reinforcement will be $\beta = A'_s/A_s \approx 95/190 = 0.5$. Hence the compression stress in the top bars at the same section may be gauged by $\gamma/\lambda_o \approx (35/135) \times 0.5 = 0.13$. This indicates that beam bars at least 35% larger than those derived with Eqs. (4.56) and (4.57c) for fully ductile frames could be used. Bond criteria for the top beam bars at column B (Fig. 4.73(g)) will be even less critical.

As a consequence of restricted ductility demand, reduced curvature ductilities in beam hinges can also be expected. Large tensile steel strains, associated with the maximum flexural overstrength of beam sections, may not be developed. In such cases, with some judgment the value of the materials overstrength factor λ_o, quantified in Section 3.2.4(e), may well be reduced.

When column-sway mechanisms, as shown in Fig. 8.2(b) to (d), are relied on, bar diameter limitation, enumerated in Section 4.8.6, must be applied to the column bars. As described in Section 8.3.3 and illustrated in Fig. 8.2, curvature ductility demands in column hinges of frames with restricted ductility may be as large as those in fully ductile structures. The development of maximum overstrength can be expected. These facts should be taken into account when using Eq. (4.56) and selecting bar diameters in columns.

(d)   Contribution of the Concrete to Joint Shear Resistance   Once the internal beam forces, such as shown in Fig. 4.48, have been derived, as described in Section 8.3.4(a), the horizontal joint shear force V_{jh} is readily evaluated from Eq. (4.47b). However, in frames of restricted ductility a plastic hinge with the bottom beam reinforcement yielding will seldom develop. Hence the horizontal joint shear force in general will be less than what Eq. (4.47c) would predict.

Because the tension force in the bottom beam reinforcement [Fig. 4.48(b)] $T = f_{s2}A_{s2}$, can be estimated, the horizontal joint shear force from Eq. (4.47b) becomes

$$V_{jh} = (1 + k)T - V_{col} \tag{8.9}$$

where from Fig. 4.48(b), $T = A_{s1}\lambda_o f_y$ and

$$k = \beta f_{s2}/(\lambda_o f_y) \geq 0 \tag{8.10}$$

quantifies the tension force in the bottom reinforcement in terms of the tension force in the top beam reinforcement at overstrength.

By following the procedure described in Section 4.8.7(a), the contributions to the strut mechanism may be obtained from Eq. (4.66). However, for common situations, for example when $\beta = A'_s/A_s < 0.75$ and $f_{s2}/(\lambda_o f_y) < 0.5$, it is found that C'_c becomes rather small, and hence the entire tension force generated in the bottom beam reinforcement may be assigned to the top beam bars to be carried in compression. With this conservative assumption the steps in the derivation of Eqs. (4.60) to (4.67) in Section 4.8.7(a) may be followed. This will lead to the estimation of the horizontal joint shear force to be resisted by horizontal joint shear reinforcement

$$V_{sh} = (1 + k)\left(0.75 - 0.85\frac{P_u}{f'_c A_g}\right)T \qquad (8.11)$$

For example using typical relevant values for the joint at column C of the gravity load dominated structure shown in Fig. 4.73(g), such as $f_{s2} = 0.5f_y$, $\beta = 0.5$ and $\lambda_o = 1.25$, from Eq. (8.10) we find that

$$k = 0.5 \times 0.5f_y/(1.25f_y) = 0.2$$

and hence from Eq. (8.11)

$$V_{sh} \approx \left(0.9 - \frac{P_u}{f'_c A_g}\right)T$$

In this case the amount of horizontal joint shear reinforcement (Eq. (4.70b) will be approximately 78% of that required in an identical fully ductile frame. In the case of two-way frames, the beneficial effect of axial compression load on the column in Eq. (8.11) must be reduced in accordance with the recommendations of Section 4.8.8 and Eq. (4.77).

(e) *Joint Shear Reinforcement*   Once the contribution of the truss mechanism [Fig. 4.53(c)] to the horizontal joint shear resistance has been determined from Eq. (8.11), the required joint shear reinforcement should be evaluated as for joints of fully ductile frames using Eqs. (4.70) and (4.72). As can be expected, all other limitations for joints set out in Section 4.8 are readily satisfied in the case of frames with restricted ductility.

(f) *Exterior Joints*   As plastic hinges, with the top beam reinforcement yielding, can be expected at exterior columns, the joint design should be as outlined in Section 4.8.11. As stated, the designer may consider the relevance of a lower material overstrength factor λ_o when computing the maximum beam tension force introduced to the column. Because, in general, requirements for the shear reinforcement at exterior joints are not critical, further refinements of Eq. (4.82) are not warranted. Thus for the design of exterior joints of frames with restricted ductility, the approach described in Section 4.8.11 is recommended.

8.4 WALLS OF RESTRICTED DUCTILITY

Wall dimensions are frequently dictated by functional rather than structural requirements of building design. Therefore, it is often found that the potential strength of structural walls is in excess of that required when considering fully ductile response to the design earthquake. In such cases, the design and detailing of walls may be based on structural behavior with restricted ductility. A justification of the elastic response design approach will not be uncommon, particularly for squat walls.

8.4.1 Walls Dominated by Flexure

It is emphasized again that if ductility is required to be available in a wall during seismic response, every attempt should be made to restrict inelastic deformations to regions of the wall controlled by flexure. Therefore, even for walls of restricted ductility, capacity design procedures will best ensure the predictable inelastic behavior during the extreme design earthquake.

For cantilever walls this approach is extremely simple. All that is required is to ensure that the shear strength of the wall is in excess of the possible shear demand. The latter is associated with the development of the flexural overstrength of the base section. This reasoning also suggests that for the sake of an economic solution, the flexural strength of the wall should not be significantly larger than that required to resist the bending moment derived from code-specified lateral forces.

It is not uncommon that in the process of arranging both the vertical and horizontal wall reinforcement, perhaps using little more than nominal quantities, and by subsequently providing additional vertical bars at wall edges and corners (because of traditional habits), the flexural strength of the wall will exceed its shear strength. The procedure may impart a false sense of safety to the designer. Such a wall may fail in shear.

However, by observing the "shear strength > flexural strength" hierarchy in walls of restricted ductility, a number of simplications of the design process can be utilized. In evaluating the ideal flexural strength, the presence of axial compression on a wall must not be overlooked.

(a) Instability of Wall Sections   In Section 5.4.3(c) a guide was given to assist the designer to select a wall thickness or a boundary element which should ensure that out-of-plane buckling of walls in the plastic hinge region will not occur before the intended displacement ductility capacity is exhausted. Figure 5.35 shows the critical wall thickness for the full range of displacement ductilities that may be encountered in design, and hence it is applicable also to walls with restricted ductility (i.e., $\mu_\Delta \leq 3$). Instability criteria are not likely to govern the dimensions of walls with restricted ductility.

(b)   Confinement of Walls   Confinement of the concrete in the flexural compression zone of wall sections, which is discussed in Section 5.4.3(e), will very seldom be required. Restricted displacement ductility also results in limited curvature ductility, principles of which were examined in Section 5.4.3(e). Curvature ductility in walls is developed primarily by inelastic strains in the tensile reinforcement rather than by large compression strains in the concrete. Therefore, it is relatively easy to achieve restricted curvature ductility, while concrete strains in the extreme compression fibers (Figs. 5.31 and 5.32) remain small. Typical relationships between displacement and curvature ductilities, particularly relevant to walls with small aspect ratios ($A_r < 3$), which are more common in structures of limited ductility, are shown in Fig. 5.33. It may thus be concluded that additional transverse reinforcement in the end regions of wall sections, to confine compressed concrete, are not required in these types of structures. However, in case of doubt, Eqs. (5.18a) or (5.18b) may be used to verify this condition. For example, for a wall with $\phi_{o,w} = 1.4$ and $\mu_\Delta = 2$, Eq. (5.18b) will indicate that confinement is not required unless at the development of this ductility the flexural compression zone c exceeds 25% of the length (l_w) of the wall section.

(c)   Prevention of Buckling of the Vertical Wall Reinforcement   Instability of compression reinforcement is another important consideration when structural behavior relies on ductility. The relevant issues were examined in considerable detail in Section 5.4.3(e). Because of low ductility demand, spalling of the cover concrete in the potential plastic hinge region is not expected. Therefore, the buckling in this region of medium-sized bars [typically, $d_b = 20$ mm (0.75 in.) and larger], even though the Bauschinger effect may still be significant, is not likely to be a critical issue. It is therefore recommended that transverse ties with tie diameter not less than one-fourth of that of the bar to be tied should be provided around vertical bars with area A_b when the reinforcement ratio is $\rho_l = \Sigma A_b/(bs_v) > 3/f_y$ (MPa) [$0.43/f_y$(ksi)] [Eq. (5.21)]. The vertical spacing s_v of such ties should not exceed the limit recommended by Eq. (8.4).

(d)   Curtailment of the Vertical Wall Reinforcement   If at all practicable, curtailment of wall reinforcement should be conservative, to ensure that if ductility demand does arise, inelastic deformations are restricted to the base of the wall, where appropriate detailing, as outlined above, has been provided. For this purpose a moment envelope of the type shown in Fig. 5.29 should be used.

(e)   Shear Resistance of Walls   The design of walls dominated by flexural response should follow the principles discussed in Section 5.4.4. The design

shear force given by Eqs. (5.22) and (5.23) is

$$V_{\text{wall}} = \omega_V \phi_{o,w} V_E$$

where ω_v is the dynamic shear magnification factor and $\phi_{o,w}$ is the overstrength factor for a wall, defined by Eq. (5.13). This simple but conservative approach will normally lead to the use of moderate horizontal shear reinforcement. The limitation of $W_{\text{wall}} \leq \mu_\Delta V_E/\phi$ means that the ideal shear strength of the wall need not be taken larger than that corresponding to elastic response [Eq. (5.24a)]. It is a conservative limit, based on the "equal displacements" concept (Section 2.3.4).

For the determination of the required amount of shear reinforcement, an estimate needs be made as to the share of the concrete, v_c, in the total shear resistance. By similarity to the approach used in the design for shear of columns of limited ductility [Eq. (8.7)], the basic equation [Eq. (3.36)] for walls may be modified so that

$$v_c \leq \frac{4 - \mu_\Delta}{3}\left(A\sqrt{f_c'} + \frac{P_e}{4A_g}\right) \geq B\sqrt{\frac{P_u}{A_g}} \tag{8.12}$$

where the constants are $A = 0.27\text{MPa})$ or $3.3(\text{psi})$ and $B = 0.6(\text{HPa})$ or $7.2(\text{psi})$, and where P_u is the minimum design axial compression load on the wall; for example, 90% of that due to dead load alone and tension due to earthquake, if applicable. No limitation of shear stress, such as given by Eq. (5.26), other than that applicable for gravity loading [Eq. (3.30)] need be applied.

(f) Coupling Beams   In the process of transferring shear forces between two walls, coupling beams are susceptible to sliding shear failure, such as shown in Fig. 5.43(b), when subjected to large reversed cyclic shear stresses, in conjunction with high ductility demand. For this reason diagonal reinforcement, as discussed in Section 5.4.5(b) and shown in Fig. 5.55, should be used whenever the nominal shear stress exceeds the limiting intensity given by Eq. (5.27). For such beams in structures of limited ductility, this limitation may be relaxed, if desired, so that

$$v_i \leq C(5 - \mu_\Delta)\frac{l_n}{h}\sqrt{f_c'} < 0.2f_c' \leq 6 \text{ MPa (870 psi)} \tag{8.13}$$

where $\mu_\Delta \leq 4$, and the constants are $C = 0.1$ (MPa) or 1.2 (psi).

When diagonally reinforced coupling beams are used, shear stress is based on the shear force derived from the elastic analysis with or without vertical shear redistribution between stories, (i.e., $V_i = V_u/\phi$, where $\phi = 0.9$). Of course, in this case no stirrups or ties, apart from those required to preserve the integrity of cracked concrete within the beam (Fig. 5.55), need be provided.

When Eq. (8.13) is satisfied, conventionally reinforced coupling beams, such as shown in Fig. 5.43(a) and (b), may be preferred. In this case shear resistance relies primarily on the stirrup reinforcement, this being an essential component of the traditional truss mechanisms. The shear strength of such beams must be based, in accordance with capacity design principles, on the flexural overstrengths of the end sections as detailed. An exception to this is when coupling beams are assured to remain elastic. This will be the case when the value of ϕ_o exceeds that of μ_Δ/ϕ [Eq. (8.6)].

8.4.2 Walls Dominated by Shear

(a) Considerations for Developing a Design Procedure As a general rule it is easy to ensure that structural walls with cross sections of the type shown in Fig. 5.5(a) to (d) will develop the necessary ductility, when inelastic response is dominated by flexural yielding. The prerequisites of this behavior have been presented in the preceding section. However, when flanged walls, such as shown in Fig. 5.5(e) to (h), are to be used, it is common that the ideal flexural strength of the base section, even with the code-specified minimum amount of vertical reinforcement, will be in excess of that required by code-specified earthquake forces. In some cases the flexural strength may even be in excess of that required to ensure elastic response.

The designer is then tempted to consider the code-specified lateral static force only to satisfy requirements of shear strength. In these cases the response of the wall to the design earthquake may lead to a shear failure and to consequent dramatic reduction of strength, while developing ductilities less than anticipated. These are the types of walls, the inelastic behavior of which is dominated by shear rather than flexure.

Diagonal tension failure due to shear need not be brittle. However, if the horizontal shear reinforcement yields, rapid deterioration of both stiffness and energy dissipation will follow. This deterioration is approximately proportional to the ductility developed by inelastic shear deformations.

An example of a two-thirds-scale test wall, which eventually failed in shear, is shown in Fig. 8.3. The crack pattern seen and the failure loads of this wall can be appreciated only if its hysteretic response is also studied. This is shown in Fig. 8.4. The wall's ideal flexural strength, based on measured material properties, was 28% in excess of its ideal shear strength. The latter was evaluated on the assumption (X3) that the contributions of concrete and horizontal shear reinforcement [Eq. (3.32)] were 55% and 45%, respectively, of the total ideal shear strength, V_i. Following a number of small-amplitude ($\Delta < \Delta_y$) displacement cycles, the ideal flexural strength of the unit was attained and even slightly exceeded in the ninth cycle in the positive direction of loading. This was associated with a displacement ductil-

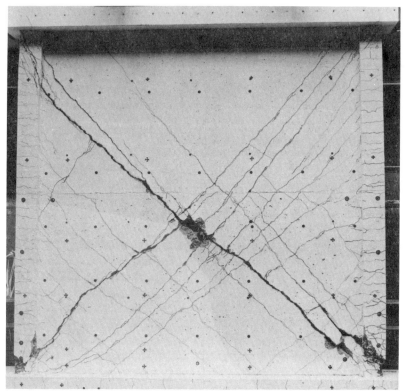

Fig. 8.3   Failure of a squat wall due to diagonal tension after reversed cyclic loading.

ity of $\mu_\Delta = 2.5$. For all subsequent load cycles, particularly in the negative direction of loading, after the development of 5-mm (0.2 in.)-wide diagonal cracks, significant reduction in both stiffness and strength was observed. Nearly total loss of resistance occurred when the displacement ductility exceeded -3.75.

Some loss of energy dissipation in such walls may well be accepted, provided that the imposed ductility demand does not result in excessive reduction of lateral force resistance. The design strategy for such walls may be formulated on the basis of a shear strength–ductility capacity relationship shown in Fig. 8.5. The approach used is similar to that proposed for the assessment of the shear performance of bridge piers [A12].

Figure 8.4 shows that the initial shear strength of the test wall deteriorated after cyclic loading with some ductility demand has been applied. Therefore, if shear strength is to control the inelastic response of a wall, either the ductility demand on the wall must be reduced, or only significantly reduced shear strength is to be relied on.

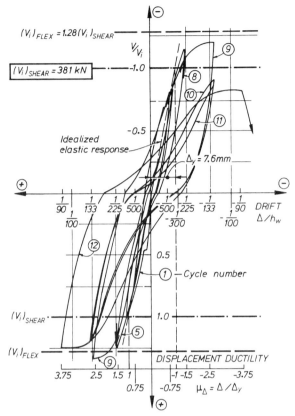

Fig. 8.4   Hysteretic response of a squat wall that eventually failed in shear.

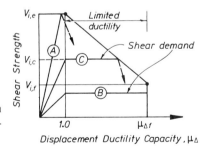

Fig. 8.5 Postulated relationship between shear strength and displacement ductility capacity.

The initial shear strength of a wall $V_{i,e}$ may be based on the models used for nonseismic situations, in which case a considerable portion of the shear resistance may be assigned to the concrete [Eq. (3.36)]. It is, however, this mechanism that deteriorates with increasing ductility demands. Therefore, this initial shear strength $V_{i,e}$ should be relied on only if the expected ductility demand is very small. Conservatively, this approach, shown by curve A in Fig. 8.5, may be used for the design of elastically responding $(\mu_\Delta \leq 1)$ walls.

If the shear strength of the wall V_i, based on the predominant contribution of the horizontal shear reinforcement [Eqs. (3.39) and (3.41)], exceeds the shear force that could possibly be generated when the flexural strength of the base section V_{if} is developed, ductile response, controlled by flexure, can be expected. This is shown in Fig. 8.5 by the relationship B. The capacity design procedure and a conservative estimate of the contribution of the concrete to shear strength with Eq. (3.39), given in Section 3.3.2, should ensure this behavior.

As the measure of shear dominance in a wall is squatness, the dependable displacement ductility capacity, $\mu_{\Delta f}$, of such walls may be assumed to be affected primarily by the aspect ratio, $A_r = h_w/l_w$. This dependable displacement ductility capacity can be conservatively estimated [X3] with

$$\mu_{\Delta f} = 0.5(3A_r + 1) \leq 5 \qquad (8.14)$$

Therefore, the lateral design force for a wall of the type shown in Fig. 8.3 $(A_r = 1)$ should correspond with a ductility capacity of 2 if this ductility is to be derived from flexural deformations. On the other hand, when $A_r \geq 3$, Eq. (8.14) would allow $\mu_\Delta = 5$ to be used.

An intermediate case of shear-dominated walls with restricted ductility is shown in Fig. 8.5 by the relationship C. As stated earlier, in this case the wall's flexural strength is well in excess of that required by the design forces. Therefore, its reduced shear strength $V_{i,c}$, after some cycles of inelastic displacement, is expected to govern its use. This may be achieved if a correspondingly reduced ductility capacity, μ_Δ, is relied on when selecting the lateral design force intensity.

(b) Application of the Design Procedure To illustrate aspects that should be considered while designing such walls, a short and simplified example is used. A structural wall, with an aspect ratio of $A_r = 2$, responding elastically to the design earthquake, would need to be designed, at dependable strength, for a base shear of 4000 kN (900 kips). If flexural ductility can be assured, according to Eq. (8.14), a displacement ductility capacity of only $\mu_{\Delta f} = 0.5 (3 \times 2 + 1) = 3.5$ could be relied on. Using the principles of inelastic response shown in Fig. 2.22 and by considering this example wall to have a short period of vibration, the equal-energy concept is found to be appropriate when estimating the required lateral force resistance. Thus with Eq. (2.15b),

we find that $R = \sqrt{2\mu - 1} = \sqrt{2 \times 3.5 - 1} = 2.45$, and hence $V_u = 4000/2.45 = 1633$ kN (367 kips). However, because of wide flanges, the dependable flexural strength of this wall with minimum amount of vertical reinforcement was found to correspond to a shear force of 2200 kN (495 kips).

It is decided to rely on limited ductility derived from shear deformations in accordance with the relationships shown in Fig. 8.5. The ideal strength corresponding with a dependable displacement ductility of $\mu_\Delta = 3.5$, derived from flexure only, would need to be at least $V_{if} = 1633/0.85 = 1921$ kN (432 kips). Using Eq. (3.39) it is found that for this wall, $V_c = 804$ kN (181 kips) and hence shear reinforcement will be provided to resist a shear force [Eq. (3.32)] of $V_s = 1921 - 804 = 1117$ kN (251 kips). With reduced ductility the contribution of the concrete V_c is going to increase and it may well be that the shear reinforcement, as provided, will be adequate. During elastic response ($\mu_\Delta = 1$) the full contribution of the concrete to shear strength is found from Eq. (3.36) to be $V_c = 1992$ kN (448 kips), and thus $V_{i,e} = 1992 + 1117 = 3109$ kN (700 kips). From interpolation between $V_{i,f}$ and $V_{i,e}$ in Fig. 8.5, assuming a reduced dependable displacement ductility of, say, $\mu_\Delta = 2.5$, we find that

$$V_{i,c} = V_{i,f} + \frac{\mu_{\Delta f} - \mu_\Delta}{\mu_{\Delta f} - 1}(V_{i,e} - V_{i,f})$$

$$= 1921 + \frac{3.5 - 2.5}{3.5 - 1.0}(3109 - 1921) = 2396 \text{ kN (539 kips)}$$

or $\phi V_{i,c} = 0.85 \times 2396 = 2037$ kN (458 kips).

From Eq. (2.15b) we find that the force reduction factor for $\mu_\Delta = 2.5$ is $R = \sqrt{2 \times 2.5 - 1.0} = 2.0$, and hence the lateral design force for reduced ductility increases from 1633 kN (367 kips) to

$$V_u = 4000/2.0 = 2000 < 2037 \text{ kN (458 kips)}$$

and thus the shear strength provided is adequate. Note that flexural yielding cannot be expected with an applied lateral force of less than $V_1 = 2200/0.9 = 2444$ kN (550 kips).

(c) Consideration of Damage   Structures of restricted ductility are expected to suffer limited and hence more easily repairable damage during the design earthquake. Restricted ductility, derived from inelastic flexural deformation of well-detailed structural walls, is likely to result in moderate crack widths. Experiments have shown, however, that walls lightly reinforced for shear develop only a few diagonal cracks with large width. The wall shown in Fig. 8.3 developed instantaneously a main diagonal crack of 1.4 mm (0.06 in.) width when subjected to a lateral force corresponding to less than one-half of

its ideal shear strength. At the imposition of a displacement ductility of only $\mu_\Delta = 1.5$, the width of the diagonal crack in the other direction increased to 2.4 mm (0.1 in.) at a top deflection of $h_w/225$ (i.e., 0.45% drift).

This aspect should be borne in mind when using lightly reinforced walls for restricted ductility. To ensure better damage control under moderate earthquakes, it may prove to be more attractive to rely on very small ductility, if any at all, when designing lightly reinforced walls, the behavior of which during the design earthquake would be dominated by shear rather than flexure. This may often be achieved with the use of more reinforcement, without, however, incurring economic penalties. A more promising approach to damage control in lightly reinforced walls of restricted ductility is to follow the capacity design procedure by providing for shear strength in excess of the flexural strength of the wall.

8.5   DUAL SYSTEMS OF RESTRICTED DUCTILITY

The principles outlined in previous sections of this chapter for distinct types of structures with restricted ductility are also applicable to relevant components of similar dual systems, such as examined in Chapter 6. For this reason a detailed procedure for the design of these structures is not presented here. In typical dual systems of this category, for example as shown in Fig. 8.1(c), the role of walls in lateral force resistance is likely to be dominant. Therefore, primarily situations reviewed in Section 8.4 will be relevant.

9 Foundation Structures

9.1 INTRODUCTION

An important criterion for the design of foundations of earthquake resisting structures is that the foundation system should be capable of supporting the design gravity loads while maintaining the chosen seismic energy-dissipating mechanisms. The foundation system in this context includes the reinforced concrete or masonry foundation structure, the piles or caissons, and the supporting soil. In this chapter we aim to highlight only some special seismic features of the foundation system.

To conceive a reliable foundation system, it is essential that all mechanisms by which earthquake-induced structural actions are transmitted to the soil be clearly defined. Subsequently, energy dissipation may be assigned to areas within the superstructure and/or the foundation structure in such a manner that the expected local ductility demands will remain within recognized capabilities of the concrete or masonry components selected. It is particularly important to ensure that any damage that may occur in the foundation system does not jeopardize gravity load-carrying capacity [T6].

In reviewing the general principles that might govern the choice of a foundation system, the possible failure mechanisms relevant to seismic actions will be considered. In this, attention is paid separately to foundation systems suited to support ductile frame and structural wall types of superstructures, details of which were presented in Chapters 4, 5, and 7. No attempt is made, however, to provide detailed recommendations for the proportioning and detailing of the components of foundation structures, as the principles are similar to those applicable to components of typical superstructures, which are covered in previous chapters.

The expected seismic response of the foundation structure will dictate the necessary detailing of the reinforcement. Where there is no possibility for inelastic deformations to develop during the seismic response, detailing of the reinforcement as for foundation components subjected to gravity and wind induced loads only should be adequate. However, where during earthquake actions, yielding is intended to occur in some components of the foundation structure, the affected components must be detailed in accordance with the principles presented in previous chapters, to enable them to sustain the imposed ductility demands. Therefore, at the conceptual stage of design, a clear decision must be made regarding the admissibility of inelastic deformations within the foundation system. Accordingly, in this chapter we give separate consideration to elastic and ductile foundation systems.

Moments and shear forces in the foundation structure may be strongly affected by the distribution of reactive pressure induced in the supporting soil. Consequently, account should be taken of the uncertainties of soil strength and stiffness, particularly under cyclic dynamic actions, by considering a range of possible values of soil properties.

9.2 CLASSIFICATION OF INTENDED FOUNDATION RESPONSE

A clear distinction must be made between elastic and inelastic response for both the superstructure and the foundation system. This distinction is a prerequisite of the deterministic seismic design philosophy postulated in previous chapters. Although there will be cases where the combined super-structure–foundation system will not readily fit into categories presented here, the principles outlined should enable designers to develop with ease approaches suitable for intermediate systems. The choice between elastic or ductile foundations response is to some extent dependent on the philosophy adopted for the design of the superstructure.

9.2.1 Ductile Superstructures

In previous chapters we described in detail the application of capacity design principles to ductile superstructures, ensuring that energy dissipation is derived from ductile mechanisms only, while other regions are provided with sufficient reserve strength to exclude in any event the possibility of brittle failures. To enable such a ductile superstructure to develop its full strength under the actions of lateral forces, and hence the intended ductility, the foundation structure must be capable of transmitting overstrength actions from the superstructure to the supporting soil or piles.

9.2.2 Elastic Superstructures

In certain cases the response of the superstructure to the largest expected earthquake will be elastic. This could be the result of a design decision, or of code requirements for minimum levels of reinforcement in the superstructure providing adequate strength for true elastic response. Foundation systems that support such elastic superstructures may then be considered in three groups:

(a) Elastic Foundation Systems When an elastic response design procedure is appropriate, the entire structure, including the foundation, is expected to respond within elastic limits. Usually, only in regions of low seismicity or in

low buildings with structural walls will it be possible to satisfy overall stability (overturning) criteria for this high level of lateral forces.

(b)   Ductile Foundation Systems   When the potential strength of the superstructure with respect to the specified lateral seismic forces is excessive, the designer might choose the foundation structure to limit lateral forces that are to be resisted. In such cases the foundation structure rather than the superstructure may be chosen to be the principal source of energy dissipation during the inelastic response. All requirements relevant to ductile performance will be applicable to the design of the components of such a foundation structure. Before choosing such a system the designer should, however, carefully weigh the consequences of possible damage during moderately strong earthquakes. Cracks, which may be large if some yielding has occurred, and spalling of concretes may be difficult to detect. Moreover, because of difficulty with access to members of the foundation structure, which may well be situated below the water table, repair work is likely to be costly.

(c)   Rocking Structural Systems   A common feature in the design of earthquake-resisting structural walls is the difficulty with which the flexural capacity of such walls, even when only moderately reinforced, can be absorbed by the foundation system without it becoming unstable (i.e., without overturning). For such situations the designer may chose rocking of parts or of the entire structure to be the principal mechanism of earthquake resistance. Consequently, rocking parts of the superstructure and their foundation members may be designed to remain elastic during the rocking motions.

9.3   FOUNDATION STRUCTURES FOR FRAMES

9.3.1   Isolated Footings

Gravity- and earthquake-induced forces in individual columns may be transmitted to the supporting soil by isolated footings, shown in Fig. 9.1. The overturning moment capacity of such a footing will depend on the axial

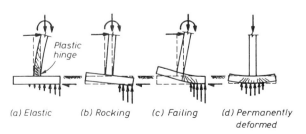

Fig. 9.1   Response of isolated footings.

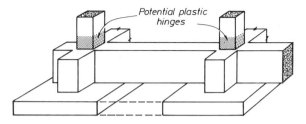

Fig. 9.2   Combined footings.

compression load on the columns acting simultaneously with the lateral force due to earthquake, and on the footing dimensions.

The common and desirable situation whereby a plastic hinge at the base of the column can develop with flexural overstrength while the footing remains elastic is shown in Fig. 9.1(*a*). If the footing is not large enough, rocking or tipping can occur, as seen in Fig. 9.1(*b*), while both the column and the footing remain elastic. Unless precautions are taken, permanent tilt due to plastic deformations in the soil could occur. When the footing is not protected by application of capacity design principles, inelastic deformations may develop in the footing only, as seen in Fig. 9.1(*c*). If these occur due to earthquake attack from the other direction also, the bearing capacity at the edges of the footing to sustain gravity loads may also be lost [Fig. 9.1(*d*)]. Particular attention must be paid to the detailings of the column–footing joint. The relevant principles are those discussed in Section 4.8. Isolated footings of the type discussed here are suitable only for one- or to two-story buildings.

9.3.2   Combined Footings

More feasible means to absorb large moments transmitted by plastic hinges at column bases involve the use of stiff tie beams between footings. Figure 9.2 shows that a high degree of elastic restraint against column rotations can be provided. The depth of the foundation beams usually enables the overstrength moments from the column bases to be resisted readily at ideal strength. Although the footings, too, may transmit some moments, it is usually sufficient to design them to transmit to the soil only axial loads from the columns due to gravity and earthquake forces. The latter are associated with the overstrength actions of the mechanism developed in the ductile frame superstructure, such as derived in Section 4.6.6. The shear forces from the foundation beams must also be included. The model of this structure is shown in Fig. 9.3(*a*).

When the ideal strength of the foundation structure, shown in Fig. 9.2, is based on load input from the superstructure at overstrength, no yielding

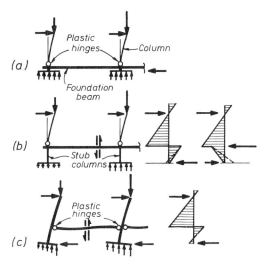

Fig. 9.3   Models of combined footings with foundation beams.

should occur and hence the special requirements of the detailing of the reinforcement for ductility, examined in detail in Chapter 4, need not apply.

Due consideration must be given to the joints between columns and foundation beams. These are similar to the types shown in Figs. 4.45(*a*) to (*c*) and 4.46(*a*) and (*c*). If it is necessary to reduce the bearing pressure under the footing pads, they may be joined, as shown by the dashed lines in Fig. 9.2, to provide a continuous footing.

When the safe bearing stratum is at a greater depth, stub columns or pedestals, extending between the footing and the foundation beam, are often used, as shown in Fig. 9.4. Stub columns require special attention if inelastic deformations and shear failure are to be avoided. It is generally preferable to restrict energy dissipation to plastic hinges in columns above the beams, as shown on the left of Fig. 9.4. The model given in Fig. 9.3(*b*) shows that the

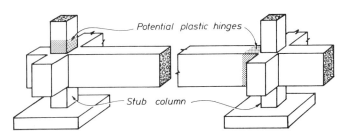

Fig. 9.4   Combined footings with foundation beams and stub columns.

moments and shear forces for the stub columns will be influenced by both the mode of horizontal earthquake shear resistance and the degree of rotational restraint provided by the footing pad. Therefore, it is important to establish whether the column shear forces are resisted at the level of the footings or at the level of the foundation beam.

In exceptional circumstances, foundation beams may be made to develop plastic hinges, as shown in Fig. 9.3(c) and on the right of Fig. 9.4. The capacity design of columns above and below the foundation beam would then follow the steps outlined in Section 4.6. Similarly, all principles discussed in Section 4.8 are relevant to the design of joints adjacent to plastic hinges. Tie beams, extending in both horizontal directions also serve the purpose of ensuring efficient interaction of all components within the foundation system, as an integral unit.

Estimates for the horizontal load transfer between footings and the soil are rather crude. Some provision should thus be made for the horizontal redistribution of lateral forces between individual column bases. Codes [A6] recommend arbitrary levels of axial forces, producing tension or compression, which should be considered together with the bending moments acting on tie or foundation beams. A typical level of the design force in such a beam is of the order of 10% of the maximum axial load to be transmitted by either of the two adjacent columns.

Consideration needs to be given to the modeling of the base of columns in ductile frames. The common assumption of full fixity at the column base may be valid only for columns supported on rigid raft foundations or on individual foundation pads supported by short stiff piles or by basement walls. Foundation pads supported on deformable soil may have considerable rotational flexibility, resulting in column moments in the bottom story quite different from those resulting from the assumption of a rigid base. The consequence can be unexpected column hinging at the top of the lower-story columns. In such cases the column base should be modeled by a rotational spring [Fig. 9.5(b)] of flexural stiffness

$$M/\theta = K_f = k_s I_f \qquad (9.1)$$

where k_s is the vertical coefficient of subgrade modulus [units: MPa/m) (psi/in.)] and I_f is the second moment of area of the foundation pad/soil

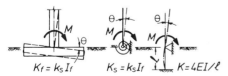

(a) Pad rotation (b) Rotational   (c) Fictitous
                      spring   column

Fig. 9.5 Modeling of column base rotational stiffness.

interface corresponding with pad rotation, as shown in Fig. 9.5(*a*). For computer programs that do not have the facility for directly inputting the characteristics of rotational springs, the foundation rotational flexibility may be modeled by use of a fictitious prismatic column member, as shown in Fig. 9.5(*c*), of length *l* and flexural rigidity *EI*, so that

$$K = 4EI/l = k_s I_f \tag{9.2}$$

Alternatively, modeling as shown in Fig. 6.7 may be used.

9.3.3   Basements

When basements extending over one or more stories below the ground floor are provided and the drift within the basement or subbasement is very small, because of the presence of basement walls, ideal foundation conditions for the ductile frames of the superstructure are available. Hence no difficulties should normally arise in providing an elastic foundation system to absorb actions readily from the superstructure at its overstrength.

9.4   FOUNDATIONS FOR STRUCTURAL WALL SYSTEMS

Often, instead of being distributed over the entire plan area of a building, seismic resistance is concentrated at a few localities where structural walls have been positioned. As a consequence, the local demand on the foundations may be very large and indeed critical [B19]. The performance of the foundation system will profoundly influence the response of the structural wall superstructure (Section 5.1). For these reasons foundation structures, supporting wall systems, are examined here in greater detail.

9.4.1   Elastic Foundations for Walls

The design of elastic foundation systems for elastically responding structures [Section 9.2.2(*a*)] does not require elaboration. The simple principles relevant to ductile superstructures (Section 9.2.1) may be stated as follows:

1. The actions transmitted to the foundation structure should be derived from the appropriate combination of the earthquake- and gravity-induced actions at the base of walls, at the development of the overstrength of the relevant flexurally yielding sections in accordance with the principles of capacity design (Sections 5.4 and 5.5). To determine the corresponding design actions on various components of the foundation structure, the appropriate "soil or pile reactions" must be determined. In this it may be necessary to make limiting assumptions (Section 9.1) to cover uncertainties in soil strength and stiffness.

When foundations are designed for ductile cantilever walls, the actions transmitted from the inelastic superstructure to the foundation structure should be as follows:

(a) The bending moment should be that corresponding to the flexural overstrength of the base section of the wall, developing concurrently with the appropriately factored gravity load. This is, from Eq. (5.13), $\phi_{o,w}M_E$, where $\phi_{o,w}$ is the wall flexural overstrength factor and M_E is the base moment derived from code-specified lateral forces.

(b) The earthquake-induced shear force, assumed to be transmitted at the base of the cantilever, should be taken as the critical shear force [Eq. (5.22)] used in the design of the plastic hinge zone of the wall, that is, $V_{wall} = \omega_v \phi_{o,w} V_E$, where ω_v is the dynamic shear magnification factor given by Eq. (5.23) or (6.12), and V_E is the shear force obtained from the code-specified lateral forces.

2. All components of the foundation structure should have ideal strengths equal to, or in excess of, the moments and forces that are derived from the seismic overstrength of the wall superstructure.

3. Bearing areas of footings, piles, or caissons should be such that negligible inelastic deformations are developed in the supporting soil under actions corresponding to overstrength of the superstructure.

4. Because yielding, and hence energy dissipation, is not expected to occur in components of a foundation structure so designed, the special requirements for seismic detailing of the reinforcement need not be satisfied. This means, for example, that reliance may be placed on the contribution of the concrete to resist shear forces [Eqs. (3.33) to (3.36)] and that transverse reinforcement for the purpose of confinement of the concrete or the stabilizing of compression bars need be provided only as in gravity-loaded reinforced concrete structures.

5. The principles outlined above apply equally to masonry walls and walls of ductile dual structural systems, examined in Chapters 6 and 7.

9.4.2 Ductile Foundations for Walls

For the type of foundation response described in Section 9.2.2(b), the major source of energy dissipation is expected to be the foundation structure. It is emphasized again that the consequence of extensive cracking in components situated below ground should be considered carefully before this system is adopted. When proceeding with the design, the following aspects should be taken into account:

1. If energy dissipation is to take place in components of the foundation structure, the designer must clearly define the areas of yielding. Moreover, when members have proportions markedly different from those encountered

in frames, the ductility capacity likely to be required in potential plastic hinges may need to be checked (Section 3.5). When the foundation element is squat, its length-to-depth ratio should be taken into account in determining the ductility that it could reliably develop, as for cantilever walls (Section 5.7). In this context the length of the foundation beam or wall should be taken as the distance from the point of zero moment to the section of maximum moment, where the plastic hinge is intended to develop.

2. Such foundations are likely to belong to the class of structures with restricted ductility capacity discussed in Chapter 8. Accordingly, increased lateral forces (Section 2.4) must be used for the design of these structures.

3. Design shear forces for components of the foundation structure should be based on capacity design procedures, evaluating the flexural overstrength of potential plastic hinges. In deep foundation members, where shear is critical, diagonal principal reinforcement, similar to the system used in coupling beams of coupled walls, may be appropriate. Because of the scarcity of experimental evidence relating to ductile response of foundation systems, caution and conservative detailing procedures should be adopted. Existing code requirements do not cover contingencies for such situations.

4. To determine the required strength of the elastic superstructure, capacity design procedure, applied in reverse when compared with that relevant for example to ductile frames, will enable the intensity of the lateral design forces resisted by the entire structural system to be estimated. From the computed overstrength of the plastic mechanism of the foundation system, the overturning moment $\phi_{o,f}M_E$, applied to the top of the foundation structure to satisfy equilibrium criteria, can readily be evaluated. The overturning moment and base shear at the top of the foundation structure, originating from the code-specified lateral forces, are M_E and V_E, respectively, and $\phi_{o,f}$ is the flexural overstrength factor for the foundation structure. Hence the ideal flexural strength of the wall base must be

$$M_i \geq \phi_{o,f}M_E \qquad (9.3)$$

and the curtailment of the wall flexural reinforcement should satisfy the moment envelope given in Fig. 5.29.

The effects on shear demands of higher modes during the dynamic response of the structure, examined in Section 5.4.4 and illustrated in Fig. 5.41, should also be taken into account to ensure adequate shear strength of walls at all levels. Accordingly, by similarity to Eq. (5.22), the ideal shear strength of the wall should be

$$V_i \geq \omega_v\phi_{o,f}V_E \qquad (9.4)$$

where the dynamic shear magnification factor ω_v is given in Eq. (5.23).

5. Wall superstructures whose ideal strength exceeds that corresponding to flexural overstrength of a ductile foundation system do not need to meet the special seismic detailing requirements of Section 5.4.

9.4.3 Rocking Wall Systems

It is now recognized that with proper study, rocking may be an acceptable mode of energy dissipation. In fact, the satisfactory response of some structures in earthquakes can only be attributed to foundation rocking. For rocking mechanisms the wall superstructure and its foundation structure should be considered as an entity. In this context rocking implies soil–structure interaction. Rocking at other levels of the building or rocking of one part of the superstructure on another part is not implied here. The design should be based on special studies, including appropriate dynamic analyses [P36], to verify viability. Some aspects of the effects of loss in base fixity for walls in dual structural systems were examined in Sections 6.2.3 and 6.2.4. The following aspects should be considered when designing the foundations:

1. The design vertical load 'on the rocking foundation structure of a wall should be determined from the factored gravity loads (Section 1.3.2), together with contributions from slabs, beams, and other elements, adjacent to walls induced by relative displacements due to rocking. Whether plastic hinges will develop, possibly with overstrength, will depend on the magnitude of deformations imposed on the affected members. A dynamic analysis will produce the magnitude of relevant displacements. In some cases a simple response spectrum approach [P36] will provide adequate prediction of displacements. Components connected to a wall may be yielding during rocking of the wall. The three dimensional nature of the behavior of the entire structure must also be considered. Transverse beams, which may extend between the rocking wall and adjacent nonrocking frames, such as shown in Fig. 6.9, must be detailed for ductility to at least preserve their integrity for carrying the intended gravity loads. Such members should be subject to capacity design procedures.

2. The total sustained lateral forces, acting simultaneously with the vertical loads derived from the considerations above, should be determined from the lateral forces that cause rocking of the walls and from the effects of linkages with other walls or frames through floor diaphragms. The total of lateral forces sustained by the entire structure may then be derived from the summation of the lateral resistance of all rocking walls and nonrocking frames, all of which are effectively interconnected by rigid floor diaphragms. Further aspects of the interaction of frames with rocking walls (Fig. 6.9) are examined in Section 6.2.4.

3. The lower limit on lateral force resistance of walls, at which rocking could be permitted to begin, should be based on considerations of damage

control. In this context rocking implies rigid body rotation of a wall about the theoretical point of overturning, involving loss of contact with the soil over most of the initial bearing area. Inelastic deformations in ductile superstructures are not expected to occur before the lateral force intensity on the buildings as a whole approaches the dependable strength of the system with respect to the design earthquake force. To ensure against premature damage in rocking systems, elastic structural walls should not begin rocking at lateral force levels less than those associated with the code-specified earthquake force.

4. Analysis should be carried out on all structural elements of the building to estimate the ductility demands implied by calculated vertical and horizontal displacements of the rocking wall or walls, to ensure that these do not exceed the ductility capacity of the elements.

5. Rocking walls may impose large forces on the supporting soil. Therefore, bearing areas within the foundation structure should preferably be so proportioned as to protect the soil against plastic deformations that might result in premature misalignment of the otherwise undamaged structural wall or the entire building. This consideration may lead to the selection of independent footings of adequate size that can distribute the total vertical load on a wall to the soil at points or lines of rocking. It is thus possible to ensure that plastic deformations are negligible or do not occur in the soil. Alternatively, oversize footings should be provided to limit soil pressure to a safe value during rocking of the superstructure. Figure 9.6 shows schematic details for a rocking wall. A small foundation beam over the length of the four-story reinforced concrete or masonry wall distributes the relatively small gravity load to the supporting soil. In the event of earthquake-induced rocking the entire gravity load W_g and additional forces W_t, mobilized in floor slabs and edge beams, transverse to the wall, by the uplift displacement Δ_v, must be transmitted to the ground at the left-hand point of rocking. Figure 9.6(b) shows details of an independent reinforced concrete block, separated from the remainder of the structure, which can transmit both the vertical and horizontal components of the resulting force R. An articulation of the foundation beam is provided to introduce the horizontal earthquake force F_E which initiates the rocking motion.

6. Rocking wall superstructures and their foundations, whose ideal strengths exceed the actions derived from capacity design procedures related to all interconnecting elements, need not meet the special seismic ductility requirements of Section 5.4.

9.4.4 Pile Foundations

(a) *Mechanisms of Earthquake Resistance* Pile foundation systems supporting structural walls may be subjected to large concentrated forces as a result of overturning moments and shear forces developed in the wall during an

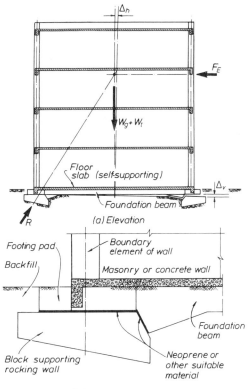

Fig. 9.6 Schematic details of a rocking wall.

earthquake. Three distinct situations, are shown schematically in Fig. 9.7 to examine some design considerations. Similar lateral force and gravity load intensities are assumed to require in each of these cases that the left-hand pile or caisson should develop a significant tensile reaction. Small arrows indicate localities where the necessary horizontal and vertical forces from the surrounding soil may be applied to a pile.

The most common and desirable situation, shown in Fig. 9.7(a), is that of a ductile cantilever wall with piles and foundation beam (pile cap) designed to resist at ideal strength the overstrength load input from the superstructure. However, in certain situations, as will be shown subsequently, a greater degree of conservatism than that used in the capacity design of superstructures will be necessary if it is intended to ensure that piles remain elastic at all times.

For an inelastic foundation system, examined in Sections 9.2.2(b) and 9.4.2, it would be possible to assign energy dissipation to the piles, while the

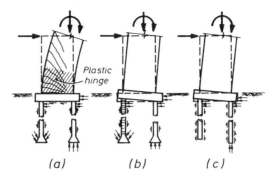

(a)   *(b)*   *(c)*

Fig. 9.7 Walls supported on piles or caissons.

wall above remains elastic. This could be effected by yielding of the longitudinal reinforcement of the tension pile [Fig. 9.7(*b*)] or by the use of friction piles [Fig. 9.7(*c*)]. Neither alternative is particularly desirable. In the first case, wide cracks may develop well below ground level, and as a result of alternate tension and compression actions on the pile, large amounts of confining reinforcement will be required. In the second case, reliance is placed on skin friction, the value of which is uncertain during seismic response, as was evidenced by the complete pullout of such a pile during the 1985 Mexico earthquake [M15].

(b) ***Effects of Lateral Forces on Piles*** The behavior of piles embedded in the ground and subjected to lateral forces and consequent bending moments, shear forces, and distortions is extremely difficult to predict with accuracy. Predictions of the dynamic response of such piles depend, among other variables, on modeling techniques used and the simulation of soil stiffness and density distribution, frequency variability of soil reactions, and damping resulting from wave radiation and internal friction [W5]. In more simplified approaches the Winklerian model of beams on elastic foundations is used, where allowances may be made for the relative position of a pile within a group with suitable variation of the modulus of subgrade reaction over the length of the pile [M16]. However, nonlinear Winkler springs will normally be needed to represent soil properties adequately under strong seismic response.

Within the soil, lateral pile displacement will be affected by the earthquake response of the superstructure (dynamic effects) and in certain cases also by movements of the surrounding soil (kinematic effects). The resulting pile–soil interaction can induce large local curvatures in piles, particularly when piles cross an interface between hard and soft layers of soil. In such cases, as shown in Fig. 9.8, it may not be possible to avoid the development of plastic hinges, even if a capacity design approach was used in an attempt to

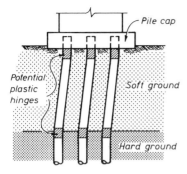

Fig. 9.8 Locations of potential plastic hinges in embedded piles.

protect the pile from damage during the inertial response of the superstructure. While the total base shear transmitted to the superstructure at the development of its overstrength can be determined with reasonable precision, the evaluation of the intensities of bending moments and shear forces along piles, such as in Fig. 9.8, is much more uncertain. This is because, apart from the variability of soil properties, the levels at which piles might be restrained against rotations may also vary considerably over a site and would not normally be verified for each pile. Therefore, conservatism tempered with engineering judgment is warranted in such situations. This implies a deliberate underestimation of the distance between peak moments in order to obtain an estimate for the maximum shear demand. Because the location of peak moments, other than that adjacent to the pile cap, is uncertain, detailing for at least limited curvature ductility should extend over a considerable length where large moments in the pile are feasible.

When pile foundations are used, effective base shear transfer by means other than piles should only be assumed if it can be shown that:

1. Shear forces can be effectively transferred from the soil in contact with an extensive foundation structure by means of friction and bearing against ribs keyed into undisturbed soil or against appropriately designed concrete faces cast directly against soil (i.e., not those supporting backfill).

2. In the process of shear transfer, soil shear strains in strata just below the pile cap will be negligible.

Suggested values of friction factors are given in Table 9.1. The numbers given are ultimate values, the development of which requires some movement before failure would occur. The effect of adhesion, where applicable, is included in the friction factor.

Buildings on sloping sites may be supported by piles, some of which may extend a considerable distance between the soil surface and the pile cap.

TABLE 9.1 **Friction Factors for Mass Concrete Placed on Different Foundation Materials [U4]**

Foundation Materials	Friction Factor
Clean sound rock	0.70
Clean gravel, gravel–sand mixtures, coarse sand	0.55–0.60
Clean fine to medium sand, silty medium to coarse sand, silty or clayey gravel	0.45–0.55
Clean fine sand, silty or clayey fine to medium sand	0.35–0.45
Fine sandy silt, nonplastic silt	0.30–0.35
Very stiff and hard residual or preconsolidated clay	0.40–0.50
Medium stiff and stiff clay and silty clay	0.30–0.35

When assigning lateral forces to piles, their shear stiffness must be taken into account. For example, a circular elastic pile at the eastern edge of the building, symmetrical in plan, as shown in Fig. 9.9, would resist only about 4% of the lateral force induced in the pile at the western edge, assuming that all piles have the same diameter and that lengths are of the proportions shown. As Fig. 9.9(c) shows the efficiency of the piles in resisting lateral forces is reduced dramatically as pile lengths increase. While the foundation system in Fig. 9.9 is symmetrical in terms of east–west seismic response, gross eccentricity, as defined in Section 1.2.3(b) and shown in Fig. 1.10(d), exists when lateral forces are generated in the north–south direction. The eccen-

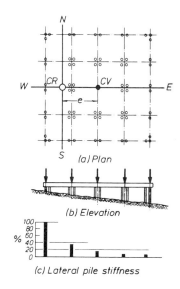

Fig. 9.9 Piles on sloping sites.

tricity e shown in Fig. 9.9(a) is drawn to scale and is based on the assumption that the ground slopes only in the east–west direction. In this example structure elastic north–south displacements of the piles at the eastern edge due to torsion alone would be approximately 1.5 times as large as that due to translation of the platform.

(c) Detailing of Piles

(i) *Reinforced Concrete Piles*: The proportioning and detailing of reinforced concrete piles should follow the methodology recommended for columns at upper levels of capacity-designed ductile frames, examined in Section 4.6. The end region of a pile over a distance l_0 should be reinforced as described in Section 4.6.11(e). However, as pointed out in Section 9.4.4(b), the region of partial confinement may need to be extended considerably so as to cover the probable locations of peak moments along a pile.

Although design computations may show that no tension load would be applied to a reinforced concrete pile with gross sectional area A_g, minimum longitudinal reinforcement, with total area $A_{st} = \rho_p A_g$, should be provided so that:

1. Requirements for transferring any tension resulting from the design actions are satisfied, or
2. $\rho_p \geq 1.4/f_y$ (MPa) $[0.2/f_y$ (ksi)]
3. $\rho_{min} \geq 0.7/f_y$ ($0.1/f_y$ ksi) when it can be shown that the reinforcement content required to resist 1.4 times the design actions does not exceed this amount.

The maximum amount of longitudinal reinforcement in reinforced concrete piles and its arrangement should be as for columns given in Section 4.6.9. To enable, if necessary, the flexural strength of the pile section to develop, it is essential that all longitudinal bars in a pile are adequately anchored within the pile cap.

In noncritical regions, nominal transverse ties or spirals at vertical spacing not exceeding 16 times the diameter of longitudinal bars should also be provided [A6].

(ii) *Prestressed Concrete Piles*: Where prestressed pretensioned concrete piles are used with or without additional nonprestressed bars, plastic hinge rotations corresponding with a member ductility factor of 8 may be developed [P48] if transverse reinforcement according to Section 3.6.1(a) and Eq. (3.62) is provided. However, the terms in parentheses in this equation, allowing for the effect of axial load intensity, should be replaced by the term

$$\left(\frac{P_u}{f'_c A_g} + f_p - 0.08 \right) \tag{9.5}$$

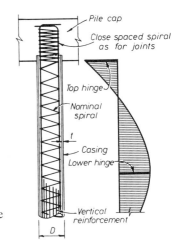

Fig. 9.10 Steel-encased reinforced concrete pile with spiral confining reinforcement.

where f_p is the compressive stress in the concrete due to prestress alone. When prestressed piles are protected from hinge formation on the basis of capacity design principles [Section 9.4.4(a)], 50% of the confining reinforcement intended for plastic hinge regions is considered to be adequate in regions of peak moments along the piles.

Because transverse (spiral) reinforcement is also required to provide lateral support to longitudinal reinforcement, to prevent premature buckling in critical regions, vertical spacing or pitch should be limited as for columns in Section 4.6.11(d). Restraint against buckling of individual prestressing strands, which have been strained beyond yield in tension, are more difficult to provide. Limited experiments indicated [P48] that to delay buckling of strands, the maximum pitch of confining transverse reinforcement be limited to $s_h \leq 3.5d_b$, where d_b is the nominal diameter of the prestressing strand used.

(iii) Steel-Encased Concrete Piles: Piles consisting of reinforced concrete cores contained within cylindrical steel casings have performed well in experiments simulating seismic response. Two characteristic situations must be considered, as illustrated in Fig. 9.10. At the top of the pile, the casing will normally be discontinuous at or somewhat above the base of the pile cap. Longitudinal reinforcement in the pile is thus necessary at this critical section to provide the required moment capacity, and must be properly anchored into the pile cap. However, the casing acts as very efficient transverse reinforcement for the plastic hinge region at the pile top, restraining the longitudinal bars from buckling and confining the concrete. Provided that the volumetric ratio of confinement $\rho_s = 4t/D$, where t = casing thickness and D = casing diameter, exceeds that required in Section 3.6.1(a) for concrete columns, only nominal spiral reinforcement is required in the core concrete

to position the longitudinal reinforcement. However, closer spacing of the spiral reinforcement within the pile cap will be required to assist in resisting joint shear forces, in accordance with the principles of Section 4.8. Experiments on pile/pile cap connections so designed have resulted in extremely ductile response, with flexural strength exceeding predictions as a result of the casing bearing against the pile cap on the compression side, thus acting as compression reinforcement.

The second critical location occurs some depth below the surface, where bending moments of opposite sign to those at the pile cap develop. At this location the state of stress in the casing is complex, as longitudinal stresses can develop as a result of flexural action, hoop tensions develop by confining action, and shear stress is developed as a result of shear forces. However, it is sufficiently accurate to assume that the casing is capable of developing its yield strength independently in flexural and confining actions.

Experiments on critical sections of steel-encased piles, where the casing is continuous through the critical region, indicate that fully composite action may be expected. The displacement ductility capacity found from simple fixed based cantilever tests is typically $\mu_\Delta \geq 4$, with the limit corresponding to onset of casing buckling on the compression side. It has been found [P48] that buckling occurs at this level regardless of the diameter-to-thickness ratio within the range $30 \leq D/t < 200$. There thus seems little reason to limit the D/t ratio as is common in many codes [A1, X3]. It should also be recognized that a plastic hinge forming at depth in soil will be distributed over a much greater length than corresponding to a fixed-base cantilever, as a result of the low moment gradient at the critical section (see Fig. 9.10). As a consequence, the true ductility capacity is likely to be more than twice that indicated from cantilever tests.

9.4.5 Example Foundation Structures

To illustrate the relevance of the design philosophy outlined in previous sections, a few examples, necessarily simplified, are introduced and discussed.

Example 9.1 A simple cantilever wall, subjected to earthquake forces and gravity loading, is shown in Fig. 9.11. Its foundation consists of a spread footing. The base shear is assumed to be transmitted by friction at the underside and by bearing at the end of the footing pad. It is evident that significant tension in the principal vertical wall reinforcement, necessary to develop a ductile plastic hinge at the wall base, can be generated only if the resultant of the reactive soil pressure is located close to or beyond the compression edge of the wall. For efficient wall base fixity the footing must therefore project beyond the edge of the wall, as shown in Fig. 9.11. Unless the footing is wide, large soil stresses are associated with a soil reaction shown in Fig. 9.11. The structure shown possesses limited base fixity unless exceptionally large gravity forces are to be transmitted. It may be necessary

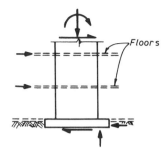

Fig. 9.11 Foundation for an isolated cantilever wall.

to consider its contribution in the rocking mode. If the wall is slender, its contribution, when rocking, to the total lateral force resistance for the building may be very small.

When piles, caissons, or rock anchors with significant tensile capacity are provided, as shown in Fig. 9.7(*a*), the flexural capacity of the cantilever wall at its base can readily be developed. The potential plastic hinge zone at the wall base, where special detailing requirements in accordance with Section 5.4 need to be satisfied, is shown by the shaded area [Fig. 9.7(*a*)]. The footing or pile cap and the piles [Fig. 9.7(*a*)] would need to be provided with ideal strengths at least equal to the flexural overstrength of the cantilever wall.

Example 9.2 Two cantilever walls are supported on a common foundation structure, consisting of piles and a deep foundation beam, as shown in Fig. 9.12. Arrows indicate qualitatively actions due to gravity and earthquake and the corresponding reactions at the foundation–soil interface. With a strong

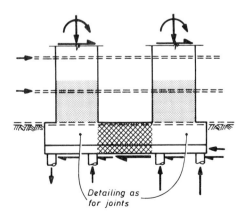

Fig. 9.12 Combined foundation for two cantilever walls.

and stiff foundation beam or wall, the major part of the moments introduced by the cantilevers through the potential plastic hinge regions, again shown shaded in Fig. 9.12, may be resisted by the portion of the foundation structure between the inner faces of the two walls, shown cross hatched.

The design for shear in this region will require special attention. When actions on the foundation are derived from capacity design consideration, yielding in the foundation structure can be prevented. Consequently, the contribution of the concrete to shear strength can be relied upon. In comparison with the structure of Fig. 9.7, with this type of foundation structure loads on piles can be reduced considerably, while the formation of the intended plastic hinges in the walls can more readily be assured. Careful detailing for continuity, shear resistance and bar anchorage, as for beam–column joints, is required in the area common to the walls and foundation beam. Detailed calculations for a similar type of foundation wall supporting a masonry structure are given in Section 7.7.4 and Fig. 7.52.

Example 9.3 It is often difficult, if not impossible, to provide base fixity for walls located adjacent to the boundary of the building, particularly under seismic forces transverse to the boundary. Service cores, accommodating lift and stair wells and consisting of two or more flanged walls, are often located in such positions and are required to transmit large overturning moments to the foundations.

Figure 9.13 shows one solution whereby a deep foundation beam connects the core with one or more adjacent columns. Thereby the internal lever arm acting on the foundation–soil interface that is required to resist the overturning moment introduced at the wall base is increased. Hence the forces to be transferred to the supporting soil are reduced. Moreover, the gravity load on the columns can be made use of in stabilizing the service core against

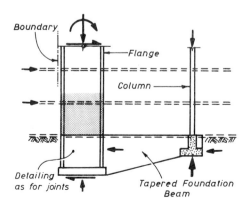

Fig. 9.13 Foundation for walls adjacent to a boundary.

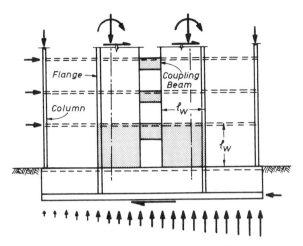

Fig. 9.14 Foundation for a coupled wall structure.

overturning when earthquake forces, opposite to those shown in Fig. 9.13, act on the building.

In designing the foundation structure, the flexural overstrength of the wall base should again be considered to determine the design forces. Particular attention needs to be paid to the junction of the wall and the foundation beam, which should be designed, using principles presented in Section 4.8, as a large knee joint subjected to reversed cyclic loading [P1].

Example 9.4 As a consequence of seismic axial forces generated in the walls, the capacity of coupled structural walls to resist the overturning moment can be considerably more than the sum of the moment of resistance of the walls [Eq. (5.3)] that are being coupled. Therefore, massive foundations may be required to enable ductile coupled walls to develop their full potential as major energy-dissipating structural systems. Figure 9.14 shows a foundation wall receiving the load from the coupled wall superstructure and two columns.

The potential plastic hinge regions within the ductile superstructure are again indicated by the shaded areas. The foundation wall shown is shallow relative to the coupled walls. Therefore, it may require considerable amounts of flexural reinforcement to resist at ideal strength the overstrength overturning moment input from the coupled walls. Of particular importance is the region of the foundation beam between flanges of the two walls, where large shear forces may need to be transferred. This is shown in Section 9.5.

Example 9.5 Cantilever or coupled walls assigned to resist the major part of the earthquake forces and placed at the ends of long buildings typically carry

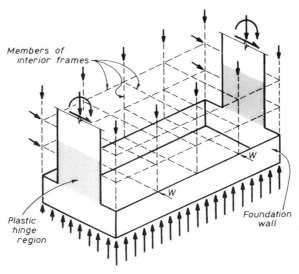

Fig. 9.15   Foundation structure for cantilever walls situated at boundaries.

relatively small gravity load. For this reason it is difficult to provide foundations for them which are large enough to ensure that these walls will not
overturn or rock prior to the development of their flexural overstrength.
Therefore, the foundations of such end walls may need to be connected to
the remainder of the structure, situated between the ends, in order to
"collect" additional gravity loads. Figure 9.15 shows such a situation. The
end walls are connected to a box-type basement structure, consisting of
peripheral and perhaps internal foundation walls, supporting a raft and a
ground-floor slab. Fixity of the ductile cantilever walls is provided by the
peripheral long foundation walls, which usually also support a row of columns.
Because the reactive pressure due to overturning moments, introduced by the
end walls, may be induced primarily under the longitudinal foundation walls,
these walls are usually subjected to very large bending moments. These
require massive flexural reinforcement in both the top and bottom of the
foundation walls (Section 9.5).

The demand for flexural reinforcement in the exterior foundation walls
may be considerably reduced if the cantilever walls are placed away from the
ends. In Fig. 9.15 a structurally, if not architecturally, more advantageous
position for such walls is marked W.

Example 9.6   When a basement is provided with deep peripheral foundation
walls, it may be more convenient to transfer the base moment due to
earthquake forces on interior walls or service cores to those exterior foundation walls. Such an interior flanged wall is shown in Fig. 9.16. The spread

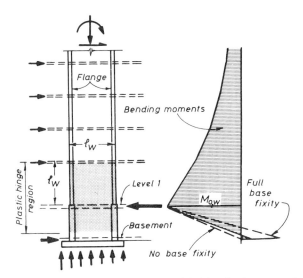

Fig. 9.16 Base fixity for a cantilever wall provided by diaphragm action of floors.

footing under the wall is provided primarily to resist vertical loading on the wall. The moment at the development of the flexural overstrength of the ductile cantilever wall $M_{o,w}$ is, however, to be transferred by means of a horizontal force couple to the level 1 floor and basement slabs, respectively. Consequently, these slabs are to be designed as diaphragms to transfer the forces to peripheral or other long foundation walls.

The degree of fixity of the wall, where it is in contact with the soil, may be difficult to evaluate and some estimate between extreme limits, indicated in the bending moment diagram in Fig. 9.16, may have to be made. In any case, some base fixity should be assumed to ensure that the required shear resistance of the wall, between basement and ground-floor level, is not underestimated. The large shear force in this relatively short region of the wall may warrant the use of some diagonal shear reinforcement.

The extent of the plastic hinge region (shown shaded) below ground-floor level is not clearly defined. Detailing of the reinforcement for ductility of this region should not be overlooked. Such detailing should be used over the length l_w below level 1 or down to the basement, whichever is the smaller distance, as well as over the obvious plastic hinge region above level 1.

Example 9.7 Whereas it would be difficult to develop in individual footings the full moment capacity of cantilever walls, this could be achieved when a massive foundation beam, connecting two or more cantilever walls, as shown in Figs. 9.12 and 9.17 is used.

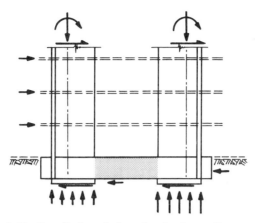

Fig. 9.17 Ductile foundations for elastic cantilever walls.

When the flexural capacity of the walls is excessive, as a result, perhaps of architectural restraints on wall geometry, and code-specified minimum reinforcement levels, the designer may choose the foundation to be the major source of energy dissipation. Accordingly, as Fig. 9.17 shows, the foundation wallbeam between two walls may be designed to develop the necessary plastic hinges. Such beams should be treated the same way as deep beams of coupled walls. Hence they are best reinforced with diagonal bars to resist fully both the moment and shear to be transferred between the two walls. The moment of resistance assigned to the footings will depend on the stiffness of the soil relative to the foundation beam. Thus some judgment will be required in assigning resistance of the wall base moments to the footing and to the foundation wall beam. Limiting situations are similar to those shown in Fig. 9.3. Moreover, it must be recognized that plastic hinge rotations at the ends of the foundation wallbeam will result in equal rigid-body rotations of the cantilever walls and footings attached to them. Hence in this plastic state additional moment resistance may develop due to soil reactions. When evaluating the flexural overstrength factor $\phi_{o,f}$, allowance for the moments introduced to the footings should be made; otherwise, the moment input in the cantilever walls will be underestimated.

Once the foundation beam is designed and its flexural overstrength is determined, it is possible to provide for the corresponding ideal strength at the base of the walls so that yielding in the walls should not develop. To keep the walls elastic may then result in saving in transverse reinforcement for shear, confinement, and bar stability, because the walls would not need to be detailed for ductility.

Because of the nonsymmetrical configuration of wall sections, or the influence on moment strengths of greatly reduced net axial forces, corresponding with opposite directions of seismic lateral forces, the flexural strength of a wall may be considerably less in one direction of earthquake attack than in the other. In such cases the designer may allow the yielding of one wall, while the other wall remains elastic when the direction of earthquake force corresponds with its large flexural strength. In this case plastic hinging will occur in the foundation beam only adjacent to the strong wall.

9.4.6 Effects of Soil Deformations

The elastic and inelastic response of structural wall systems is sensitive to deformations that originate in the foundation systems. Usually, it is soil deformations, rather than component distortions within the foundation structure, which significantly affect the stiffness of walls. For the assessment of foundation compliance the approach described in Section 9.3.2 for footings resting on deformable soils [Eq. (9.1)] is relevant. Because of the difficulty in the determination of the coefficient of subgrade modulus, k_s, accuracy in the prediction of soil stiffness is generally not comparable with that accepted in the analyses of reinforced concrete superstructures. Recommendations have also been formulated to assess the effects of soil–structure interaction on the dynamic response of the structure [A6].

When the elastic deformations of the soil are estimated, their contribution may be included in the total deflection of the structure for the purpose of estimating the fundamental period of vibration. Equations (9.1) and (9.2) may be used for this purpose. Inelastic deformations, required to develop the intended displacement ductility, should originate entirely in the plastic hinges of the superstructure, such as at the base of the ductile cantilever wall or in the inelastic foundation structure, not in the ground. In the design, permanent inelastic soil deformations are not anticipated. Consequently, for a given displacement ductility demand, much larger curvature ductility will be required in plastic hinges. This is because the yield displacement results from structural and soil deformations, as shown in Fig. 7.3(c), but subsequent inelastic displacements should originate from plastic distortions of the structure only. A quantitative evaluation of the effect of soil deformations on ductility demands for the superstructure is given in Section 7.2.4(a).

9.5 DESIGN EXAMPLE FOR A FOUNDATION STRUCTURE

9.5.1 Specifications

The entire earthquake resistance of a 10-story reinforced concrete building has been assigned to eight structural walls, symmetrically arranged around the periphery of a square building. The elevation of the prototype exterior framing is shown in Fig. 9.18(a).

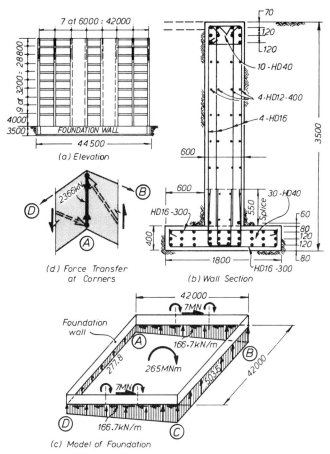

Fig. 9.18 Details of example foundation structure. (1000 mm = 3.28 ft; 1 kN = 0.225 kip; 1 kN/m = 0.0686 kip/ft; 1 MNm = 738 kip-ft.).

The floor system consists of a waffle slab supported by the peripheral frame and by interior columns. A central service core is separated from the floors and is not assumed to participate in lateral force resistance. All exterior columns and the boundary elements of walls are 500 mm square. No lateral force resistance has been assigned to any of the columns shown.

Only highlights of the design of the perimeter foundation walls are given here. Gravity loads on the exterior framing at the top of the foundation wall are listed in Table 9.2. Each of the cantilever walls, as designed and detailed, is estimated to introduce at flexural overstrength a moment M_{Eo} = 54,000 kNm (39,700 kip-ft) and a shear, with dynamic magnification [Section 5.4.4(a)] of 3500 kN (784 kips) at the top of the foundation wall. The system overstrength factor [Section 1.3.3(g)] was ψ_o = 1.53.

TABLE 9.2   Gravity Loads on the Foundation Wall

	Dead Load		Reduced Live Load	
Element	kN	kips	kN	kips
Corner column[a]	400	90	100	22
Interior column	1800	403	320	72
Wall	6000	1344	750	168

[a] Only half of the total load on the corner column, assigned to one foundation wall, is recorded here.

The strength properties to be used are $f_c' = 30$ MPa (4350 psi), $f_y = 400$ MPa (58 ksi). Unit weights of concrete and soil may be assumed 23.5 (150 lb/ft^3) and 16.0 kN/m^3 (100 lb/ft^3), respectively. Linear distribution of soil reactions, as shown in Fig. 9.18(c), may be assumed. The maximum soil pressure should not exceed 500 kPa (10.4 kips/ft^2). Dimensions of the foundation wall may be assumed as shown in Fig. 9.18(b).

9.5.2   Load Combinations for Foundation Walls

1. Forces due to dead and live load and corresponding uniform soil pressures are shown for one-half of the symmetrical wall in Figs. 9.19(a) and (b). For convenience in analysis, distributed loads and soil reactions on the walls have been replaced by equivalent components at 3-m (9.87-ft) centers. With some approximations, the soil reaction due to dead load from the superstructure alone is

$$2(400 + 1800 + 6000)/42 = 390.5 \text{ kN/m } (26.7 \text{ kips/ft})$$

2. Earthquake actions from each of the walls at overstrength are replaced by force couples of $54,000/6 = 9000$ kN (2016 kips) and a shear of 3500 kN (784 kips), as shown in Fig. 9.19(c). The total overturning moment introduced by four cantilever walls to the foundation system at the underside of the footings is

$$M_{o,t} = 4(9000 \times 6 + 3500 \times 3.5) = 265,000 \text{ kNm } (194,800 \text{ kip-ft})$$

With the assumption of linear distribution of soil reactions the maximum soil reaction due to this overturning moment is, when

$$I = 2(1 \times 42^3/12 + 1 \times 42 \times 21^2) = 49,392 \text{ m}^3$$
$$p_{max} = 265,000 \times 21/49,392 = 112.7 \text{ kN/m } (7.7 \text{ kips/ft})$$

By assuming that shear is transmitted by uniformly distributed friction at the

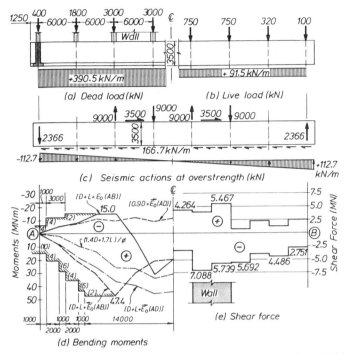

Fig. 9.19 Loading of the foundation wall and resulting design actions. (1 kN = 0.225 kip; 1 kN/m = 0.0686 kip/ft; 1 MNm = 738 kip-ft.)

underside of the footing, this shear is [Figs. 9.18(c) and 9.19(c)]

$$2 \times 3500/42 = 166.7 \text{ kN/m } (11.4 \text{ kips/ft})$$

3. Soil pressures must be checked. For the combination of actions due to dead load and the development of the overstrength of the superstructure, (i.e., $U = D + E_o$), soil reactions as shown in Fig. 9.18(c) will result.

From Fig. 9.18(b) the weight of the foundation wall is estimated as 2.6 m² × 23.5 = 61 kN/m (4.2 kips/ft) and that of the soil as 51 kN/m (3.5 kips/ft), so that soil pressures for various load combinations may be derived from

$$U = 1.4D + 1.7L = 1.4(390 + 61 + 51) + 1.7 \times 92$$
$$= 859 \text{ kN/m } (59 \text{ kips/ft})$$
$$U = D + 1.3L + E = (390 + 61 + 51) + 1.3 \times 92 + 113/1.53$$
$$= 695 \text{ kN/m } (47 \text{ kips/ft})$$
$$U = D + L + E_o = 502 + 92 + 113 = 707 \text{ kN/m } (48 \text{ kips/ft})$$

Hence the footing needs to be $859/500 = 1.718$ m (5.64 ft) wide. As seen in Fig. 9.18(b), this dimension is 1.8 m (5.90 ft).

4. Moments and shear forces for the foundation walls are to be determined. The weight of the foundation wall and soil does not affect design actions. For actions from the superstructure, moments and shear are determined for each case and critical values are obtained from appropriate superpositions. The results of the static analysis are presented in Fig. 9.19(d) and (e). While bending moment diagrams were constructed for each case, only the shear force envelopes have been provided. The directions of lateral forces applied to the superstructure along wall AB are shown as $\vec{E}_o$ and $\overleftarrow{E}_o$.

Positive and negative earthquake action in the direction shown have been considered separately. The various cases, as they affect walls AB and AD [Fig. 9.18(c)], have been recorded in Fig. 9.19(d) and (e) and the corresponding moment envelopes were determined.

9.5.3   Reinforcement of the Foundation Wall

(*a*) *Footings* From Fig. 9.19(a) and (b) the maximum pressure due to superstructure actions is

$$[1.4 \times 390 + 1.7 \times 92]/1.8 = 390.2 \text{ kPa } (8.1 \text{ kips/ft}^2)$$

Pressure due to the weight of the wall above the footing is

$$1.4 \times 0.6 \times 3.1 \times 23.5/1.8 = \underline{34.0 \text{ kPa } (0.7 \text{ kip/ft}^2)}$$

$$\text{Total pressure} = 424.2 \text{ kPa } (8.9 \text{ kips/ft}^2)$$

The moment on the 600-mm-long cantilever footing in Fig. 9.18(b) is

$$424 \times 0.6^2/2 = 76.3 \text{ kNm } (56.3 \text{ kip-ft})$$

50 mm (2 in.) of cover has been provided to the bottom reinforcement in the footing. Hence $d \approx 400 - 50 - 8 = 342$ mm (13.5 in.) and

$$A_s \approx 76.3 \times 10^6/(0.9 \times 400 \times 0.95 \times 342) = 652 \text{ mm}^2/\text{m } (0.31 \text{ m}^2/\text{ft})$$

This is a rather small amount ($\rho = 0.0019$). Provide HD16 (0.63-in.-diameter) bars on 300-mm (11.8-in.) centers ($\rho = 0.00196$) in the bottom of the footing [Fig. 9.18(b)].

(*b*) *Flexural Reinforcement* After some preliminary calculations the reinforcement shown in Fig. 9.18(b) has been chosen and its adequacy will be checked here. To resist a moment of $M = -15$ MNm (11,000 kip-ft) [Fig. 9.19(d)], 10 HD40 (1.57-in.diameter) top bars with $A_b = 1256$ mm² (1.95

in.^{2}) are used. Note that $\phi = 1.0$. Thus

$$T = A_s f_y = 12{,}560 \times 400 \times 10^{-3} = 5024 \text{ kN (1125 kips)}$$

Neglecting the contribution of bottom bars in compression gives

$$a = 5024 \times 10^3/(0.85 \times 30 \times 1800) = 109 \text{ mm (4.3 in.)}$$
$$jd = 3500 - 166 - 0.5 \times 109 \approx 3280 \text{ mm (129 in.)}$$
$$M_i = 5024 \times 3.28 = 16.48 > 15.0 \text{ MNm   satisfactory}$$

For the maximum positive moment of $M_u = 47.4$ MNm (34,840 kip-ft) 32 HD40 bars (1.57 in. diameter) are considered. $d = 3500 - 220 = 3280$ mm (129 in.).

Assuming that the top bars will yield in compression, we obtain from:

Eq. (3.21): C_s $= 5{,}024$ kN (1125 kips)

   $T = 32 \times 1256 \times 400 \times 10^{-3}$ $= 16{,}077$ kN (3601 kips)

Eq. (3.21): $T - C_s = C_c$ $= 11{,}053$ kN (2476 kips)

Eq. (3.25): $a = 11{,}053 \times 10^3/(0.85 \times 30 \times 600)$ $= 722$ mm (28 in.)

Eq. (3.22): $M_i = 11.053 (3.28 - 0.5 \times 0.722)$ $= 32.26$ MNm (23,700 kip-ft)

    $+ 5024(3.28 - 0.166)$ $= 15.64$ MNm (11,490 kip-ft)

              $M_i = 47.90$ MNm (35,190 kip-ft)

This is satisfactory, as $M_u = 47.4$ MNm (34,830 kip-ft) $\approx M_i$. Check the steel compression strains. From Eq. (3.23),

$$c = 722/0.85 = 850 \text{ mm (33 in.)}$$

and from Fig. 9.18(b),

$$\epsilon_{s,\,min} = (850 - 310)0.003/850 = 0.0019 \approx 0.002 = \epsilon_y$$

*(c) **Shear Reinforcement*** From Fig. 9.19(e) at 12 m from the corner,

$$v_i = 5692 \times 10^3/(600 \times 3280) = 2.89 \text{ MPa (419 psi)}$$

With $\rho_w = 32 \times 1256/(600 \times 3280) = 0.0204$, Eq. (3.33) gives

$$v_b = (0.07 + 10 \times 0.0204)\sqrt{f_c'} = 0.274\sqrt{f_c'} > 0.2\sqrt{f_c'} = 1.09 \text{ MPa (158 psi)}$$

From Eq. (3.40),

$$A_v/s = (2.89 - 1.09)600/400 = 2.70 \text{ mm}^2/\text{mm } (0.11 \text{ in.}^2/\text{in.})$$

Use four HD16 (0.63-in.-diameter) legs $= 804 \text{ mm}^2 (1.24 \text{ in.}^2)$. Then

$$s = 804/2.7 = 298 \approx 300 \text{ mm } (12 \text{ in.})$$

At 6 m from the corner, $V_u = 4486$ kN (1005 kips); hence $v_i = 2.28$ MPa (331 psi) and

$$A_v/s = (2.28 - 1.09) \times 600/400 = 1.80 \text{ mm}^2/\text{mm } (0.071 \text{ in.}^2/\text{in.})$$

Use four HD16 (0.63-in.-diameter) legs at $s = 804/1.80 = 447 > 400$ mm (16 in.).

(d) ***Shear Reinforcement in the Tension Flange*** From Fig. 9.19(*d*) the average shear force (moment gradient) over a 12-m length from the corner of the wall is $V_{\text{avg}} = 47,400/12 \approx 4000$ kN (896 kips). The tension reinforcement in one flange is approximately $9 \times 100/32 = 28\%$ of the total. Hence the shear flow at the flange-web junction is approximately

$$v_o \approx 0.28 \times 4000/3.28 = 341 \text{ kN/m } (23.3 \text{ kips/ft})$$
$$v_i = 341,000/(400 \times 1000) = 0.85 \text{ MPa } (123 \text{ psi})$$

As the flange is subjected to axial tension, assume [Eq. (3.35)] that $v_c = 0$. Hence $A_v/s = 0.85 \times 400/400 = 0.85 \text{ mm}^2/\text{mm } (0.033 \text{ in.}^2/\text{in.})$.

As some excess flexural reinforcement has been provided in the bottom of the footing, place additional HD16 (0.63-in.-diameter) transverse bars in the top of the footing (flange) on 300-mm (12-in.) centers. Thus $A_v/s = 201/300 = 0.67 < 0.85 \text{ mm}^2/\text{mm}$.

(e) ***Joint Shear Reinforcement*** From the analysis of actions due the development of plastic hinges at the base of the walls, shown in Fig. 9.19(*c*), the forces around the cantilever wall–foundation wall joint were extracted and these are shown in Fig. 9.20. The moment about the center of the joint in

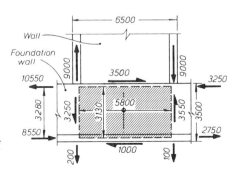

Fig. 9.20 Forces at the boundaries of the foundation–wall joint.

equilibrium due to these forces is

$$\Sigma M = [9000 \times 6.0 + (3500 + 1000)0.5 \times 3.5]$$
$$- [(10{,}550 + 3250 + 8550 + 2750)0.5 \times 3.28 - 3450 \times 6]$$
$$= 61{,}875 - 61{,}864 \approx 0$$

The vertical joint shear is

$$V_{jv} = 9000 - 3250 - 200 = 5550 \text{ kN } (1243 \text{ kips})$$

and the nominal joint shear stress is, from Eq. (4.68),

$$v_{jv} \approx 5550 \times 10^3/(600 \times 3500)$$
$$= 2.64 \text{ MPa } (383 \text{ psi}) < 0.25 f_c' = 7.5 \text{ MPa } (1090 \text{ psi})$$

This joint is similar to an exterior beam–column joint, such as shown in Fig. 4.66, with the foundation beam taking the role of the column. However, anchorage conditions in this situation are much better, with bar diameter-to-joint length ratios of $40/6500 = 0.006$ horizontally and $28/3500 = 0.008$ vertically. These ratios are much less than the most severe criteria for a typical beam–column joint required by Eq. (4.56). Hence it is justified to assume that undiminished bar forces will readily be transmitted by bond within the joint, and that little or no reliance need be placed on strut action, illustrated in Figs. 4.44(a) and 4.66. The mechanism of shear transfer may be based on the model shown in Fig. 4.62 with, say, a 45° compression field. If no allowance is made for the contribution of concrete to joint shear strength, from Fig. 4.62 and Eq. (4.78b) the required horizontal or vertical joint reinforcement per unit length would be

$$\frac{A_j}{s} \approx \frac{v_j b_c}{f_y}$$

where in this example v_j is based on core dimensions of approximately 5800×3130 mm, as shown in Fig. 9.20. Hence, by considering both vertical and horizontal joint shear stresses,

$$v_j = \frac{5550 \times 10^3}{600 \times 3130} \approx \frac{10{,}300 \times 10^3}{600 \times 5800} = 2.96 \text{ MPa } (429 \text{ psi})$$

Therefore,

$$A_j/s = 2.96 \times 600/400 = 4.44 \text{ mm}^2/\text{mm } (0.17 \text{ in}^2/\text{in.})$$

To compensate for this conservative approach, the effects of addition shear due to gravity loads, from Fig. 9.19(e) approximately

$$\Delta V = 0.5(7088 + 5739) - 5550 = 864 \text{ kN } (193 \text{ kips})$$

will be neglected.

Vertically, provide four HD20 (0.79-in.-diameter) bars on 300-mm (11.8-in.) centers; that is,

$$4 \times 314/300 = 4.19 \approx 4.44 \text{ mm}^2/\text{mm } (0.17 \text{ in}^2./\text{in.})$$

The nominal horizontal reinforcement extending over the entire length of the foundation wall is HD12 at 400 m (0.47 in. diameter at 15.7 in.):

$4 \times 113/400$ $= 1.13 \text{ mm}^2/\text{mm } (0.044 \text{ in.}^2/\text{in.})$

Provide additional four HD20 bars at 400 m $= \underline{3.14 \text{ mm}^2/\text{mm } (0.124 \text{ in.}^2/\text{in.})}$

$A_j = 4.44$ $\approx 4.27 \text{ mm}^2/\text{mm } (0.168 \text{ in.}^2/\text{in.})$

These horizontal HD20 (0.79-in.-diameter) bars, approximately 7 m (23 ft) long, should be provided with a standard horizontal hook anchored immediately beyond the vertical HD28 (1.1-in.-diameter) bars extending from the boundary elements of the cantilever walls.

9.5.4 Detailing

(a) *Anchorage and Curtailment* From Eq. (3.63), assuming that $m_{db} = 1.0$, and from Fig. 9.18(b), with $c \approx 80$, the development length for HD40 bottom bars is

$$l_d = \frac{1.0 \times 1.38 \times 1256 \times 400}{80\sqrt{30}} = 1582 \text{ mm } (5.0 \text{ ft})$$

Hence bars could be curtailed (Section 3.6.3) at a distance $d + l_d \approx 3300 + 1582 \approx 5000$ mm (16.4 ft) past the section at which they are required to contribute fully to flexural resistance. The stepped shaded line in Fig. 9.19(d) shows how the curtailment of these HD40 (1.57-in.-diameter) bars, with numbers of curtailed bars given in brackets, could be done. Because lapped splices for such large bars are impractical and are not recommended [A1], mechanical connections at staggered locations should be used to ensure continuity of the 40-mm bars over a length of 44.5 m (146 ft).

(b) *Detailing of Wall Corners* As Figs. 9.19(c) and 9.18(d) show, a shear force of 2366 kN (530 kips) needs to be transmitted from one foundation wall to another at wall corners. This necessitates a vertical tension member,

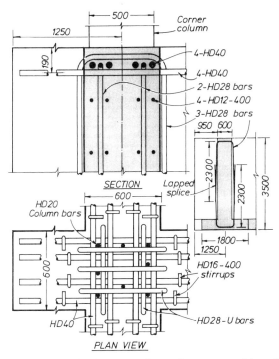

Fig. 9.21 Details to enable shear transfer at the corners of foundation walls.

illustrated in Fig. 9.18(d), even when the wall shear forces are reversed. Hence provide five HD28 (1.1-in.-diameter) U-shaped bars as detailed in Fig. 9.21, with a tension capacity of

$$2 \times 5 \times 616 \times 400 \times 10^{-3} = 2464 > 2366 \text{ kN (530 kips)}$$

These bars, with a lapped splice at midheight of the wall, must be carefully positioned to allow starter bars for the lightly reinforced corner column, (not shown in this figure) to be placed.

APPENDIX A
Approximate Elastic Analysis of Frames Subjected to Lateral Forces

From among a number of approximate analyses used to derive the internal actions for lateral forces, one, developed by Muto [M1], is examined here in some detail but in a slightly modified form. This procedure will give sufficiently accurate results for a reasonably regular framing system, subjected to lateral forces, for which a number of simplifying assumptions can be made. Equilibrium will always be satisfied in this approximation. However, elastic displacement compatibility may be violated to various degrees, as a result of assumptions with regards to member deformations.

(a) *Lateral earthquake forces on the structure* are simulated by a set of horizontal static forces, each acting at the level of a floor as shown in Fig. 1.9(b). The derivation of these forces was given in Section 2.4.3. Because the building as a whole is a cantilever, the total shear forces in each story (story shear), as well as the overturning moment at each level, can readily be derived. These are qualitatively shown in Fig. 1.9(c) and (d). The sum of all the horizontal forces acting on the upper four floors, for example, gives rise to the story shear force V_j. The aim of this approximate analysis is to determine the share of each of the columns in that story in resisting the total story shear force V_j. Subsequently, the column and beam moments will also be found.

(b) *Prismatic elastic columns*, when subjected to relative end displacements Δ as shown in Fig. A.1(a), will develop a shear force

$$V_f = \frac{12 E_c I_c}{l_c^3} \Delta \tag{A.1}$$

when both ends are fully restrained against rotations.

As the modulus of elasticity E_c and the column height h are common values for all columns in a normal story, the simple expression

$$V_f = \alpha k_c \Delta \tag{A.2}$$

may be used to relate displacement Δ to the induced shear, where $\alpha =$

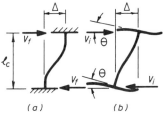

(a)   (b)   **Fig. A.1** Displacement induced shear forces in a
Full restraint Partial restraint column.

$12E_c/l_c^2$ is a common constant. The relative stiffness of the column is $k_c = I_c/l_c$, where I_c is the second moment of area of the section of the prismatic column relevant to the direction of bending.

In real frames only partial restraints to column end rotation θ will be provided by the beams above and below the story in question, as shown in Fig. A.1(b). Consequently, the induced shear force across the column will be less for the same drift Δ than given by Eq. (A.2). However, the column shear may be expressed thus:

$$V_i = \alpha(ak_c)\Delta \tag{A.3}$$

where the value of a will depend on the degree of rotational restraint offered by the adjacent beams. Obviously, $0 < a < 1$. The first aim will be to establish the magnitude of coefficient a.

A simplifying assumption to be made is that all joint rotations, shown as θ in Fig. A.1(b), in adjacent stories are the same. This enables the boundary conditions for all beams and columns, and hence their absolute flexural stiffness, to be uniquely defined.

(c) *Isolated subframes*, such as those used for gravity load analysis, can be further simplified by assuming that the stiffness of each beam is equally utilized by the column above and that below a floor, where the beams restrain the end rotations of these columns. Hence beams can be split into hypothetical halves, each half possessing 50% of the stiffness of the original beam while interacting with one column only. Thus for the purpose of this approximate lateral force analysis, stories of the frame of Fig. 1.9(a) can be separated as shown in Fig. A.2. The subframe drawn by full lines can then be studied separately. After the evaluation of actions, they can be superimposed upon each other by reassembly of adjacent hypothetical subframes.

(d) *Restraints provided by beams* against column end rotations can be evaluated from the study of isolated columns together with the beams that frame into them. For example, column i of the subframe of Fig. A.2, together with the appropriate beams, can be isolated as shown in Fig. A.3(a). The pinned ends of beams represent points of contraflexure resulting from the

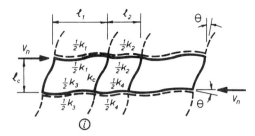

Fig. A.2   Simplified subframe used for lateral force analysis.

previously made assumption of equal rotations at adjacent beam–column joints.

Because of reasonable uniformity in member sizes, for the purpose of approximation it may be assumed that beam stiffnesses at the top and the bottom of the story are similar and are approximately equal to the average stiffness: that is,

$$k_1 + k_2 = k_3 + k_4 = (k_1 + k_2 + k_3 + k_4)/2 \qquad (A.4)$$

If a relative displacement of the ends of the column Δ is now imposed, as shown in Fig. A.3(b), the resulting column shear can readily be assessed. From the distorted shape of the subframe it is evident that only the three pin-ended members, shown by full lines in Fig. A.3(b), need be considered.

The moment induced by the displacement Δ in the fully restrained column is, from Eq. (A.1),

$$M_F = \frac{V_F l_c}{2} = \frac{6 E_c I_c}{l_c^2} \Delta = \frac{12 E_c}{l_c^2} \frac{l_c}{2} \frac{I_c}{l_c} \Delta = \alpha l_c \frac{k_c}{2} \Delta \qquad (A.5)$$

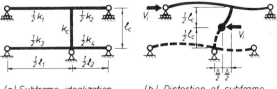

(a) Subframe idealization   (b) Distortion of subframe

Fig. A.3   Isolated column in a story restrained by beams.

The absolute flexural stiffnesses of the beams and columns* are $3E_c k_1$, $3E_c k_2$, and $6E_c k_c$, respectively, and the distribution factor d_c with respect to the column, to be used in the next step, becomes

$$d_c = \frac{6E_c k_c}{3E_c(k_1 + k_2 + 2k_c)} = \frac{2}{2 + \bar{k}} \qquad (A.6)$$

where from Eq. (A.4) the following substitution is made:

$$\bar{k} = \frac{k_1 + k_2 + k_3 + k_4}{2k_c} \qquad (A.7)$$

Using a single cycle moment distribution and the expressions above, the final end moment in the column becomes

$$M_c = (M_F - d_c M_F) = \left(\alpha l_c \frac{k_c}{2}\right) \frac{\bar{k}}{\bar{k} + 2} \Delta \qquad (A.8)$$

The column shear force resulting from the displacement Δ is thus

$$V_i = \frac{2M_c}{l_c} = \alpha \frac{\bar{k}}{\bar{k} + 2} k_c \Delta \qquad (A.9)$$

When this is compared with Eq. (A.3), it is seen that the parameter expressing the degree of restraint provided by the beams is

$$a = \frac{\bar{k}}{\bar{k} + 2} \qquad (A.10)$$

The relative shear stiffness** of the column, to be used in subsequent analyses and termed by Muto as the D value of the column, is therefore

$$D_i = \frac{\bar{k}}{\bar{k} + 2} k_c \qquad (A.11)$$

---

*The absolute flexural stiffness of a member is defined as the magnitude of the bending moment, which when applied in a specified form to a member of given geometric properties and boundary conditions will produce unit rotation at the point or points of application.
**The absolute shear stiffness αD_i is defined as the magnitude of the shear force that is required to cause unit transverse displacement of one end of a member, with given geometric and elastic properties and boundary conditions, relative to the other end.

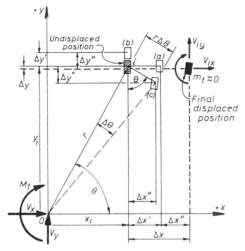

Fig. A.4   Displacement of a column due to story translation and story twist.

With the use of the D value, the displacement-induced column shear is determined in accordance with Eq. (A.3).

$$V_i = \alpha D_i \Delta \tag{A.12}$$

provided that the value of Δ is known. Its evaluation is given in the next section.

(e) *Displacement compatibility* for all columns of a story is assured by rigid diaphragm action of the floor slab, which has been assumed. The displacement of one end of a column, relative to the other end, results from possible translations of the floor in either of the principal directions of the framing, x and y, and from its in-plane rotation (twist), as shown in Fig. 1.10(a) to (c). It is seen that the displacement of column i due to torsion in the building about a vertical axis can be resolved into two displacement components, $\Delta x''$ and $\Delta y''$, as shown in Fig. 1.10(c).

The position of the undisplaced and hence unloaded column in relation to a specially chosen center of a coordinate system O, is shown by its shaded cross section in Fig. A.4. As a result of story drifts $\Delta x'$ and $\Delta y'$ in the x and y directions, respectively, the column is also displaced by these amounts, as shown by positions (a) and (b) in Fig. A.4. The displacement resulting from twisting of the building moves one end of the column in position (c) in Fig. A.4. By superimposing the displacements resulting from the three kinds of relative floor movements, the final position of column i, as shown by its blackened cross section in Fig. A.4, is established. Using the notation and sign convention of Fig. A.4, the total column displacement can be expressed as follows:

$$\Delta x = \Delta x' + \Delta x'' \tag{A.13a}$$

$$\Delta y = \Delta y' - \Delta y'' \tag{A.13b}$$

However,

$$\Delta x'' = r\,\Delta\theta\,\sin\theta = y_i\,\Delta\theta; \qquad \Delta y'' = r\,\Delta\theta\,\cos\theta = x_i\,\Delta\theta$$

Hence

$$\Delta x = \Delta x' + y_i\,\Delta\theta \tag{A.14a}$$

$$\Delta y = \Delta y' - x_i\,\Delta\theta \tag{A.14b}$$

where all symbols are as defined in Fig. A.4. The actions induced by these column displacement components can now be expressed using Eq. (A.12) thus:

$$V_{ix} = \alpha D_{ix}\,\Delta x \tag{A.15a}$$

$$V_{iy} = \alpha D_{iy}\,\Delta y \tag{A.15b}$$

Because the angle of twist $\Delta\theta$, as well as the torsional stiffness of a column, is small, the torsional moment m_t induced in the column, and shown for completeness in Fig. A.4, can be neglected. The two D values for each column in Eq. (A.15) are directional quantities. They are evaluated from the column and beam stiffness, relevant to imposed displacements in the x or y directions, respectively. Beams framing perpendicular to the plane of any subframe are assumed to make no contribution to the in-plane lateral force resistance of that frame.

(f) *Equilibrium criteria for story shears* require that

$$V_x = \sum V_{ix} \tag{A.16a}$$

$$V_y = \sum V_{iy} \tag{A.16b}$$

$$M_t = \sum y_i V_{ix} - \sum x_i V_{iy} \tag{A.16c}$$

where V_x and V_y are the total applied story shear forces in the two principal directions, respectively, and M_t is the torsional moment generated by these forces about the vertical axis passing through the center of the coordinate system O. When Eqs. (A.15) and (A.16) are combined and the displacement components of Eq. (A.14) are used, it is seen that

$$V_x = \sum \alpha D_{ix} \, \Delta x = \alpha \sum D_{ix}(\Delta x' + y_i \, \Delta \theta) = \alpha \left(\Delta x' \sum D_{ix} + \Delta \theta \sum y_i D_{iy} \right)$$
$$(A.17a)$$

$$V_y = \sum \alpha D_{iy} \, \Delta y = \alpha \sum D_{iy}(\Delta y' - x_i \, \Delta \theta) = \alpha \left(\Delta y' \sum D_{iy} - \Delta \theta \sum x_i D_{iy} \right)$$
$$(A.17b)$$

$$M_t = \sum y_i(\alpha D_{ix} \, \Delta x) - \sum x_i(\alpha D_{iy} \, \Delta y)$$
$$= \alpha \left(\Delta x' \sum y_i D_{ix} + \Delta \theta \sum y_i^2 D_{ix} - \Delta y' \sum x_i D_{iy} + \Delta \theta \sum x_i^2 D_{iy} \right) \quad (A.17c)$$

When the coordinate system is chosen so that $\sum y_i D_{ix} = \sum x_i D_{iy} = 0$, the expressions above simplify. The point O, about which the first moment of the D values is zero, is defined as the center of rigidity or the center of stiffness of the framing system in the story considered [Section 1.2.2(b)]. Whenever the story shear is applied at this center, only story translations will occur. With this selection of the coordinate system, the displacement components resulting separately from the application of each story action are readily expressed from Eq. (A.17) as follows:

$$\Delta x' = \frac{V_x}{\alpha \sum D_{ix}} \qquad (A.18a)$$

$$\Delta y' = \frac{V_y}{\alpha \sum D_{iy}} \qquad (A.18b)$$

$$\Delta \theta = \frac{M_t}{\alpha \left(\sum y_i^2 D_{ix} + \sum x_i^2 D_{iy} \right)} \qquad (A.18c)$$

The torsional response depends on the polar second moment of D values and this is abbreviated for convenience as

$$I_p = \sum y_i^2 D_{ix} + \sum x_i^2 D_{iy} \qquad (A.19)$$

(g) *Final column shear forces* can now be determined by substituting the story displacements expressed by Eq. (A.18) into the general relationship,

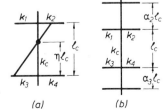

Fig. A.5   Locating the point of contraflexure along a column.

TABLE A.1 Value of η_0 for Multistory Frames with m Stories

m	n	$\bar{k}$ 0.1	0.2	0.3	0.4	0.6	0.8	1.0	3.0	5.0
1	1	0.80	0.75	0.70	0.65	0.60	0.50	0.55	0.55	0.55
2	2	0.50	0.45	0.40	0.40	0.40	0.40	0.45	0.45	0.50
	1	1.00	0.85	0.75	0.70	0.65	0.65	0.60	0.55	0.55
	3	0.25	0.25	0.25	0.30	0.35	0.35	0.40	0.45	0.50
3	2	0.60	0.50	0.50	0.50	0.45	0.45	0.45	0.50	0.50
	1	1.15	0.90	0.80	0.75	0.70	0.65	0.65	0.55	0.55
	4	0.10	0.15	0.20	0.25	0.30	0.35	0.40	0.45	0.45
	3	0.35	0.35	0.35	0.40	0.40	0.45	0.45	0.50	0.50
4	2	0.70	0.60	0.55	0.50	0.50	0.50	0.50	0.50	0.50
	1	1.20	0.95	0.85	0.80	0.70	0.70	0.65	0.55	0.55
	6	− 0.15	0.05	0.15	0.20	0.30	0.35	0.35	0.45	0.45
	5	0.10	0.25	0.30	0.35	0.40	0.40	0.45	0.50	0.50
6	4	0.20	0.35	0.40	0.40	0.45	0.45	0.45	0.50	0.50
	3	0.50	0.45	0.45	0.45	0.45	0.45	0.50	0.50	0.50
	2	0.80	0.65	0.55	0.55	0.55	0.50	0.50	0.50	0.50
	1	1.30	1.00	0.85	0.80	0.70	0.65	0.65	0.55	0.55
	8	− 0.20	0.05	0.15	0.20	0.30	0.35	0.35	0.45	0.45
	7	0.00	0.20	0.30	0.35	0.40	0.40	0.45	0.50	0.50
	6	0.15	0.30	0.35	0.40	0.45	0.45	0.45	0.50	0.50
8	5	0.30	0.45	0.40	0.45	0.45	0.45	0.45	0.50	0.50
	4	0.40	0.45	0.45	0.45	0.45	0.50	0.50	0.50	0.50
	3	0.60	0.50	0.50	0.50	0.50	0.50	0.50	0.50	0.50
	2	0.85	0.65	0.60	0.55	0.55	0.50	0.50	0.50	0.50
	1	1.30	1.00	0.90	0.80	0.70	0.70	0.65	0.55	0.55
	10	− 0.25	0.00	0.15	0.20	0.30	0.35	0.40	0.45	0.45
	9	− 0.05	0.20	0.30	0.35	0.40	0.40	0.45	0.50	0.50
	8	0.10	0.30	0.35	0.40	0.40	0.45	0.45	0.50	0.50
	7	0.20	0.35	0.40	0.40	0.45	0.45	0.50	0.50	0.50
10	5	0.40	0.45	0.45	0.45	0.45	0.50	0.50	0.50	0.50
	3	0.60	0.55	0.50	0.50	0.50	0.50	0.50	0.50	0.50
	2	0.85	0.65	0.60	0.55	0.55	0.50	0.50	0.50	0.50
	1	1.35	1.00	0.90	0.80	0.75	0.70	0.65	0.55	0.55
	$m - 1$	− 0.30	0.00	0.15	0.20	0.30	0.30	0.35	0.45	0.45
	$m - 2$	− 0.10	0.20	0.25	0.30	0.40	0.40	0.40	0.45	0.50
	$m - 3$	0.05	0.25	0.35	0.40	0.40	0.45	0.45	0.50	0.50
	$m - 4$	0.15	0.30	0.40	0.40	0.45	0.45	0.45	0.50	0.50
	$m - 5$	0.25	0.35	0.40	0.45	0.45	0.45	0.45	0.50	0.50
≥ 12	$m - 6$	0.30	0.40	0.40	0.45	0.45	0.50	0.50	0.50	0.50
		0.45	0.45	0.45	0.45	0.50	0.50	0.50	0.50	0.50
	4	0.55	0.50	0.50	0.50	0.50	0.50	0.50	0.50	0.50
	3	0.65	0.55	0.50	0.50	0.50	0.50	0.50	0.50	0.50
	2	0.70	0.70	0.60	0.55	0.55	0.50	0.50	0.50	0.50
	1	1.35	1.05	0.90	0.80	0.70	0.70	0.65	0.55	0.55

Eq. (A.15), thus:

$$V_{ix} = \alpha D_{ix}\, \Delta x = \alpha D_{ix}(\Delta x' + y_i\, \Delta\theta) = \frac{D_{ix}}{\sum D_{ix}} V_x + \frac{y_i D_{ix}}{I_p} M_t \quad (A.20a)$$

$$V_{iy} = \alpha D_{iy}\, \Delta y = \alpha D_{iy}(\Delta y' - x_i\, \Delta\theta) = \frac{D_{iy}}{\sum D_{iy}} V_y - \frac{x_i D_{iy}}{I_p} M_t \quad (A.20b)$$

(h) *Determination of column moments* is easily completed after the column shear forces have been determined. According to the initial assumption [Fig. A.3(b)], points of contraflexure are at midheights and thus the column end moments could be obtained as $M_i = 0.5 l_c V_i$.

Muto [M1] studied the parameters that affect the position of the point of contraflexure in columns of multistory frames of different heights. He gave the distance of the point of contraflexure ηl_c measured from the bottom of the column, as shown in Fig. A.5(a), where

$$\eta = \eta_0 + \eta_1 + \eta_2 + \eta_3 \qquad (A.21)$$

in which η_0 gives (Table A.1) the position of this point in each story as a function of $\bar{k}$, the ratio of the stiffnesses of beams to that of the column [Eq. (A.7)], the total number of stories m, and the story under consideration n. η_1, given in Table A.2, is a correction factor allowing for beams with different

TABLE A.2   Correction Coefficient η_1 for Different Beam Stiffnesses[a]

α_1	$\bar{k}$								
	0.1	0.2	0.3	0.4	0.6	0.8	1.0	3.0	5.0
0.4	0.55	0.40	0.30	0.25	0.20	0.15	0.15	0.05	0.05
0.5	0.45	0.30	0.20	0.20	0.15	0.10	0.10	0.05	0.05
0.6	0.30	0.20	0.15	0.15	0.10	0.10	0.05	0.05	0.00
0.7	0.20	0.15	0.10	0.10	0.05	0.05	0.05	0.00	0.00
0.8	0.15	0.10	0.05	0.05	0.05	0.05	0.00	0.00	0.00
0.9	0.05	0.05	0.05	0.05	0.00	0.00	0.00	0.00	0.00

[a] $\alpha_1 = (k_1 + k_2)/(k_3 + k_4) < 1$ [see Fig. A.5(a)]. When $\alpha_1 > 1$, use $1/\alpha_1$ and take value of η_1 as negative.

TABLE A.3   Correction Coefficient η_2 and η_3 for Different Story Heights[a]

		$\bar{k}$								
α_2	α_3	0.1	0.2	0.3	0.4	0.6	0.8	1.0	3.0	5.0
2.0		0.25	0.15	0.15	0.10	0.10	0.10	0.05	0.05	0.00
1.8		0.20	0.15	0.10	0.10	0.05	0.05	0.05	0.00	0.00
1.6	0.4	0.15	0.10	0.10	0.05	0.05	0.05	0.05	0.00	0.00
1.4	0.6	0.10	0.05	0.05	0.05	0.05	0.05	0.00	0.00	0.00
1.2	0.8	0.05	0.05	0.05	0.00	0.00	0.00	0.00	0.00	0.00
1.0	1.0	0.00	0.00	0.00	0.00	0.00	0.00	0.00	0.00	0.00
0.8	1.2	−0.05	−0.05	−0.05	0.00	0.00	0.00	0.00	0.00	0.00
0.6	1.4	−0.10	−0.05	−0.05	−0.05	−0.05	−0.05	0.00	0.00	0.00
0.4	1.6	−0.15	−0.10	−0.10	−0.05	−0.05	−0.05	−0.05	0.00	0.00
	1.8	−0.20	−0.15	−0.10	−0.10	−0.05	−0.05	−0.05	0.00	0.00
	2.0	−0.25	−0.15	−0.15	−0.10	−0.10	−0.10	−0.05	−0.05	0.00

[a]From Fig. A.5(b) evaluate η_2 with α_2 and η_3 with α_3. These corrections do not apply in the bottom and the top stories.

stiffnesses above and below the column, and η_2 and η_3 (Table A.3) correct the location of the point of contraflexure when the story height above or below is different from the height of the story that is being considered. These factors are applicable when lateral forces are applied in form of an inverted triangle. Muto has also provided values for other lateral force patterns [M1]. The application of the technique is shown in Section 4.11.6.

APPENDIX B
Modified Mercalli Intensity Scale

The Mercalli (1902) scale, modified subsequently by Wood and Neumann (1931) and Richter (1958), is based on the observed response of buildings constructed in the first decades of this century. Because it is based on subjective observations and is primarily applicable to masonry construction, to avoid ambiguity in language and perceived quality of construction, the following lettering [R4] is used:

Masonry A: good workmanship, mortar, and design; reinforced, especially laterally, and bound together by using steel, concrete, and so on; designed to resist lateral forces

Masonry B: good workmanship and mortar; reinforced, but not designed in detail to resist lateral forces

Masonry C: ordinary workmanship and mortar; no extreme weaknesses like failing to tie in at corners, but neither reinforced nor designed against horizontal forces

Masonry D: weak materials, such as adobe; poor mortar; low standards of workmanship; weak horizontally

Modified Mercalli Intensity Scale of 1931 (abridged and rewritten by C. F. Richter)

   I. Not felt. Marginal and long period of large earthquakes.

  II. Felt by persons at rest, on upper floors, or favorably placed.

 III. Felt indoors. Hanging objects swing. Vibration like passing of light trucks. Duration estimated. May not be recognized as an earthquake.

 IV. Hanging objects swing. Vibration like passing of heavy trucks; or sensation of a jolt like a heavy ball striking the walls. Standing motor cars rock. Windows, dishes, doors rattle. Glasses clink. Crockery clashes. In the upper range, wooden walls and frames crack.

  V. Felt outdoors; direction estimated. Sleepers wakened. Liquids disturbed, some spilled. Small unstable objects displaced or upset.

Doors swing, close, open. Shutters, pictures move. Pendulum clocks stop, start, change rate.

VI. Felt by all. Many frightened and run outdoors. Persons walk unsteadily. Windows, dishes, glassware broken. Knickknacks, books, and so on, off shelves. Pictures off walls. Furniture moved or overturned. Weak plaster and masonry D cracked. Small bells ring (church, school). Trees, bushes shaken visibly, or heard to rustle.

VII. Difficult to stand. Noticed by drivers of motor cars. Hanging objects quiver. Furniture broken. Damage to masonry D, including cracks. Weak chimneys broken at roof line. Fall of plaster, loose bricks, stones, tiles, cornices, unbraced parapets, and architectural ornaments. Some cracks in masonry C. Waves on ponds; water turbid with mud. Small slides and caving in along sand or gravel banks. Large bells ring. Concrete irrigation ditches damaged.

VIII. Steering of motor cars affected. Damage to masonry C; partial collapse. Some damage to masonry B; none to masonry A. Fall of stucco and some masonry walls. Twisting, fall of chimneys, factory stacks, monuments, towers, elevated tanks. Frame houses moved on foundations if not bolted down; loose panel walls thrown out. Decayed piling broken off. Branches broken from trees. Changes in flow or temperature of springs and walls. Cracks in wet ground and on steep slopes.

IX. General panic. Masonry D destroyed; masonry C heavily damaged, sometimes with complete collapse; masonry B seriously damaged. General damage to foundations. Frame structures, if not bolted, shifted off foundations. Frames racked. Conspicuous cracks in ground. In alluviated areas sand and mud ejected, earthquake fountains, sand craters.

X. Most masonry and frame structures destroyed with their foundations. Some well-built wooden structures and bridges destroyed. Serious damage to dams, dikes, embankments. Large mudslides. Water thrown on banks of canals, rivers, lakes, and so on. Sand and mud shifted horizontally on beaches and flat land. Rails bent slightly.

XI. Rails bent greatly. Underground pipelines completely out of service.

XII. Damage nearly total. Large rock masses displaced. Lines of sight and level distorted. Objects thrown into the air.

Symbols

Symbols defined in the text that are used only once, and those which are clearly defined in a relevant figure, are in general not listed here.

A	= factor defined by Eq. (5.4), or tributary floor area
A_b	= area of an individual bar, mm^2 ($in.^2$)
A_c^*	= area of concrete core within wall area A_g^*, measured to outside of peripheral hoop legs, mm^2 ($in.^2$)
A_e	= effective area of section, mm^2 ($in.^2$)
A_{ev}	= effective shear area, mm^2 ($in.^2$)
A_f	= flue area in a masonry block
	= effective shear area of flexural compression zone in a wall
A_g	= gross area of section, mm^2 ($in.^2$)
A_g^*	= gross area of the wall section that is to be confined in accordance with Eq. (5.20), mm^2 ($in.^2$)
A_{jh}	= total area of effective horizontal joint shear reinforcement, mm^2 ($in.^2$)
A_{jv}	= total area of effective vertical joint shear reinforcement, mm^2 ($in.^2$)
A_p	= effective area of confining plate used with masonry
A_r	= aspect ratio of a wall
A_s, A_{s1}, A_{si}	= area of tension reinforcement, mm^2 ($in.^2$)
A_s'	= area of compression reinforcement, mm^2 ($in.^2$)
A_{sd}	= area of diagonal reinforcement, mm^2 ($in.^2$)
A_{sf}	= area of reinforcement in a wall flange
A_{sh}	= total effective area of hoops and supplementary cross-ties in direction under consideration within spacing s_h, mm^2 ($in.^2$)
A_{si}	= steel area in layer i
A_{ss}	= area of slab reinforcement contributing to flexural strength of beam
A_{st}	= total area of longitudinal reinforcement, mm^2 ($in.^2$)
A_{sw}	= area of web wall reinforcement, mm^2 ($in.^2$)
A_{te}, A_{tr}	= area of one leg of a stirrup tie or hoop, mm^2 ($in.^2$)
A_v	= area of shear reinforcement within distance s, mm^2 ($in.^2$)
A_{vf}	= area of shear friction reinforcement, mm^2 ($in.^2$)

A_{wb}	= area of a boundary element of a wall
a	= coefficient quantifying the degree of restraint
	= depth of flexural compression zone at a section
a_g, a_i, a_u	= accelerations
a_o	= peak ground acceleration
a_r	= response acceleration
B_b	= overall plan dimension of building
b	= width of compression face of member or width of a flange, or thickness of wall element, mm (in.)
b_c	= overall width of column, critical width of a wall, mm (in.)
b_e	= effective width of tension flange, mm (in.)
b_f	= maximum width of ungrouted flue, mm (in.)
b_j	= effective width of joint, mm (in.)
b_w	= width of web or wall thickness, mm (in.)
C_c, C'_c, C_{c1}	= internal concrete compression forces
C_j	= factor expressing ratio of joint shear forces
C_m	= compression force at a masonry section
C_s, C'_s, C_{s1}	= steel compression forces
$C_{T, S, \mu, p}$	= inelastic seismic base shear coefficient
C_x	= compression force introduced to beams due to tension flange action
c, c_1, c_2, c^*	= neutral-axis depth measured from extreme compression fiber of section
c_{av}	= average neutral-axis depth
c_c	= critical neutral-axis depth
D	= dead load
	= diameter of a pile
	= diameter of deformed reinforcing bar with yield strength $f_y = 275$ MPa (40 ksi)
D_c	= diagonal compression force in a joint core due to strut action
D_i	= relative shear stiffness of column i
D_{ix}, D_{iy}	= the value of D_i in the x and y directions, respectively
D_s	= diagonal compression force in a joint sustained by joint shear reinforcement
D_1, D_2	= diagonal compression forces in a joint
D20	= designation of a 20-mm (0.79-in.)-diameter deformed bar with specified yield strength of $f_y = 275$ MPa (40 ksi)
d, d_c	= distance from extreme compression fiber to centroid of tension reinforcement
d'	= distance from extreme compression fiber to centroid of compression reinforcement
d_b	= nominal diameter of bar, mm (in.)
d_c	= distribution factor applicable to a column
d_m	= length of diagonal strut developed in masonry-infill panel
d_s	= diameter of spiral, mm (in.)

$\overleftarrow{E}, \overrightarrow{E}$	= code-specified earthquake force applied to the building in the direction shown by the arrow
E_c	= modulus of elasticity of concrete
E_m	= modulus of elasticity for masonry
$\overleftarrow{E}_o, \overrightarrow{E}_o$	= earthquake action corresponding with given direction due to the development of overstrength in an adjacent member or section
E_s	= modulus of elasticity of steel
e	= eccentricity
e_d	= design eccentricity
e_{dx}, e_{dy}	= design eccentricity in x and y directions, respectively
e_s, e_x, e_y	= static torsional eccentricity
e_v	= tension shift
F	= a factor defined by Eq. (5.9b)
	= force in general
F_a	= permitted stress for axial load
F_b	= permitted stress for pure bending
F_j, F_n	= total lateral force acting at level j or n
f	= constant related to foundation flexibility
f_a	= computed stress due to axial load
f_b	= computed stress due to bending moments
f_c	= compression stress in extreme fibre
f'_c	= specified compressive strength of concrete; the term $\sqrt{f'_c}$ is expressed in MPa (psi)
f_{cb}	= compression strength of stack-bonded masonry unit, MPa (psi)
f'_{cc}	= strength of confined concrete in compression, MPa (psi)
f_{cj}	= compression strength of confined mortar, MPa (psi)
f'_g	= compression strength of grout, MPa (psi)
f'_j	= compression strength of mortar, MPa (psi)
f'_m	= compression strength of masonry, MPa (psi)
f_p	= compression stress in concrete due to prestress alone, MPa (psi)
f'_p	= compression strength of masonry prism, MPa (psi)
f_r	= modulus of rupture of concrete, MPa (psi)
f_s, f_{si}	= steel tension stress, MPa (ksi)
f'_s	= steel compression stress, MPa (ksi)
f'_t	= tensile strength of concrete MPa (ksi)
f_y	= specified yield strength of nonprestressed reinforcement, MPa (ksi)
f_{yd}	= specified yield strength of diagonal reinforcement, MPa (ksi)
f_{yh}	= specified yield strength of hoop or supplementary cross-tie reinforcement, MPa (ksi)
f_{yt}	= specified yield strength of transverse reinforcement, MPa (ksi)

G	= shear modulus
G_m	= shear modulus for masonry
g	= acceleration due to gravity
H	= total height of structure
HD20	= designation of 20-mm (0.79-in.)-diameter deformed bar with specified yield strength of $f_y = 380$ MPa (55 ksi)
H_u	= total factored lateral force
h	= overall thickness of member, mm (in.)
	= overall depth of member, mm (in.)
h_i, h_n	= height of floors i and n above base
h, h_1, h'_1	= story heights
h''	= dimension of concrete core of section measured perpendicular to the direction of confining hoop bars, mm (in.)
h_b	= overall depth of beam, mm (in.)
h_c	= overall depth of column, mm (in.)
h_j	= length of joint, mm (in.)
h_m	= height of masonry-infill panel in a story
h_s	= thickness of slab, mm (in.)
h_w	= total height of wall from base to top
I	= importance factor
	= second moment of area (moment of inertia)
	= modified Mercalli intensity
I_a, I_b, I_c, I_i	= second moment of area of members a, b, c, and i
I_c	= second moment of area of a column section
I_{cr}	= second moment of area of cracked section transformed to concrete
I_{cry}	= second moment of area of cracked section corresponding to M_y
I_e	= equivalent second moment of area of section
I_f	= second moment of bearing area of a footing pad
I_g	= second moment of area of the gross concrete section ignoring the reinforcement
I_{ix}, I_{iy}	= second moment of area with respect to the x and y axes, respectively
I_p	= polar moment of D values or frame stiffnesses
I_w	= equivalent second moment of wall area allowing for shear deformation and cracking
K	= absolute stiffness
K_{xi}, K_{yi}	= horizontal force factor
k	= relative stiffness of a member
$\bar{k}$	= ratio of beam to column stiffnesses
k_c	= relative stiffness of column
k_f	= flexural stiffness of footing
k_s	= coefficient of subgrade modulus, MPa/m (kips/ft^3)
k_1, k_2	= relative stiffness of beams or piers 1, 2, and so on

L	= live load
L_r	= reduced live load
l, l_1, l_2	= span of beam measured between center lines of supporting columns or walls
l^*	= distance between plastic hinges in a beam or column
	= buckling length
l_{AB}	= clear span of beam $A-B$
l_c, l'_c	= story height, length of column
l_d	= development length of straight bar, mm (in.)
l_{db}	= basic development length in tension, mm (in.)
l_{dh}	= development length for hooked bars, mm (in.)
	= moment arm for diagonal bars
l_e	= effective embedment length, mm (in.)
l_{hb}	= basic development length for a hooked bar, mm (in.)
l_m	= length of masonry-infill panel between adjacent columns
l_n, l_{1n}, l_{2n}	= clear length of member measured from face of supports
l_{ny}	= clear distance to next web
l_0	= length of plastic region
l_p	= plastic hinge length
l_s	= length of lapped splice
l_w	= horizontal length of wall
l_x, l_y	= length and width of a rectangular slab panel
M	= bending moment
	= Richter magnitude of an earthquake
	= mass
M_{AB}	= moment applied to member $A-B$ at A
M_c	= column moment of a node point
M_{cr}	= moment causing cracking
M_d	= flexural resistance of diagonal reinforcement
M_E	= moment due to the code-specified earthquake forces alone
M_E^*	= moment at the base of a column due to code-specified earthquake forces
M_f	= midspan moment for a simply supported beam
M_G	= moment due to gravity load
M_i	= ideal flexural strength
	= moment assigned to element i
M_{ot}	= total overturning moment at the base of a cantilever wall or coupled walls
$M_{o,\text{col}}$	= flexural overstrength of a column section
M_o, M_{io}, M_{xo}	= flexural overstrength of a section in general or at location i or at x
M_{1o}, M_{2o}	= flexural overstrength of the base section of walls 1 and 2, respectively
M_p	= moment sustained by axial load
M_s	= moment sustained by reinforcement

M_t	= torsional moment developed in a story
M_u	= moment due to the combined actions of factored loads and forces
	= required flexural strength
$M_{u,r}$	= reduced design moment for a column
M_y	= moment at the onset of yielding
M_1, M_2	= design moment assigned to walls 1 and 2
M_1', M_2'	= design moments assigned to walls 1 and 2 after lateral force redistribution
m	= design load magnifier
	= mass per unit length
	= torsional moment induced in a column by relative story twist
N	= vertical compression force
	= total number of floors in a building
n	= number of floors above the level considered
P	= axial force
P_D	= axial force due to dead load
P_E	= axial force induced by the code-specified lateral earthquake forces only
P_{ed}	= elastic design load
P_{Eo}	= axial force induced in a column or wall by earthquake at the development of the overstrength of the structure
P_g	= gravity-load-induced axial compression force
P_i	= axial seismic force at level i
	= axial force associated with ideal strength
P_{Lr}	= axial force due to reduced live load
P_u	= design axial force at given eccentricity, derived from factored loads and forces
P_{1o}, P_{2o}	= total axial force on walls 1 and 2, respectively, at the development of the overstrength of the structure
p	= probability of exceedence
	= depth of pointing in a mortar bed
Q_u	= shear force in coupling beam
Q_o, Q_{io}	= shear force developed in a coupling beam at overstrength
Q_r	= stability index
Q_r^*	= modified stability index
q_1, q_2, q_b	= theoretical shear forces per unit story height in coupling beams
R	= seismic force reduction factor on account of ductility
	= diameter of plain round bar
R_d	= parameter controlling sliding shear
R_e	= epicentral distance
R_m	= moment reduction factor
R_s	= diagonal compression force developed in masonry infill panel
R_v	= axial load reduction factor for columns
	= lateral force reduction factor applicable to coupled walls

R16	= round plain bar 16 mm (0.63 in.) in diameter
r	= polar coordinate
	= ratio of negative to positive shear force at a section
	= live-load reduction factor
r_{dy}	= radius of gyration of story stiffness
S_E	= code-required strength for earthquake forces
S_i	= ideal strength
S_{ix}, S_{iy}	= shear in column or wall due to unit story shear force
S_o	= overstrength
S_p	= probable strength
S_u	= required strength to resist combined actions due to factored loads and forces
S_y	= yield strength
s	= spacing of sets of stirrups or ties along a member, mm (in.)
s_h	= vertical center-to-center spacing of horizontal hoops, ties, spirals or confining plates, mm (in.)
s_v	= horizontal center-to-center spacing of vertical bars, mm (in.)
	= spacing of horizontal or vertical shear reinforcement, mm (in.)
T	= tension force
	= axial force induced in coupled walls
T, T', T_1	= tension force in a beam or column
T_h, T'_h	= tension force in a bar at a hook
T_s	= steel tension force
T_w, T_s	= period of vibration due to out-of-plane and in-plane response, respectively
T_x, T_y	= tensile membrane force in a slab acting as a flange
T_1	= fundamental period of vibration, seconds
$T_{1,o}$	= steel tension force at overstrength
t	= wall thickness of a steel tube
	= thickness of masonry wall
t_f	= thickness of wall flange
t_w	= thickness of web of a flanged wall
$\Delta T, \Delta T_s, \Delta T_c$	= force increment transmitted from reinforcement to surrounding concrete by bond
ΔT	= time interval
U	= sum of actions due to factored loads and forces
u	= bond stress, MPa (psi)
u_o, u'_o	= average bond force per unit length
V	= shear force
V_b	= total base shear applied to a building
V_{beam}, V_b	= shear force across a beam
V_{bf}	= frame base shear
V_c	= shear resistance assigned to the concrete
	= shear across a column

V_{ch}	= ideal horizontal joint shear strength assigned to concrete shear resisting mechanism only
V_{col}	= design shear force for a column
V_{cv}	= ideal vertical joint shear strength assigned to concrete
V_{dh}	= shear resistance of diagonal reinforcement in controlling diagonal tension failure
V_{di}	= shear resistance of diagonal reinforcement in squat walls
V_{do}	= shear resistance of dowel mechanisms
V_E	= shear force derived from code-specified lateral earthquake forces
V_{Eo}	= shear force due to the development of plastic hinges at flexural overstrength
$V_{E,x}, V_{E,y}$	= shear force derived from the code-specified lateral forces in the x and y directions, respectively
V_f	= shear force induced in a column by flexural deformations
	= shear resistance of flexural compression zone in a wall
V_i, V_i'	= shear force associated with ideal strength
	= shear assigned to element or section i
V_{ix}, V_{iy}	= shear force induced in column or wall i in the x and y directions, respectively
V_{ix}', V_{iy}'	= total lateral force to be resisted by one frame in the x and y directions due to translation of the entire framing system
V_{ix}'', V_{iy}''	= total lateral force induced in a frame in the x and y directions by torsion only
V_{jh}	= horizontal joint shear force
V_{jv}	= vertical joint shear force
V_{jx}, V_{jy}	= horizontal joint shear forces in the x and y directions, respectively
V_m	= strength of shear mechanisms in unreinforced masonry construction
V_{mh}	= horizontally joint shear force assigned to masonry
V_{mv}	= vertical joint shear force assigned to masonry
V_r	= total factored lateral story shear force
V_s	= shear resistance assigned to web reinforcement
V_{sh}	= ideal horizontal joint shear strength assigned to horizontal joint shear reinforcement
V_{sv}	= ideal vertical joint shear strength assigned to the joint truss mechanism
V_u	= required shear strength due to combined action of factored loads and forces
V_{wall}	= design wall shear force derived from capacity design principles
V_x, V_y	= total applied story shear force in the x and y directions, respectively

V_1, V_2	= beam shear forces at 1, 2, etc.
v	= shear stress, MPa (psi)
v_b	= basic shear stress, MPa (psi)
v_c	= ideal shear stress provided by the concrete, MPa (psi)
v_i	= total shear stress, MPa (psi)
$v_{i,\,max}$	= maximum total shear stress, MPa (psi)
v_{jh}	= nominal horizontal shear stress in joint core, MPa (psi)
v_m	= nominal shear strength for unreinforced masonry, MPa (psi)
v_p	= characteristic velocity of P waves
v_s	= characteristic velocity of S waves
W_D, W_L	= weight due to dead and live load, respectively
W_t	= total weight
W_{tr}	= total weight at level r
w	= effective width of diagonal strut action provided by masonry infill panel
	= unit weight, kg/m^3 (lb/ft^3)
x, x_i, x_m	= coordinate or displacement
x_f	= focal distance
$\Delta x, \Delta y$	= total displacement in x and y directions, respectively
$\Delta x', \Delta y'$	= displacements due to story translation
$\Delta x'', \Delta y''$	= displacements due to relative story twist
y, y_i	= coordinate
y_t	= distance from centroid of gross uncracked concrete section to extreme fiber in tension
Z	= zone factor
	= factor allowing for squatness of walls
z_b, z_c	= internal moment arm in a beam and column, respectively
α	= stress block parameter
	= $12E_c/l_c^2$
α, α^*	= inclination of diagonal compression in a joint core
	= inclination of diagonal reinforcement
β	= reinforcement ratio, A_s'/A_s
β_1	= factor defining the depth of a equivalent rectangular stress block
$\gamma_D, \gamma_L, \gamma_E$	= load factors
Δ	= displacement
Δ_e, Δ_e'	= elastic displacement
Δ_f	= displacement due to foundation deformation
Δ_m	= maximum displacement
Δ_{mc}	= displacement of center of mass
Δ_p	= plastic displacement
Δ_s	= total structural displacement
Δ_y	= displacement at first yield
$\Delta\theta$	= angle of relative story twist
δ	= relative displacement of floors
	= story drift

δ_e	= story drift in elastic structure
ϵ	= strain
ϵ_c	= concrete compression strain in the extreme fiber
ϵ_{cc}	= compression strain in confined concrete at maximum stress
ϵ_{ce}	= elastic concrete compression strain
ϵ_{cm}	= maximum concrete compression strain
ϵ_{cu}	= ultimate concrete compression strain
ϵ_h, ϵ_v	= tensile strains in the horizontal and vertical joint shear reinforcement, respectively
ϵ_s, ϵ_s'	= steel strains in tension and compression, respectively
ϵ_{sm}	= steel strain at maximum tensile stress
ϵ_u	= concrete compression strain at the ultimate state
ϵ_y	= yield strain of steel
η, η_0, η_1	= coefficients locating position of point of contraflexure in a column
η_v	= wall shear ratio in dual systems
θ	= polar coordinate
$\theta, \theta', \theta'', \theta_1$	= plastic hinge rotation
θ_Δ	= stability coefficient
κ	= constant related to load distribution
λ	= story displacement magnification factor
$\lambda_o, \lambda_1, \lambda_o'$	= overstrength factors for steel
μ	= coefficient of friction
	= ductility factor, ductility ratio
μ_b	= displacement ductility capacity of a beam
μ_ϵ	= strain ductility
μ_f	= displacement ductility capacity of a frame unit
μ_m	= maximum ductility demand
μ_u	= ductility capacity
μ_Δ	= displacement ductility ratio
μ_ϕ	= curvature ductility ratio
$\xi_f, \xi_m, \xi_p, \xi_t$	= factors affecting bar anchorage in a joint
ρ	= ratio of tension reinforcement $= A_s/bd$
ρ_h	= ratio of horizontal wall reinforcement
ρ_l	= ratio of vertical wall reinforcement
$\rho_{\max}, \rho_{\min}$	= maximum and minimum values of the ratio tension reinforcement computed using width of web
$\rho_s, \rho_{s,r}$	= ratio of volume of spiral or circular hoop reinforcement to total volume of concrete core, measured to outside of spirals or hoops
ρ_t	= ratio of total reinforcement in columns $= A_{st}/A_g$
ρ_{tm}	= mechanical reinforcement ratio $= (f_y/f_c')\rho_t$
ρ_w	$= A_s/b_w d$
$\rho_x \rho_y$	= effective reinforcement ratios to confine concrete
ρ'	= ratio of compressional reinforcement $= A_s'/(bd)$
ρ^*	= modified reinforcement ratio for masonry walls

τ	= shear stress in masonry construction
τ_f	= shear friction stress
ϕ	= strength reduction factor
ϕ_b, ϕ_c	= strength reduction factor for beams and columns, respectively
ϕ_m	= maximum curvature
$\phi_o, \phi_{o,i}$	= flexural overstrength factor
ϕ_o^*	= flexural overstrength factor relevant to the base section of a column
$\phi_{o,f}$	= flexural overstrength factor relevant to a foundation structure
$\phi_{o,w}$	= flexural overstrength factor for a wall
ϕ_u	= ultimate curvature
ϕ_y	= yield curvature
ψ_o	= system overstrength factor
ω	= circular frequency
ω, ω_p	= dynamic moment magnification factor for columns
ω_c	= dynamic shear magnification factor for columns of dual systems
ω_v, ω_v^*	= dynamic shear magnification factor for walls

References

A1   ACI Committee 318, *Building Code Requirements for Reinforced Concrete* (*ACI 318-89*) *and Commentary*, American Concrete Institute, Detroit, 1989, 111 p.

A2   Andrews, A. L., "Slenderness Effects in Earthquake Resisting Frames," *Bulletin of the New Zealand National Society for Earthquake Engineering*, Vol. 10, No. 3, September 1977, pp. 154–158.

A3   ACI-ASCE Committee 352, "Recommendations for Design of Beam-Column Joints in Monolithic Reinforced Concrete Structures," *Journal ACI*, Vol. 73, No. 7, July 1976, pp. 365–436.

A4   Arnold, C., and Reitherman, R., *Building Configuration and Seismic Design*, John Wiley & Sons, New York, 1982, 296 p.

A5   Abrams, D. P., and Sozen, M., *Experimental Study of Frame–Wall Interaction in Reinforced Concrete Structures Subjected to Strong Earthquake Motions*, University of Illinois at Urbana–Champaign Civil Engineering Studies Structural Research Series, No. 460, 1979, 386 p.

A6   Applied Technology Council, *Tentative Provisions for the Development of Seismic Regulation for Buildings*, Publication ACT3-06, June 1978, 505 p.

A7   Aoyama, H., "Outline of Earthquake Provisions in the Recently Revised Japanese Building Code," *Bulletin of the New Zealand National Society for Earthquake Engineering*, Vol. 14, No. 2, 1981, pp. 63–80.

A8   Architectural Institute of Japan, *Design Guidelines for Earthquake Resistant Reinforced Concrete Buildings Based on the Ultimate Strength Concept* (draft), October 1988.

A9   Aktan, A. E., and Bertero, V. V., "Conceptual Seismic Design of Frame–Wall Structures," *Journal of Structural Engineering*, *ASCE*, Vol. 110, No. 11, November 1984, pp. 2778–2797.

A10   Architectural Institue of Japan, *AIJ Standard for Structural Calculations of Reinforced Concrete Structures* (*1982*), September 1983, p. 8.

A11   Aoyama, H., "Earthquake Resistant Design of Reinforced Concrete Frame Building with 'Flexural' Walls," *Journal of the Faculty of Engineering, University of Tokyo* (*B*), Vol. XXXIX, No. 2, 1987, pp. 87–109.

A12   Ang, Beng Ghee, Priestley, M. J. N., and Paulay, T., "Seismic Shear Strength of Circular Reinforced Concrete Columns," *ACI Structural Journal*, Vol. 86, No. 1, January–February 1988, pp. 45–59.

A13   Ang, Beng Ghee, *Seismic Shear Strength of Circular Bridge Piers*, Research Report 85-5, Department of Civil Engineering, University of Canterbury, Christchurch, New Zealand, 1985, 439 p.

A14   ACI-ASCE Committee 442, *Response of Concrete Buildings to Lateral Forces*, American Concrete Institute, Detroit, 1988, 36 p.

A15   Adriono, T., and Park, R., "Seismic Design Considerations of the Properties of New Zealand Manufactured Steel Reinforcing Bars," *Proceedings of the*

Pacific Conference on Earthquake Engineering, Wairakei, New Zealand, August 1987, Vol. 1, pp. 13–24.

B1   Beck, H., "Contribution to the Analysis of Coupled Shear Walls," *Journal ACI*, Vol. 59, August 1962, pp. 1055–1070.

B2   Bertero, V. V., and Popov, E. P., "Seismic Behaviour of Ductile Moment-Resisting Reinforced Concrete Frames," in *Reinforced Concrete Structures in Seismic Zones*, ACI Publication SP-53, American Concrete Institute, Detroit, 1977, pp. 247–291.

B3   Birss, G. R., *The Elastic Behaviour of Earthquake Resistant Reinforced Concrete Interior Beam-Column Joints*, Research Report 78-13, Department of Civil Engineering, University of Canterbury, New Zealand, February 1978, 96 p.

B4   Bertero, V. V., *Proceedings of a Workshop on Earthquake-Resistant Reinforced Concrete Building Construction*, University of California, Berkeley, 1977, Vol. I, 115 p., Vols. II and III, 1940 p.

B5   Beckingsale, C. W., *Post-Elastic Behavior of Reinforced Concrete Beam-Column Joints*, Research Report 80-20, Department of Civil Engineering, University of Canterbury, Christchurch, New Zealand, August 1980, 398 p.

B6   Blakeley, R. W. G., Megget, L. M., and Priestley, M. J. N., "Seismic Performance of Two Full Size Reinforced Concrete Beam–Column Joint Units," *Bulletin of the New Zealand National Society for Earthquake Engineering*, Vol. 8, No. 1, March 1975, pp. 38–69.

B7   Buirguieres, S. T., Jirsa, J. O., and Longwell, J. E., *The Behaviour of Beam–Column Joints Under Bidirectional Load Reversals*, Bulletin D'Information 132, Comité Euro-International du Béton, April 1979, pp. 221–228.

B8   Buchanan, A. H., "Diagonal Beam Reinforcing for Ductile Frames," *Bulletin of the New Zealand National Society for Earthquake Engineering*, Vol. 12, No. 4, December 1979, pp. 346–356.

B9   Barnard, P. R., and Schwaighofer, J., "Interaction of Shear Walls Connected Solely Through Slabs," *Proceedings of the Symposium on Tall Buildings*, University of Southampton, April 1966, Pergamon Press, Elmsford, N.Y., 1967.

B10   Berrill, J. B., Priestley, M. J. N., and Chapman, H. E., "Design Earthquake Loading and Ductility Demand," *Bulletin of the New Zealand National Society for Earthquake Engineering*, Vol. 13, No. 3, September 1980, pp. 232–241.

B11   Blakeley, R. W. G., Cooney, R. C., and Megget, L. M., "Seismic Shear Loading at Flexural Capacity in Cantilever Wall Structures," *Bulletin of the New Zealand National Society for Earthquake Engineering*, Vol. 8, No. 4, December 1975, pp. 278–290.

B12   Bertero, V. V., Popov, E. P., Wang, T. Y., and Vallenas, J., "Seismic Design Implications of Hysteretic Behaviour of Reinforced Concrete Structural Walls," *6th World Conference on Earthquake Engineering*, New Delhi, 1977, Vol. 5, pp. 159–165.

B13   Barda, F., "Shear Strength of Low-Rise Walls with Boundary Elements," Ph.D. dissertation, Lehigh University, Bethlehem, Pa., 1972, 278 p.

B14   Bertero, V. V., "Seismic Behaviour of Reinforced Concrete Wall Structural Systems," *Proceedings of the 7th World Conference on Earthquake Engineering*, Istanbul, 1980, Vol. 6, pp. 323–330.

B15   Bertero, V. V., Aktan, A. E., Charney, F., and Sause, R., "Earthquake Simulator Tests and Associated Experiments, Analytical and Correlation Studies of One-Fifth Scale Model," in *Earthquake Effects on Reinforced Concrete Structures*, U.S.–Japan Research, ACI Publication SP-84, American Concrete Institute, Detroit, 1985, pp. 375–424.

B16   Bertero, V. V., and Shadh, H., *El Asnam, Algeria Earthquake, October 10, 1980*, Earthquake Engineering Research Institute, Oakland, Calif., January 1983, 190 p.

B17   Bertero, V. V., "Lessons Learned from Recent Earthquakes and Research and Implications for Earthquake Resistant Design of Building Structures in the United States," *Earthquake Spectra*, Vol. 2, No. 4, 1986, pp. 825–858.

B18   Bertero, V. V., "State of the Art Practice in Seismic Resistant Design of Reinforced Concrete Frame–Wall Structural Systems," *Proceedings of the 8th World Conference on Earthquake Engineering*, San Francisco, 1984, Vol. V, pp. 613–620.

B19   Binney, J. R., and Paulay, T., "Foundations for Shear Wall Structures," *Bulletin of the New Zealand National Society for Earthquake Engineering*, Vol. 13, No. 2, June 1980, pp. 171–181.

B20   Bazant, Z. P., and Bhat, P. D., "Endochronic Theory of Elasticity and Failure of Concrete," *Proceedings ASCE*, Vol. 102, No. EM4, April 1976, pp. 701–702.

B21   Butler, R., Stewart, G., and Kanamori, K., "The July 27, 1976 Tangshan, China Earthquake—A Complex Sequence of Interplate Events," *Bulletin of the Seismological Society of America*, Vol. 69, No. 1, 1979, pp. 207–220.

B22   Berg, V. B., and Stratta, J. L., *Anchorage and the Alaska Earthquake of March 27, 1964*, American Iron and Steel Institute, New York, 1964, 63 p.

C1   Council of Tall Buildings and Urban Habitat, "Tall Building Criteria and Loading," *Monograph of Planning and Design of Tall Buildings*, Vol. CL, 1980, 888 p.

C2   Council of Tall Buildings and Urban Habitat, "Structural Design of Tall Concrete and Masonry Buildings," *Monograph of Planning and Design of Tall Buildings*, Vol. CB, 1978, 938 p.

C3   Collins, M. P., and Mitchell, D., "Shear and Torsion Design of Prestressed and Non-prestressed Concrete Beams," *Prestressed Concrete Institute Journal*, Vol. 25, No. 5, September–October 1980, pp. 32–101.

C4   Clough, R. W., and Penzien, J., *Dynamics of Structures*, McGraw-Hill, New York, 1975, 652 p.

C5   Celebi, M. (Ed.), *Seismic Site-Response Experiments Following the March 3, 1985 Central Chile Earthquake*, U.S. Department of Interior Geophysical Survey Report 86-90, Menlo Park, Calif., 1986.

C6   Coull, A., and Choudhury, J. R., "Analysis of Coupled Shear Walls," *Journal ACI*, Vol. 64, September 1967, pp. 587–593.

C7   Collins, M. P., and Mitchell, D., *Prestressed Concrete Structures*, Prentice Hall, Englewood Cliffs, N.J., 1991, 766 p.

C8   Clough, R. W., "A Replacement for the SRSS Method of Seismic Analysis," *Earthquake Engineering and Structural Dynamics*, Vol. 9, 1988, pp. 187–194.

C9   Coull, A., and Elhag, A. A., "Effective Coupling of Shear Walls by Floor Slabs," *Journal ACI*, Vol. 72, No. 8, August 1975, pp. 429–431.

C10 Ciampi, V., Eligehausen, R., Bertero, V. V., and Popov, E. P., "Hysteretic Behaviour of Deformed Reinforcing Bars Under Seismic Excitation," *Proceedings of the 7th European Conference on Earthquake Engineering*, Athens, September 1982, Vol. 4, pp. 179–187.

C11 Comité Euro-International du Béton, *Seismic Design of Concrete Structures*, Technical Press, New York, 1987, 298 p.

C12 Corley, W. G., Fiorato, A. E., and Oesterle, R. G., *Structural Walls*, ACI Publication SP-72, American Concrete Institute, Detroit, 1981, pp. 77–131.

C13 Charney, F. A., and Bertero, V. V., *An Analytical Evaluation of the Design and Analytical Seismic Response of a Seven Storey Reinforced Concrete Frame–Wall Structure*, Report UCB/EERC 8208, Earthquake Engineering Research Center, University of California, Berkeley, 1982, 176 p.

C14 Cheung, P. C., Paulay, T., and Park, R., *Interior and Exterior Reinforced Concrete Beam-Column Joints of a Prototype Two-Way Frame with Floor Slab Designed for Earthquake Resistance*, Research Report 89-2. Department of Civil Engineering, University of Canterbury, Christchurch, New Zealand, 165 p.

E1   Eligehausen, R., Popov, E. P., and Bertero, V. V., *Local Bond Stress–Slip Relationships of Deformed Bars Under Generalized Excitation*, Report UCB/EERC-83/23, Earthquake Engineering Research Center, University of California, Berkeley, 1983, 178 p.

E2   Esteva, L., "Bases para la Formulación de Decisiones de Diseño Sísmico," Doctoral thesis, National University of Mexico, 1968.

E3   Esteva, L., and Villaverde, R., "Seismic Risk, Design Spectra, and Structural Reliability," *Proceedings of the 5th World Conference on Earthquake Engineering*, Rome, 1972, Vol. 2, pp. 2586–2596.

F1   Fenwick, R. C., and Irvine, H. M., "Reinforced Concrete Beam-Column Joints for Seismic Loading," *Bulletin of the New Zealand National Society for Earthquake Engineering*, Part I: Vol. 10, No. 3, September 1977, pp. 121–128; Part II: Vol. 10, No. 4, December 1977, pp. 174–185.

F2   Fintel, M., Derecho, A. T., Freskasis, G. N., Fugelso, L. E., and Gosh, S. K., *Structural Walls in Earthquake Resistant Structures*, Progress Report to the National Science Foundation (RANN), Portland Cement Association, Skokie, Ill., AUgust 1975, 261 pp.

F3   Fenwick, R. C., "Strength Degradation of Concrete Beams Under Cyclic Loading," *Bulletin of the New Zealand National Society for Earthquake Engineering*, Vol. 16, No. 1, March 1983, pp. 25–38.

F4   Freeman, S. A., Czarnecki, R. M., and Honda, K. K., *Significance of Stiffness Assumptions on Lateral Force Criteria*, ACI Publication SP-63, American Concrete Institute, Detroit, 1980, pp. 437–457.

F5   Filippou, F. C., Popov, E. V., and Bertero, V. V., *Effects of Bond Deterioration on Hysteretic Behaviour of Reinforced Concrete Joints*, Report UCB/EERC-83/19, Earthquake Engineering Research Center, University of California, Berkeley, 1983, 191 p.

G1   Goodsir, W. J., *The Design of Coupled Frame–Wall Structures for Seismic Actions*, Research Report 85-8, Department of Civil Engineering, University of Canterbury, Christchurch, New Zealand, 1985, 383 p.

G2   Goodsir, W. J., Paulay, T., and Carr, A. J., "A Study of the Inelastic Seismic Response of Reinforced Concrete Coupled Frame–Shear Wall Structures," *Bulletin of the New Zealand National Society for Earthquake Engineering*, Vol. 16, No. 3, September 1983, pp. 185–200.

G3   Gulkan, P., and Sozen, M. A., "Inelastic Responses of Reinforced Concrete Structures to Earthquake Motions," in *Reinforced Concrete Structures in Seismic Zones*, N. M. Hawkins, Ed., ACI Publication SP-53, American Concrete Institute, Detroit, 1977, pp. 109–116.

G4   Gill, W. D., Park, R., and Priestley, M. J. N., *Ductility of Rectangular Reinforced Concrete Columns with Axial Load*, Research Report 79-1, Department of Civil Engineering, University of Canterbury, Christchurch, New Zealand, February 1979.

H1   Hilsdorf, H. K., "An Investigation into the Failure Mechanism of Brick Masonry Under Axial Compression in Designing," in *Engineering and Constructing with Masonry Products*, F. B. Johnson, Ed., Gulf Publishing, Houston, May 1969, pp. 34–41.

H2   He, L., and Priestley, M. J. N., *Seismic Behaviour of Flanged Masonry Shear Walls: Preliminary Studies*, Structural Systems Research Project Report SSRP-88/-1, University of California, San Diego, May 1988, 119 p.

I1   Iqbal, M., and Derecho, A. T., "Inertia Forces over Height of Reinforced Concrete Structural Walls During Earthquakes," in *Reinforced Concrete Structures Subjected to Wind and Earthquake Forces*, ACI Publication SP-63, American Concrete Institute, Detroit, 1980, pp. 173–196.

I2   Iliya, R., and Bertero, V. V., *Effects of Amount and Arrangement of Wall-Panel Reinforcement on Hysteretic Behaviour of Reinforced Concrete Walls*, Report UCB/EER-80/04, Earthquake Engineering Research Center, University of California, Berkeley, February 1980, 175 p.

J1   Jury, R. D., *Seismic Load Demands on Columns of Reinforced Concrete Multistorey Frames*, Research Report 78-12, Department of Civil Engineering, University of Canterbury, Christchurch, New Zealand, February 1978.

J2   Jennings, P. C., *Engineering Features of the San Fernando Earthquake, February 9, 1971*, Report EERL 71-02, Earthquake Engineering Research Laboratory, California Institute of Technology, Pasadena, Calif., June 1971, 512 p.

K1   Kelly, T. E., "Some Comments on Reinforced Concrete Structures Forming Column Hinge Mechanisms," *Bulletin of the New Zealand National Society for Earthquake Engineering*, Vol. 10, No. 4, December 1977, pp. 186–195.

K2   Kent, D. C., and Park, R., "Flexural Members with Confined Concrete," *Proceedings ASCE*, Vol. 97, No. ST7, July 1971, pp. 1969–1990.

K3   Kanada, K., Kondon, G., Fujii, S., and Morita, S., "Relation Between Beam Bar Anchorage and Shear Resistance at Exterior Beam–Column Joints," *Transactions of the Japan Concrete Institute*, Vol. 6, 1984, pp. 433–440.

K4   Kitayama, K., Asami, S., Otani, S., and Aoyama, H., "Behaviour of Reinforced Concrete Three-Dimensional Beam–Column Connections with Slabs," *Transactions of the Japan Concrete Institute*, Vol. 8, 1986, pp. 38–388.

K5   Klingner, R. E., and Bertero, V. V., "Infilled Frames Aseismic Construction," *Proceedings of the 6th World Conference on Earthquake Engineering*, New Delhi, 1977, Vol. 5, pp. 159–190.

K6   Kitayama, K., Otani, S., and Aoyama, H., "Earthquake Resistant Design Criteria for Reinforced Concrete Interior Beam–Column Joints," *Proceedings of the Pacific Conference on Earthquake Engineering*, Wairakei, New Zealand, August 1987, Vol. 1, pp. 315–326.

K7   Kurose, Y., *Recent Studies of Reinforced Concrete Beam–Column Joints in Japan*, PMFSL Report 87-8, Department of Civil Engineering, University of Texas, December 1987, 164 p.

K8   Keintzel, E., "Seismic Design Shear Forces in RC Cantilever Shear Wall Structures," *European Earthquake Engineering*, 3, 1990, pp. 7–16.

L1   Lukose, K., Gergely, P., and White, R. N., "Behaviour and Design of R/C Lapped Splices for Inelastic Cyclic Loading," *Journal ACI*, Vol. 79, No. 5, September–October 1982, pp. 366–372.

L2   Leuchars, J. M., and Scrivener, J. C., "Masonry Infill Panels Subjected to Cyclic In-Plane Loading," *Bulletin of the New Zealand National Society for Earthquake Engineering*, Vol. 9, No. 2, 1976, pp. 122–131.

L3   Leslie, P. D., "Ductility of Reinforced Concrete Bridge Piers," Master of Engineering report, Department of Civil Engineering, University of Canterbury, Christchurch, New Zealand, 1974, 147 p.

M1   Muto, K., *A Seismic Design Analysis of Buildings*, Maruzen, Tokyo, 1974, 361 p.

M2   Ma, S. Y., Bertero, V. V., and Popov, D. P., *Experimental and Analytical Studies on the Hysteretic Behaviour of Reinforced Concrete Rectangular and T-Beams*, Report EERC 76-2, Earthquake Engineering Research Center, University of California, Berkeley, 1976.

M3   Montgomery, C. J., "Influence of *P*–Delta Effects on Seismic Design," *Canadian Journal of Civil Engineering*, Vol. 8, 1981, pp. 31–43.

M4   Moss, P. J., and Carr, A. J., "The Effects of Large Displacements on the Earthquake Response of Tall Concrete Frame Structures," *Bulletin of the New Zealand National Society for Earthquake Engineering*, Vol. 13, No. 4, December 1980, pp. 317–328.

M5   Mander, J. B., Priestley, M. J. N., and Park, R., "Observed Stress–Strain Behaviour of Confined Concrete," *Journal of Structural Engineering, ASCE*, Vol. 114, No. 8, August 1988, pp. 1827–1849.

M6   MacGregor, J. G., and Hage, S. E., "Stability Analysis and Design of Concrete Frames," *Journal of the Structural Division, ASCE*, Vol. 103, No. ST10, October 1977, pp. 1953–1970.

M7   Mattock, A. H., *Shear Transfer Under Monotonic Loading, Across an Interface Between Concrete Cast at Different Times*, Structures and Mechanics Report SM 76-3, University of Washington, Seattle, September 1976, 68 p.

M8   Mattock, A. H., *Shear Transfer Under Cyclically Reversing Loading, Across an Interface Between Concretes Cast at Different Times*, Structures and Mechanics Report SM 77-1, University of Washington, Seattle, June 1977, 97 p.

M9   Mahin, S. A., and Bertero, V. V., "Nonlinear Seismic Response of a Coupled Wall System," *Journal of the Structural Division, ASCE*, Vol. 102, No. ST9, September 1976, pp. 1759–1780.

M10   Milburn, J. R., and Park, R., *Behaviour of Reinforced Concrete Beam–Column Joints Designed to NZS 3101*, Research Report S2-7, Department of Civil Engineering, University of Canterbury, Christchurch, New Zealand, 1982, 107 p.

M11   Megget, L. M., and Fenwick, R. C., "Seismic Behaviour of a Reinforced Concrete Portal Frame Sustaining Gravity Loads," *Proceedings of the Pacific Concrete Conference*, Auckland, New Zealand, 1988, Vol. 1, pp. 41–52.

M12   Morgan, B., Hiraishi, H., and Corley, W. G., "Medium Scale Wall Assemblies: Comparison of Analysis and Test Results," in *Earthquake Effects on Reinforced Concrete Structures*, U.S.–Japan Research, ACI Publication SP-84, American Concrete Institute, Detroit, 1985, pp. 241–269.

M13   Mander, J. B., Priestley, M. J. N., and Park, R., "Theoretical Stress–Strain Model for Confined Concrete," *Journal of Structural Engineering, ASCE*, Vol. 114, No. 8, August 1988, pp. 1804–1826.

M14   Meinheit, D. F., and Jirsa, J. O., "Shear Strength of Reinforced Concrete Beam–Column Connections," *Journal of the Structural Division, ASCE*, Vol. 107, No. ST11, November 1983, pp. 2227–2244.

M15   Mitchell, D., "Structural Damage Due to the 1985 Mexico Earthquake," *Proceedings of the 5th Canadian Conference on Earthquake Engineering*, Ottawa, 1987, A. A. Balkema, Rotterdam, pp. 87–111.

M16   Müller, F. P., and Keintzel, E., *Erdbebensicherung von Hochbauten*, 2nd ed., Wilhelm Ernst, Berlin, 1984, 249 p.

M17   Markevicius, V. P., and Gosh, S. K., *Required Shear Strength of Earthquake Resistant Shear Walls*, Research Report 87-1, Department of Civil Engineering and Engineering Mechanics, University of Illinois at Chicago, February 1987.

M18   Mayes, R. C., Omoto, Y., and Clough, R. W., *Cyclic Shear Tests of Masonry Piers*, Report EERC 76-8, University of California, Berkeley, May 1976, 84 p.

M19   Matsumura, A., "Effectiveness of Shear Reinforcement in Fully Grouted Masonry Walls," *Proceedings of the 4th Meeting of the Joint Technical Coordinating Committee on Masonry Research UJNR*, San Diego, October 1988, 10 p.

N1   Newmark, N. M., and Rosenblueth, E., *Fundamentals of Earthquake Engineering*, Prentice Hall, Englewood Cliffs, N.J., 1971, 640 p.

N2   Noland, J. L., *U.S. Research Plan: U.S.–Japan Coordinated Program for Masonry Building Research*, Atkinson-Noland Associates, Boulder, Colo., 1984, 22 p.

N3   *New Zealand Reinforced Concrete Design Handbook*, New Zealand Portland Cement Association, Wellington, New Zealand, 1978.

N4   Newmark, N., and Hall, W., *Earthquake Spectra and Design*, Monograph, Earthquake Engineering Research Institute, Oakland, Calif., 1982, 103 p.

N5   Neville, A. M., *Properties of Concrete*, Pitman Publishing, New York, 1971, 532 p.

O1   Oesterle, R. G., Fiorato, A. E., and Corley, W. G., "Reinforcement Details for Earthquake-Resistant Structural Walls," *Concrete International Design and Construction*, Vol. 2, No. 12, 1980, pp. 55–66.

O2   Oesterle, R. G., Fiorato, A. E., Aristazabal-Ochoa, J. D., and Corley, W. G., "Hysteretic Response of Reinforced Concrete Structural Walls," in *Reinforced Concrete Structures Subjected to Wind and Earthquake Forces*, ACI Publication SP-63, American Concrete Institue, Detroit, 1980, pp. 243–273.

O3   Otani, S., Li, S., and Aoyama, H., "Moment-Redistribution in Earthquake Resistant Design of Ductile Reinforced Concrete Frames," *Transactions of the Japan Concrete Institute*, Vol. 9, 1987, pp. 581–588.

P1   Park, R., and Paulay, T., *Reinforced Concrete Structures*, John Wiley & Sons, New York, 1975, 769 p.

P2   Park, R., and Gamble, W., *Reinforced Concrete Slabs*, John Wiley & Sons, New York, 1980, 618 p.

P3   Paulay, T., "Seismic Design of Ductile Moment Resisting Reinforced Concrete Frames, Columns: Evaluation of Actions," *Bulletin of the New Zealand National Society for Earthquake Engineering*, Vol. 10, No. 2, June 1977, pp. 85–94.

P4   Paulay, T., *Deterministic Design Procedure for Ductile Frames in Seismic Areas*, ACI Publication SP-63, American Concrete Institute, Detroit, 1980, pp. 357–381.

P5   Paulay, T., Carr, A. J., and Tompkins, D. N., "Response of Ductile Reinforced Concrete Frames Located in Zone C," *Bulletin of the New Zealand National Society for Earthquake Engineering*, Vol. 13, No. 3, September 1980.

P6   Paulay, T., "Capacity Design of Earthquake Resisting Ductile Multistorey Reinforced Concrete Frames," *Proceedings of the 3rd Canadian Conference on Earthquake Engineering*, Montreal, June 1979, Vol. 2, pp. 917–948.

P7   Paulay, T., "Developments in the Design of Ductile Reinforced Concrete Frames," *Bulletin of the New Zealand National Society for Earthquake Engineering*, Vol. 12, No. 1, March 1979, pp. 35–48.

P8   Priestley, M. J. N., Park, R., and Potangaroa, R. T., "Ductility of Spirally Reinforced Concrete Columns," *Journal of the Structural Division, Proceedings ASCE*, Vol. 107, No. ST1, January 1981, pp. 181–202.

P9   Paulay, T., "A Consideration of P–Delta Effects in Ductile Reinforced Concrete Frames," *Bulletin of the New Zealand National Society for Earthquake Engineering*, Vol. 11, No. 3, September 1978, pp. 151–160; Vol. 12, No. 4, December 1979, pp. 358–361.

P10   Powell, G. H., and Row, D. G., *Influence of Analysis and Design Assumptions on Computed Inelastic Response of Moderately Tall Frames*, Report EERC 76-11, Earthquake Engineering Research Center, University of California, Berkeley, April 1976, 111 p.

P11   Park, R., Priestley, M. J. N., and Gill, W. G., "Ductility of Square-Confined Concrete Columns," *Journal of the Structural Division, Proceedings ASCE*, Vol. 108, No. ST4, April 1982, pp. 929–950.

P12   Park, R., and Hopkins, D. C., "United States/New Zealand/Japan/China Collaborative Research Project on the Seismic Design of Reinforced Concrete

Beam–Column–Slab Joints," *Bulletin of the New Zealand National Society for Earthquake Engineering*, Vol. 22, No. 2, June 1989, pp. 122–126.

P13   Paulay, T., Park, R., and Birss, G. R., "Elastic Beam–Column Joints for Ductile Frames," *Proceedings of the 7th World Conference on Earthquake Engineering*, Istanbul, 1980, Vol. 6, pp. 331–338.

P14   Paulay, T., "Seismic Design in Reinforced Concrete: The State of the Art in New Zealand," *Bulletin of the New Zealand National Society for Earthquake Engineering*, Vol. 21, No. 3, September 1988, pp. 208–232.

P15   Paulay, T., "An Application of Capacity Design Philosophy to Gravity Load Dominated Ductile Reinforced Concrete Frames," *Bulletin of the New Zealand National Society for Earthquake Engineering*, Vol. 11, No. 1, March 1978, pp. 50–61.

P16   Paulay, T., and Bull, I. N., *Shear Effect on Plastic Hinges of Earthquake Resisting Reinforced Concrete Frames*, Bulletin D'Information 132, Comité Euro-International du Béton, April 1979, pp. 165–172.

P17   Park, R., and Dai, Ruitong, "A Comparison of the Behaviour of Reinforced Concrete Beam–Column Joints Designed for Ductility and Limited Ductility," *Bulletin of the New Zealand National Society for Earthquake Engineering*, Vol. 21, No. 4, December 1988, pp. 255–278.

P18   Paulay, R., Park, R., and Priestley, M. J. N., "Reinforced Concrete Beam–Column Joints Under Seismic Actions," *Journal ACI*, Vol. 75, No. 11, November 1978, pp. 585–593.

P19   Paulay, T., "Lapped Splices in Earthquake Resisting Columns," *Journal ACI*, Vol. 79, No. 6, November–December 1982, pp. 458–469.

P20   Portland Cement Association Team, "Caracas Earthquake Damage," *Journal ACI*, Vol. 65, No. 4, April 1969, pp. 292–294.

P21   Paulay, T., and Santhakumar, A. R., "Ductile Behaviour of Coupled Shear Walls," *Journal of the Structural Division*, ASCE, Vol. 102, No. ST1, January 1976, pp. 93–108.

P22   Paulay, T., "Coupling Beams of Reinforced Concrete Shear Walls," *Journal of the Structural Division*, ASCE, Vol. 97, No. ST3, March 1971, pp. 843–862.

P23   Paulay, T., "Simulated Seismic Loading of Spandrel Beams," *Journal of the Structural Division*, ASCE, Vol. 97, No. ST9, September 1971, pp. 2407–2419.

P24   Paulay, T., and Taylor, R. G., "Slab Coupling of Earthquake Resisting Shear Walls," *Journal ACI*, Vol. 78, No. 2, March–April 1981, pp. 130–140.

P25   Petersen, H. B., Popov, D. P., and Bertero, V. V., "Practical Design of Reinforced Concrete Structural Walls Using Finite Elements," *Proceedings of the IASS World Congress on Space Enclosures*, Building Research Centre, Concordia University, Montreal, July 1976, pp. 771–780.

P26   Priestley, M. J. N., and Hart, G. C., *Design Recommendations for the Period of Vibration of Masonry Wall Buildings*, Research Report SSRP 89/05, University of California, San Diego and Los Angeles, 1989, 46 p.

P27   Paulay, T., and Spurr, D. D., "Simulated Seismic Loading on Reinforced Concrete Frame–Shear Wall Structures," *6th World Conference on Earthquake Engineering*, New Delhi, 1977, Preprints 3, pp. 221–226.

P28 Paulay, T., and Uzumeri, S. M., "A Critical Review of the Seismic Design Provisions for Ductile Shear Walls of the Canadian Code and Commentary," *Canadian Journal of Civil Engineering*, Vol. 2, No. 4, 1975, pp. 592–601.

P29 Priestley, M. N. J., and He Limin, "Seismic Response of *T*-Section Masonry Shear Walls," *Masonry Society Journal*, Vol. 9, No. 1, August 1990, pp. 10–19.

P30 Paulay, T., "Developments in the Seismic Design of Reinforced Concrete Frames in New Zealand," *Canadian Journal of Civil Engineering*, Vol. 8, No. 2, 1981, pp. 91–113.

P31 Paulay, T., and Santhakumar, A. R., "Ductile Behaviour of Coupled Shear Walls Subjected to Reversed Cyclic Loading," *6th World Conference on Earthquake Engineering*, New Delhi, 1977, Preprints 3, pp. 227–232.

P32 Paulay, T., and Binney, J. R., "Diagonally Reinforced Coupling Beams of Shear Walls," in *Shear in Reinforced Concrete*, ACI Publication SP-42, American Concrete Institute, Detroit, 1974, Vol. I, pp. 579–598.

P33 Paulay, T., "The Ductility of Reinforced Concrete Shear Walls for Seismic Areas," in *Reinforced Concrete Structures in Seismic Zones*, ACI Publication SP-53, American Concrete Institute, Detroit, 1977, pp. 127–147.

P34 Paulay, T., "Coupling Beams of Reinforced Concrete Shear Walls," *Proceedings of a Workshop on Earthquake-Resistant Reinforced Concrete Building Construction*, University of California, Berkeley, 1977, Vol. 3, pp. 1452–1460.

P35 Paulay, T., Priestley, M. J. N., and Synge, A. J., "Ductility in Earthquake Resisting Squat Shearwalls," *Journal ACI*, Vol. 79, No. 4, July–August 1982, pp. 257–269.

P36 Priestley, M. J. N., Evison, R. J., and Carr, A. J., "Seismic Response of Structures Free to Rock on Their Foundations," *Bulletin of the New Zealand National Society for Earthquake Engineering*, Vol. 11, No. 3, September 1978, pp. 141–150.

P37 Poland, C. D., "Practical Application of Computer Analysis to the Design of Reinforced Concrete Structures for Earthquake Forces," in *Reinforced Concrete Structures Subjected to Wind and Earthquake Forces*, ACI Publication SP-63, American Concrete Institute, Detroit, 1980, pp. 409–436.

P38 Priestley, M. J. N., "Ductility of Unconfined and Confined Concrete Masonry Shear Walls," *Masonry Society Journal*, Vol. 1, No. 2, July–December 1981, pp. T28–T39.

P39 Paulay, T., "Earthquake Resisting Shear Walls: New Zealand Design Trends," *Journal ACI*, Vol. 77, No. 3, May–June 1980, pp. 144–152.

P40 Park, R., and Milburn, J. R., "Comparison of Recent New Zealand and United States Seismic Design Provisions for Reinforced Concrete Beam–Column Joints and Test Results for Four Units Designed According to the New Zealand Code," *Bulletin of the New Zealand National Society for Earthquake Engineering*, Vol. 16, No. 1, March 1983, pp. 21–42.

P41 Paulay, T., and Scarpas, A., "The Behaviour of Exterior Beam–Column Joints," *Bulletin of the New Zealand National Society for Earthquake Engineering*, Vol. 14, No. 3, September 1981, pp. 131–144.

P42 Paulay, T., "The Design of Ductile Reinforced Concrete Structural Walls for Earthquake Resistance," *Earthquake Spectra*, Vol. 2, No. 4, October 1986, pp. 783–824.

P43   Priestley, M. J. N., and Wood, J. H., "Behaviour of a Complex Prototype Box Girder Bridge," *Proceedings of the RILEM International Symposium on Testing In-Situ of Concrete Structures*, Budapest, September 1977, Vol. 1, pp. 140–153.

P44   Paulay, T., and Goodsir, W. J., "The Ductility of Structural Walls," *Bulletin of the New Zealand National Society for Earthquake Engineering*, Vol. 18, No. 3, September 1985, pp. 250–269.

P45   Park, R., "Ductile Design Approach for Reinforced Concrete Frames," *Earthquake Spectra*, Vol. 2, No. 3, May 1980, pp. 565–620.

P46   Popovics, S., "A Numerical Approach to the Complete Stress–Strain Curves for Concrete," *Cement and Concrete Research*, Vol. 3, No. 5, September 1973, pp. 583–599.

P47   Pussegoda, L. N., "Strain Age Embrittlement in Reinforcing Steel," Ph.D. thesis report, Department of Mechanical Engineering, University of Canterbury, Christchurch, New Zealand, 1978.

P48   Priestley, M. J. N., and Park, R., *Strength and Ductility of Bridge Substructures*, RRU Bulletin 71, National Roads Board, Wellington, New Zealand, 1984, 120 p.

P49   Paulay, T., "A Seismic Design Strategy for Hybrid Structures," *Proceedings of the 5th Canadian Conference on Earthquake Engineering*, Ottawa, 1987, A. A. Balkema, Rotterdam, pp. 3–25.

P50   Priestley, M. J. N., "Masonry Wall-Frame Joint Test," Report No. 90-5, Seqad Consulting Engineers, Solana Beach, Calif., Nov. 1990.

P51   Park, R., and Dai Ruitong, "Ductility of Doubly Reinforced Concrete Beam Sections," *Structural Journal ACI*, Vol. 85, No. 2, March–April 1988, pp. 217–225.

P52   Paulay, T., Bachmann, H., and Moser, K., *Erdbebenmessung von Stahlbetonhochbauten*, Birkhauser Verlag, Basel, 1990, 579 p.

P53   Priestley, M. J. N., and Chai, Y. H., "Prediction of Masonry Compression Strength," *New Zealand Concrete Construction*, Part 1: Vol. 28, No. 3, March 1984, pp. 11–14; Part 2: Vol. 28, No. 4, April 1984, pp. 21–25.

P54   Popov, E. P., "Bond and Anchorage of Reinforcing Bars Under Cyclic Loading," *Journal ACI*, Vol. 81, No. 4, July–August 1984, pp. 340–349.

P55   Priestley, M. J. N., and Elder, D. M., "Cyclic Loading Tests of Slender Concrete Masonry Shear Walls," *Bulletin of the New Zealand National Society for Earthquake Engineering*, Vol. 15, No. 1, March 1982, pp. 3–21.

P56   Priestley, M. J. N., "Seismic Resistance of Reinforced Concrete Masonry Shear Walls with High Steel Percentages," *Bulletin of the New Zealand National Society for Earthquake Engineering*, Vol. 10, No. 1, March 1977, pp. 1–16.

P57   Priestley, M. J. N., and Bridgeman, D. O., "Seismic Resistance of Brick Masonry Walls," *Bulletin of the New Zealand National Society for Earthquake Engineering*, Vol. 7, No. 4, December 1974, pp. 149–166.

P58   Priestley, M. J. N., "Seismic Design of Concrete Masonry Shear Walls," *Journal ACI*, Vol. 83, No. 1, January–February 1986, pp. 58–68.

P59   Priestley, M. J. N., "Ductility of Unconfined and Confined Concrete Masonry Shear Walls," *Masonry Society Journal*, Vol. 1, No. 2, July–December 1981, pp. T28–T39.

P60   Priestley, M. J. N., and Chai, Y. H., "Seismic Design of Reinforced Concrete Masonry Moment Resisting Frames," *Masonry Society Journal*, Vol. 5, No. 1, January–June 1985, pp. T1–T17.

P61   Priestley, M. J. N., and Elder, D. M., "Stress–Strain Curves for Unconfined and Confined Concrete Masonry," *Journal ACI*, Vol. 80, No. 3, May–June 1983, pp. 192–201.

P62   Powell, G. H., *Drain-2D User's Guide*, Report EERC 73-22, Earthquake Engineering Research Center, University of California, Berkeley, October 1973.

R1   Richter, C. R., *Elementary Seismology*, W. H. Freeman, San Francisco, 1958.

R2   Rosman, R., "Approximate Analysis of Shear Walls Subjected to Lateral Loads," *Journal ACI*, Vol. 61, June 1964, pp. 717–733.

R3   Rosman, R., *Die statische Berechnung von Hochhauswänden mit Öffnungsreihen*, Bauingenieur Praxis Heft 65, Wilhelm Ernst, Berlin, 1965, 64 p.

R4   Rosenblueth, E., Ed., *Design of Earthquake Resistant Structures*, Pentech Press, London, 1973, 295 p.

S1   Scott, B. D., Park, R., and Priestley, M. J. N., "Stress–Strain Behaviour of Concrete Confined by Overlapping Hoops at Low and High Strain Rates," *Journal ACI*, Vol. 79, No. 1, January–February 1982, pp. 13–27.

S2   Santhakumar, A. R., "The Ductility of Coupled Shear Walls," Ph.D. thesis, University of Canterbury, Christchurch, New Zealand, 1974, 412 p.

S3   Soesianawati, M. T., and Park, R., "Flexural Strength and Ductility of Reinforced Concrete Columns with Various Quantities of Transverse Reinforcement," *Proceedings of the Pacific Concrete Conference*, Auckland, New Zealand, November 1988, Vol. 1, pp. 65–76.

S4   Stratta, J. L., *Manual of Seismic Design*, Prentice Hall, Englewood Cliffs, N.J., 1987, 272 p.

S5   Suzuki, N., Otani, S., and Aoyama, H., "The Effective Width of Slabs in Reinforced Concrete Structures," *Transactions of the Japan Concrete Institute*, Vol. 5, 1983, pp. 309–316.

S6   Suzuki, N., Otani, S., and Kobayashi, H., "Three-Dimensional Beam–Column Subassemblages Under Bidirectional Earthquake Loadings," *Proceedings of the 8th World Conference on Earthquake Engineering*, San Francisco, 1984, Vol. 6 (9.4), pp. 453–460.

S7   Schneider, R. R., and Dickey, W. L., *Reinforced Masonry Design*, Prentice-Hall, Englewood Cliffs, N.J., 1980, 612 p.

S8   Saiidi, M., and Hodson, K. E., *Analytical Study of Irregular Reinforced Concrete Structures Subjected to In-Plane Earthquake Loads*, Report 59, College of Engineering, University of Nevada, Reno, May 1982, 160 p.

S9   Soleimani, D., Popov, E. P., and Bertero, V. V., "Hysteretic Behaviour of Reinforced Concrete Beam–Column Subassemblages," *Journal ACI*, Vol. 76, No. 11, November 1979, pp. 1179–1195.

S10   Steinbrugge, K. V., *Earthquakes, Volcanoes, and Tsunamis: An Anatomy of Hazards*, Skaudia America Group, New York, 1982, 392 p.

S11   Sahlin, W., *Structural Masonry*, Prentice Hall, Englewood Cliffs, N.J., 1971.

S12   Stafford-Smith, B., and Carter, C., *A Method of Analysis for Infilled Frames*, *Proceedings ICE*, Vol. 44, 1969, pp. 31–44.

S13   Schlaich, J., Schaefer, K., and Jennewein, M., "Toward a Consistent Design of Structural Concrete," *Journal of the Prestressed Concrete Institute*, Vol. 32, No. 3, May–June 1987, pp. 74–150.

S14   Schueller, W., *High-Rise Building Structures*, John Wiley & Sons, New York, 1977, 274 p.

S15   Sinha, B. P., Gerstle, K. H., and Tulin, L. G., "Stress-Strain Behaviour for Concrete Under Cyclic Loading," *Journal ACI*, Vol. 61, No. 2, February 1964, pp. 195–211.

S16   Sheikh, S. A., and Uzumeri, S. M., "Strength and Ductility of Confined Concrete Columns," *Proceedings ASCE*, Vol. 106, No. ST5, May 1980, pp. 1079–1102.

S17   Shing, P. B., Noland, J. L., Klamerus, E., and Spaeh, H., "Inelastic Behaviour of Concrete Masonry Shear Walls," *Journal of Structural Engineering, ASCE*, Vol. 115, No. 9, September 1989, pp. 2204–2225.

S18   Stevens, N. J., Uzumeri, S. M., and Collins, M. P., "Reinforced Concrete Subjected to Reversed Cyclic Shear-Experiments and Constitutive Model," *ACI Structural Journal*, Vol. 88, No. 2, March–April 1991, pp. 135–146.

T1   Tompkins, D. M., *The Seismic Response of Reinforced Concrete Multistorey Frames*, Research Report 80-5, Department of Civil Engineering, University of Canterbury, Christchurch, New Zealand, 1980, 152 p.

T2   Tanaka, H., and Park, R., "Effectiveness of Transverse Reinforcement with Alternative Anchorage Details in Reinforced Concrete Columns," *Proceedings of the Pacific Conference on Earthquake Engineering*, Wairakei, New Zealand, August 1987, Vol. 1, pp. 225–235.

T3   Taylor, R. G., "The Nonlinear Seismic Response of Tall Shear Wall Structures," Ph.D. thesis, Research Report 77/12, Department of Civil Engineering, University of Canterbury, Christchurch, New Zealand, 1977, 234 pp.

T4   Tomii, M., Sueoka, T., and Hiraishi, H., "Elastic Analysis of Framed Shear Walls by Assuming Their Infilled Panel Walls to be 45 Degree Orthotropic Plates," *Transactions of the Architectural Institute of Japan*, Part I: No. 280, June 1980; Part II: No. 284, October 1979.

T5   Tomii, M., and Hiraishi, H., "Elastic Analysis of Framed Shear Walls by Considering Shearing Deformation of the Beams and Columns of Their Boundary Frames," *Transactions of the Architectural Institue of Japan*, Part I: No. 273, November 1978; Part II: No. 274, December 1978; Part III: No. 275, January 1979; Part IV: No. 276, February 1979.

T6   Taylor, P. W., and Williams, R. C., "Foundations for Capacity Designed Structures," *Bulletin of the New Zealand National Society for Earthquake Engineering*, Vol. 12, No. 2, June 1979, pp. 101–113.

T7   Trigo, J. d'A., *Estruturas de Paneis sob Accao de Solicitacoes Horizontais*, Laboratorio Nacional De Engenharia, Civil, Lisbon, 1968.

U1 Uzumeri, S. M., "Strength and Ductility of Cast-in-Place Beam–Column Joints," in *Reinforced Concrete in Seismic Zones*, Publication SP-53, American Concrete Institute, Detroit, 1977, pp. 293–350.

U2 U.S. Department of Commerce, "The Prince William Sound, Alaska, Earthquake of 1964 and Aftershocks," *Coast and Geodetic Survey*, Vol. II, Part A, Washington, D.C., 1967, 392 p.

U3 Unemori, A. L., Roesset, J. M., and Becker, J. M., "Effect of In-Plane Floor Slab Flexibility on the Response of Crosswall Building Systems," in *Reinforced Concrete Structures Subjected to Wind and Earthquake Forces*, ACI Publication SP-63, American Concrete Institute, Detroit, 1980, pp. 113–134.

U4 U.S. Department of the Navy, *Soil Mechanics, Foundations and Earth Structures*, Design Manual DM-7, Alexandria, Va., 1971, p. 7-10-7.

V1 Vallenas, J. M., Bertero, V. V., and Popov, E. P., *Hysteretic Behaviour of Reinforced Concrete Structural Walls*, Report UCB/EERC-79/20, Earthquake Engineering Research Center, College of Engineering, University of California, Berkeley, August 1979, 234 p.

V2 Vallenas, J., Bertero, V. V., and Popov, E. P., *Concrete Confined by Rectangular Hoops Subjected to Axial Loads*, Report EERC-77/13, Earthquake Engineering Research Center, University of California, Berkeley, August 1977, 114 p.

W1 Wilson, E. L., Hollings, J. P., and Dovey, H. H., *Three Dimensional Analysis of Building Systems*, Report EERC-75/13, Earthquake Engineering Research Center, University of California, Berkeley, April 1975.

W2 Wilson, E. L., and Dovey, H. H., *Three Dimensional Analysis of Building Systems: Tabs*, Report EERC-72/8, Earthquake Engineering Research Center, University of California, Berkeley, December 1972.

W3 Wakabayashi, M., Minami, K., Nishimura, Y., and Imanaka, N., "Anchorage of Bent Bar in Reinforced Concrete Exterior Joints," *Transactions of the Japan Concrete Institute*, Vol. 5, 1983, pp. 317–324.

W4 Wakabayashi, M., *Design of Earthquake Resistant Buildings*, McGraw-Hill, New York, 1986, 309 p.

W5 Whitman, R. V., and Bialek, J., "Foundations," in *Design of Earthquake Structures*, E. Rosenblueth, Ed., Pentech Press, London, 1973, pp. 223–260.

W6 Wood, S. L., "Minimum Tensile Reinforcement Requirements in Walls," *ACI Structural Journal*, Vol. 86, No. 5, September–October 1989, pp. 582–591.

W7 Wong, P. K. C., Priestley, M. J. N., and Park, R., "Seismic Resistance of Frames with Vertically Distributed Longitudinal Reinforcement in Beams," *ACI Structural Journal*, Vol. 87, No. 4, July–August 1990, pp. 488–498.

X1 "The Chile Earthquake of March 3, 1985," *Earthquake Spectra*, Vol. 2, No. 2, February 1986, pp. 249–513.

X2 *Loma Prieta Earthquake October 17, 1989*, Preliminary Reconnaissance Report, Earthquake Engineering Research Institute, Oakland, Calif., 50 p.

X3 *New Zealand Standard Code of Practice for the Design of Concrete Structures*, NZS 3101: Part 1, 127 p.; Commentary NZS 3101: Part 2, 156 p.; Standard Association of New Zealand, Wellington, New Zealand, 1982.

X4   Seismology Committee, Structural Engineers Association of California, *Recommended Lateral Force Requirement and Tentative Commentary*, 1988, 123 p.

X5   *Design of Concrete Structures for Buildings*, A National Standard of Canada, CAN-A23.3-M84, Canadian Standards Association, 1984, 281 p.

X6   *Final Review Draft of Recommended Comprehensive Seismic Design Provisions for Buildings*, ATC 3-05, prepared by Applied Technology Council, Palo Alto, Calif., January 1977.

X7   *NEHRP Recommended Provisions for the Development of Seismic Regulations for New Buildings*, Building Seismic Safety Council, 1988, Part 1, *Provisions*, 158 p.; Part 2, *Commentary*, 281 p.

X8   *Code of Practice for General Structural Design and Design Loadings for Buildings*, *NZS 4203:1984*, New Zealand Standard, Standards Association of New Zealand.

X9   *Code of Practice for General Structural Design and Design Loadings for Buildings*, *2nd Draft NZS 4203:1989*, New Zealand Standard, Standards Association of New Zealand, 242 p.

X10   *Uniform Building Code*, International Conference of Building Officials, Whittier, Calif., 1988, 926 p.

X11   "The San Salvadore Earthquake of October 10, 1986," *Earthquake Spectra*, Vol. 3, No. 3, August 1987, pp. 415–635.

X12   *Earthquake Resistant Regulations, A World List—1988*, International Association for Earthquake Engineering, Tokyo, 1988, 1026 p.

X13   *Code of Practice for Masonry Design*, NZS 4230:1985, New Zealand Standard, Standards Association of New Zealand, Wellington, 1985, p. 130.

X14   *Caltrans Seismic Design References*, California Department for Transportation, Sacramento, Calif., June 1990.

Y1   Yamaguchi, I., Sugano, S., Higashibata, Y., Nagashima, T., and Kishida, T., *Seismic Behaviour of Reinforced Concrete Exterior Beam–Column Joints Which Used Special Anchorages*, Takenaka Technical Research Report 25, 1981, pp. 23–30.

Y2   Yoshimura, M., and Kurose, Y., "Inelastic Behaviour of the Building," in *Earthquake Effects on Reinforced Concrete Structures*, U.S.–Japan Research, ACI Publication SP-84, American Concrete Institute, Detroit, 1985, pp. 163–201.

Y3   Yanez, F., Park, R., and Paulay, T., "Strut and Tie Models for Reinforced Concrete Design and Analysis," Technical Papers TR9, New Zealand Concrete Society Silver Jubilee Conference, Wairakei, New Zealand, 1989, pp. 43–55.

Z1   Zahn, F. A., Park, R., Priestley, M. J. N., and Chapman, H. E., "Development of Design Procedures for the Flexural Strength and Ductility of Reinforced Concrete Bridge Columns," *Bulletin of the New Zealand National Society for Earthquake Engineering*, Vol. 19, No. 3, September 1986, pp. 200–212.

Index